Lecture Notes in Electrical Engineering

Volume 965

The book series *Lecture Notes in Electrical Engineering* (LNEE) publishes the latest developments in Electrical Engineering—quickly, informally and in high quality. While original research reported in proceedings and monographs has traditionally formed the core of LNEE, we also encourage authors to submit books devoted to supporting student education and professional training in the various fields and applications areas of electrical engineering. The series cover classical and emerging topics concerning:

- Communication Engineering, Information Theory and Networks
- Electronics Engineering and Microelectronics
- Signal, Image and Speech Processing
- Wireless and Mobile Communication
- Circuits and Systems
- Energy Systems, Power Electronics and Electrical Machines
- Electro-optical Engineering
- Instrumentation Engineering
- Avionics Engineering
- Control Systems
- Internet-of-Things and Cybersecurity
- Biomedical Devices, MEMS and NEMS

For general information about this book series, comments or suggestions, please contact leontina.dicecco@springer.com.

To submit a proposal or request further information, please contact the Publishing Editor in your country:

China

Jasmine Dou, Editor (jasmine.dou@springer.com)

India, Japan, Rest of Asia

Swati Meherishi, Editorial Director (Swati.Meherishi@springer.com)

Southeast Asia, Australia, New Zealand

Ramesh Nath Premnath, Editor (ramesh.premnath@springernature.com)

USA, Canada

Michael Luby, Senior Editor (michael.luby@springer.com)

All other Countries

Leontina Di Cecco, Senior Editor (leontina.dicecco@springer.com)

**** This series is indexed by EI Compendex and Scopus databases. ****

More information about this series at https://link.springer.com/bookseries/7818

Mikhailo Klymash · Andriy Luntovskyy ·
Mykola Beshley · Igor Melnyk ·
Alexander Schill
Editors

Emerging Networking in the Digital Transformation Age

Approaches, Protocols, Platforms, Best Practices, and Energy Efficiency

Springer

Editors
Mikhailo Klymash
Lviv Polytechnic National University
Lviv, Ukraine

Mykola Beshley
Lviv Polytechnic National University
Lviv, Ukraine

Alexander Schill
Dresden University of Technology
Dresden, Germany

Andriy Luntovskyy
Saxon Study Academy
BA Dresden University of Cooperative Education
Dresden, Germany

Igor Melnyk
Igor Sikorsky Kyiv Polytechnic Institute
National Technical University of Ukraine
Kyiv, Ukraine

ISSN 1876-1100 ISSN 1876-1119 (electronic)
Lecture Notes in Electrical Engineering
ISBN 978-3-031-24962-4 ISBN 978-3-031-24963-1 (eBook)
https://doi.org/10.1007/978-3-031-24963-1

This Springer imprint is published by the registered company Springer Nature Switzerland AG
The registered company address is: Gewerbestrasse 11, 6330 Cham, Switzerland

Foreword of Prof. Dr. habil. Dr. h. c. Alexander Schill, Chairman of Chair for Computer Networks at TU Dresden

This Lecture Notes book covers a broad range of leading-edge topics in the area of Emerging Networking (EmN). It is suitable as a textbook for advanced lectures in the domains of systems architecture, networking, and distributed platforms. Moreover, it can serve as a basis for research and development seminars and as an inspiration for the interested IT and Telco specialists looking for new challenges and developments. The authors have many years of experience as professors and lecturers in computer science and have already published a significant number of textbooks, journal contributions, and conference papers.

The book is based on the latest advanced research and close cooperation of the scientists from Germany, Ukraine, Israel, Switzerland, Slovak Republic, Poland, Czech Republic, South Korea, China, Italy, North Macedonia, Azerbaijan, Kazakhstan, France, Latvia, Greece, Romania, USA, Finland, Morocco, Ireland, and UK.

The book starts with an introduction into the principles of EmN with multiple smart applications based on 5G and Beyond technologies. Under the category Emerging Networking, we also cover the steadily growing diversity of unmanned scenarios, where the process automation grade is nearly 100%. Operative disaster networking (emergency networking), which appears at sites, aiding rescue troops, belongs to the above-mentioned problems, too.

The future EmN architecture includes the components of artificial intelligence, Blockchain, software-based networking with efficient network slicing, quality of service management, cloud computing support, and advanced security approaches.

Furthermore, in the age of Digital Transformations Smart Applications like Smart Shopping, supported via Internet of Things (IoT), contactless systems (RFID, NFC) and emerging combined networks (mobile, wireless, and fixed) become a fundamental challenge both for urban and rural environments. The book discusses the software and networking design of smart applications in dynamically changing domains, such as online shopping, smart shopping, and social shopping. Further aspects of EmN support for smart applications will be examined in the book, too, including smart energy, smart office, smart home, robotics, unmanned (aerial) vehicles and flying objects, as well as context-sensitive intelligent applications, which are aimed at domains such as traffic telematics, health care, and entertainment.

Moreover, EmN supports distributed intelligence across the combined mobile, wireless and fixed networks in the Edge-to-Cloud Continuum. Therefore, the book also examines 6G driving factors, features, and potentials in the mid-term. Topics include antenna and cell structures, as well as network software based on further network function virtualization (NFV), artificial intelligence (AI), and advanced cyber-security approaches like Blockchain (BC).

Altogether, the content of this book is surely technically interesting while still being rather practically oriented and therefore straightforward to understand. We would like to wish all readers successful studies and lots of inspiration by this useful textbook!

September 2022 Alexander Schill

Preface

This volume is a collection of the latest research on Emerging Networking deployment and appeared in close cooperation of scientists from Germany, Ukraine, Israel, Switzerland, Slovak Republic, Poland, Czech Republic, South Korea, China, Italy, North Macedonia, Azerbaijan, Kazakhstan, France, Latvia, Greece, Romania, USA, Finland, Morocco, Ireland, and UK. The authors of the chapters from this collection present their in-depth extended research results in the above-mentioned scientific area.

The volume consists of 37 chapters, a foreword, an introduction, conclusions, and this preface.

Chapter 1, presented by **M. Beshley, M. Klymash, I. Scherm, H. Beshley, and Y. Shkoropad,** is titled *Emerging Network Technologies for Digital Transformation: 5G/6G, IoT, SDN/IBN, Cloud Computing, and Blockchain.* The chapter starts with a discussion of the state of the art in the development of network technologies for the global digitalization of society, including new trends and future research directions. The importance and necessity of active development of network technologies are also emphasized. In particular, the top ten trending network technologies that will play an important role in a comprehensive digital transformation are highlighted. The authors have developed unique concepts and concise explanations of such emerging technologies as follows: the Internet of Things, software-defined networking, network function virtualization, (Docker) container networking, cloud computing, edge computing and AI, information-centric networking and services, time-sensitive and deterministic networking, 5G/6G (generation) mobile networks, AI-based networking, and many other contexts related to advanced communications. Furthermore, it is emphasized that all emerging network technologies discussed in this chapter should be combined to organize new digital infrastructures in different deployment scenarios and use cases. Only then will we be able to achieve the necessary capabilities to optimize the performance and efficiency of different infrastructures in the new era of digital transformation.

In Chapter 2, titled *A Framework for Context-sensitive Control-Loops with Roles* by **I. Schmelkin, A.Schill, and A. Kropp,** it has been established that the increasing complexity of modern information and communication systems in the context of digital transformation requires self-adaptive software that can autonomously adapt to changing business needs of users, changing work environments, and system failures. Control loops are one of the popular approaches to the design and implementation of self-adaptation in such systems, but existing solutions are characterized by unification for specific applications without flexibility and universality of usage. For this reason, the authors developed the framework as a basis for flexibly adding or adjusting functions in existing code when small changes to the managed system occur, or for replacing entire parts of control functions when large parts of the managed system are changed. The benefits of using the framework for various information technology tools are explained and demonstrated in detail, namely for monitoring distributed server architectures, for network switches and routers, embedded systems, and Internet of Things sensors. In the future, the authors' proposed solutions will allow software developers of control logic for information technology systems to ensure the universality of control coordination between different types of device elements by adjusting existing roles and introducing new ones.

Chapter 3 titled *The Impact of Smart Economy and Digitalization on Market Efficiency* and presented by **J. Smettan** is dedicated to the study of the current stage of global economic and social development, characterized by a significant impact of digitalization. The rapid development of information and network technologies has aroused an in-depth interest in the digital economy, as well as its impact on the evolution of the labor market. The digitalization era has opened new horizons for the creation of new forms of employment and new forms of enterprises. The above causes the need for research to study the impact on the economic efficiency of significant transformations of the labor market caused by digitalization. That is why the author emphasizes that the digitalization of the economy entails the use of artificial intelligence, robots, and cloud technologies in the production process, increasing the demand for workers with digital skills. However, the public administration system is not ready for such global changes. The rapid pace of digital technologies implementation in management and production processes increases the imbalance between the development of the national labor market and the digital economy. The formation of the global digital segment of the labor market is accompanied by economic and social contradictions, which are especially acute in institutionally underdeveloped countries. From the scientific point of view of the author of the section, the positive consequences of the digitalization of the labor market include the reduction of labor costs for companies and the emergence of new jobs and new professions that will be associated with information and communication systems. The negative ones include the lack of knowledge of how to benefit from automation, the problem of ensuring the confidentiality of digitized data transmission, the loss of the country's competitive position in the market of goods and services, and the increase in income asymmetries between the rich and the poor. Thus, taking into account the strengths of the massive introduction of the latest information technologies in society, informatization is becoming such an important factor in increasing productivity and improving the quality of life that the changes that are taking place are considered by researchers as the onset of a new era of economic development, which in the literature is characterized by the term "Smart Economy".

Chapter 4, titled *Self-Driving Cars* by **A. Luntovskyy, D. Guetter and A. Masiuk**, discusses the importance of the construction of unmanned vehicles in smart automotive industry, which is rapidly evolving from technological proprietary development to digital transformation based on the use of Emerging Networking. The consumers' demands on the functionality of their cars are growing steady. Therefore, digital transformation in the automotive industry is becoming increasingly more important and market attractive. The authors emphasize that the cars of the future will be autonomous and electric, driven by emerging technologies such as artificial intelligence (AI), Internet of Things (IoT), and 5G functionalities. Thus, vehicles will be able to learn and predict the behavior of their customers, recognize signs, and make fast self-driving decisions. The chapter discusses important challenges that may arise when innovating the digitalization of transport in the era of digital transformation. In particular, it is argued that modern smart cars collect information about traffic patterns, drivers, their usual location, etc., through technologies such as IEEE 802.11p/ pWLAN/ pWi-Fi, ITS-G5/Car2Car, LoRA WAN, LTE CatM, and NB-IoT. This data is useful for interaction and communication with drivers and traffic management information and analytical centers based on Emerging Networking including proven fog and cloud technologies. However, to protect driver information (security) and personal data (privacy), the connected systems must be safe. That is why manufacturers and scientists are always looking forward to new methods in the area of cyber-security and network data defense.

Chapter 5, under the title *Smart Home: Protocols, Platforms and Best Practices* by **D. Guetter and A. Luntovskyy,** is dedicated to the development of modern digital platforms for smart homes. Scientific and technological progress and digitalization are moving at a very fast manner, making life easier, more convenient and better. One of such innovations is the construction of a smart home based on advanced network technologies. Wireless data transmission technologies have become widespread: Bluetooth, Wi-Fi, ZigBee, LoRa, NB-IoT, and other. In this chapter, the basic concepts for so-called smart home systems are investigated, existing protocols for the construction of "smart home" are considered, as well as design schemes are presented. The authors solve some practical problems of compatibility of network and IoT components as well as consider pros and cons for conformity and independence from the vendor's and cloud functionalities. Some issues of bottlenecks in existing platforms in terms of network security and data privacy are presented.

In Chapter 6, titled *Construction and Methods for Solving Problems at the SDN Control Level* by **O. Romanov, M. Nesterenko, G. Boggia, and D. Striccoli,** the authors have focused on improving the efficiency of SDN functionality, which is a key technology for the implementation of the overall concept of digital transformation by telecom operators. Transferring the management to a central controller, which understands the full topology of the network, allows to increase the efficiency of utilization of all connections and optimally utilize the resources of hardware devices. To obtain optimal solutions in the network resource management process, the authors formalize a number of mathematical management models. The solution of mathematical models allows obtaining numerical values of throughput and quality of service indicators in different modes of network operation. Such models provide an opportunity to compare the effectiveness of various methods of resource management using input data of real indicators of telecommunication network functioning. In particular, the comparison of these methods on criteria of quality of service in various scenarios of service and traffic transfer, taking into account such aspects as routing of flows, load balancing, and distribution on priorities, is carried out.

In Chapter 7 under the title: *Research and Implementation Directions of the Fast ReRoute Facility Backup Scheme in Programmable Networks* by **O. Lemeshko, O. Yeremenko, M. Yevdokymenko, S. Harkusha, and Y. Dobryshkin,** noted that in the context of digitalization of society, in addition to the requirements for quality of service (QoS), which have already become classic functions in modern digital infrastructures, the functionality to ensure a high level of fault tolerance and network security comes to the fore. This is especially important in the functioning of information and communication networks in conditions of constant external and internal factors of influence, which lead to a significant change in both structural and functional parameters and properties of the telecommunication network: its topology, throughput, manageability, stability, etc. Protocol routing and traffic management tools in digital infrastructures are the key technological means of ensuring QoS, fault tolerance, and network security. The authors proposed the newest fault-tolerant routing for the emerging technology of software-defined networks, which in the near future will be the key to digital transformation for Internet service providers, cloud operators, and enterprises in the era of digitalization. The novelty of routing lies in the updated form of the objective function in the optimality criterion, which is formed as a route metric that aims to minimize the use of additional network resources associated with the protection of network elements and the implementation of backup routes. In the course of the study, it is proved that the use of the proposed routing allows to reduce the resource consumption of the solution by 23% compared to the implementation without the use of the object redundancy scheme. It is shown that the implementation of the FRR solution, which prevents the inclusion of the protected network element in both the main and backup routes, can lead to a significant reduction in the throughput of the main route (up to 75%), which is used before the failure of the protected network element.

Chapter 8, titled *Ecosystem for the Development and Testing of Emerging Network Applications in 5G and Beyond Environment* by **R. Odarchenko and E. Shapira,** is devoted to applying the 5G wireless standard in modern ecosystems. Really, among the different novel wireless technologies, the decisive role for critical applications (automotive and PPDR) is given to the fifth generation mobile technologies. 5G is a key driving force in the development of the ICT industry and other related vertical industries. In these conditions, when a large number of commercial 5G networks are already deployed, software developers begin to develop a large number of network applications for the needs and with the use of 5G. Therefore, it is very important to support developers, small and medium enterprises in order to automatically test, certify, and bring to the end customer already debugged applications that can be deployed and used in commercial projects. The chapter presents the ecosystem for the development and testing of emerging network applications in the 5G and Beyond environment. This ecosystem combines software developers, test sites in European cities, automated testing and certification tools, end customers (users) of network applications, and other tools and systems. And it is mandatory to use NetApp Store, a specially developed Marketplace, to bring fully tested and certified network applications to the end user or customer. This paper also demonstrates how the methodology developed in the project will be applied in practice on the example of remote human driving NetApp. In particular, the architecture of the application, the principles of interaction with the test platform, the requirements for the application, the test plan, and metrics to be collected were shown. The chapter has a practical focus and can be of great interest to designers of environmental control systems. All aspects of the problem of designing ecology systems on the base of 5G standard are analyzed in the chapter scrupulously, and corresponding conclusions have been made.

Chapter 9 under the title: *Comprehensive Approach for Radio Equipment Design to Ensure Reliable UAVs Operations* by **L. Uryvsky, A. Gnashchuk, S. Osypchuk, and A. Moshynska,** is addressed to the peculiarities of providing a comprehensive approach for radio equipment design for modern unmanned aerial vehicles (UAVs). The functioning of UAVs is needed as a modern and promising technological object in various society applications and requires not only the means of direct flight support (aerodynamic properties, fuel resources, etc.), but also does require comprehensive radio equipment. The following tasks are expanded in this work and need qualitative solutions for building comprehensive radio equipment, such as reliable channel for UAV position registration on UAV, UAV position registration channel at the ground control center, transmission channel for flight control commands, and channel for collecting data for which the UAV is authorized. The chapter analyzes the needs of modern UAVs in providing them with radio and telecommunication systems that monitor the state of UAV parameters, determine its location, control the UAV operations, and exchange information with UAVs via duplex channel, information protection during information transmission, electronic warfare, and others. A comprehensive approach to the design of a multifunctional radio communication system to ensure the reliable UAV operations, algorithms for its operation to ensure the UAV functioning, depending on the type of tasks for which it is intended.

Chapter 10, titled *Networked Simulation with Compact Visualization of Complex Graphics and Interpolation Results* by **I. Melnyk and A. Luntovskyy,** is devoted to the methods of presentation of complex 2D and 3D graphics in Emerging Networks. The researchers and engineers of Emerging Networks provide efficient networked simulation and use visualization tools. Modern design and simulation systems as well as multiple practical scenarios frequently use the complex plane and spatial graphics, which is oriented on compact presentation of figures with complex geometry. It is especially important for CAD/CAM systems, where the complex constructions of elaborated machines or devices are simulated. Presentation in compact form of details with complex geometry is necessary for defining mechanical stresses, temperature distribution, or electromagnetic fields distribution by solving the sets of differential equations. In the theory of communication systems, interpolation of digital signals is necessary for defining the time of its transmission through a communication channel. In network cloud simulation systems, suitable interpolation functions for the figures with complex geometry are also necessary for reducing the amount of necessary digital data and its fast transmission through communication channels. In this chapter, interpolation of complex plane and spatial curves with using Bezier curves, Levitan polynomials, and Rvachov functions is considered. Possibility of applying arithmetic-logic functions for describing the figures with complex plane geometry is also considered. Corresponded examples for different interpolation functions are also given.

Chapter 11 under the title: *Digitalisation in Shopping: An IoT and Smart Applications Perspective* by **O. Cvetkovski, P. Gkikopoulos, and J. Spillner** demonstrates the development of a smart and cost-effective shopping application using Internet of Things (IoT) innovations. Such a system is suitable for use in different stores, providing customers with better shopping knowledge and time saving. The chapter provides a broad overview of ideas for using smart shopping in everyday life. It also describes unique concepts of evolution of architectural solutions together with research on software, hardware, and the latest network technologies. The information system of smart shopping proposed by the authors allows one to find the nearest route to get the listed products on different shelves in real time. The developed system was tested in the field of smart shopping. The experiment was conducted to ensure accurate determination of the proximity between the consumer and the product. The conducted experience describes how to apply known methods relying on wireless signal power to the field of smart shopping, where beacon signals are available from electronic shelf labels (ESL). Technical recommendations for the use of the proposed shopping information system in the conditions of various external and internal factors of influence, based on real research, are formed. In the future, it is expected that the innovative, scientifically based solutions of the authors of this chapter on smart stores can be transferred to smarter cities and regions.

Chapter 12 is titled: *Increasing Functional Stability of Telecommunications Network in the Depressed Zone of HPS Reservoir* by **P. Anakhov**, **V. Zhebka**, **L. Berkman**, and **V. Koretska.** The authors have analyzed the possibilities of applying HPS for defense of communication systems. The sphere of influence of the reservoirs of large HPS involves colossal massifs of rocks. The complex of geophysical fields and processes, mechanical and electrical transformations caused changes in the geophysical situation of the local environment in the depressed zone, which determined the need to formulate recommendations for the protection of telecommunications. A method for developing measures to protect the Emerging Telecommunication Networks from the effects of destructive influences, which includes data collecting of their impact on hardware resources, their analysis, and development of appropriate countermeasures. The conditions of functional stability of the Emerging Telecommunication Networks are formulated. They are represented by the resistance of the network infrastructure components to the impact of hazards, the ability to reconfigure the operational systems, and the transmission network. To verify network protection measures, a matrix of compliance with the threats has been developed, the occurrence of which may be due to processes in the depressed zone.

Chapter 13, titled *Discrete Vehicle Automation Algorithm Based on the Theory of Finite State Machine* by **I. Melnyk, S. Tuhai, M. Surzhykov, and I. Shved,** considers the model of vehicle control for the linear law of change of its speed in time. Corresponded mathematical models are created using analytical methods of discrete mathematics, set theory, finite state machines theory, and mathematical statistics. Based on the analysis of a simple law of motion of a vehicle between two given points in a straight line with a constant decrease in speed, it is theoretically substantiated that the mathematical model of such a law of motion can be realized in space of finite states over time. Such a simple algorithm is very suitable for smart-driving cars. An advantage of the proposed mathematical model of the vehicle control system is its simplicity, as well as the lack of corrective action at those times when the vehicle movement corresponds to the specified law with a slight error. This avoids unnecessary release of vehicle speed and oscillating processes that can occur in the case of a continuous control action if the control system parameters are selected incorrectly. Mathematical models for cases of quasi-stationary and random perturbing action are considered separately, and Student's distribution law is used to model random perturbation.

Chapter 14 is titled: *A Modified Federated Singular Value Decomposition Method for Big Data and ML optimization in IIoT Systems* by **O. Hordiichuk-Bublivska, H. Beshley, M. Kyryk, Y. Pyrih, O. Urikova, and M. Beshley.** The chapter is devoted to developing new digitalization methods for industrial systems based on big data, IoT, and machine learning technologies, which significantly contribute to the current stage of economic and social development. Data storage, processing, and transmission in digital format have become important attributes of the processes of production, distribution, exchange, and consumption of industrial products. The main conditions for successful digital modernization of industrial systems can be considered the effective use of the main advantages and minimization of risks and possible negative side effects of the implementation of information and digital technologies in the processes of production and sale of industrial products. The authors proposed a method of federated machine learning, which is implemented on industrial end devices. This approach allows the increase of industrial systems efficiency in the context of digital transformation. For data optimization and fast operation of machine learning, it is proposed to use the SVD algorithm, which allows the detection of redundant data and does not take them into account in further processing.

To increase the confidentiality of data transmission, an improved federated method of Singular Value Decomposition is proposed. This method involves the generation of all parameters for masking directly on local devices. The study results showed high accuracy of calculations by the proposed method and ensures the reliability of the data obtained. Based on simulations, it is proven that the proposed method allows optimization of big data and ML processing in industrial systems, which reduces the duration of calculations and makes it possible to make important decisions quickly. The authors also propose to use the Funk-SVD algorithm to increase the efficiency of sparse data processing in intelligent industrial systems.

Chapter 15 under the title: *Smart House Management System* by **B. Zhurakovskyi, O. Nedashkivskiy, M. Klymash, O. Pliushch, and M. Moshenchenko** describes in detail the process of developing a smart home control system using automation devices from different developers, solving the problem of interaction and data exchange in a heterogeneous digital environment. The authors have created their own Telegram bot for such an information system, which will allow the user to communicate with the IFTTT automation service, which communicates with the services of home automation device developers. An algorithm for combining the interaction of end systems into one interface with a detailed explanation of the functioning has been developed. The system is created using the Telegram messenger, which acts as an interface and protects the system using TelegramID. The system includes a Chatbot, a database, IFTTT system, and Web services of developers. At the same time, the developed system will have a large extension in the form of a new interface and several additional automation scenarios by adding triggers based on user interaction with the mobile device and the environment.

In Chapter 16, titled: *Searching Extreme Paths Based on Travelling Salesman's Problem for Wireless Emerging Networking* by **O. Pavlenko, A. Tymoshenko, O. Tymoshenko, A. Luntovskyy, Yar. Pyrih, and I. Melnyk,** the problems of finding the optimal route in the network with fully connected topology by solving the Traveling Salesman's Problem (TSP) are considered. Nowadays, the world is characterized by Emerging Networking in various areas of human activity and industries. Among the combined networks of today (mobile, wireless, fixed), the networks with a fully connected topology occupy a special place. They play an essential role in the software agents interacting with each other according to the point-to-point (P2P) model, the machine-to-machine (M2M) communication model for robotics and IoT with a specified gateway, energy-efficient clusters, or data centers with a large number of central processing units and radio channels between them, as well as Multiple Input Multiple Output (MIMO) antennas for the 5G and Beyond standards. Typically, the analysis of network systems with a fully connected topology leads to solving the TSP. A solution, however, requires combinatorial enumeration and needs a massive amount of computing time, even on powerful modern computers (NP-complexity). The given work provides a new TSP algorithm, which is based on the analysis of the digital series by unique sums. That allows one to find the best route between network blocks with a fully meshed topology. The obtained simulation results and computational experiments have shown the very high efficiency of the proposed algorithm. Furthermore, an attempt for a specific TSP-based MAC protocol is herewith offered (OSI layer 2), which distinguishes it from existing routing protocols (layer 3) such as OSPF+BGP or RIP

+BGP and above (with QoS considering). The considered TSP-based MAC protocol for wireless sensor network (WSN) provides access from a start node (cluster head, gateway) to all other nodes within the limited small net (cluster). The start node can be selected dynamically for future reconfiguration for energy efficiency and long life without frequent battery exhaustion. The start node provides TSP-based polling to all other networking energy-autarky nodes within the cluster. The obtained results are very important both theoretically and practically for forming novel network protocols.

Chapter 17, under the title: *A Survey of Intrusion Detection Methods in Wireless Networks* by **A. Babaryka, I. Katerynchuk, and I. Chesanovskyi,** focuses on the fact that the use of wireless networks for digital transformation is one of the most popular and, at the same time, sensitive to various types of cyber-attacks. If information is transmitted through wires in traditional networks, the radio waves used for wireless solutions are quite easy to intercept with the appropriate equipment. The principle of wireless network operation leads to many possible attacks and intrusion vulnerabilities. The paper describes the characteristic vulnerabilities of wireless networks and, based on these properties, identifies several possible types of attacks on wireless networks. Based on the analysis of modern research in the field of security of information systems, the classification of intrusion detection systems by monitoring methods, the nature of the response to threats, the nature of the data processing process, security levels, time characteristics, data processing location, and the type of anomaly detection is proposed. In particular, the authors describe that intrusion detection system (IDP) provides an additional level of protection for wireless networks together with the intrusion prevention system (IPS). IDS can notify about the beginning of an attack on the network, and some can detect previously unknown attacks. IPS is not limited to alerting but also takea various measures to block the attack (e.g., breaking the connection or executing a script specified by the administrator). Most intrusion prevention systems use one of three detection methods: signature, a statistical anomaly, and protocol analysis state. Studies have shown that at the present stage, models based on neural networks can be used quite effectively to block hybrid attacks. Such systems are quite easily implemented in existing tools, can adapt training, and can work with large amounts of data, which is important when processing network traffic.

Chapter 18, titled: *Applied Steganographic System for Hiding Textual Information on Audio Files* by **S. Buchyk, S. Toliupa, N. Lukova-Chuiko, O. Khomenko, and Y. Serpinskyi,** is devoted to the development of effective methods of digital information protection, in particular methods of computer steganography and steganalysis, which are relevant and important in the era of digital transformation of the state and society. The subject area is analyzed, the existing methods of audio steganography are considered, their advantages and disadvantages are analyzed, and their comparison is carried out. The authors analyzed the reduction of significant changes to the container files and to increase the resistance of containers to attacks. Based on the analysis, it was decided to develop a modification of the least significant bit method because this method provides greater security and is an effective way to hide secret information from hackers and send it to the destination safely and safely. The LSB steganographic algorithm for hiding textual information in audio files has been enhanced by the additional use of a cryptographic method and pseudo-random numbers to create a stable crypto key. Also, the algorithm ensures that the file size does not change even after encoding and is also suitable for any type of audio file format. This algorithm also allows to hide much more secret information in container files compared to other algorithms. The proposed improved algorithm can be used in next-generation networks. It makes sense to use the method to build covert network channels in the new network environment, for example, covert channels based on streaming media networks or covert channels based on IPv6 networks.

Chapter 19 is titled: *Reliability Assessment of UAV Fleets* by **E. Zaitseva, V. Levashenko, N. Brinzei, A. Kovalenko, M. Yelis, V. Gopejenko, and R. Mukhamediev.** The authors proposed the method of simulation unmanned aerial vehicles (UAVs) fleets, based on an approach of binary-state system. UAVs are widely used in various industries for monitoring, searching, communication, etc., which is a very interesting and constantly improving direction in the era of digital transformation. The proposed method can be used to analyze homogeneous and heterogeneous fleets that have different types of management. The proposed readiness definitions for different types of UAV fleets allow to analyze the reliability of fleets with reduced computational complexity. This chapter presents models for assessing the reliability of a UAV fleet managed centrally or decentralized. The following topologies are considered: homogeneous redundant UAV fleet, homogeneous hot redundant UAV fleet, heterogeneous redundant UAV fleet, heterogeneous hot redundant UAV fleet, and heterogeneous hot redundant UAV fleet. For the listed topologies, reliability estimates are obtained depending on the number of primary and backup UAVs.

In Chapter 20 under the title: *Method of Cyber-Resilience Information-Control System Synthesis of Mosaic Structure* by **Yu. Danyk** and **V. Shestakov,** authors make a significant contribution to the further development of cyber-physical systems, which are a key technological component of the Industry 4.0 concept. Cyber-physical systems allow to receive and control information for automation and virtualization of processes, to influence production systems, and to determine the number of employees required to meet the needs of production within the concept of Industry 4.0. The section presents the results of the development of the model and method of synthesis of cyber-resilience of the information control system of the mosaic structure. The model and method of ICS synthesis allow to automate the process of determining the rational structure and characteristics of automated workplaces of cyber-resistant ICS. At the stage of structural and parametric synthesis of the method, the mathematical apparatus of multicriteria optimization by a nonlinear trade-off scheme is used. According to the results of the research carried out by the authors during the deployment of a cyber-physical specialized center for the operational control of forces and means, using the developed method, an increase in its cyber-resilience was obtained in the range of up to 95-98%.

In Chapter 21 under the title: *Neural Network Architecture with Oscillatory Synaptic Blocks for Signal Processing of Spatially Spaced Multiband Radar Complex* by **R. Peleshchak, V. Lytvyn, I. Peleshchak, and O. Veres,** the authors develop a neural network method of radar target recognition based on a spatially distributed multi-band radar complex, which is an important area of information technology. The proposed complex consists of spatially distributed radar stations that conduct joint radar target detection. The main idea of a multi-band radar complex is that radar systems operate at different frequencies with different polarization and are located in different places. This allows to simultaneously detect dozens of radar targets and recognize them in real-time using a neural network with parallel blocks of oscillatory synaptic connections. The input features for recognition are target scattering diagrams, which are integral characteristics of effective target scattering zones. Based on the modeling, it was found that air objects with the largest effective target scattering area have the highest probability of correct target recognition. As the signal-to-noise ratio decreases, the probability of correct target recognition decreases.

Chapter 22, titled *Modeling, Simulation and Visualization of Acoustic Coherent Images* by **I. Hvozdeva and V. Myrhorod,** is devoted to the consideration of methods for creating mathematical models of acoustic coherent images in relation to the tasks of monitoring the underwater environment of the World Ocean using modern distributed information and telecommunication networks. The combination of the theoretical approach to constructing an acoustic image as a section of the secondary field at the aperture of the receiving antenna and the experimental method for obtaining such images with instrumental measuring instruments of a hydroacoustic tank was chosen as the main research method. Comparison of theoretical and experimental results makes it possible to verify the proposed mathematical models of acoustic coherent images. The proposed models, confirmed experimentally, make it possible to perform computer simulation of the processes of formation and processing of acoustic coherent images (ACI) to solve the problems of creating and improving methods for processing such images for their recognition and classification, as well as for testing the technical means of underwater monitoring networks during their tests.

Chapter 23 under the title: *5G and Beyond: Perspectives of 6G* by **A. Luntovskyy** examines 6G driving factors, features, and potentials in the mid-term. Further development will concern better coding and spectral efficiency, new antenna, and dense heterogenous cell structures. A special meaning has the use of network softwarization based on further network function virtualization (NFV), interoperability with existing terrestrial and non-terrestrial solutions, as well as deployment of advanced cyber-security approaches like Blockchain (BC). Convergence with non-terrestrial services like Starlink in the giga-cell area is possible in mid-term too. Emerging mobile and wireless networks provide herewith the best QoS (higher data rates and lower latencies). The significant growth of channel power and spectral efficiency promise approaching to the known Nyquist–Shannon's limits (post-Shannon metaphor). Mid-term 6G networks include, moreover, further "enabling technologies" with better than for 5G sensing and passive localization techniques, when the distance and their derivations (velocity and acceleration) can be efficient determined via evaluation of the propagation delays due to reflections on the objects (radar techniques, WLANp, VLC, UWB, LiFi) in multiple smart applications for unmanned PHY objects and vehicles.

In Chapter 24, titled *Method and Means of Synthesis of Barker-Like Codes for Data Transmission Coding* by **O. Riznyk, I. Tsmots, and V. Teslyuk,** the authors investigated possibility of increasing the noise immunity of data transmission in communication networks by using barker-like codes. Improving noise immunity at fixed data rates is an urgent problem for emerging wireless networks in the era of digital transformation. The method of synthesis of noise immune barker-like code sequences using ideal ring beams is improved. The method of a fast finding of such noise immune barker-like code sequences, which can find and correct errors to the greatest extent according to the length of the received code sequence, is improved. The algorithm for fast finding of noise immune barcode sequences, which are able to find and correct errors in the greatest number according to the length of the received barcode sequence, is implemented. The simulation model of noise immune barker-like coding using ideal ring beams is developed. The simulation model of noise immune barker-like coding for search and correction of errors in the obtained noise immune barker-like code sequences was implemented. The presented noise immune barker-like code sequences have practical value, as the resultant barker-like code sequence contains up to 50% and corrects up to 25% of distorted symbols from the length of the noise immune barker-like code sequence.

In Chapter 25 under the title: *Processing Marker Arrays of Clustered Transformants for Image Segments* by **V. Barannik, A. Krasnorutsky, V. Barannik, Yu. Babenko, S. Shulgin, O. Chernenko, O. Slobodyanyuk, and M. Bondarchuk,** the authors proposed an innovative technological approach to the preservation of the information component of video data in the aerospace segment of the remote video service. The imbalance between the reliability of the retrieved video image and the efficiency of its delivery by onboard data transmission channels is substantiated. On the basis of the structural characteristics of the binary representation of the clustered component, a method of forming markers and a marker array of clustered transforms of the image segment are developed. It is substantiated that in order to reduce the volume of clustered transforms, it is necessary to reduce the volume of marker arrays. This approach allows to reduce the structural description of the linearized marker array of clustered transformants by reducing the number of repeating markers. A method of forming a sequence of two-component tuples containing the length of the cluster trend that precedes the corresponding marker is developed. This approach allows to reveal structural regularities in arrays of markers due to the presence of cluster trends. It is substantiated that there is a relationship between the components of tuples, which is due to the positions of the markers in the linearized array of markers, and hence the correlated lengths of the marker chains. For a structural approach to the processing of tuples, a method of forming domains for the marker matrix is developed, which takes into account the correlation of the lengths of the marker chains.

Chapter 26, titled: *A Method of Scrambling for The System of Cryptocompression of Codograms Service Components* by **V. Barannik, S. Sidchenko,D. Barannik, M. Babenko, V. Manakov, O. Kulitsa, V. Kryshtal, V. Kolesnyk,and O. Kuzmin,** is devoted to the advanced method of scrambling for a service components system in cryptocompression codograms formed under the condition of discarding the pixel brightness values in RGB space least significant. Such approaches are useful for Emerging Networking with the increased modern demands on content security for smart applications. The method provides an increase in the availability of video information due to further cryptocompression representation of image volume reduction; increase in crypto resistance by changing the values of the service elements of the data system, breaking the correlation between the elements and changing the pixel frequency. The method of crypto-resistant compression of codograms of service components of the scrambling system, formed under the condition of discarding the least significant bit in the values of pixel brightness in RGB space, is developed. The difference of this method from the known ones is that before performing scrambling transformations, the merging of service data presented in a reduced dynamic range into 8-bit combined elements is organized. At the stage of permutation of the merged 8-bit data, not only the location of the values of the original 7-bit elements of the service components is organized, but also the change of their values. This allows to increase cryptographic characteristics of known permutation transformations. The software implementation of the method of scrambling the system of service components in codogram cryptocompression provides formation of a secure design of codogram of crypto-resistant compression with scrambled service components.

Chapter 27, titled *Smart Ergonomic Lighting for Circadian Rhythm* by **Ye. Zheliazkov and J. Jamnenko,** addresses the important scientific and applied problem of lighting characteristics for the human visual and the circadian systems for designing light intensity and color for various healthcare and medical applications. An algorithm of defining human circadian rhythm has been elaborated as well as electrical scheme of corresponded adaptive device based is proposed.

In Chapter 28 under the title: *The Approach to Assessment of Technical Condition of Microprocessor Systems that are Implemented on Integrated Circuits with a Programmable Structure* by **S. Shtanenko, Yu. Samokhvalov, S. Toliupa, and O. Silko,** the problem of diagnostics of programmable logic device (PLD) microprocessor systems is the basis for building complex technical systems operating in adverse conditions, including cyber-attacks. A significant advantage of field-programmable gate array (FPGAs) is their versatility and the fact that they can be quickly programmed to perform the functions of almost any digital device. The use of FPGAs as an element base for the construction of modern microprocessor systems today is a promising direction in terms of designing highly reliable, fault-tolerant complex technical digital systems that operate in adverse conditions, including cyber-attacks. The analysis of existing methods of testing integrated circuits with programmable structures is carried out, and their advantages and disadvantages are noted. It is proposed to use a service processor as a diagnostic device as an integral element of a multiprocessor system. This processor will not only collect diagnostic information but also make decisions about the reconfiguration of the system to restore its functioning in automatic mode by reprogramming the PLC. In general, the proposed approach to diagnostics of microprocessor systems based on FPGA is a further development of the self-diagnostic method. If a complex technical system is given the property of adaptation to changing operating conditions, then such a system can be considered as a system with elements of artificial intelligence.

Chapter 29, titled *Robust Approach to the Signal and Data Processing* by **I. Prokopenko**, is devoted to the development of the theory of robust methods for signal and data processing construction of space distributed information-measuring systems in their distortion by non-Gaussian impulse interference and failures of the recording equipment, purposed for Emerging Networking. The problem of robust estimation of regression parameters of a random process is considered, the recurrent algorithm of gradient search of the solution of plausibility equations for non-Gaussian perturbations with density of distribution of Tukey type is synthesized, and their efficiency is investigated.

In Chapter 30 under the title: *Automation of Vehicle Control Systems and Emerging Networking* by **Yu. Zaychenko, S. Melnykov, G. Hamidov, A. Gasanov, and P. Malezhyk,** the issues of organization of intelligent vehicle control systems, including those using the capabilities of Emerging Networks, are considered. Directions for the organization of automated and intelligent control systems are shown. The main stages of automation of control systems for dynamic objects and the evaluation of the effectiveness of control functioning are determined. As a result of the provided review, the characteristic levels of development of the intelligence of control systems are presented, the features of emerging networking are highlighted, and the effectiveness of the above ideas for solving practical problems is shown.

Chapter 31 is entitled: "*Metaverse of Things in 6G Era: An Emerging Fusion of IoT, XR, Edge AI and Blockchain Technologies*" by **T. Maksymyuk, J. Gazda, B. Shubyn, O. Karpin, O. Kapshii, O. Urikova, El-M. Amhoud, M. Liyanage, M. Jo, and M. Dohler**. The authors describe an interesting paradigm of synchronized physical and virtual realities within the Metaverse of Things (MoT). The prerequisites and demand for the user interoperability across real and virtual world are described. Authors describe the key elements of the MoT, namely immersive vision, pervasive sensing, ubiquitous computing, comprehensive intelligence, and decentralized trust. The relation among those elements and the corresponding underlying state-of-the-art technologies within the MoT are clearly outlined. The chapter also pays a reasonable attention to the underlying 5G/6G networks architecture, which should orchestrate a large number of data flows associated with the MoT applications. The interesting insights are given toward the fact that emerging 6G networks should be able to synchronize humans, things, and digital twins across real and virtual worlds in real time. This can be achieved by augmenting visual and sensory information from the real world and transferring it to the digital twins in the metaverse via multiple synchronized data flows. In the opposite direction, virtual actions in the metaverse should be translated to the real IoT devices in the physical world. To tackle this challenge, authors define the 6G architecture with decentralized service management via Blockchain. Particular feature of the proposed vision is that each user is able to adjust his quality requirements for each device and select the best network operator for each data flow independently. In this case, the synchronization and timely scheduling of data flows are ensured by the smart contracts and non-fungible tokens, according to the specified preferences of each user. The vision, provided in the chapter, opens a wide range of previously unimaginable use cases in the emerging 6G era.

Chapter 32 under the title: *Data Management in IoT: from Sensor Data to Intelligent Applications* by **V. Vasyutynskyy and A. Luntovskyy** is devoted to new solutions in Emerging Networks, including cloud computing and data mesh. Globalization processes and digitalization, which lead to the use of emerging network technologies IoT, artificial intelligence, big data, data analytics, cloud technologies, and robotics, provided an opportunity to analyze large amounts of data in a short period of time, which radically modifies approaches to digital transformation. Emerging networks allow the smart devices and applications to connect efficiently and everywhere, opening a lot of opportunities and use cases for getting and using data. The resulting huge amount of heterogeneous data leads to new challenges in data management, like data storage, processing, privacy, life cycle, etc. This chapter provides an overview of these challenges and modern solutions such as cloud computing and mesh data networks.

Chapter 33 is titled: *A Neuron Network Learning Algorithm for the Recognition of Fractal Image and the Restoration of their Quality* by **O. Yunak, B. Strykhaliuk, M. Klymash, Y. Pyrih, and O. Shpur.** The chapter focuses on one of the most up-to-date tasks in the field of information technology in the context of global digitalization of society, namely intelligent image recognition. Fractal images are often used for graphical representation of data during computer modeling of some processes, for automatic generation of abstract images, in particular entertainment applications. Fractal graphics are contained in scientific visualization packages for building both simple structures and complex illustrations that simulate natural processes and three-dimensional objects. Recognition of fractal structures is one of the most difficult tasks, and this is due to the fact that it is very hard to collect data, since the construction of fractal structures requires significant information computing and resources. To solve this problem, the authors used the RSIF algorithm. The proposed algorithm does not use recursive functions and looping, which in turn requires fewer computing resources. The results of the algorithm showed that the use of neural networks allows to effectively recognize fractal images and to convert them into RSIF. The resulting RSIF matrix allows you to improve the quality of the fractal image and converts raster graphics into vector graphics, which in turn will allow you to resize the image without losing quality. In the future, this neural network can be integrated into pattern recognition systems and graphic editors in the form of a library.

Chapter 34, titled *Advances in Data Reduction Techniques to Solve Power Spectrum Estimation Problems for Emerging Wireless Networks* by **A. Lozynsky, I. Romanyshyn, B. Rusyn, M. Beshley, M. Medvetskyi, and D. Ivantyshyn,** discusses new trends and research directions for reducing redundancy in the spectrum analysis process for emerging wireless technologies, in particular 5G/6G. This research direction will allow future wireless digital infrastructures to improve coding resilience and ensure the efficient use of the spectrum of licensed signals, the lack of which is observed today. The authors proposed a technique for estimating the autocorrelation function (and power spectrum) of a stationary random signal for a finite sample under different variants of non-uniform data sampling. The main difference between the methodology and the known ones is the focus on minimizing the number of sampling data by reducing redundancy. For this purpose, the authors applied a reduction ruler to estimate the autocorrelation function and the power spectrum. New construction techniques based on the known Golomb ruler, based on irregular sampling with sampling frequency switching of two close frequencies, are presented. The approaches for implementation in practice are described in detail. Modeling of the estimation accuracy based on the calculation of the standard deviation, which tends to increase as the amount of data decreases, has been performed. It is proved that the proposed solutions will reduce the number of calculations in the implementation of digital signal processing for wireless networks, which are emerging in the new era of digital transformation.

Chapter 35 is titled: *Investigation of High-Speed Methods for Determining the Equilibrium State of a Network Based on the Principle of Maximum Entropy* by **R. Politanskyi, A. Samila, L. Politanskyi, V. Vlasenko, V. Popa, Yu. Bobalo, and I. Tchaikovsky.** In this chapter, the authors propose to use the entropy method to determine the most probable state of the network, which is subject to restrictions on the total traffic generated and received in each node of the network. It is also assumed that the network is closed, that is, the sum of all generated information flows is equal to the sum of all received flows. The possibilities of applying the method in networks with several decentralization planes are considered. The method of building a model of a decentralized network in the form of a two-dimensional sparse matrix of information flows, which is structured in a special way, is substantiated. Accordingly, the purpose of the analysis of a decentralized network is to determine the optimal value of the entropy of the system represented by such a matrix. A method for estimating the range of values of the entropy of a decentralized network by analyzing significantly unbalanced networks with a significant number of zero values in the matrix of information flows is developed. Calculations are presented for elementary configurations of the network represented by a flat graph and a decentralized network with the lowest topological complexity. The advantage of this method is the significantly lower computing power consumption for the analysis of the network state, even if the network has a complex decentralized structure.

Chapter 36 under the title: *Elliptical Polarized Radio Wave Decay in Land Communication Links* by **Y. Ben-Shimol, N. Blaunstein, and C. Christodoulou** is devoted to the topic of depolarization in terrestrial wireless communication channels, which is insufficiently studied even for technologies beyond 4G. An innovative analytical approach based on a multi-parameter stochastic model as a combination of the vision of radio wave propagation over the built-up area and a statistical description of the main characteristics of the terrain is presented. An approach is developed that provides a sufficient way to predict the depolarization attenuation of radio waves and allows to reduce of polarization losses in two-way terminal antennas, thereby achieving cheaper data transmission in wireless networks. The expression of the total wave intensity is proposed, and the probability distributions for the parameters describing the spatial ellipse of polarization, the distribution of energy along and across the propagation directions, as well as the change in the shape and spatial orientation of an elliptically polarized wave depending on the terrain features are obtained. These equations are based on a new physical view of the Stokes parameters and on test results that take into account real data of mixed radio communications in the residential sector and suburban/urban radio lines that are widely used.

Chapter 37, tilted *Fault Identification in Linear Dynamic Systems by the Method of Locally Optimal Separate Estimation* by **A. Volovyk, V. Kychak, A. Osadchuk, and B.Zhurakovskyi,** provides a new method of synthesis of two-stage structures for recursive filters. Due to the rapid development of so-called Emerging Networks of new generations, the problem of improving the stability and reliability of communication equipment has become very important. Emerging Networks support nowadays multiple smart applications, including industries, automotive, robotics, unmanned vehicles, and flying objects. They enable automation and "Distributed Intelligence" across the combined mobile, wireless, and fixed networks in the Edge-to-Cloud Continuum. Therefore, advanced efficient solutions for the synthesis problems for optimal filters are necessary. An advanced method is proposed in the chapter, which provides a successful realization of such filters. The presented synthesis method is based on the assumption that there is no a priori information about the dynamics of operating faults and disturbances. The provided synthesis method is based on assumption of the lack of prior data about fault dynamics and perturbations. In this case, it is assumed that faults affect both the state of the system and the output variables, and disturbances affect only the state variables. Synthesis of the recursive filter was made in two options. In the first case, it was assumed that the distribution matrix of faults in the observation channel is a full rank matrix, and then the result obtained was generalized to the case of an arbitrary rank matrix. The proposed filters are locally optimal in the sense of forming separate unbiased fault estimates and states with minimal variance. For these purposes, a matrix version of the Lagrange multiplier method was used. The operability of the considered method was tested on the example of an aircraft landing system.

October 2022

Mikhailo Klymash
Andriy Luntovskyy
Mykola Beshley
Igor Melnyk
Alexander Schill

Editorial Team

	Mikhailo Klymash	Prof. Dr. habil.	Lviv National Polytechnic University, Main Building, Stepan Bandera Street 12, 79000 Lviv, Ukraine
	Andriy Luntovskyy	Prof. Dr. habil.	BA Dresden Univ. of Coop. Education, Saxon Study Academy, Hans-Grundig-Str. 25, 01307 Dresden, Germany

(continued)

(continued)

	Mykola Beshley	Ass. Prof. Dr. habil	National Polytechnic University, ITRE, Profesorska Street 2, 79000 Lviv, Ukraine
	Igor Melnyk	Prof. Dr. habil.	National Technical University of Ukraine "Igor Sikorsky Kyiv Polytechnic Institute", Main Building, Peremogy Avenue 37, 03056 Kyiv, Ukraine
	Alexander Schill	Prof. Dr. rer. nat. habil. Dr. h. c.	Dresden University of Technology, Andreas-Pfitzmann-Bau, Faculty for Computer Sciences, Noethnitzer Str. 46, 01187 Dresden, Germany

Introduction

Protocols, Platforms, Best Practices, and Energy Efficiency

Emerging Networking in the Digital Age: What does it mean and what is inside? Which paradigms are commonly used for Emerging Networking?

Emerging Networking (EmN) appears nowadays spontaneously and accompanies us in our everyday life as well as supports modern industrial processes, ongoing digitalization of workflows, and multiple smart office scenarios.

Such kinds of networks are based on up-to-date heterogeneous combinations of mobile, wireless, and fixed networks, which are augmented via ad hoc and PAN as well as IoT devices. The other important part of them is virtualization on the basis of well-known software-defined networking (SDN), which represents one of the most important paradigms for emerging networking.

Paradigms, per definition, are a set of concepts or design patterns, including theoretical and empirical models, research methods, practical tools, platforms, and standards, structure opportunities, what means a significant contribution to a field. We would like to study such paradigms for EmN in detail.

EmN supports distributed intelligence across the combined mobile, wireless, and fixed networks in Edge-to-Cloud Continuum.

Under category EmN, we also mean steady growing diversity of unmanned scenarios, where the process automation grade is nearly 100%. Operative disaster networking (emergency networking), which appears at sites, aiding rescue troops, belongs to the above-mentioned problems too. This is the next paradigm.

Intelligent vehicle automation algorithms based on analytical methods of discrete mathematics, set theory, finite state machines (FSM) theory, and mathematical statistics are developed. Such algorithms can be implemented on the sensors and onboard computers of the vehicles. They define the best time intervals, change the velocity and acceleration corresponding to the current situation on the road, and communicate to the central server, which generally controls the situation on the road. EmN can realize communication between the central server and onboard computers with built-in M2M (machine-2-machine) style, i.e., not only by Wi-Fi but also by LoRA WAN, LTE-CatM, NB-IoT, etc.

Together with the widespread Internet of Things (IoT) and digitalization challenges, EmN begins a new era in the history of networking, where the network itself is part of the digital environment, and IT translates business intentions into appropriate network configurations for all devices.

The EmN provides flexible network monitoring, performs network analytics, and demonstrates the full advantage of machine learning (ML). The EmN solutions enable better process automation and proactive problem solving as well as decentralized operation of the network infrastructure.

Furthermore, within the work, we would like to pay attention to the embedded systems in industries, entertainment, medicine, and smart office, where EmN offers many operations that can be performed automatically using intelligent control algorithms (artificial intelligence and machine learning).

As a result, network operation costs (OPEX), application response times (even up to real-time solutions), and energy consumption can be significantly reduced, network reliability and performance are improved, and network security and flexibility are enhanced. This will be a benefit for existing networks as well as evolved 5G and Beyond mobile networks, emerging Internet of Things (IoT) in cooperation with fog and cloud systems, and soon for the future 6G networks. The future mobile networks will reach a whole new level of self-awareness, self-configuration, self-optimization, self-recovery, and self-protection based on terrestrial and non-terrestrial (SAT-based) hierarchical and heterogeneous cellular infrastructures.

Embedded systems integration, modern electronic devices, and IoT deployment require an interdisciplinary approach as multiple technologies come together to the networked control algorithms, automation routines, and monitored workflows. Hence, an important part is the networked supporting tools for CAD/ CAM, modeling, simulation, and visualization of electronic signals and data processing.

The future EmN architecture includes the components of artificial intelligence (AI), Blockchain (BC), SDN, and network functions virtualization (NFV) with efficient network slicing, quality of service (QoS) management, supported via cloud computing, and advanced security approaches. So that is the next paradigm.

Virtualization, creating of "Digital Twins" based on SDN and NFV, on the other hand, can be considered as key technologies for EmN. SDN provides a programmability feature that is essential to facilitate 5G Beyond networks' operation while reducing capital and operational costs (CAPEX/ OPEX). NFV allows network resources to be virtualized and decoupled from hardware platforms, facilitating programmatic enhancement of network functionality, thereby making the network easier to deploy and more adaptable to possible changes.

We would like to develop our knowledge not only in theoretical areas of EmN, but also deepen these in the fascinating practical scenarios: from industrial robotics to current environment protection under the use of AI-based systems to the robotic space management program. EmN means applying, testing, developing, and teaching using new "Smart Applications", networked "Virtual Testbeds" for electronic devices, IoT, networked simulation and modeling tools. Do you have an interest in our work? Then you are welcome!

Furthermore, in the age of Digital Transformations Smart Applications like Smart Shopping, supported via IoT, contactless systems (RFID, NFC) and emerging combined networks (mobile, wireless, and fixed), become a fundamental challenge for urban and rural environments both. The work discusses the software and networking design of smart applications in such dynamically changing domains, such as the attractive apps for online shopping, apps for so-called smart shopping, and social shopping. Further aspects of EmN supported smart applications will be examined in the book too, inter alia, smart energy, smart office, smart home, robotics, unmanned (aerial) vehicles and flying objects, and context-sensitive intelligent applications, which are aimed at traffic telematics, health care, and entertainment.

Digitalization forces the markets to be more transparent and manageable. The psychological basis for decision-making by market participants is also being significantly changed through digitalization and smartness. The work examines, in particular, the characteristics of well-functioning markets and the possibilities of digital tools to generate or influence these characteristics.

With the emergence of new services and applications in the digital age, the networks are forced to satisfy a wide range of users and meet their needs in terms of end-to-end latency, reliability, and scalability, to promote the use of mobile devices and provide flexible and efficient network connectivity.

And, last but the least, EmN can support so-called Networked Ergonomics based on appropriate lighting and HVAC monitoring, aimed at the optimization of working environments and office communication.

With the development of EmN, customer needs and behavior have changed. The focus shifts from improving network performance to improving the perception of Quality of Experience (QoE). Providing according to the intentions of users of a given level of QoE for services and applications becomes a fundamental task in the implementation of end-to-end resource management in the known concept of intent-based networking (IBN).

The book examines multiple case studies and gives the answers to the critical questions about secured and Blockchain-supported as well as energy-efficient EmN and design paradigms. Summarizing, the following EmN features are considered:

- Small energy consumption and energy efficiency
- Wide interoperability to 4G, 5G and Beyond, WSN, RFID, NFC, Robotics, Wearable
- New efficient communications models with decentralization, i.e., M2M, fog-based models, P2P instead of convenient C-S or cloud-centric, but not only....

The work also examines 6G driving factors, features, and potentials in the mid-term. Further development of 6G in the frame of EmN will concern better coding, new antenna, and cell structures. A special meaning has the use of network softwarization based on further network function virtualization (NFV), artificial intelligence (AI), and advanced cyber-security approaches like Blockchain (BC).

The emerging mobile and wireless networks provide the best QoS (higher DR and lower latencies). The significant growth of channel power and spectral efficiency EmN promises leading to approaching known Shannon limits. The above-mentioned growth is based on:

- New radio access technologies (RATs)
- Dense heterogeneous cell structures
- Advanced spectrum spreading
- New modulation schemes
- Massive 3D-MIMO antennas and beamforming.

Future 6G networks include, moreover, further enabling technologies with better than for 5G sensing and passive localization techniques, when the distance and their derivations (velocity and acceleration) can be efficiently determined via evaluation of the propagation delays due to reflections on the objects (radar techniques, WLANp, VLC, UWB, LiFi) in multiple smart applications for unmanned PHY objects and vehicles.

Conventional network security approaches, which mean centralized solutions with service orchestration and 1:1 and 1:n paradigms and are filtered by the convenient firewalls, IDS/ IPS as well as use of public key method and PKI, are to be substituted via advanced approaches which prefer decentralized solutions, more flexible service choreography for P2P and n:n, as well as Blockchain technology and so-called Cooperative Intrusion Detection P2P Networking (CIDN).

EmN networking raises new challenges for researchers to investigate, redesign, and develop future intelligent network-based algorithms for Embedded Electronics.

The Distinguishing Features of the Book That We Feel Sets Our Book Apart From Others in the Field

- Multiple considerations of the EmN aspects based on modern approaches like IoT, IBN, SDN, NFV, smart office, and unmanned vehicles energy efficiency
- EmN can provide robust QoS and usual QoE as well as attractive, up-to-date content.
- EmN is deployed under the wide use of Blockchain technology.
- Any modern (fixed, radio, SAT-based, wireless, or mobile) vitalization technology can be configured and used as well as networking heterogeneity can be consolidated.
- EmN is energy efficient and economizing of CAPEX/OPEX, in mid-term also CO2 footprint minimizing.

- EmN content is, per definition, intrusion protected as well as workflow and transactions are secured via PKI and Blockchain as well as Cooperative Intrusion Detection P2P Networking.
- EmN is mainly used to transform users' business intent into a network configuration, operation, and maintenance strategies, which are prominent for designing the AI-enabled 6G networks.

We will provide you with a clearer demarcation of our own concepts for EmN, which are built on SDN, NFV, BC, and AI concepts with subsequent software implementation of network functions with a significant improvement in service quality, which feels like the usual Quality of Experience (QoE), which can be every time monetized. The protection against normal risks for networked data from intrusions is guaranteed. Such networks already create convenient conditions for reducing their cost and the cost of their operation (the expenditures, namely CAPEX and OPEX). In the future, EmN guarantees energy efficiency, optimal Power Usage Effectiveness (PUE), and minimizing CO2 emissions (CO2 footprint).

The analysis of modern literature highlights demonstrated a trend in the development of heterogeneous IPv4 and IPv6 networks (heterogeneous networks), which include sensor networks and the routes of IoT (e.g., NB-IoT, LTE-CatM, LoRa WAN, 6LoWPAN, EnOcean), DSL, ATM, MPLS, Wi-Fi-6, WLANp, VLN, LiFi as well as hierarchical cells 5G and Beyond combined with SpaceX/ Starlink and radio relay areas.

When distinguishing the categories of OoS, QoE, and the introduction of a specific quality parameter "Q" by the method of expert assessments, we consider it necessary to determine the parameter of network security (e.g., as the intrusion probability, the percentage of intrusion detection and intrusion prevention, etc.).

Contents

Emerging Network Technologies for Digital Transformation: 5G/6G, IoT, SDN/IBN, Cloud Computing, and Blockchain

Mykola Beshley[1](✉), Mikhailo Klymash[1], Ilona Scherm[2], Halyna Beshley[1], and Yuriy Shkoropad[1]

[1] Lviv Polytechnic National University, Lviv 79013, Ukraine
{mykola.i.beshlei,mykhailo.m.klymash,halyna.v.beshlei, yurii.i.shkoropad}@lpnu.ua

[2] BA Dresden University of Cooperative Education, Saxon Study Academy, Hans-Grundig-Str. 25, 01307 Dresden, Germany
Ilona.Scherm@ba-sachsen.de

Abstract. This chapter is one of the first approaches to systematize knowledge and describe in an accessible form emerging network technologies that are commonly associated with digital transformation. The main focus is on emerging network technologies such as the Internet of Things (IoT), Software-Defined Networking (SDN), Network Function Virtualization (NFV), (Docker) Container Networking, Cloud computing, Edge Computing, Information-centric Networking and Services, Time-Sensitive and Deterministic Networking, 5G/6G, AI-based Networking and many other contexts related to advanced communications. Practical and innovative applications of such networks are also discussed, including network security, smart cities, e-health, and smart systems. A number of key issues in the areas of Internet of Things (IoT) and intent-based networking (IBN) smart systems are discussed. Fundamental concepts and architectures for new developments in communications are proposed. The chapter looks at the impact of AI on various networking products and solutions, including embedded hardware, components, and software platforms (automation, optimization, and network transformation). It also discusses the role of SDN, edge computing, NFV and augmented intelligence in shaping and supporting AI-driven network ecosystems. A more detailed consideration of each of the outlined visions of network technologies on the essence of the term "digital transformation" makes it possible to highlight common features in the concept of future development.

Keywords: IoT · SDN · NFV · Cloud computing · 5G/6G · Artificial intelligence · IBN

1 Introduction

In today's digitized world, information is one of the main assets that determine the efficiency of public administration's economic, social and socio-political spheres [1].

M. Klymash et al. (Eds.): TCSET 2022, LNEE 965, pp. 1–20, 2023.
https://doi.org/10.1007/978-3-031-24963-1_1

Digital transformation is a continuous process of using information technology to meet customers' ever-changing needs [2]. The essence of digital transformation is to simplify routine processes by automating them [1]. In fact, digital transformation is the process of integrating new and emerging digital technologies into all aspects of the business [3]. The goal is to take advantage of these technologies to improve safety and working conditions for employees, operational efficiency, business flexibility, and manageability. Digital transformation is seen as a way to optimize operations and maintain their own competitiveness in a global and open marketplace, it is especially relevant. But the drive toward innovative digital transformation is not at all easy to understand in the face of emerging networking technologies. Innovative networking technologies such as 5G/6G, IoT, SDN/IBN reduce deployment costs and improve Quality of service (QoS) by modernizing traffic engineering processes through integrating advanced software, artificial intelligence, big data analytics, and cloud technologies [4–8]. This chapter looks at emerging network technologies that are innovative and evolving in the context of digital transformation.

Emerging network technologies in the digital transformation age are shown in Fig. 1.

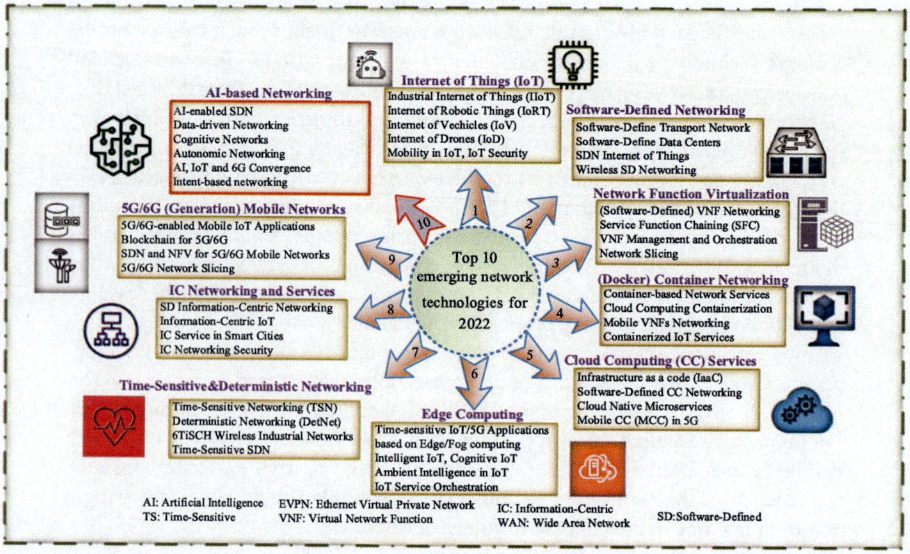

Fig. 1. Top 10 Trends of emerging network technologies for digital transformation

New emerging technologies are changing the architectures of networks and services. Key benefits of the future networks include intelligent network management, a faster process of innovation, which contributes primarily to the fast pace of technological progress and competition, increased network reliability and security, quality of experience, and harmonized access control policies.

2 Emerging Network Technologies: Concepts, Challenges, and Next Frontier for Digital Transformations

2.1 Internet of Things

The IoT is an emerging technology that creates a global network of machines and devices that are able to communicate and exchange data with each other via the Internet. IoT technology is significantly transforming the public sector into a digital form of management, fundamentally changing the way government agencies collect data and information and integrating key technical and business trends, including mobility, automation, and data analytics. The IoT connects physical objects into a network using embedded sensors, actuators, and other devices that collect and transmit real-time information about network activity. A typical smart city network consists of users, devices, and network infrastructure that connects users to multiple applications in the data center. The digital transformation of infrastructure using IoT is depicted in Fig. 2. This is followed by connected IoT devices such as sensors, cameras, and telemetry devices that support each of the groups within the smart city deployment, including utilities, healthcare facilities, educational institutions, security companies, and the like. The limited computing power of the connected facilities prevents the devices from having sophisticated security features built in. This creates two major problems: the devices are difficult to configure and easy to hack.

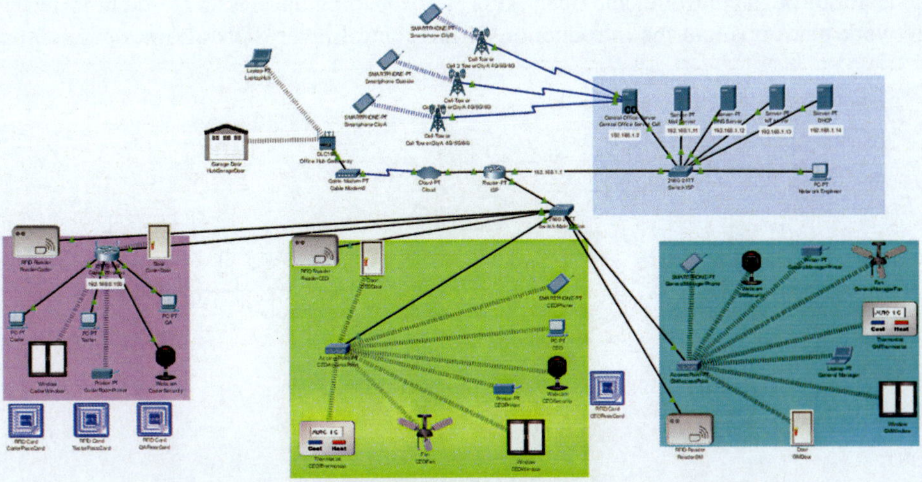

Fig. 2. An example of digital transformation of infrastructure using IoT.

Several technologies need to be used to network household items. The unique identification system allows to collect and accumulate information about objects. This can be provided by Radio-Frequency IDentification (RFID) chips. They are able to transmit information to reading devices without their own power source. Each chip has an individual number. Global Positioning System (GPS) technology is used to determine the exact location of the device. To track changes in the state of the element or the environment, objects are equipped with sensors. To process and accumulate data from the sensors, an embedded computer (for example, Raspberry Pi, Intel Edison, etc.) is used. Wireless network technologies (Wi-Fi, Bluetooth, ZigBee, 6LoWPAN, 4G, 5G, 6G) can be used to exchange information between devices [9–11].

IoT will allow the development of applications in many different domains, such as home automation, industrial automation, medical aids, traffic management, and many others. Accordingly, IoT can be classified as follows: Industrial Internet of Things (IIoT), Internet of Robotic Things (IoRT), Internet of Vehicles (IoV), Internet of Drones (IoD) [12–14].

2.2 Software-Defined Networking

Several competing forces surround enterprises, operators, and service providers. The significant growth of multimedia content, the explosion of cloud computing, the increased use of mobile devices, and the requirement to reduce costs while maintaining stable revenues are disrupting traditional business models. The key enabler that businesses will need to unlock the full potential of their digital transformation initiatives is software-defined networking (SDN) [15].

To stay on-trend, providers often consider SDN technology to fundamentally transform the structure and routine operations of the network in order to provide a high-quality digitalization of the enterprise [16]. SDN allows centralized programming of network behavior through controller and open APIs [17]. Opening the classic closed proprietary network platforms and the introduction of the control layer SDN allows operators to

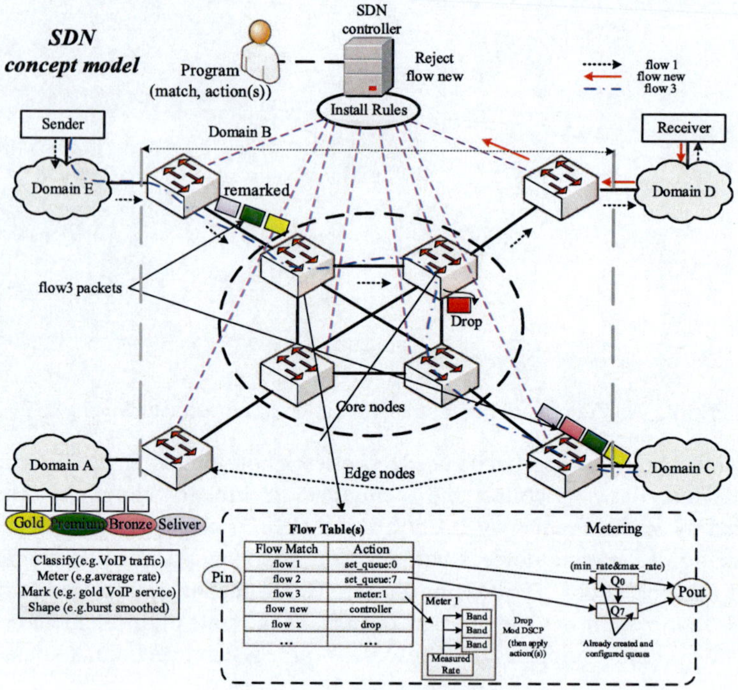

Fig. 3. The fundamental model of QoS management in SDN.

manage the entire network in a coordinated manner [18]. The fundamental model of QoS management in SDN is depicted in Fig. 3.

SDN functionality is based on four basic principles.

1. The programmability of the network. SDN allows to manage the behavior of the entire network using software installed on a central network controller. As a result, operators can customize the behavior of their networks to support new services and even specific customers. The separation of hardware from software allows operators to quickly introduce innovative, differentiated services without the constraints imposed by closed proprietary platforms [19].
2. Logically centralized intelligent management. SDN technology is based on logically centralized network topologies, providing intelligent management of network resources. Traditional methods of network management are distributed. Devices operate autonomously and are characterized by limited information about the state of the network. The centralized control provided by SDN technology can make the management of bandwidth, recovery, protection, and QoS policies highly intelligent and optimized so that the organization will have a complete view of the networks [20].
3. Network abstraction. SDN-based services and programs are separated by core technology, and the equipment that provides the physical connectivity is taken out of network management. Instead of hardware-bound management interfaces, applications interact with the network through APIs that enable general network maintenance, including routing, security, access control, bandwidth management, traffic management, quality of service, processor optimization and data storage, energy efficiency, and all forms of individual policy management to achieve business goals. For example, the SDN architecture makes it easy to define and enforce consistent policies for wired and wireless communications across campus. Similarly, SDN allows the entire network to be managed with intelligent orchestration and pre-provisioning systems.
4. Openness. The SDN architecture enables interoperability among various vendors and promotes a commercially neutral ecosystem, thereby heralding a new era of openness. Openness is at the core of SDN. Open APIs support a variety of applications, including cloud computing and network regulation solutions, as well as business-critical services. In addition, intelligent software allows the management of hardware from different vendors using open programmable interfaces such as OpenFlow. Thanks to SDN, advanced network services and applications can work in a single software environment. With SDN, IT teams can test new processes without impacting the network, making it easier to introduce new decisions as part of a digital transformation strategy.

In addition, it is possible to create their own programs in the form of additional control modules on the network controller, instead of waiting for the implementation of a function in proprietary and closed software environments in traditional networks. SDN allows users to develop programs based on the network bandwidth, monitor its status, and automatically adjust the configuration to specific conditions. SDN technology based on OpenFlow allows system administrators to solve the problems of high bandwidth, the dynamic nature of modern services, network adaptation to the ever-changing business needs, as well as the complexity of operation and management. In particular, the article gives an overview of recent research in the direction of modification, quality of service, standards and features of software-defined networks in more detail, where the authors have done a tremendous amount of review work. Currently, SDN technology is used for the development of various infrastructures and is classified into the following areas: Software-Defined Transport Network, Software-Defined Data Centers, SDN Internet of Things [21], and Software-Defined Wireless Networking.

2.3 Network Function Virtualization

The goal of digital transformation of telecom operators is to become more customer-oriented, more flexible, quickly create products and services, provide customers with a wide range of deeply customized and automated digital services. The transformation of telecom operators into digital service providers (transformation of the network and operations in accordance with new requirements) begins with SDN and NFV. A large number of highly specialized and expensive to maintain equipment prevents operators from quickly implementing and monetizing innovative services with acceptable capital and operating costs. The idea of NFV involves the transfer of network services from specialized devices to standard computer platforms with virtualization [21]. The operator can create such virtual devices at significant capacities at any point in the network. The NFV deployment model can be centralized or distributed, in which the virtualization of network functions is carried out anywhere in the network—in the data center of the operator, at the communication node and even on the user side, in the client network equipment (Fig. 4). Network functions are placed where it is most beneficial. The orchestrator is a tool for distribution across different sites of data centers and other manipulations (activation, cloning, deletion and restoration of instances of virtual functions). This system allows you to organize packages of individual services based on a combination of different virtual network functions (VNFs, virtual network functions).

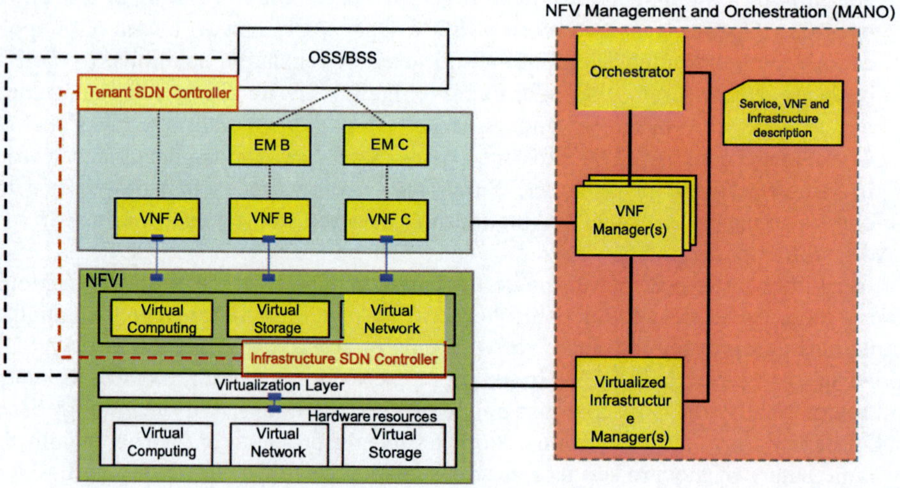

Fig. 4. NFV deployment model.

The value of this approach is assumed to be that subscribers have the opportunity to order a list of necessary network services in a specific period of time. The set of services is determined by the chain of network services. The orchestration platform allows to form chains of network services that affect the way traffic flows and is processed by network elements under the control of the SDN controller. The orchestrator and the controller interact in order to synchronize the specified set of services and network services. And here is the synergy of SDN and NFV. Network architecture, based on virtualization of network functions and supported by SDN, becomes more open, more flexible, and provides tools for automation and data analysis, which are not available in legacy networks. Transformation of the network and business model of the operator on the basis of SDN/NFV is aimed at solving tasks that traditional infrastructure cannot cope with [22]. Such tasks include the transfer of traffic growing due to the increasing number of users, devices (IoT) and data volume (video), as well as rapid deployment of new services with minimal costs. NFV does not require the abandonment of already deployed network infrastructure when creating new services, so that the "new" can coexist peacefully with the "old". Moreover, NFV makes it possible to return to the old network infrastructure if necessary, which reduces the operator's risks. Future scenarios of NFV development are associated with the transition to 5G networks and the introduction of IoT. The communications industry is confident that such networks will be built using NFV.

2.4 (Docker) Container Networking

Container networking is a virtualization method that divides applications into independent blocks. Containers are similar to virtual machines, but have several key differences [23].

Containers can run large distributed applications with low overhead. This is achieved by sharing a stripped-down operating system kernel (usually Linux-based), which makes them more efficient than virtual machines (VMs). Containers are also simpler than virtual machines.

The container is created from a prepared image assembled from a special registry, such as the Docker Hub or another private registry. To perform its functions, the platform uses a Docker daemon, which is a server or a long-running process that runs on the operating system and accepts requests from the Docker engine API.

As the daemon instances a container, it assigns it a unique network address, connecting it to a virtual Ethernet bridge to communicate between containers. In the case of Docker, the bridge is called docker0. All containers in the system communicate with each other by directing packets to docker0, which then automatically forwards those packets across the subnet.

When creating new types of applications, developers have the option of using a containerized software system built from microservices linked together using APIs.

Containerized systems take up less storage space and run faster. With microservices, each container behaves as if it were its own Web site with its own Web server. Containers can access each other's workloads through API calls. The APIs are parsed by the DNS server assigned by the container daemon.

Container orchestration systems, such as Kubernetes and Apache Mesos, are network operations centers used to manage container subnets. As a virtual subnet, a Docker or Rocket container network is a kind of SDN.

Separating the Docker subnet from the underlying network hardware is extremely important. This allows entire container networks to migrate between platforms, including from private clouds to public clouds, without change. Container networks can span multiple platforms - for example, core services can run on private premises, while overflow containers can be installed in a public cloud such as Amazon Web Services (AWS). This improves efficiency and ensures consistent bandwidth and application performance even during times of peak network load. Containers offer many advantages in managing IoT data, deploying and supporting applications and operating systems.

2.5 Cloud Computing

Cloud computing is a trending technology that provides users with access to remote computing resources via the global Internet. With this technology, the user has access to personal data, but cannot control it and does not have to take care of the infrastructure, operating system, and software with which he works. Resources can be both computing power and data storage, various services, software, and even data networks.

Cloud technologies have emerged as a response to business needs in reducing the time to deploy and scale IT and have opened up a whole range of new opportunities for businesses. The cloud model is also a means of reducing the chain of intermediaries: instead of building infrastructure, the need to communicate with dozens of hardware and software vendors, integrators and infrastructure support companies, the client needs to communicate only with the cloud service provider. In practice, of course, there may be more than one cloud service provider, and integration of cloud services from different providers may be required, but the general trend towards reducing the chain of intermediaries is obvious. For its part, the owner of a cloud platform that provides self-service tools can attract a large pool of customers with minimal service maintenance costs. In addition to the fact that the cloud can be public and private, it can be implemented on the customer's site or provided from the provider's data center. Inside the public cloud, a more secure premium delivery option is possible - the so-called virtual private cloud. Another option is when the cloud is built by a cloud provider at the customer's site and managed by this provider, in this case it is a managed private cloud. There is also an option when the integrator has built a cloud at the customer's site and finished its services, and after that the cloud is managed by the customer's internal IT services. Cloud services, in terms of the delivery model, are also divided into Software-as-a-Service (SaaS), Platform-as-a-Service (PaaS) and Infrastructure-as-a-Service (IaaS) [24].

If when using a traditional IT infrastructure, its maintenance falls entirely on the shoulders of the user, in the case of SaaS, the provider is responsible for managing all elements of the infrastructure. The SaaS service provider controls all elements of the complex system, the cloud infrastructure, including networks, servers, operating systems, middleware, etc. The provider owns the infrastructure, as well as the application itself, and leases it to the user. The PaaS model is mostly in demand by developers who receive tools for creating, testing and executing application software within the platform.

In this scheme, the user controls the application and data layer. In the case of IaaS, the customer only buys the "infrastructure from the cloud".

Mobile cloud computing (MAC) is a technology that uses the cloud to develop mobile applications. Today's sophisticated mobile applications perform tasks such as authentication, user location, providing targeted content and communication to end users. To do this, they require large computing resources such as storage capacity, memory and processing power. With the help of cloud computing, we can reduce the load from mobile devices due to the power of the cloud infrastructure. Using cloud services, developers create and update feature-rich mobile applications and then deploy them for remote access from any device. With cloud technologies, mobile applications store and process data in such a way that they can be used on all types of old and new mobile devices.

The main advantage of cloud technology is precisely the ability to accommodate unpredictable uneven resource consumption while optimizing infrastructure utilization. By consolidating servers in the cloud, individual consumption patterns from different users are balanced and capacity utilization is optimized.

2.6 Edge Computing and AI

The purpose of Edge computing is to move computing resources from a hyper-scaled cloud data center, which may be located at a considerable distance (in the "core" of the network) closer to the user or device, to the "edge" of the network.

Edge Computing enables the distribution of computing resources and application services over the channel through a decentralized computing infrastructure. Computing needs are met more efficiently when Edge Computing is used [25]. Wherever there is a need to collect data or when a user performs a certain action, it can be done in real time. In general, the two main benefits associated with edge computing are increased productivity and reduced operating costs, which are briefly described below.

Benefits of using Edge Computing:

- Improved performance
  In addition to collecting data for transfer to the cloud, edge computing also processes, analyzes, and performs necessary actions on the collected data locally. Because these processes are completed in milliseconds, it becomes essential to optimize technical data, no matter what the operations may be.

  Transferring large amounts of data in real time in a cost-effective way can be a problem, primarily when it is done from remote industrial sites. This problem is eliminated by adding smart data to devices at the edge of the network. Marginal computing brings analytical capabilities closer to the intermediary eliminator machine. This system provides less costly asset performance optimization options.
- Reduced operational costs
  The cloud computing model has connectivity, data migration, bandwidth and latency options that are quite expensive. Using edge computing requires significantly less bandwidth and lower latency. Using edge computing creates a valuable continuum from device to cloud that can handle huge amounts of data. More expensive bandwidth additions are no longer required because there is no need to transfer gigabytes of data to the cloud. It also analyzes sensitive data on the local network, thereby protecting

sensitive data. Enterprises are now opting for modern computing. This is because of their optimized performance, compliance and security protocols, and lower costs.

Edge computing helps reduce reliance on the cloud and, as a result, improve data processing speeds. In addition, there are already many advanced devices with processing power and available memory. Moving to advanced computing power allows these devices to be used to their full potential. A multi-level system of edge and cloud computing for future network is depicted in Fig. 5.

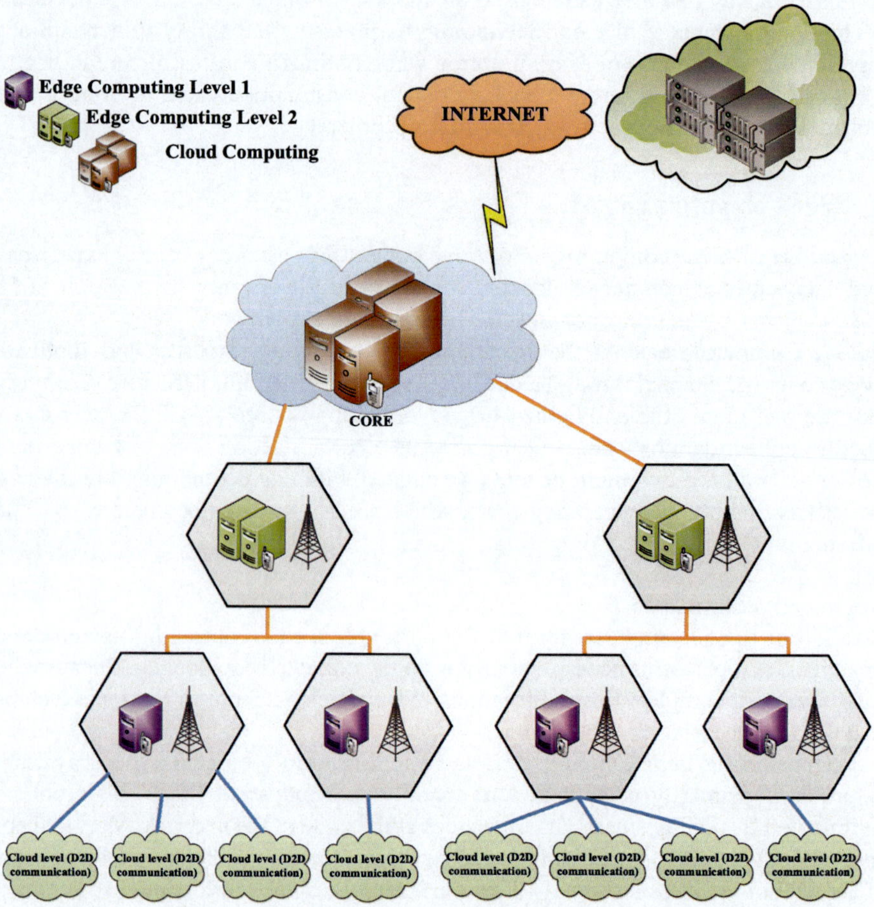

Fig. 5. A multi-level system of edge and cloud computing for future network.

Edge computing is establishing a new century computing paradigm that brings AI and machine learning to where data generation and computation take place: at the edge of the network. Combining edge computing and AI has led to a new direction: Edge AI.

Edge AI provides faster computation and insight, better data security, and effective control over continuous operation. As a result, it can improve the performance of AI-enabled applications and reduce operational costs. Edge AI can also help AI overcome related technology challenges.

Edge AI simplifies machine learning, autonomous application of deep learning models and advanced algorithms on Internet of Things (IoT) devices themselves, away from cloud services.

The best way to demonstrate the use of Edge Computing is through a few key examples. Here are some scenarios where edge computing is most useful:

– Autonomous vehicles

Autonomously driven or AI-integrated cars and other vehicles need a huge amount of data from their surroundings to work properly in real time. Delay will occur if cloud computing is used.

– Streaming services

Services such as Netflix, Hulu, Amazon Prime and Disney + put a heavy load on network infrastructure. Edge computing helps to create a more optimal approach by using edge caching: popular content is cached on objects located closer to end users for faster access.

– Smart homes

Similar to streaming services, the growing popularity of smart homes poses certain challenges. There is a very heavy load on the network that does not allow relying on cloud computing alone. Processing information closer to the source means less latency and faster response time in emergency situations. Examples are medical teams, firefighters or police.

Note that organizations can lose control over their data if the cloud is located in multiple locations around the world. This system can create problems for certain institutions, such as banks, which are required by law to store data only in their own country. Although efforts are being made to find solutions, cloud computing has clear drawbacks when it comes to cloud data security.

2.7 Information-Centric Networking and Services

Information-Centric Networks (ICN) is an approach to developing Internet infrastructure to support such use directly by introducing unique named data as a core principle of the Internet [26]. Data becomes independent of location, application, storage and means of transport, enabling intranet caching and replication. The expected benefits are increased efficiency, scalability based on information/bandwidth demand, and increased resiliency in complex communication scenarios. These concepts are known by various terms, including but not limited to: Information Network (NetInf), Named Data Network (NDN), and Publication/Subscription Network.

New security models are needed for ICN. Today's host-centric trust model - getting data from a trusted server through a secure connection - is no longer applicable. Instead, security/trust/identity functions are tied to the information objects themselves, using signed objects and ensuring the integrity of names and data. Moreover, not all objects will be universally accessible, requiring authorization and boundary-defining mechanisms.

The ICN paradigm is also expected to require new interfaces for applications to interact with the network. For example, the new API should allow application developers to take advantage of the location-independent naming, caching and multiple-access functionality of ICN.

An ICN approach would potentially have better energy efficiency than existing approaches because data is transported over shorter distances on average. Although the network needs to power more cache memory, even this can be energy efficient because the cooling load can be better distributed over a larger but smaller number of installations. The ICN has the capability to improve relevant metrics, such as invested energy per user perceived latency.

2.8 Time-Sensitive and Deterministic Networking

Time-sensitive and deterministic networks have become a future-proof technology for achieving stringent QoS guarantees, such as limited end-to-end latency and jitter, as well as higher reliability [27]. As the mass deployment of time-sensitive and deterministic networks has progressed, however, they have also presented many challenges, such as synchronous and asynchronous scheduling and shaping mechanisms, and so on. One of the most important applications for 5G/6G networks is the tactile Internet. Tactile Internet requires a 1ms end-to-end delay from future networks, which is the most challenging task in terms of creating a fifth-generation communication network. Therefore, fifth-generation communication networks are also called ultra-low latency networks.

2.9 5G/6G (Generation) Mobile Networks

Today we are at the peak of a huge digital era where innovation and new opportunities are constantly growing. The transformation process will touch virtually everything from every person, their family, and car to entire industries in many countries, transforming the way we live, work, learn and stay healthy. The global deployment of 5G networks is gaining momentum and opening up incredible opportunities. We see how 5G is transforming every aspect of our lives, industry, and society as a whole. Therefore, in this part of the chapter, we will describe what functional characteristics the emerging 6G networks will have in 2030 and what new services that are not implemented today will become a reality. Observing the development of previous generations of mobile communication systems, we can predict the directions and opportunities of the next generations. In 2G and 3G networks, the main driving force is the provision of voice services. When the penetration of mobile phones and the use of voice services reached a maximum. Aspects of creating wireless networks capable of providing multimedia services began to be considered. The dominant technology was 4G, which provided broadband communication. Over the past 10 years, significant achievements in the field of mobile data transmission and the creation of various applications have radically changed the way of life of people, directing it to the maximum digitalization of society. Global traffic growth and the lack of capacity of 4G networks led to the emergence of 5G technology [28]. This development was a step forward to provide ultra-low latency for critical services, enabling a variety of IoT applications.

The evolutionary process of communication systems, which allows for the ever-increasing volume of data transmitted on mobile networks, has gone through five generations, and the world is very close to their sixth generation (6G), which is currently being actively developed and tested [29]. Approximately every 10 years, a new generation of cellular technology appears in the world. Each new generation provides new features and services (Fig. 6).

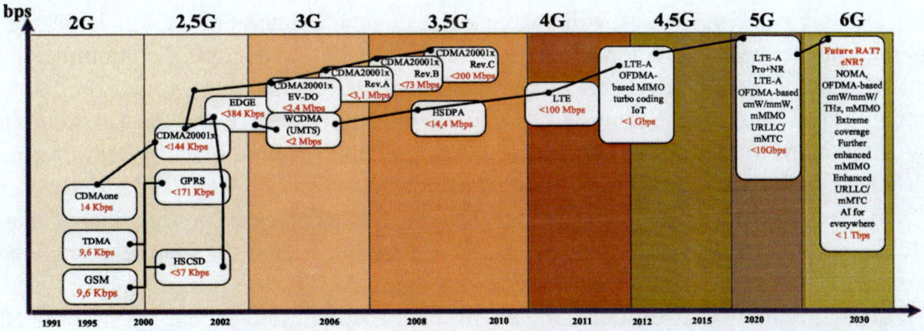

Fig. 6. Technological evolution up to 6G in mobile communications.

With the evolution of the 5G architecture, the 6G wireless network has advanced the goal of a ubiquitous intelligent revolution. In essence, 6G will serve as a human-scale neural network and a link between the physical and digital worlds. Artificial Intelligence (AI), based on machine learning, will be the foundation of 6G. In this area, our society will completely shift from connected people and connected things to connected intelligence. In other words, the 6G wireless network aims to provide an AI service to every person, home, and business, which, in turn, will create universal intelligence (intelligence of everything). In terms of wireless technology, for the first time, we can sense our environment and objects using the radio waves of wireless communication. In addition to transmitting bits of data, the 6G wireless network can serve as networked senses that extract knowledge and big data from the physical world in real-time. This extracted information will not only make a big difference in improving data transmission but will also contribute to machine learning for AI services.

Using AI and machine learning (ML) technology, 6G networks will be fully autonomous, requiring no manual operation. In this sense, 6G networks will be customized rather than proprietary, leading to the concept of "Individual mobile network on demand".

Ultra-high data rates of up to 1 Tbps are required to support new applications such as AR/VR/MR and holographic communications. The sub-THz and mm Wave bands will be the baseband spectrum for 6G cellular networks, while the lower THz band (0.3–1.0 THz) will be a prime candidate for short-haul data transmission, such as indoor or device-to-device (D2D) applications. The THz band provides ultra-wide bandwidths exceeding tens of GHz. The THz band will enable a wide range of data-intensive and delay-sensitive applications. In addition, the THz band can be used for wireless scanning. THz communication is a new wireless technology, the implementation of which is associated with many challenges. Currently, research is underway on amplifier design, new antenna materials, power RF transistors, THz receiver-on-chip architecture, channel modeling, and signal array processing. The possibility of successful application of terahertz technologies in 6G depends on technical progress in the development and production of THz-related components such as electronic, photonic, and hybrid receivers and antenna arrays on a crystal.

Compared to traditional radar technology, 6G scanning, which uses broadband spectrum and large antenna arrays, will enable technological innovations such as large-scale interaction between base stations and user devices, common design of communication and scanning devices, advanced interference suppression techniques, and the use of scanning data to aid AI. Scanning has the potential to be one of the most revolutionary innovations of 6G, on the basis of which numerous real-time machine learning and artificial intelligence applications can be created.

As the cost of manufacturing and launching satellites decreases, a huge fleet of small satellites in low earth orbit (LEO) or VLEO will become a reality. On the other hand, the use of UAVs will mean that the coverage provided by the future mobile system will no longer be flat (two-dimensional). Instead, a three-dimensional hybrid network architecture containing multiple layers and many mobile 6G base stations will enable communication and navigation services to be provided anywhere, anytime. This will require a radical change in communication protocols in terms of connection planning, resource allocation and handover of user connections, as well as wireless backhaul from the base station to the network control center. Currently, UAVs are designed and operated separately, but it is expected that in future 6G networks their functions and operations, as well as resource and mobility management, will be tightly integrated. Such an integrated system will identify each user device with a unique identifier, unify billing processes for individual requirements and consistently deliver high-quality services through optimal access points. To easily integrate a new UAV or LEO satellite without the need for manual configuration, the integrated network must have a self-organizing function. With an intelligent radio interface device, adding and removing a base station will be transparent to user devices in terms of physical layer procedures (such as beamforming, measurement and feedback). Given that the deployment, maintenance, and power supply of UAVs are completely different from terrestrial networks, new operational and business models can be expected to emerge with the ability to organize green solar-powered networks. However, as AI spreads into data centers, edge nodes, and mobile devices, the energy consumption associated with AI learning and mastery becomes a key issue that needs to be addressed now.

By leveraging the flexibility of software-defined networking (SDN) and network functions virtualization (NFV), a user-centric core network enables each client device to obtain its own "private" virtual instance of the core network through a decentralized structure of virtual instances implemented as virtual machines or containers. For a decentralized network, nodes can communicate with each other using certain algorithms, such as distributed hash tables or blockchain. To achieve high efficiency, 6G networks must have simple management and operation mechanisms. At the same time, availability and reliability should not only be maintained but also improved. In this regard, availability can be improved by implementing blockchain technology at all levels of the network, the so-called Metaverse blockchain. That is, the 6G network should be able to withstand a certain level of failures or malicious attacks without requiring significant human intervention. The blockchain, combined with smart contracts, allows business logic and business process mechanisms to be integrated into the deployment of individual 6G networks [30].

The convergence of AI, blockchain, IoT, SDN, NFV for 6G is depicted in Fig. 7.

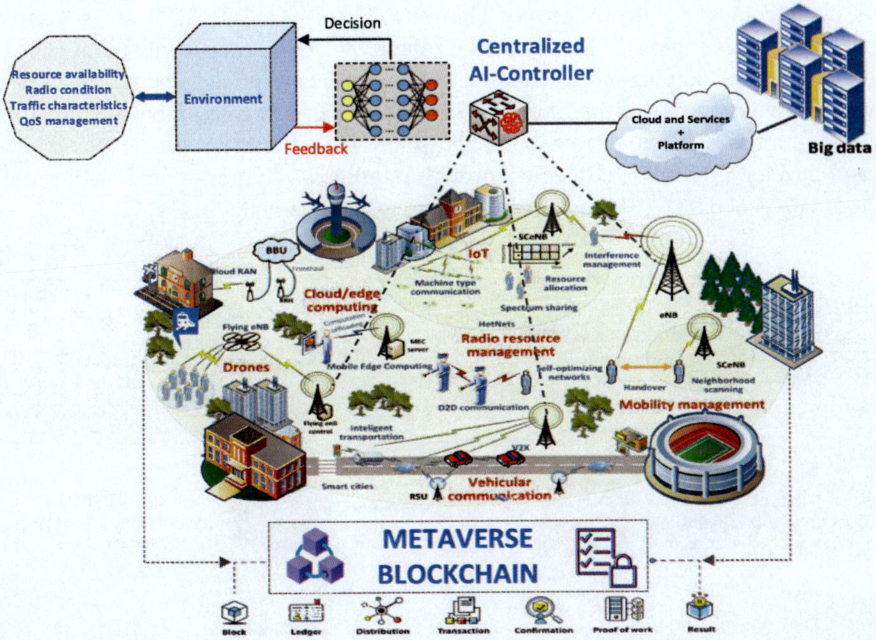

Fig. 7. The convergence of AI, blockchain, IoT, SDN/NFV, Cloud/Edge computing, Big Data for 6G.

In the next decade we will enter an era in which humans, things, and artificial intelligence will be connected. The communication system will perceive the physical world by analyzing the propagation of its own signals and form a digital vision of the environment for use by artificial intelligence. To shape the vision of 6G and define the technologies that will make 6G a reality, it is essential that industry, academic and system actors work together as they have done before in 4G and 5G. The standardization of 6G worldwide will undoubtedly lay the foundation for successful development for decades to come and will make significant progress in the new era of digital transformation.

2.10 AI-Based Networking

Artificial intelligence has become an essential part of modern digital technology, embedding intelligent software into the control logic of network infrastructure. AI is expected to become one of the main applications for emerging network infrastructure as providers look for ways to reduce costs and improve operational efficiency while reducing the number of errors and adverse network events. AI for emerging networks is an innovative research area that mainly focuses on intelligent management to optimize and improve resource utilization in wireless, wired, and heterogeneous information and communication systems.

Today, exciting research approaches are the development of AI for UAV communications, cloud computing, the Internet of Things (IoT), 5G/6G, SDN/IBN, and security. The emerging AI technology will play a vital role in the digital transformation process, providing automated intelligent management of business enterprise resources by creating the cognitive networks of the future [31]. In particular, using cognitive networks in business will allow enterprises to expand and improve their current products and services and develop innovative strategies through intelligent monitoring and analysis of collected statistical data. AI-based networking concept is depicted in Fig. 8.

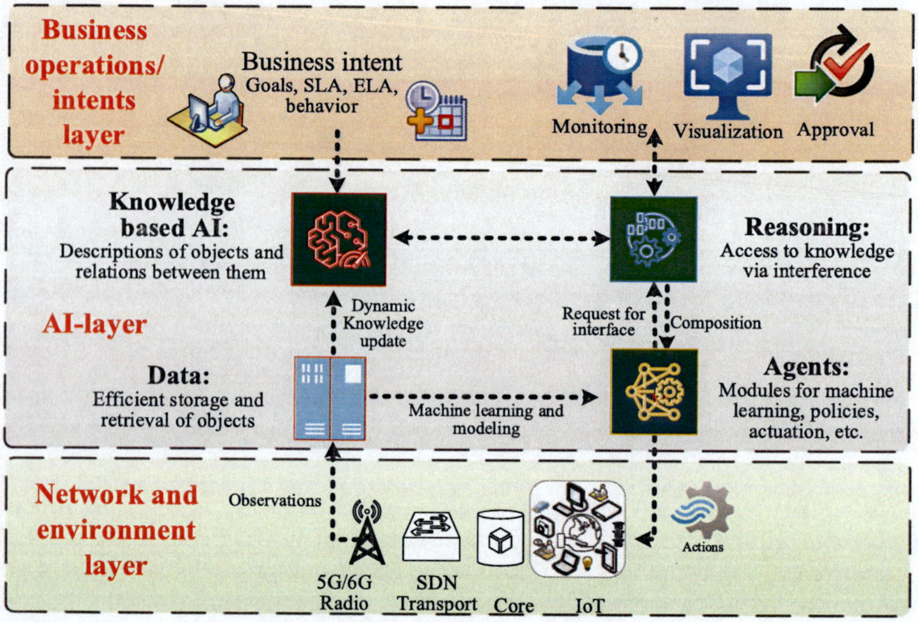

Fig. 8. AI-based networking concept [11]

For networks to be autonomous, a certain level of freedom must be allowed to choose their actions. One such approach is to use intents to manage network resources. The intent is defined as a specific logical utility function that is understandable to end users focused on performing specific actions on the network. In their individual accounts, the operator and users can form the necessary intent or business goal. The network translates the intents into the configuration parameters necessary to achieve those goals using specialized software. In doing so, unlike traditional solutions, there is no more coding and manual configuration of network devices [32].

In AI-based networks, the system automatically generates specific device configurations in the form of specific scenarios based on intelligent intent analysis. This network then continuously checks the current state against the target state by monitoring configuration parameters using self-regulated processes. For example, within an autonomous system, there may also be conflicting situations where one intent is responsible for keeping the service with high QoE and another intent is to minimize energy consumption. With a specialized knowledge base AI, such conflicts can be resolved either explicitly based on relative importance weights or implicitly based on the predictive properties of the outcome defined in utility-level goals. Ensuring the quality of service is derived from SLA/ELA (Service Level Agreements/Experience Level Agreements) contracts, or business strategies and an orderly level of QoS in the form of user intent is guaranteed under changing load fluctuations. Machine learning plays a key role in processing intents due to its ability to understand abstract concepts from various fields and provide accurate conclusions based on statistical data collected by the monitoring system.

According to the ideology described above, this trending technology is called intent-based network (IBN), which is able to compute application and user behavior, learn and reconfigure on-the-fly to minimize operational costs, and improve service quality and network security. It is known that SDN will play a special role in the development of IBN through a combination of automation, centralized management, and artificial intelligence. In academia, IBN is still under development and is closed for use by enterprises in the digital transformation process. Thus, IBN is among the latest emerging trends of the top 10 networking technologies, which are of extraordinary interest to the academic community in the era of digital transformation.

3 Conclusions

The world is undergoing profound changes with the advent of the digital era of intellectual transformation. Digitization is related to the organization of intelligent transport, entertainment, healthcare, and "smart city" and will contribute to improving the quality of service. The chapter describes how emerging network technologies, such as blockchain, IoT, Cloud computing, data analytics, AI and 6G can digitalize society everywhere. Combined with existing 5G technologies, some key radical 6G technologies will guarantee high-quality services to achieve ubiquitous wireless connectivity, from the telecommunications industry to digital smart industries. The 6G networks that are evolving are set to be a game changer for the next generation of wireless and wired communication systems that will address the limited data rates that are growing along with the billions of data applications that traditional networks face. The chapter explains that the IoT technology together with the 6G architecture will reform the current industry into an intelligent infrastructure. The use of blockchain technology is relevant for the networks of the future due to decentralization, transparency, lack of spectrum resources, privacy, and security.

It is established that networks will gradually become cognitive systems using artificial intelligence. Such networks will be able to feel, think, acquire new knowledge and act autonomously. They will be fully controlled by intent-driven technologies, simplifying the human task of defining services and operational goals, while becoming the supporting infrastructure for a truly digital society.

Big data-driven operations, distributed intelligence, continuous learning, and intent-based automation will be key drivers of future digital transformation.

Acknowledgement. This work was supported by the Ukrainian project № 0120U100674 "Development of the novel decentralized mobile network based on blockchain-architecture and artificial intelligence for 5G/6G development in Ukraine". This work was supported by the Ukrainian project No. 0120U102201 "Development the methods and unified software-hardware means for the deployment of the energy-efficient intent-based multi-purpose information and communication networks".

References

1. Tran-Dang, H., Kim, D.-S.: The physical internet in the era of digital transformation: perspectives and open issues. IEEE Access **9**, 164613–164631 (2021). https://doi.org/10.1109/ACCESS.2021.3131562
2. Baslyman, M.: Digital transformation from the industry perspective: definitions, goals, conceptual model, and processes. IEEE Access **10**, 42961–42970 (2022). https://doi.org/10.1109/ACCESS.2022.3166937
3. Klymash, M., Beshley, M., Luntovskyy, A. (eds.): Future Intent-Based Networking. LNEE, vol. 831. Springer, Cham (2022). https://doi.org/10.1007/978-3-030-92435-5
4. Selmchenko, M., Beshley, M., Panchenko, O., Klymash, M.: Development of monitoring system for end-to-end packet delay measurement in software-defined networks. In: 2016 13th International Conference on Modern Problems of Radio Engineering, Telecommunications and Computer Science (TCSET), pp. 667–670 (2016). https://doi.org/10.1109/TCSET.2016.7452147
5. Klymash, M., Seliuchenko, M., Beshley, M., Redchuk, S.: Increasing wavelengths utilization efficiency in OTNoDWDM network based on local resource distribution method. In: 2015 Second International Scientific-Practical Conference Problems of Infocommunications Science and Technology (PIC S&T), pp. 157–160 (2015). https://doi.org/10.1109/INFOCOMMST.2015.7357300
6. Kryvinska, N., Strauss, C.: Conceptual model of business services availability vs. interoperability on collaborative IoT-enabled eBusiness platforms. In: Bessis, N., Xhafa, F., Varvarigou, D., Hill, R., Li, M. (eds.) Internet of Things and Inter-cooperative Computational Technologies for Collective Intelligence. Studies in Computational Intelligence, vol. 460, pp. 167–187. Springer, Heidelberg (2013). https://doi.org/10.1007/978-3-642-34952-2_7
7. Klymash, M., Savchuk, R., Pozdnyakov, P., Beshley, M.: The researching and modeling of structures of mobile networks for providing of multiservice radio access. In: Proceedings of International Conference on Modern Problem of Radio Engineering, Telecommunications and Computer Science, pp. 281–282 (2012)
8. Kryvinska, N., Lepaja, S., Nguyen, H.M.: Service and personal mobility in Next Generation Networks. Mob. Wirel. Commun. Netw. (2003)

9. Beshley, M., Kryvinska, N., Seliuchenko, M., Beshley, H., Shakshuki, E., Yasar, A.: End-to-End QoS "Smart Queue" management algorithms and traffic prioritization mechanisms for narrow-band internet of things services in 4G/5G networks. Sensors **20**(8), 2324 (2020). https://doi.org/10.3390/s20082324
10. Chettri, L., Bera, R.: A Comprehensive survey on internet of things (IoT) toward 5G wireless systems. IEEE Internet Things J. **7**(1), 16–32 (2020). https://doi.org/10.1109/JIOT.2019.2948888
11. Beshley, M., Kryvinska, N., Beshley, H., Medvetskyi, M., Barolli, L.: Centralized QoS routing model for delay/loss sensitive flows at the SDN-IoT infrastructure. Comput. Mater. Continua **69**(3), 3727–3748 (2021)
12. Wang, L.-L., Gui, J.-S., Deng, X.-H., Zeng, F., Kuang, Z.-F.: Routing algorithm based on vehicle position analysis for internet of vehicles. IEEE Internet Things J. **7**(12), 11701–11712 (2020). https://doi.org/10.1109/JIOT.2020.2999469
13. Pu, C., Wall, A., Choo, K.-K.R., Ahmed, I., Lim, S.: A lightweight and privacy-preserving mutual authentication and key agreement protocol for internet of drones environment. IEEE Internet Things J. **9**(12), 9918–9933 (2022). https://doi.org/10.1109/JIOT.2022.3163367
14. Beshley, H., Klymash, M., Beshley, M., Kahalo, I.: Improving the efficiency of LTE spectral resources use by introducing the new of M2M/IoT multi-service gateway. In: 2019 IEEE 15th International Conference on the Experience of Designing and Application of CAD Systems (CADSM), pp. 114–117 (2019). https://doi.org/10.1109/CADSM.2019.8779270
15. Huang, X., Cheng, S., Cao, K., Cong, P., Wei, T., Hu, S.: A survey of deployment solutions and optimization strategies for hybrid SDN networks. IEEE Commun. Surv. Tutorials **21**(2), 1483–1507 (2019). https://doi.org/10.1109/COMST.2018.2871061
16. Romanchuk, V., Beshley, M., Polishuk, A., Seliuchenko, M.: Method for processing multi-service traffic in network node based on adaptive management of buffer resource. In: 2018 14th International Conference on Advanced Trends in Radioelecrtronics, Telecommunications and Computer Engineering (TCSET), pp. 1118–1122 (2018). https://doi.org/10.1109/TCSET.2018.8336390
17. Beshley, M., Panchenko, O., Demydov, I., Seliuchenko, M.: Method of dynamic quality management of services in the integrated SDN/CLOUD network. In: Physical and Technological Problems of Information Transmission, Processing and Storage in Materials: Materials V International Scientific and Practical Conference, 3–5 November 2016 in Chernivtsi, pp. 74–75 (2016)
18. Jun, S., et al.: A cost-efficient software based router and traffic generator for simulation and testing of IP network. Electronics **9**(1), 40 (2020). https://doi.org/10.3390/electronics9010040
19. Amin, R., Reisslein, M., Shah, N.: Hybrid SDN networks: a survey of existing approaches. IEEE Commun. Surv. Tutorials **20**(4), 3259–3306 (2018). https://doi.org/10.1109/COMST.2018.2837161
20. Casellas, R., Vilalta, R., Martínez, R., Muñoz, R.: Highly available SDN control of flexi-grid networks with network function virtualization-enabled replication. J. Opt. Commun. Network. **9**(2), A207–A215 (2017). https://doi.org/10.1364/JOCN.9.00A207
21. Beshley, M., Kryvinska, N., Beshley, H., Kochan, O., Barolli, L.: Measuring end-to-end delay in low energy SDN IoT platform. Comput. Mater. Continua **70**(1), 19–41 (2021)
22. Xu, Z., Liang, W., Huang, M., Jia, M., Guo, S., Galis, A.: Efficient NFV-enabled multicasting in SDNs. IEEE Trans. Commun. **67**(3), 2052–2070 (2019). https://doi.org/10.1109/TCOMM.2018.2881438
23. Qi, S., Kulkarni, S.G., Ramakrishnan, K.K.: Assessing container network interface plugins: functionality, performance, and scalability. IEEE Trans. Netw. Serv. Manage. **18**(1), 656–671 (2021). https://doi.org/10.1109/TNSM.2020.3047545

24. Nadeem, F.: Evaluating and ranking cloud IaaS, PaaS and SaaS models based on functional and non-functional key performance indicators. IEEE Access **10**, 63245–63257 (2022). https://doi.org/10.1109/ACCESS.2022.3182688
25. Beshley, M., Klymash, M., Hamal, M., Shkoropad, Y., Branytskyy, A.: Method for estimating service delay in edge and cloud computing architecture. In: 2020 IEEE 15th International Conference on Advanced Trends in Radioelectronics, Telecommunications and Computer Engineering (TCSET), pp. 915–919 (2020). https://doi.org/10.1109/TCSET49122.2020.235570
26. Qin, J., Xue, K., Li, J., Sun, Q., Lu, J.: Service prioritization in information centric networking with heterogeneous content providers. IEEE Trans. Netw. Serv. Manage. **18**(4), 4476–4488 (2021). https://doi.org/10.1109/TNSM.2021.3105198
27. Chahed, H., Kassler, A.J.: Software-defined time sensitive networks configuration and management. In: 2021 IEEE Conference on Network Function Virtualization and Software Defined Networks (NFV-SDN), pp. 124–128 (2021). https://doi.org/10.1109/NFV-SDN53031.2021.9665120
28. Beshley, H., Beshley, M., Medvetskyi, M., Pyrih, J.: QoS-aware optimal radio resource allocation method for machine-type communications in 5G LTE and beyond cellular networks. Wirel. Commun. Mob. Comput. **2021**, 9966366-1–9966366-18 (2021)
29. Alwis, C.D., et al.: Survey on 6G frontiers: trends, applications, requirements, technologies and future research. IEEE Open J. Commun. Soc. **2**, 836–886 (2021). https://doi.org/10.1109/OJCOMS.2021.3071496
30. Maksymyuk, T., Klymash, M., et al.: Blockchain-empowered framework for decentralized network management in 6G. IEEE Commun. Mag. **58**(9), 86–92 (2020). https://doi.org/10.1109/MCOM.001.2000175
31. Beshley, M., Klymash, M., Beshley, H., Urikova, O., Bobalo, Y.: Future intent-based networking for QoE-driven business models. In: Klymash, M., Beshley, M., Luntovskyy, A. (eds.) Future Intent-Based Networking. LNEE, vol. 831, pp. 1–18. Springer, Cham (2022). https://doi.org/10.1007/978-3-030-92435-5_1
32. Abbas, K., Khan, T.A., Afaq, M., Song, W.-C.: Network slice lifecycle management for 5G mobile networks: an intent-based networking approach. IEEE Access **9**, 80128–80146 (2021). https://doi.org/10.1109/ACCESS.2021.3084834

A Framework for Context-Sensitive Control-Loops with Roles

Ilja Shmelkin(✉), Alexander Schill, and Alexander Kropp

Technische Universität Dresden, 01062 Dresden, Germany
ilja.shmelkin@tu-dresden.de

Abstract. Control loops for self-adaptation are one approach to reduce the complexity of information technology systems by splitting a system into a managing and a managed part. While the managing part comprises the actual domain-specific application logic, the managed part acts as its overseer by monitoring sensors, analyzing operational data, as well as planning and executing corrections in case the application logic does not function as expected. To date, two challenges remain in this area. First, in most cases, the managing part and the managed part were specifically designed to work together, leaving little room for replacing the managed part without redesigning and reimplementing the managing part. As a result, the managing functions are tied to single application domains and require a large amount of research and development to apply self-adaptation to new application domains. Second, even within the same application domain, small changes in application logic often prohibit the reuse of the same management logic. In this chapter, we propose a solution to both challenges. First, we introduce a framework for creating a control-loop based managing system that is capable of managing systems from different application domains. Second, by applying the concept of roles in our framework, we provide a basis for developers to flexibly add or adjust features in the existing code when small changes occur in the managed system, or to replace entire parts of the management functions when large parts of the managed system change. We discuss the implementation of our framework in different scenarios as well as the application of our framework to different use cases within different application domains.

Keywords: Context sensitivity · Control loops · MAPE-K · Flexible software

1 Introduction

One challenge in creating and maintaining information technology (IT) systems and networks of IT systems is their ever growing complexity. Today, highly specialized and complex IT systems in many application domains are still managed

This work is funded by the German Research Foundation (DFG) within the Research Training Group Role-based Software Infrastructures for continuous-context-sensitive Systems (GRK 1907).

M. Klymash et al. (Eds.): TCSET 2022, LNEE 965, pp. 21–48, 2023.
https://doi.org/10.1007/978-3-031-24963-1_2

and maintained by individual domain experts. To face this challenge and reduce the effort of maintaining IT systems, the field of self-adaptive systems emerged. With self-adaptation, the IT system itself collects additional data about its operational environment, dynamics in the availability of resources, changing system goals, and malfunctions to manage itself [20]. Control loops are one popular approach to design and implement self-adaptivity in IT systems.

1.1 Problem Analysis

With control loops for self-adaptation, the most popular of which is the MAPE-K (control loop) [4,15,21], where a system is divided into a managing and a managed part. The managing part consists of a five-stage control loop (described in Sect. 2.2) while the managed part comprises the domain-specific application logic. To date, two main challenges have arisen from using such system architecture.

1. As most of the time, both the managing and managed part were specifically developed to function together, re-using the managing logic together with the application logic from other domains is difficult [15]. This results in the necessity to re-design and re-implement the managing control loop for every new application domain that can profit from self-adaptivity.
2. Also, small differences (e.g., data formats, transport protocols, etc.) between managed IT systems from the same application domain (e.g., different monitoring systems) already prohibit re-using the same managing logic. Additionally, adding support for a large variety of very similar managed IT systems results in large, complex, and incomprehensible software.

1.2 Chapter Contributions

With this chapter, we want to propose a solution for both challenges. First, by introducing a framework for creating a control-loop based managing system that is able to manage IT systems from a variety of application domains, we remove the strong cohesion that exists between current managing and managed systems.

Second, by applying the role concept to our framework, we create a basis for developers to flexibly add or adjust functionality in existing code in the case when small changes occur in the managed system or replace whole parts of the management capabilities when large portions of the managed system change.

The remainder of this chapter is structured as follows: In Sect. 2, we present the necessary background information for understanding our contributions. This includes an explanation of control loops for self-adaptive systems and the role concept. Following that, we briefly discuss related approaches for engineering self-adaptive systems as well as research on creating flexible software systems with the use of design patterns in Sect. 3. In Sect. 4, we present our framework in detail. We discuss the application of our approach in Sect. 5 in a case study. For this, we apply our framework to different use cases from two different application domains and briefly discuss performance data from an upstream evaluation of our framework applied to one use case. Finally, we conclude this chapter in Sect. 6 and discuss opportunities for future work.

2 Background and Basics

In this section we want to introduce two main concepts to the reader. First, we will provide general background information on self-adaptive systems and further explain the use of control loops to engineer *external* self-adaptive systems. Next, we will provide a general idea of the notion of roles for modeling and implementing software. As we explain those concepts, we will use a running example to better illustrate them to the reader. Over the course of this chapter, we will build up the running example further to apply our framework to it in Sect. 5.1.

2.1 Running Example

Our running example consists of a monitoring system that is capable of monitoring the processing load that is applied to an infrastructure of servers. Within the monitoring system, we distinguish different components which are affiliated with certain tasks that are covered by state-of-the-art monitoring systems [15]. Those tasks are:

1. Acquiring, filtering, and saving data (e.g., CPU load) from the monitored system(s) (one or multiple servers) to a persistent database.
2. Analysis of queried data in real-time via simple threshold analysis.
3. When thresholds are exceeded or subceeded, a predefined action is triggered that notifies an administrator or web-based API.

In modern-day monitoring systems, an additional web-based API is often used to execute changes in the monitored infrastructure (e.g., change the availability of resources, enable or disable functions, etc.). Such API is not usually part of the monitoring system but acts as a necessary addition.

Consider Fig. 1 for a visual representation of our running example. From Fig. 1, we can derive additional properties of this generic monitoring system that apply to most modern-day monitoring systems. First, we see that the acquisition of data from the monitored infrastructure is performed via one specific protocol (i.e., *COMPROTa*) and that the data is acquired via a log file, from which one line at a time is transferred in a fixed re-occurring interval (e.g., every second) from the monitored server to the acquisition component of the monitoring system. Then, the acquisition component filters and transmits valid lines of the log file towards a database for persistence and analysis component for real-time evaluation. The evaluation of gathered data occurs via thresholds that are defined by the user of the monitoring system. If a threshold is exceeded or subceeded, two notifications are sent, one towards an administrator via E-Mail over SMTP and another towards a web-based API via HTTP/S. Other than the properties of the monitoring system, the functionality of the API can be implemented freely

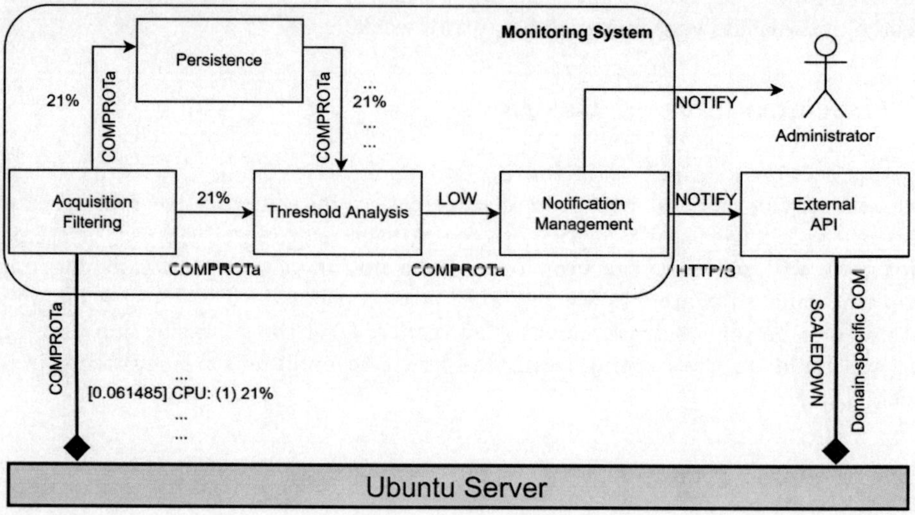

Fig. 1. Generic monitoring system with three components for data acquisition, data analysis, and action planning with one additional web-based API for action execution.

by the user of the system. That means, while the user can decide freely how to react on received notifications (e.g., adding or removing computing resources, changing datacenters, shutting down the system, etc.), the user has no control over:

- The communication protocol used for data transport (e.g., possible protocols are HTTP/S, message queues, gRPC, SNMP, etc.).
- The format of representation used for monitored data (e.g., possible formats are metrics, events, log files, checks, binary formats).
- The interval at which data are queried, the amount of data that are queried within one re-occurring interval, or how the filtering mechanisms function exactly.

This reveals the main shortcoming of modern-day monitoring systems, namely that they are inflexible and tailored towards specific application domains or even specific applications [15].

In the upcoming sections, we will introduce control loops and the notion of roles into this running example to frame the contributions of this chapter that will solve this issue.

2.2 Control Loops

Control loops are one popular approach to designing and implementing self-adaptive systems. Self-adaptive systems that use dedicated control loops for their management are often referred to as external self-adaptive systems as opposed to internal self-adaptive systems whose adaptation logic is scattered around the program code of the managed system [20]. The main benefit of control loops for external self-adaptive systems is the use of multiple clear cut components, each dedicated to performing a certain management task. The most prominent example for such control loop is the MAPE-K loop, which was first described by IBM [4], based on the idea of an autonomic manager by Kephart and Chess [10].

The MAPE-K Control Loop consists of five components that, by its original definition, work together in a fixed sequence of control tasks on a high level of abstraction to manage an underlying *managed subsystem* [20]. Those components are:

- **Monitoring**, for gathering data from the managed subsystem, filtering of those data and transporting them to a knowledge- and analysis component.
- **Analysis**, for performing evaluation (e.g., quantitative or statistical) of monitored data against a set of system goals.
- **Planning**, for calculating a sequence of actions to solve potential problems in the system based on the results of the upstream analysis.
- **Execution**, for performing this sequence of actions on the managed subsystem.
- And finally, **Knowledge**, for providing a persistent database to the other components for saving data that are important for future loop-iterations.

Such a division can often be found in software design without being explicitly labeled as such. Returning to the running example of this chapter, we can distinguish all five MAPE-K components within the parts of a monitoring system that we described earlier, as visualized in Fig. 2.

The mapping of the functions which a monitoring system usually provides onto the components of a MAPE-K control loop is intuitive. The only difference to our earlier description is that the execution of actions based on received notifications is now natively included within the execution component instead of being outsourced to a web-based API.

The MAPE-K control loop is defined so that one loop is responsible for managing one system. In monitoring, however, one monitoring system often covers the surveillance of multiple similar nodes (e.g., a set of servers). Adding to the idea of a MAPE-K based monitoring system, we need to extend our visualization (Fig. 2) to support monitoring of multiple managed subsystems.

MAPE-K Patterns add two important concepts to the base definition of the control loop which ease that task.

First, Weyns et al. recognized the need for having a framework for describing re-occurring patterns in self-adaptive systems. Most importantly, the authors recognized that in practice, MAPE-K based systems often use multiple control loops instead of only one. Additionally, nodes often do not include all five components but rather focus on specific tasks like monitoring and execution, while

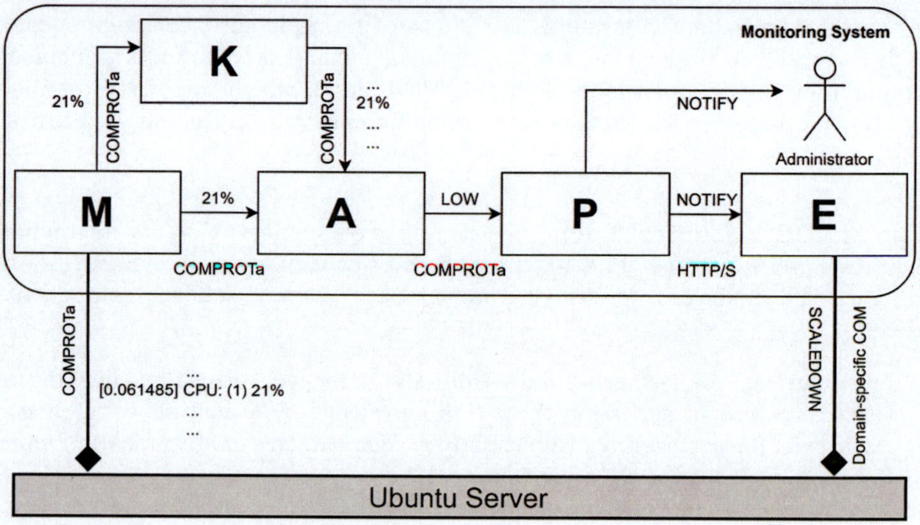

Fig. 2. Generic monitoring system from Fig. 1 modeled with MAPE-K control loop components.

another node may perform centralized analysis and planning for the whole system of control loops. This includes describing components of the same type that communicate and operate towards one common goal [21].

Second, Shmelkin and Springer identified that regardless of the grade of distribution and actual placements of MAPE-K components in a distributed system of control loops, the communication paths between components of the same type (e.g., monitoring to monitoring) and between components in the original sequence of the control loop invariably stay the same. Therefore, the authors defined interface descriptions for MAPE-K patterns [16].

By adding those concepts to our running example, we can now distinguish two[1] control loops running across several nodes (see Fig. 3).

- The **main loop**, that includes an aggregating monitoring component, centralized analysis- and planning components and one knowledge component.
- One (or more) **secondary loop(s)**, that include(s) an agent[2] monitoring component and an execution environment.

There can only be one main loop, which provides the functions of a monitoring system, whereas there is no limit to the number of secondary loops that can be present, i.e., one per monitored system. The secondary (sub) loops contain an agent, which measures properties of the monitored system and transfers them

[1] This particular infrastructure consist of one main loop and any number of secondary loops.

[2] In the monitoring domain, software that collects data from a system and provides them to a central component is referred to as an *agent*.

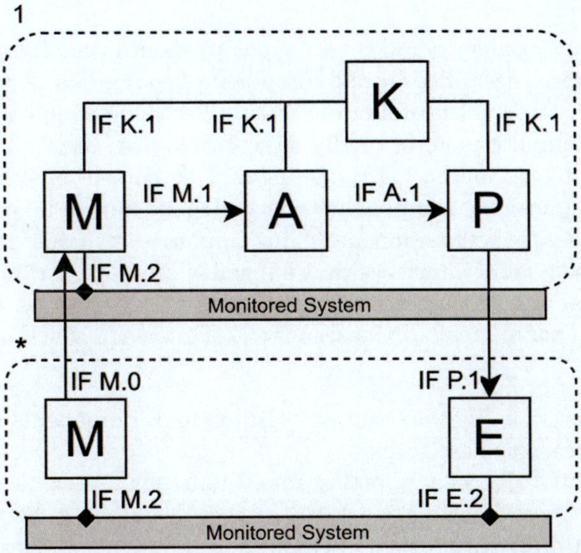

Fig. 3. Generic monitoring system from Fig. 1 and Fig. 2 modeled with one main- and sub loop and MAPE-K interfaces.

to the monitoring system (i.e., the main loop) over an interface (IF M.0) and an execution environment, that allows receiving notifications (IF P.1) from the monitoring system and acting upon those accordingly.

The interfaces are named systematically. Communication interfaces are always assigned the first letter of the respective senders component type and a suffix that determines the interface type. Interfaces for communication between components of the same type carry the suffix .0 (M.0, A.0, P.0, E.0), whereas the suffix .1 (M.1, A.1, P.1, K.1) is used for communication in sequence of the MAPE-K control loop, the suffix .2 (M.2, E.2) is used for communication between monitoring- and monitored system, and .3 (M.3) is used for communication between the system and its environment.

With that, our running example now includes the following advantages compared to its earliest iteration:

- It has a defined architecture of abstract components that separate the concerns of a monitoring system.
- Due to having five distinguished component types whose copies can be arranged almost arbitrarily, the running example can be used seamlessly in any distributed infrastructure.
- As communication happens over a maximum of eleven different abstract interfaces with clear-cut separation of concerns, all possible communication paths are predefined [16].

However, as framed in our introduction, the archetype of MAPE-K based systems also has downsides.

Our running example includes two types of monitoring components. While their main purpose (described by the component type) is the same, i.e., handling monitored data, their actual functionalities differ. One acquires data from the monitored nodes and one additionally aggregates those data in the main node. Currently, we have no method of modeling and implementing different behavior of the generic monitoring component type in dependence of its *context* (i.e., the concrete use case of the component). This implies that, while having a framework of components and interfaces that allow any kind of distributed monitoring infrastructure, all functions need to be specifically tailored towards the concrete use case of each component of the resulting architecture. Such approach would be:

- **inflexible**, as changing the function of individual components has to be investigated on a case to case basis,
- **incomprehensible**, as supporting many different use cases at once implies large and incomprehensible software (further described in Sect. 3),
- and **hard to maintain**, as small changes in the monitored system imply the need to re-implement or change code that was created on a case-to-case basis.

In the following section, we will introduce a method for modeling and implementing *behavior of MAPE-K components and interfaces under context* with roles that solves those issues.

2.3 Role-Based Systems

The concept of roles under context has been recently used to naturally model and implement complex, context-dependent and dynamic domains [11,12]. The original definition of roles under context by Bachmann and Daya dates back to 1977 and has been revised and updated by many researchers since then [1]. Sowa first distinguished between *natural types* and *role types* and later strengthened his new modeling concept by asserting that role types are a subtype of natural types [17,18]. Guarino separated these concepts even more by identifying that natural types are characterized by semantic rigidity and in contrast, role types are not [7].

Within our running example, we can distinguish several natural types. That includes, first and foremost, the five components of the MAPE-K control loop. Each of the components is semantically rigid in their respective task, i.e., the monitoring component is always concerned with receiving data, filtering it, and transmitting it towards connected components. Furthermore, an infrastructure of such rigid components directly implies the communication paths that are present in the system, as depicted in Fig. 3. Consequently, each component includes a subset of eleven different interfaces that are also semantically rigid.

While the general semantics of a system architecture based on this framework are determined by its composition of rigid components and their connections via rigid interfaces, the actual functionality of each part of the system is left open. This encourages the usage of roles to provide context-sensitive implementation of concrete functionality. With roles, each part of the system (components and interfaces) can be provided with new or adjusted functionality that is tailored towards a specific use case (the context). Roles are *played by* the rigid system parts and, with that, the system parts gain the roles' properties and methods. If the context of the infrastructure changes, so do the roles and, with that, the functionality of the whole system.

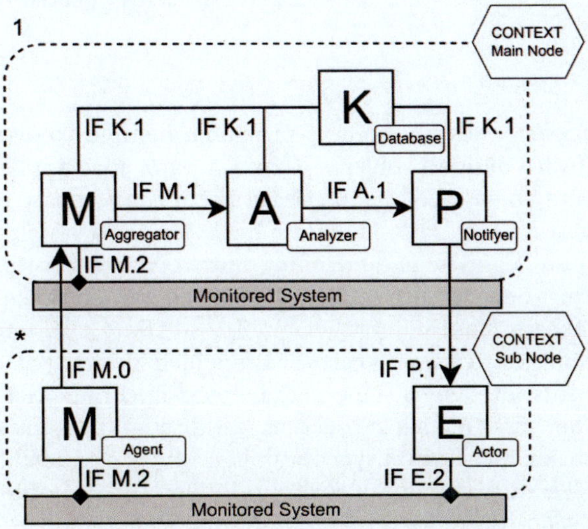

Fig. 4. Generic monitoring system from Fig. 1, Fig. 2, and Fig. 3 with rigid natural components and interfaces that play roles under context. Roles that are played by interfaces are for better representation.

With this extension of our running example, we can solve the three issues that were identified earlier (i.e., inflexibility, incomprehensibility, and maintainability issues). Since it can change context and provide fitting roles at any time, our running example provides separation of concerns and a flexible, context-specific implementation of functions that allow the monitoring of a variety of application domains. Furthermore, the resulting monitoring system is far more comprehensible than individually developed systems that support multiple domains, as the system's functions are bound to context, which is a concept that is easy to understand for humans. Finally, due to separating individual implementations in a two stage hierarchy (i.e., roles that are played in a context), the effort of maintaining the system and reacting to small and larger changes in the monitored system (i.e., the application domain) is greatly reduced compared to traditional individually developed systems.

A complete framework for modeling and implementing software systems with this concept will be discussed in detail in Sect. 4. Furthermore, we discuss how the framework can be applied to the running example of this chapter and different application domains in Sect. 5.1.

3 Related Work

Before moving to an in-depth discussion of our framework, we want to highlight relevant related work. This includes a brief review of alternative concepts for creating flexible software systems and their implications on flexibility, comprehensibility and maintainability. Finally, we highlight how the work of Danny Weyns on MAPE-K patterns for self-adaptive systems affected our framework.

3.1 Design Patterns

In software technology, design patterns provide a method to overcome different re-occurring software design problems. Given a particular application, different design patterns may be applied to the structure of the software to boost certain aspects like: re-usability of code, dynamic behavior, software flexibility, decomposition of complex algorithms, and many more. While those patterns usually focus on providing benefits for certain aspects of the software, they are also detrimental to others (e.g., comprehensibility, software size, etc.). Thus, it is necessary to decide on a case-to-case basis, which design patterns should be used for specific software and if some may be used in conjunction with others to provide further benefits. In this subsection, we discuss different design patterns and their implications on flexibility, comprehensibility and maintainability.

First of all, design patterns are roughly divided into inheritance-based patterns and delegation-based patterns. While inheritance models an "is-a" relationship, delegation models the behavior of an entity.

Inheritance-Based Patterns allow a child class (also: derived class) to inherit the attributes and methods of its parent class (also: base class or super class). A child class can also define new attributes and methods as well as overload (i.e., overwrite) the inherited properties. Using inheritance is beneficial when a child class is a natural extension of its parent class (e.g., a computer is-an electronic device). Applied to software systems, we can create more generic classes that define all common attributes and functions of a system. The case-specific implementation is then created by an inheriting child class. Naturally, this inheritance chain can become complex very quickly. Combinatorial explosion of inheriting classes can occur particularly when small changes occur between children on the same level of inheritance.

The **strategy pattern** [6] (also: policy pattern) enables the decoupling of the client and the algorithm used by the client [5]. This is achieved by encapsulating different strategies for aspects of a class in behavior interfaces. In this way, a monitoring system can have multiple behavioral interfaces for all methods, which can change their strategy (i.e., their behavior) from case to case. Instead of creating a new child class for a new use case, we can use the strategy pattern to choose between different strategies that fit the use case. When a new strategy is needed, it is created as a child class of the respective behavior interface. This leads to massively improved code reusability and maintainability.

This is even more true the higher the level of specialization of each behavior interface, which means that creating one interface for the entire monitoring process provides less reusability and maintainability than creating many interfaces for many smaller parts of the respective algorithm. This phenomenon enables the intuitive application of the template method pattern.

The **template method pattern** [5,6] allows the methods of an abstract class to be divided into several smaller helper methods, each of which performs a particular step of the overall algorithm. For example, the monitoring process as such can be divided into several equally important tasks such as data collection, data analysis, alert and notification management, and action execution. Combined with the strategy pattern, we can create a behavioral interface for each of the helper methods. In this way, strategies are now provided for a fraction of the entire algorithm. The more helper methods we use, the smaller each of them becomes. A large number of helper methods allows the creation of many, very specialized strategies. This increases the flexibility of a system, as support for new use cases can be added by using existing strategies or implementing new ones if no suitable strategy exists. Developers must balance the number of helper methods (i.e., the granularity) because the users of the system must be aware of all of them. A system that is too fine-grained may overwhelm users, while one that is too coarse-grained will not provide the same level of flexibility and code reusability, as large portions of the system will need to be re-implemented to support a new use case. However, with inheritance-based patterns, a large number of fine-grained strategies on a multilevel inheritance chain almost certainly result in incomprehensible models and large software. Monitoring systems, for example, have the peculiarity that small changes in the monitored system (e.g., supported transport protocol, changed scraping intervals, change in data format, etc.) already imply that the current configuration of the monitoring system is no longer functional. Over the long term, this implies that a flexible inheritance-based software system has to maintain many different strategies, thus, it has to be large and incomprehensible.

Delegation-Based Patterns provide another option to achieve code reuse as well as flexibility in software that is as powerful as inheritance. With delegation, we can define delegate classes, that have a reference to their original class by using composition [9]. During run time, given specific circumstances (i.e., a context), the delegate classes perform actions that fit the purpose of the original class in the given context. Thus, with delegation, two objects are involved in handling a request: a receiving object delegates operations to its delegate [6]. With the concept of delegation, we can once again define an abstract interface for the components of a system. When this interface is implemented, the original class holds a reference to a delegate, which also implements the same interface. Different delegates of the same abstract interface model different *behavior* of the interface, very similar to what we achieved with roles in Sect. 2.3. Therefore, this relationship between the original class and the delegates can be understood as a role-playing relationship, in which the delegate is played by the original class in a given context. Since delegates encapsulate context-specific behavior, we achieve the same flexibility compared to a combination of the inheritance-based strategy and template method patterns. However, the size of the respective software models is smaller with delegation, thus, comprehensibility is increased as the behavior of a class is bound to a given context.

To emphasize the role-playing relationship that occurs between an original class and its delegates (and to make the resulting models more intuitive and natural), the **role-object pattern** was introduced [3], a combination of the decorator pattern [6] and the product trader pattern [2]. The role-object pattern makes the role-playing relationship explicit within an object-oriented model using delegation. With that, abstract interfaces do not define any implementation state but only component-unspecific communication operations as well as a minimal protocol for managing roles. Roles are modeled as child classes of a common abstract parent class. The common abstract parent class is never instantiated, instead, a role is chosen that fits the provided context and is instantiated during run time. Each role can define additional operations and attributes as needed in a specific context, e.g., the monitoring of an application via different protocols, similar to what we achieved with pure delegation. With this pattern, we achieve a context-specific view on the components of our monitoring system and are able to change roles at run time that are associated with our core object if the context of our system changes.

Unfortunately, the role-object pattern offers no method to make context explicit within models. This is a downside that all object-oriented models have in common. Pure role-based technologies, as introduced in Sect. 2, enable context to be made explicit, hence, they are superior compared to an object-oriented representation of roles. While our framework can be designed and implemented with object-oriented technology, there are benefits regarding flexibility, comprehensibility, and maintainability in using role-based modeling and implementation technologies that natively support context.

3.2 MAPE-K Patterns for Self-adaptive Systems

Contemporary systems generally tend to make use of modern *distributed* infrastructures. When introducing distribution to the MAPE-K control loop, suddenly, one single managing subsystem does not suffice most of the time, which is why distributed self-adaptive systems feature multiple MAPE-K control loops. Having multiple control loops within a system also raises several challenges. One main challenge is how to coordinate the operation of multiple control loops in one system. This challenge has been addressed by Weyns et al. with the introduction of patterns for decentralized control loops in self-adaptive systems [21].

According to the authors, a MAPE-K pattern is a simple graphical representation of a systems architecture. Generally, the use of patterns should ease the re-usability of software artifacts. The pattern itself describes how individual control loop components or whole control loops interact with each other. Furthermore, the authors distinguish between the pattern as an abstract representation and its derived instance. Although MAPE-K was originally introduced as a hierarchical approach, where each level of the hierarchy contains instances of all five MAPE-K components, in practice, numerous different approaches exist. Their complexity and associated challenges grow with the grade of distribution and decentralization.

Over the course of this chapter, we have already used multiple visualizations of MAPE-K control loops (i.e., Figs. fig:InterfacesspsPatterns spsrunningspsexample and fig:Rolesspsrunningspsexample) that use the notation of MAPE-K patterns from Weyns et al. In the literature, it was stated that "*research should focus on [...] the communication paradigms used for the various interactions in the patterns*" and underlines "*the need for standardizing the interactions between the MAPE loop components.*" [21]. As an addition to their work, we added eleven possible interfaces as already sketched in Sect. 2 of this chapter [16] to approach those challenges.

Furthermore, one important aspect that the work from Weyns et al. misses is the consideration of the modeling, implementation, and actual deployment of the pattern-based systems. With our framework, we aim to provide a general and flexible solution to this.

4 Context-Sensitive Control Loops with Roles

Over the course of this chapter we have framed our main contribution by introducing and extending an inflexible generic monitoring system towards a flexible role-based monitoring system that operates in different contexts. In this section, we want to establish a deeper understanding of the framework by providing in-depth insight in the structure of natural components and interfaces. We will discuss the rigid properties and functions connected to them and how they can be augmented or replaced by roles which are played in a given context. Using the monitoring component from our running example, we will show how we use the programming model ObjectTeams/Java (OT/J) for the implementation of a system based on our framework [8].

4.1 Concept and Framework

Unlike our running example, we want to generalize our concept towards being applicable to a variety of application domains that allow operation as an infrastructure of MAPE-K components and interfaces. For this, the components are designed as containers for their respective interfaces with a minimal protocol for managing the tasks that are determined by their component type (e.g., monitoring component handles data acquisition, filtering, and data shipment). The interfaces are designed as ephemeral storage for data with a minimal protocol for receiving, storing, and sending data. The natural components provide a framework for possible combinations of MAPE-K components and how they interact through interfaces, however, their specific functions are added through roles to provide maintainability and comprehensibility through separation of concerns and flexibility in implementation. Roles can only exist within a context, thus, we use an objectified representation of context to delimit different use cases from each other with OT/J.

Our concept envisions a three phase protocol for each component (we use the monitoring component as an example):

1. The **retrieval phase**. When triggered, the component starts collecting data from interfaces that allow for incoming data (i.e., M.0, M.2, M.3, K.1). All received data are temporarily saved within the respective interface objects.
2. The **processing phase**. When triggered, all temporarily saved data within the interface objects are pulled to the component and processed, e.g., the monitoring component takes care of organizing and filtering monitored values as well as labeling them for further transmission.
3. The **finalization phase**. When triggered, all processed data that are labeled for transmission are sent to the next component in sequence over the respective interface (i.e., M.0, M.1, K.1). After successful transmission, all non-persisted data are deleted from the components' internals so that the component is ready for the next iteration of the loop.

The three phases are visualized in Fig. 5 using the monitoring component as an example. The phases are triggered by a callback method. The callback, in turn, can be triggered by a re-occurring interval or by external action.

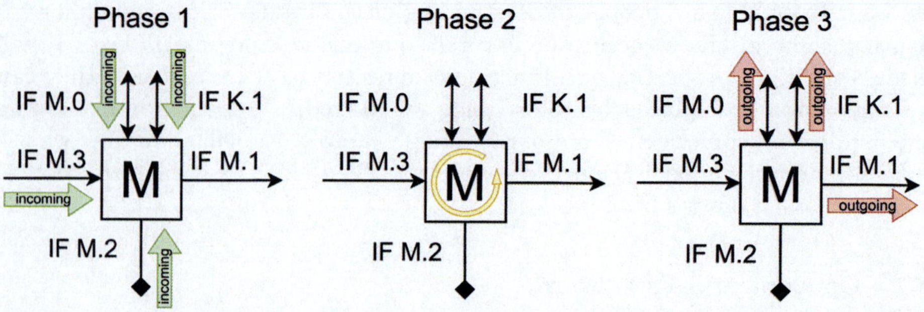

Fig. 5. Visual representation of the retrieval-, the processing-, and the finalization phase of a monitoring component from a role-based context-sensitive MAPE-K control loop.

Each interface consists of three fields (sender information, receiver information, and a *dataBuffer* for saving temporary data) as well as one or two methods, depending on the interface, one that handles receiving data and/or one that handles sending data. Combined, the concept envisions a total of 16 different naturals, five for the components of the loop and eleven for all possible interfaces.

4.2 Implementation

For the implementation of our concept, we used the programming model Object-Teams Java (OT/J), which originated from the issue that systems built from isolated objects suffer from increasingly poor structure, the more complex they get. In OT/J, this problem is lifted by natively allowing objects to play roles within contexts. The language definition of OT/J provides three main concepts:

1. Base classes, which are the classes to which a role is bound. Base objects (instances of a base class) have attributes and functions that describe their very nature, similar to naturals. Those attributes and functions may be inherited or overwritten by roles that are played during run time.
2. Role classes, whose instances bind to base objects by the *playedBy* keyword. Roles can declare *callout* and *callin* bindings. A callout binding allows a method call to a role object to be forwarded to its bound base method. A callin binding, on the other hand, allows a method call to a base object to be intercepted by its bound role method to be either executed *before*, *after*, or *instead* of the base method.
3. Finally, team classes, which are meant as containers for role classes. Every direct inner class of a team class is a role class. Teams and their objects encapsulate context in OT/J. Teams can be activated, enabling callout and callin bindings from their contained roles to respective base objects.

In the following, we want to discuss an exemplary implementation of a monitoring component (see Listing 1.1) that includes its rigid properties and functions and explain how it can be used in different application domains by adding context-specific functions with roles.

Listing 1.1. Program code of the MonitoringBase class.

```
public class MonitoringBase {
  InterfaceM0Base m0;
  InterfaceM1Base m1;
  InterfaceM2Base m2;
  InterfaceM3Base m3;
  InterfaceK1Base k1;

    public MonitoringBase() {
      this.m0 = new InterfaceM0Base();
      this.m1 = new InterfaceM1Base();
      this.m2 = new InterfaceM2Base();
      this.m3 = new InterfaceM3Base();
      this.k1 = new InterfaceK1Base();
    }

    public void callback() {
      retrieval();
      processing();
```

```
19         finalization();
20     }
21     protected void retrieval() {
22         System.out.println("In retrieval();");
23     }
24     protected void processing() {
25         System.out.println("In processing();");
26     }
27     protected void finalization() {
28         System.out.println("In finalization();");
29     }
30 }
```

The class contains all interfaces of the monitoring component, which are represented by the respective interface base objects (lines 2–6 and 9–13). Also, the base class defines the methods of the monitoring component (lines 16–29), which shape the semantics of the class. The same concept also applies to the base classes of interfaces (see Listing 1.2 as an example), where the attributes (lines 3–5) and semantics (lines 11–16) of the interface are defined. Base classes should always contain a complete representation of the attributes and semantics that are defined for the natural, even if not all of them are used in every context.

Listing 1.2. Program code of the M.0 interface base class.

```
1  public class InterfaceM0Base {
2
3      public ArrayList dataBuffer;
4      public String sender;
5      public String receiver;
6
7      public InterfaceM0Base() {
8          this.dataBuffer = new ArrayList();
9      }
10
11   public void receive() {
12       System.out.println("M0 Interface receive()");
13   }
14   public void send() {
15       System.out.println("M0 Interface send()");
16   }
17 }
```

Within both listings, we highlighted specific parts of the code that are replaced with role functions if the software is executed within a specific use case. To achieve this replacement, we need to introduce context into the software. With OT/J, this is achieved via a team class that encapsulates context-specific role classes.

Listing 1.3 shows an example for that, the team class *ExampleMasterTeam* and its direct inner role classes *MonitoringRole* (lines 5–18) and *M0Role* (lines

19–32). When this team is activated during run time (line 3), it allows binding the (highlighted) functions of the base classes MonitoringBase and InterfaceM0Base to the concrete implementations within the inner roles of the team, i.e., MonitoringRole and M0Role. This is achieved by using *callin*-bindings (lines 15–17 and line 31). As long as the team is activated, whenever a bound function is executed within one of the base classes, the callin-bindings take effect and execute the associated callin-function (lines 11–13 and lines 27–29) *instead.* This causes the software to activate the receiving of data through a *SocketServer* over the InterfaceM0 when the ExampleMasterTeam is active instead of printing a message on the console when the ExampleMasterTeam is inactive.

Listing 1.3. Program code of the ExampleMasterTeam and its inner role-class MonitoringRole.

```
1 public team class ExampleMasterTeam {
2
3   public ExampleMasterTeam(MonitoringBase as
      MonitoringRole mBase) {activate(ALL_THREADS);}
4
5   protected class MonitoringRole playedBy MonitoringBase {
6
7     InterfaceM0Base getM0() -> get InterfaceM0Base m0;
8
9     public MonitoringRole(MonitoringBase mBase) {...}
10
11    callin void retrieval()    { getM0().receive() }
12    callin void processing()   { ... }
13    callin void finalization() { ... }
14
15    retrieval <- replace retrieval;
16    processing <- replace processing;
17    finalization <- replace finalization;
18  }
19  protected class M0Role playedBy InterfaceM0Base {
20
21    ArrayList getDataBuffer() -> get ArrayList dataBuffer;
22    SocketServer m0Server;
23
24    public M0Role(InterfaceM0Base m0Base) {
25      m0Server = new SocketServer();
26    }
27    callin void receive() {
28        getDataBuffer().fillWithData(m0Server.receiveData());
29    }
30
31    receive <- replace receive;
32  }
33 }
```

Again, we highlighted the specific methods in the role classes that are executed instead of the base class functions. Between different use cases, the implementation of base classes always remains the same, however, team classes and role classes have to be adjusted or re-implemented.

In a real world scenario, one team class should contain role classes for all base classes that exist in the scenario. To implement our running example, this requires having base classes for all five MAPE-K components and their interfaces as well as two team classes, one for the main node and one for the sub nodes. While the main node team class contains direct inner role classes that have callin-bindings for monitoring, analysis, planning, knowledge and their respective interfaces, the sub node team class only contains direct inner role classes that have callin-bindings for monitoring and execution with their respective interfaces.

With this framework, while base classes remain rigid, we have full flexibility regarding the implementation of team- and role classes. This solves the issues we identified in the beginning of this chapter:

1. For each new use case (also across application domains), we can create one team class and a set of role classes to support self-adaptivity and autonomic management within this domain without re-designing and re-implementing a whole new system each time but rather using a framework of naturals (base classes) and adjust their behavior with teams and role classes that fit the use case.
2. Small differences between managed IT systems no longer prohibit the usage of existing logic, as semantics and functions are clearly separated between different concepts (i.e., base classes, teams, and roles) to support separation of concerns, good maintainability, good comprehensibility and good code re-use.

5 Case Study

A quantitative evaluation of this framework was already conducted in the monitoring application domain in a recent publication [14]. In their work, the authors compared a role-based monitoring system that was created with our framework with two state-of-the-art monitoring systems while keeping all three versions as semantically equal as possible to provide a realistic comparison. The results indicated that the performance of all three versions was close with the role-based framework having an overall slightly better performance than its state-of-the-art counterparts. This showed that our framework can provide a more general and flexible *monitoring solution* while maintaining reasonable performance. Whether this applies to other application domains as well has to be investigated and evaluated in the future. We want to continue this chapter by discussing two further subjects:

1. First, we want to show how our framework can be applied to different use cases from the monitoring application domain. For this, we further develop our running example to vividly present how our framework supports changing environments.

2. Second, we want to show how our framework can be applied to other application domains than monitoring. For this, we discuss the application of our framework to the embedded systems domain.

5.1 Flexible Monitoring Systems

When using our framework, the first step always is to determine the naturals that are needed for the desired architecture. This means that we need to determine how many dedicated control loops our architecture uses, which natural components each of them includes, and which of the components communicate with each other (this indicates the natural interfaces that are needed). Having the individual architecture set, we can derive the roles that are needed in a particular use case. While the naturals are provided by the framework, the roles and their encapsulating team have to be implemented on a case to case basis.

Monitoring Servers. Over the course of this chapter, we've already executed those steps considering our running example of a monitoring system for servers (i.e., an Ubuntu server). The architecture (see Fig. 4) consist of a main control loop that includes the monitoring service capabilities for data aggregation, saving data in a database, data analysis as well as sending out notifications to monitored servers. Furthermore, the architecture includes a sub control loop on each monitored server that contains an agent for sensing CPU load on its host machine and an actor that performs actions that are received from the main control loop.

In the main loop, the system uses four interfaces for internal communication (i.e., IF M.1, IF M.2, IF A.1, and IF K.1). Internal communication (color red) is achieved within the three phase protocol of the framework by using getter and setter methods on the dataBuffers of the respective incoming and outgoing interface objects, as shown in Listing 1.4 for the monitoring component (lines 17–18 and 21–22). Before accessing objects that are located within a natural from a role, they have to be retrieved via the -> operator of OT/J (see lines 7–9). The same strategy can be used to achieve internal communication in the sub loop (i.e., IF M.2 and IF E.2).

Listing 1.4. Program code that shows internal and external communication for the monitoring component.

```
1 public team class ExampleMasterTeam {
2
3   public ExampleMasterTeam(MonitoringBase as
       MonitoringRole mBase) {activate(ALL_THREADS);}
4
5   protected class MonitoringRole playedBy MonitoringBase {
6
7     InterfaceM0Base getM0() -> get InterfaceM0Base m0;
8     InterfaceM1Base getM1() -> get InterfaceM1Base m1;
```

```
    InterfaceM2Base getM2() -> get InterfaceM2Base m2;

    public MonitoringRole(MonitoringBase mBase) {...}

    process(ArrayList toProcess) {...}

    callin void retrieval()      { getM0().receive();
                                   ArrayList incoming =
                                   getM0().getDataBuffer().join(
                                   getM2().getDataBuffer()); }
    callin void processing()     { ArrayList processed =
                                   process(incoming); }
    callin void finalization() { getM1()
                                   .setDataBuffer(processed); }
    retrieval <- replace retrieval;
    processing <- replace processing;
    finalization <- replace finalization;
  }

  protected class M0Role playedBy InterfaceM0Base {

    ArrayList getDataBuffer() -> get ArrayList dataBuffer;
    SocketServer m0Server;

    public M0Role(InterfaceM0Base m0Base) {
      m0Server = new SocketServer();;
    }
    callin void receive() {
        getDataBuffer().fillWithData(m0Server.receiveData());
    }
    receive <- replace receive;
  }

  protected class M1Role playedBy InterfaceM1Base {

    ArrayList getDataBuffer() -> get ArrayList dataBuffer;
  }

  protected class M2Role playedBy InterfaceM2Base {

    ArrayList getDataBuffer() -> get ArrayList dataBuffer;

    callin void receive() { osBean.getCpuLoad() * 100) + "%"; }

    receive <- replace receive;
  }
}
```

Furthermore, the architecture uses external communication (color blue) across control loop borders (i.e., IF M.0 and P.1). External communication is also achieved within the three phase protocol of the framework, however, instead of just using getters and setters on the respective dataBuffers, the data also has to be transmitted over the network. This is achieved by calling the interface's receive() and send() methods, as shown in Listing 1.4 (line 15 and 37). The transmission (receiving and sending) is handled by an asynchronous SocketServer (lines 34 and 37) that is included as a class in the codebase of the framework.

Listing 1.4 includes code that handles the functionality of the monitoring component within the main control loop. To achieve correct function, the roles for all other needed components and interfaces have to be added. Furthermore, a second team class with all components and interfaces of the sub control loop have to be added. An in-depth description of the functions that all the components provide, a description of the deployment of the system, and a performance evaluation of this use case is included in an earlier publication [14]. In this work, we explained how such role-based monitoring system can be used in an autoscaling environment for an infrastructure of web-servers. In this chapter, we focus on the semantics and architectural parts of the framework and how it can be applied to different use cases and application domains, which is why we omitted further details. However, we still want to briefly highlight one experiment that was executed to measure the performance of the role-based monitoring approach compared with two state-of-the-art solutions in this specific use case.

Table 1. Comparison of the average performance of Elastic Stack, Prometheus, and the role-based monitoring approach (RBMA) over ten experiment iterations.

Metric	Elastic Stack	Prometheus	RBMA
Avg. CPU% (Manager)	9% ($\sigma = 0.3\%$)	4% ($\sigma = 0.7\%$)	3% ($\sigma = 0.4\%$)
Avg. CPU% (Workers)	39% ($\sigma = 1.6\%$)	52% ($\sigma = 1.1\%$)	39% ($\sigma = 2.7\%$)
Peak CPU% (Manager)	35% ($\sigma = 3.8\%$)	18% ($\sigma = 16\%$)	13% ($\sigma = 4.8\%$)
Peak CPU% (Workers)	82% ($\sigma = 9.9\%$)	97% ($\sigma = 0.5\%$)	98% ($\sigma = 1.1\%$)
Avg. first >30% to scale-up	31, 1s ($\sigma = 22.8s$)	110, 1s ($\sigma = 17.9s$)	27, 8s ($\sigma = 3.1s$)
Avg. alert to new node avail	5, 3s ($\sigma = 0.5s$)	4, 2s ($\sigma = 0.4s$)	6, 7s ($\sigma = 0.6s$)
Avg. alert to node removed	3, 5s ($\sigma = 0.5s$)	4, 6s ($\sigma = 0.5s$)	5, 5s ($\sigma = 0.5s$)
Avg. scaling operations	10 ($\sigma = 0$)	4 ($\sigma = 0$)	7, 6 ($\sigma = 1.5$)

For the experiment, we ran a total of 10 iterations of a load test. During the tests we focused on measuring the longterm performance of the three versions of the architecture (role-based, Prometheus-based[3], and Elastic-based[4]). We've used additional metrics for evaluating the three infrastructures. All metrics are calculated as an average over all 10 iterations of the experiment and are listed in Table 1. The standard deviation of each metric is included in parentheses. During the experiments, the working nodes running Prometheus generated the

[3] Prometheus website - https://prometheus.io.
[4] Elastic website - https://www.elastic.co.

highest load on average, followed by Elastic Stack and the role-based monitoring approach (RBMA), which both generated the same load. Load peaks were highest on RBMA, followed closely by Prometheus. Elastic Stack generated the lowest load peaks during the experiment. High peak loads are a result of the booting process of new nodes. As the load measurements begin as soon as a new machine is online, usually only the first one or two measurements are highly elevated. Next, we measured the average time between the first load measurement exceeding the 30% threshold and the availability of a new working node. RBMA performed best on this metric, followed closely by Elastic Stack. Prometheus took more than three times as long on average to provision a new node. This is because all three versions were configured to only raise an alert if the threshold was exceeded for more than 10 s. If the threshold of 30% was not surpassed once during this time, the 10 s period was reset. The results that are indicated by this metric do not tell that Prometheus is far worse in recognizing high load, however, they tell that with the exact same configuration, Prometheus scales less aggressive than Elastic Stack and RBMA. Next, we measured the time between when an alert was triggered and when a new working node became available. Prometheus performed best, with an average time to scale of 4.2 s, while Elastic Stack took 5.3 s and RMBA took 6.7 s. We also measured the time to scale down, with Elastic Stack performing best, followed by Prometheus and then RBMA. Finally, we measured the average number of scaling operations (upscaling and downscaling) during the experiments. The Elastic Stack scaled more frequently (10 times per iteration) than RBMA (7.6 times) and Prometheus (4 times).

In summary, the performance of all three versions was close. All three versions were able to scale appropriately to offload an unsustainable load from one machine by dividing it among a maximum of three machines. This shows that RBMA can provide a more general and flexible monitoring solution while maintaining reasonable performance.

Monitoring Network Devices. Network devices, also known as network components, are physical devices required for communication and interaction between devices in a computer network. In particular, they regulate the transfer of data in a computer network. Different active network devices (e.g., gateways, routers, bridges, access points, hubs, etc.) nowadays support common standardized interfaces for data communication and device management. For example, network device management is usually performed with SNMP. For that, SNMP allows the communication and modification of a variety of device parameters, including parameters that indicate network device performance such as applied load, memory utilization, bandwidth, and many more. Consequently, SNMP was used to monitor the performance of network devices for a long time. While SNMP also was successfully used in a control loop architecture for monitoring and managing servers [16], its main application domain remains the monitoring and management of network devices. With our framework, adding support for monitoring and managing network devices through SNMP is an simple task. First, we need to determine the desired architecture and naturals that are needed. As modern network devices already provide an SNMP interface that allows querying data and performing adjustments, there is no need to use a dedicated control loop on the network device, as opposed to our first use case, where a sub control loop was located on each monitored device to provide an actor and agent. The resulting architecture (see Fig. 6) contains one control loop which includes

- a monitoring component, that communicates with the SNMP interface on the network device,
- an analysis component, that reasons about the state of the network device,
- a planning component, that communicates adjustments to the execution component,
- an execution component, that handles the execution of adjustments with the SNMP interface on the network device,
- and a knowledge component, that acts as the database of the system.

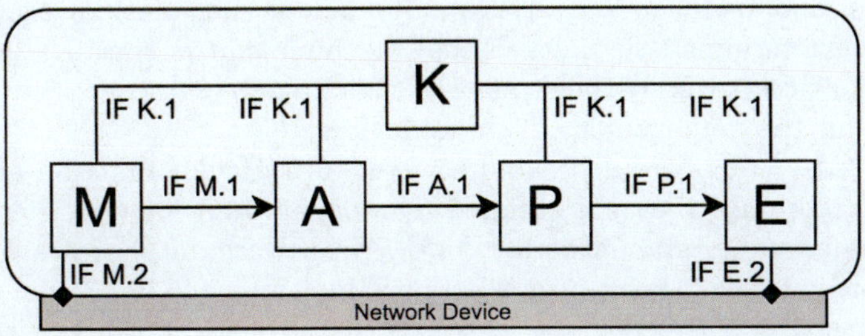

Fig. 6. Architecture for a network device monitoring system based on the framework for context-sensitive control loops with roles.

Furthermore, the architecture contains two interfaces for external communication (i.e., M.2 and E.2) and four interfaces for internal communication (i.e., IF M.1, IF A.1, IF P.1, and IF K.1). As this architecture only uses one control loop that is located on one device, we have to create one team class for its implementation. The team class in turn contains the inner role classes for the five components and six interfaces.

The implementation of the basic communication primitives of SNMP has to be included in the roles for M.2 and E.2 and can be provided by the third party library SNMP4J[5]. The library includes ready-to-use implementations for querying parameters from the network device(i.e., SNMPGet, SNMPGetBulk) as well as changing parameters of the network device (i.e., SNMPSet). The analysis, planning, and knowledge components do not need to support SNMP specific primitives, yet their role-based implementation has to respect the data formats that are provided by the respective network device as those are usually carried through the whole control loop. In this case, the data representation format

[5] SNMP4J Website - https://www.SNMP4J.org/.

for network performance metrics is usually associated with either pure numerical values or with an SNMP Management Information Base (MIB) that formally describes the SNMP objects and their provided values. While support for numerical data is trivial, the SNMP MIB is usually provided by the network device in question.

Adding support for monitoring and managing devices that use SNMP with our concept is simple and shows that our framework supports niches that are not supported by most modern monitoring toolkits [15].

5.2 Flexible Embedded Systems Management

We want to highlight another strength of our concept, which lies in creating support for the management of devices from the embedded system application domain. Embedded systems are nowadays included in close to every day-to-day electrical device (e.g., refrigerators, washing machines, kitchen appliances, headphones, and many more). A large number of such devices executes Java programs on a minimalistic Java virtual machine that is specifically made for embedded devices to fulfil their purpose which enables the application of our framework for the management of those devices.

Embedded systems are dedicated towards a particular function and are difficult to update once they are shipped. Developers and manufacturers have to maintain high quality standards for deployed software and hardware. By using our concept, it is easy to support whole product lines of similar devices as the framework separates different use case specific implementations in roles under context. This allows developers of embedded device software to maintain an established and rather generic management toolkit based on our framework, while adding support for new devices of the same product line through adjusting existent and implementing new roles. This results in less programming overhead, better maintainability, and better quality as developers operate on a common code base for different device iterations within the same product line.

Embedded devices and systems of embedded devices are very diverse, however, we can determine a general control loop architecture for systems of embedded devices. Such architecture (see Fig. 7) is inspired by the master/slave pattern from Weyns et al. [21] in which a centralized master control loop is responsible for the analysis and planning part and multiple slaves are responsible for monitoring and execution. Using a washing machine as an example, such system of embedded devices has many different sensors (for e.g., water temperature, water level, valve status) and actors (e.g., water heater, valve actuator) that have to be operated correctly in terms of the chosen washing program. With the proposed architecture, such sensors and actuators are located within the slaves of the system and communicate with the master on a regular basis. The master decides whether an action is required within an actor based on the output data of the respective sensors. The goal of the master is the correct execution of the chosen washing program. With such architecture, a manufacturer can easily adjust the implementation of individual system parts, e.g., adding washing programs, adjusting thresholds and timings, adding/removing sensors and

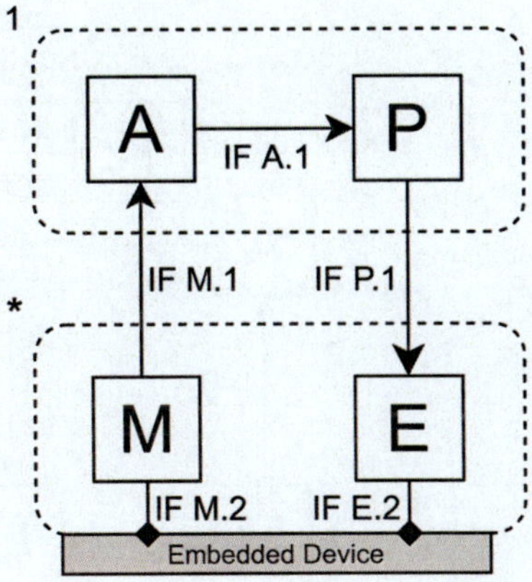

Fig. 7. Master/slave architecture [21] for an embedded device management system based on the framework for context-sensitive control loops with roles.

actors by adding or adjusting the respective roles. Such changes may occur over the course of years, every time a new iteration of a washing machine from this specific product line is released.

The Internet of Things Domain is an extension of the embedded systems domain and became a widespread phenomenon in recent years. Its main difference to the domain of embedded systems lies in devices having networking capabilities. With our framework, such connectivity can be included in the system via allowing sensing components to interact with the environment of the system and having a dedicated knowledge component for keeping information about connected devices available (see Fig. 8). This enables, e.g., rolling out updates, cloud communication, device to device communication, wireless communication with the user of the system via smartphone applications and many more.

In the IoT domain, recent efforts have been made to apply self-adaptivity to an application in an IoT management domain [19]. Similar to our approach, the authors used a control loop-based approach, but the control loop of Oreizy et al. is more specialized to self-adaptive applications, while our approach allows a more general and fine-grained view of architectures [13]. In their work, the authors emphasize the need to perform updates for the communication primitives of an IoT management domain consistently and concurrently to avoid data loss. Data loss could result from two different IoT devices communicating with each other while the versions of their communication primitives are incompatible because one was updated earlier than the other. The authors describe a three-layer architecture for their approach. The first layer (*adaptation management*) takes care of triggering and distributing updates and contains knowledge about distributed service provisioning. The second layer (*coordination middleware*) contains a set of local update managers that collect data and collaborate to coordinate their respective local updates to ensure consistent updates to the sys-

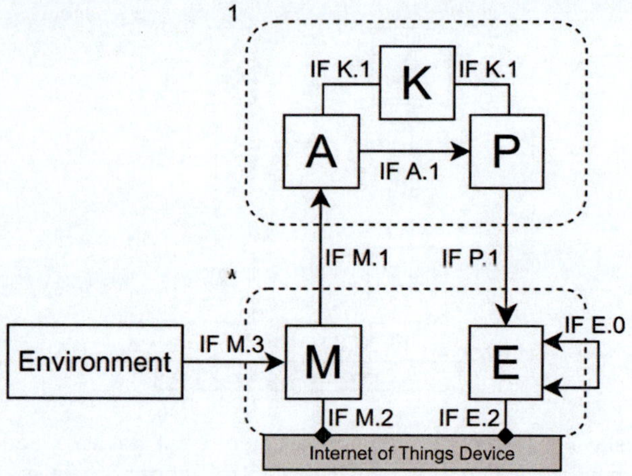

Fig. 8. Architecture for an IoT device management system [19] based on the framework for context-sensitive control loops with roles.

tem. The third layer (*application layer*) represents the actual application logic. Our framework makes it easy to replicate this architecture and functionality. Consider Fig. 8 for a representation of this IoT application using our framework. The adaptation management layer is embodied by the main control loop, which contains analysis, planning, and knowledge. The coordination middleware is embodied by an arbitrary number of sub loops, one on each IoT device, that handle data collection (monitoring) and application of updates (execution). The authors emphasize that the application of updates must be executed in a consistent manner where either all or none of the participating IoT devices successfully apply the update. With our framework, execution components can interact through the E.0 interface to coordinate adjustments initiated by the planning component of the main loop and execute them consistently through the E.2 interface on the actual application logic. The actual implementation of those functions should be contained within dedicated roles for each of the involved interfaces and components.

The combination of role-based technology with our flexible framework of MAPE-K components can provide extensive support for various distributed IoT applications in the future, with the one given here as an example covering only one single application area. Therefore, we strongly encourage other researchers to use our framework in their IoT infrastructures in the future.

6 Conclusion and Future Work

Within this chapter, we have provided an solution to two major challenges for the use of external self-adaptive systems that consist of a managing- and managed part. Firstly, we introduced a framework for creating a control-loop based managing system that is capable of managing systems from different application domains. Second, by applying the concept of roles in our framework, we created a basis for developers to flexibly add or adjust features in existing code that is based on our framework when small changes occur in the managed system, or to replace entire parts of the management functions when large parts of the managed system change. Furthermore, we discussed the implementation of our framework in different use cases as well as the application of our framework to different use cases within different application domains.

Going forward, we opened the opportunity to apply our concept to new application domains which we encourage other researchers to use. In the future, it is necessary to research the performance implications on other domains than monitoring (for which we already conducted a performance evaluation in an earlier publication). Moving forward from that, an overview of the application of our framework to different domains and the performance of the framework in those domains compared to established state-of-the-art solutions is an interesting research opportunity.

References

1. Bachman, C.W., Daya, M.: The role concept in data models. In: Proceedings of the Third International Conference on Very Large Data Bases, vol. 3, pp. 464–476 (1977)
2. Bäumer, D., Riehle, D.: Product Trader. Pattern languages of program design **3**, 29–46 (1997)
3. Bäumer, D., Riehle, D., Siberski, W., Wulf, M.: The role object pattern. In: Washington University Department of Computer Science. Citeseer (1998)
4. IBM Corporation: An architectural blueprint for autonomic computing. IBM White Pap. **31**(2006), 1–6 (2006)
5. Freeman, E., Robson, E., Bates, B., Sierra, K.: Head first design patterns. O'Reilly Media, Inc. (2008)
6. Gamma, E., Helm, R., Johnson, R., Vlissides, J., Patterns, D.: Elements of Reusable Object-Oriented Software, vol. 99. Addison-Wesley Reading, Massachusetts (1995)
7. Guarino, N.: Concepts, attributes and arbitrary relations: Some linguistic and ontological criteria for structuring knowledge bases. Data Knowl. Eng. **8**(3), 249–261 (1992)
8. Herrmann, S.: A precise model for contextual roles: The programming language objectteams/java. Appl. Ontol. **2**(2), 181–207 (2007)
9. Johnson, R.E., Zweig, J.M.: Delegation in c++. J. Object-Orient. Program. **4**(7), 31–34 (1991)
10. Kephart, J.O., Chess, D.M.: The vision of autonomic computing. Computer **36**(1), 41–50 (2003)
11. Kühn, T.: A family of role-based languages. Technische Universität Dresden (2017). https://nbn-resolving.org/urn:nbn:de:bsz:14-qucosa-228027
12. Leuthäuser, M.: A pure embedding of roles. Technische Universität Dresden (2017). https://nbn-resolving.org/urn:nbn:de:bsz:14-qucosa-227624
13. Oreizy, P., Medvidovic, N., Taylor, R.N.: Architecture-based runtime software evolution. In: Proceedings of the 20th International Conference on Software Engineering, pp. 177–186. IEEE (1998)

14. Shmelkin, I., Matusek, D., Schill, A.: Towards role-based context-aware monitoring systems. In: Advanced Approaches to Intelligent Data Processing and Smart Networking. Springer, Cham (2022). https://doi.org/10.1007/978-3-031-16368-5_1
15. Shmelkin, I., Schütze, L., Kluge, T.: Modeling flexible monitoring systems with a role-based control-loop. In: Proceedings of the 14th ACM International Workshop on Context-Oriented Programming and Advanced Modularity (COP2022). ACM (2022)
16. Shmelkin, I., Springer, T.: On adapting SNMP as communication protocol in distributed control loops for self-adaptive systems. In: 2021 IEEE International Conference on Autonomic Computing and Self-Organizing Systems (ACSOS), pp. 61–70. IEEE (2021)
17. Sowa, J.F.: Conceptual Structures: Information Processing in Mind and Machine. Addison-Wesley Longman Publishing Co., Inc. (1984)
18. Steimann, F.: On the representation of roles in object-oriented and conceptual modelling. Data Knowl. Eng. **35**(1), 83–106 (2000)
19. Weißbach, M., Taing, N., Wutzler, M., Springer, T., Schill, A., Clarke, S.: Decentralized coordination of dynamic software updates in the internet of things. In: 2016 IEEE 3rd World Forum on Internet of Things (WF-IoT), pp. 171–176. IEEE (2016)
20. Weyns, D.: An Introduction to Self-Adaptive Systems: A Contemporary Software Engineering Perspective. John Wiley & Sons (2020)
21. Weyns, D., et al.: On patterns for decentralized control in self-adaptive systems. In: de Lemos, R., Giese, H., Müller, H.A., Shaw, M. (eds.) Software Engineering for Self-Adaptive Systems II. LNCS, vol. 7475, pp. 76–107. Springer, Heidelberg (2013). https://doi.org/10.1007/978-3-642-35813-5_4

The Impact of Smart Economy and Digitalization on Market Efficiency

Jürgen Smettan(✉)

BA Dresden University of Cooperative Education, Saxon Academy of Studies, Hans-Grundig-Str. 25, 01307 Dresden, Germany
Juergen.Smettan@ba-sachsen.de

Abstract. The chapter about the effect of smart economy and digitization on market efficiency and functionality is based on the question what properties markets must have in order to good operate. What requirements must be deployed for markets to serve the society well? This includes many features that can be strongly influenced by digitalization (i.e. Digital Transformations in so-called Emerging Networks) and smart tools, such as transparency of the relevant market information, the quick reactions of market players and the punctuality of markets. Market efficiency goods and price signals. Information processing in emerging networks is intensified by multiple smart applications and offers for their networking, by use of software platforms and market simulations. Digitalization forces the markets to be more transparent and manageable. The psychological basis for decision-making by market participants is also being significantly changed through digitalization and smartness. The chapter examines the characteristics of well-functioning markets and the possibilities of digital tools to generate or influence these characteristics.

Keywords: Market efficiency · Conditions of market efficiency · Digital transformation · Smart tools to enhance market efficiency

1 Digital Transformation and Politics

Everywhere in our society we meet technologies of digitalization. Every citizen has a smartphone and an email account. Everybody is chatting several hours a day, is surfing through websites and buys tickets for events and buys foods and products with his mobile. And we are looking for the right way using rout planners. Hotels use booking systems. Every company, every business and every firm use digital platforms in every office. Everywhere we meet digital technology. But the valuation of these techniques is very different. Many politicians estimate and assess digital technologies very little. When in February 2022 additional equipment for the German forces was promise, as Olaf Scholz announced to the German parliament, no cent was planned to combat cyberattacks (cp. Schröder 2022).

Up to now we meet this discrepancy between real importance on the one side and valuation of digital technologies on the other side. Today we face a level of very complex development. Many parts of our society or nearly all fields and spheres of our society

M. Klymash et al. (Eds.): TCSET 2022, LNEE 965, pp. 49–61, 2023.
https://doi.org/10.1007/978-3-031-24963-1_3

don't work at all without digital equipment. All the predictable coming crisis cannot be managed without digital support.

One of the early digital transformation processes took place within the economy: Production processes and consumption, trade and commerce, traffic and logistics are fields of business were experiments with new software and new tools happened from the very beginning of digitalization.

In many other fields of the German society we see a substantial lack of digital transformation on the one side. But on the other side, there is a lack of political performance to increase activities managing these deficits. One main lack of equipment is the broadband support to guarantee the digital traffic in our networks. Critics about these facts can be heard since many years (cp. Sächsische 2017).

2 Digital Transformation and Economy

The use of digital tools and smart solutions produced a lot of advantages in competitive markets. At the beginning we saw software from Apple and Microsoft, little later freeware and shareware of databanks and writing tools, calculation tools and painting software. The advantages could be seen clearly in an early stage of economic software development. The reason is that the use of digital support leads to quick gains and utility. Investment in digital tools lead to quick return of investments. Already during the 80th and former 90th digital tools generated a large increase and improvement of efficiency and productivity even on offline based hardware structures. Several quick and easy software solutions lead to a large increase of efficiency despite of low capability of processors and digital stores. Many offline solutions enhanced production processes and improved production precision not only during manufacturing process but also in trading, marketing, sales and delivery. A very important impact of smart solutions on transformation of the market processes could be seen in capital markets. Stock exchanges in the whole world changed to complete digital driven markets (cp. Laudon et al. 2016).

We have a lot of examples for former use of digital software solutions in the field of economy and trading, business and management. Economy today does not work without smart tools. First considerable use of all kinds of software outside universities was given in the field of economy.

The number of email accounts increased in the beginning 80th very slowly. But in the end of the 80th an beginning 90th in the United States and in Europe the number of accounts began to grow exponential (c.p. Laudon et al. 2016).

When the internet spread in a higher degree all around the world and the number of internet access points increased exponentially, a lot of new kinds of tools were developed. For example, during the early 20th hotels began to organize their booking transactions with new platforms like booking.com or Hrs.com. Hotel services, staff organization, room cleaning plans, organization of breakfast and meal, control of cost development, purchase of energy, water, gas, food, cars and so on were organized using digital tools. Without digital tools and digital control no one is able to run such a business today.

Another field, especially online trade, could grow and develop in a significant dimension, when an increasing number of customers got access to internet and digital trading at all. It took some years in the new century, until higher developed touch screens and smartphones opened the door to millions of customers who are looking for their products in the internet. This development of smart solutions was the main condition for the growth of digital trading and digital business. Along with this growth some special enterprises and companies emerged like Amazon, Alibaba or ebay and others. Some others who did not recognize the massive change in business structures and did not recognize the digital transformation disappeared and played furthermore no important role in the markets. In Germany we lost companies like "Neckermann", "Ottoversand" or "Quelle" and others who did not rely on digital instruments and tools.

Digital transformation had a very important impact on payment and transaction of money. Terms of payment changed to a high degree with digitalisation. Steps of transaction became more and more safe and secure. The development from the beginning with credit cards to money cards up to online banking lead to more complex versions of payment with special internet currencies and monetary systems like bitcoin, pay pal, ethereum or solana and others. Cashless money transfer first reduced cash payment in the stationary commerce. With growing data traffic in the world wide web, digital payment became a basic function to run the markets. Today every company and every enterprise depends to a high degree on network solutions for payment and money transfer and furthermore depends on network solutions for enterprise control.

A more and more threatening development is the problem of attacks against enterprises and companies. Criminal attacks, cybercrime, blackmailing, destruction and fraud lead to higher and higher damage and millions of losses in the business. There is an increasing demand for security measurements against cyber criminality (c.p. Stirnimann 2018).

A deep change happened when digital transformation reached the capital markets and stock exchange. A lot of software tools for different tasks were constructed in short times. Since the 80th a very quick and strong development pushed the science of economic informatics into a position of a game changer. The old setting of stock exchanges was completely replaced by digital solutions. Not only the number of transfers but also the volume of transactions increased a lot. And furthermore, the speed of trading and trading frequency raised up steeply. The implementation of digital trading platforms like XETRA or tradegate in Germany replaced the traditional trading on the floor completely (c.p. Sillescu 1997).

But aside the many different tools and applications in special markets there is to mention, that we have a lot of knowledge about the market functions that forms the theoretical basis for creating new tools and useful instruments to get efficient and powerful markets. Now we should have a look on this theoretical basis.

All markets have common principles and dynamics. No matter what kind of goods or product is traded. The main reason why digital transformation very early took place in economic field is that digital tools are able to enhance important functions of well running markets. Trading is much easier when market participants use digital help. They can act much quicker, more efficient, more productive on the basis of more and better information. Markets are quicker, more rapid and speedy and include more participants. These changes in market processes can be seen in a systematic connection to the features of efficient markets.

3 Economy and Efficient Markets

Digitalisation has an important impact on market functions. Without digital tools a lot of markets would run worse. Markets would generate low efficient results. But what is an efficient market? What is the difference between efficient and nonefficient or lower efficient markets? What is the contribution of smart solution and digital tools? (c.p. Diedrich 1992, S.52).

The Idea of Efficiency: Efficiency of a market can be seen as a realization of economic rationality. Efficient markets lead to rational results of production and consumption in the society. Rational in a sense of "Pareto optimal" are the results of human production and consumption if no resources are wasted or squandered (cp. for ex. Varian 2014, S.15; Lüppertz 2007, S. 8, 25). Two principles can be seen as rational: Maximum principal and minimum principal. The use of resources in a way that we get maximum results (maximum principal) is in our view one kind of behaviour that we call "rational". To reach defined targets by using a minimum of input (minimum principal) is in our view as well as the maximum principal a kind of behaviour that we call "rational". This is reasonable behaviour.

In an economic view every society must find answers to three principal questions to manage limited resources: Questions of coordination, of allocation and distribution. The answers must be rational and reasonable. "Rational" is a kind of behaviour that leads to a maximum of utility in the society using given limited resources. To get optimal solutions every society should trust on well running markets. The markets should answer the questions. This leads to optimal solutions. But normally there exist some restrictions. Digital tools help us to get well running markets and to remove these restrictions.

a) Coordination: Every society has to establish communication channels between the producing forces on the one side and the needs and demands of the people on the other side. The needs and the producing forces must come together. Production must be oriented on the demand and along to the wishes of the people. If this matching is not guaranteed the result is too much or too low production. Supply and demand must be coordinated and adjusted.
b) Allocation: The producing participants of the economy must know how much products in which number and quality they have to produce. Every society has to decide how many resources are used to produce these special numbers of goods and products. To get optimal solutions every society should trust on well running markets because this decision has a lot of consequences. To produce a special number of goods we need a lot of input factors: labour, capital and ground. And know-how is necessary to combine these three factors of production. The quantity of labour, capital and ground that is necessary for production depends on the number of products the society decided to produce.
c) Distribution: A third question that every society has to answer in managing limited resources is the question of distribution: Who gets the produced goods and products? Every society must distinguish between persons, who get parts of the production and others who do not get one or more parts of the production. As a result of the market processes we get four groups of persons in every (product-)market. First group is the group of participants in the market who try to sell their goods and products and

they can sell their products. Second group are the market participants who try to sell their goods and products but they cannot sell their products because they try to realize prices higher than the going price. Third group is the group of participants who want to buy goods and who get the demanded goods. The fourth group is the group of market participants who want to buy goods but don't get the goods they wish to buy. We have different answers to the question, which distribution is a fair and justified distribution of income, products or assets. Markets give an answer in terms of ability and willingness to pay. In terms of poverty and richness this must not be a fair solution (cp. Diedrichs 1992). But it is a market result as a consequence of payment.

4 Complete Markets, Perfect Markets and Ideal Markets

4.1 Complete/Uncomplete Markets

a) Complete markets are markets in which very many protagonists on the supply-side of the market as well as very many protagonists on the demand-side of the market take part in the market process. Theoretically endless numbers of suppliers and endless numbers of demanders offer and demand products. This market situation and these conditions we call it a "polypoly".
b) Uncomplete markets have less than endless numbers of participants on both sides of the market. Usually subdivisions of uncomplete markets show 8 more types of markets. Combinations of numbers of participants on the supply-side and participants on the demand-side can be subdivided into three steps: "endless", "some" or "one" participants in the market. These steps lead to 3x3 types of markets. Following these steps we can distinguish nine types of markets. But only one combination "endless by endless" means the realisation of a complete market. Every other combination is not complete (vgl. Pindyck et al. 2018). This means that 8 of these types of markets are restricted in the number of participants. Within the field of microeconomic we call this grid (3x3) the "market shape scheme". If a complete market in addition generates one price for all traded products, this market is called the "ideal market".

4.2 Perfect Markets

a) Perfect/unperfect markets

Perfect markets include a special ability. These markets are able to generate one single price. All the participants in these markets behave in a special way. They make rational decisions. They act quickly and they have full information. They have no preferences and all products in these markets are homogeneous. All the negotiations in this market lead to one price. In this single-price situation all buyers in this market can buy all products from all suppliers to the same price. All the homogeneous goods in this market have the same price (c.p. Herdzina et al. 2015).

As an example of a split price market we have a beermarket. A person who wants to drink a beer in a city normally pays a different price depending on the special location, a restaurant, bar or a beer garden or a pub and so on. The same kind of beer in different locations usually has a different price. This is an unperfect market. In different locations we see different prices for the same product. Perfect markets lead to a single-price-situation. Every product which is sold in a perfect market is sold to the same price. But there are several conditions which must be given so that a market is able to generate one single price.

Attention! A perfect market must not be an ideal market. The ideal market is both, perfect and complete. The ideal market is the "perfect polypoly". This is the situation of full competition (completeness) with single-price. All the uncomplete markets can be perfect in the sense that they produce single-price-situations. Take especially the two-side-monopoly. This is the situation in which two persons make a deal and their negotiations generate a single price. This leads to the following question: What are the conditions that must be given to generate a single-price situation?

a) Transparency: Market conditions must be transparent for market participants. Market participants must have a complete overview on decision relevant signals. And all participants must know everything about alternative decisions. Participants in the market must have full information about every relevant element of the market, the goods and other participants. Transparency is necessary for common conclusion. Every participant must be able to come to the same conclusion.
b) Homogeneous products: Homogeneous products means that the products have the same quality and same features. For example, this is the case in shares. Every share of an enterprise has the same value. The same part of the enterprise is represented in every share. The actual number of the share does not count. All products in the market must have the same value.
c) Rational behaving protagonists: Every participant in the market must think and act in a rational way. Economic principles must be realized 1. Maximizing utility or profit with given resources. 2. Reaching given targets with minimum resources. Nonrational behaviour leads to waste and squander.
d) No preferences: Participants in the market evaluate differences between the products only with respect on the price. No further features make differences between the goods. The evaluation of every product in the market must be independent from other circumstances. Not the place or other persons or circumstances may influence the valuation of the products.
e) Point market: Ideal markets react very quickly on modifications. Changes in quantity or price cause immediate reactions of protagonist who demand or supply goods. We have immediate reactions when all participants come together in one place. There is no time-lag in reaction. Changes cause immediate reactions. If the participants meet at different places it is possible that different prices can be negotiated.

4.3 Ideal Markets

Ideal markets are efficient markets. These markets are able to create optimized results. These markets are able to provide society with products and bring demand and supply together. Solutions for the questions of allocation, coordination and distribution can be generated in optimized form (vgl. Diedrichs 1992, S. 50f).

Ideal markets are perfect and complete. Because in the market shape scheme only the type "polypoly" is a complete market, so we can only get one combination of an ideal market, this is the perfect polypoly. In this ideal market we have both features realized in one market. Ideal markets are perfect and complete. Endless numbers of buyers meet endless numbers of suppliers in these perfect markets. At the end of all negotiations we have one price as a result of the price mechanism.

We call ideal markets „ideal" because these markets lead to optimum solutions in the questions of allocation, coordination and distribution. Ideal markets lead to optimum production because we get the situation that as many products are produced as consumed (production optimum). Nothing is wasted and there is no lack in production. As many products as needed are produced. Additional ideal markets lead to optimum coordination (exchange optimum). There exist no more possibilities to enhance the wealth of a society by an alternative coordination. And ideal market generates a solution for the distribution problem (socio-economic optimum). Those people who are willing and able to pay the price get the products. Those who are not able or not willing to pay the price do not get any of the products.

If well going markets do not exist and answers on these three questions cannot be generated by the markets, society must maintain an organisation that answers to these questions and makes decisions. That organisation must be a planning unit. That unit must decide on the number of products to produce, must decide the allocation of input factors into the production an must decide how much products each single member of the society receives. But if the markets are working complete and perfect and if the markets are able to find solutions for all these questions: allocation, distribution and coordination, then we have a situation in which optimum results are given for production and consumption. Market solutions in fully functioning markets are called pareto-optimum-solutions (Pareto-Optimum). In this situation under these conditions is no further exchange of products possible which leads to different allocation, different distribution or coordination with a better use of resources. No further increase of wealth is possible (c.p. Diedrichs 1992, S. 52).

5 Market Failure and Smart Solutions

The idea of an efficient and ideal market is a scientific vision. The idea is a model that tells us the way to optimum economic solutions for social questions. But within our reality we can hardly find such ideal markets. Only a few markets like stock market fulfil nearly all conditions of these ideal markets. Everywhere we find some restrictions and we find some conditions that do not fit to the well working model of an ideal market. The model includes a lot of theoretical assumptions that must be realized. The number of features that must be fulfilled is a number of theoretical assumptions which guarantee that the model works well. This is not a description of the reality. But we have to notice that not every market that does not fit to the model is a bad market. Even strong restricted markets can still be useful to generate social solutions and to provide society with necessary goods and products. But the model shows what must be done if markets do not run perfect. The model shows the way to perfect and complete markets. If a market doesn't meet the conditions we can think about measurements to enhance the market performance. And this is the point where smart solutions and digital tools can be useful in many different ways.

Complete market failure is given if markets do not have any social utility left. But complete failure of a market is not widespread. Market failure is not a question of yes or no. Restrictions are given less or more, are coming and going. Final market failure is given if no further exchange of goods and products exist. In the most cases restrictions lead to split prices, to excessive prices for consumers or to poor prices for producers.

Well running markets are very valuable for every society. Perfect and complete markets are important institutions for every society and it is important to keep them running. To keep the markets running, we have a lot of smart digital tools and intelligent solutions. Terms of „smart solution“ or „digital transformation in the economy“ enclose the manifold digital solutions in soft- and hardware to solve problems of production and consumption, investment and saving, finance and money and so on (vgl. Pindyck et al. 2018).

5.1 Restriction of Completeness

Markets are not complete when the number of participants is low. When there are only a few protagonists involved on the supply-side or a few protagonists are involved on the demand-side or only few persons on both sides take part in the market processes. Less participants in the market means low competition and restricted price mechanism. Restriction of completeness causes restriction of competition and malfunction of the price mechanism.

a) Few participants on every side: The question of completeness was one of the first economic questions which were tackled by digital tools. The main task of digital tools and smart solutions concerning this question is to get more and more participants into the market. Smart tools have to enhance the number of protagonists. In many markets these tools have already been implemented successfully. International networks, world wide web, international information platforms open the international markets for every player. Different local or national markets are combined together to one worldwide market.

   Groups with few participants on the one or other side of the market grow up to large groups and competition is growing. It took some years of development from isolated offline solutions up to business networks. IT-networks of enterprises raised up in the middle of 1990th. Capable hardware was necessary to provide enterprises with sufficient data traffic (Laudon 2016 11f).

   Meanwhile we have a lot of more efficient network technologies available like SDN/IBN (c.p. Klymasch 2022). Completeness of markets will be reached in ever higher degrees. The higher the number of participants in the markets the higher is the chance that price dynamics are working efficiently.

b) Restriction of competition based on price agreements

Every company tries to make income safe. Every enterprise wants to make safe profits. And most of the enterprises try to enhance their profits. Pressure caused through competition is unpleasant. Competition forces the companies to undertake measurements against the pressure of the market. Many of them use illegal behavior to reject competition. Many businessmen and businesswomen are involved into price agreements.

But in some cases, mainly observed in oligopoly situation, digital tools enhance the price mechanism in a way that seems to be a price agreement but is not such an agreement. Example fuel price: Federal Cartel Office looks for circumstantial evidence of price agreements in fuel market. This market is a classic oligopoly. But the proof of price agreements was not found up to now. From an economic view it is enough to explain the price dynamics with digital tools in the market. There are a lot of applications which are used to watch fuel prices. Not only customers but also the suppliers use these apps and react with their own prices on the dynamic in the markets. It looks like to be arranged. But this is not the case. Digital tools make it possible to get the full information of a region, to react quickly and react along the given market dynamics. Smart solutions make it possible to react quickly on the market without illegal agreements.

c) Fusions, Cartels, Takeovers

Main task of Federal Cartel Office is the observation of cartels and fusions and to fight against illegal fusions. Fusions are illegal if the markets are disturbed by these fusions, if competition is eliminated. It is the responsibility of the office to find out and to eliminate the illegal restrictions of the markets. Digital tools are used in many ways to gain information about suspect organizations. A lot of digital tools can be used to do this job and to collect information about the enterprises which are under control.

5.2 Restriction of Perfection

Each of the listed restrictions can be the reason for split prices. If it is strong enough every restriction causes split prices. If a producer is able to split price that means, he is in a position in which he can get different prices from different buyers. This split causes an increase of the producer's surplus.

b) Restriction of transparency: If a buyer does not have enough information on products or on alternative decisions or on other producers and further important market information, he is not able to calculate an appropriate price. A lack of information leads normally to higher prices. But the producer can also run a lack of information which normally leads to lower prices because he has no complete information about the value of his product.

Information support is a main achievement of different smart solutions. A magnitude of information search programs, databanks and translation applications support customers as well as suppliers with information that makes markets transparent to a high degree. Visibility of market information increased in many markets during the last years along with the process of digitalization. Furthermore the velocity of comparisons in product markets is intensified by smart tools. Smart tools enhance transparency in terms of availability of information about products, prices and market conditions. Information systems enhance transparency concerning velocity of orientation in the market.

b) Restriction of homogeneity of products: In normal market situations it is the intention of producers and suppliers to present their goods in a special light with special utility and special features. Products with a unique selling point are more valuable than products which are highly exchangeable. A special selling argument makes it easier to sell a good. But the maintained differences make it difficult for the buyers to see the different utility in equal goods or to see the same utility in different goods. The more information the buyers have the more qualified evaluation of products is possible. Here we have another example how information systems provide participants in the markets with important knowledge to improve their market position. Buyers can use smart tools for a higher understanding and better description of the products they intend to buy. Digitalization and digital transformation provide the participants in the markets with information that enhances the precision of comparison of products. Participants in the markets are enabled to make better decisions.

c) Restriction of rationality: Rational behavior is a precondition for an efficient use of resources. Irrational behavior in the most situations is emotional driven. Spontaneous decisions are often very expensive. In an optimistic mood people accept higher risk in their economic decisions as in pessimistic mood. In pessimistic mood people often avoid risk and avoid decisions and they avoid spending money at all. Digital tools can help to get more rationality into personal decisions. Decision algorithms help to avoid bad individual decisions. Comparisons of products, suppliers, enterprises, market places and quick information on alternatives make decisions less emotional driven and make it to a higher degree target oriented. If one takes additional information systems into account before he or she decides to buy a product, spontaneous emotions are reduced, mental distance to the product is enhanced and rationality has a better chance to determine decisions.

d) Preferences of market participants: If buyers or producers have special preferences, this means that buyers do not buy a special product from a special supplier with regard only to the price. They also have special interest to meet the special supplier on other reasons, not only product and price of the product. Take the example that a wheelchair patient buys bread and cake at a special baker because this baker has a ramp for barrier free driving into the baker's shop. Another baker not far away doesn't have such a ramp. The wheelchair patient will buy his bread at the bakery with a ramp even when the bread and cake is on a higher price. Preferences lead to split price. If the other baker would be informed about the effect of the ramp, he will also buy such a ramp for his baker's shop and this baker would get more customers. Digital tools and networks enhance knowledge about the markets and can improve market decisions.

e) Restriction of point markets/speed of reaction: In our market models there is usually no factor 'time' included. Market reactions are immediate. Market reactions have to happen suddenly. See the rule of supply: "A rising price leads to increasing supply." Within the market models there is no time lag between the change of price and the change of magnitude of supply. In real situations the impact of price signals on producing behavior needs time. But digital solutions in a lot of cases are able to generate the punctuality of markets. Important examples for nearly ideal markets in the sense of punctuality are the stock markets of the world. Without electronic trading platforms the huge numbers of buyers and sellers in the international capital markets and the huge number of trades with securities and stocks, loans and assets could not be managed. Trading platforms like XETRA or 'tradegate' in Germany punctuality of the markets can be enhanced a lot and the traders can immediately react on market signals. All participants in the markets are brought into a virtual neighborhood. Very quick and sudden reactions become possible. Smart tools are able to bring markets and participants closer together and they can meet in complete markets (vgl. Diedrichs 1992, S. 48f).

On the background of stronger networks and more capable laptops and notebooks a very important change of work organization has happened during the last two or three years. In times of corona many enterprises changed their work organization and began to restructure their relationship to their employees. Many companies changed their work organization into home office or mobile work. New and stronger solutions are to expect when new strong solutions like 5G and 6G are used to run these networks.

5.3 Restrictive Conditions

Market inefficiency can be caused by several other external restrictions. These external conditions can restrict market functions up to total breakdown. This includes the danger that the three basic social questions get no answer. It is possible that the market fails completely. Three reasons are mainly relevant: External effects (externalities), asymmetric information (information lacks) and specific public goods (common goods) (cp. Varia, 2014, 663f, 714f, 737f).

a) External effects are given if costs or utilities that appear during production processes exist outside the market. The existence of these externalities is not part of the price mechanism. The price of the product is usually too low. The differences have to be taken into consideration to solve the problem. Take the case of forestry: Enterprises which earn their income in producing wood. An aside result of the production of wood is that in the forest a lot of animals find a place to live and people find nice places for hiking and furthermore the forest reduces CO2 in the air and so on. But the enterprise is only payed for producing wood. No one pays for the incremental utility. The other side is the case if we have damage that is not included in the price of the products. Then we have costs of the public sector. Normally for instance we do not pay for the use of streets. A well-known example for including external effects in market mechanism is the international trading with CO2-certificates. The international community has established a worldwide trading system with CO2-certificates. Nations which plan to produce products with emission of CO2 will pay for this emission. Other nations which are able to produce without CO2 emission or can reduce CO2 from the atmosphere can sell their certificates. The market for these certificates is a very highly abstract market and can only be driven by using smart tools and trading platforms (eg. Eex 2022).
b) Specific public goods: The problem of public goods (common goods) lies in the fact that these goods have specific features which make them market-incompatible. The whole community receives the utility from these goods. Single users cannot be excluded. Well running markets are a very powerful instrument to solve social

problems and public questions. But some goods are not manageable by markets. If the society needs these goods it is necessary to organize production and consumption of these goods by planning units. Such a specific public good is for example internal and external security. Another category of public goods is the public infrastructure of streets, public places or the public sewage system. Here we see an important task for smart administration and digital government to enhance the capability of the public systems to keep society running.

c) Asymmetric information: This means a restriction of market information on one side, on the side of the buyers or the side of sellers. Take the example of trading with new or used houses. The former owner of a house normally knows very well which parts of the house are in a bad stage and have to be repaired. Somebody who is interested in buying the house normally does not know anything about the offer until the former owner gives him information on the object. The result of asymmetric information normally is a price which is too high and bad for the less informed participants in the market.

In summary there exist a lot of tools which are useful to enhance the markets completeness and perfection. But it is still the single individuum which decides to buy or sell, which decides to invest or save, to realize production or consumption. We face the danger, that algorithms and machines take over the markets. But human beings are still responsible for the results and consequences of their decisions (cp. Huber 2022). Smart economy, smart tools, artificial intelligence and algorithms should be used to increase the functionality of our markets. As a result we receive optimized answers to the questions of allocation, distribution and coordination.

References

Diedrichs, D.: Mikroökonomik. WRW-Verlag Wirtschafts- und Rechtswissenschaften, Köln (1992)

European Energy Exchange AG (2022): https://www.eex.com/de/. Augustusplatz 9, 04109 Leipzig, Germany

Herdzina, K., Seiter, S.: Einführung in die Mikroökonomik. Verlag Franz Vahlen München, QC 100 H541 (12) (2015)

Huber, W.: Menschen, Götter und Maschinen. Eine Ethik der Digitalisierung. Verlag C.H. Beck (2022)

Klymash, M.: Wissenschaftliche und wissenschaftstechnische Tätigkeit an dem Lehrstuhl für Telekommunikation der Nationalen Universität "Lviver Politechnik", interner Bericht, ppt (2022)

Laudon, K.C., Laudon, J.P., Schoder, D.: Wirtschaftsinformatik. Eine Einführung. Pearson Deutschland GmbH, 3. Auflage, ST 205 L372 (3) (2016)

Lüppertz, V.: Problemorientierte Einführung in die Volkswirtschaftslehre. Lehr- und Aufgabenbuch. Winklers GmbH, Braunschweig (2007)

Pindyck, R.S., Rubinfeld, D.L.: Mikroökonomie. Pearson Deutschland GmbH, 9. Aktualisierte Auflage, QC 100 P648 (9) (2018)

Zeitung, S.: Digitalisierung? Deutschland lädt noch ... Sächsische Zeitung, 14. Juni 2017, S.2 (2017)

Schröder, M.: Fürs Digitale fehlt es an politischem Willen. Sächsische Zeitung, 9.6.22, S.1 (2022)

Sedgewick, R., Wayne, K.: Algorithmen. Algorithmen und Datenstrukturen. Pearson Deutschland GmbH, 4. Auflage, ST 134 S448 (4) (2014)

Sillescu, D.: Per Knopfdruck an die Börse. Fischer Taschenbuch Verlag, Frankfurt am Main, Aktiengeschäfte mit PC, Fax und Telefon (1997)

Stirnimann, S.: Der Mensch als Risikofaktor bei Wirtschaftskriminalität: Handlungsfähig bei Non-Compliance und Cyberkriminallität. Springer Gabler (2018)

Varian, H.R.: Intermediate Microeconomics. A Modern Approach. 9th Edn. Norton and Company, New York (2014)

Wagner, P.: Aufbruch Wirtschaft. Wie die digitale Transformation gelingt. Google LLC Anzeigen-sonderveröffentlichung. SZ Scala GmbH (2016)

Self-driving Cars

Andriy Luntovskyy[1], Dietbert Guetter[1](✉), and Andrii Masiuk[2]

[1] Computer Science Department, BA Dresden University of Cooperative Education, Saxon Academy of Studies, Hans-Grundig-Str. 25, 01307 Dresden, Germany
Andriy.Luntovskyy@ba-dresden.de, Dietbert.Guetter@t-online.de
[2] Lviv Polytechnic National University, S. Bandery Str., 12, Lviv 79013, Ukraine

Abstract. Self-Driving Cars, also so-called Unmanned Vehicles (UmV): what is about their appearance in mid-term? In February 2018, UNECE's Inland Transport Committee (ITC) acknowledged the importance of the Working Party 29 (GRVA) activities related to automated, autonomous and connected vehicles: Automated, Autonomous and Connected Vehicles. We would like to discuss technical, ethical and legal aspects of the technology which domineering and bold emerge in the era of Digital Transformation. The work contains a brief survey and some case studies.

Keywords: Unmanned Vehicles (UmV) · Self-Driving Cars (SDC) · Emerging Networking (EmN) · IoT · Telematics · Connectivity · Automation · Reliability and control · Cybersecurity · Smart applications

1 Motivation

Self-Driving Cars (SDC), also so-called Unmanned Vehicles (UmV), are a part of the above-mentioned GRVA and, besides, a part of the Emerging Networking in the contemporizing Digital Age. Which paradigms are normally used for Emerging Networking?

Such kind of networks is based on up-to-date heterogenous combination of mobile, wireless and fixed networks which are augmented via ad hoc and PAN as well as IoT devices. The further important part of them is virtualization on the basis of well-known SDN (Software-Defined Networking) which represents one of the most important paradigms for emerging networking [1, 2].

Paradigms per definition are a set of concepts or design patterns, including theoretical and empirical models, research methods, practical tools, platforms and standards, structure opportunities what means a significant contribution to a field. We would like to study such paradigms for EmN in detail.

Under category EmN we mean also steady growing diversity of unmanned scenarios, where the process automation grade is nearly of 100%. Operative disaster networking (emergency networking), which appear at sites, aiding rescue troops and robotics belongs to the above-mentioned problematics too. This is the next paradigm [3].

Together with wide-spread Internet of Things (IoT) and digitalization challenges, EmN begins a new era in the history of networking, where the network itself is part of

M. Klymash et al. (Eds.): TCSET 2022, LNEE 965, pp. 62–72, 2023.
https://doi.org/10.1007/978-3-031-24963-1_4

the digital environment and IT translates business intentions into appropriate network configurations for all included devices [4–6]. This is the next paradigm.

EmN support unmanned scenarios via highly-decentralized and distributed apps and services for adaptation, coordination and choreography [2, 3, 7].

2 State-of-the-Arts

A good survey on the given subject is provided by [9]. The state-of-the-arts for Car Control Systems is depicted below (Fig. 1).

They include the embedded 4G/5G SIM cards to communicate with service centers, braking and running gear control systems, ABS, tire condition and pressure, battery and ignition in the car, fluid and motor oil, condition suspension and lighting controls, cockpit and steering assistance [2].

Modern automotive technology and cars (refer Fig. 1) use also complex IT with multiple integrated Embedded Systems and IoT devices, which are aimed to, for example, HVAC (Heating, Ventilation, Air-Conditioning, windows regulators) as well as to:

- Seating mechanism,
- Lighting,
- Embedded electronics,
- Braking systems,
- Steering systems etc.

IoT devices are also used for traffic safety control functions like:

- Monitoring of safety-relevant components (brakes, …),
- Load securing for trucks,
- Condition of maintenance (tires, oil level),
- Optimal operation (fuel consumption, amortization or wear and tear),
- Driver status monitoring (fatigue, illness, alcohol, drugs, reaction time, driving errors, qualifications, driving times, braking actions),
- Security checks: OK, ALERT
    - o Multi-locks
    - o Manipulation attempts
    - o Individual recognition of all components
    - o Recognition must only be able to be manipulated if the components are destroyed, e.g. RFID circuits integrated windows, engine blocks.
- Theft detection,
- Are the original parts installed?

Furthermore, there are different systems that help the self-driving car control the car, including:

- car navigation system,

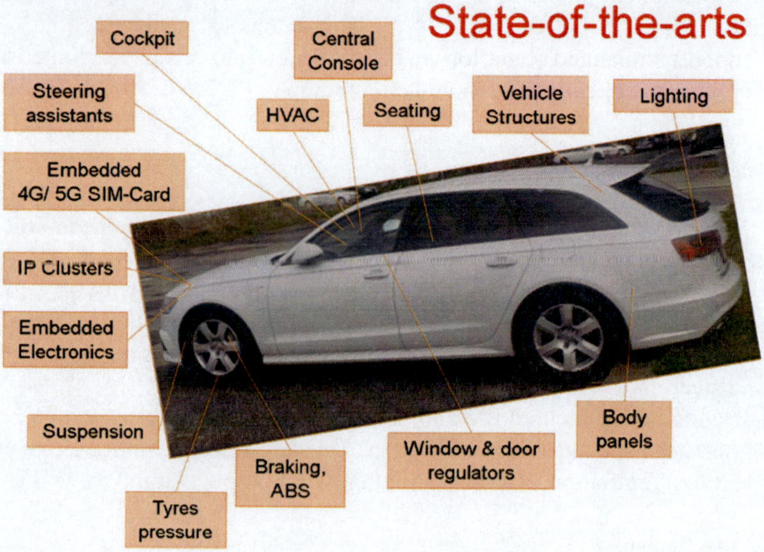

Fig. 1. Convenient car control systems [2].

- location system,
- electronic map,
- map matching,
- global path planning,
- environment perception,
- laser perception,
- radar perception,
- visual perception,
- vehicle control,
- perception of vehicle speed and direction,
- vehicle control method.

The most important assistance systems with intelligence at a glance are as follows [8–10]:

- Anti-lock braking system (ABS),
- Electronic Stability Program (ESP),
- Anti-slip regulation (ASR),
- Brake Assist (BAS),
- Hill Descent Control (HDC),
- Adaptive Cruise Control (ACC),
- Automatic Emergency Braking Systems (AEBS).

Nowadays, advanced wireless protocols, such as IEEE 802.11p/pWLAN/pWi-Fi, ITS-G5/Car2Car, LoRA WAN, LTE Cat-M, NB-IoT are used for data transfer between self-driving cars and cloud analytics [1–4].

3 Foundations

The Society of Automotive Engineers (SAE: https://www.sae.org) was founded in 1905 (USA, PA). Nowadays SAE provides further development of UV. To other foundations belongs Waymo (https://www.waymo.com) as a daughter of Google (2009–2016), then Alphabet (2016-present).

A prototype of Waymo's self-driving car, is navigating public streets in Mountain View, California since 2017. The Tesla Autopilot system is classified even as an SAE Level 2 system. A short list of leading international vendors for SDC/UmV is represented below (alphabetically) [8–10]:

- Apple,
- Baidu (China),
- BMW,
- Bosch,
- Ford,
- General Motors,
- Mercedes-Benz,
- Nissan,
- Tesla,
- Toyota,
- Uber,
- Waymo.

The following Overall Objectives for UmV are strength persecuted [10] considering the survey of AVL Software and Functions GmbH in Regensburg (Bavaria):

- Stable development and growth of complex automated mobility vehicles with higher quality, safer and secured (i.e. harmless to yourself and other road users), an in a cost-efficient way.

  These overall objectives for UmV are persecuted under use of the three concepts: Connectivity + Automation + Cybersecurity.
  Driving **Connectivity** means based on [10]:

- The ability to link to and communicate with other Embedded, PHYs, Wearable, IoT and electronic devices, software services and agents via Internet. Connectivity is provided between:

  o vehicles and infrastructure (V2I, vehicle-to-infrastructure),
  o vehicles and centralized back-end systems i.e. cloud (V2C, vehicle-to-cloud),
  o the vehicles (V2V, vehicle-to-vehicle).

  Driving **Automation** (Autonomy) means based on [10]:

- Moving the responsibility from the human drivers to an electronic system (UmV) for the fulfilling of the following tasks:

- o 2D-control of the vehicle (steering, braking, acceleration, deceleration),
- o monitoring of driving environments,
- o performance of dynamic driving tasks.

Cybersecurity means based on [10] protection of a data and telecommunication system from theft or damage to the hardware, the software, and to the information on them, as well as from disruption or misdirection of the services that have to be provided.

Connectivity and Automation are technical areas for UmV, which are continuously merging together and lead to more intelligent and sustainable solutions. "The market drivers for the introduction of connectivity and automation in automotive industries (for unmanned cars, trucks, tractors, buses) are different, but from the point of view on technical and legal processes, the synergies are clearly visible" – by [10]. Data security (cybersecurity) aspects with regard to privacy, integrity and intrusion detection are to be accurately considered because they lead towards higher degrees of connectivity and automation of UmV.

To the most important products of Society of Automotive Engineers belong multiple industry standards tighter with conferences and publications for professional development in the automotive branch. One of them is the standard SAE (J3016) for Automation Levels (Fig. 2).

4 Anticipated Levels

Between manually driven vehicles (SAE Level 0) and fully autonomous vehicles (SAE Level 5), there are a variety of vehicle types that can be described to have some degree of automation. These are collectively known as semi-automated vehicles. As it could be a while before the technology and infrastructure are developed for full automation, it is likely that vehicles will have increasing levels of automation. These semi-automated

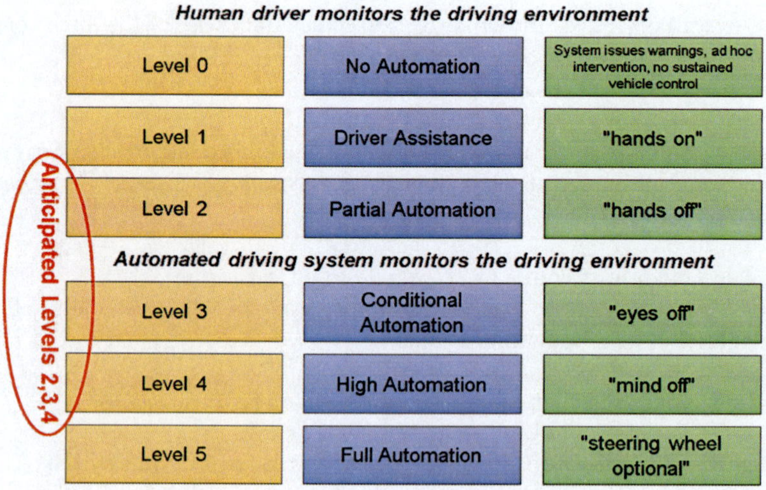

Fig. 2. SAE (J3016) automation levels [8].

vehicles could potentially harness many of the advantages of fully automated vehicles, while still keeping the driver in charge of the vehicle.

Nowadays SAE Levels 2,3,4 are so-called "Anticipated Levels 2, 3, 4" (refer Fig. 2). As SAE Level 0 a human is considered. The leading vendors like Tesla, Waymo (Google/Alphabet), Toyota provide in their products the elements of Levels 2, 3 and, partially 4, according to the given classification (refer Table 1):

Table 1. Levels of automated driving features [10].

Level	Status, modus operandi	Content
Level 0 – Warning/Indication	Standard	Haptic, acoustic or optical display
Level 1 - Driver Assistance	Standard	Execution by one system of either steering, braking, acceleration/deceleration
Level 2 – Partial Automation	Standard	Execution by more driver assistance systems e.g. steering & braking
Level 3 – Conditional Automation	Largely implemented	Limited self-driving with all-time driver fallback
Level 4 – High Automation	Experimental	Full automation within certain environment without driver fallback
Level 5 – Full Automation	Experimental	All-time automated driving at all traffic/road conditions without driver interaction

5 IoT in Traffic Telematics

A current trend is the increasing support for autonomous driving. The classic services are integrated herewith [2]:

- Navigation (via e.g. GPS, Galileo, Google, NavGear),
- Road condition reports, traffic jam reports via radio TC (traffic channel),
- Technical support:
  - o Measuring instruments for the driving situation (speed, cruise control),
  - o Problems (lack of fuel, higher oil pressure, cooling water absence),
  - o Alarms (engine oil pressure, brake oil pressure, used brakes),
  - o Active aids and assistance (ABS,…).

In addition, there are solutions for environment recognition too, such as position / value of important objects):

- Edge of the road, lane detection,
- Traffic signs, traffic lights,

- Dangerous road surface condition (oil spill, ice, aquaplaning…),
- Information about other traffic participants like,
- Cars and trucks (predecessor, successor, neighbors),
- Pedestrians,
- Bikes and moto-rollers,
- Relative speed (amount / direction),
- Expected intentions of the traffic participants (blinking,…),
- Special situations,
- Fire brigades, police, First Aid…,
- Children, disabled persons,
- Modern systems also support active driver actions,
- Driver information (visual / acoustic),
- automatic speed limit,
- automatic braking (traffic signs, dangers,…),
- Choice of traffic route,
- Choice of traffic speed,
- Decision to change lanes (navigation, overtaking).

6 Legislation and Social Acceptance

In addition to the technical problems, there are also problems of social acceptance, e.g. attribution of blame in the event of accidents (Table 2).

Table 2. Legislation and social acceptance [2].

Who is responsible?	There are some possible cases
Driver	Then it is not autonomous driving!
Vendor	High financial risk by multiple precedents and law cases on prominent car vendors are already known
Car insurance	High cost of policies as a sequence!
State	Makes only sense for a test period almost

In general, extreme real-time requirements and very high traffic safety requirements are applied to autonomous driving. Use of the navigation algorithms for route planning is already state-of-the-art. The timing is not critical, as the routes are planned in advance and, in the worst case, driving errors lead to detours.

On the other hand, the driving itself is time-critical. At a vehicle, velocity of approx. 100 km/h is equal to ~28 m/s. It means that 1 m is covered in 36 ms. Such comprehension leads to the fact that the detection of the traffic situation must be steady updated to a cm-accuracy in time intervals of at least 1 ms and all this with absolute reliability (higher radio network QoS requirements).

7 Case Studies

7.1 Case Study 1: Recognizing Traffic Signs and Detection of Traffic Lanes

Let us consider some sub-problems, e.g. the problem of recognizing traffic signs. This must be done with extreme security, even if the traffic signs are not always in optimal condition (dirt, corrosion, damage, overridden by hand-regulating police officers).

The traffic signs can be covered by trees or trucks in side lanes etc.

A related problem is the detection of traffic lanes. The borders must be reliably detected, even in problematic situations (no colour borders, yellow temporary repair tracks, locking cones, delimitation by leaves, snow as well as not easily recognizable, damages).

Both problems require complex, intelligent sensors that reliably guarantee measurement and transmission operations in the millisecond-time range.

The intentions of other traffic participants must also be recognized in real time, for instance:

- A car in the back in the same lane starts to change its lane, possibly without blinking,
- Playing children,
- Unsafe traffic participants and drivers (beginners, seniors, disabled persons),
- Vehicles with technical defects,
- Extra-long vehicles with out-swinging,
- Extremely powerful vehicles and sport cars (with macho drivers at the steering wheel).

7.2 Case Study 2: Warning Messaging

Due to the risk of rear-end collisions, traffic jam detection is important. In urban areas, limited control is possible through communicating induction loops in the road surface or light barriers on the road.

These are relatively sluggish because only counts are made and queue lengths are not measured directly.

Much more powerful is wireless communication between control systems and traffic participants/ road users who know the position of each participant and, if possible, their intended destination and, in principle, enable optimal control.

One of the important research areas is decentralized support for autonomous driving, which can be implemented, for example, under use of 5G radio technology and beyond. On this basis, it is possible to realize a comfortable detection of traffic jams for suburb highways and country roads. Traffic jams in curved areas are particularly dangerous (see Fig. 3) in such cases.

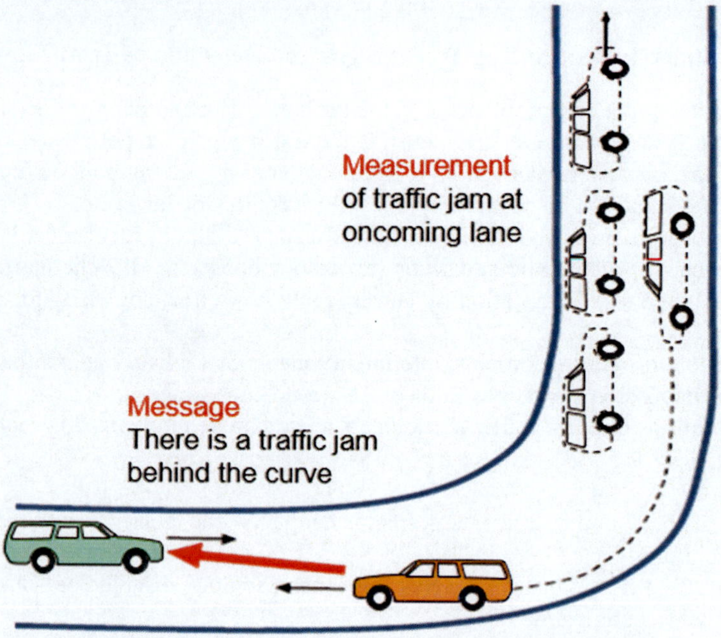

Fig. 3. Sensors for traffic: warning messaging [2].

In the given example, vehicle A wants to take a left turn without having a view of the near end of the traffic jam. The example makes use of the fact that vehicle B saw the traffic jam in the opposite direction while driving past. Vehicle B informs vehicle A that there is a risk of congestion in its direction. Vehicle A can then reduce its speed punctual.

7.3 Case Study 3: Truck Platooning

At present, motorways are heavily loaded by truck traffic. There is practically a permanent truck lane. Despite the large number of drivers, there is a high risk of rear-end collisions and jams.

With the help of the so-called electronic drawbar, columns of trucks can be formed and shaped under distance approx. 2 m (so-called Platooning). A driver drives in the first truck of the convoy (platoon), the following trucks are controlled by the lead vehicle. You drive at a short distance without a driver (see Fig. 4). The advantage of such solution is that there are no rear-end collisions and jams, less traffic space is required, only one driver drives and the others can relax and rest.

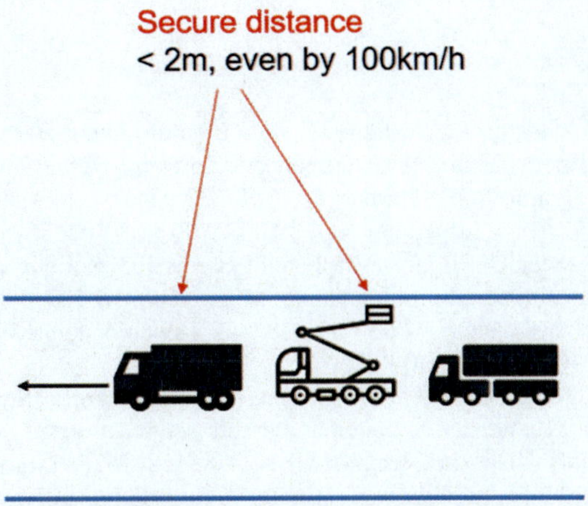

Fig. 4. Sensors for traffic: platooning [2].

The prerequisite is that all participants in the column are equipped with automatic driving systems that communicate via 5G and beyond radio technology under extreme real-time conditions (latency < 1ms). With individually assembled trains, variable train lengths would be possible. However, this would require a fundamental change in the safety concept of the railways.

The greatest benefits can be achieved through integrated transport systems.

8 Conclusions

Some technical and legal aspects of the SDC or UmV technology which domineering and bold emerge in the era of Digital Transformation are discussed.

The mostly important Self-Driving Cars Foundations are examined.

The following case studies are considered:

1. The problem of recognizing traffic signs and detection of traffic lanes,
2. The problem of Warning Messaging,
3. The problem of Truck Platooning.

The data security aspects remain out of scope of this work because of they require a separate and accurate consideration what is surely provided in other chapters.

The work does not claim to be complete in terms of the consideration of the above-mentioned subject and is merely a position paper.

The work contains a brief survey and some case studies.

Acknowledgment. The authors' kindly acknowledgment belongs to the editorial team and BA Dresden students and fellows for standing support, useful ideas, inspiration and challenges by fulfilling of this work.

References

1. Klymash, M., Luntovskyy, A., Beshley, M., (eds.): Future Intent-Based Networking: on the QoS Robust and Energy Efficient Heterogeneous Software-Defined Networks. LNEE 831, 28 chapters, monograph, Book, Hardcover, XXI, 530 pages, 1st ed. 2022, Springer, Cham. ISBN 978-3-030-92433-1. https://doi.org/10.1007/978-3-030-92435-5
2. Luntovskyy, A., Guetter, D.: Highly-Distributed Systems: IoT, Robotics, Mobile Apps, Energy Efficiency, Security, 1st edn. Springer, Cham, Monograph, XXXII, 321 pages, March 2022. ISBN-10: 3030928284, ISBN: 978-3-030-92828-5, 189 color figures (Foreword: A Schill) https://doi.org/10.1007/978-3-030-92829-2
3. Luntovskyy, A., Gütter, D.: Moderne Rechnernetze: Protokolle, Standards und Apps in kombinierten drahtgebundenen, mobilen und drahtlosen Netzwerken, 481 Seiten, 263 Abb., Springer Nature, July 2020. (in German). ISBN: 9783658256166 https://www.springer.com/gp/book/9783658256166. https://doi.org/10.1007/978-3-658-25617-3
4. Luntovskyy, A., Gütter, D.: Moderne Rechnernetze - Übungsbuch: Aufgaben und Musterlösungen zu Protokollen, Standards und Apps in kombinierten Netzwerken, 145 Seiten, 44 Abb., Springer, July 2020. (in German). ISBN: 9783658256180 https://www.springer.com/gp/book/9783658256180. https://doi.org/10.1007/978-3-658-25619-7
5. Luntovskyy, A., Spillner, J.: Architectural Transformations in Network Services and Distributed Systems: Current Technologies, Standards and Research Results in Advanced (Mobile) Networks, 344p. Springer, Wiesbaden (2017). ISBN: 9783658148409. https://www.springer.com/gp/book/9783658148409#otherversion=9783658148423. https://doi.org/10.1007/978-3-658-14842-3
6. Luntovskyy, A., Gütter, D., Melnik, I.: Planung und Optimierung von Rechnernetzen. Methoden, Modelle, Tools für Entwurf, Diagnose und Management im Lebenszyklus von drahtgebundenen und drahtlosen Rechnernetzen: Planung von Rechnernetzwerke theoretisch anspruchsvoll und praxisnah, 415 Seiten, 245 Abb.. Springer, Wiesbaden (2012). ISBN: 9783834814586. https://www.springer.com/gp/book/9783834814586#otherversion=9783834882424. https://doi.org/10.1007/978-3-8348-8242-4
7. Luntovskyy, A., Klymash, M.: Software Technologies for Mobile Apps, Apps for Fog Computing, Robotics and Cryptoapps, Lviv, 247 p. (2019). (Monograph, Ukrainian, ISBN 978-617-642-399-7)
8. Society of Automotive Engineers (2022). https://www.sae.org
9. ADAC Autonomous Driving (2022) (in German). https://www.adac.de/rund-ums-fahrzeug/ausstattung-technik-zubehoer/autonomes-fahren/
10. Assmayr, H., Geyer, D., Schwab, G.: Autonomes Fahren ist der Trend der Zukunft: Synergien zwischen Automotive und Offroad, Agrartechnik, Survey by Company AVL Software and Functions GmbH (2022). https://core.ac.uk/download/pdf/236374514.pdf

Smart Home: Protocols, Platforms and Best Practices

Dietbert Guetter(✉) and Andriy Luntovskyy

Computer Science Department, BA Dresden University of Cooperative Education, Saxon Academy of Studies, Hans-Grundig-Str. 25, 01307 Dresden, Germany
Dietbert.Guetter@t-online.de, Andriy.Luntovskyy@ba-dresden.de

Abstract. Emerging Networking (EmN) appears nowadays spontaneous and accompanies us in our everyday life as well as supports modern industry processes, ongoing digitalization of workflows and multiple Smart Home scenarios. This work is devoted to the Smart Home design: a brief smart survey of the used protocols, platforms and best practices under considering of price mirror, easy configuring and manageability, data security and privacy aspects as well as energy-efficiency. The practical problems for network and IoT components compatibility as well as cloud-independence are solved. The authors favor the open source and cloud-free solutions as it was shown in the presented case studies. Smart Applications become not only cloud-centric, but also container- and microservice-based.

Keywords: Emerging Networking (EmN) · IoT and IIoT · Sensing technologies · Smart home · HVAC (Heating · Ventilation · Air-Conditioning) · ZigBee · Bluetooth · Energy efficiency · Sensor GW · Process automation · Reliability and control · Smart applications · Smart office · Wireless sensor networks

1 Motivation

The concepts for Smart Home like cross-layered design, Low-Duty-Cycle, energy efficiency and minimal CO2 footprint are considered as the most important goals for IoT solution development [1–6]. The following technologies or their combinations are usually deployed (Table 1):

M. Klymash et al. (Eds.): TCSET 2022, LNEE 965, pp. 73–84, 2023.
https://doi.org/10.1007/978-3-031-24963-1_5

Table 1. IoT sensing technologies (radio/contactless) for smart home.

Type	Title	Frequency, MHz	Purpose
Radio sensing			
WAN	SigFox	868 MHz (EU), 902 MHz (USA)	Low-power wide-area network
	LoRa	2,4 GHz, 868/915 MHz, 433 MHz, 169 MHz	Low-power wide-area network
	NarrowBand-IoT	700–2100 MHz	4G mobile network
	LTE-CatM	700–2100 MHz	4G mobile network
LAN	WiFi5 IEEE 802.11ac	ISM 2400/5000 MHz	Wireless LAN
	WiFi6 IEEE 802.11ax	ISM 2400/5000 MHz	Wireless LAN
PAN	ZigBee IEEE 802.15.4	900/2400 MHz	Low-power personal-area network
	6LoWPAN	ISM 2400/5000 MHz	Low-power personal-area network
	Bluetooth IEEE 802.15.1	ISM 2400/5000 MHz	Low-power personal-area network
	EnOcean	868 MHz (EU), 928 MHz (Japan), 902 MHz (USA)	Low-power personal-area network (Energy Harvesting)
Contactless sensing			
Tags, labels, chipcards	RFID	LW 125–134 kHz, SW 13,56 MHz; UHF 865–869 MHz (EU), UHF 950 MHz (US/Asia), SHF 2,45 GHz/ 5,8 GHz	Multiple standards
	NFC	13,56 MHz	Standardized RFID, ISO 14443 / ISO 15693

2 System Integrators for IoT and Smart Home

In the world of IoT and IIoT there are multiple cloud platforms to support the integration of the existing IoT sensor hardware.

The system integrators for IoT and Smart Home possess a proven knowhow and as a rule use the following cloud-based or cloud-centric platforms [1–3]:

- IBM Watson IoT Platform,
- Microsoft Azure IoT Hub and Win for IoT,
- Google Cloud IoT,

- Amazon AWS IoT,
- SAP Internet of Things, SAP Leonardo IoT, SAP Edge Service, as well as a lot of further cloud-free commercial and freeware integrators.

They guarantee integrated connectivity and interoperability (refer Fig. 1):

- Nowadays there is always a technology available which can connect a device.
- For IoT devices NB-IoT, EnOcean and LoRa WAN are the most interesting technologies with low power and long-range communication.
- On the edge of a modern IoT network you can still use 4G, 5G and Beyond, Wi-Fi, Ethernet and LTE-CatM.
- These technologies are augmented via RFID/ NFC, refer [1–5].

The discussed platforms support new efficient and secured solutions (based on: https://www.imbuildings.com).

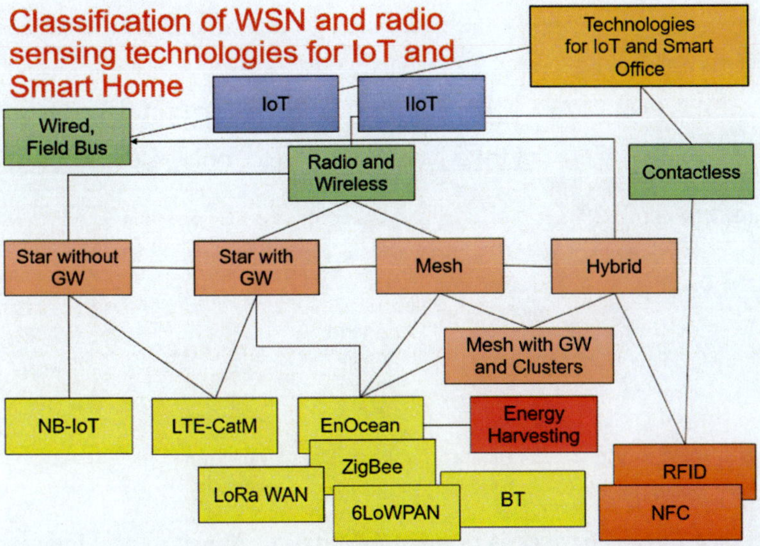

Fig. 1. Smart Home: integrated connectivity and interoperability [1, 3].

3 Challenges for IoT and Smart Home

The works [1–6] examine multiple case studies and gives the answers to the key questions about secured and blockchained as well as energy efficient IoT/ Smart Home and design paradigms. Summarizing, the following challenges are be considered:

- Small energy consumption and energy efficiency,
- Wide interoperability to 4G, 5G and Beyond, WSN, RFID, NFC, Robotics, Wearable,

- New efficient communications models with decentralization, i.e. M2M, fog-based models, P2P instead of convenient C-S or cloud-centric, but not only….

Conventional network security approaches, which mean centralized solutions with service orchestration and 1:1 and 1:n paradigm an are filtered by the convenient firewalls, IDS/IPS as well as use of Public Key Method and PKI, are to be substituted via advanced approaches which prefer decentralized solutions, more flexible service choreography for P2P and n:n, as well as Blockchain technology and so-called Cooperative Intrusion Detection P2P Networking (CIDN) [2, 5].

The next challenge is an any cases so-called Big Data problem [1, 3].

4 Big Data Problematic

However, the IoT data accumulating can lead to so-called Big Data [1, 3, 4, 6] problematics practically inevitably (Fig. 2):

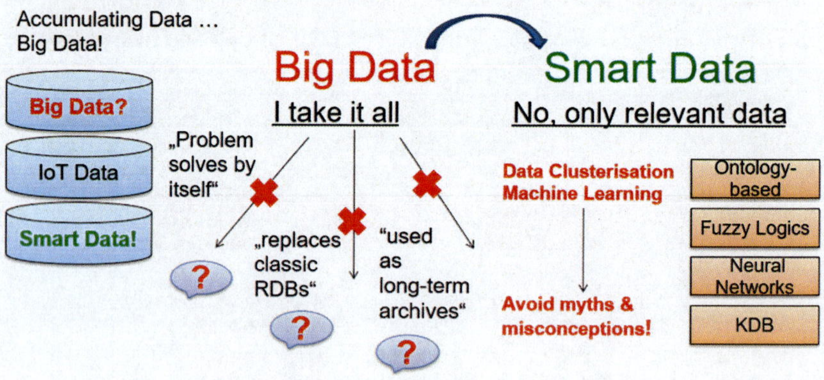

Fig. 2. "Big Data" to "Smart Data" transition [1].

Smart data for IoT devices is of great importance. Aimed to avoiding myths and misconceptions such as [1]:

- The problem solves itself,
- Such data replace only the classic relational databases,
- Such data can be used as long-term archives,

the data must be converted into smart data, which is also a challenge here. The wrong developer position "I take it all" must be replaced by "No, only relevant data". This can be achieved through a preliminary "data clusterization" and, then ML processes. From this point of view, the following methods are usually considered [1, 3, 4, 6]: ontologies, fuzzy logic, neural networks, knowledge DB.

5 Platform for Easy Integration and Best Practices

There is a wide spectra of communication technologies are practically relevant for IoT device integration, e.g. via WSN (Wireless Sensor Networks) and WPAN (Wireless Personal Area Networks).

LAN and Wi-Fi deployment is relevant for use of data flows based on "communication gateways-2-routers", "gateways-2-complex devices" connectivity, e.g. PC, notebooks, MFP, NAS, Smart TVs, radios etc. The Wi-Fi mesh networks use as a rule 3D-roaming protocol IEEE 802.11s (refer Fig. 3). The presented building with multiple floors provides a "smart" Wi-Fi mesh network with 3D-roaming for the listed networking devices.

Furthermore, the following standards can be mostly used for the mobile IoT boards in intranet (refer Sect. 1, Table 1) for small and simple devices' connectivity: ZigBee (e.g. for Alexa Gateways for LED-lamps, subwoofers, mini-cameras, digital cinemas…) and its multiple clones like e.g. ConBee II [8], Bluetooth (BT) as well as established on the market Z-Wave, EnOcean, also Homematik IP etc. (refer Fig. 1).

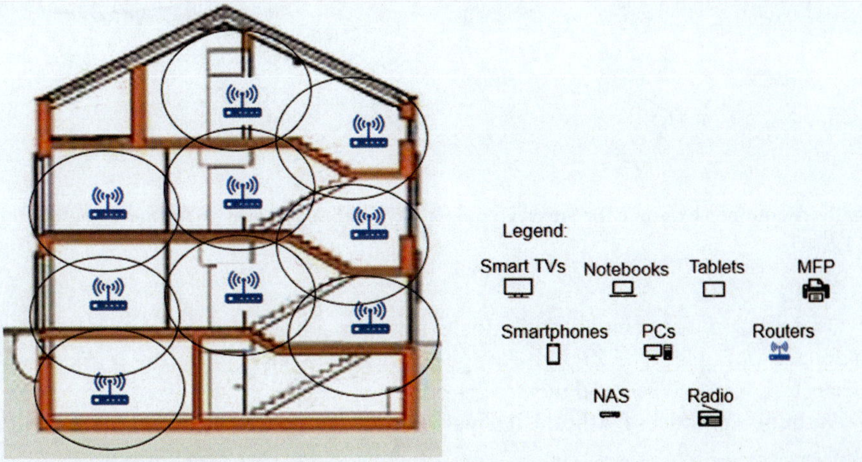

Fig. 3. A generalized example for Smart Home based on mesh Wi-Fi with 3D-roaming [1–7].

A gateway (GW) supports the protocols: Z-Wave, Wi-Fi, ZigBee and Bluetooth. The gateways should be compatible with many other vendors. The voice control can be provided via Amazon Alexa or Google Home. A Philips Hue Bridge enables simple control by voice via Alexa too.

However, use of license-free ISM-bands lead rather to harmony problems with Wi-Fi, ZigBee and Bluetooth [14]. Anyway, the compatibility guarantees between diverse network technologies are a hardcore problem [7–14]. For instance, the choice of a concrete technology and supporting platform [1–6] is problematically because of multiple companies do not fully implement the up-to-date ZigBee 3.0 standard and therefore the integrators work on a lot of compatibility problems (refer Sect. 2).

The Smart Home, based on further sensing protocols, i.e. BT, controls HVAC (Heating, Ventilation, Air-Conditioning) as well as lighting is shown in Fig. 4. The designers face frequently to the following problems:

- System interoperability,
- Confusion and obtrusiveness of different user hardware and software interfaces,
- Lack of data security and privacy due to proprietary software or external data storage effects based on the clouds (theft of data, outflow of data to companies),
- Requirement of IoT connections for reliable control.

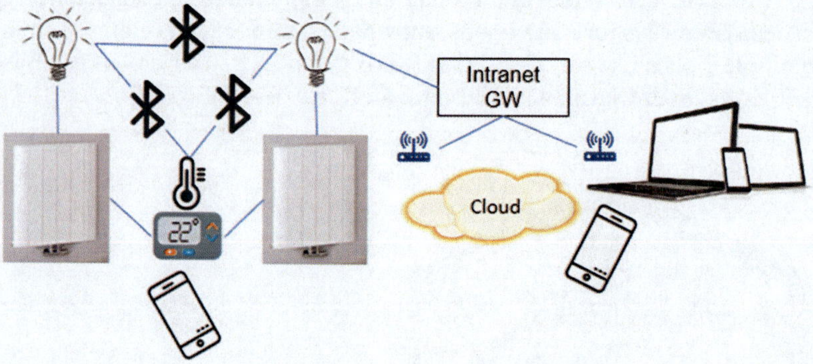

Fig. 4. A common example for Smart Home (HVAC, lighting) based on further sensing protocols [2, 12].

Own concept for a mobile IoT-Board for testing and learning aims using ZigBee is depicted in Fig. 5. The ZigBee Bridge provides the passage for the traffic from IoT devices to LAN or Wi-Fi and then to Internet.

As the best practices for the design platform alternatives, three variants are considered below:

- Tuya IoT Development Platform [9],
- Home Assistant V5 [10],
- Azure IoT Hub [11].

The discussed platforms provide pros and cons regarding to:

- Cloud-centricity,
- Data security,
- Language support and manageability.

Let's consider and provide details.

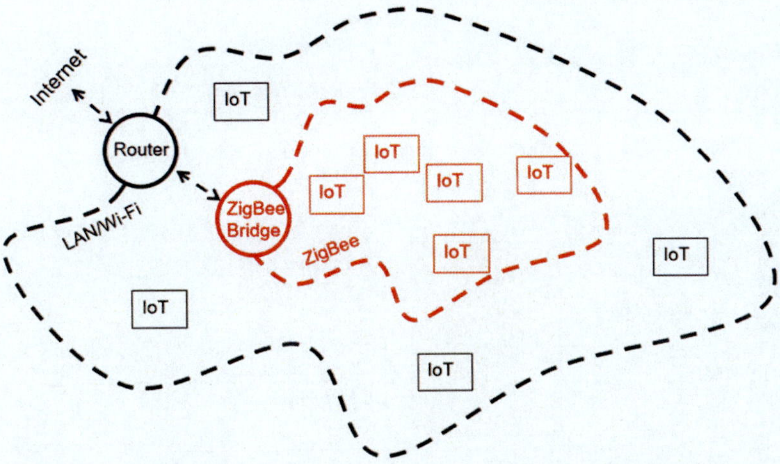

Fig. 5. An example for mobile IoT-Board for testing and learning aims [2, 12].

5.1 Tuya IoT Development Platform

Tuya Inc. is a Chinese software company [9] founded in 2014 and focused on the development of AI and IoT solutions. The Tuya Smart platform provides many solutions related to Smart Home too. Tuya provides a cloud-based platform that connects multiple IoT devices.

The Tuya vendor company offers various inexpensive solutions in a wide range of Smart Home areas and, therefore, as hardware basis is selected here. The products of other companies are often compatible too and can also be integrated, e.g. LED-lamps from Philips. However, the compatibility must be tested separately for each type of IoT device.

You can develop embedded Zigbee with Tuya products as required. For instance, ZSLCB Module Datasheet is considered. ZSLCB is a Zigbee module that Tuya has developed. It consists of a highly integrated wireless RF processor chip (EFR32MG21A020F768IM32-B), a few peripherals, a built-in 802.15.4 PHY/MAC Zigbee network protocol stack, and rich library functions.

ZSLCB is embedded with a low-power 32-bit ARM Cortex-M33 core, 768-KB flash program memory, 64-KB RAM, and rich peripheral resources. ZSLCB integrates all function libraries of the Zigbee MAC and TCP/ IP protocols.

In Fig. 6 the workflow for a Tuya cloud-based solution is depicted. Unfortunately, a row of such solutions is unsecure and privacy critical. The most important details of the hack of the Smart Home and IoT platform from the Chinese vendor Tuya have been published by [12]. Provisioning as a process means the initialization as well as preparation of IoT network to provide the set of services to its users. During provisioning, an IoT device listens to broadcast packets without being on the Wi-Fi itself. Subsequently it communicates unencrypted with Tuya servers in the cloud. Then, it's easy to intercept this traffic.

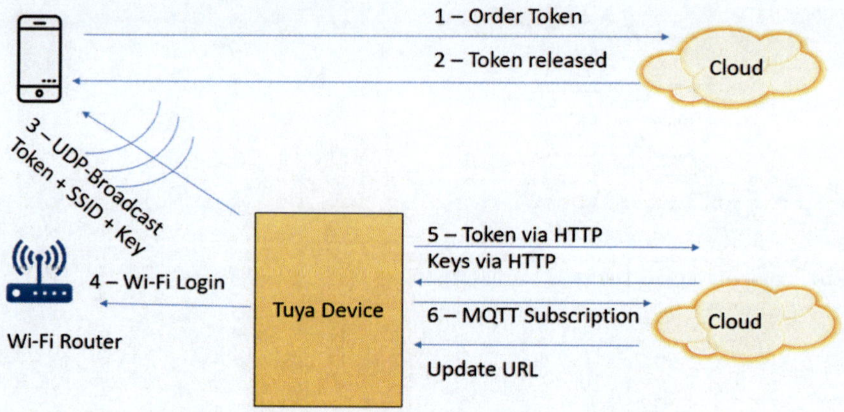

Fig. 6. Tuya Cloud-based solution (based on [12]).

5.2 Home Assistant v5

The author and founder of the concept [10] is Paulus Schoutsen together with Home Assistant (HA) Core Team and Community (2013). According to his own clever statement: "A good home automation never annoys, but is missed when it is not working." The integration platform works under the operating system Software Appliance & Virtual Appliance (Linux) and the programming language Python 3.9. As a category for use Smart Home, IoT under the Free License of Apache Software Foundation must be mentioned.

HA v5 is nowadays a free and open source home automation software designed to be the central control system in a Smart Home. Written in Python too, it focuses on local control and privacy within an intranet. It has a wide range of IoT devices supported with modular plug-ins to various IoT sensing technologies, systems and integrators' services.

5.3 Azure IoT Hub

Azure IoT Hub [11] acts a part of Microsoft's Azure portal and managed service hosted in the cloud. Its most important advantage is as follows: simple and fast scalability. PaaS (Platform as a Service) is used as the cloud service model. Azure IoT Hub offers reliable and secure two-way communication between IoT devices and applications with a wide range of functions. The contract, or service level agreement (SLA), corresponds to at least 99.9%. The costs reach approx. €21.08 per month (400,000 messages daily). The clients are divided into devices and services. The implementation can be done with the specialized SDKs for Java, NodeJS, Python, .NET, C# as well as under use of the standard protocols like [1–6]:

- MQTT (Message Queuing Telemetry Transport),
- AMQP (Advanced Message Queuing Protocol),
- REST (Representational State Transfer) over HTTPS.

6 Case Studies

Paradigms are a set of concepts or design patterns, including theoretical and empirical models, research methods, practical tools, platforms and standards, what means a significant contribution to a field.

Osram, Amazon, Bosch, ConBee/ ZigBee and CC25351 (sticks from Dresden Elektronik), Samsung, Tuya, Xiaomi, Aquar and certainly a few more. Still problems are available with the compatibility of the different manufacturers, despite the same protocols [1–7, 13].

The given example is aimed for learning aims. A concept for a Mobile IoT-Board for Testing and Learning Aims is depicted in Fig. 7. The hardware is deployed and used for learning aims for an appropriate IoT course at Saxon Academy of Studies, BA Dresden University of Coop. Education [14]. The IoT device programming is easily: simple menu-driven interface with a smartphone and based only on "IF-THEN-ELSE" rules [2, 8–14]. Some Python-similar languages can be used alternately. A preliminary inventory list for selected compatible components is shown in Table 2. The list for students' use and learning goals was created advisedly price-sensitive.

Table 2. Inventory list for selected compatible components.

Vendor	Trade mark	Functionality
Pearl	7Links	ZigBee Gateway (Tuya compatible)
	7Links	Wi-Fi Security Camera
	Revolt	ZigBee Radiator Thermostat
	VisorTech	RF Gateway
	VisorTech	RF Remote Control
Lidl	SilverCrest	ZigBee Intermediate Plug
	SilverCrest	ZigBee Movement Sensor
	SilverCrest	ZigBee Door / Window Closing Contact + Alarm
	Livarno	ZigBee LED-Lamp (color)
TP-Link		Router (TL-WR710N)
ELV		Siren (122772)
Ikea	TradFri	ZigBEE LED-Lamp

A box with multiple built-in IoT devices is depicted in Fig. 7. The top for the mobile IoT devices (battery-based) is given in Fig. 8.

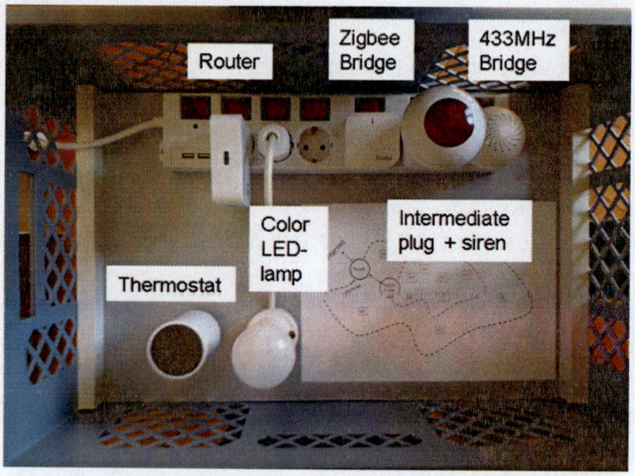

Fig. 7. A box with multiple built-in IoT devices (own construction of Dietbert Guetter [2, 14]).

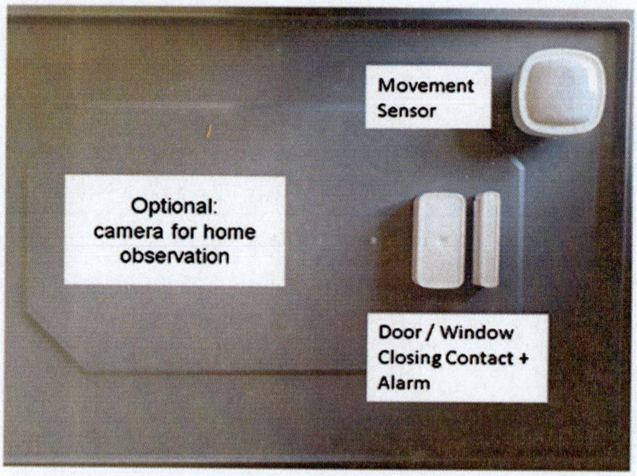

Fig. 8. The top for the mobile IoT devices (battery-based, own construction of Dietbert Guetter [2, 14]).

7 Conclusions

The work contains a brief smart survey of the used protocols, platforms and is illustrated with some best practices. The practical problems for network and IoT components compatibility as well as cloud-independence are solved. This work can be positioned as a Work-in-Progress.

The choice of a concrete technology and supporting platform is problematically because of multiple companies do not fully implement the up-to-date standards. The cloud solutions provide a comfortably way to integration of IoT devices but are problematic regarding to data security; sometimes, they provoke such disadvantageous like privacy legislation conformity.

Furthermore, the authors favor mostly the open source and cloud-free solutions for IoT and Smart Home applications. For instance, company Dresden Elektronik Ingenieurtechnik GmbH is situated on the right path with its Zigbee Gateway ConBee II for multiple IoT devices. The gateways are not cloud-based and are better secured regarding to the traffic within the ZigBee piconet. Therefore, Smart Applications become not only cloud-centric, but also container- and microservice-based.

Acknowledgment. The authors' kindly acknowledgment belongs to the BA Dresden students and graduates Tom Schwabe, Tim Zobjack, Alexander Spielberg, Clemens Gawer, Richard Everth, Christin Flischikowski, Marco Hebestadt, Martin Funke, Willi Knabe for technical support, useful ideas, inspiration and challenges by fulfilling of this work.

References

1. Luntovskyy, A., Guetter, D.: Highly-Distributed Systems: IoT, Robotics, Mobile Apps, Energy Efficiency, Security, 1st edn. Springer, Cham, Monograph, March 2022. ISBN-10: 3030928284, ISBN: 978-3-030-92828-5, XXXII, 321 pages (2022). 189 Color Figures (Foreword: A. Schill). https://doi.org/10.1007/978-3-030-92829-2
2. Homepage of Dietbert Guetter (online). http://www.guetter-web.de/
3. Klymash, M., Beshley, M., Luntovskyy, A. (eds.): Future Intent-Based Networking. LNEE, vol. 831. Springer, Cham (2022). https://doi.org/10.1007/978-3-030-92435-5
4. Luntovskyy, A., Gütter, D.: Moderne Rechnernetze: Protokolle, Standards und Apps in kombinierten drahtgebundenen, mobilen und drahtlosen Netzwerken, 481 Seiten, 263 Abb., Springer, Wiesbaden, Juli 2020. ISBN: 9783658256166. https://www.springer.com/gp/book/9783658256166. https://doi.org/10.1007/978-3-658-25617-3
5. Luntovskyy, A., Gütter, D.: Moderne Rechnernetze - Übungsbuch: Aufgaben und Musterlösungen zu Protokollen, Standards und Apps in kombinierten Netzwerken, 145 Seiten, 44 Abb., Springer, Wiesbaden, July 2020. ISBN: 9783658256180. https://www.springer.com/gp/book/9783658256180. https://doi.org/10.1007/978-3-658-25619-7
6. Luntovskyy, A, Spillner, J.: Architectural Transformations in Network Services and Distributed Systems: Current technologies, Standards and Research Results in Advanced (Mobile) Networks, 344 p. Springer, Wiesbaden (2017). ISBN: 9783658148409 https://www.springer.com/gp/book/9783658148409#otherversion=9783658148423. https://doi.org/10.1007/978-3-658-14842-3
7. Luntovskyy, A., Gütter, D., Melnik, I.: Planung und Optimierung von Rechnernetzen. Methoden, Modelle, Tools für Entwurf, Diagnose und Management im Lebenszyklus von drahtgebundenen und drahtlosen Rechnernetzen: Planung von Rechnernetzwerke theoretisch anspruchsvoll und praxisnah, 415 Seiten, 245 Abb., Springer, Wiesbaden (2012). ISBN: 9783834814586. https://www.springer.com/gp/book/9783834814586#otherversion=9783834882424. https://doi.org/10.1007/978-3-8348-8242-4
8. Dresden Elektronik Ingenieurtechnik GmbH. https://www.dresden-elektronik.de/produkte.html
9. Tuya Smart: Global IoT Development Platform Service Provider. https://www.tuya.com/
10. Home Assistant Platform. https://www.home-assistant.io/
11. Azure IoT Hub. https://azure.microsoft.com/de-de/services/iot-hub/
12. Merlin Schumacher. Der Tuya-Smart-Home-Hack: IoT-Komponenten von unsicherer Tuya-Firmware befreien (in German). https://www.heise.de/select/ct/2019/4/1550226849835235

13. IoT System Integration. https://www.imbuildings.com/
14. Dietbert Guetter. Vorlesungsskript und Praktikum "Internet der Dinge" (2022). (in German). https://bildungsportal.sachsen.de/opal/auth/repository/catalog/3183116302;jsessionid=DDE92063C03990E03632F376619D1F55.opalN7/

Construction and Methods for Solving Problems at the SDN Control Level

Oleksander Romanov[1](✉), Mykola Nesterenko[1], Gennaro Boggia[2], and Domenico Striccoli[2]

[1] National Technical University of Ukraine "Igor Sikorsky Kyiv Polytechnic Institute", Peremoga Ave. 37, Kyiv 03056, Ukraine
a_i_romanov@ukr.net

[2] Polytechnic University of Bari, Via Edoardo Orabona, 4, 70126 Bari, BA, Italy
{gennaro.boggia,domenico.striccoli}@poliba.it

Abstract. Today, many leading telecom operators are beginning to actively implement SDN technology. This allows them to increase network agility, quickly introduce new services, and break free from OEM dependency. The optimality of the decisions made largely depends on the capabilities of the control plane in the SDN architecture. The chapter analyzes possible principles for constructing a logically centralized control plane for SDN networks, the purpose and functions of the main blocks of the controller and the open operating system of the network, the process of interaction of system elements in the process of solving control problems. To obtain optimal solutions in the control process, a mathematical formulation of control problems is required using real indicators of the functioning of telecommunication networks as input data. The chapter analyzes several methods for solving control problems, gives a mathematical description of each method and conducts their comparative evaluation. The results obtained made it possible to determine the limitations of the application of management methods and the areas of their optimal use. The mathematical apparatus makes it possible to obtain numerical values of throughput and quality of service indicators in various network operation modes.

Keywords: Controller · Software-defined networking · Open network operating system · Control plane disaggregation · Message distribution system · OpenFlow · Datapath model

1 Introduction

Existing telecommunication networks are a set of specialized network devices that were created on the basis of software and hardware devices from various manufacturers. At the same time, any equipment upgrade cannot be carried out without the participation of the manufacturer. Therefore, the introduction of new modern services on networks, as a rule, requires the replacement of the old equipment fleet with a new one. This approach leads to long cycles of design, purchase of new equipment and commissioning. This negatively affects the effectiveness of the introduction of new services.

M. Klymash et al. (Eds.): TCSET 2022, LNEE 965, pp. 85–101, 2023.
https://doi.org/10.1007/978-3-031-24963-1_6

The analysis shows that at present the networks of telecom operators consist mainly of "monolithic physical" network elements, where the functions of management, administration and data transmission are performed by physical devices. Often the network of the telecom operator is built using network elements of the same manufacturer, since in this case it is easier to ensure compatibility. The introduction of new services, modification of equipment or services is usually carried out in turn at each element of the network and requires coordination of the available resources of the operator. This approach makes the operator's network inflexible, makes it difficult to introduce new services, and also increases the operator's dependence on specific vendor solutions. Therefore, many leading telecom operators are beginning to actively implement SDN technology in their networks.

A greater contribution to the development of SDN is made by representatives of the Open Networking Foundation (ONF) consortium, who have developed a number of documents describing the principles of building and operating SDN networks [1, 2]. In [3–6], the general requirements, system approaches and generalized architecture of SDN networks are considered. The papers [7–10] consider the tasks and features of the protocols used to solve various problems in SDN networks. The works [10, 11] describe the features of building SDN network elements and the order of their interaction in the process of servicing information flows. In [12–16], the principles of building an optical transport network are considered and recommendations are given to ensure the safety of their operation. Quite complete and systematized material is presented in [17, 18]. Articles [19–22] explore various aspects of servicing information flows and focus on the need to meet network security requirements.

It should be noted that most of the works deal with the practical implementation of SDN networks based on open source platforms. At the same time, such issues as the mathematical description of methods for solving control problems in various operating conditions, ensuring the optimal use of network resources, and improving the quality of user service are not considered very often. And the solution of these tasks is usually assigned to the controller, network operating system or control applications located above the control level. Optimal control algorithms are the basis of the software of these functional elements. In this chapter, possible formulations of control problems, their formalized representation in the form of mathematical models will be analyzed, and a comparative assessment of control methods in terms of throughput will be carried out.

2 SDN Control Plane Architecture

Let's consider the SDN architecture and analyze the functions of its elements in the process of serving user requests. In doing so, we will focus on the elements that are directly involved in the process of managing load servicing. It should be noted that in the near future SDN networks will work with networks of the old fleet. Therefore, there must be elements that will ensure the interaction of SDN networks with networks using outdated technologies in automatic and semi-automatic modes of operation. In addition, it should be possible to include a human operator in the control loop for manually entering commands related to setting up service policies, assigning priorities in the distribution of network resources, and other tasks that do not have a formalized description. The architecture of such a system and its components are shown in Fig. 1. This architecture takes into account the recommendations of ONF TR-502 [6, 23, 26].

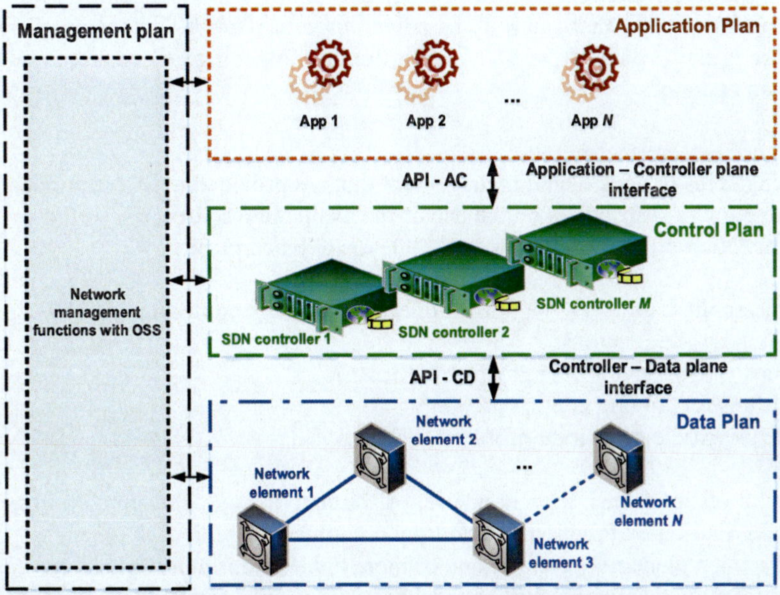

Fig. 1. SDN with function management OSS.

An SDN architecture may include the following main components:

- data plane;
- control plane;
- application plane;
- management plane.

At the same time, the controller plane, application plane, management plane can take direct part in the management process.

The management plan ensures the interaction of SDN networks with networks that incorporate legacy control systems built on the basis of Operations support systems (OSS). In addition, this functional block provides for the possibility of human intervention in the control process - the operator. Generally, it can be used for:

- ensuring the interaction of control systems for supporting operations of OSS networks of the old park with the SDN controller;
- connection of a human operator to the SDN controller;
- initial configuration of network elements;
- assigning to SDN controllers the network segments that will be under their control.

It should be noted that the initial setup of the network management system without the participation of a human operator is practically impossible. At this stage, there is a fairly large amount of poorly structured tasks that cannot be automated. In addition, there are management tasks that can only be solved by organizational measures. Therefore, during the initial launch of the SDN controller, the following tasks can be assigned to the human operator:

- configuring user service policies;
- defining areas of control and management that are provided to SDN applications;
- introduction of restrictions on the use of the available resource;
- determination of the list of system parameters to be controlled.

At the application level, the human operator typically configures:

- contracts and service level agreements (SLAs);
- algorithms for solving control problems;
- device priority level in joint problem solving.

At all levels of the SDN architecture, the human operator sets up a security policy that allows distributed functions to securely communicate with each other.

Application plane may contain one or more applications and applications can be of various types. First, these are applications that are designed to provide services to clients of the SDN network architecture. Such applications do not participate in the management process. They are the source of increasing the number of users on the network. And as a result, an increase in the income of the operator. Secondly, these are applications that are designed to solve individual specific control problems. These are applications that have a modular horizontal structure and their number increases as new methods for solving control problems are developed. These applications are directly involved in the management process and are launched by the SDN controller as needed. Thirdly, applications that themselves can act as an SDN controller. Such applications can take on some of the control functions. Can be used as a backup to the primary SDN controller. But they must always run under the control of the SDN controller, which is designated as the main (leader).

Each application has its own set of resources over which it has exclusive control. However, an application can only be able to use this set of resources under the control of one or more SDN controllers. Applications can:

1. Form requests for the use of network resources to the controller plane.
2. To interact with each other using the SDN controller as a control.
3. Submit network requirements to the controller plane in the process of solving control problems.
4. To act as an SDN controller.

Control plane. It should be noted that the solution of all network management tasks is carried out in cooperation with the controller plane, which may contain one or more SDN controllers. Each SDN controller is the authority for a set of network resources of one or more elements in the data plane or application plane. The functionality of the SDN controller should provide the following network tasks:

- collection of initial data for solving management problems;
- control of the state of the main network elements;
- processing requests for applications that are in the area of responsibility of its domain and are assigned to it;
- isolation of applications from each other;
- control of network operation parameters;
- human access - operator with the required priority level for manual input of control commands;
- timely response to changes in the structure, load and quality of service in the network;
- providing a solution to control problems to restore normal operation after a failure.

3 Functional Blocks for Collecting Information for Solving Control Problems

Management of a system of any level of complexity involves collecting information about the state of network elements, identifying deviations of system parameters from normalized values, making decisions on methods for bringing network characteristics to the required standards, and generating control commands for control bodies. In [6, 8] for solving the problems of managing SDN networks, it is pro-posed to use the following functional blocks:

1. Agents.
2. Coordinators.
3. Virtualizers.
4. Resource databases (RDB).
5. Managers.

The analysis of the recommendations [8, 13, 16] made it possible to detail the functions of these elements, clarify their locations and the order of interaction in the process of solving control problems. Taking into account the results obtained, the structure of the modular control plane was proposed, which is shown in Fig. 2.

Consider the functions of the "Agent" block. Usually this is the control object. The purpose of this function block is as follows:

1. Monitoring of failures in network elements located in the agent's area of responsibility.
2. Collection of data on the performance of controlled network elements.
3. Collection of statistical data on network operation (loading ports, queue length, service delays).
4. Collection of initial data for solving problems of optimizing the parameters of the network operation.
5. Determination of deviations of controlled parameters from normalized values.

6. Timely transfer of information to the control body about changes in the parameters of the controlled object.

It should be noted that depending on the location of the agent, the list of tasks assigned to him may vary significantly. Therefore, you should immediately detail the location of the agent by assigning the appropriate index. And accordingly, to clarify the list of tasks to be solved by him. Based on the SDN architecture for placing agents and coordinators in Fig. 2. at least two groups can be distinguished:

1. Agents that are located at the Application level, and designate them as agent – $A_{APi}, i \in \overline{(1, N)}$.
2. Agents that are located at the Data plane level, and designate them as agent–$A_{APl}, l \in \overline{(1, N)}$.

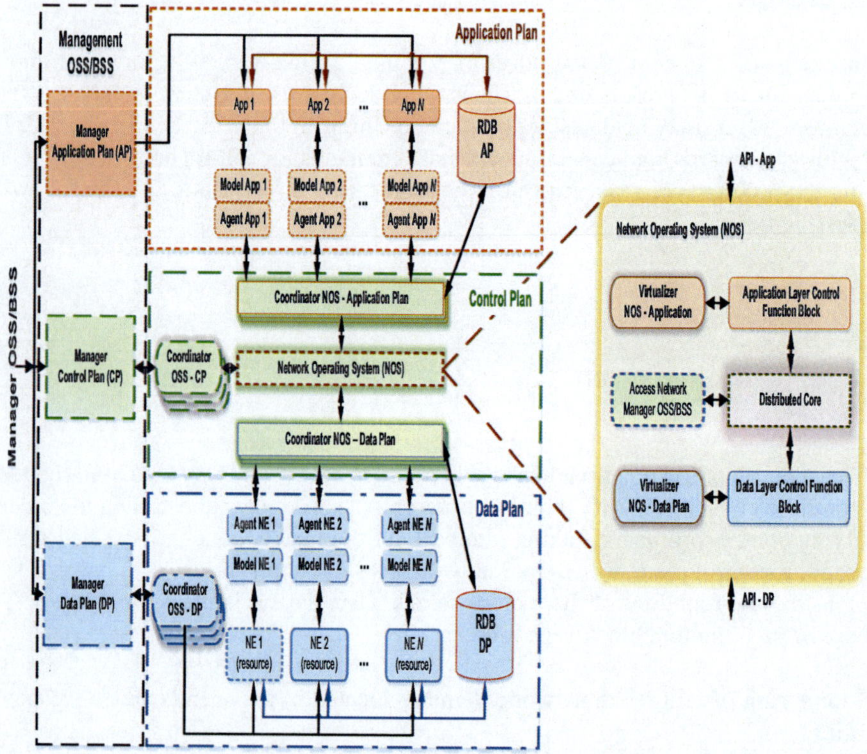

Fig. 2. Functional structure of the control modules plane.

An important feature of SDN networks is that agents must control the network at different levels of abstraction. At the same time, the quality and level of network control should not depend on what type of elements the agent serves. It can be both physical and virtual elements. Therefore, it is advisable that the agent includes a model of a controlled object in its composition. In this case, the process of servicing both virtual and physical network resources will be the same and will not depend on the equipment manufacturer.

On Fig. 2. The proposed functional structure of agents, recommended places for their placement and the order of interaction in the process of solving control problems are presented. In addition, here is a detailed diagram of the ONOS functional block, the location of coordinators, virtualizers and other elements.

The coordinator is a functional component with the help of which a human operator can influence the order of the network operation. It acts on behalf of a human operator and has the ability to provide and modify network service policy requirements. The coordinator may communicate control information to the network elements in the form of data models from the control plane and the application plane. Therefore, the functional blocks of the coordinator are located everywhere.

An important function of coordinators is to ensure that the SDN control layer interacts with legacy technology networks that use OSS/BSS control systems. All coordinators are directly connected to the OSS/BSS Management layer, that is, the layer that can transmit control commands from a human operator with a high priority level.

There are four types of coordinators:

1. Coordinators are placed, on the Control Plan level, for the interaction of NOS and Application plan – $C_{NOS\_APb}, b \in \overline{(1, N)}$
2. Coordinators are placed, on the Control Plan level, for the interaction of OSS/BSS and Control plane – $C_{OOS/BBS\_CPd}, d \in \overline{(1, N)}$.
3. Coordinators are placed, on the Control Plan level, for the interaction of NOS and Data plan – $C_{NOS\_DPw}, w \in \overline{(1, N)}$.
4. Coordinators are placed, on the Data plane level, for the interaction of OSS/BSS and Data plan $C_{OOS/BBS\_DPy}, y \in \overline{(1, N)}$.

The main contribution to the solution of control problems is made by the coordinators of the Control plane level.

An important element of the control plane is the virtualizer. In the SDN architecture, virtualization solves the problem of forming and allocating virtual resources for specific applications. The SDN controller offers application services as an instance of an information model that abstractly describes the resources available and the policy for their use. A virtualizer is a functional object that stores an instance of the information model of resources and policies for their use.

The virtualizer is created by the OSS/coordinator for each client application. The OSS/Coordinator allocates resources and defines the policy that the virtualizer should use when providing services to applications through the API. Next, the virtualizer creates an agent, in which its own copy of the information model of available resources and the policy of its use is written to provide a service for a specific program.

Management in such a complex system as SDN is impossible without monitoring the parameters of the system functioning and the state of its elements. These data are used as initial data in solving control problems. To do this, the state of the system elements is monitored both at the Data plane and Application plan levels. The monitoring data is stored in a Distributed Resource Database (RDB). Based on this data, models of virtual network resources are created and the tasks of reshaping the load distribution plan are solved in cases where the network parameters deviate from the required norms.

4 Architecture of an Open Network Operating System

ONOS development is a separate open source ONF project distributed under the Apache 2.0 license [15]. In addition, the system description of ONOS is published in the public domain [20].

ONOS provides the flexibility to create and deploy new dynamic network services with simplified APIs. ONOS supports the required real-time network configuration and management. This approach eliminates the need to run routing and switching control protocols within the operator's network infrastructure.

Currently, a search is being made for possible ways to optimize and improve the efficiency of ONOS functioning. One of these ways can be the transfer of the ONOS functional block to the cloud. This will allow end users to create new network applications without having to change the data plane.

A possible version of the ONOS architecture is shown in Fig. 3. The design of the ONOS architecture is modular. This approach allows you to ensure the optimal consumption of resources, since each specific deployment will determine the optimal number of required subset of modules.

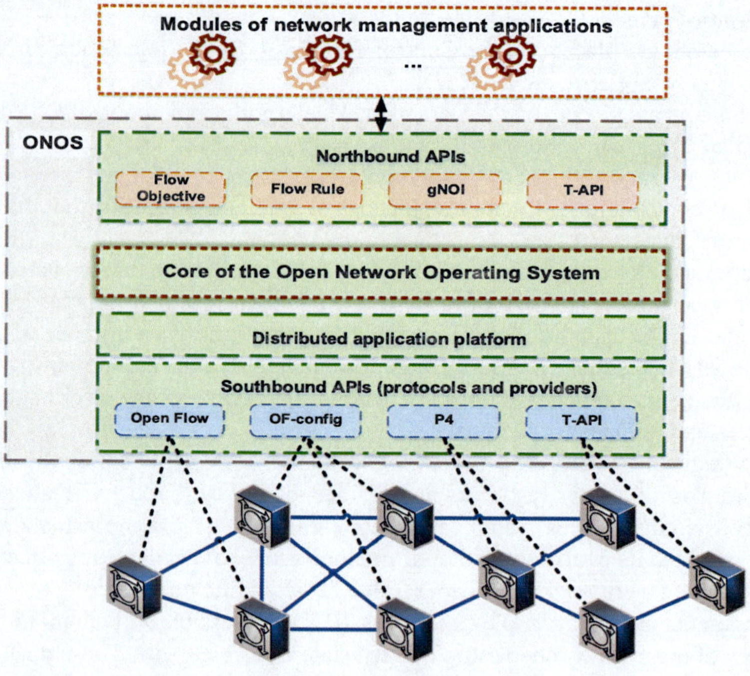

Fig. 3. Architecture ONOS.

ONOS architecture includes:

1. The core of an open operating system, which is responsible for solving all the intellectual tasks of network management.
2. Platform of distributed applications, which is a set of applications for solving network management problems. This platform is built as an extensible modular system. In addition, each modular functional block solves a specific control task. As the list of management tasks to be solved increases, the number of modules in the platform of distributed applications will increase.
3. A set of open network interfaces. At the same time, just like in the SDN architecture, it is proposed to use two types of interfaces:
- open southern interface (SBI), which is a set of modules for interacting with the data layer.
- open northern interface (NBI), which is a set of modules for interacting with functional blocks of the application layer.

This ONOS architecture will allow:

1. Simplify the process of reinstalling and upgrading software on network elements.
2. Reduce the cost of introducing new technologies and services throughout the operator's network.
3. Step-by-step increase in the number of control tasks to be solved by adding new functional blocks as part of the platform of distributed applications.
4. Create a horizontally scalable system that provides a high level of fault tolerance, which is very important to ensure that the specified requirements for the reliability of the functioning of centralized control systems are met.

Based on the above, it is possible to formulate requirements for the ONOS architecture. First, the horizontal plane of the SBI must be large enough. This is because any access to the underlying hardware will be through ONOS. Therefore, the totality of all southern APIs should be sufficient to set up, maintain, and manage the network.

In addition, a multi-layer system of applications and services running on ONOS should be provided. Depending on the priority of the task solved by the application and the frequency of its use, the level of efficiency of task processing should be determined. For example, applications that define the main control functional block in the event of failure of network elements should have a direct connection with ONOS. And applications that provide the installation of new software, certificates, configuration parameters and use the OF-CONFIG protocol in their work can be located at the level with a large delay.

Another requirement for ONOS is to provide simultaneous two-way exchange of information through NBI and SBI. As can be seen from Fig. 3, through NBI, applications use ONOS to manage network elements, and through SBI, southbound modules transmit information about the state of the core network to the ONOS core.

The interaction between the ONOS core and network devices is provided by a set of protocols and interfaces, such as OpenFlow, OF-CONFIG, P4, T-API, which take care of the details of interaction with devices, thereby isolating the ONOS core and applications running on top of it from the details of a variety of network devices.

The core of ONOS consists of a number of subsystems, each of which is responsible for a certain aspect of the functioning of the network. Each subsystem maintains its own service abstraction, which is responsible for propagating network state parameters across the cluster.

5 SDN Mathematical Description of SDN Network Management Methods

So far, we have considered the principles of building the control plane as a set of functional blocks. Their possible structure and order of interaction are considered. Another important issue in SDN networks is to ensure the optimal solution of control problems in various operating conditions. The optimal use of SDN network resources, its bandwidth and quality of service depend on what methods for solving control problems will be implemented in the control plane.

Let us consider possible methods for solving problems of managing load distribution in a network. These methods in the form of algorithmic or software can be written in the control plane, and the solution of control problems can be assigned both to the controller or network operating system, and to control applications located above the control level. Let's analyse control methods, give their mathematical description and define the boundaries of their optimal use.

One of the possible ways to increase network throughput is to use workarounds in the process of servicing incoming requests. This is due to the fact that when the load increases in some information directions, in other directions at this time a reserve of free channels is possible due to the lack of load. Using unloaded sections of the network as workarounds, it is possible to reduce overall network losses and, as a result, increase the overall throughput of the communication network.

However, the use of workarounds can also lead to a decrease in network throughput. When the load entering the network increases to a certain value, an overload occurs in the network. At the same time, denials increases sharply, as well as the quality of service decreases than in a networks of direct connections. This is explained by the fact that at high loads, a significant part of client requests use bypass connections, the maintenance of which requires more network equipment.

Let us consider the process of network transition from an area where detours are profitable to an area where detours are not advantageous using the example of the functioning of the simplest three-node network (Fig. 4.), the branches of which contain one channel each.

Let the load Z_{ij} supplied for servicing in different information directions be equal in magnitude and independent of the direction:

$$(Z_{12} = Z_{21}) = (Z_{23} = Z_{32}) = (Z_{13} = Z_{31}) = Z, \tag{1}$$

where: Z – load in the information direction is measured in Erlangs.

We accept the restriction that the flows of requests arriving for service are Poisson flows.

As noted above, servicing incoming requests using workarounds does not always lead to an increase in the efficiency of using network equipment. There are areas where connection management not only fails to reduce overhead, but also leads to an increase in overhead compared to serving claims over direct paths.

Let's analyse the operation of this network for two algorithms:

1. Using only direct paths (network of direct connections).
2. Using bypass paths (network with bypass connections).

For both algorithms, it is possible to obtain analytical dependences of losses in the network on the magnitude of the incoming load.

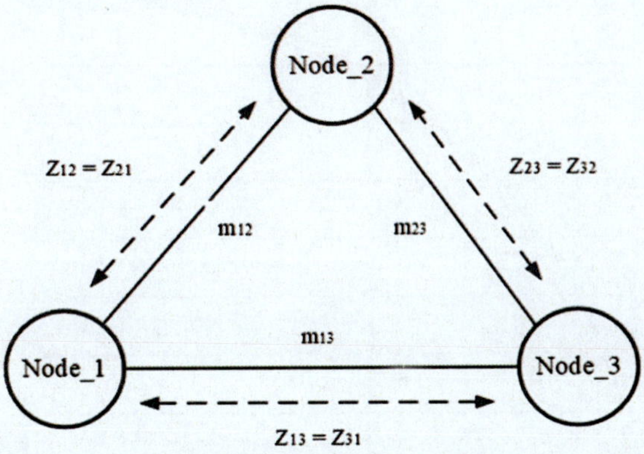

Fig. 4. Three-node topology with one virtual channel in network branches.

For a direct connection network (bypasses are prohibited), the dependence of the quality of service $P_{ij\_network\_DC}$ on the value of the incoming load Z can be described by the following relationship:

$$P_{ij\_network\_DC} = \frac{Z}{Z+1}. \tag{2}$$

A similar dependence $P_{ij\_network\_BC}$ in a communication network with bypasses, can be described by the following relationship:

$$P_{ij\_network\_BC} = \frac{Z^2 \cdot (Z^2 + 7 \cdot Z + 9)}{(Z^4 + 8 \cdot Z^3 + 15 \cdot Z^2 + 8 \cdot Z + 2)}. \tag{3}$$

For a comparative assessment of dependences (2) and (3), their plots $P_{ij\_network\_DC} = f(Z)$ and $P_{ij\_network\_BC} = f(Z)$ are shown in Fig. 5 (curves Network_DC and Network_BC, respectively). The curves intersect at load $Z = 1$Erl. At this point, the probability of denial of service $P_{ij\_network\_DC} = P_{ij\_network\_BC} = 0, 5$.

Thus, the use of workarounds for this network is reasonable when the load is less than 1 Erl. When the load increases above 1 Erl., the use of workarounds becomes unprofitable.

The load Z at which $P_{ij\_network\_DC} = P_{ij\_network\_BC}$ is called the critical load $Z_{critical\_load}$, i.e. $Z_{critical\_load} = Z = 1Erl$.

As the number of virtual channels in the branches of the network shown in Fig. 4, the value of the loss corresponding to the critical point decreases.

The dependence of the coordinates of critical points on the number of channels in the network branches obtained in [24, 25] is shown in Fig. 6.

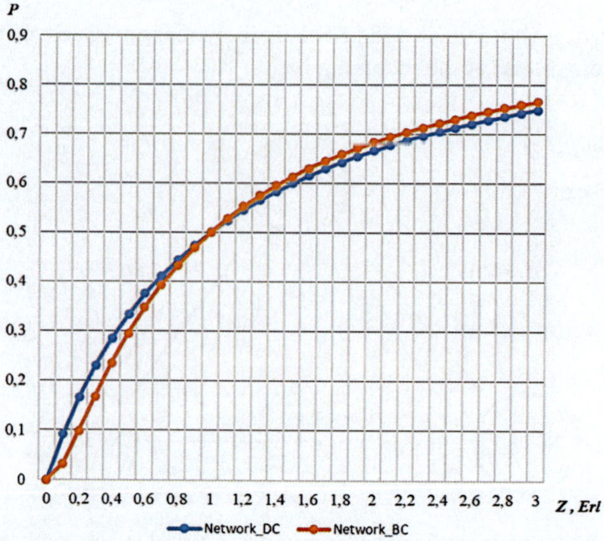

Fig. 5. Probability of denial of service in operation with direct and bypass routes.

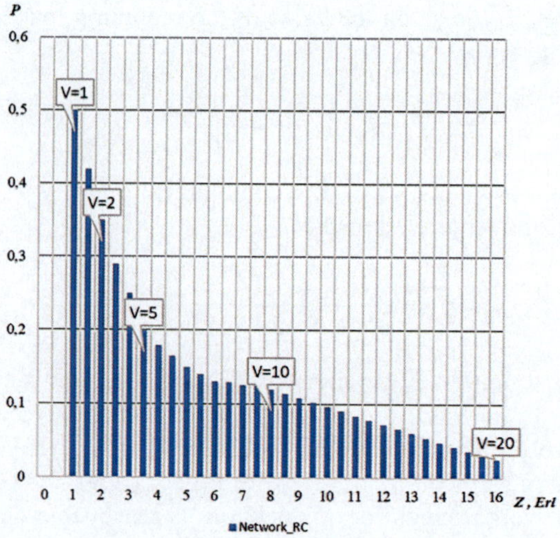

Fig. 6. Graph of the rational use of bypass routes from the input load and the count of virtual channels.

The shaded part of the graph shows the area of reasonable use of bypass routes. The analysis of this dependence shows that if at $V = 1$ the critical point is in the area of losses corresponding to a strong overload of the communication network, then at $V > 20$ the point shifts to the area of normalized values of the quality of service in the information directions.

Study of the influence of multiple bypasses on the quality of service in a communication network shows that at low losses, the use of multiple bypasses (more than one bypass) can increase the network throughput. However, as the load grows, the moment of needing to disable bypasses comes much earlier than in a network with simple bypasses (only one bypass).

Therefore, one of the main management strategies should be to control the state of the network and the magnitude of the load entering it. Based on an appropriate analysis, multiple bypasses may be allowed at first, then, as the load increases, it is necessary to switch to simple bypasses, and finally, bypasses should be completely prohibited and limited to direct routes only.

However, managing the use of bypasses based on the magnitude of the critical load on the communication network is fraught with a number of difficulties. Firstly, determining the values of the critical load is a significant difficulty, since it depends on the structure of the network, its channel capacity, the composition of information directions, etc. Secondly, the collection of reliable statistical data on the magnitude of the load entering the network for servicing is also quite complicated and time-consuming.

Therefore, it is necessary to search for methods that allow efficient use of network resources under congestion conditions that do not require complete information about the magnitude of the load entering the communication network for service.

One of the possible ways to take into account the current state of the network in the process of servicing requests can be the use of the rebuilding method. The essence of the functioning of the communication network with changes is as follows. Let in the communication network, the structure of which is shown in Fig. 4, a service request has been received from the switching centre 1 (Node_1) to the Node_3, and let's assume that the m_{13} branch connecting these nodes is busy servicing the request received earlier.

In this case, the received request is placed for service along the bypass route, i.e. Node_1–Node_2–Node_3. However, as soon as the m_{13} branch is released, the service of the request from the bypass path is transferred to the direct path, and the virtual channels in the bypass path are released.

Let us determine the dependence of the probability of losses on the value of the load received for service in network with rebuilds connections. To determine it, we construct a Markov process for the functioning of the network.

In this case, for the network, the structure of which is shown in Fig. 4, the following five states can be distinguished:

a) 0, 1, 2, 3 – the number of conversations on direct routes;
b) 4 – two conversations, one of which is on a direct route, and the second is on a bypass route.

The transition diagram will have the form shown in Fig. 7.

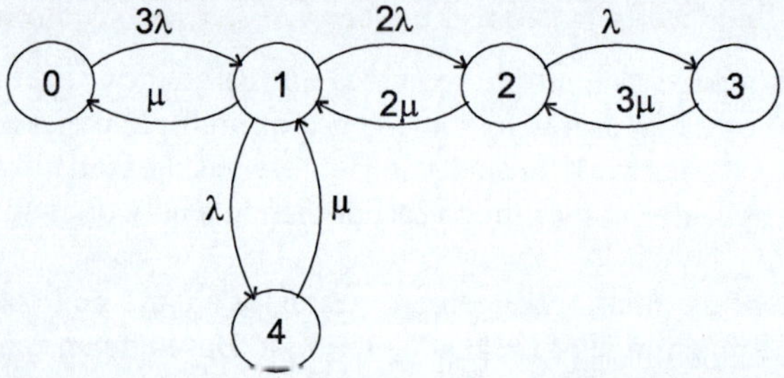

Fig. 7. Description of the functioning of a three-nodal network with rebuilds connections used Markov chain.

According to the transition diagram, we can write down the system of equations for the stationary state:

$$\begin{cases} 3 \cdot \lambda \cdot P_0 = \mu \cdot P_1 \\ (3 \cdot \lambda + \mu) \cdot P_1 = 3 \cdot \lambda \cdot P_0 + 2 \cdot \mu \cdot P_2 + 2 \cdot \mu \cdot P_4 \\ (\lambda + 2 \cdot \mu) \cdot P_2 = 2 \cdot \lambda \cdot P_1 + 3 \cdot \mu \cdot P_3 \\ 3 \cdot \mu \cdot P_3 = \lambda \cdot P_2 \\ 2 \cdot \mu \cdot P_4 = \lambda \cdot P_1 \end{cases}, \tag{4}$$

under the normalizing condition

$$\sum\nolimits_{i=0}^{4} P_i = 1. \tag{5}$$

Conditional loss probabilities γ_i in each individual state are defined as follows [24]:

$$\gamma_i = \frac{Q(i) - q(i)}{Q(i)}, \tag{6}$$

where: $Q(i)$ is the total intensity of the flow of applications arriving in the i-th state;

$q(i)$ – the intensity of the flow of applications accepted for service.

Then the probability of losses $P_{ij\_network\_RC}$ in the communication network with rebuilds connections is determined as follows:

$$P_{ij_{networkRC}} = \sum\nolimits_{i=1}^{k} \gamma_i \cdot P_i, \tag{7}$$

where: k is the number of states of the Markov process;

P_i is the probability of losses in the i-th state.

Having solved the system of Eqs. (4) taking into account (5), having determined from (6) γ_i and substituting into (7) we get:

$$P_{ij\_network\_RC} = \frac{2 \cdot Z^3 + 7 \cdot Z^2}{2 \cdot Z^3 + 9 \cdot Z^2 + 6 \cdot Z + 2}, \tag{8}$$

where: Z – the load coming into service ($Z = {}^{\lambda}/{}_{\mu}$);

λ – intensity of incoming requests;

μ – intensity of service requests.

To assess the effectiveness of the functioning of communication networks with rebuilding, Fig. 8 shows the dependence of the probability of losses on the magnitude of the incoming load:

a) curve 1 – for a communication network of direct connections;
b) curve 2 – for a communication network with bypass connections;
c) curve 3 – for a communication network with rebuilds connections.

The obtained results show that in the area of losses up to 0.67 networks with rebuilds provide a higher quality of service than the other two types of networks. In the loss area of more than 0.67, networks with rebuilds are inferior in quality of service to networks direct connections.

Thus, when designing the SDN network management level, it is possible to take into account the requirements for the quality of service and calculate strategies to optimize the process of servicing network subscribers.

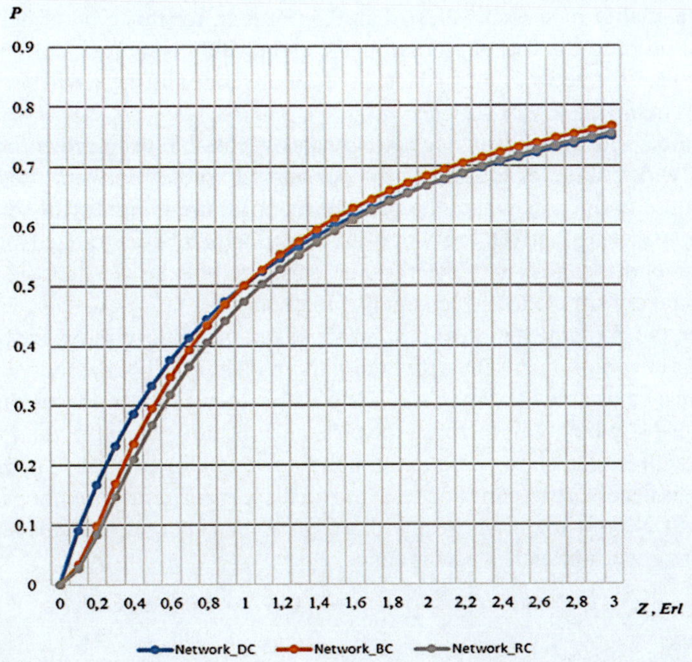

Fig. 8. Probability of denial of service in operation with direct, bypass and rebuilds route.

Implementing the rebuild method on SDN networks is relatively simple. Since the principles of centralized control are used here, there is complete information about the state of all virtual channels and paths of the network. As soon as a request is placed on a bypass route, the control plane takes control of that request. Further, the monitoring system determines the moment of freeing virtual channels in the direct path and re-establishes the connection to the shortest path, and releases the resources of the bypass path to service newly incoming requests.

6 Conclusion

The chapter mathematically analyzes possible options for making decisions by the SDN controller (or control applications) when forming load distribution in a telecommunications network. The first strategy (network of direct connections) involves the use of the shortest paths for information transfer in terms of the number of intermediate nodes between the source and recipient of information.

However, as mathematical analysis shows, such a strategy has high performance when the network operates under congestion conditions, but in the field of standard network performance indicators, it leads to a decrease in the quality of service.

The use of bypass connections improves the quality of service on the network. Moreover, this is observed in a fairly wide range of loads, where the probability of failure in the quality of service is less than 0,5. Further, for this type of networks, it is necessary to prohibit the use of workarounds, and switch to service along the shortest paths. Otherwise, the network may go into congestion mode and there will be a significant deterioration in the quality of service.

The method with rebuilds connections makes it possible to improve the quality of service in a wider range of loads. For the considered type of network, this method is advisable to use in the area where the probability of failure in quality of service is less than 0,67. Considering that SDN networks use the principles of centralized management, and the SDN controller has complete information about the state of all network elements, it will be relatively simple to implement this method.

However, the implementation of new management methods requires further research in this area. In our opinion, in SDN networks with a high level of connectivity of nodes and high performance of network branches, it is possible to organize many virtual channels and virtual paths, which will increase the efficiency of using network equipment. And this, in turn, will lead to an increase in throughput and quality of service in the network. Therefore, it is necessary to continue research on these management methods in the field of complex SDN network structures, with a high level of connectivity of nodes with a high level of network branches performance.

References

1. O'Connor, C.C., Vachuska, T., Davie, B.: Software-defined networks: a systems approach, p. 152 (2021)
2. Romanov, O., Nesterenko, M., Mankivskyi, V.: The method of redistributing traffic in mobile network. In: Ageyev, D., Radivilova, T., Kryvinska, N. (eds.) Data-Centric Business and Applications. LNDECT, vol. 69, pp. 159–182. Springer, Cham (2021). https://doi.org/10.1007/978-3-030-71892-3_7
3. ONF: Accelerating the Adoption of SDN & NFV. Issue 1, June 2016

4. Lam, J., Lee, S.-G., Lee, H.-J., Oktian, Y.E.: Securing SDN Southbound and Data Plane Communication with IBC. Hindawi Publishing Corporation Mobile Information Systems Volume 2016. https://doi.org/10.1155/2016/1708970
5. Romanov, O., Nesterenko, M., Veres, L., Kamarali, R., Saychenko, I.: Methods for calculating the performance indicators of IP multimedia subsystem (IMS). In: Ilchenko, M., Uryvsky, L., Globa, L. (eds.) MCT 2019. LNNS, vol. 152, pp. 229–256. Springer, Cham (2021). https://doi.org/10.1007/978-3-030-58359-0_13
6. ONF TR-502: SDN architecture. Issue 1, June 2014
7. ONF TR-537: Negotiable Datapath Model and Table Type Pattern Signing Version 1.0. ONF TR-537 (2016)
8. ONF TR-522: SDN Architecture for Transport Networks, March 2016. ONF TR-522
9. Romanov, O., Korniienko, N., Obod, I, Svyd, I.: Construction of the SDN control level based on ONOS UkrMiCo 2021. In: 2021 IEEE International Conference on Information and Telecommunication Technologies and Radio Electronics, Proceedings, pp. 127–132 (2021)
10. ONF TS-029: MPLS-TP OpenFlow Protocol Extensions for SPTN. Version 1.0, 16 June 2017
11. ONF TR-539: OpenFlow Controller Benchmarking Methodologies Version 1.0, November 2016
12. Optical Transport Protocol Extensions Version1.0 March 15, 2015 ONF TS-022
13. ONF TR-508. Requirements Analysis for Transport OpenFlow/SDN. V1.0 August 20, 2014. ONF TR-508
14. Romanov, O., Korniienko, N., Burlaka, H.: Construction of the SDN transport network model using the T-API interface. In: 2021 IEEE 4th International Conference on Advanced Information and Communication Technologies, AICT 2021, pp. 220–224 (2021)
15. ONOS. Security and Performance. Analysis: Report No. 1. 19 September 2017
16. ONF specification: OpenFlow Table Type Patterns. Version No. 1.0, 15 August 2014
17. Comer, D., Rastegarnia, A.: Externalization of packet processing in software defined networking. All content following this page was uploaded by Adib Rastegarnia on 11 January 2019
18. Romanov, O., Siemens, E., Nesterenko, M., Mankivskyi, V.: Mathematical description of control problems in SDN networks. In: Proceedings of International Conference on Applied Innovation in IT, vol. 9, no. 1, pp. 33–39 (2021)
19. ONF TR-525 SDN Interoperability Event Technical Issues Report AppFest 2015
20. ONOS Security and Performance Analysis (Report No. 2), 2 November 2018
21. Phemius, K., Bouet, M., Leguay, J.: DISCO: distributed multi-domain SDN controllers. Thales Commun. Secur. (2013)
22. Yeremenko, O., Lemeshko, O., Persikov, A.: Secure routing in reliable networks: proactive and reactive approach. In: Shakhovska, N., Stepashko, V. (eds.) CSIT 2017. AISC, vol. 689, pp. 631–655. Springer, Cham (2018). https://doi.org/10.1007/978-3-319-70581-1_44
23. IETF RFC 7426. Request for Comments: 7426. ISSN: 2070-1721 EICT. Category: Informational Pentikousis, K. (ed.). ISSN: 2070-1721 EICT
24. Feinberg, E.A., Shwartz, A.: Handbook of Markov Decision Processes: Methods and Applications. Springer, Netherlands, January 2002. ISBN: 0792374592. https://doi.org/10.1007/978-1-4615-0805-2
25. Romanov, O., Nesterenko, M., Marinov, A., Skolets, S., Burlaka, H.: SDN network modeling using the GUI MiniEdit. In: Proceedings - 16th International Conference on Advanced Trends in Radioelectronics, TCSET 2022, pp. 637–642 (2022)
26. Romanov, O., Mankivskyi, V., Veres, L., Saychenko, I.: Analysis of performance in Docker net deployed in AWS cloud. In: 2021 IEEE 8th International Conference on Problems of Infocommunications, Science and Technology, PIC S and T 2021 - Proceedings, pp. 285–290 (2021)

Check for updates

Research and Implementation Directions of the Fast ReRoute Facility Backup Scheme in Programmable Networks

Oleksandr Lemeshko[1], Oleksandra Yeremenko[1](✉), Maryna Yevdokymenko[1], Serhii Harkusha[2], and Yurii Dobryshkin[3]

[1] Kharkiv National University of Radio Electronics, Nauky Avenue 14, Kharkiv 61166, Ukraine
{oleksandr.lemeshko.ua,oleksandra.yeremenko.ua, maryna.yevdokymenko}@ieee.org
[2] Poltava University of Economics and Trade, Koval Street 3, Poltava 36014, Ukraine
[3] State Scientific Research Institute of Armament and Military Equipment Testing and Certification, Strilets'ka Street 1, Chernihiv 14033, Ukraine

Abstract. The work is devoted to research and implementation of the main directions of the Fast ReRoute (FRR) facility backup approach in programmable networks. The basic FRR flow-based model has been described under link, node, and bandwidth protection schemes during the primary and backup routes calculation within the data plane communication. Additionally, the facility backup feature of the studied approach allowed improvement of network resource consumption, which is vital in implementing redundancy technics such as Fast ReRoute. Finally, the numerical research for different configurations of the FRR facility backup scheme in optimization form proved its high efficiency and better network fault tolerance, performance, and scalability. The optimal implementation of the facility backup, link, and node protection schemes minimizes the routing table size by providing the best match of the primary and backup routes, increasing the network throughput available for reuse. Practical recommendations have been formulated to utilize the FRR facility backup approach in programmable networks.

Keywords: Network resilience · Fast ReRoute · Facility backup · Performance · Programmable networks

1 Introduction

The main requirement for modern infocommunication networks (ICN) was and remains the provision of a high level of reliability (fault tolerance), security, and Quality of Service (QoS) [1–7]. At the same time, it is worth noting that improving the efficiency of ICN according to one type of indicators, for example, reliability or security, can have a negative impact on performance and QoS level in general [4, 8–11]. This is due to the fact that ICN has a limited network (link, buffer, and computing) resource, which is constantly redistributed by a set of network protocols, each of which is individually responsible for a certain aspect of network functioning [4, 8].

M. Klymash et al. (Eds.): TCSET 2022, LNEE 965, pp. 102–114, 2023.
https://doi.org/10.1007/978-3-031-24963-1_7

The described situation is especially vividly observed in the example of the application of fault-tolerant routing protocols when solving Fast ReRouting (FRR) problems in programmable networks [12–24]. The FRR protocols should ensure the prompt network response to the denial of service by one or another element (router and/or link) based on a pre-calculated and involved backup routing solution [12]. Similar local and global protection against failures caused by various internal and external factors is implemented based on reserving an additional network resource. The introduction of redundancy in the network resource utilization has a positive effect on the reliability (fault tolerance) and security level but, as a rule, provokes a decrease in the current QoS indicators. Therefore, the tasks of effective use of the available network, first of all, link resource, with the control of reliability and QoS indicators when organizing fast rerouting, come to the fore [14–18].

As the analysis [11–14] showed, an inevitable compromise in ensuring proactive protection of network elements against failures is the implementation of multipath routing when defining both primary and backup routes, which allows for more efficient use of the available network resource. In addition, a significant potential for increasing the ICN efficiency lies in using the *facility backup* scheme functionality [12]. It aims at the optimization and coordinated use of primary and backup routes resource by different packet flows. Providing FRR with the correct setting of the *facility backup* scheme will minimize the use of additional network resources associated with the organization of the network elements protection, which must necessarily affect the QoS level. In this regard, an urgent scientific and practical task is the development of adequate mathematical models and methods of fast rerouting in ICN, which would implement the facility backup scheme and become the algorithmic basis of improved or new fault-tolerant routing protocols in programmable networks.

Within the presented work, a generalized architecture of the Fast ReRoute facility backup scheme in programmable networks has the following components, as shown in Fig. 1. Firstly, it contains two planes: the data plane and the control plane with the central SDN controller. Here, the control plane optimizes fast rerouting processes with a facility backup scheme. In addition, the control plane is responsible for weighting coefficients adjustment during facility backup scheme implementation under primary and backup paths calculation. The data plane consists of SDN routers, e.g., SD-WAN routers.

Therefore, the remainder of this chapter is structured as follows: Sect. 2 proposes the Fast reroute facility backup flow-based model in a communication network. Section 3 contains the numerical research of the Fast ReRoute facility backup scheme. Subsection 3.1 provides Fast ReRoute solution without facility backup scheme analysis. Then Subsects. 3.2 and 3.3 investigate the facility backup scheme implementation cases during modeling. Finally, the conclusion discusses the results obtained and gives the Fast ReRoute facility backup approach application directions.

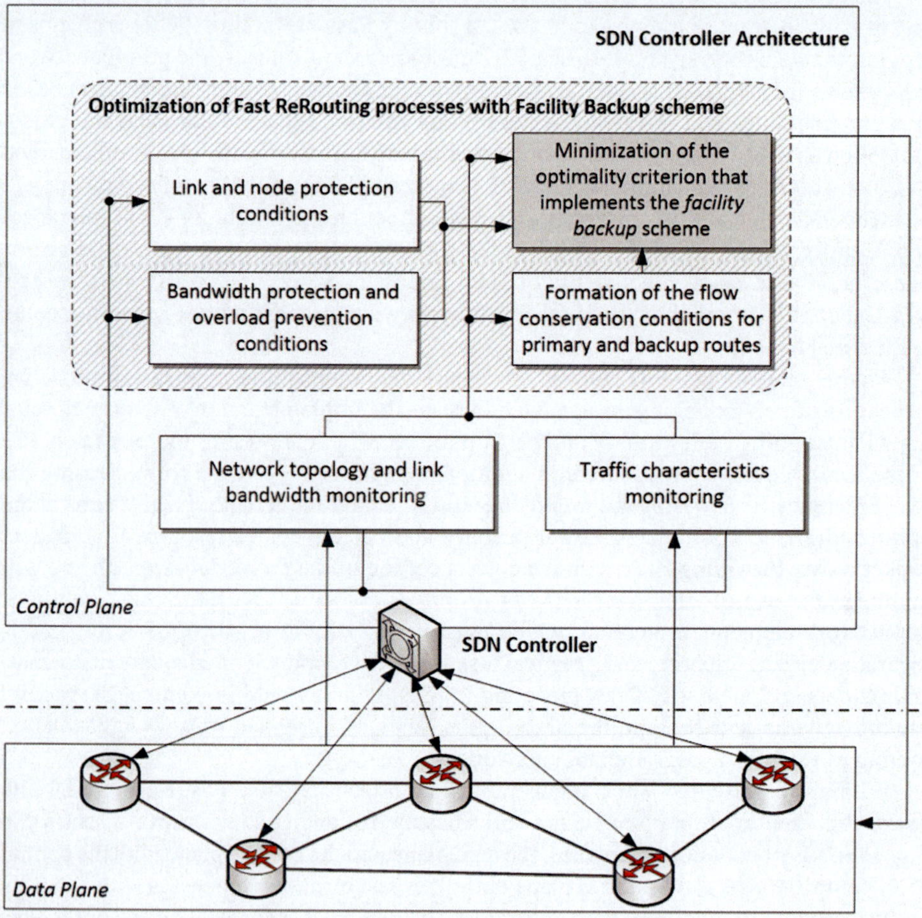

Fig. 1. Generalized architecture of the Fast ReRoute Facility Backup Scheme in Programmable Networks.

2 Fast ReRoute Facility Backup Flow-Based Model

Suppose that the graph $G = (R, E)$ describes the network structure with the following characteristics [11, 12]:

- $R = \{R_i;\ i = \overline{1, m}\}$ is a set of vertices modeling routers;
- $E = \{E_{i,j};\ i, j = \overline{1, m};\ i \neq j\}$ is a set of edges representing links ($|E| = n$);
- $R_i^* = \{R_j : \exists E_{i,j} \in E;\ j = \overline{1, m};\ i \neq j\}$ is a subset of routers incident to the R_i router;
- $\varphi_{i,j}$ is a link $E_{i,j} \in E$ capacity.

Then every kth unicast flow ($k \in K$) is connected with functional parameters:

- λ^k is the average intensity of the kth flow measured in packets per second (pps);

- s_k is the source node;
- d_k is the destination node.

Under the Fast ReRoute facility backup problem solving, it is necessary to calculate a set of variables $x_{i,j}^k$ describing the fraction of the kth flow intensity in the link $E_{i,j} \in E$ of the primary route. The product $|K| \cdot |E|$ defines the total number of routing variables $x_{i,j}^k$.

The following constraints are imposed on the control variables $x_{i,j}^k$ according to single-path routing:

$$x_{i,j}^k \in \{0; 1\}. \tag{1}$$

The flow conservation conditions are introduced in the model for the primary path [12]:

$$\begin{cases} \sum\limits_{j:E_{i,j}\in E} x_{i,j}^k - \sum\limits_{j:E_{j,i}\in E} x_{j,i}^k = 0; k \in K, R_i \neq s_k, d_k; \\ \sum\limits_{j:E_{i,j}\in E} x_{i,j}^k - \sum\limits_{j:E_{j,i}\in E} x_{j,i}^k = 1; k \in K, R_i = s_k; \\ \sum\limits_{j:E_{i,j}\in E} x_{i,j}^k - \sum\limits_{j:E_{j,i}\in E} x_{j,i}^k = -1; k \in K, R_i = d_k. \end{cases} \tag{2}$$

The backup paths must be calculated together with the primary routes when solving the problem of fast rerouting in the network. Therefore, as shown in [11–14] additional routing variables $\overline{x}_{i,j}^k$ are involved in characterizing the fraction of the kth flow intensity in the link $E_{i,j} \in E$ of the backup route and imposed with similar flow conservation conditions:

$$\begin{cases} \sum\limits_{j:E_{i,j}\in E} \overline{x}_{i,j}^k - \sum\limits_{j:E_{j,i}\in E} \overline{x}_{j,i}^k = 0; k \in K, R_i \neq s_k, d_k; \\ \sum\limits_{j:E_{i,j}\in E} \overline{x}_{i,j}^k - \sum\limits_{j:E_{j,i}\in E} \overline{x}_{j,i}^k = 1; k \in K, R_i = s_k; \\ \sum\limits_{j:E_{i,j}\in E} \overline{x}_{i,j}^k - \sum\limits_{j:E_{j,i}\in E} \overline{x}_{j,i}^k = -1; k \in K, R_i = d_k. \end{cases} \tag{3}$$

In turn, the following condition implements the link $E_{i,j} \in E$ protection scheme for the single path routing over the backup paths [12]:

$$\overline{x}_{i,j}^k \in \left\{0;\ \delta_{i,j}^k\right\} \tag{4}$$

where

$$\delta_{i,j}^k = \begin{cases} 0, \text{ under the link } E_{i,j} \text{ protection;} \\ 1, \text{ otherwise.} \end{cases} \tag{5}$$

Fulfilling conditions (4) and (5) guarantees that the protected link $E_{i,j} \in E$ will not be included in a backup route. Their linearity also becomes a valuable advantage of the considered model.

In this work, the generalization of (4) and (5) are also used for the node $R_i \in R$ protection scheme under the single path fast rerouting [12]:

$$\overline{x}_{i,j}^k \in \left\{0;\ \delta_{i,j}^k\right\} \text{ under } R_j \in R_i^*,\ j = \overline{1, m} \tag{6}$$

where a selection of values $\delta_{i,j}^k$ is subject to the condition (5).

Finally, the following condition is used to implement the bandwidth protection scheme and overload prevention [14]:

$$\sum_{k \in K} \lambda^k (x_{i,j}^k + \overline{x}_{i,j}^k - x_{i,j}^k \cdot \overline{x}_{i,j}^k) \le \varphi_{i,j},\ E_{i,j} \in E. \tag{7}$$

Thus, the conditions (7) prevent overload of network links, even when some flows switch from the primary to the backup route. In this case, some of the links' bandwidth of backup paths for these flows will always be untapped, thereby implementing a bandwidth protection scheme in fault-tolerant routing.

An essential step in developing and improving the model is the selection of the Fast ReRoute optimality criterion. Typically, these criteria aim to minimize or maximize the objective function selected, reflecting, for example, the total cost of the network resource utilization by the primary and backup routes or the overall effectiveness of the problem-solution. While the following criterion strived to minimize the objective function [12]:

$$F = \sum_{k \in K} \sum_{E_{i,j} \in E} c_{i,j}^k x_{i,j}^k + \sum_{k \in K} \sum_{E_{i,j} \in E} \overline{c}_{i,j}^k \overline{x}_{i,j}^k - b \sum_{k \in K} \sum_{E_{i,j} \in E} x_{i,j}^k \overline{x}_{i,j}^k, \tag{8}$$

where $c_{i,j}^k$ and $\overline{c}_{i,j}^k$ are routing metrics of the primary and backup paths' links; b is weighting coefficient. The following discussion is devoted to routing metrics and weighting coefficients' range and numerical values.

The first and second terms in (8) determine the value of the conditional cost of formation and use of primary and backup path/multipath. As shown in works [6, 7], if the metrics and are chosen inversely proportional to the link bandwidth $\varphi_{i,j}$, as, for example, in the OSPF (Open Shortest Path First) protocol, then considering the linear and additive form of the objective function (8), optimality primary and backup routes will be based on the fact that they should have the maximum bandwidth and the minimum number of links. Therefore, the following ratios will take place further:

$$c_{i,j}^k = \frac{a}{\varphi_{i,j}} \text{ and } \overline{c}_{i,j}^k = \frac{\overline{a}}{\varphi_{i,j}} \tag{9}$$

where a and $\overline{a}$ are some weighting coefficients. Their counterparts in OSPF and EIGRP (Enhanced Interior Gateway Routing Protocol) are equal to 10^8 and 10^7, accordingly.

The third term in optimality criterion (8) determines the *facility backup* scheme implementation to improve routing decisions' scalability. Therefore, the facility backup scheme allows for minimizing routing table size under the implementation of the link and node protection schemes when the primary and backup routes have the best match in terms of their unprotected constituent network elements. Additionally, such an approach helps reserve a minimum volume of network resources (nodes and links), increasing the network throughput available to arriving flows.

Given that the objective function (8) is subject to minimization, the third term is introduced with a "minus" because the degree of coincidence of primary and backup paths should be maximized. In contrast, the first two terms characterize the conditional cost of utilizing primary and backup paths that should be as low as possible. The highest efficiency of the claimed capabilities can be achieved through the proper selection of ratios between weighting coefficients a, $\overline{a}$, and b, which is the subject of further investigation.

Therefore, in the framework of the mathematical model improvement, the fast reroute technical task was formulated in the optimization form, and the optimality criterion serves the minimum of the objective function (8) under constraints and conditions (1)-(7). Following single path routing due to the Boolean type of routing variables (1) and (4), it is necessary to solve the optimization problem of the Mixed Integer NonLinear Programming (MINLP) class.

3 Numerical Research of the Fast ReRoute Facility Backup Scheme

The study aims to substantiate the relationship between the weighting coefficients a, $\overline{a}$, and b for the effective implementation of FRR with the support of the *facility backup* scheme when protecting routers and network links. For this purpose, several network structures and protection schemes were analyzed. The research methodology will be demonstrated on a specific numerical example when the network was analyzed, which structure is presented in Fig. 2. The network contained twelve routers and seventeen communication links, in the gaps of which their bandwidth (pps) is given (Fig. 2).

Let packets of two flows, the characteristics of which are presented in Table 1, be transmitted in the network.

In work, a comparative analysis of the resource consumption of routing solutions, which corresponded to different variants of the weighting coefficients a, $\overline{a}$, and b was carried out. The resource capacity of an arbitrary routing solution was estimated through the value W, which characterized the total volume of the link resource, namely, the network link bandwidth, which was used (reserved) along both the primary and backup routes:

$$W = \sum_{k \in K} \lambda^k \left| P_{pr}^k \bigcup P_{bc}^k \right|, \tag{10}$$

where P_{pr}^k and P_{bc}^k are the subsets of the network link set E, which are included according to the primary and backup paths used in transmitting the kth packet flow; $|P|$ is the set's P power.

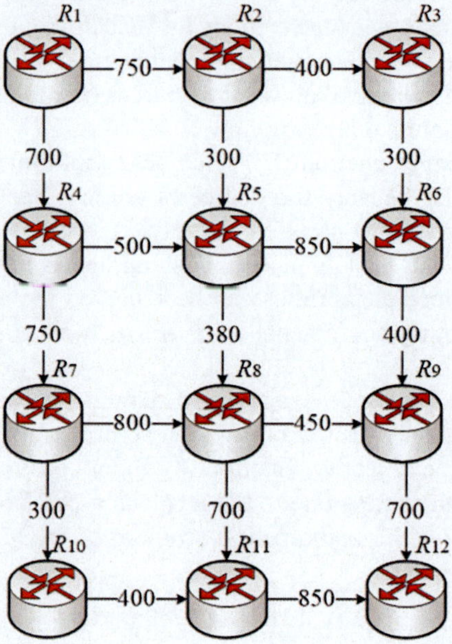

Fig. 2. Network structure under the numerical study.

Table 1. Packet flows characteristics.

Flow #	Flow rate, λ^i, pps	Source	Destination
1	200	R_1	R_{12}
2	100	R_4	

3.1 Fast ReRoute Solution Without Facility Backup Scheme

Then Fig. 3 shows the FRR problem solution without facility backup with link $E_{4,7}$ protection, i.e., when the objective function lacked the third term ($b = 0$). Figure 2 notations are the following:

- *a blue color line* indicates the first flow's primary route;
- *a blue dotted line* corresponds to the first flow's backup route;
- *a pink color* line indicates the second flow's primary route;
- *a pink dotted* line corresponds to the second flow's backup route.

In this case, by analogy with the OSPF protocol, the weighting coefficients a and $\overline{a}$ took the value 10^8. When choosing such a combination of weighting coefficients, the goal of FRR is to protect the selected network element while ensuring the maximum throughput of the primary and backup routes while using the minimum number of communication links.

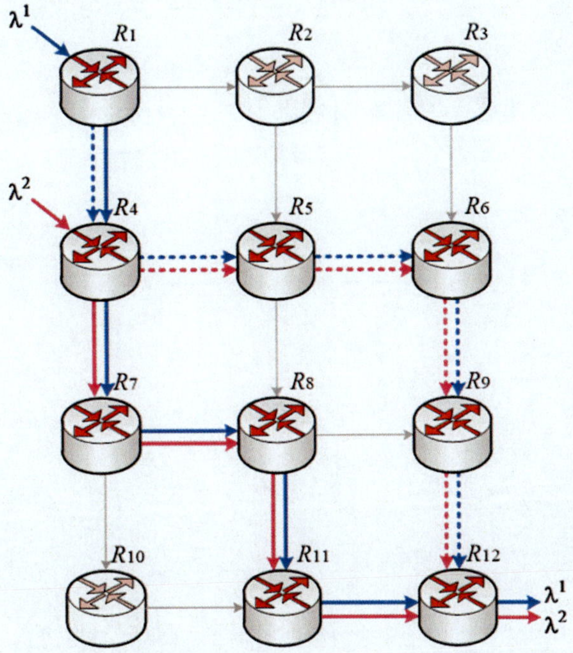

Fig. 3. FRR solution without *facility backup.*

Based on the results of the calculations, Table 2 presents the primary and backup routes for the first and second packet flow, as well as their main characteristics, including bandwidth and the number of used communication links. In addition, Table 2 shows the resource consumption of the presented FRR solution.

Table 2. FRR solution without *facility backup* ($a = \bar{a} = 10^8;\ b = 0$).

Flow #	Path	Routers	Links number	Throughput, pps	W, pps
1	Primary	$R_1 \to R_4 \to R_7 \to R_8 \to R_{11} \to R_{12}$	5	700	2600
	Backup	$R_1 \to R_4 \to R_5 \to R_6 \to R_9 \to R_{12}$	5	400	
2	Primary	$R_4 \to R_7 \to R_8 \to R_{11} \to R_{12}$	4	700	
	Backup	$R_4 \to R_5 \to R_6 \to R_9 \to R_{12}$	4	400	

3.2 Fast ReRoute Solution with Facility Backup Scheme

Figure 4 shows the FRR problem solution with facility backup and link $E_{4,7}$ protection when the influence of the third term in the objective function (8) was predominant ($b >> a$ and $b >> \bar{a}$).

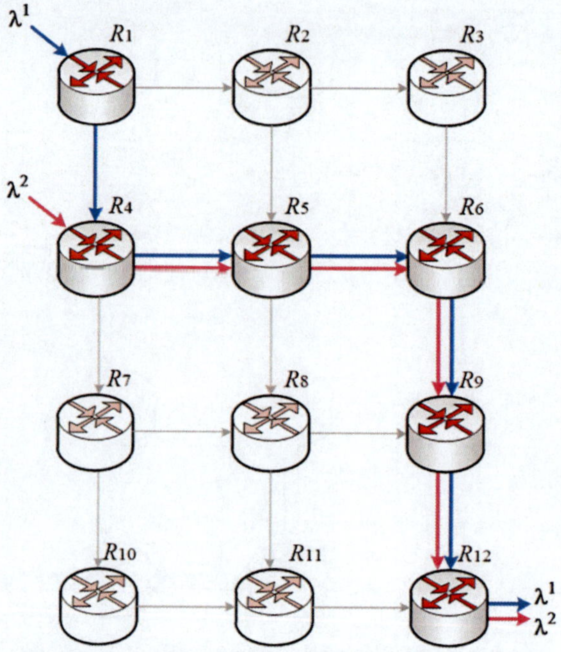

Fig. 4. FRR solution with *facility backup* ($a = 10^2$; $\overline{a} = 10^2$;$b = 10^8$).

In this variant, the weighting coefficients a and $\overline{a}$ took the value of 10^2 at $b = 10^8$. When choosing such a combination of weighting coefficients, the main goal of FRR is, first of all, to ensure the minimum difference between the primary and backup routes concerning the set of communication links included in them. Then, during optimization, attention was paid to the bandwidth of these routes and the number of communication links that comprise them.

Based on the results of the calculations, Table 3 presents the characteristics of the obtained FRR solution. Choosing a large value of b compared to a and $\overline{a}$ resulted in a complete match (coincidence) between the primary and backup routes, neither of which included the $E_{4,7}$ link to be protected (Fig. 4). That is, in this case, the FRR solution with

Table 3. FRR solution with *facility backup* ($a = 10^2$; $\overline{a} = 10^2$;$b = 10^8$).

Flow #	Path	Routers	Links number	Throughput, pps	W, pps
1	Primary	$R_1 \to R_4 \to R_5 \to R_6 \to R_9 \to R_{12}$	5	400	1500
	Backup	$R_1 \to R_4 \to R_5 \to R_6 \to R_9 \to R_{12}$	5	400	
2	Primary	$R_4 \to R_5 \to R_6 \to R_9 \to R_{12}$	4	400	
	Backup	$R_4 \to R_5 \to R_6 \to R_9 \to R_{12}$	4	400	

facility backup ignored the most productive routes $R_1 \to R_4 \to R_7 \to R_8 \to R_{11} \to R_{12}$ and $R_4 \to R_7 \to R_8 \to R_{11} \to R_{12}$ for the first and second flows, which will have a negative impact on the QoS level.

3.3 Fast ReRoute Solution with Facility Backup Scheme (Special Case)

In the third case, we chose such a variant of the ratio between weighting coefficients when the condition was fulfilled

$$a > b > \overline{a}. \tag{11}$$

Figure 5 shows the solution to the FRR problem with *facility backup* and the protection of the $E_{4,7}$ link, when in the objective function (8), the weighting coefficients under condition (11) took the values $a = 10^8, \overline{a} = 10^2$, and $b = 10^4$.

Here the primary routes had the maximum bandwidth allowed for the selected network (Fig. 1) – 700 pps, which is 75% higher compared to the second FRR solution. The backup routes for both flows differed minimally from the first FRR solution, i.e., by two instead of four communication links. This made it possible to reduce the volume of the used link resource from 2600 pps to 2000 pps (Table 4), i.e., by 23% on average. At the same time, the backup routes used had a bandwidth of 380 pps (Table 4), which is only 5% less compared to the results obtained within the second FRR solution.

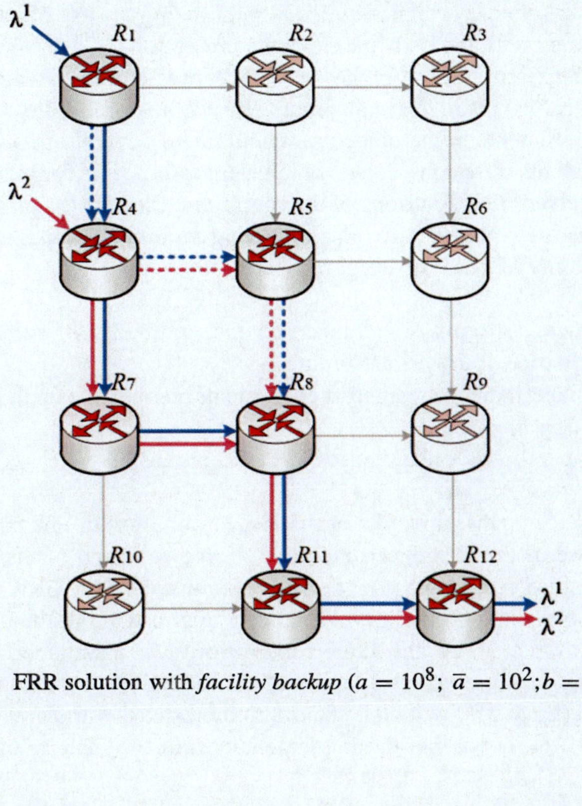

Fig. 5. FRR solution with *facility backup* ($a = 10^8$; $\overline{a} = 10^2$; $b = 10^4$).

Table 4. FRR solution with *facility backup* ($a = 10^8$; $\bar{a} = 10^2$;$b = 10^4$).

Flow #	Path	Routers	Links number	Throughput, pps	W, pps
1	Primary	$R_1 \to R_4 \to R_7 \to R_8 \to R_{11} \to R_{12}$	5	700	2000
	Backup	$R_1 \to R_4 \to R_5 \to R_8 \to R_{11} \to R_{12}$	5	380	
2	Primary	$R_4 \to R_7 \to R_8 \to R_{11} \to R_{12}$	4	700	
	Backup	$R_4 \to R_5 \to R_8 \to R_{11} \to R_{12}$	4	380	

4 Conclusion

The paper proposed an optimization model of fault-tolerant routing based on fast rerouting (1)–(8), in which expressions (1)–(7) served as constraints on routing variables, and the optimality criterion was the minimum of the objective function (8). The formulated optimization problem, which is related to the calculation of a set of primary and backup paths with the protection of network links and routers (4)–(6), belongs to the MINLP class of problems because the routing variables are Boolean (1), (4), and the conditions of network bandwidth protection (7) are non-linear. The model's novelty is the updated form of the objective function in the optimality criterion (8) due to the need to implement the facility backup scheme, which is aimed at minimizing the use of additional network resources associated with the network elements protection.

A study of the fast rerouting processes in ICN was carried out to substantiate the choice of numerical values of the main parameters of the proposed model, in particular, the weighting coefficients in the objective function (8). Several options for the ratio of weighting coefficients were considered, which were responsible for influencing the optimality of the received FRR solutions of the routing metrics for the primary and backup routes, as well as the conditions for implementing facility backup. An experimentally established hierarchy of ratios of weighting factors:

- in the first place, in terms of influence on the optimality of solutions, should be precisely the metrics of the primary routes;
- in the second place, there are weighting coefficients responsible for the implementation of the *facility backup* scheme;
- in the third place, there are the metrics of backup paths.

The analysis of numerical results of the fast rerouting with link protection (Fig. 1) shows that the use of the introduced hierarchy in the selection of the specified model parameters will allow reducing the resource consumption of the FRR solution by 23% (Table 4) in comparison with the implementation option without the use of the *facility backup* scheme (Table 2). At the same time, priority is maintained in choosing the high capacity of the primary routes, and a slight reduction in the capacity of backup routes is ensured (up to 5%), which is the minimum price for implementing the *facility backup* scheme. It is shown that the implementation of the FRR solution, which bans the inclusion of the network element that is protected, both to the primary and backup routes (Table 3), can lead to a significant decrease in the primary route bandwidth (up to 75%), which is used before the failure of the network element being protected.

The proposed mathematical model (1)–(8) with reasonably selected parameters can be applied as the algorithmic basis of prospective fault-tolerant routing protocols with the provision of fast rerouting of flows while protecting routers and network communication links according to the *facility backup* scheme.

References

1. Dodd, A.Z.: The Essential Guide to Telecommunications (Essential Guide Series). Pearson. (2019)
2. Tealab, M., Hassebo, A., Dabour, A., AbdelAziz, M.: Smart cities digital transformation and 5G–ICT Architecture. In: 2020 11th IEEE Annual Ubiquitous Computing, Electronics & Mobile Communication Conference (UEMCON) Proceedings, pp. 0421–0425. IEEE (2020). https://doi.org/10.1109/UEMCON51285.2020.9298156
3. Tran-Dang, H., Kim, D.S.: The physical internet in the era of digital transformation: perspectives and open issues. IEEE Access. **9**, 164613–164631 (2021). https://doi.org/10.1109/ACCESS.2021.3131562
4. Gooley, J., Yanch, D., Schuemann, D., Curran, J.: Cisco software-defined wide area networks: designing, Deploying and Securing Your Next Generation WAN with Cisco SD-WAN. Cisco Press (2020)
5. Strelkovskaya, I., Solovskaya, I., Makoganiuk, A.: Forecasting 5G network multimedia traffic characteristics. In: 2020 IEEE 15th International Conference on Advanced Trends in Radioelectronics, Telecommunications and Computer Engineering (TCSET) Proceedings, pp. 982–987. IEEE (2020). https://doi.org/10.1109/TCSET49122.2020.235585
6. Smelyakov, K., Chupryna, A., Hvozdiev, M., Sandrkin, D.: Gradational correction models efficiency analysis of low-light digital image. In: 2019 Open Conference of Electrical, Electronic and Information Sciences (eStream) Proceedings, pp. 1–6. IEEE (2019). https://doi.org/10.1109/eStream.2019.8732174
7. Ahmad, S., Mir, A.H.: SDN interfaces: protocols, taxonomy and challenges. Int. J. Wirel. Microwave Technol. (IJWMT) **12**(2), 11–32 (2022). https://doi.org/10.5815/ijwmt.2022.02.02
8. Medhi, D., Ramasamy, K.: Network Routing: Algorithms, Protocols, and Architectures. Morgan Kaufmann, Burlington (2017)
9. Strelkovskaya, I., Solovskaya, I., Makoganiuk, A.: Predicting self-similar traffic using cubic B-splines. In: 2019 3rd International Conference on Advanced Information and Communications Technologies (AICT) Proceedings, pp. 153–156. IEEE (2019). https://doi.org/10.1109/AIACT.2019.8847761
10. Strelkovskaya, I.V., Solovskaya, I.N.: Tensor model of multiservice network with different classes of traffic service. Radioelectron. Commun. Syst. **56**, 296–303 (2013). https://doi.org/10.3103/S0735272713060058
11. Lemeshko, O., Yeremenko, O., Yevdokymenko, M., Shapovalova, A., Hailan, A.M., Mersni, A.: Cyber resilience approach based on traffic engineering fast reroute with policing. In: 2019 10th IEEE International Conference on Intelligent Data Acquisition and Advanced Computing Systems: Technology and Applications (IDAACS) Proceedings, vol. 1, pp. 117–122. IEEE (2019). https://doi.org/10.1109/IDAACS.2019.8924294

12. Yeremenko, O.S., Lemeshko, O.V., Tariki, N.: Fast ReRoute scalable solution with protection schemes of network elements. In: 2017 IEEE First Ukraine Conference on Electrical and Computer Engineering (UKRCON) Proceedings, pp. 783–788. IEEE (2017). https://doi.org/10.1109/UKRCON.2017.8100353
13. Lemeshko, O., Yeremenko, O., Hailan, A.M.: Two-level method of fast ReRouting in software-defined networks. In: 2017 4th International Scientific-Practical Conference Problems of Infocommunications. Science and Technology (PIC S&T) Proceedings, pp. 376–379. IEEE (2017). https://doi.org/10.1109/INFOCOMMST.2017.8246420
14. Lemeshko, O., Yevdokymenko, M., Yeremenko, O., Hailan, A.M., Segeč, P., Papán, J.: Design of the fast reroute QoS protection scheme for bandwidth and probability of packet loss in software-defined WAN. In: 2019 IEEE 15th International Conference on the Experience of Designing and Application of CAD Systems (CADSM) Proceedings, pp. 1–5. IEEE (2019). https://doi.org/10.1109/CADSM.2019.8779321
15. Rhamdani, F., Suwastika, N.A., Nugroho, M.A.: Equal-cost multipath routing in data center network based on software defined network. In: 2018 6th International Conference on Information and Communication Technology (ICoICT) Proceedings, pp. 222–226. IEEE (2018). https://doi.org/10.1109/ICoICT.2018.8528730
16. Zhang, X., Cheng, Z., Lin, R., He, L., Yu, S., Luo, H.: Local fast reroute with flow aggregation in software defined networks. IEEE Commun. Lett. **21**(4), 785–788 (2016). https://doi.org/10.1109/LCOMM.2016.2638430
17. Qiu, K., Zhao, J., Wang, X., Fu, X., Secci, S.: Efficient recovery path computation for fast reroute in large-scale software-defined networks. IEEE J. Sel. Areas Commun. **37**(8), 1755–1768 (2019). https://doi.org/10.1109/JSAC.2019.2927098
18. He, F., Oki, E.: Backup allocation model with probabilistic protection for virtual networks against multiple facility node failures. IEEE Trans. Netw. Serv. Manage. **18**(3), 2943–2959 (2021). https://doi.org/10.1109/TNSM.2021.3075458
19. Islam, N., Rahman, H., Nasir, M.K.: A Comprehensive analysis of QoS in wired and wireless SDN based on mobile IP. Int. J. Comput. Netw. Inf. Secur. **13**(5), 18–28 (2021). https://doi.org/10.5815/ijcnis.2021.05.02
20. Rohitaksha, K., Rajendra, A.B., Mohan, J.: Routing in hybrid software defined networking using hop-count approach. Int. J. Wirel. Microwave Technol. (IJWMT) **12**(3), 54–60 (2022). https://doi.org/10.5815/ijwmt.2022.03.04
21. Omumbo, N., Muhambe, T.M., Ratemo, C.M.: Evaluation of routing performance using OSPF and multi-controller based network architecture. Int. J. Comput. Netw. Inf. Secur. (IJCNIS) **13**(4), 45–61 (2021). https://doi.org/10.5815/ijcnis.2021.04.05
22. Mohammadinejad, H., Mohammadhoseini, F.: Proposing a method for enhancing the reliability of RPL routing protocol in the smart grid neighborhood area networks. Int. J. Comput. Netw. Inf. Secur. **11**(7), 21–28 (2019). https://doi.org/10.5815/ijcnis.2019.07.04
23. Jayalakshmi, D.S., Hemanand, D., Kumar, G.M., Rani, M.M.: An efficient route failure detection mechanism with energy efficient routing (EER) protocol in MANET. Int. J. Comput. Netw. Inf. Secur. **13**(2), 16–28 (2021). https://doi.org/10.5815/ijcnis.2021.02.02
24. Kamamura, S., Shimazaki, D., Hiramatsu, A., Nakazato, H.: Autonomous IP fast rerouting with compressed backup flow entries using OpenFlow. IEICE Trans. Inf. Syst. **96**(2), 184–192 (2013). https://doi.org/10.1587/transinf.E96.D.184

Ecosystem for the Development and Testing of Emerging Network Applications in 5G and Beyond Environment

Roman Odarchenko[1,2(✉)] and Eli Shapira[3]

[1] National Aviation University, 1 Liubomyra Gusara Ave., Kyiv, Ukraine
odarchenko.r.s@ukr.net
[2] Bundleslab KFT, Budapest, Hungary
[3] DriveU, Kefar Sava, Israel
eli@driveu.auto

Abstract. There are a lot of different wireless technologies nowadays. And among them the decisive role for critical applications (Automotive and PPDR) is given to the 5-th generation mobile technologies. 5G is a key driving force in the development of the ICT industry and other related vertical industries. In these conditions, when a large number of commercial 5G networks are already deployed, software developers begin to develop a large number of network applications for the needs and with the use of 5G. Therefore, it is very important to support developers, small and medium enterprises in order to automatically test, certify and bring to the end customer already debugged applications that can be deployed and used in commercial projects. Therefore, the paper presents the ecosystem for the development and testing of emerging network applications in 5G and beyond environment. This ecosystem combines software developers, test sites in European cities, automated testing and certification tools, end customers (users) of network applications and other tools and systems. And it is mandatory to use NetApp Store, specially developed Marketplace, to bring fully tested and certified network application to the end user or customer. This paper also demonstrates how the methodology developed in the project will be applied in practice on the example of Remote human driving NetApp. In particular, the architecture of the application, the principles of interaction with the test platform, the requirements for the application, the test plan and metrics to be collected were shown.

Keywords: 5G · Network application · Software development · Ecosystem · Testing · Network optimization · Automotive · PPDR

1 Introduction

The number of devices connected to the Internet and the requirements of subscribers to the data rate of mobile access are increasing every year [1]. In 2021, the number of cellular subscribers reached 5.3 billion, which is 67% of the world's population, and 4.2 billion of them are mobile Internet users [2].

M. Klymash et al. (Eds.): TCSET 2022, LNEE 965, pp. 115–139, 2023.
https://doi.org/10.1007/978-3-031-24963-1_8

Telecom equipment developers and telecom operators are updating network architectures and interoperability in response to new challenges. Thus appeared the fifth generation (5G) networks, which are the next stage in the evolution of the fourth generation networks (see LTE, 4G). They will obviously be followed by the next generation networks - 6G [3].

5G is a new generation of mobile wireless that will replace the current 4G LTE connection in the coming years [4]. The 5G wireless connection assumes a higher loading/unloading throughput, less latency and increased capacity. About 200 cellular operators have already launched commercial 5G services. Between 2022 and 2025, companies will invest $ 620 billion in their networks, of which 85% will be in the fifth generation. The mobile industry is starting to develop a roadmap for 6G.

In general, this will lead to the following benefits for business and consumers [5]:

- **Shorter delays**: 5G provides greater data transfer and minimizes latency since data is sent/received. This provides the ability to connect in real time to exchange IoT data, connected cars and other "smart" objects.
- **Expanding connectivity**: the 5G mobile network architecture significantly increases the capacity to provide resources. More people and devices will be able to communicate at the same time without overloading the network.
- **Higher data rate**. As already mentioned, the 5G data rate will be comparable (or even higher in some cases) with fiber-optic wired networks.
- **Mobility**: 5G allows base stations to maintain traffic from 0 to 310 miles per hour, which means stable operations for mobile objects (for example, on a high-speed train or helicopter).
- **Improved connection density**: 5G is already starting to accommodate more connected devices than LTE, and support up to 1 million connected objects per square kilometer. This is important given the growth of IoT devices.

Thus, the demand for 5G mobile and wireless communications technologies will be driven mainly by increasing demands on digital technology, cloud computing and security in many industries, including automotive, logistics, retail, entertainment and manufacturing.

In particular, the release of 5G will accelerate technical advances in the following areas of business (Fig. 1):

1. Connected cars, autonomous driving and independent driving.
2. AR, VR and MR, including 3D video and UHD screens.
3. Industry 4.0 – Industrial IoT and enhanced automation.
4. Smart Cities – transport, infrastructure, logistics.
5. Connected Healthcare - robotics, wearable telemetry, HD image transmission and the use of blockchain in healthcare.
6. Smart home.

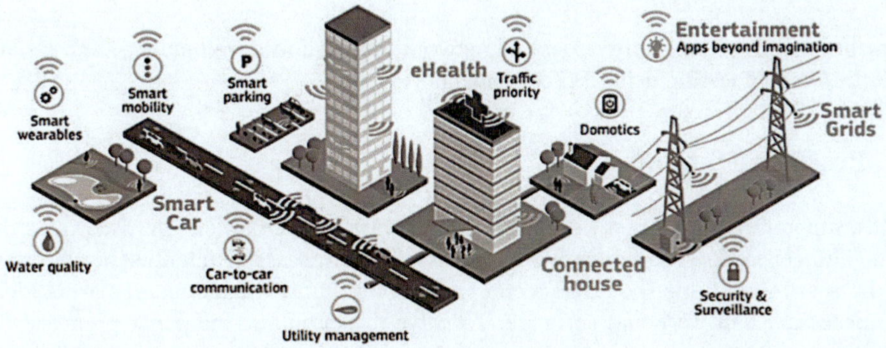

Fig. 1. 5G use cases.

Under these conditions, software vendors are beginning to produce more and more network applications specifically for 5G networks. Thus, as 5G approaches a very high maturity level, testing and validation of 5G solutions has become of utmost importance. Since verticals have very different needs such as the case of the Automotive industry and Public Protection and Disaster Relief (PPDR), it will require different levels of support into taking their applications and services from concepts, to prototypes and finally to products. However, it is also more and more frequent that services in different verticals need to exchange data, particularly mobility services, such as those found in connected automobiles and PPDR [6].

Under these 5GPPP conditions, the 5GASP (5G Application & Services experimentation and certification Platform) project was launched [7]. The main technical objective of 5GASP is to build and operate an Open, and Inter-Domain 5G NFV-based Reference (Open5G-NFV) ecosystem of Experimental Facilities [8].

The rest of the paper is organized as follows: Sect. 2 contains related literature review, Sect. 3 presents the problem statement and the aim of paper, 5GASP verticals use cases analysis was made in Sect. 4, Sect. 5 describes the 5GASP developed ecosystem with the definition of all its parts, an analysis of 5GASP methodology use for Remote human driving NetApp, is presented in Sect. 6, and, finally, Sect. 7 describes a conclusion.

2 Background Analysis

Considering the 5G networks benefits and prospects [9–11], can be stated that this type of networks will continue take the leading role in the ICT sector. In these conditions, the development of network applications also will be a new trend for software developers in different business sectors [12–15]. Particular attention is paid to the development of applications for critical sectors of the economy and social life. For example, a large number of publications are devoted to the development and research of approaches to application development for the Automotive [16–18] and PPDR [19–21] industries. Also, a large number of publications are devoted to testing the developed software [22, 23]. But it is the development and testing of network applications in 5G is devoted to a very small number of works [24–26], therefore, a large number of unexplored issues remain related to the improvement of network application development methods and their automated testing using 5G equipment.

3 Problem Statement

Thus, supporting the process of developing software for the needs of vertical industries using 5th generation cellular networks is a very important issue that must be addressed in the near future. This should be a comprehensive solution involving all stakeholders, the necessary hardware and software. To solve this problem, the paper proposed the following:

- analysis of 5GASP vertical use cases (Automotive and PPDR);
- 5GASP ecosystem analyses with the definition of the development lifecycle, automated testing, validation and certification lifecycle;
- definition of the involved to the ecosystem entities;
- presentation of an example of 5GASP methodology use for Remote human driving NetApp.

4 5GASP Verticals and Use Cases

4.1 Automotive Industry

The Internet of Vehicles (IoV) is one of the established use cases for 5G in the vertical industry. According to statistics, in 2020 the global market for C-V2X (vehicle communication with other objects via cellular communication) exceeded $86.7 billion.

Level 5 fully autonomous vehicles (the car doesn't require a human to drive it) will be popular in at least a decade.

However, to capitalize on this innovation of tomorrow, experienced organizations are already actively investing in supporting the autonomous car ecosystem. Eventually, self-driving cars will act as mobile computing platforms and transfer huge amounts of data - approx. 4 TB every 90 min. This data will need to be quickly uploaded to a cloud platform, triggered by AI algorithms, and transferred back to the car in the form of driving instructions. These exchanges must occur within milliseconds to ensure smooth and safe driving. The features of 5G technology described earlier can make this fast data exchange a reality.

5G is the ideal candidate to provide the next level connectivity necessary to fulfil the stringent requirements of automated driving applications. 5GASP will capitalise on this factor and innovate in the following technologies and enablers, NR-SA, NR-NSA, C-RAN, mMTC, eMBB, uRRLC, Slicing, SDN/NFV, MEC, SFC, etc. to develop, test and validate in heterogeneous test sites the CCAM functionalities.

Innovation in telecommunications cannot happen without standards. Standardisation of 5G and spectrum allocation issues and their effect on automotive applications are one of the main barriers to innovation in 5G applications.

In addition to providing vehicles with connectivity to the cloud, new evolutions of 5G are challenging the way vehicles are communicating with other vehicles and the roadside infrastructure. Time-critical services such as "emergency braking" (Cooperative ADAS) or remove driving for Level-5 automated vehicles require extremely low latencies that can only be provided by localized communications means between vehicles and with the roadside infrastructure (V2X).

From all of this it is clear the 5G is considered for all types of connectivity services, whether they are contemplated by automotive OEM to implement the "extended vehicle" concept or by the C-ITS community whom needs a collection of access technologies with short, medium and long range characteristics in order to achieve the long-term goal of zero fatalities on the road and the fully automated vehicle.

5GASP therefore ambitions to develop NetApps either compatible with the "extended vehicle" standards or with "C-ITS" standards – possibly both at once – in order to facilitate the adoption of 5G and offer a framework for the development of innovative services that meet the constraints of these new standards.

4.2 Public Protection and Disaster Relief (PPDR) Industry

In today's world, public safety has turned out to be the main problem of every country around the world. Thus, in the event of an emergency, forecasting the situation, quickly responding to it and acting after the emergency are key in order to avoid or reduce losses. Of the 198 countries worldwide, no more than 20 countries have reasonably complete public safety systems, including the United States, Japan, developed countries in the European Union, and some large developing countries.

The emergency response system is responsible for conducting all types of public protection events, such as:

- Natural disasters: including floods, meteorological disasters, earthquakes, geological disasters, maritime disasters, biological disasters, steppe and forest fires, etc.
- Natural disasters in public health: including outbreaks of infectious diseases, the emergence of unknown diseases, food safety and occupational hazards, animal epidemics and other serious events that affect public health and life.
- Serious accidents: including all types of accidents, traffic accidents, accidents in public institutions and equipment accidents, environmental pollution and environmentally damaging accidents, etc.
- Social security: including terrorist attacks, economic security incidents and emergencies related to events abroad, etc.

For all the types of public protection events, an integrated ICT solutions have to be applied and provide early warning and event prediction, big data mining, cloud storage, integrated control and dispatch, emergency management, mobile applications, etc. to meet the needs of peacetime and wartime, to fully implement the concept of use in wartime and peacetime.

With the standardization of dedicated PPDR capabilities in 3GPP, including MCX (MC Voice, Video and Data), IOPS and high-power UE, 5G is staring to gain traction with the prospective stakeholder organizations, which incentivises intensification of research and real-world piloting.

The ambition of 5GASP is to support development of specialized PPDR NetApps in response to such expectations and demonstrate their value through verification, certification and demonstration of 5G network resilience, privacy, trust, security and ad-hoc extensibility during massive disasters. Core research and development will pursue service resilience and self-healing properties through NetApp provisioning and orchestration exploiting cloud-native and MANO technologies and adaptive 5G PPDR architecture augmentation with private and hybrid deployment options, supported with novel connectivity modes between UE, RAN and core network segments.

4.3 5GASP Use Cases

In order to provide opportunities for SMEs and developers to experiment their applications in the context of specific vertical use cases we have identified Automotive and PPDR vertical as those of most relevance to showcase our innovation with a set of use cases resulting from different NetApps orchestrations, as can be seen summarized in Fig. 2. We will target 3 use-cases: Automotive, PPDR and a Cross Vertical. The use cases will be performed near the end of the project to show case in public events the 5GASP result: How easy is to deploy a NetApp in minutes with Zero-Touch Orchestration.

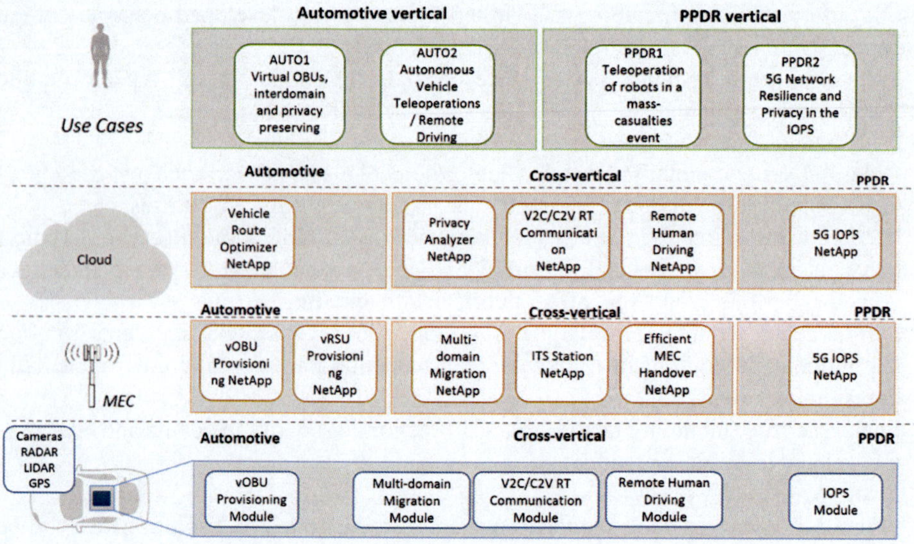

Fig. 2. 5GASP NetApps and orchestrated use cases.

5 5GASP Ecosystem

As the softwarization in 5G opened the door to a huge number of new players, mostly SME's, they are able to provide innovative products in specific verticals. The capability of such SME's to innovate is closely linked to the low-cost barrier of developing software for the virtualized 5G environment. Nonetheless 5G still poses a considerable barrier to these same SME's in the form of a deeply entrenched community with large amounts of standards, definitions and knowledge dating back to the beginning of cellular networks. For SME's coming from an IT background this can become a daunting task which can be only overcome through training and education of Human Resources.

Thus, 5GASP ecosystem (Fig. 3) is a set of own project services and testbeds united around one goal: support of the development and fully automated testing of emerging network applications in 5G and beyond era.

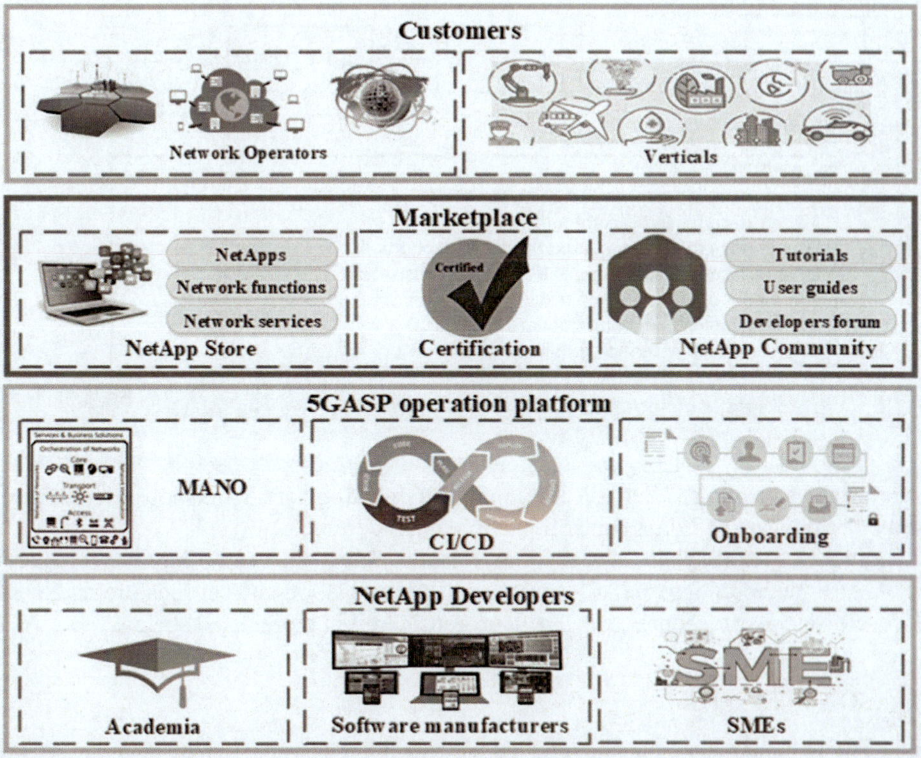

Fig. 3. 5GASP ecosystem.

5GASP ecosystem includes: NetApp developers (academia institutions, SMEs, software manufacturers), NetApp potential customers (mobile network operators, CDN providers, end users, etc.), still developing 5GASP operation platform (6 connected EU testbeds with all the required equipment and software tools), and, of course, the platform for placing all this together – Marketplace (NetApp Store, certification center, Community, etc.). The interaction of different ecosystem actors is shown in the Fig. 4 [27].

To support the goals of the project, 5GASP is promoting the development of a 5G NetApps developers' community (NetApp Community) [28] which will promote the

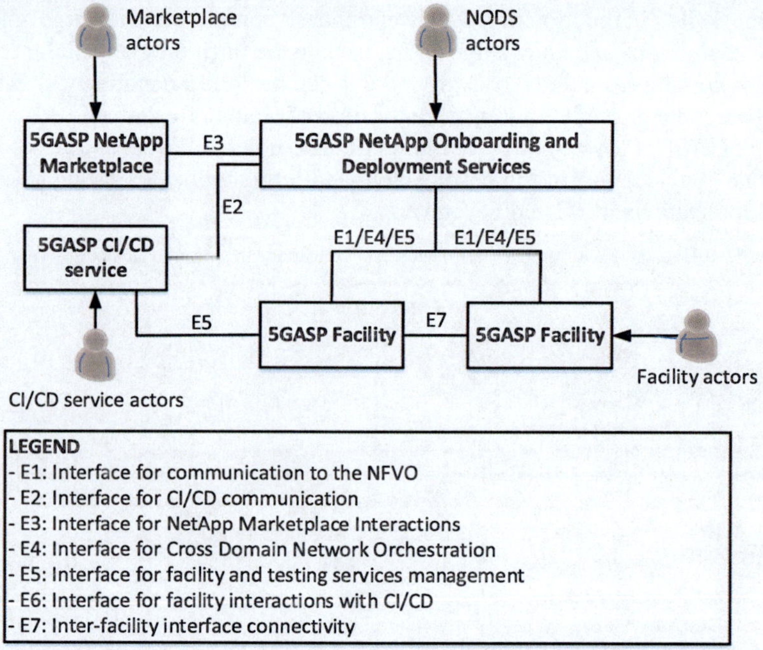

Fig. 4. The interaction scheme of different ecosystem actors.

training and education of NetApp developers through Webinars, Workshops, Hackathons and other relevant activities for SME's Human Resources.

According to the proposed ecosystem, were developed the main comprehensive parts of 5GASP methodology: 5GASP lifecycle, experimentation and certification readiness lifecycle, testing procedures, etc. They are considered in more details below.

5.1 5GASP Lifecycle

5GASP Lifecycle is a process used by the NetApp developers to design, develop and test high quality NetApps. This lifecycle aims to produce a high-quality NetApp that meets or exceeds customer (for example, MNO) expectations, reaches completion within times and cost estimates. 5GASP lifecycle includes four main phases: design, development, testing, and validation (see Fig. 5) [29].

Design. The design of a NetApp consists of three main steps, as they are depicted in Fig. 6.These steps are necessary to be completed by the programmer so that their existing code can be onboarded as a NetApp through the 5GASP platform and tested through 5GASP's testing framework.

Development. Following the design phase, the developer can start developing the NetApp. The development consists of materializing the design considerations that have been studied in the previous step, resulting at the end in a NetApp ready to be onboarded

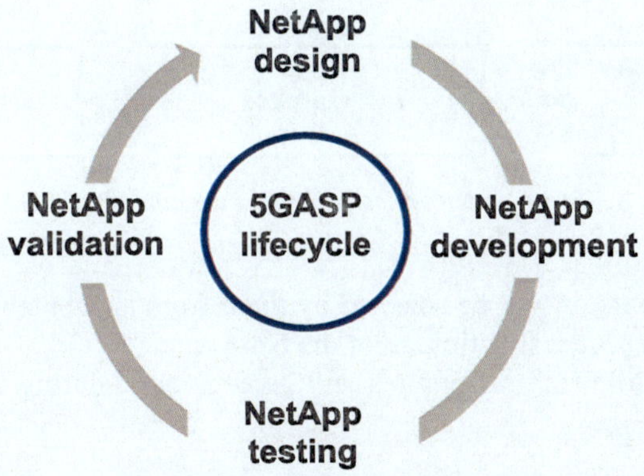

Fig. 5. 5GASP lifecycle.

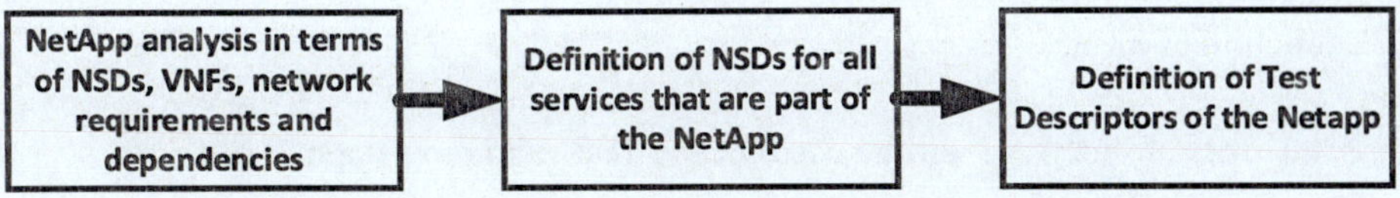

Fig. 6. Phases of the design of a NetApp in the context of 5GASP.

to the 5GASP portal. Consequently, the development of the NetApp is also divided into three steps (see Fig. 7), one for each component of the needed triplet: the VNFDs/NSDs, the network slice and the test descriptors.

Fig. 7. Phases of the development of a NetApp in the context of 5GASP.

Testing. Testing is a very important and required phase of the lifecycle, which aims to check whether the developed NetApp matches expected requirements and to ensure that NetApp is defect-free. It involves the execution of NetApp components using manual or automated tools to evaluate one or more properties of interest. The purpose of NetApp testing is to identify errors, gaps, or missing requirements in contrast to actual requirements, which are pre-defined for each NetApp on the first stage of the overall 5GASP methodology. In 5GASP, the testing phase will include the next stages (see Fig. 8).

NetApp and test VNFs deployment in test environment stage includes:

- Scheduling of the tests.
- Deployment of NetApps and test VNFs.

During the Test Execution stage, the system will run the included, selected and/or necessary tests to ensure the well-functioning of the NetApp. These tests will be created

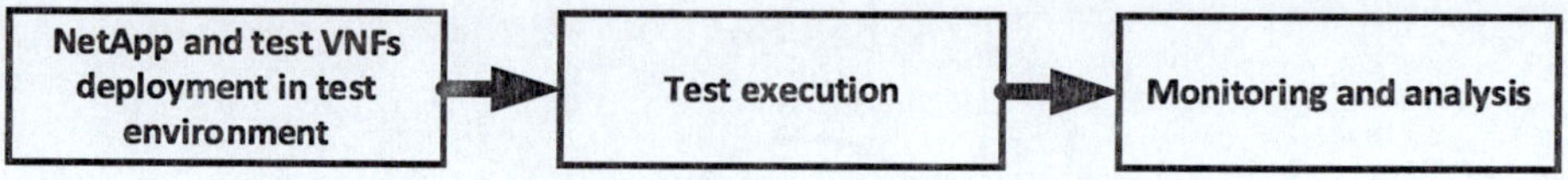

Fig. 8. Testing stages of a NetApp in the context of 5GASP.

by the NetApp owners or will be selected by them from an available repository in the platform, to test the proper functioning of the NetApp.

After this, the following procedures will be performed during the Monitoring and analysis stage:

- Monitor infrastructure, NetApps and test VNFs,
- Analysis of gathered data,
- Evaluation of metrics,
- Extraction of KPIs.

Outcomes of this stage will be used for the Test validation stage.

Validation. One of the biggest challenges in upcoming 5G DevOps environments is the validation and verification (V&V) of virtual network functions (VNFs) and network services (NSs) against different execution platforms so that service operators can be sure that VNFs and NSs behave as expected immediately after they are deployed and put into production [30].

After Testing phase Validation process takes place. Validation can be defined as the process of evaluating the developed NetApp during or at the end of the development process to determine whether it satisfies specified requirements (defined KPIs) [31].

Several KPI could be extracted from 3GPP/ETSI/others covering:

- Functional Suitability (Functional Completeness, Functional Correctness, Functional Appropriateness);
- Performance efficiency (Time-behavior, Resource utilization, Capacity);
- Compatibility (Co-existence, interoperability);
- Reliability (Maturity, availability, Fault tolerance, recoverability);
- Security (Confidentiality, integrity, non-repudiation, accountability, authenticity);
- Maintainability (reusability, analyzability, Modifiability, testability);
- Portability (adaptability, instability, replaceability).

As the issue of validation is critical, ETSI NFV ISG has released a number of documents approaching the need and the framework for pre-deployment validation and testing, as well as interoperability and portability [32].

5.2 Experimentation and Certification Readiness Lifecycle

The proposed 5GASP approach on DevOps experimentation and certification readiness Lifecycle is depicted in Fig. 9.

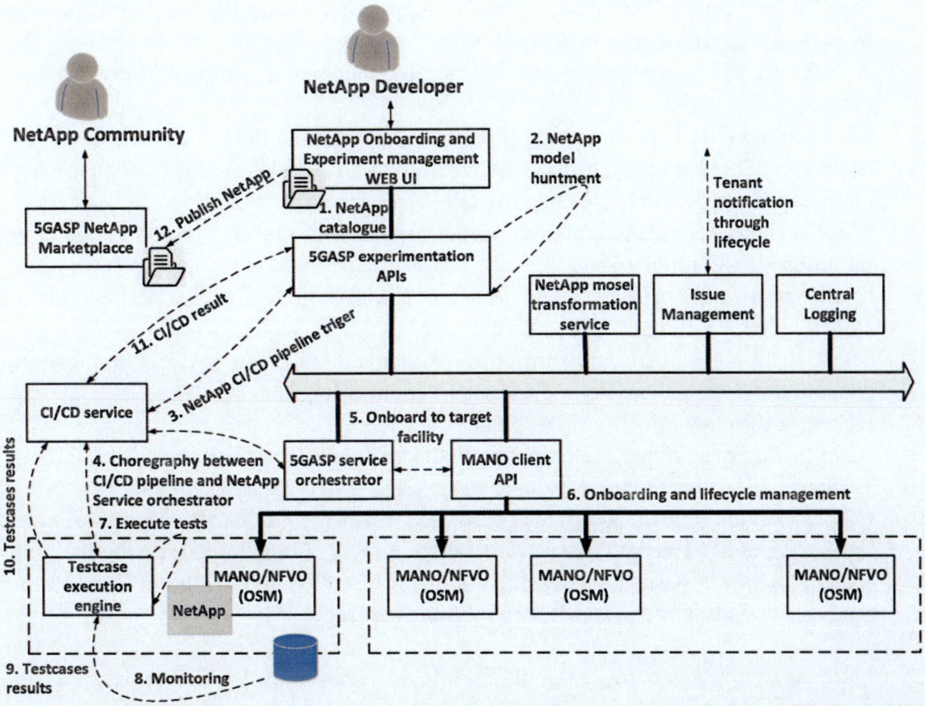

Fig. 9. 5GASP approach on DevOps experimentation and certification readiness Lifecyle [33].

The 5GASP approach defines the following functionalities and services:

- NetApp onboarding and Experiment Management UI,
- 5GASP Experimentation APIs service,
- NetApp model transformation service,
- Issue Management,
- Central Logging,
- 5GASP Service Orchestrator,
- MANO Client API (SOL005) service,
- 5GASP facilities:
  - Testcase Execution Engine,
  - Monitoring repository per facility,
- CI/CD service,
- 5GASP NetApp marketplace.

The experimentation and certification-readiness Lifecyle is envisaged and will be realized by the following interactions:

1. The NetApp developer (5GASP tenant), via the NetApp onboarding and Experiment Management UI Onboards the NetApp under test to the NetApp Catalogue in 5GASP Experimentation APIs service
2. A NetApp Model enhancement follows by the NetApp model transformation service.
3. The NetApp CI/CD main testing pipeline is triggered to the CI/CD service.
4. A Service Choreography follows between the CI/CD service and NetApp Service Orchestrator to properly manage the deployment to target facilities
5. The Service Orchestrator resolves dependencies and Onboard NetApps to target facilities NFVO repositories
6. The Onboarding & Lifecycle Management towards the facilities is realized via the MANO Client API (SOL005) service
7. The CI/CD service from information exchanged with the Service Orchestrator execute tests via the facility's Testcase Execution Engine
8. During testing Monitoring takes place
9. After testing execution, Testcases results from facility are available
10. Testcases results are made available to CI/CD service for assessment
11. CI/CD service makes available Testcases results to 5GASP Experimentation Service so 5GASP tenants are aware for the testing results and assessment
12. If the NetApp is passed successfully the full lifecycle and all the defined testing pipelines is Publish to 5GASP NetApp marketplace.

Figure 10 attempts to capture the methodology followed by 5GASP as derived from the reference 5GASP DevOps experimentation and certification readiness lifecycle process presented in the previous section. It discusses how the major activities will be followed and interact with each other during the lifetime of the project. The V-Model demonstrates the relationships between each phase of the development and onboarding life cycle and its associated phase of testing. The V-model approach supports flexibility and agility throughout the development process. In addition to be a highly disciplined approach, it promotes meticulous design, development, and documentation necessary to build stable software product [35].

In general, the 5GASP methodology follows an iterative approach, while the following phases and their relationships are identified in [36].

5.3 5GASP Facilities and Their Interaction

There are 6 European testbeds, which are been used for the deployment and testing the developed NetApps. They are situated in Aveiro, Bucharest, Mursia, Patras, Bristol and Ljubljana. All the facilities are looking forward towards the establishment of the multi-domain fabric and the provision of hosting network slices in each facility. Also, some envisioned plans are presented with aspects on how to tailor 5GASP facilities to vertical NetApps requirements. The full description of the testbeds can be found in [36].

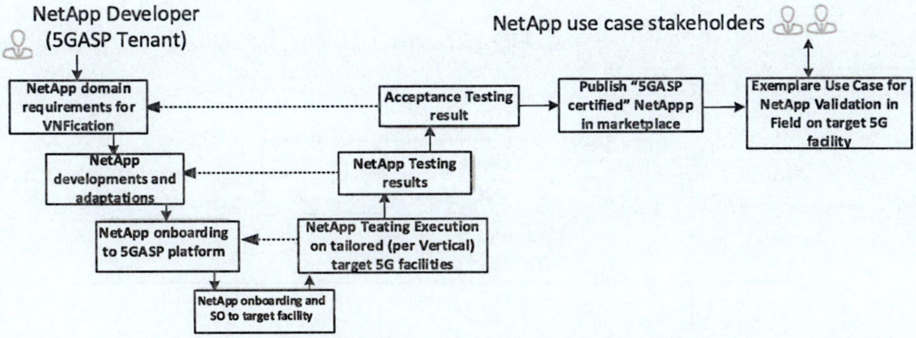

Fig. 10. 5GASP NetApps V- Model methodology [34].

Figure 11 shows the representation of the multidomain scenario, when different NetApps may interact and make more complex scenarios for NetApps usage and testing accordingly.

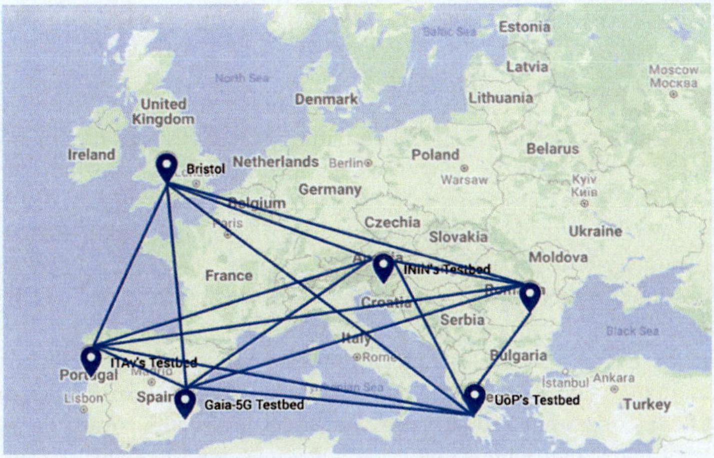

Fig. 11. Representation of the multidomain scenario.

To provide the above-mentioned scenario, NetOr creates an overlay of [37] VPN Tunnels. Profound details on the above option can be found in [38], thus are omitted. The majority of the facilities employed OSM as its selected NFVO, thus exposing a SOL005 compliant Northbound Interface (NBI). NODS incorporates a relative SOL005 compliant client, thus rendering the direct exposure of facilities' OSMs as the nominated scenario, as Fig. 12 illustrates.

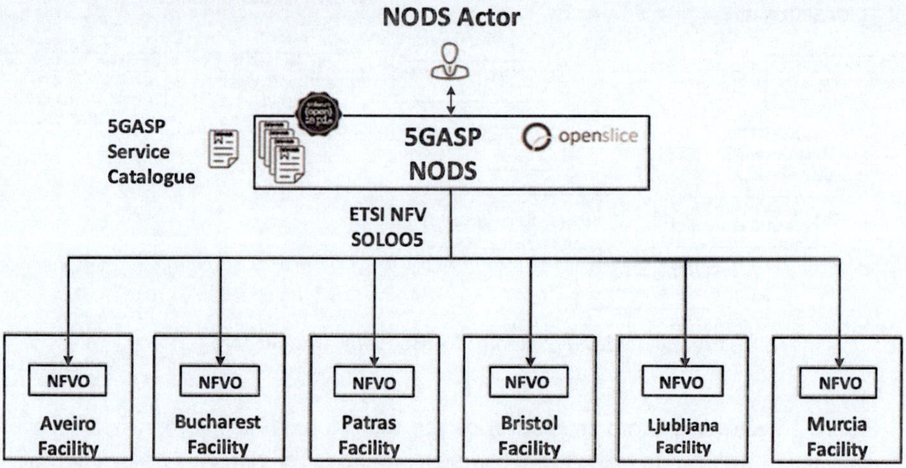

Fig. 12. NODS - facilities interaction.

6 5GASP Methodology Use for Remote Human Driving NetApp

6.1 Remote Human Driving Use Case

There is an industry consensus today that autonomous vehicles will need help in making decisions, especially in unusual, dangerous situations that can happen on the road and may require violating the traffic laws (e.g. crossing double yellow lines). For these cases, the industry has started to adopt teleoperation/remote driving solutions that enable a remote human operator to monitor and take control over the car if needed. In addition, some countries have issued regulations that mandate the ability of systems that enable a remote human to take control over the vehicle if needed.

The Teleoperations use case enables a remote operator to take full/partial control over an autonomous vehicle in unusual/dangerous situations that can happen on the road. Ensuring safe Teleoperation and human remote assistance entails reliable and secured transmission of high-quality real-time video with minimum latency. The use case will use various NetApps provided by different 5GASP SME partners (Fig. 13).

The figure below illustrates the use case architecture [36].

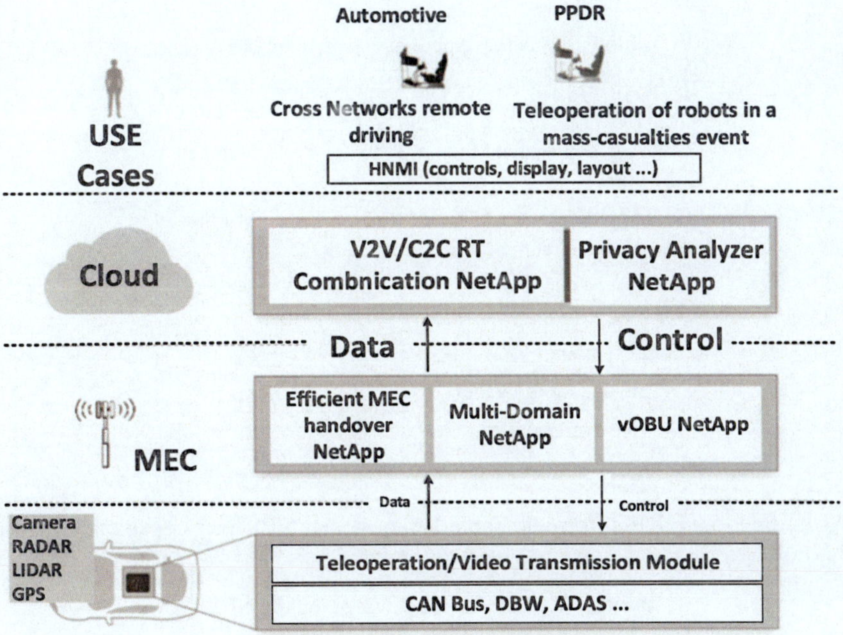

Fig. 13. Cross-vertical Use Case.

6.2 Remote Human Driving NetApp Architecture and Initial Requirements

A virtual vehicle instance is equipped with one or more 5G modems and a software (SW) module that enables the transmission of HD video to the tele-operation centre (Fig. 14).

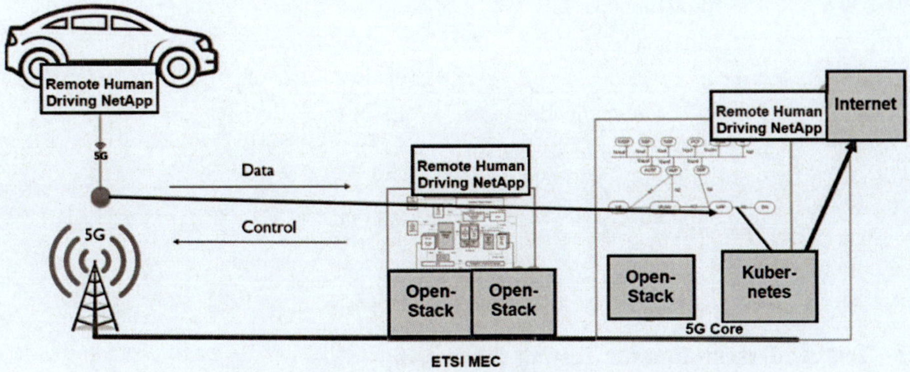

Fig. 14. Remote human driving NetApp architecture.

This NetApp has strict connectivity requirements, which have to be achieved for uninterrupted work of the application (Table 1) [36].

Table 1. KPIs for remote human driving NetApp.

Generic KPI name	KPI value	KPI unit
NetApp deployment time at the edge	<5	Minutes
NetApp deployment time in the cloud	<5	Minutes
Contineous control messages	<1/100.000	Messages
Contorl latency	<5	Ms
Frame decoding latency	<15	Ms
Service availability	>99	%

According to the KPIs from the Table 1, the special NEST was developed (Table 2).

Table 2. Remote human driving NetApp NEST.

Characteristics	Specific value
Area of service	Spain
Area of service: Region specification	Murcia
Downlink maximum throughput per UE	100000 kbps
Uplink maximum throughput per UE	1000 kbps
Isolation level	Virtual resources isolation
V2X communication mode	YES-NR
Slice quality of service parameters: 3GPP 5QI	9
Supported device velocity	120 km/h

6.3 Testbed Description for Testing the NetApp

Mursia's testbed was chose to showcase the full cycle for the development, testing, validation and certification for the Remote human driving NetApp. Murcia site is a laboratory for experimentation located in the Computer Science Faculty in the University of Murcia. Its research purposes are focused on the area of 5G technologies with virtualization and backhaul infrastructure for the Wireless connectivity, including 5G, LoRaWAN and 802.11p self-managed infrastructure that allows experimentation in different fields.

Mursia's testbed provides the experimenters with all the infrastructure and tools required by developers to test and validate their NetApps in a real-world 5G network.

The Murcia facility relies on Free5GC [39] as 5G core and uses Amarisoft's [40 5G gNB stack with SDR radios and outdoor RRH. Murcia Site architecture is presented on Fig. 15.

It has a network infrastructure based on Coarse Wavelength-Division Multiplexing (CWDM) which provides a bandwidth of 10 Gbps fully available and customizable with QinQ/802.1ad allowing the isolation of different traffic. MEC functionality, including the virtualisation of edge devices. SDN is provided by two Delta AG7648-R with the latest PiCOS operating system in the core and HPE 2920 for leaf nodes switches supporting OpenFlow technology; in addition network programmability is granted by two barefoot Tofino-powered switches EdgeCore WEDGE100BF-32X-O-AC-F and Stordis BF2556X-1T-A1F. Connectivity between OpenFlow enabled switches and Programmable switches (P4) works at 6×40 Gbps while programmable switches can work up to 100 Gbps.

The site is equipped with a cloud platform which hosts core network components, as well as SDN, NFV and MEC deployments. Two OpenStack are deployed; rocky, a full-fledged deployment offering 160 vcpus (Intel(R) Xeon(R) Gold 6138 CPU based) and 512 GB RAM splitted into two compute nodes one on each side and Ussuri, a lightweight deployment offering $12 \times 86\_64$ vcpus (Intel(R) Xeon(R) CPU E5–2603 v3) based with 48 GB RAM and some ARM RPI 4 nodes with 4 vcpu each (Broadcom BCM2711B0 quad-core A72 based) and 8 GB RAM each. Hyper-Converged Infrastruture with a 4-node cluster with 128vCores and 4 TB RAM and two edge clusters providing 24vCores and 512 GB RAM each is also available for extending the VIM capabilities. OSM is used as NFV orchestrator.

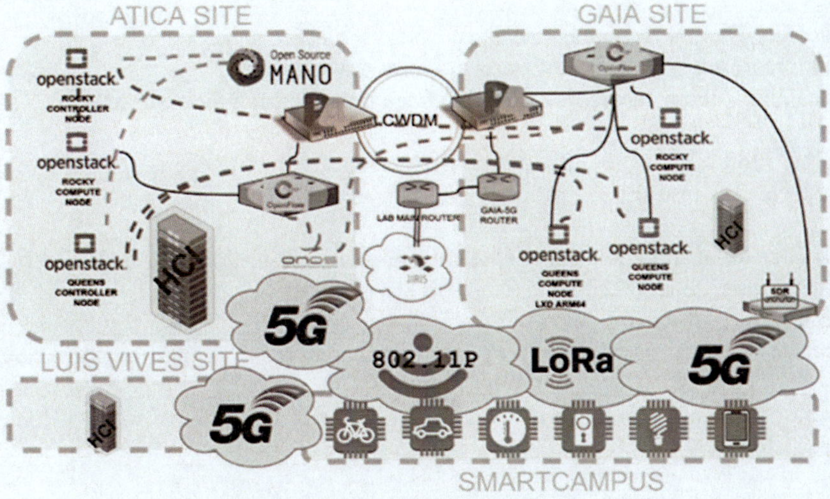

Fig. 15. Murcia site architecture.

Regarding wireless connectivity, 5G and LTE provision is done by using SDR technology, in particular, 2 ETTUS B210 and one ETTUS N310 are available for experimentation. These SDR devices are accompanied with a 4 Core (I7–8700 based) with 8 GB RAM the B210s and a 16 Core (AMD EPYC 7302P based) with 32 GB RAM server the N310. Besides, a commercial core from Amarisoft with the associated RAN hardware and both systems are being fully integrated for having a completely functional and multi-site 5G solution. The deployment consists of two sites located in the Campus of Espinardo of the University of Murcia, and two cores are available, one turnkey solution by Amarisoft, and an experimental one which can implement multiple open source 5G cores.

Murcia Testbed also has LoRaWAN capabilities based on the 868 MHz band. Currently, the infrastructure relies on a Kerlink iStation gateway, located in the Luis Vives bulding. This gateway is connected to our own deployed LoRaWAN network server (Chirpstack), the received data is also forwarded to The Things Network LoRaWAN network servers. In the future, it is planned to deploy two more gateways in the campus, in order to achieve full coverage of the Espinardo campus and also to cover Murcia area. Also, NB-IoT could be integrated into the platform provided by external providers.

6.4 Test Cases

5GASP defines the test cases based on the test references in Sect. 3.1 [27]. The following subsections present the lists of the predefined test cases, which can be used for the Remote human driving NetApp. In each of 5GASP test case list, the well-defined fields are expected as follows. Some other fields (e.g., Expected Value) could be added and defined by the test designer.

- Test Case Name - the name of the test case.
- Test Description - the detailed description of the test case.
- Test Points - the measurable technical check points during the testing:
  - Test Method - it should be either a) Automation or b) Manual. 5GASP aims to execute the testing automatically as much as possible.
- Test Format - it should be either a) testing scripts) or b) testing tool xxx). Hereby, xxx is a tool name, e.g., iPerf.
- Test Priority - it should be one of the low 4 priorities: (i) P0 basic test cases, (ii) P1 recommended test cases, (iii) P2 optional test cases, (iv) P3 conditional test cases.

Table 3 presents a template of the test case list with the example value of the well-defined fields.

Table 3. 5GASP cross-vertical service test cases for remote human driving NetApp.

Test description	Test points	Expected value	Test method	Test priority
Verify the system initiation - the modules are up and reporting as running	Control Plane	PASS/FAIL	Automation	P0
The streamer and connector are connected to relay and node and reporting as connected	Control Plane	PASS/FAIL	Automation	P0
The system is synchronized using the cellular network	Control Plane	PASS/FAIL	Automation	P1
Connectivity to reporting service is established	Control Plane	PASS/FAIL	Automation	P1
Beginning of the video stream – the streamer is generating a video pattern pushed through the streamer and 5G system to the relay and node (terminated and measured)	End-2-End User Plane	PASS/FAIL	Automation	P1
Verify End-2-End Packet Data Rate of both Uplink and Downlink from the view of NetApp	End-2-End User Plane	1. Uplink >50 Mbps 2. Downlink >100 Mbps	Automation	P1
Verify End-2-End Packet Latency of both User Plane and Control Plane from the view of NetApp	End-2-End User Plane	1. User Plane <4 ms 2. Control Plane <20 ms	Automation	P1
Verify End-2-End Packet Loss of both Uplink and Downlink from the view of NetApp	End-2-End User Plane	1. Uplink <0.025% 2. Downlink <0.025%	Automation	P1

(*continued*)

Table 3. (*continued*)

Test description	Test points	Expected value	Test method	Test priority
Verify End-2-End Packet Jitter of both User Plane and Control Plane from the view of NetApp	End-2-End User Plane	1. User Plane <1 ms 2. Control Plane <10 ms	Automation	P1
Verify End-2-End Frame Loss of both Uplink and Downlink from the view of NetApp	End-2-End User Plane	<50 frame skips per hour	Automation	P1
To verify the stability of the service while video BW is being changed according to the predefined pattern (between 2 Mbps to 20 Mbps)	End-2-End User Plane	PASS/FAIL	Automation (under investigation)	P1
This test validates the NetApp has REST interfaces of the EMHO NetApp that are accessible for the supported/enhanced NetApp using the 5G network	Control Plane	PASS/FAIL	Automation (under investigation)	P1
This test validates the availability of the interface of NetApp to receive 5G monitoring data	Control Plane	PASS/FAIL	Automation (under investigation)	P1
Verify the system initiation - the modules are up and reporting as running	Control Plane	PASS/FAIL	Automation	P0

6.5 NetApp Testing Results

At the initial stages, the NetApp was tested using the LTE systems. The initial tests were designed to prove the system is working and capable of provide the required service. In addition, the NetApp should have the capability to gather the statistical data and verify a proper operation according to the table in the previous section. Bandwidth (BW) tests were chosen to verify the NetApp operation (Fig. 16).

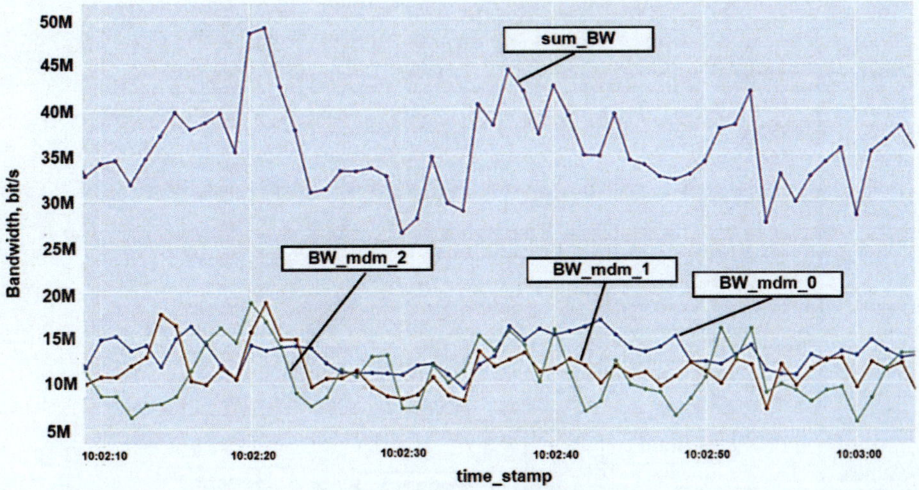

Fig. 16. NetApp bandwidth tests.

Following the initial tests, the system was deployed in a hybrid LTE and 5G system to verify the jitter, latency and loss measuring system capabilities (Fig. 17).

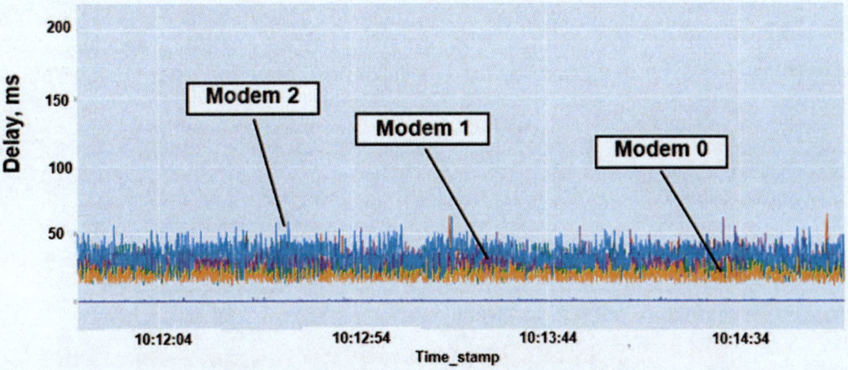

Fig. 17. Packet received latency per modem.

As it can be seen from the test results, 5G modem performance is much better than the LTE system (Fig. 18).

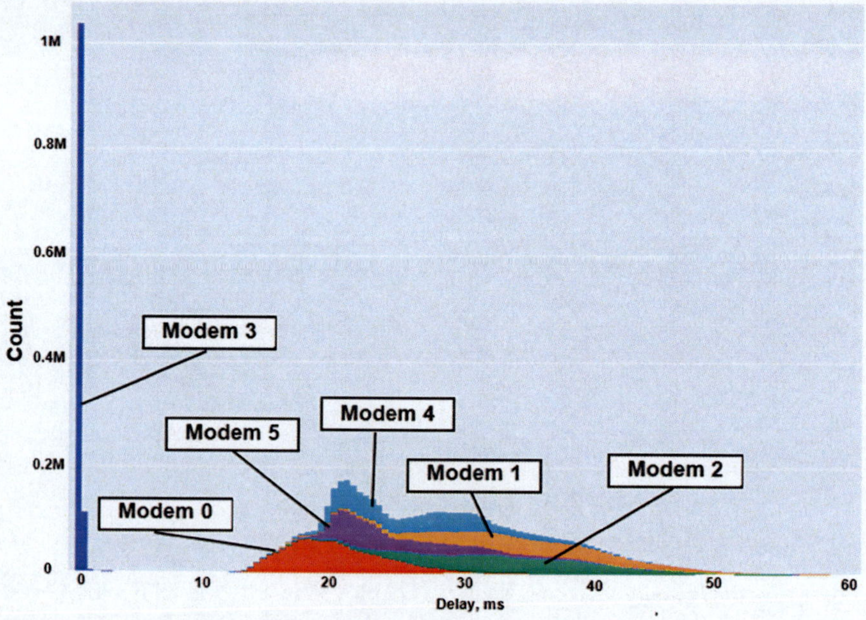

Fig. 18. Packet received latency histogram per modem.

To verify the proper system operation, the whole video pipeline was analyzed using a test system. The system was tested for video frames delay and jitter (Fig. 19).

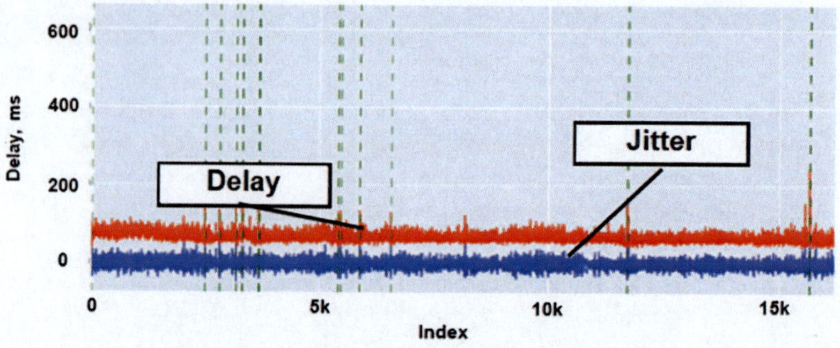

Fig. 19. Video delay and jitter test results.

7 Conclusions

Mobile networks have long been part of our lives. At the same time, 5G is a key driving force in the development of the ICT industry and other related vertical industries. In these conditions, when a large number of commercial 5G networks are already deployed, software developers begin to develop a large number of network applications for the needs and with the use of 5G. Therefore, it is very important to support developers, small and medium enterprises in order to automatically test, certify and bring to the end customer already debugged applications that can be deployed and used in commercial projects.

Therefore, the 5GASP project has developed an ecosystem for the development and testing of emerging network applications in 5G and beyond environment. This ecosystem combines software developers, test sites in 6 European cities, automated testing and certification tools, end customers (users) of network applications and other tools and systems. And it is mandatory to bring the tested and certified network application to the end user or customer through the Marketplace, which was called the NetApp Store.

This paper also demonstrates how the methodology developed in the project will be applied in practice on the example of Remote human driving NetApp. In particular, the architecture of the application, the principles of interaction with the test platform, the requirements for the application, the test plan and metrics to be collected were shown.

Acknowledgement. This work was supported in part by the European Commission under the 5GASP: 5G Application & Services experimentation and certification Platform (H2020 – ICT-2020, grant agreement ID: 101016448). The views expressed in this contribution are those of the author and do not necessarily represent the project.

References

1. The state of mobile internet connectivity 2021, GSMA report
2. GSMA * The Mobile Economy 2022 report
3. David, K., Berndt, H.: 6G vision and requirements: Is there any need for beyond 5G? IEEE Veh. Technol. Mag. **13**, 72–80 (2018)
4. Andrews, S., et al.: What will 5G be?, IEEE J. Sel. Areas Commun. **32**, 1065–1082 (2014)
5. https://www.t-mobile.com/business/resources/articles/benefits-of-5G-for-innovation
6. 5GASP project website. https://www.5gasp.eu/project/project.html
7. CORDIS EU research results: 5G Application & Services experimentation and certification Platform – Access. https://cordis.europa.eu/project/id/101016448
8. 5GPPP: about 5GASP. https://5g-ppp.eu/5gasp/
9. Chochliouros, I.P., et al.: Business and market perspectives in 5G networks. In: 2017 Internet of Things Business Models, Users, and Networks. IEEE, pp. 1–6 (2017)
10. Bulashenko, A., et al.: New traffic model of M2M Technology in 5G wireless sensor networks. In: 2020 IEEE 2nd International Conference on Advanced Trends in Information Theory (ATIT), pp. 125–131. IEEE (2020)
11. Mi, D., Odarchenko, R., et al.: Demonstrating immersive media delivery on 5G broadcast and multicast testing networks. In: IEEE Transactions on Broadcasting, https://doi.org/10.1109/TBC.2020.2977546
12. Liu, E., Effiok, E., Hitchcock, J.: Survey on health care applications in 5G networks. IET Commun. **14**(7), 1073–1080 (2020)
13. Shaohua, S.W., et al.: Toward offloading internet of vehicles applications in 5G networks. IEEE Trans. Intell. Transp. Syst. **22**(7), 4151–4159 (2020)

14. Odarchenko, R., Tran, T., Navrátil, D., Sanders, P., Burdinat, C., Gomez-Barquero, D.: Enabling multicast and broadcast in the 5G Core for converged fixed and mobile networks. IEEE Trans. Broadcast. **66**(2), 9099058, 428–439. https://doi.org/10.1109/TBC.2020.2991548 https://ieeexplore.ieee.org/document/9099058
15. Odarchenko, R., Altman, B., Aguiar, R., Sulema, Y.: Multilink approach for the content delivery in 5G networks. In: 5th International Scientific-Practical Conference Problems of Infocommunications Science and Technology, PIC S and T 2018 - Conference Proceedings, pp 140–144 (2018)
16. Ding, A.Y., Janssen, M.: Opportunities for applications using 5G networks: requirements, challenges, and outlook. In: Proceedings of the Seventh International Conference on Telecommunications and Remote Sensing, pp. 27–34 (2018)
17. Rabe, M., Milz, S., Mader, P.: Development methodologies for safety critical machine learning applications in the automotive domain: a survey. In: Proceedings of the IEEE/CVF Conference on Computer Vision and Pattern Recognition, pp. 129–141 (2021)
18. Rahim, M.A., et al.: Evolution of IoT-enabled connectivity and applications in a aut omotive ind ustry: a review. Veh. Commun. **27**, 100285 (2021)
19. Bai, F., et al.: Towards characterising and classifying communication-based automotive applications from a wireless networking perspective. Int. J. Veh. Autonom. Syst. **10**(3), 165–197 (2012)
20. Apostolakis, K.C., et al.: Cloud-Native 5G infrastructure and network applications (NetApps) for public protection and disaster relief: the 5G-EPICENTRE Project. In: 2021 Joint European Conference on Networks and Communications & 6G Summit (EuCNC/6G Summit), pp. 235–240. IEEE (2021)
21. Ergul, O., et al.: Adaptive and cognitive communication architecture for next-generation PPDR systems. IEEE Commun. Mag. **54**(4), 92–100 (2016)
22. Arampatzis, D., et al.: Unification architecture of cross-site 5G testbed resources for PPDR verticals. In: 2021 IEEE International Mediterranean Conference on Communications and Networking (MeditCom), pp. 13–19. IEEE (2021)
23. Kuliamin, V.V., Pakoulin, N.V., Petrenko, A.K.: Practical approach to specification and conformance testing of distributed network applications. In: Malek, M., Nett, E., Suri, N. (eds.) Service Availability. ISAS 2005. Lecture Notes in Computer Science, vol. 3694. Springer, Berlin, pp. 68–83 (2005). https://doi.org/10.1007/11560333_6
24. Jorgensen, P.C.: Software Testing: a Craftsman's Approach. Auerbach Publications, Boca Raton (2013)
25. Zhang, P., et al.: A survey of testing for 5G: Solutions, opportunities, and challenges. China Commun. **16**(1), 69–85 (2019)
26. Trantzas, K., et al.: An automated CI/CD process for testing and deployment of network applications over 5G infrastructure. In: 2021 IEEE International Mediterranean Conference on Communications and Networking (MeditCom), pp. 156–161. IEEE (2021)
27. 5GASP deliverable 5.1 initial report on test plan creation and testing methodologies
28. 5GASP NetApp Community portal – Access. https://community.5gasp.eu/
29. 5GASP deliverable 4.1 initial methodology for designing and developing NetApps – Access. https://www.5gasp.eu/assets/documents/deliverables/D4.1%20Initial%20Methodology%20for%20Designing%20and%20Developing%20NetApps.pdf
30. Engel, A.: Verification, Validation, and Testing of Engineered Systems. Wiley, Hoboken (2010)
31. Dasso, A. (ed.). Verification, Validation and Testing in Software Engineering. IGI Global, Hershey (2006)
32. ETSI NFV Management and orchestration - an overview
33. 5GASP objectives. https://www.5gasp.eu/project/objectives.html

34. 5GASP Certification Guidelines – Access. https://community.5gasp.eu/index.php/certification-guidelines/
35. Turner, R.: Toward Agile systems engineering processes. Crosstalk. J. Defense Softw. Eng. 11–15 (2007)
36. 5GASP Deliverable 2.1 Architecture, Model Entities Specification and Design
37. Donenfeld, J.A.: Wireguard: next generation kernel network tunnel. In: NDSS, pp. 1–12 (2017)
38. 5GASP Deliverable 3.1 5GASP experimentation services, middleware and multi-domain facilities continuous integration
39. Neto, F.J.D.S., et al.: Analysis for comparison of framework for 5G core implementation. In: 2021 International Conference on Information Science and Communications Technologies (ICISCT), pp. 1–5. IEEE (2021)
40. Amarisoft official website. https://www.amarisoft.com/

Comprehensive Approach for Radio Equipment Design to Ensure Reliable UAVs Operations

Leonid Uryvsky[1(✉)], Andrii Gnashchuk[2], Serhii Osypchuk[1], and Alina Moshynska[1]

[1] Igor Sikorsky Kyiv Polytechnic Institute, Industrialnyi Lane, 2 (Campus #30), Kyiv 03056, Ukraine
leonid_uic@ukr.net
[2] National Aviation University, Kyiv, Ukraine

Abstract. The work is devoted to the peculiarities of providing a comprehensive approach for radio equipment design for modern unmanned aerial vehicles (UAVs). The functioning of UAVs is needed as a modern and promising technological object in various society applications and requires not only the means of direct flight support (aerodynamic properties, fuel resources, etc.), but also does require comprehensive radio equipment. The following tasks are expanded in this work and need qualitative solutions for building comprehensive radio equipment:

– Reliable channel for UAV position registration on UAV.
– UAV position registration channel at the ground control center.
– Transmission channel for flight control commands.
– Channel for collecting data for which the UAV is authorized.
– Channel for transmission of collected data to the ground control center.
– Means of radio electronic protection.
– Means of wireless telecommunication channels protection.

This chapter analyzes the needs of modern UAVs in providing them with radio and telecommunication systems that monitor the state of UAV parameters, determine its location, control the UAV operations, exchange information with UAVs via duplex channel, information protection during information transmission, electronic warfare, and others. A comprehensive approach to the design of a multifunctional radio communication system to ensure the reliable UAV operations, algorithms for its operation to ensure the UAV functioning, depending on the type of tasks for which it is intended.

Keywords: UAV · Reliable communication · Communication approach · Radio · GPS · IEEE 802.11

M. Klymash et al. (Eds.): TCSET 2022, LNEE 965, pp. 140–174, 2023.
https://doi.org/10.1007/978-3-031-24963-1_9

1 Introduction

World trends of modern information and telecommunication technologies development are expanded in [1, 2]. One of the main trends is communication for **unmanned aerial vehicles (UAVs)**. UAVs, also commonly known as **drones**, or **remotely piloted aircrafts**, with their high mobility and low cost have found a wide range of applications during the past few decades [3].

The functioning of UAVs is needed as a modern and promising technological object in various society applications and requires not only the means of direct flight support (aerodynamic properties, fuel resources, etc.), but also **does require comprehensive radio and communication equipment**.

The **object** of study is **UAVs**. The **subject** of research is a comprehensive **radio and communication equipment design** for UAVs. The **purpose** of paper is a **comprehensive analysis** for reliable communication and radio equipment design for modern UAVs.

The following **tasks** are expanded in this work for building comprehensive radio equipment for UAVs:

- Reliable channel for UAV position registration on UAV.
- UAV position registration channel at the ground control center.
- Transmission channel for flight control commands.
- Channel for collecting data for which the UAV is authorized.
- Channel for transmission of collected data to the ground control center.
- Means of radio electronic protection.
- Means of wireless telecommunication channels protection.

1.1 UAV Applications

Historically, UAVs have been primarily used in the military, deployed in hostile territory to reduce pilot losses [4]. With continuous cost reduction and device miniaturization, small UAVs (typically with weight not exceeding 25 kg) are now more easily accessible to the public.

UAVs have been successfully used for the delivery of goods in humanitarian, medical, agriculture and commercial contexts. UAVs are also considered critical to support public safety missions in different domains, including search and rescue operations and disaster management [5]. UAVs have numerous applications (Table 1): military, traffic control, cargo delivery, precise agriculture, video streaming, aerial inspection, rescue, and search, and many others [6].

Table 1. Commercial and governmental UAVs applications.

Role	Commercial	Role	Governmental	
			Federal	State & Local
Media	- Event filming - Aerial photography - Information services	Science	*Earth & Environment* - Pollution and land monitoring - Meteorological services - Biological services - Research	
Transport	- Cargo planes			
Monitoring	*Security* - Pipeline *Inspection* - Power and rail line	Security	- Coastline - Border patrol - Anti-terror fight - Search & Rescue - Disaster and catastrophe management	- Police support - Fire fighting - Traffic spotting - Forest fire avoidance
Communications	- Relay - Remote sensing - Disaster eNBs			
Agriculture	- Crop spraying/dusting - Forestry operations	Humanitarian Support	- Famine relief - Medical support - Emergency relief	

1.2 UAV Classification

UAVs can be broadly classified (Fig. 1) into **two** categories, **fixed** wing, and **rotary** wing (Fig. 2), each with their own strengths and weaknesses. For example, fixed-wing UAVs usually have high **speed** and heavy **payload**, but they must maintain continuous forward motion to remain aloft, and thus are not suitable for **stationary** applications like close inspection. In contrast, rotary-wing UAVs such as **quadcopters**, while having limited mobility and payload, can move in any direction as well as to stay stationary in the air. Thus, the choice of UAVs critically **depends on the applications** [7].

Currently, **military** branches as one of the main operators of UAVs use different kind of UAVs. To differentiate UAVs, the most common factors for UAV classification in the literature are takeoff **weight**, operational **altitude**, level of control **autonomy** as well as **military** based and **ownership**-based UAV distinctions [6]. Another approach for **UAVs classification is shown in** Table 2 [8].

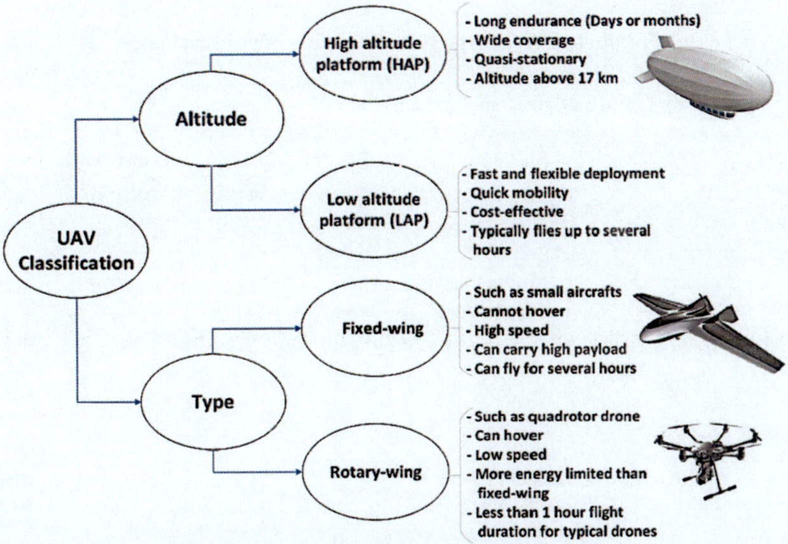

Fig. 1. UAVs classification by type and altitude.

Fig. 2. Rotary-wing (left) and fixed-wing (right) UAVs [15].

1.3 UAV Communication Paradigms

UAV Communication Paradigms. Multiple UAVs connected in a wireless ad hoc manner form a so-called **flying ad hoc network (FANET)**, offering multihop connectivity to each other. UAVs can be equipped with different networking technologies for communication with multiple UAVs, and between UAVs and other systems such as **ground control stations (GCSs)**, wireless sensor networks (**WSNs**), and on-ground robots.

Specifically, in a FANET, we can distinguish two main types of communication paradigms (Fig. 3):

- **UAV-to-UAV (U2U)** communications, where UAVs communicate with each other via line of sight (LoS) propagation links in a multihop manner in order to extend the communications range.

Table 2. UAVs classes and respective metrics values.

Class	Category	Normal employment	Normal operating altitude	Normal mission radius	Primary supported commander	Example platform
CLASS I (less than 150 kg)	SMALL >20 kg	Tactical Unit (employs launch system)	Up to 1524m AGL	50 km (LOS)	BN/Regt, BG	Luna, Hermes 90
	MINI 2–20 kg	Tactical Sub-unit (manual launch)	Up to 914m AGL	25 km (LOS)	Coy/Sqn	Scan Eagle, Skylark, Raven, DH3, Aladin, Strix
	MICRO <2 kg	Tactical PI, Sect, Individual (single operator)	Up to 61m AGL	5 km (LOS)	PI, Sect	Black Widow
CLASS II (150 kg to 600 kg)	TACTICAL	Tactical Formation	Up to 3048m AGL	200 km (LOS)	Bde Comd	Sperwer, Iview 250, Hermes 450, Aerostar, Ranger
CLASS III (more than 600 kg)	Strike/Combat	Strategic/National	Up to 19812m	Unlimited (BLOS)	Theatre COM	
	HALE	Strategic/National	Up to 19812m	Unlimited (BLOS)	Theatre COM	Global Hawk
	MALE	Operational/Theatre	Up to 13716m MSL	Unlimited (BLOS)	JTF COM	Predator B, Predator A, Heron, Heron TP, Hermes 900

- **UAV-to-GCS** communications **(U2GCS)** where a UAV communicates directly with the GCS.

Mixing these two communication paradigms is the basis of **hybrid approaches**, which represent **combination between U2U and U2GCS** communications, and where the UAV can send its data directly to the GCS in a one-hop or multihop fashion via the different UAVs in the mission area. UAVs can then function as relay nodes for data forwarding in a FANET.

UAV-to-Ground Control Station Communications. In general, U2GCS communication mode allows UAVs to communicate with the fixed central control center, such as a ground station, and are deployed to provide wireless services in case of terrestrial base

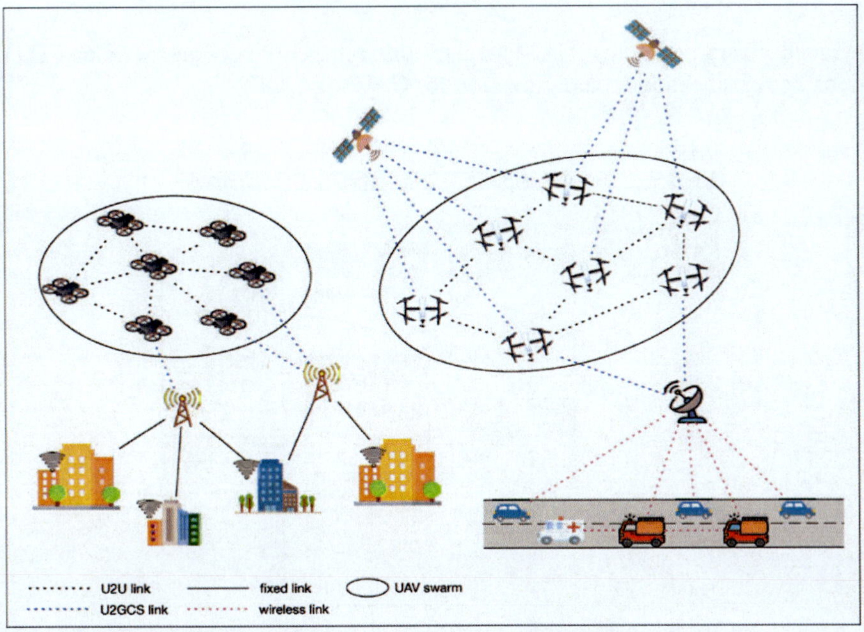

Fig. 3. UAV main communication paradigms (**U2U** and **U2GCS**) included in a typical example for aerial surveillance and imaging (left), search and rescue (**SAR**) applications (right). All the links are bidirectional.

station failure. U2GCS connectivity can be exploited to fill the gap between ground networks and satellites, thus representing an intermediate layer. In this scenario, UAVs can be exploited to cover connectivity holes, provide services in harsh environments, enhance network performance, and more.

Direct U2GCS communications exploit the **unlicensed spectrum** or reuse the spectrum bands that have been assigned for other applications, thus **resulting in limited data rates, unreliable connections, and insecure communications**. Other issues are related to signal propagation in free space since high mobility of UAVs **cannot always guarantee LoS propagation**. In such scenarios, **large-scale fading occurs**, as well as **multiple reflections, scattering**, and diffractions due to obstacles or unexpected **strong interference** (i.e., jamming signals). Furthermore, considering the high mobility of UAVs and the coverage requirements of swarm applications, the **U2GCS mode is not stable**, with the **possibility that the entire communication network becomes paralyzed once the ground station or the satellite breaks down, loses the radio link, or is attacked**.

1.4 UAV Communications Standardization

The Third Generation Partnership Project (**3GPP**), an industry-driven consortium that is **standardizing cellular networks**, has been active in identifying the **requirements, technologies**, and **protocols** for **aerial communications**. As a result, the 3GPP Technical Specification TS 22.125, *and many others*, establishes the requirements for providing

Unmanned Aircraft Systems (**UAS**) services through 3GPP networks. **IEEE, ITU** also work on communications standardization for UAV (Fig. 4, 5).

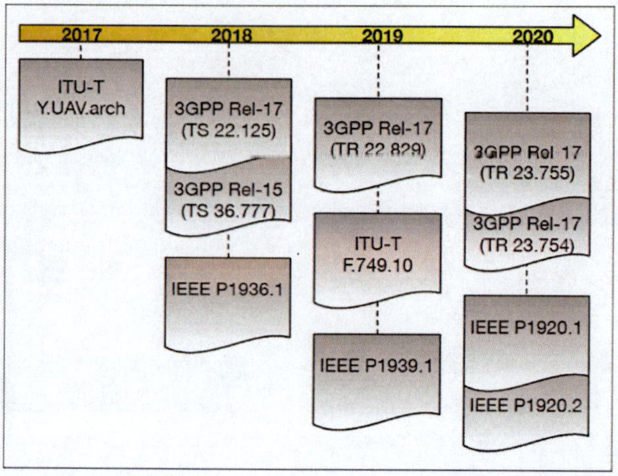

Fig. 4. Overview of the main initiatives of UAV standardization.

The 3GPP recommends using video feedback for supporting direct stick steering with the following video parameters [5]:

- **Visual** line of sight (**VLOS**) operation: 2 Mbps data rate for a 480p video with 30 fps and 1 s latency
- **Beyond** VLOS (**BVLOS**) operation: 4 Mbps data rate for a 720p video with 30 fps and 140 ms latency

The BVLOS operation has more stringent video feedback requirements because the video is the only way for the pilot to follow the UAV flight. The KPIs for UAV-assisted wireless services in accordance with 3GPP TS 22.125, V17.1.0, are shown in Table 3.

Recent developments in the UAVs field have made evident the **need for a standardization process of the communication technologies** supporting **direct information exchange, thus enabling U2U networking** [9]. This is necessary to achieve all sorts of cooperative tasks requiring real-time (or near-real-time) synchronization, including swarm formation and collision avoidance.

The **UAVs sector** has been experiencing unprecedented **growth** in recent years and was estimated to generate a total of US$22.5 billion at the end of 2020. Such estimations are even more promising in coming years, where a growth rate of 13.8% is foreseen, reaching US$42.8 billion by 2025. This outstanding growth is to be expected when considering the ever-growing number of application areas where the contribution of UAVs is significant, including site surveys, industrial inspection, surveillance, smart agriculture, search, and rescue (**SAR**) missions, topography, parcel delivery, human transportation, and so on.

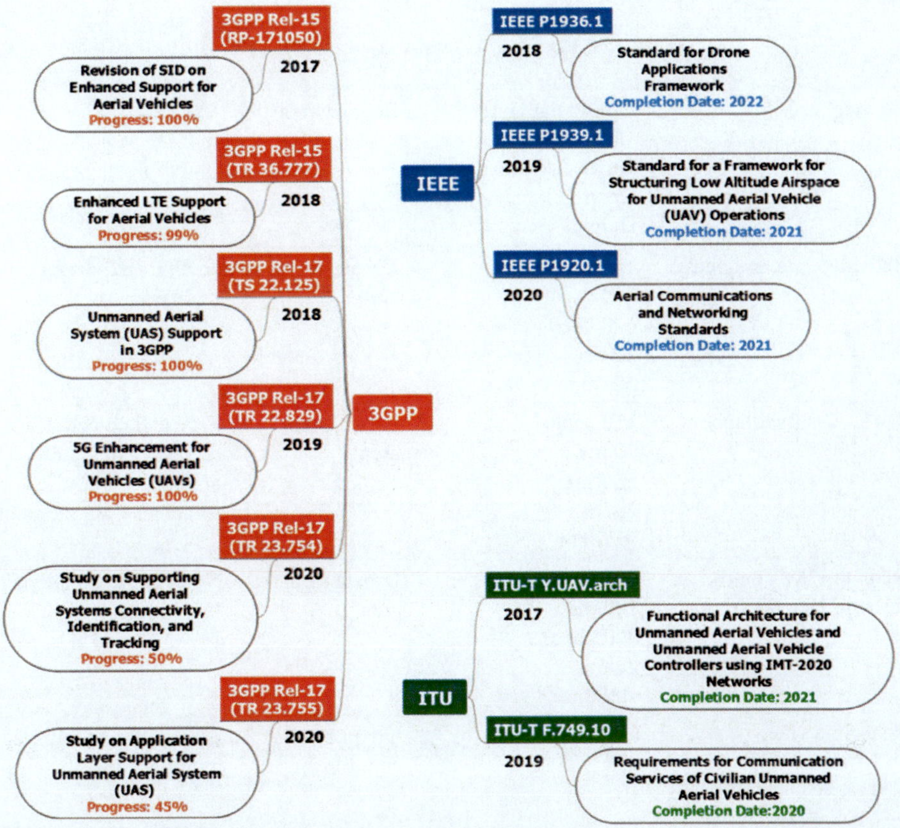

Fig. 5. Recent **standardization efforts of 3GPP, IEEE, and ITU for UAS communications** with completion percentages as of November 2020.

Modernization of Aerial Communication for Future Growth. According to a report issued by Eurocontrol [10], it is predicted that Europe will have 14.4 million flights in 2035 under the most likely scenario "Regulated Growth." With this continuously growing demand for air travel, it is likely that the current **air traffic management** (**ATM**) systems in the United States and Europe will reach their capacity limits in the coming years. Two large-scale projects have been started to develop a next-generation air traffic management system that can cope with this demand:

– Single European Sky ATM Research (Europe);
– Next Generation National Airspace System (United States of America).

To enable these new air traffic management systems, it is necessary to develop improved communication, surveillance, and navigation techniques.

Digital Communication for ATM. ATM modernization requires a paradigm shift away **from voice towards digital data communications**. Almost all communication

Table 3. KPIs for UAV-assisted wireless services.

Service	Uplink data rate Uplink latency Downlink data rate Downlink latency Height	Service area	Use case
8K video live broadcast	100 Mbps 200 ms 600 kbps 20 ms <100 m	Urban, scenic area	Healthcare, virtual reality
4*4K AI surveillance	120 Mbps 20 ms 50 Mbps 20 ms <200 m	Urban, rural area	Face and object recognition, environment awareness, autonomous driving
Remote UAV controller through HD video	25 Mbps 100 ms 300 kbps 20 ms <300 m	Urban, rural area	Non-LoS bidirectional UAV control
Real-time video	0.06 Mbps 100 ms – – –	Urban, rural, countryside	Entertainment with multiple camera views
Video streaming	4 Mbps → 720p, 9 Mbps → 1080p 100 ms – – –	Urban, rural, countryside	Public safety, emergency surveillance, law enforcement
Periodic still photos	1 Mbps 1s – – <120 m	Urban, rural, countryside	Exploration, inspections, search, and rescue

between air traffic control and the pilot is performed using a voice channel. Voice has a high overhead, making it unsuitable for efficiently transferring the information needed for future operational procedures. While voice works reliably, it also has an extremely low spectral efficiency, as the content of the message could typically be represented in just a few data bits. To resolve these problems, it is necessary to develop a digital data link. Ideally, it would work in the **VHF** spectrum band to obtain optimal coverage.

In fact, **International Civil Aviation Organization (ICAO)** has already standardized a VHF data link two decades ago: **VDL**. The link has a very low throughput and as such is not suited for future ATM scenarios. Therefore, it was decided to develop a new data link standard, operating in an **L-band** portion (**960 MHz** to **1164 MHz**), which is already allocated on a secondary basis to aeronautical communications. Two systems are under development to cover these needs: **LDACS1** and **LDACS2. LDACS** stands for ***L-band Digital Aeronautical Communications System***. The **communication range** of both systems is designed to be 200 nautical miles **(370 km)**.

LDACS1 is the most promising candidate for the future air-to-ground communications standard LDACS. The LDACS1 mode aims to be deployed in the inlay scenario. Therefore, the total **bandwidth** is limited to around **500 kHz**. Of course, it is also possible to deploy the system in a non-inlay scenario. LDACS1 is an **FDD** mode based on **OFDM** modulation. LDACS1 parameters are shown in Table 4.

Table 4. LDACS1 system parameters.

Name	Value
Total bandwidth (B)	625 kHz
FFT size (N_{fft})	64
Subcarrier spacing ($f_{\mathrm{sc}} = B/N_{\mathrm{fft}}$)	9.77 kHz
Active subcarriers (N_{a})	50
Active bandwidth ($N_{\mathrm{a}} \cdot (f_{\mathrm{sc}} + 1)$)	498.1 kHz
OFDM symbol duration ($t_{\mathrm{fft}} = 1/f_{\mathrm{sc}}$)	102.4 μs
Cyclic prefix duration (t_{cp})	17.6 μs
Total symbol duration ($t_{\mathrm{s}} = t_{\mathrm{fft}} + t_{\mathrm{cp}}$)	120 μs

LDACS2 is the second standard for the described **ground-to-air** data link. Its physical layer is actually very similar to the second-generation GSM system. Since it is not able to cope with DME interference, it can only be used in the 960 MHz to 975 MHz frequency range. Table 5 summarizes the most important parameters of LDACS1 and LDACS2.

Control and Telemetry. The wireless link used for **controlling** and **monitoring** the UAV is typically designed to have a **long range** and **high reliability. Bandwidth** requirements are typically **low**. If the trajectory of the UAV is controlled by the pilot, the **latency** should also be as **low** as possible.

Network Communication Standards. The network standards implemented by the lower layers in a multi-layer system have an **impact** on the **achievable communication range**. Currently there are two widely accepted network standards for **900** MHz and **2.4 GHz** that provide specifications for the **physical** and **data link** layers. For this research project, the key aspects that differentiate these standards are the achievable data rates and the communication range.

Table 5. Comparison of LDACS systems proposed.

Parameter	LDACS 1	LDACS2
Duplex method	FDD	TDD
Modulation	OFDM	GMSK
Throughput	561kbps–2.6 mbps	270.8 kbps
Frequency range	960–1164 MHz	960–975 MHz
Coexistence	Spectral inlay and pulse blanking	None

2 UAV Networks

2.1 Basic UAV Network Architecture

Basic networking architecture of UAV-aided wireless communications is shown on Fig. 6 with primary and secondary Control and Non-Payload Communications (**CNPC**) links, data link, U2U and U2GCS communication directions, satellite, GCS, and ground terminals [3].

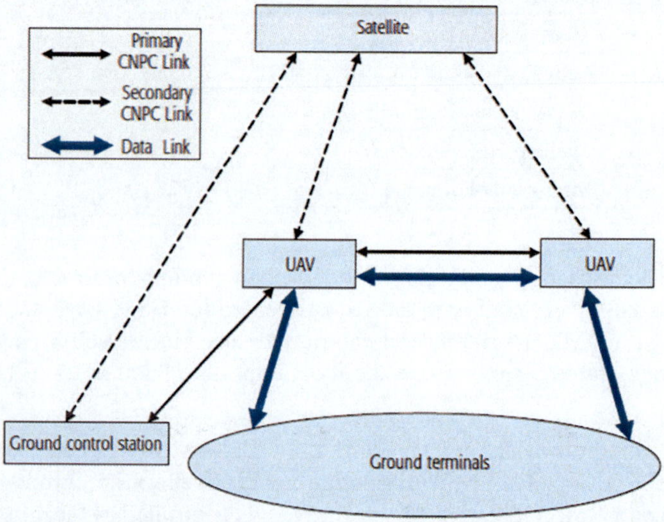

Fig. 6. Basic UAV network architecture.

Due to the **critical functions** to be supported, CNPC links should in general operate in **protected** spectrum. Currently two such bands have been allocated: the **L-band** (960–977 MHz) and the **C-band** (5030–5091 MHz) [11].

Furthermore, although the **direct links between GCS and UAVs** (primary CNPC links) are always **preferred** for delay reasons, **secondary** CNPC links via **satellite** could also be exploited as a **backup** to **enhance reliability and robustness**.

2.2 UAV Airborne Environment

Airborne networks are a class of **mobile wireless** networks (**MWN**): **wireless** due to the untethered communication links between aircraft, and **mobile** due to the movement of aircraft (Fig. 7). Depending on the communication paradigm, they **may or may not need to be ad hoc networks that self-organize.**

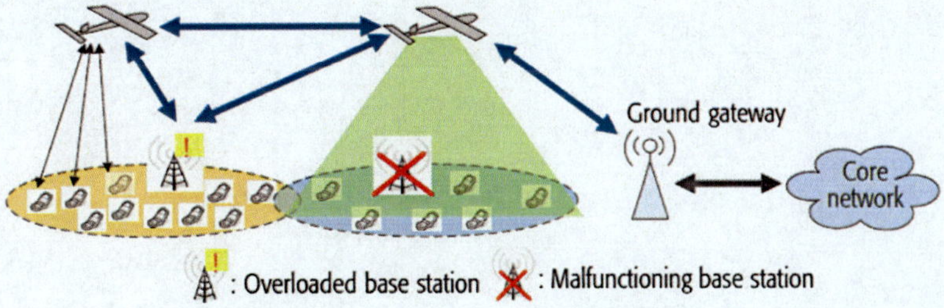

Fig. 7. UAV airborne environment [26].

A typical **airborne tactical network** as depicted in Fig. 8 consists of three types of nodes: **airborne nodes** (AN), **ground stations** (GS), and **relay nodes** (RN). The airborne nodes span the variety of classes: commercial, civilian, drones (shown with a quadcopter icon), and military. At the top, satellite links connect commercial aircraft, some of which are also connected to ground stations.

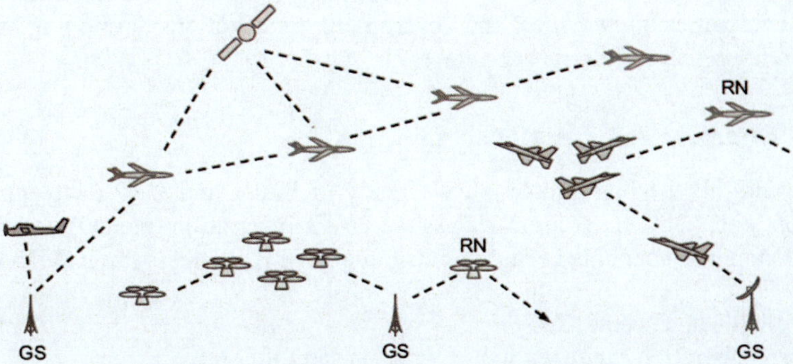

Fig. 8. Airborne network environment.

2.3 UAVs Integration into National Airspace (NAS)

There is an increasing need to fly Unmanned Aircraft Systems (**UAS**) in the National Airspace System (**NAS**) to perform missions of vital importance to national security and defense, emergency management, science, and to enable commercial applications (Fig. 9). However, routine access by UAS to the NAS remains unrealized [12].

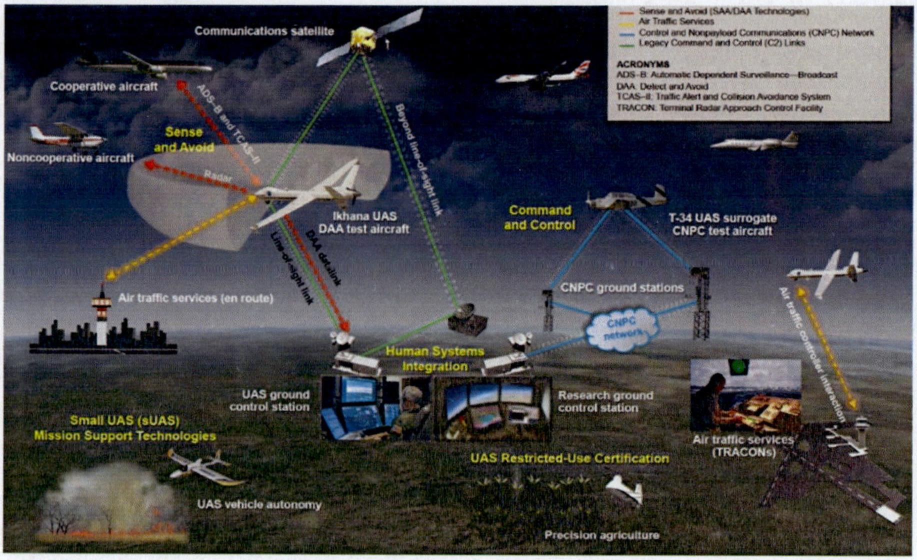

Fig. 9. UAS integration in the **NAS.**

UAS promise to dramatically change the face of commercial aviation, enabling whole new markets, and potentially spurring economic growth and job creation. According to industry forecasts, UAS operations will increase exponentially once they are fully integrated within the national airspace system. Today, in the United States, there are about 50 companies, universities, and government organizations developing more than 150 different unmanned aircraft designs [12].

2.4 UAVs Integration into Cellular Networks

Together with all the highlighted ways already of UAVs and UAS development and integration, one of the most promising is the UAVs integration into cellular networks. There are numerous benefits because of such merge, e.g., benefits for **UAV industry:**

- More reliable and secure CNPC.
- Unlimited operation range without relying on satellite.
- High capacity enabled by advanced cellular techniques.
- Cost effective: avoid building new infrastructure for UAV communication.
- Enhanced air traffic management: legitimate control takeover when needed.

Benefits for **cellular industry**:

- New business opportunities by incorporating aerial users.
- More robust and cost-effective cellular network with aerial communication platform.

UAVs integration into cellular networks with all the actors, communication links and components is shown on Fig. 10.

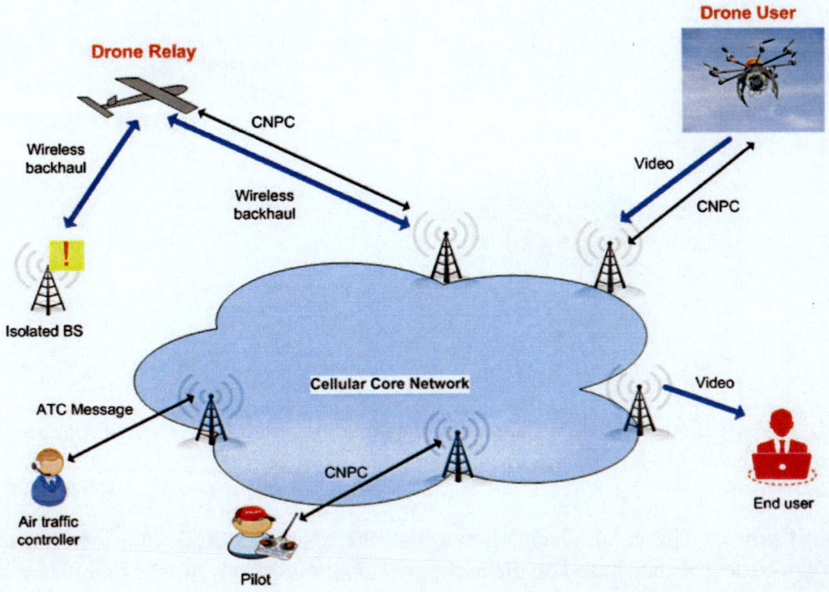

Fig. 10. UAVs integration into cellular networks.

However, there are the following **challenges** for cellular-connected UAVs:

- High altitude:
  - o 3D coverage requirement: existing BS antenna tilted downwards (Fig. 11),
  - o Unique channel characteristics: strong LoS link,
- High 3D mobility,
- Both CNPC and payload communications,
- Severe aerial interference: uplink and downlink,
- Additional design degree of freedom with UAV mobility control.

With all the pros and cons, challenges, – integrating UAVs into 5G and beyond is clearly a new and promising paradigm to embrace forthcoming IoD (**internet-of-drones**) era.

2.5 UAVs Optimal Wireless Coverage

The **optimal deployment of multiple UAVs equipped with directional antennas used as aerial base stations is presented in** [13, 14]. First, the downlink coverage probability was derived based on the probabilistic **LoS/NLoS links and considering the shadow fading** (Fig. 12). Next, given a desired geographical area which needs to be covered by multiple UAVs, an efficient deployment approach was proposed based on the circle packing theory that leads to a **maximum coverage** while each UAV uses a **minimum**

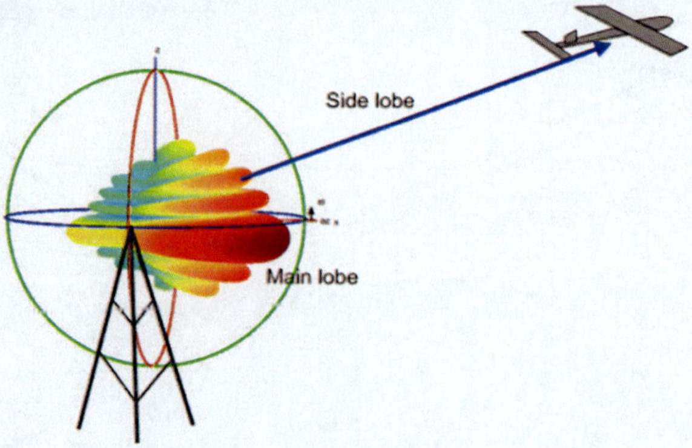

Fig. 11. Cellular-Connected UAVs served by side-lobe or reflected signals with current base station.

transmit power. The results have shown that, the optimal altitude and location of the UAVs can be determined based on the number of available UAVs and the gain/beamwidth of the **directional antennas**.

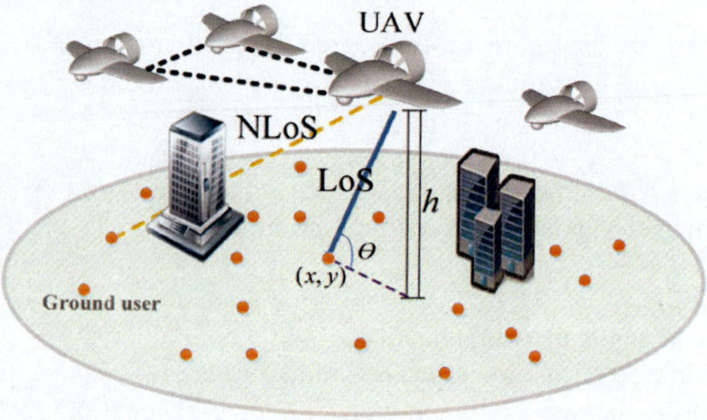

Fig. 12. LoS and NLoS links considered for UAVs optimal wireless coverage.

3 UAVs Construction

3.1 Communication Architecture

ITU defines UAS in [6] as the **aggregation of multiple systems**:

- Unmanned aircraft subsystem,
- Ground control station (GCS) subsystem,
- Air traffic control (ATC) subsystem,
- Sense and avoid (S&A) subsystem,

– Payload subsystem (e.g., video camera, etc.).

The general information flow of a UAS divided in internal and external exchange of data is shown in Fig. 13. Internally, command and control (C2) data is transferred among GCS and UAV. Externally, ATC is involved in the exchange of NavAID, air traffic services (ATS), ATC voice, S&A and non-payload data. The direct transfer of throughput-intensive payload data (e.g., HD video) is typically an internal downlink (UAV to GCS) information flow. Depending on the location of GCS and responsible ATC, information from GCS to ATC or vice versa need to be relayed through the UAV. CNPC is utilized in command and control, ATC relay and S&A information exchange between UAV and GCS only.

An unmanned aircraft system (**UAS**) consists of an unmanned aerial vehicle (**UAV**) and its **controller**, which use **radios to communicate**. While the **remote controller (RC)** is traditionally operated by a person who is maintaining visual **line of sight** with the UAV it controls, the **trend** is **moving toward long-range control and autonomous operation**. To enable this, reliable and widely available wireless connectivity is needed because it is the only way to remotely control a UAV or take control of an autonomous UAV flight.

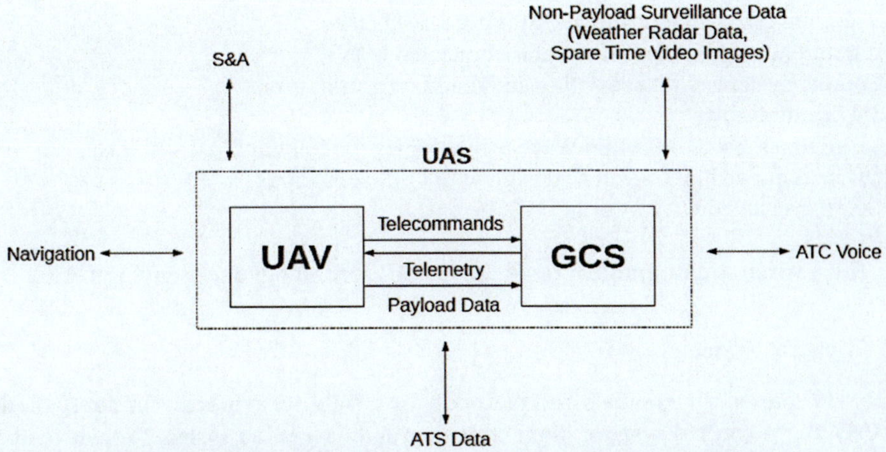

Fig. 13. Internal and external information flow of UAS as per ITU.

3.2 Complexities

Federal Aviation Administration released **582 new reports of incidents** involving unmanned aircraft in the U.S. airspace system [15]. These reports date from August 21, 2015, to January 31, 2016. Cases:

– **Close Encounters** (Near MidAir Collision): drone within 152m of a manned aircraft.
– **Sightings**: drone within aircraft flight paths but no immediate potential threat.

So, the question is coming – **are we ready for the UAV era?**

At the same time, more **drone crashes are caused by technical glitches, not human error** [16]. While pilots may take a brunt of the public's worries on drone safety, a new study suggests that most of the time, they are not the ones at fault. About **64% of drone crashes are caused by technical issues**, a study released today by the Royal Melbourne Institute of Technology (RMIT).

The researchers looked at **150 reported drone crashes** worldwide, occurring between 2006 and 2016. The study showed that **technical glitches** were the **guilty party** in **more than half the cases**. Dividing those causes down even further, the team said that the **loss of communication between the drone and the operator was the most common error (!)**, what clearly shows the point where the research and design efforts for UAVs should be spent.

Promising Research Directions for UAVs communication:

- New architectures/protocols for cellular-connected UAVs,
- Field measurement and prototype for LTE-connected UAVs,
- Channel measurement and modelling for UAV-BS communications,
- Performance analysis for aerial coverage,
- Security and safety issues of UAV systems,
- Spectrum management and multiple access schemes,
- Interference mitigation for cellular-connected UAVs,
- Cellular systems with coexisting aerial and terrestrial users,
- 3D beamforming,
- MIMO/massive MIMO/mm Wave technologies for cellular UAV systems,
- Offline/online UAV trajectory design with QoS constraints,
- Cellular-enabled UAV swarm.

Thus, **wireless communication is the CORE technology for securing UAVs.**

3.3 Control System

Paper [17] presents the **models and methods for reliability synthesis for components of UAV flight control system**: flight computer and navigation system. The developed reliability models depict different variants of fault tolerant designs including design for systems with complex reliability behavior.

Flight control computers and **control systems** are traditionally designed as custom-built, which results in long development cycle, high full life-cycle costs, and inconvenient maintenance. **Commercial off-the-shelf (COTS)** technologies are showing outstanding performance, reliability and developing period in various applications.

UAV autopilot systems (APS) allow UAV to perform entire missions autonomously without the need for manual remote control. These missions may include cargo delivery, mapping, surveillance, and many other applications. Operators use ground control stations to set the parameters of the mission and the UAV autopilot directs the drone or other unmanned craft to complete the task. There are many manufacturers for UAV APS (e.g., PC/104, Pixhawk and many others). **UAV APS** is designed to work with sensor and communication systems for monitoring and control UAV from GCS.

3.4 Position Registration Systems

A modern UAVs use **global positioning system** (**GPS**) to determine coordinates and the speed with respect to Earth.

The **spatial orientation** and the **acceleration** of a UAV is determined using **gyroscopes** and **accelerometers**. These are the key sensors used in controlling a UAV's movement in space. The **control of a UAV** is implemented using specialized **digital signal processing microcontrollers** called **APS**.

With the advancement of GPS, as well as Galileo and GLONASS, **UAVs** are provided with **highly accurate absolute position information** [18]. While the availability of such solutions has led to readily available and accurate navigation solutions, reliance on them can be problematic as they are **easily jammed**.

Many **GPS-denied** navigation solutions resolve depth using **light detecting and ranging** (**LiDAR**) devices. These sensors perform **accurate range measurements**, though long-range LiDARs often have **large size, weight, and power**, rendering them a **poor solution for small UAVs. LiDARs** actively transmit an **optical**, or **infrared**, **laser** and measure the **reflection** of the environment to determine range. **Optical sensors** measure the relative orientation of features as observed by one or more cameras.

In contrast to optical sensors which are hampered by poor weather conditions, such as rain, dust, and fog, **radar remains unaffected**. Additionally, as **radar** is an active sensor, it remains unaffected by **night-time navigation**. Radar, as a range-only sensor, may be viewed as the dual to optical sensors. Radar is an active sensor that transmits a succession of electromagnetic pulses. As the pulses are reflected by the surrounding environment, many features, called point scatterers, reflect the transmitted pulses straight back to the radar. Using the delay between the transmitted and received pulses, the radar is able to detect the range to these features.

A variety of navigation systems use radar to communicate with **beacons**, used as **landmarks**, that are distributed throughout the navigation environment. Each beacon transmits a radio signal, which is received, decoded, and converted to a range measurement, as is done on a global scale with GPS. These sensors are often very accurate, though typically have low update rates.

Sonar is a sensor that uses transmitted sound waves, and the echo from environment features, to detect range.

3.5 Communication Systems

LoS and BLoS. UAS missions typically operate within RF **line-of-sight** (**LoS**) or **beyond line-of-sight** (**BLoS**).

ITU defines **LoS** as the *direct radio line of sight radiocommunication between the UAV and GCS*, and BLoS as the *indirect radio communication between the UAV and a GCS using satellite communication services* [18]. In this regard, the difference between **non-line-of-sight** (**NLoS**) and **BLoS** is that the latter refers to lack of visibility is caused by the curvature of the Earth's surface (Fig. 14).

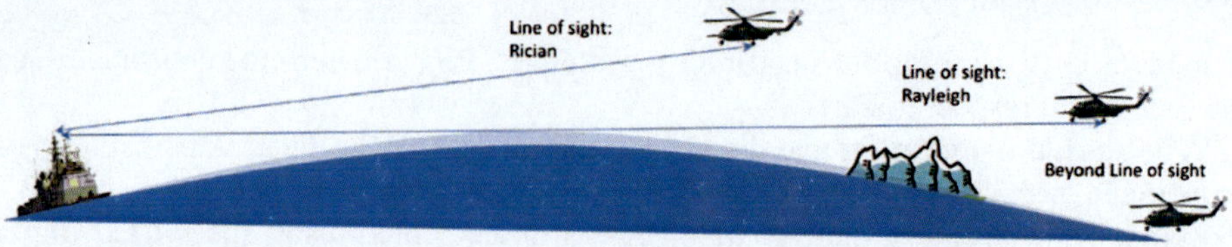

Fig. 14. Difference between LoS and BLoS for ship-based communications.

LoS communication is limited by several factors, amongst others, link design. The **maximum achievable distance**, however, is impaired primarily by the **radio horizon** leading to **typical value of 370 km** depending on the operational altitude of the UAV. Nevertheless, the **actual distance** through LoS can be **considerably less than 370 km.**

LoS communication is applicable to three types of UAVs: low, medium, and high endurance. In this context, it is important to note that low endurance UAVs almost exclusively operate under LoS. Medium and high endurance UAVs, however, are usually operable in both LoS and BLoS.

BLoS links as an alternative to LoS communication use **geostationary satellites** that have a turnaround link **latency of 0.48 s or more**. In military terminology, BLoS is often used to characterize links with a range greater than 600 miles, or 965 km. The footprint of satellite based BLoS links depend for instance on deployed beamwidth (spot beam, wide beam, etc.), link margin and propagation impairments (ionospheric/atmospheric attenuation such as fading through rain).

Basic requirements for control and non-payload communications (CNPC):

- Ensure safe, reliable, and effective flight operation,
- Low data rate, high reliability, high security, low latency,
- Telemetry (UAV status reporting),
- Command and control,
- Navigation aids,
- Sense and avoid (S&A),
- Air traffic control (ATC) relay.

Basic Requirements for Payload Communications (PC):

- Application specific information,
- Typically, higher rate than CNPC.

The Fig. 15 depicts the command and control, sense and avoid and ATC relay information flows between the GCS and UAV [19]. It shows not only the basic exchange of control data between GCS and UAV, but also the relay of ATC voice and ATS data messages using the UAV as a relay, and the downlink of navigational data, non-payload target tracks (from the S&A system and other sources), weather radar outputs, and video images from the UAV to the GCS.

Quantitative Communication Requirements. The Table 6 summarizes the required data exchange **frequency**, **throughput**, **delay**, and **traffic type** for expected applications of UAV networks. The **types of traffic** are real-time, periodic, and delay tolerant (DT). Coordination traffic may include telemetry, some sensed traffic, and decision making and task-assignment commands. Sensed data include any traffic generated by the sensors on-board the UAVs. The control data that consist of telemetry data on the downlink and control commands on the uplink are common to all applications. The standard **frequency of telemetry** data exchange is **4–5 Hz** or less.

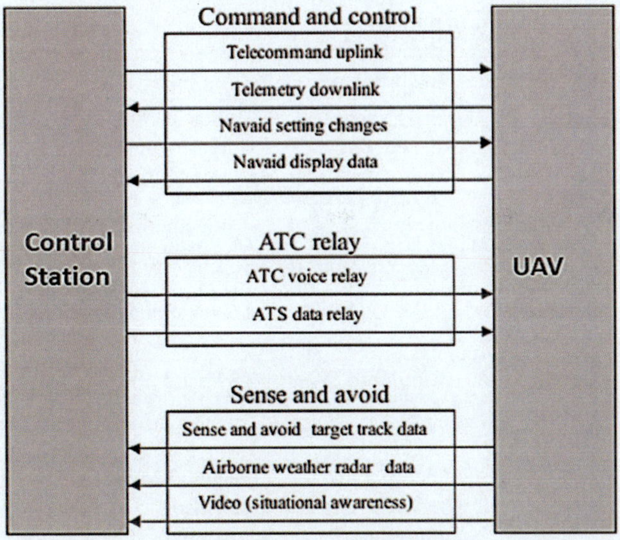

Fig. 15. CNPC information flows.

The **applications** are provided in Table 6 are very diverse in terms of the number of nodes, the area sizes the UAVs need to span, the terrain the aerial network is deployed for, etc. Therefore, the QoS requirements vary significantly, which imposes constraints on the communication technology that can be used. The provided values in Table 6 can help in designing a robust network for the application scenario at hand.

The **CNPC** links are **essential to ensure the safe operation** of all UAV systems. **Highly reliable**, **low-latency**, and **secure two-way communications**, usually with **low data rate** requirements, must be supported by these links for exchanging **safety-critical information among UAVs**, as well as between the **UAV** and ground control stations (**GCS**), such as dedicated mobile terminals mounted on ground vehicles. The main CNPC information flow can be broadly categorized into three types:

- Command and control from GCS to UAVs,
- Aircraft status report from UAVs to ground,
- Sense-and-avoid information among UAVs.

Table 6. Quantitative communication requirements for UAV applications.

Application	Data type	Frequency	Throughput	Traffic type	QoS (delay)
Search and rescue	Coordination	depends on plan-ahead timesteps, >1.7 Hz with 2 nodes for required network throughput of 1Mbps	4.8 kbaud	Real-time	50–100 ms
	Sensed	image sensor dependent: >20 fps, 30fps, laser sensor: 10Hz	2Mbps in case of video streaming, > = 64 kbaud	Real-time	50–100 ms
Monitoring and mapping	Coordination	0.1Hz, 1Hz, 4–5Hz	4.8 kbaud	Periodic or DTN	–
	Sensed	1Hz, 12Hz	9.6–64 kbaud	DTN	Can be offline
Visual tracking and surveillance	Coordination	Similar to SAR	4.8 kbaud	Real-time	<3s
	Sensed	>10Hz to incorporate mobility and noise in urban area, >15Hz, 300Hz	1 Mbps for images, 2 Mbps for video streaming, > = 64 kbaud	Real-time	50–100 ms
Network coverage provisioning	Coordination	Depends on plan-ahead timesteps	–	Periodic	–
	Sensed	50Hz for voice traffic and 30Hz for video (H.264)	12.2 kbps for voice, 384 kbps for video, 9.6 kbaud	Real-time	50–100 ms
Construction	Coordination	Control Command: 100 Hz–50 Hz, Motion Tracking: 150 Hz	<250 Kbps	Real-time	–

(*continued*)

Table 6. (*continued*)

Application	Data type	Frequency	Throughput	Traffic type	QoS (delay)
Delivery of goods	Coordination	Control Command: 100 Hz–20 Hz, Motion Tracking: 100 Hz	<250 Kbps	Periodic	–

4 Modern UAVs Review

4.1 UAVs Over 25 Years (2000 to 2025)

Document [20] presents the Department of Defense's (DoD) roadmap for developing and employing unmanned aerial vehicles (UAVs) over 25 years (2000 to 2025). It describes the missions identified by theater warfighters to which UAVs could be applied, and couples them to emerging capabilities to conduct these missions.

As presented in Table 7, the two UAVs at the two ends of the spectrum, the RQ4 Global Hawk and the Nano Hummingbird *could not be more different*. Their wingspans range from 40m to 0.16m, weight spans from 14,628 kg to 0.179 kg, and flight time from 28 h to 11 min. Accordingly, the *technical and research challenges posed by them are quite different* and a **one size fits all type of approach is unlikely to be effective** in such a diverse environment.

4.2 Communication Design of Other Modern UAVs

There is a plenty of UAVs as of today already designed, evaluated, and put to mass production, while even more are under development and improvements. Let us review some of those.

Bayraktar TB2S [21]. Bayraktar **TB2** is a **medium-altitude long-endurance** (MALE) **unmanned combat aerial vehicle** (UCAV) capable of remotely controlled or autonomous flight operations. In October 2020, Baykar presented new, improved version of TB2, named TB2S. TB2S has a protrusion on its body and an antenna on its nose for **satellite communication** (SATCOM). In the **basic model of TB2**, **communication** between the UAV and GCS was via **ground-based antenna**. Communication via the **TÜRKSAT satellite** provides a much bigger control range than the **150–300 km** range of the **basic model**. The **satellite communication** also makes the **TB2S more resistant** to the jamming of communication by the enemy. Communication range: LoS propagation, <300 kms, endurance: 27 h.

EOS C VTOL [22] is a highly capable electric VTOL (vertical take-off and landing) mini-UAV with best-in-class flight performance and **FullHD high performance payload**, GPS/GLONASS/GALILEO **navigation**, **Barometer** and **Radar** altimeter, **Inertial Navigation** System. Communication range is 50 km with RLOS (Radio Line Of

Table 7. Some UAVs characteristics summaries over 25 years (2000 to 2025).

UAV name	RQ-4 global hawk	MQ-9 reaper	MQ-1 predator	MQ-8 fire scout
Manufacturer	Northrop Grumman	General Atomics	General Atomics	Northrop Grumman
Wingspan	40 m	20.12 m	16.76 m	8.53 m
Gross Weight	14628 kg	4763 kg	1021 kg	1429 kg
Flight Time	28 h	24 h	24 + h	6 + h
Range	22780 km	1852 km	1250 km	204 km
Communication Range	–	–	–	–
Primary Users	Armed Forces	Armed Forces	Armed Forces	Armed Forces
Primary Use	Surveillance	Surveillance and combat	Surveillance and combat	Situational awareness
UAV Name	***RQ-7 Shadow***	***Phantom 2 Vision +***	***U818A***	***Nano Hummingbird***
Manufacturer	AAI	DJI	UDI	AeroVironment
Wingspan	4.23 m	0.45 m	0.4 m	0.16 m
Gross Weight	170 kg	4.9 kg	1.36 kg	19 g
Flight Time	6 h	25 min	10 min	11 min
Range	50 km	0.7 km	30 m	–
Communication Range	–	–	–	–
Primary Users	Armed Forces	Amateur and professional photographers	Flight enthusiasts, amateur photographers	Research and Development personnel

Sight), operating on 2.2–2.5 GHz frequency, with 5/10/20 MHz bandwidth, using 2 × 2 MIMO MANET/MESH datalink, AES256 encryption. It has been designed primarily to be used for **video surveillance** without any additional load to UAV.

5 Modern UAV Communication Technologies

5.1 Analog Communication Systems

Analog communication system is mostly applicable to the transmission of **video** signals. Many UAVs in the hobby and light commercial category actually transmit the camera images as a simple analog signal.

The systems are based on legacy **PAL** or **NTSC** video signals. This signal is then **FM modulated** on an RF carrier and transmitted. There is no real standard, and the signal bandwidth and video equalization parameters are manufacturer dependent. A bandwidth of **20 MHz** is, however, typical.

Analog transmission has many **disadvantages**: poor image quality, low resolution, no possibility for secure encryption, susceptibility to noise, and very bad spectral efficiency. Nevertheless, there are **two** clear **advantages** that *are not easily achieved with modern digital communication systems.*

The **first advantage** is that since the pixel stream from the image sensor directly modulates the RF signal, the **latency is extremely low** (less than 15ms is achievable). This is often used in hobby systems to pilot the UAV using the video feed (FPV) while flying fully under manual control. **Another advantage** is the **graceful degradation.** When the distance increases more and more noise appear on the image, telling the user that the radio system is getting near its limits. This contrasts with most **digital systems** that tend to work perfectly until a certain point, and then not at all.

Of course, as with **DVB-based** technologies, this system will be transmitting with a duty cycle of 100% prohibiting any coexistence with other communication technologies. In many countries, this will imply **strong transmit power limitations** or the **need for licensed spectrum.**

Today it is still possible to find remaining civilian UAV models employing **analog communication** techniques, called modulation in wireless communication, working on **35 MHz or 40 MHz** frequencies. In the **analog communication** the information is transmitted by changing one of the time-varying properties of an electrical signal, such as the amplitude, frequency, or phase. From the wireless ad hoc network point of view, as the analog signal is re-transmitted, maintaining the originally sent signal becomes harder. Each radio node introduces a noise degrading the original signal. There are no practical techniques that would allow to recover the original signal; therefore, **analog communication techniques are not used in a multi-UAV environment**.

5.2 Features of Analog and Digital Communication Systems

Digital communication techniques allow to recover part of the corrupted information or at least minimize the noise effects as the signal is re-transmitted. In the case of the digital communication the main indicator of errors is the bit error rate. When data is transmitted over-the-air, the **digital signal processing techniques provide more immunity to noise and interference compared to the analog signal processing methods**. Also, the electronic circuitry for digital signal processing is **more economical** solution compared to the analog variants. **Therefore, all the modern radio control is implemented using the digital signal processing techniques.**

The **other solution** used for increasing the range between a UAV and the GCS is using the **900 MHz** band **with high-gain antennas**. *The 900 MHz band is not part of the 802.11 standard,* and it is widely used with personal communication standards at a **lower data rate**. Some of the developed ***civilian UAVs** use **both bands**, the **900 MHz** for the long-range **telemetry** acquisition and the **movement control**, and the **2.4 GHz** for the **close-range multimedia streaming**.*

Radio Frequency for the Civil UAV Control. The **72 MHz** is a radio frequency dedicated specifically to **control civil UAVs** (Fig. 16). This is a **globally accepted standard.**

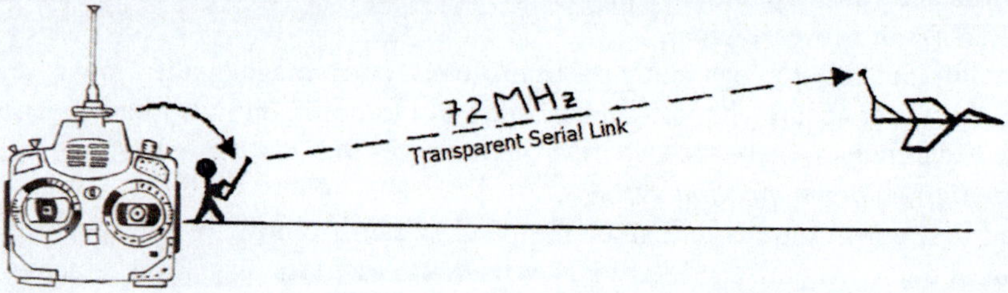

Fig. 16. Typical radio control for civil UAV [15].

Using radio controllers, such as the one shown on Fig. 16, operators can control the movement of the UAV in space. When an operator moves the knobs on the controller, a sequence of electrical pulses representing the command is generated. This sequence is then mapped onto a **72 MHz** wave via a **pulse code modulation (PCM)**. The 72 MHz transmitter then transmits the sequence of pulses into the air. It is like a broadcast message in the network. Any receiver on the **same radio channel as the transmitter will receive the transmitted sequence and decode it as a control command**. Having two controllers and two UAVs in a close vicinity creates **interference** to each other.

To avoid the described case, the **72 MHz** band is broken into **60 channels**, each having a bandwidth of **10 kHz**. Note that there is a **20 kHz safety margin** between any two adjacent channels to account for non-ideal signal processing circuits like filters. Every radio controller needs to use several channels, with each channel controlling one function of the UAV. A typical radio controller (e.g., the ones manufactured by Futaba) can support up to 8–10 channels. **If the UAV uses all the supported channels this would allow the operation of no more than 7–10 UAVs in a close vicinity**.

The **advantage of the 72 MHz** transparent link is the UAV's **fast response** to the movement **control commands**. The link is **transparent** in a sense that there is **no security, error correction**, or routing used. Once the control signal reaches the receiver it is demodulated and sent to the appropriate motor controller. Because there are **no error checking or security protocols, there are no additional delays**.

The **disadvantage** of the **72 MHz transparent link** is having **no data encoding against interference**. The link is designed to **transmit control signals over a small bandwidth**.

Based on the type of control used for a civilian UAV they can be categorized into three different groups:

- **Manual**: during its entire flight, the UAV is controlled by the operator.
- **Semi-autonomous**: only some aspects of the UAV's flight is controlled by the operator (e.g. take-off and landing).
- **Autonomous**: UAV can perform the entire assignment without any input from the operator.

The simplest way to control a UAV while streaming sensory data from it can be achieved by using the **72 MHz controllers for the movement control**, with **900 MHz and/or 2.4 GHz radios for the data transmission**. The UAV communication systems show that a combination of 900 MHz and 2.4 GHz is a **field-proven implementation** to work for data links on civilian UAVs.

Regulations for ISM Bands. The amount of **power** allowed for **wireless transmission** of data inside the ISM bands is limited by **government regulations**. Because the communication distance directly depends on the transmission power, it is a **challenge in achieving the long distance** using the ISM bands with limited power. One way to overcome this challenge is **using high-gain directional antennas** at the GCS and keep them pointing at the UAV. **Lower frequency** is preferable for the **long-range communication**, as range will be higher under the same all other conditions (power, bandwidth, etc.). Friis transmission equation, which relates the transmitted and the received powers (1):

$$P_{receive} = P_{transmit} G_{transmit} G_{receive} \left(\frac{\lambda}{4\pi R} \right)^2, \tag{1}$$

where R is the separation distance between a transmitter and a receiver in meters, $G_{receive}$ is the gain of the receiving antenna, and λ is the wavelength in meters.

Antenna Types in UAV Communication. Directional antennas are preferred in a long-range communication because they decrease the interference by focusing the received or transmitted power in a specific direction. **Omni-directional** antennas are subjected to greater interfering signals since they receive signals from all directions around them. There are designs which use an **omni-directional** antenna on the fixed-wing model airplane and a tracking high-gain **directional** antenna on the GCS (Fig. 17).

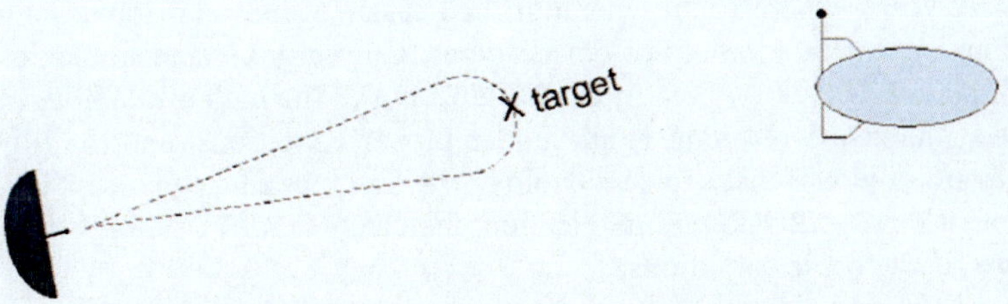

Fig. 17. UAV tracking high-gain directional antenna scenario.

In such system design, the rotating mechanical platform works with **a 2.4 GHz** reflective antenna to keep a track of the UAV in space and point the antenna at it. But there are several challenges associated with such approach. The main challenge is keeping the antenna's signal beam always pointing at flying UAV, as illustrated on Fig. 17. Experiments show that at a distance between **50 to 700 m their tracking was successful**. At longer distances than **700 m**, the tracking mechanism **kept losing the UAV** due to the weak received signal. This suggests that this tracking system is well suited for an open-field type of environment with a constant **line-of-sight**. There are not enough results to show the performance of the system when the signal from the UAV is lost. With such design we intend to restore the broken communication by introducing a **new intermediate node** either on the ground or by lunching **another UAV**.

Another important issue to consider with a tracking directional antenna is the **stabilization of the rotating platform**. With an unstable rotating platform, the data communication with a UAV will keep breaking because the directional antenna's signal beam cannot constantly point at the target.

Beyond Line-of-sight Communication (BLOS). Limited transmit **power** and the fixed **frequency** bands are the main **limiting** factors in achieving **beyond** line-of-sight communication links. Commercial solutions offer **satellite** communication equipment. Satellites are used as relay stations, transmitting data between two points on the ground. This allows achieving beyond LOS communication between a GCS and a UAV. However, communication with a satellite depends on satellite's location for a reliable communication and a clear view of the sky. It also requires frequency **license** in **each country** for operation and carries monthly costs for transmitted data. **Alternative to satellite communication**, beyond LOS communication can be achieved using **multiple UAVs** for **relaying data among themselves**. Multiple UAVs cooperate to relay the transmitted data to the destination using the 900 MHz or 2.4 GHz links. Such networks can be referred to as multi-hop ad hoc networks, wireless sensor networks, or wireless embedded networks. The achievable distance depends on the ad hoc routing protocols used, the power management configuration, and type of antennas used on the individual wireless nodes.

5.3 IEEE 802.11 Technologies

Wi-Fi technology, based on the **IEEE 802.11** standard, has continuously improved during the past years, where each generation had faster speeds and lower latency than the previous one, with a variety of device types. Currently used technologies include **IEEE 802.11n, IEEE 802.11ac, and IEEE 802.11ax**. The high availability of devices and drivers, the **relatively long range** under proper conditions, and the **high bandwidth** they off er would make this technology the first immediate choice. 802.11 allow to implement direct U2GCS communication channels (Fig. 17). **Main limitations of 802.11 are** (under some conditions):

- Unreliable,
- Insecure,
- Vulnerable to interference,
- Limited data rate,
- Visual line of sight (LoS) operation,
- Difficult for legitimate monitoring and management.

This standard is most known for wireless networks in the home and enterprise. Popular versions of the standard such as IEEE 802.11a/n operate using an OFDM physical layer in the **2.4 and 5G Hz range**. The channel **bandwidth** is most commonly **20 MHz** or **40 MHz**. Two narrowband modes are defined, typically used in professional applications, offering a bandwidth of **5 MHz and 10 MHz**. The latest version of the standard (IEEE 802.11ac) allows up to **160 MHz** channel width on the **5GHz** band. A major **benefit** of this is that due to its wide application in consumer products, chips and wireless modems are available at very **low cost** (Fig. 18).

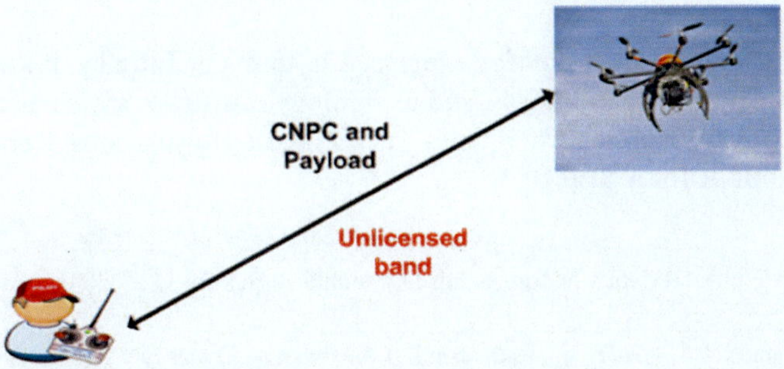

Fig. 18. Unlicensed band usage of IEEE 802.11 for CNPC and PC.

IEEE 802.11 a/b/g/n standard-based Wi-Fi technology is limited to **70 and 300 m** in an indoor and outdoor environment, respectively [15]. To increase the range up to **1–1.5 km, even up to 30–40 km, a high-gain directional antenna can be used on both transmitter and receiver sides**, but in a mobile environment it would require antenna tracking mechanism.

Another **disadvantage** is the **energy cost** of this technology, which would **reduce** the overall duration of the UAV **batteries**. The latest **IEEE 802.11ah** standard focuses on this limitation to a certain extent by combining the **advantages of Wi-Fi and low-power sensor network** communication technologies, achieving communication over longer distances among many low-power devices. IEEE 802.11ah **uses 900 MHz license-exempt** bands to provide **extended range** compared to conventional Wi-Fi networks operating at **2.4 GHz and 5 GHz**. It also benefits from **lower energy consumption**, allowing the creation of large groups of stations or sensors that cooperate to share signals. Thanks to its low power consumption, it is a good competitor of Bluetooth, and has the added benefit of off erring **higher data rates and wider coverage range**. *Unfortunately, commercial **IEEE 802.11ah chipsets remain unavailable yet***, and their possible adoption in the UAV field will take time.

5.4 LoRa

Last, the surge of the **LPWAN** technologies, like **SigFox**, narrowband Internet of Things (**NB-IoT**), and **LoRa**, can open new possibilities for UAV interconnection. Specifically, **LoRa** uses license-free sub-gigahertz radio frequency bands like **433 MHz**, **868 MHz** (in Europe), and **915 MHz** (in Australia and North America). It enables **long-range transmissions up** to **various kilometers** with **low power consumption**, achieving **data rates from 0.3 kbps to 27 kbps** depending on the different modulation configurations. On the **negative** side, **LoRa** has a very simple medium access control (MAC) protocol that would make it **inadequate** for **reliable transmissions** without a proper transport layer protocol to make it more robust, but its physical layer characteristic and low power requirements make it interesting. Nevertheless, the ***slow transmission rates offered fail to meet the low-delay and throughput requirements of many UAV applications***.

Table 8 collects different metrics expressed in terms of **latency**, **bandwidth**, radio **coverage**, **power** usage, **robustness**, and technology **maturity**, which allow us to identify some potential technology candidates **for U2U communications**. Robustness refers to the required SNR for a BER of 10^{-4}.

Table 8. Characteristics of four candidate technologies for U2U communications.

Technology	Radio range	Power usage	Radio band	Bandwidth	Latency	Robustness	Technology maturity (TRL)
Wi-Fi	A few kilometers	High	2.4/5.0 GHz	Very high	50 ms, high	15.2 to 17 dB (medium)	9, high
Bluetooth	100 m	Moderate to very low	2.4 GHz	Moderate to low	100 ms, high	9.5 to 14 dB (medium)	9, high
LoRa	Several kilometers	Very low	Sub GHz	Low	Low latency Class-C	14 dB (medium)	9, high
IEEE 802.11ah	A few kilometers	Moderate to very low	900 MHz	Low	50 to 100 ms, high	15 dB (medium)	6, medium

6 Ensuring Reliable UAV Communication

6.1 Functional and Non-functional Requirements for UAVs Communications Ability

Reliability is the quality of being able to be trusted or believed because of working or behaving well.

In paper [23] the ultrareliable and low latency communications in UAV communication systems is described. The framework for enabling ultrareliable and low-latency communications in the **control** and **non-payload communications** (CNPC) links of the unmanned aerial vehicle (UAV) communication systems is presented. The available range of the CNPC links between UAVs and a ground control station. The available range is defined as the maximal horizontal communication distance within which the round-trip delay and the overall packet loss probability can be ensured with a required probability.

The set of **preventive** or **corrective** methods and tools can be considered for ensuring reliable UAV communications with GCS, other UAVs, or satellites, when performing architecture design for UAV communication system. **Known risks occurrence** can lead to UAVs hacking or even to irrevocable UAV loss. The Table 9 expands the key **functional requirements (FR)** and **non-functional (NFR) requirements** for UAVs communications ability, with some communication channel characteristics description.

6.2 UAVs Communication Channels Security Recommendations

In terms of **security**, ITU states that the "***impact of latency on UAS command and control systems*** *is a prime factor when considering the safety of operations* [6]. For that reason, irrespective of the UAV type, all unmanned aircraft should ideally be capable of LoS communication.

Table 9. Key FRs and NFRs for UAVs communication channels.

#	FRs and NFRs for UAVs communications ability	Communication channel characteristics
1	**Control** channel: - Two-way communication between UAV and GCS to control UAV operations	- **Duplex** mode - Exceptionally **low latency** - **Low throughput** - Lower carrier frequency is preferable for **longer communication range**
2	**Telemetry** channel: - One-way communication from UAV to GCS - Data from UAV sensors - Global positioning UAV data	- **Simplex** mode - **Medium** latency - **Low throughput** - Lower carrier frequency is preferable for **longer communication range**
3	**Payload** channel: - One-way communication from UAV to GCS - Data from UAV video, photo cameras	- **Simplex** mode - **Low** or **medium** latency - **High throughput** - **Longer communication** requires **high frequency band** what is usually a limitation that can be resolved by using unlicenced spectrum (0.9, 2.4, 5 GHz), or needs licenced spectrum what is challenging
4	**Retransmission channels** if UAV is designed to work as a repeater in UAV-to-UAV communication networks	- **Duplex** mode - **Low latency** - **High throughput** - **LoS** channel conditions
5	**Alternative** or **redundant** control, telemetry, and payload communication channels for ensuring connection availability with UAV, with reliable and on time data transmission	- **Simplex** or **Duplex** mode channels - UAV to GCS, UAV to UAV, UAV to Air Platform, UAV to Satellite **channel types** - **Licenced** and **unlicenced** frequencies band use
6	**Proper security** of communication channels and data encryption	- **Channel** and **data encoding** on physical, channel and presentation OSI levels - **Physical security** of GCS - **Authentication** and **authorization** mechanisms for managing GCS and UAV
7	**Radio electronic protection** and **warfare**	- Channels design to resist to warfare attacks

UAVs operations are inevitably confronted with **threats** [23, 25]: loss of radio communication, loss of transmitted information, or control loss, or deliberate external intervention. The threats come in **two types**: those **independent** of humans, and those **related** to humans. **To mitigate** the threats, there are the following security primitives and recommendations:

1. **Reliability of radio channel**. Radio quality on aerial vehicles is influenced by several factors: the **frequency band** used, the **distance**, the **weather** conditions, the intensity of **external noise**, the **position** of the aircraft. To ensure the quality and reliability of radio communication, it is necessary to **use duplicating transmitters** with perpendicular antennas. When there is a communication problem with the main device, automatically turns the second transmitter and continues to maintain the link. Several different frequencies (e.g., 35 MHz, 2.4 GHz) can be used. The second most crucial factor is a **material** from which the aircraft body is made. If the carbon is used, – it is a conductor and shields the electrical equipment that is inside of body. Hence, all communications antennas must be mounted on the UAV surface, or "windows" should be made from other materials (e.g., glass fibre).
2. **Integrity**. Communication channel must be protected from **unauthorized** modification of data by an adversary attempting to hide, for example, a change in controls commands or telemetry data.
3. **Authenticity**. Furthermore, the UAV must also be protected from the injection of **misleading data** sent by unauthorized ground stations. To prevent external adversarial attacks, when receiving data, the UAV must be able to **verify the validity** of both the source and the message. For defending against compromised ground stations, the UAV can employ distributed solutions, such as a majority voting scheme that can jointly determine validity.
4. **Confidentiality**. Not all information, transferred to/from a UAV, are public: control commands, telemetry data.
5. **Mitigation of Channel Jamming**. The adversary can, for example, employ jamming attacks to block or delay critical fault detection from propagating towards the ground station. Therefore, channel-jamming attacks must be detected as soon as possible and mitigated in the UAV.
6. **Secure Routing**. The UAV and ground station need to route their readings timely and reliably even when **under attack**. The UAV and ground station routing protocol must be **robust** enough to withstand jamming attacks that induce long and energy–inefficient routes.
7. **Early and Correct Detection of Manipulation**. Any manipulation of transferred information must be **detected as soon as possible**, while false alarms must be avoided. For this threat, we can use the intrusion detection system.

So, there is several preventive and corrective methods and tools that can be designed and used for UAV communication system architecture to improve reliability of UAV communication channels.

6.3 Network and Application Layer Protocols for Reliable Communications

Error Control and QoS-Based Transfer Modes. The **existing TCP/IP-based** Internet protocol architecture **is not well suited to airborne networks**. Traditional MANET routing protocols such as AODV, DSDV, and OLSR are not designed for highly dynamic topologies with high velocity and unpredictable mobility, as is the case for emerging drone and tactical military scenarios. A **new protocol architecture** that addresses these issues with domain-specific transport and network layers is presented. It is observed that the exchange of information across layers provides significant benefit in the highly mobile environment. The domain specific **transport** (AeroTP), **network** (AeroNP), and **routing** (AeroRP) protocols leverage cross-layer information in optimizing end-to-end performance.

AeroTP provides flexible and composable disruption-tolerant mechanisms that can deliver information end-to-end between airborne nodes in challenging scenarios. By predicting when links will be available based on trajectory information, as well as actively listening for nearby nodes, AeroRP can send data opportunistically towards its destination and make much more efficient use of available network capacity.

A simulation analysis shows significant improvement over traditional Internet and MANET protocols, as well as providing insight into the cost/benefit trade-offs of various operating modes.

AeroTP: TCP-Friendly Transport Protocol. AeroTP is a domain-specific transport protocol designed to meet the needs of the highly dynamic network environment while being TCP friendly to allow efficient splicing with conventional TCP at the AeroGWs in the GS and on the AN.

Based on the application requirements, there will be several classes of data being transmitted over the airborne network. For this reason, **AeroTP** supports multiple transfer modes that are mapped to different traffic classes: reliable connection, near-reliable connection, quasi-reliable connection, unreliable connection, and unreliable datagram.

All modes except unreliable datagram are **connection-oriented for TCP-friendliness** and use sequence numbers so that packets may follow varying or multiple paths and be reordered at the AeroTP receiver.

6.4 Algorithms for UAV Operation in Case of Lost Communication Line

Flight Termination in Absence of Network Control Commands [10]**.** A basic level of flight safety is implemented in all the UAV platforms. Should UAV lose the ability to communicate with the ground station, the autopilot will navigate the airplane to a predetermined GPS waypoint and loiter until the communications link is restored. Should the engine also fail while in this state, the aircraft will continue to loiter, but will start losing altitude while trying to maintain the commanded airspeed. If the airspeed set by the ground station operator a priori is chosen properly, this will result in a semi-controlled descent at a specific location. Furthermore, the loss of both the communication and GPS links will result in the autopilot attempting to hold its specified course using its inertial navigation aids and air-data sensors. However, as time elapses, inertial sensor drift and varying winds will result in the airplane deviating from its intended path. The actual amount of deviation in the flight path could not be quantified realistically, since influencing variables, such as weather and other disturbances, were difficult to characterize. Furthermore, owing to the limited accuracy of the GPS system, and the lack of a sonic or laser altimeter on-board, the UAV testbed cannot safely autoland itself if the radio link cannot be restored.

Beyond the Radio Line of Sight (BRLoS). BRLOS [24] (Beyond Radio Line of Sight, defined by ICAO), and Radio Line of Sight (RLOS) refer to the method of operation of the command-and-control link between the GCS and the UAV. ICAO defines RLOS as a situation in which the transmitter(s) and receiver(s) are within mutual radio link coverage and thus able to communicate directly or through a terrestrial network.

When UAV is beyond the radio line of sight, satellite or airborne links are necessary for UAV and ATC to maintain communication. Absence of surveillance takes place when the aircraft is more than 400 km away from an ATC. While satellite links offer a solution to BRLoS communications, they introduce operational challenges that include signal delays caused by the unpredictable characteristics of communication links between RPS and the satellite, including the number of earth-to-satellite signal hops. Regulation and oversight of satellite communication service also pose a challenge to relying on satellite communications. The number of air-to-air hops and signal delays can be predicted better in airborne networks than satellite communications.

6.5 Radio Electronic and Telecommunication Channels Protection

Radio electronic and telecommunication channels protection equipment on UAVs are supposed to resist to electronic warfare systems that aim to harm or break UAV operation by influencing the UAV to GCS microwave links. Respectively, the defence and UAVs communication systems protection should be designed based on the assumption of attacks to UAV in the following forms, but not limited to:

- UAVs data transmission and control systems detection and direction.
- UAVs communication links signal parameters tracking.
- Statistical processing of signal parameters, formulation of classification attributes, signal intelligence database support and signal classification.
- Electronic suppression of UAVs data transmission channels, control channels and ground control stations transmitters power.
- UAVs Satellite navigation receiver jamming.

7 Conclusion

1. The chapter analyses the needs of modern UAVs in providing them with radio and telecommunication systems that monitor the state of UAV parameters, determine its location, control the UAV operator, exchange information with the UAV on the duplex channel, information protection in information transmission, electronic warfare.
2. A comprehensive approach to the design of a multifunctional radio communication system to ensure the reliable operation of the UAV is proposed. Algorithms for reliable UAV operation depending on the type of tasks for which it is intended are reviewed.

References

1. Ilchenko M., Uryvsky L., Osypchuk S.: World trends of modern information and telecommunication technologies development. In: International Conference Radio Electronics & Info Communications (UkrMiCo). - IEEE Xplore Digital Library. - https://ieeexplore.ieee.org/document/9165461/10.1109/UkrMiCo47782.2019.9165461 (Scopus)
2. Ilchenko, M., Uryvsky, L., Osypchuk, S.: The main directions of improving information and communication technologies in the global trends. In: Ilchenko, M., Uryvsky, L., Globa, L. (eds.) Advances in Information and Communication Technology and Systems. MCT 2019. Lecture Notes in Networks and Systems, vol. 152, pp. 3–22. Springer, Cham (2021). https://doi.org/10.1007/978-3-030-58359-0_1
3. Zeng, Y., Zhang, R., Lim, T.J.: Wireless communications with unmanned aerial vehicles: opportunities and challenges/wireless communications, networking, and positioning with UAV. IEEE Commun. Mag. **54**, 36–42 (2016)
4. Palik, Mátyás and Nagy, Máté (2019) Brief history of UAV development. REPÜLÉSTUDOMÁNYI KÖZLEMÉNYEK (1997-TŐL), 31 (1). pp. 155–166. ISSN 1417–0604
5. Abdalla, A.S., Marojevic, V.: Communications standards for unmanned aircraft systems: the 3GPP perspective and research drivers. IEEE Commun. Stand. Mag. **5**, 70–71 (2021)
6. Kakar, J.A.: UAV Communications: spectral requirements, MAV and SUAV channel modeling, OFDM waveform parameters, performance and spectrum management/Thesis submitted to the Faculty of the Virginia Polytechnic Institute and State University in partial fulfillment of the requirements for the degree of Master of Science in Electrical Engineering. Blacksburg, Virginia. – 110 p. 5 May 2015
7. Analysis of new drone incident reports. https://dronecenter.bard.edu/analysis-3-25-faa-incidents/ 28 March 2016 Arthur Holland Michel, Close Encounters, Dan Gettinger, FAA
8. Strategic concept of employment for unmanned aircraft systems in NATO. Jan 2010. 36 p.
9. Vegni, A.M., Loscri, V., Calafate, C.T., Manzoni, P.: Communication technologies enabling effective UAV networks: a standards perspective. IEEE Commun. Stand. Mag. **5**(4), 33–40 (2021)
10. Namuduri, K., Chaumette, S., Kim, J.H., Sterbenz, J.P.G.: UAV Networks and Communications. Cambridge University Press, Cambridge, 255 p. (2018). https://doi.org/10.1017/9781316335765
11. Matolak, D.W., Sun, R.: Unmanned aircraft systems: air-ground channel characterization for future applications. IEEE Vehic. Tech. Mag. **10**(2), 79–85 (2015)
12. UAS Integration in the NAS. https://www.nasa.gov/aeroresearch/programs/iasp/uas/about-us/
13. Mozaffari, M.: Wireless communications and networking with unmanned aerial vehicles: fundamentals, deployment, and optimization. Dissertation submitted to the faculty of the Virginia polytechnic institute and state university in partial fulfillment of the requirements for the degree of doctor of philosophy in electrical engineering. Blacksburg, Virginia, 16 May 2018
14. Mozaffari, M., Saad, W., Bennis, M., Debbah, M.: Efficient deployment of multiple unmanned aerial vehicles for optimal wireless coverage. IEEE Commun. Lett. **20**, 1647–1650 (2016)
15. Mammadov, E.: Long-range communication framework for autonomous UAVs. Thesis submitted to the Faculty of Graduate and Postdoctoral Studies for the M.A.Sc. degree in Electrical and Computer Engineering School of Electrical Engineering and Computer Science. Faculty of Engineering University of Ottawa, Canada (2013)
16. Grigonis, H.K.: More drone crashes caused by technical glitches, not human error, study shows. 24 Aug 2016. https://www.digitaltrends.com/cool-tech/drone-crashes-caused-by-technical-glitches

17. Pashchuk, Y., Salnyk, Y., Volochiy, S.: Reliability synthesis for UAV flight control system. Published in ICTERI 2017, Computer Science, 14 p. (2017)
18. Quist, E.B.: UAV Navigation and Radar Odometry. Brigham Young University, Dissertation,126 p. (2015)
19. Series, M.: Report ITU-R M.2171. Characteristics of unmanned aircraft systems and spectrum requirements to support their safe operation in non-segregated airspace. Mobile, Radiodetermination, amateur and related satellites services, December 2009
20. Unmanned Aerial Vehicles Roadmap: 2000–2025/Office of the Secretary of Defense, 130 p. April 2001
21. Baykar, TB2 UAV. https://www.baykartech.com/en/uav/bayraktar-tb2/
22. EOS C VTOL. https://threod.com/products/eos-c-vtol/
23. She, C., Liu, C., Quek, T.Q., Yang, C., Li, Y.: Ultra-reliable and low-latency communications in unmanned aerial vehicle communication systems. Comput. Sci. IEEE Trans. Commun. (2019)
24. Beyond visual line of sight in non-segregated airspace fundamental principles & Terminology, UK Civil Aviation Authority, October 2020
25. Rudinskas, D., Goraj, Z., Stankunas, J.: Security analysis of UAV radio communication system/ISSN1648-7788 print/ISSN 1822-4180 online. **13**(4), 116–121 (2009). https://doi.org/10.3846/1648-7788.2009.13.116-121. AVIATION, http://www.aviation.vgtu.lt
26. Zhang, R., Zeng, Y.: Keynote: UAV meets wireless communication: opportunities and challenges. In: IEEE GLOBECOM 2017, Singapore, 4–8 December 2017
27. Balzano, Q., Rzasa, J., Milner, S., Davis, C.: High capacity tactical networks with reconfigurable, steerable, narrow-beam agile point-to-point rf links. In: IEEE Military Communications Conference (2007)
28. Saad, W.: ISWCS 2018 – Tutorial. Wireless Communications and Networking with Unmanned Aerial Vehicles ISWCS 2018 – Electrical and Computer Engineering Department, Network Science, Wireless, and Security (NetSciWiS) Group Virginia Tech
29. Zhang, R.: UAV meets wireless communication in 5G and beyond: main research challenges and key enabling techniques. In: IEEE Wireless Communications and Networking Conference (WCNC), Barcelona, Spain, 15 April 2018

Networked Simulation with Compact Visualization of Complex Graphics and Interpolation Results

Igor Melnyk[1(✉)] and Andriy Luntovskyy[2]

[1] National Technical University of Ukraine "Igor Sikorsky Kyiv Polytechnic Institute", Peremogy Avenue 37, Kyiv 03056, Ukraine
imelnik@phbme.kpi.ua

[2] BA Dresden University of Cooperative Education, Saxon Academy of Studies, Hans-Grundig-Street 25, 01307 Dresden, Germany
Andriy.Luntovskyy@ba-dresden.de

Abstract. The researchers and engineers of Emerging Networks provide efficient networked simulation and use visualization tools. Therefore, they would like to evolve their knowledge and knowhow both: not only in theoretical areas of Emerging Networks, but also deepen these in the fascinating practical Smart Scenarios. Modern design and simulation systems as well as multiple practical scenarios frequently use the complex plane and spatial graphics, which need compact presentation of figures with complex geometry. It is especially important for CAD/CAM, where the complex constructions of elaborated machines or devices are simulated. Presentation in compact form of details with complex geometry is necessary for defining mechanical stresses, temperature distribution or electromagnetic fields distribution by solving the sets of differential equations. In the theory of communication systems interpolation of digital signals is necessary for defining the time of its transmission through a communication channel. In network cloud simulation systems suitable interpolation functions for the figures with complex geometry are also necessary for reducing the amount of necessary digital data and its fast transmission through communication channels. In this chapter interpolation of complex plane and spatial curves with using Bezier curves, Levitan polynomials and Rvachov functions are considered. Possibility of applying arithmetic-logic functions for describing the figures with complex plane geometry is also considered. Corresponded examples for different interpolation functions are also given.

Keywords: Emerging networking · Networked simulation and visualization · PHY modelling · Digital transformations · 2D/3D graphic simulation software · Plane curve simulation · Spatial curve simulation · Interpolation · Arithmetic-logic functions · Bezier curves · Levitan polynomials · Rvachov functions

M. Klymash et al. (Eds.): TCSET 2022, LNEE 965, pp. 175–196, 2023.
https://doi.org/10.1007/978-3-031-24963-1_10

1 Motivation and Demarcation

Emerging Networking appears spontaneous and accompanies us in our everyday life as well as supports modern industry processes [1–4], ongoing digitalization of workflows and multiple scenarios of R&D (Research and Design).

Emerging Networking (EmN) in the Digital Age provides Digital Transformation, Smart Modelling of different complex PHY objects and their compact visualization.

Embedded Systems integration, modern electronic devices and IoT deployment require an interdisciplinary approach as multiple technologies come together to the networked control algorithms, automation routines and monitored workflows. An important part belongs to the networked supporting tools for CAD/CAM, modelling, simulation and visualization of electronic signal and data processing.

Virtualization, creation of "Digital Twins" based on facilitating and programmable enhancements of PHY world, and network functionality, making the networks easier to deploy and more adaptable to possible changes [1–4].

The researchers and engineers of EmN provide efficient networked simulation tools and would like to develop knowledge and knowhow both: not only in theoretical areas of EmN, but also deepen these in the fascinating practical scenarios. It means: from industrial robotics to current environment protection under use of AI-based systems to the robotic space management program. EmN means: apply, test, develop and teach using new "Smart Applications", networked "Virtual Testbeds" for electronic devices, digital signals, IoT and Industrial IoT, networked simulation and PHY modelling tools. Do you have an interest in our work? Then you are welcome!

Demarcation: we don't discuss the standard software, algorithms and notations for virtual and augmented reality (VR and AR) like VRML, X3D, 3DMLW, O3D or 3DStudio, Haptic/Tactile IoT, Wearable Simulators, Rendering and Textures.

The General Aim of this Work is as Follows: Use of Different Interpolation Functions for Compact Presentation of Complex Plane and Spatial Graphics and Electronic Signals in Computer-Aided-Manufacture Systems and in Network Simulation Software.

The following important topics remain without of scope of this chapter but not without of scope of the both researchers:

- Embedded Systems and Networked Automation,
- Networked Control for Embedded and Electronics,
- Modelling, Simulation, Visualization in Networks.

2 Introduction and Foundations

Nowadays, the network cloud systems of complex plane and spatial graphics simulation are widely used for elaboration and production of complex machines and devices. In modern CAD and CAM systems, such as Solidworks, simulation of constructions with complex spatial geometry is also realized [5, 6]. In such advanced computer systems usually not only the suitable means of designing the constructions with complex graphics, but the necessary tasks of simulation through solving the sophisticated sets of differential equations are also realized. For example, such complex problems of mechanical, thermodynamic and electrotechnical engineering can be solved in the Solidworks CAM system [5, 6].

1. Defining the mechanical stresses on details of constructions.
2. Defining the temperature operation regimes for details of constructions.
3. Defining the distribution of electric field in the designed item.
4. Defining the distribution of magnetic field in the designed item.

It is well-known from the theory of numerical methods, that for solving such complex simulation tasks the accuracy of interpolation or approximation of the plane and spatial curves, which define the boundary conditions, is generally necessary [7, 8].

In the theory of electronic signals suitable interpolation of the complex signal function is also necessary for investigation of the signal properties, for its fast transmission throw communication systems and further treatment without significant errors [9].

Basic theory of interpolation and approximation is presented in book [10].

In this chapter the suitable polynomials for interpolation of complex plane and spatial curves by the given set of base points are considered. Main interpolation functions, have been considered in this chapter, are as follows:

1. Bezier curves for interpolation of plane and spatial curves by 4 basis points [9].
2. Levitan polynomials for interpolation signal functions [12].
3. Arithmetic-logic relations for describing and visualization the region of geometry figures, located between 2 or more plane curves [11].
4. Rvachov functions for interpolation the dependencies of mechanical stresses, temperature, electric or magnetic field distribution between 2 plane or more spatial curves as boundary conditions [12].

Pierre Etienne Bezier worked in Corporation "Renault", an outstanding French mathematician and engineer. He became a leader in the modernization of design and manufacturing by applying the mathematics and computing tools into computer-aided design systems and three-dimensional modeling [13]. Borys Levitan [14] and Volodymyr Rvachev [15] were outstanding Ukrainian mathematicians, they studied and were employed in Karazin National University, Kharkiv, Ukraine. Since 1992 Borys Levitan lived in the USA and was employed at the University of Minnesota [14].

Using correct polynomials for presentation of complex plane and spatial graphics, generally, leads to the reducing the amount of necessary basis data points. Usually amount of necessary basic data points for polynomial interpolation is smaller, than for spline interpolation [7, 10–12]. As for the accuracy of polynomial interpolation, in the case of suitable choosing of polynomial functions it can be in the range of a few percent or smaller. For example, in the theory of signals by the Nyquist – Kotelnikov theorem is proved, that for the number of basic points in the time N_{bp}, which correspond to value:

$$N_{bp} > \left[\frac{1}{2F_s} \right], \tag{1}$$

where F_s – cutoff signal frequency, the signal function $S(t)$ can theoretically be approximated by the Nyquist function without error [9]. It is well-known, that corresponded functional row is written as [9]:

$$S(t) = \sum_{k=-\infty}^{\infty} f\left(\frac{\pi k}{\omega_s}\right) \frac{\sin \omega_s \left(t - \frac{\pi k}{\omega_s}\right)}{\omega_s \left(t - \frac{\pi k}{\omega_s}\right)}; \; \omega_s = 2\pi F_s, \tag{2}$$

where ω_s – cyclic cutoff signal frequency.

But it should be pointed out, that Levitan polynomials, will be considered at the Sect. 3 of this chapter, have some advantages to signal interpolation relatively to Nyquist functions (2) [9].

By this reason polynomial interpolation of complex plane and spatial curves is one of the best ways to obtain compact graphic data for solving complex engineering designing problems at the up-to-time cloud network systems. But the important requirement for such presentation of graphic data is that the used method of interpolation, as well as the means of writing the polynomial coefficients, have to be known on the client side of clouds CAD or CAM system [8].

As for using Rvachov R-functions, they generally have more wide and interesting applications at simulation tasks for systems with complex geometry. With describing the basic geometry of complex designed system by arithmetic-logic relations and applying obtained logic relation to Rvachov R-function, values of R-function usually give the suitable approach for solving of complex engineering simulation tasks, such as distribution of mechanical stresses, temperature fields, electric fields or magnetic fields, without numerical solving the complex sets of differential equations [15]. In general, the application of this approach leads to a decrease in the time for solving the complex modeling problems and, ultimately, to a reduction in the required time for designing of modern machines, devices and systems.

Such approaches are especially important today, for effective presentation of complex spatial graphics objects in emerging networking in the digital transformation systems.

Therefore, the subject of this chapter is describing the possibilities of complex plane and spatial graphics visualization by using different means of polynomial interpolation of simulated curves. Some examples of solving such simulation tasks are also given.

3 Interpolation of Plane and Spatial Sets of Discrete Points by Bezier Curves

In general, presentation the equation of Bezier curves is based on Bernstein polynomials, which in the general form written as [16]:

$$b_{k,n} = C_k^n x^k (1 - x)^{n-k}; \; B_n(x) = \sum_{k=0}^{n} f\left(\frac{n}{k}\right) b_{k,n}(x), \tag{3}$$

where C_k^n– number of combinations from n to k [20], $f\left(\frac{n}{k}\right)$ – Bernstein coefficients.

With using Bernstein polynomials coefficients (3) analytical relation for Bezier curves is written as [8, 16]:

$$\mathbf{B}(t) = \sum_{i=0}^{n} b_{i,n}(t)\mathbf{P_i}, \tag{4}$$

where $b_{i,n}$ – Bernstein polynomials, $\mathbf{P}_i$ – basic point for construction of Bezier curve.

For example, analytical relation for third order Bezier curve, corresponding to general Eq. (4), is written as follows [8, 16]:

$$R(t) = P_0(1-t)^3 + 3P_1t(1-t)^2 + 3P_2t^2(1-t) + P_3t^3, 0 \leq t \leq 1, \tag{5}$$

where $P_0 - P_4$ – the coordinates x of basic discrete points, which are located in the range [0, 1] for variable t and evenly spaced.

The main property of Bezier curves, that they are usually located inside the polygon, which is formed by lines connecting the basic points. Therefore, Bezier curves always pass through the first and last basic points, and intermediate points are simply considered as anchor points for constructing the interpolation curve [8, 16].

Let's consider the example, which shows how to construct the simple, one-dimensional Bezier curves using relation (5).

Example 1. Using the relation (5), to construct Bezier curves for evenly located on the interval [0, 1] basis points $\mathbf{P}_0 - \mathbf{P}_3$ with such given coordinates x.

1. $\mathbf{P}_0 = 2, \mathbf{P}_1 = 7, \mathbf{P}_2 = 5, \mathbf{P}_3 = 3$; 2. $\mathbf{P}_0 = 7, \mathbf{P}_1 = 5, \mathbf{P}_2 = 2, \mathbf{P}_3 = 3$;
3. $\mathbf{P}_0 = 5, \mathbf{P}_1 = 2, \mathbf{P}_2 = 7, \mathbf{P}_3 = 3$; 4. $\mathbf{P}_0 = 7, \mathbf{P}_1 = 5, \mathbf{P}_2 = 2, \mathbf{P}_3 = 7$.

The results of the calculated with using relation (5) and constructed third-order Bezier curves for the different sets of given coordinates x of the basic points are shown in Fig. 1. To obtain these results the graphic means of the MATLAB scientific and technical calculation system were used [8].

It is clear that Eq. (5) can be parametric and it is usually used for N coordinates of spatial points $\mathbf{P}_0$ - $\mathbf{P}_3$. For example, in the case of $N = 2$, the points $\mathbf{P}_0$ - $\mathbf{P}_3$ have two coordinates on the plane. Then, two equations, similar to (5), for the coordinates $R_x(t)$ and $R_y(t)$, describe a two-dimensional Bezier curve in the plane through the following system of independent equations:

$$R_x(t) = P_{0x}(1-t)^3 + 3P_{1x}t(1-t)^2 + 3P_{2x}t^2(1-t) + P_{3x}t^3, \tag{6}$$

$$R_y(t) = P_{0y}(1-t)^3 + 3P_{1y}t(1-t)^2 + 3P_{2y}t^2(1-t) + P_{3y}t^3.$$

By the similar way, for $N = 3$ and spatial coordinates x, y and z, the following relations for three-dimensional Bezier curve are correct:

$$R_x(t) = P_{0x}(1-t)^3 + 3P_{1x}t(1-t)^2 + 3P_{2x}t^2(1-t) + P_{3x}t^3,$$

$$R_y(t) = P_{0y}(1-t)^3 + 3P_{1y}t(1-t)^2 + 3P_{2y}t^2(1-t) + P_{3y}t^3, \tag{7}$$

$$R_z(t) = P_{0z}(1-t)^3 + 3P_{1z}t(1-t)^2 + 3P_{2z}t^2(1-t) + P_{3z}t^3.$$

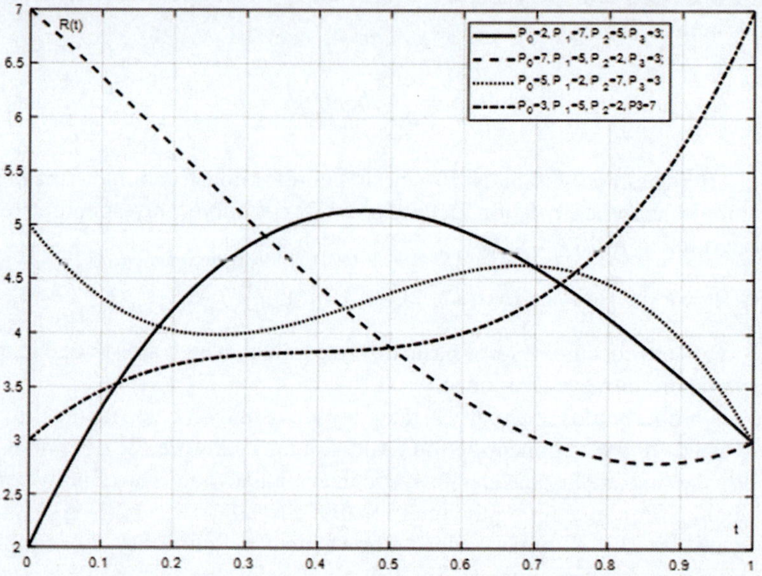

Fig. 1. One-dimension four order Bezier curves for data sets, presented at Example 1.

Consider the examples, which show how to construct two-dimensional Bezier curves using relations (6) and three-dimensional Bezier curves using relations (7).

Example 2. Using the relation (6), to construct Bezier curves for basis points $\mathbf{P}_0 - \mathbf{P}_3$ with such given coordinates (*x*, mm, *y*, mm).

1. $\mathbf{P}_0$ (0 mm, 0 mm), $\mathbf{P}_1$ (0 mm, 1 mm), $\mathbf{P}_2$ (1 mm, 0 mm), $\mathbf{P}_3$ (1 mm, 2 mm);
2. $\mathbf{P}_0$ (–1 mm, 0 mm), $\mathbf{P}_1$ (–1 mm, 1 mm), $\mathbf{P}_2$ (1 mm, 1 mm), $\mathbf{P}_3$ (1 mm, 0 mm);
3. $\mathbf{P}_0$ (2 mm, 0 mm), $\mathbf{P}_1$ (2 mm, 1 mm), $\mathbf{P}_2$ (1 mm, 2 mm), $\mathbf{P}_3$ (1 mm, 2 mm);
4. $\mathbf{P}_0$ (0 mm, 0 mm), $\mathbf{P}_1$ (1 mm, 1 mm), $\mathbf{P}_2$ (1 mm, 2 mm), $\mathbf{P}_3$ (1 mm, 0 mm).

The results of the calculated with using relations (6) and constructed two-dimensional third-order Bezier curves for the different sets of given *x* and *y* coordinates of the basic points are shown in Fig. 2.

Example 3. Using the relation (7), to construct Bezier curves for basis points $\mathbf{P}_0 - \mathbf{P}_3$ with such given coordinates (*x*, mm, *y*, mm, *z*, mm).

1. $\mathbf{P}_0$ = (0 mm, 0 mm, 0 mm), $\mathbf{P}_1$ = (0 mm, 1 mm, 1 mm), $\mathbf{P}_2$ = (1 mm, 0 mm, –1 mm), $\mathbf{P}_3$ = (1 mm, 2 mm, 0 mm);
2. $\mathbf{P}_0$ = (1 mm, 0 mm, 0 mm), $\mathbf{P}_1$ = (1 mm, 1 mm, 1 mm), $\mathbf{P}_2$ = (1 mm, 0 mm, 1 mm), $\mathbf{P}_3$ = (1 mm, 0 mm,2 mm);
3. $\mathbf{P}_0$ = (0 mm, 0 mm, 0 mm), $\mathbf{P}_1$ = (1 mm, 1 mm, 1 mm), $\mathbf{P}_2$ = (1 mm, 2 mm, 1 mm), $\mathbf{P}_3$ = (2 mm, 1 mm, 2 mm);
4. $\mathbf{P}_0$ = (0 mm, 0 mm,1 mm), $\mathbf{P}_1$ = (1 mm, 2 mm, 1 mm),

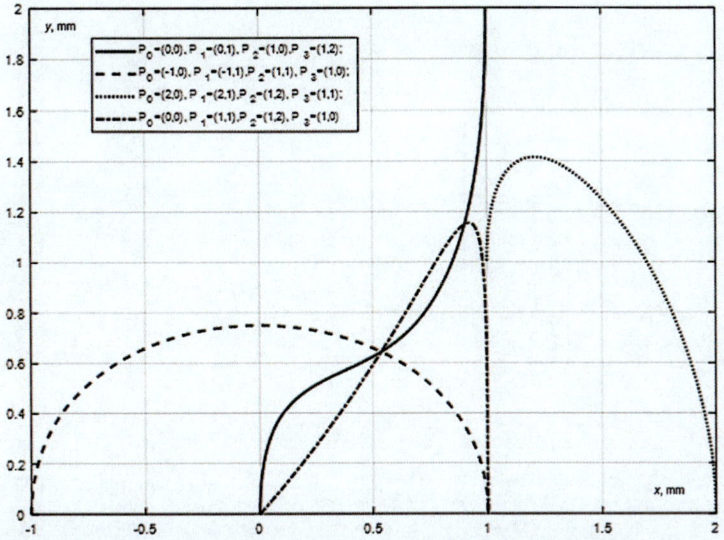

Fig. 2. Two-dimension four order Bezier curves for data sets, presented at example 2.

$\mathbf{P}_2 = (0\text{ mm}, 2\text{ mm}, 0\text{ mm})$, $\mathbf{P}_3 = (1\text{ mm}, 1\text{ mm}, 1\text{ mm})$.

The results of the calculated with using relations (7) and constructed three-dimensional third-order Bezier curves for the different sets of given *x*, *y* and *z* coordinates of the basic points are shown in Fig. 3.

Presentation of complex plane and spatial graphics with using Bezier curves in graphics simulation of designed constructions in cloud computing, which are provided in emerging networking in the digital transformation systems, allows to significantly reduce the amount of necessary digital data for describing the complex geometry of simulated details and items. Such an approach can be effectively used in the modern mechanical engineering, automotive, aviation, space and instrument-making industry [8, 16, 21]. For example, it can be applied in advanced technologies of 3D printing of complex spatial models from metals, ceramic and plastic [22, 23]. Really, only 4 basis points for correct presentation of complex plane and spatial curves with using relations (5 – 7) are necessary.

4 Levitan Polynomials for Description of Digital Signals in Electronics, Embedded and Networks

When Nyquist – Kotelnikov theorem, sense of which is described by elations (1, 2), was formulated and proved, and it was successfully applied to describing the digital signals in electronics and communication systems, the scientists started to investigate it for finding the novel, more effective approaches, to interpolation of complex signal functions.

An interesting and extremely paradoxical result on the properties of functions with finite frequency band was obtained by the famous American mathematician

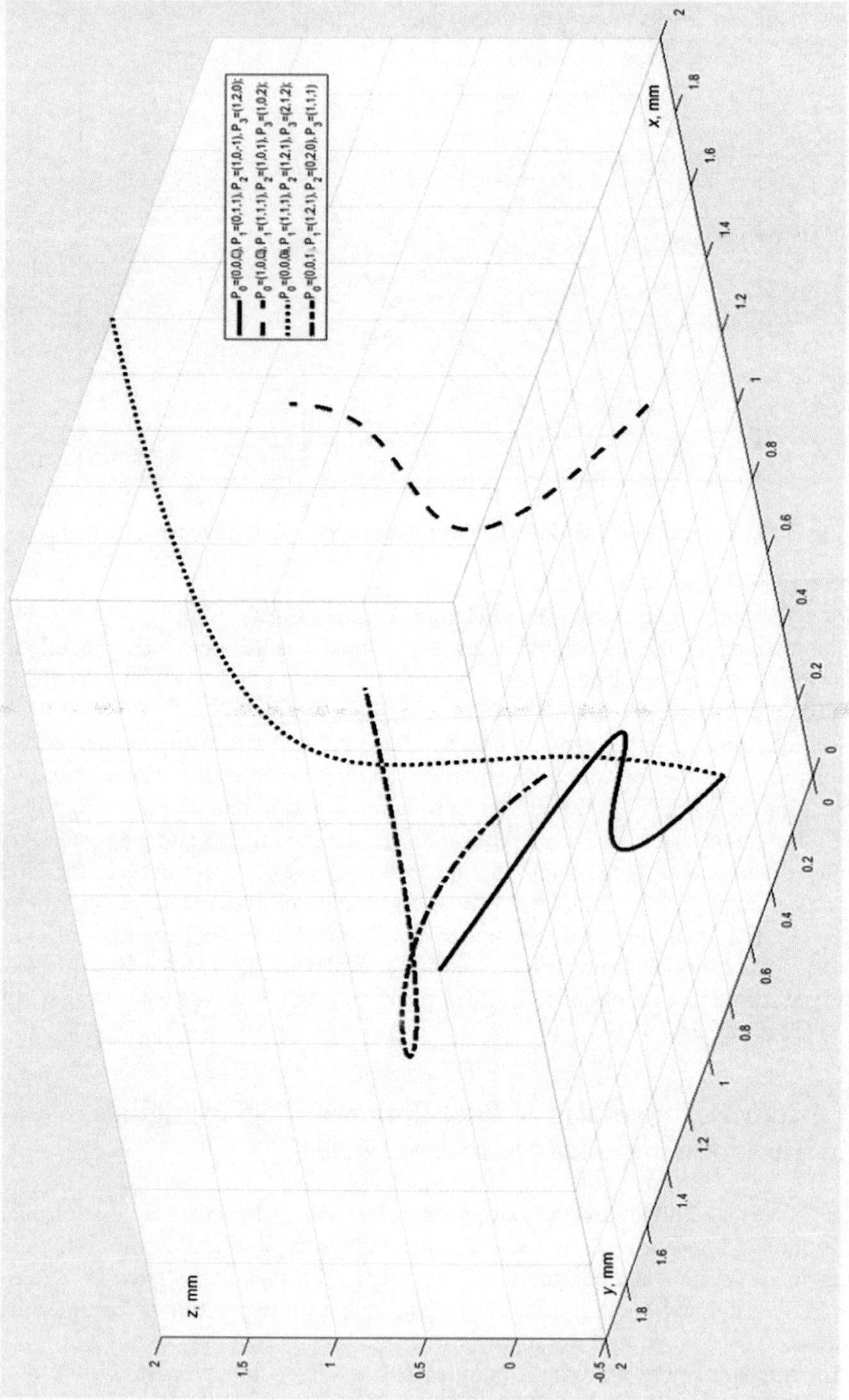

Fig. 3. Three-dimension four order Bezier curves for data sets, presented at Example 3.

D. Slepyan [24]. He, from a purely theoretical point of view, considered the problem of detecting a signal with a limited frequency band, given, and time interval [-*T*; *T*] in the presence of additive noise [5]. It was theoretically substantiated, that such a problem can be reduced to the problem of detecting the same signal under the condition $T \to \infty$.

The result of solving this problem showed that the probability of detecting a signal with a finite frequency band, in the case of proper construction of electronic equipment, can be extremely high, theoretically, it is close to 1 [9].

Another step to study the interpolation of signal functions has been made by outstanding Ukrainian mathematics B. Levitan and N. Akhiezer.

Denote by $\mathbf{W}_\sigma$ ($\sigma \geq 0$) the space of all integer functions $f(z)$ of exponential type with exponent $\alpha < \sigma$. We will also assume that through $\mathbf{B}_\sigma$ the space of functions $f(z)$ is bounded for all real numbers. B. Levitan first proved the theorem, that for any function $f(z)$, which belong to the class $\mathbf{B}_\sigma$, one can construct the periodic trigonometric sums $S_n(f; z)$ ($n = 1 \ldots N$), such, that they are bounded on the axis of real numbers by the same constant as $f(z)$, and uniformly coincide with the function $f(z)$ in each part of the plane of complex numbers. The fundamental theoretical results of B. Levitan was first systematized and published in the fundamental tutorial book on the theory of approximation by N. Akhiezer.

Let's assume, that [12]:

$$h = \frac{\sigma}{n}, \quad n = 1, 2, \ldots, N; \quad \Delta = \frac{2\pi}{h} = \frac{2\pi n}{\sigma}. \tag{8}$$

Then, Levitan polynomials is written as follows [12]:

$$S_n(f; z) = \sum_{k=-\infty}^{\infty} f(z + k\Delta)\mathrm{sinc}^2\left(\frac{\pi}{\Delta}(z + k\Delta)\right), \ \mathrm{sinc}(x) = \frac{\sin(x)}{x}. \tag{9}$$

It should be noted that the asymptotic convergence to 0 in Levitan polynomials is slightly better than in the Nyquist – Kotelnikov series, because the Levitan polynomial's function $y_2(x) = \mathrm{sinc}^2(x)$ decreases to its 0 values much faster and on a smaller numerical interval, than the Nyquist function (2) $y_1(x) = \mathrm{sinc}(x)$ [7, 8]. Accordingly, for Levitan polynomials, provided that the same calculation error is obtained as in the Nyquist – Kotelnikov series, fewer terms can be taken. On the other hand, the function $y_2(x)$ is somewhat more complex from a computational point of view, as it requires an arithmetic operation of taking a square. Therefore, these two methods of interpolation are, to a large extent, equivalent, and the choice of the optimal method in terms of computational time and obtained accuracy usually depends on the type and performance of the CPU [3, 4]. Such estimation is very important for minimizing the time of calculations in clouds in the emerging networking in the digital transformation systems [10].

Example 4. Using the relation (9), to construct Levitan polynomials for interpolation of next sets of discrete rational numbers, given by the function $f(x)$, starting from the value $x_0 = 0$ with the step $h = 0.01$:

1. $\mathbf{F_{in1}} = [1.2, 3.5, 3.7, 2.3, 4.8, 3.2, 2.1, 1.5]$;
2. $\mathbf{F_{in2}} = [1.8, 1.5, 2.5, 2.3, 1.8]$;

3. $\mathbf{F_{in3}} = [2.8, 1.1, 2.5, 1.3, 2.1, 1.6, 0.8]$.

The result of the calculation using relations (7) and constructed Levitan polynomials is presented at Fig. 4. For 8 basic points it was taking only 8 terms in the sum of row (9) for obtaining this simulation results.

A generalization of the Levitan polynomial (9) for a broader class of functions $f(z) \in \mathbf{W}_{\sigma}^{(r)}$, for which the condition $\frac{f(x)}{|x-i|^{2r+1}} \in \mathbf{L}_2(\mathbb{R})$ holds, it also existed [11–13]. The corresponded analytical relation, generally similar to (2) and (9), is written as follows [11–13]:

$$S_n^{(r)}(f; z) = \sum_{k=-\infty}^{\infty} f(z + k\Delta)\operatorname{sinc}^{2(r+1)}\left(\frac{\pi}{\Delta}(z + k\Delta)\right), \tag{10}$$

where r – the order of polynomials.

Therefore, the large extended set of signal interpolation functions (10), based on Nyquist function (2), was proposed by B. Levitan and N. Akhiezer.

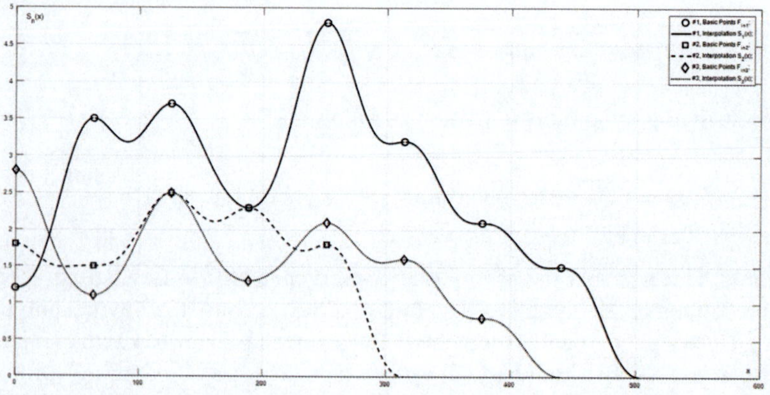

Fig. 4. Use of Levitan polynomials (9) for interpolation data sets, presented at Example 4.

Let's consider another example, in which compares Kotelnikov-Nyquist and Levitan interpolation for a low slope signal, specified by segments of a linear function.

Example 5. Using the relation (2) and (9, 10), to construct Nyquist and Levitan polynomials for interpolation of a set of discrete rational numbers, given by the 2 segments of a linear signal function $f(x)$. The corresponded signal data starting from the value $x_0 = 0$ are located with the step $h = 0.01$: $\mathbf{F_{in}} = [0, 0.2, 0.4, 0.6, 0.8, 1, 1.2, 0.6, 0]$.

The result of the calculation using relations (2, 9, 10) and constructed Nyquist and Levitan polynomials, as well as signal function $f(x)$, are presented at Fig. 5. It is clear from dependencies, presented at Fig. 5, that in the case of low slope signal the finest precision of interpolation is given, namely by Nyquist function (2). The relative difference between the signal function and Nyquist interpolation is a fraction of percent.

And with growing the order r of Levitan polynomials the error of interpolation becomes greater. Furthermore, in this case outliers of interpolation polynomials and maxima are observed, that are not characteristic of the interpolated signal function. Therefore, for the quasi-linear signal functions with the low slope Nyquist interpolation is generally the better method, than Levitan polynomials.

Therefore, let's consider another example, in which comparing Kotelnikov-Nyquist and Levitan interpolation for a high slope signal with quasi-random deviation of basis points.

Example 6. Using the relation (2) and (9, 10), to construct Nyquist and Levitan polynomials for interpolation of a set of discrete quasi-random rational numbers, which are considered as values of signal function. The corresponded signal data starting from the value $x_0 = 0$ are located with the step $h = 0.01$: $\mathbf{F_{in}} = [0, 1.2, 0.4, 1.6, 0.6, 2, 1, 2, 0]$.

The result of calculation using relations (2, 9, 10) and constructed Nyquist and Levitan polynomials for this example are presented at Fig. 6.

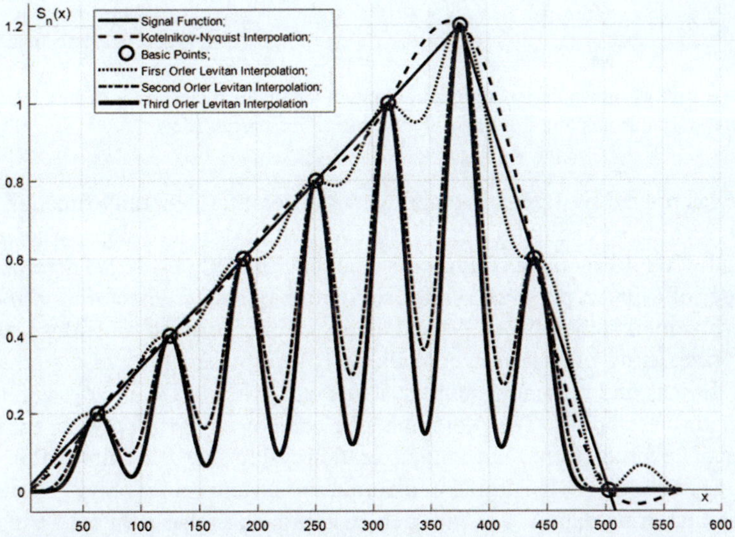

Fig. 5. The results of interpolation signal function $f(x)$, given by the set at example 5, by Nyquist polynomials and Levitan polynomials of different order.

As it is clear from graphic data, presented at Fig. 6, in the case of quasi-random data with the high slope of signal function the best precision of interpolation is given the first order Levitan polynomials, described by the relation (9). For Nyquist functions (2) the error of interpolation at the initial section of the interpolated function is nearly 50%, but for first order Levitan polynomials over the entire range of the function definition area, the error is in the range of fraction of percent. As for Levitan polynomials of second and third order, described by the Eq. (10), for them the error is significantly greater and outliers and maxima, that are not characteristic of the interpolated signal function, are also observed.

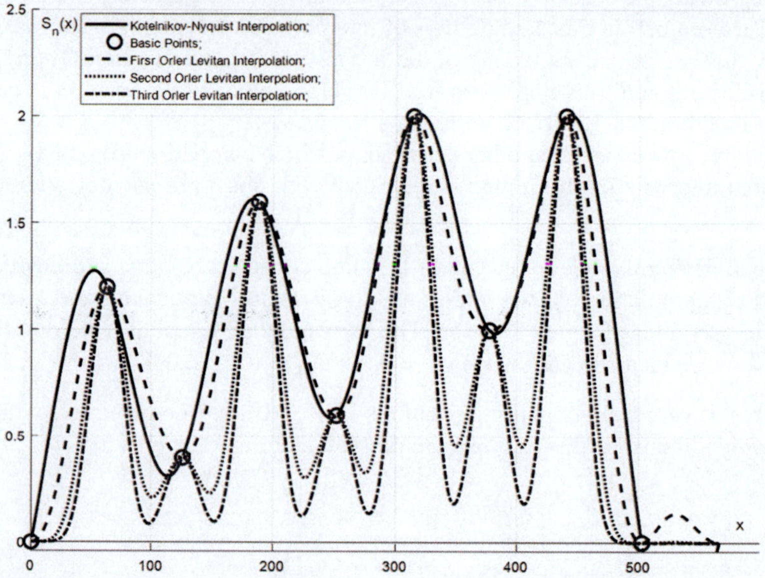

Fig. 6. The results of interpolation signal function, given by the quasi-random data set in example 6, by Nyquist polynomials and Levitan polynomials of different order.

But it was proved by B. Levitan and N. Akhiezer, that generally function (10) can give the correct interpolation for digital sets even with extra-high slope, when the range of function by frequency band is limited by the relations (8) [11, 12]. Therefore, choosing the order of Levitan polynomials for simulation quantized electronic signals is the separate problem of signal theory [9] and theory of interpolation functions [10–12].

In any case, using Levitan polynomials (9, 10) for presentation of digital signals in electronic devices and communication systems can give high accuracy with the small set of basic discrete points. Therefore, the efficiency of applying such an approach for elaboration of novel electronic and communication systems by using simulation in clouds in emerging networking in the digital transformation systems is very high. The amount of necessary transferring data with using such an approach is also reduced. Using Levitan polynomials is especially effective for analyzing the particularities of operation of novel high-frequency communication digital equipment for emerging networking.

5 Arithmetic-Logic Relations for Description the Graphic Objects with Complex Geometry

General definition of arithmetic-logic relation was given in paper [14]. It was pointed out, that arithmetic-logic relations are generally written as follows [14]:

$$\begin{aligned} \mathrm{AL}(x, y) &= F_1(x, y) \cdot L_1(x, y) + F_2(x, y) \\ &\cdot L_2(x, y) + F_3(x, y) \cdot L_3(x, y) + \ldots + F_{\mathrm{n}}(x, y) \cdot L_n(x, y) \end{aligned} \tag{11}$$

where arithmetical functions are $F_1(x, y) \ldots F_{\mathrm{n}}(x, y)$, logical functions are $L_1(x, y) \ldots L_n(x, y)$, and $\mathrm{AL}(x, y)$ – arithmetic-logic relation.

In any case, if the complex plain or spatial geometry of a considered region between a few curves is described as the set of algebraic inequality, it can be simply rewritten in the form of arithmetic-logic relation (11). Considering the simple corresponding example.

Example 7. The geometry of simulated closed region G is described as the plain in coordinates (x, y) by the following set of algebraic inequalities:

$$G(x, y) = \begin{cases} y \geq x^2; \\ y \leq x - x^2 + 4. \end{cases} \tag{12}$$

Rewritten the set of algebraic inequalities (12) as the arithmetic-logic relation (11). Corresponded arithmetic-logic relation is written as follows.

$$\begin{aligned} G(x, y) = & \left(\left(y - x^2 \geq 0\right) \wedge \left(x - y - x^2 + 4 \geq 0\right)\right) \cdot 1 \\ & + \left(\left(y - x^2 < 0\right) \vee \left(x - y - x^2 + 4 < 0\right)\right) \cdot 0. \end{aligned} \tag{13}$$

In relation (13) 1 is corresponded to the points, which are located inside the region $G(x, y)$, and 0 – to the points, which are lied outside this region. The geometry of the region, described by the relation (13), is presented at Fig. 7.

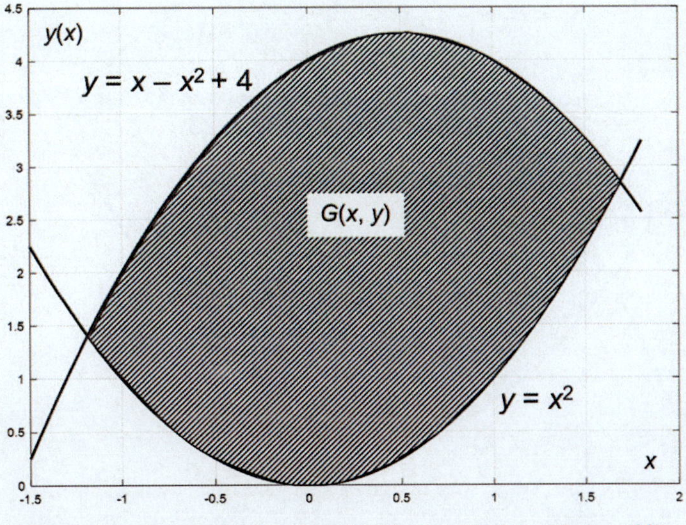

Fig. 7. The geometry of region $G(x, y)$, described by relation (13).

The main advantage of using arithmetic-logic relations, written in form (12), is simplicity of changing the function by adding or removing the new elements into the considered region. By such way the geometry of complex simulated objects is usually described. Consider the next example.

Example 8. The geometry of simulated closed region G is described at the plain in coordinates (x, y) by the following set of algebraic inequalities:

$$G(x, y) = \begin{cases} y \geq x^2; \\ y \leq x - x^2 + 4; \\ (y-2)^2 \geq 0.8 - x^2. \end{cases} \tag{14}$$

Rewritten the arithmetic-logic relation (13) corresponding to a set of algebraic inequalities (14).

Clear, that for set of inequalities (14), which is different from (12) only by the last, third inequality, the arithmetic-logic relation (13) is rewritten as follows:

$$G(x, y) = \left(\left(\left(y - x^2 \geq 0\right) \wedge \left(-y - x^2 + x + 4 \geq 0\right)\right) \wedge \left((y-2)^2 + x^2 - 0.8 \geq 0\right)\right)$$
$$\cdot 1 + \left(\left(y - x^2 < 0\right) \vee \left(-y - x^2 + x + 4 < 0\right) \vee \left((y-2)^2 + x^2 - 0.8 < 0\right)\right) \cdot 0. \tag{15}$$

The geometry of region $G(x, y)$, described by the relation (15), is presented at Fig. 8.

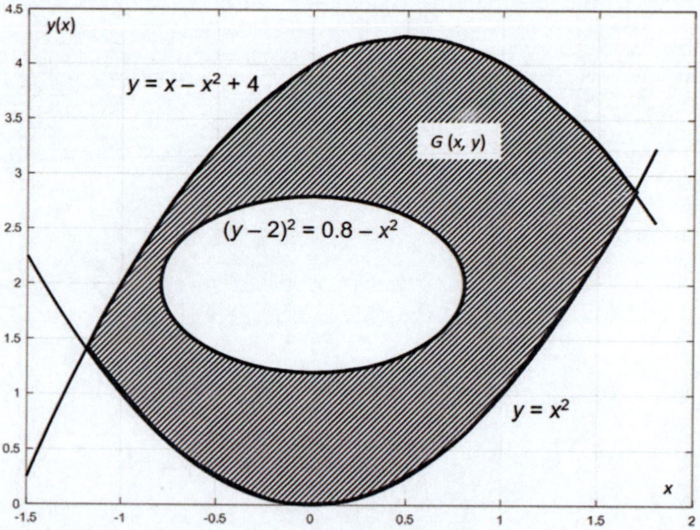

Fig. 8. The geometry of region $G(x, y)$, described by relation (15).

Example 9. The geometry of simulated close region G is described at the plain in coordinates (x, t) by the following set of algebraic inequalities:

$$G(x, t) = \begin{cases} x \leq P_1^{(1)}(t),\ P_{0,1} = 0, P_{1,1} = 2,5, P_{2,1} = 2, P_{3,1} = 1; \\ x \geq P_2^{(1)}(t),\ \ P_{0,2} = 0, P_{1,2} = 0, P_{2,2} = 1, P_{3,2} = 1, \end{cases} \tag{16}$$

where $P_1^{(1)}(t)$, $P_2^{(1)}(t)$ – correspondingly the first and second third order Bezier polynomials for space coordinate x, given by the relation (5), $\mathbf{P}_{0,1}$, $\mathbf{P}_{1,1}$, $\mathbf{P}_{2,1}$ and $\mathbf{P}_{3,1}$ are the coordinates of base points for construction polynomial $P_1^{(1)}(t)$, $\mathbf{P}_{0,2}$, $\mathbf{P}_{1,2}$, $\mathbf{P}_{2,2}$ and $\mathbf{P}_{3,2}$ are the coordinates of the base points for construction the polynomial $P_2^{(1)}(t)$.. Rewritten the set of algebraic inequalities (16) as the arithmetic-logic relation (11).

Corresponded arithmetic-logic relation is written as follows.

$$\begin{aligned} G(x,t) = {} & (((x - x_1(t)) \leq 0) \wedge ((x - x_2(t)) \geq 0)) \\ & \cdot 1 + (((x - x_1(t)) > 0) \vee ((x - x_2(t)) < 0)) \cdot 0. \end{aligned} \tag{17}$$

The geometry of region $G(x, t)$, described by the relation (17), is presented at Fig. 9.

Presentation of graphic objects using arithmetic-logic relations is usually compact themselves. Furthermore, the use of logic functions is more suitable for presentation of computer data, than the sets of inequalities. Main advantages of using such an approach in the programming systems, in simulation of physical effects in designed constructions, as well as in cloud computing in emerging networking, were considered in the paper [14].

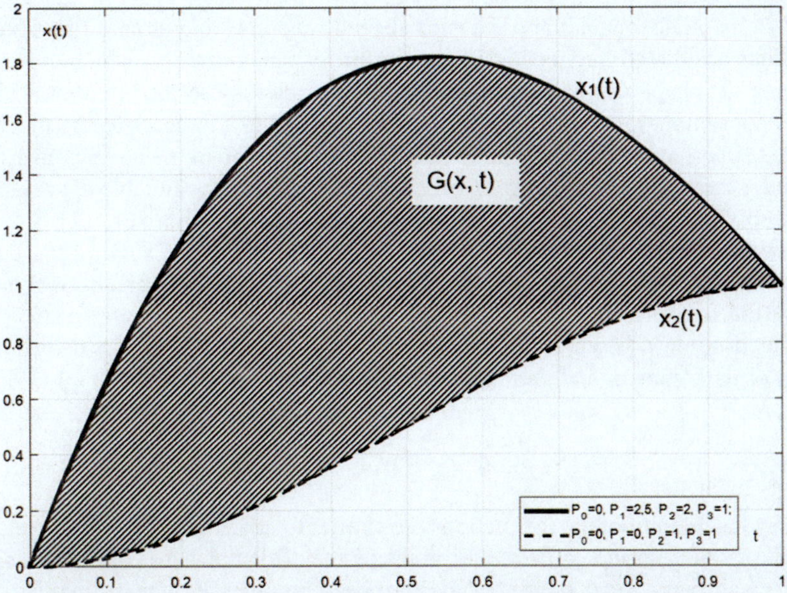

Fig. 9. The geometry of region $G(x, t)$, described by relation (17).

But it is important also, that this approach is the theoretical background to forming and applying of Rvachov *R*-functions, which are considered today as one of most advanced tools for simulation of complex graphic objects [15]. The theory of Rvachov *R*-functions and examples of its application will be considered at the next section of this chapter.

6 Rvachov R-Functions and Application Examples

The theory of Rvachev's *R*-functions, which is related to the mathematical description of the geometry of objects of complex shape, is now used both directly for the recognition and description of geometric objects and for image compression, and in applied fields of modern science and technology. The main application of *R*-functions is such areas of science and engineering, as thermodynamics, applied optics and laser technology [21], electrodynamics of superconducting structures, gyroscopes, theory of analysis and synthesis of ultrahigh-frequency transmitting and receiving antennas, including fractal antennas, space technology, communication theory, digital multidimensional signal processing, pattern recognition, boundary value problems, nanotechnology and 3D-printing [22, 23], mathematical programming, discrete mathematics and neural networks. It is clear that all these areas are extremely important for the further development of scientific and technological progress. Solving of enumerated problems, in the case of the mathematical description of the geometry of objects of complex shape, is generally associated with the use of methods of algebra of logic, set theory, predicate theory, as well as the mathematical apparatus of special functions and atomic functions [19]. *R*-functions are directly connected with arithmetic-logic relations, which are used for describing the shape of geometrical objects in a plane is described in Sect. 4. Really, using of approximation by *R*-functions in many cases allow to describe the mechanical stresses, temperature, electric or magnetic fields distribution in the novel constructions, which have the complex spatial geometry, only by applying the suitable approximation, without solving the complex sets of partial differential equation [11, 12]. Therefore, using R-functions is one of the best ways to reduce the time of solving the complex tasks of theoretical physics and applied mathematics in the clouds and networks and to optimize the software algorithms in emerging networking.

A function $z = f\ (x, y)$ is called an Rvachev *R*-function, if its sign is determined by the sign of its argument and there exists a function Φ, for which $S[z(x, y)] = \Phi[(S(x), S)]$, where

$$S(x) = 0 \cdot (\mathrm{x} < 0) + 1 \cdot (\mathrm{x} \geq 0). \tag{18}$$

Really, by the definition, the *R*-functions are similar to arithmetic logic relations. Inside the boundary of described object, the values of such functions are equal 1, outside the boundary – are equal 0 and, possible, have the negative values. But the distinguishing feature of *R*-functions is that near the boundary the value of these functions is changed from positive to negative values, and generally corresponds to solving the partial differential equations [15].

Construction of *R*-function for describing the shaping of geometrical objects is based on Boolean algebra and the theory of arithmetic-logic relations. Some examples of *R*-functions, as well as its parameters, are presented at Table 1 [15].

For the system of functions R_α, presented at Table 1, the parameter α takes values from the range] –1, 1] and can be a function of the coordinates x_1 and x_2. Usually, the value of this parameter is chosen, corresponding to solving simulation task [15].

Table 1. Methods of construction of *R*-functions from arithmetic-logic relations and corresponding parameters of *R*-functions.

Accompanying function of Boolean algebra	System of functions R_α	System of functions R_0^m
1. $F_1(x_1, x_2) = x_1 \wedge x_2$	$y_1(x_1, x_2) = x_1 \wedge_\alpha x_2 = \frac{1}{1+\alpha} \times \times\left(x_1 + x_2 - \sqrt{x_1^2 + x_2^2 - 2\alpha x_1 x_2}\right)$	$y_1(x_1, x_2) = x_1 \wedge_0^m x_2 = = \left(x_1 + x_2 - \sqrt{x_1^2 + x_2^2}\right) \times \times (x_1 + x_2)^{\frac{m}{2}}$
2. $F_2(x_1, x_2) = x_1 \vee x_2$	$y_2(x_1, x_2) = x_1 \vee_\alpha x_2 = \frac{1}{1+\alpha} \times \times\left(x_1 + x_2 + \sqrt{x_1^2 + x_2^2 - 2\alpha x_1 x_2}\right)$	$y_2(x_1, x_2) = x_1 \vee_0^m x_2 = = \left(x_1 + x_2 + \sqrt{x_1^2 + x_2^2}\right) \times \times (x_1 + x_2)^{\frac{m}{2}}$
3. $F_3(x_1) = \overline{x_1}$	$y_3(x_1) = \overline{x_1} = -x_1$	$y_3(x_1) = \overline{x_1} = -x_1$

The method of construction R_α-functions is follows [15].

1) Forming of arithmetic-logic relation for considered geometrical object. As in relations (13, 15, 17), in the right side of all inequalities in arithmetic relations should to be 0.
2) Forming from obtained Boolean functions Rvachov functions R_α or R_0^m by using corresponding analytical relations, presented in the Table 1.

Consider some examples of construction *R*-function on the basis of arithmetic-logic relations.

Example 10. Construct function R_α for the region $G(x, y)$, considered at Example 8 and presented at Fig. 8. Calculate the values of R_α function for $\alpha = 0.9$ and plot this function.

Let's solve this task for complex geometry of the considered region consequently.

1. Firstly, rewritten arithmetic-logic relation (13) for simple geometry $G_1(x, y)$, presented at Fig. 7, as R_α-function.

$$
\begin{aligned}
&G_1(x, y) = \left(y - x^2 \geq 0\right) \wedge_\alpha \left(-y - x^2 + x + 4 \geq 0\right) \\
&= (w_1(x, y) \geq 0) \wedge_\alpha (w_2(x, y) \geq 0), \\
&w_1(x, y) = \left(y - x^2\right), w_2(x, y) = \left(-y - x^2 + x + 4\right).
\end{aligned} \tag{19}
$$

2. Taking into account, that, corresponding to Table 1,

$$
R_{\alpha 1}(w_1, w_2) = w_1 \wedge_\alpha w_2 = \frac{1}{1+\alpha}\left(w_1 + w_2 - \sqrt{w_1^2 + w_2^2 - 2\alpha w_1 w_2}\right). \tag{20}
$$

Hance, the relation for R_α-function (20), with taking into account relations (19) for $w_1(x, y)$ and $w_2(x, y)$, can be rewritten in coordinates (x, y) as follows:

$$R_{\alpha 1}(x, y) = \frac{1}{1+\alpha}\Big(\big(y - x^2\big) + \big(-y - x^2 + x + 4\big)$$
$$-\sqrt{\big(y - x^2\big)^2 + \big(-y - x^2 + x + 4\big)^2 - 2\alpha\big(y - x^2\big)\big(-y - x^2 + x + 4\big)}\Big)$$
$$= \frac{1}{1+\alpha}\Big(-2x^2 + x + 4 - \sqrt{\big(y - x^2\big)^2 + \big(-y - x^2 + x + 4\big)^2 - 2\alpha\big(y - x^2\big)\big(-y - x^2 + x + 4\big)}\Big). \tag{21}$$

3. Since the geometry of the region, presented at Fig. 8 is different from the region, presented at Fig. 7 only by the excluded circle, R_α-function (20, 21) can be used directly for describing complicated figures, presented at Fig. 8. Corresponding to Table 1, the following relations are written as follows:

$$R_{\alpha,2}\big(R_{\alpha,1}(x, y), w_3(x, y)\big) = R_{\alpha,1}(x, y) \wedge_\alpha w_3(x, y) = \frac{1}{1+\alpha}\Big(R_{\alpha,1}(x, y) + w_3(x, y) -$$
$$-\sqrt{R^2_{\alpha,1}(x, y) + w_3^2(x, y) - 2\alpha R_{\alpha,1}(x, y) w_3(x, y)}\Big), w_3(x, y) = (y-2)^2 + x^2 - 0,8. \tag{22}$$

Painted mesh plot for $R_{\alpha 2}$-function (22) for value $\alpha = 0.9$ is presented at Fig. 10, a, and the corresponding contour plot of this function – at Fig. 10, b.

From a theoretical point of view in the considered Example 10 is interesting that there exists the general rule for the formation of *R*-functions for complicated geometric figures. If an analytical expression for an *R*-function has already been obtained for a simple geometric figure that is part of a more complex figure, this expression can be used directly for further calculations. Really, formula (22) is generally based on (21), and only one necessary logic relation $w_3(x, y)$ is added. This greatly simplifies analytical transformations and allows the use of previously obtained results for further analysis of complex geometric figures. Therefore, further analytical transformations of the system of Eqs. (21) through the substitution of the algebraic function $w_3(x, y)$, describing the circle, into the complicated analytical expression $R_{\alpha 2}(x, y)$ are generally superfluous. The resulting system of Eqs. (21, 22) is closed and consistent, and it can be directly used to calculate the values of the $R_{\alpha 2}$-function.

Example 11. Construct function R_α for the region $G(x, t)$, considered at Example 9 and presented at Fig. 9. Calculate the values of R_α-function for $\alpha = 0.9$ and plot this function.

Arithmetic-logic relation for graphic region $G(x, t)$, considered at Example 9 and presented at Fig. 9, is given by formula (17). R_α-function, which corresponded to this formula, by the reference to data, presented at Table 1, is written as follows:

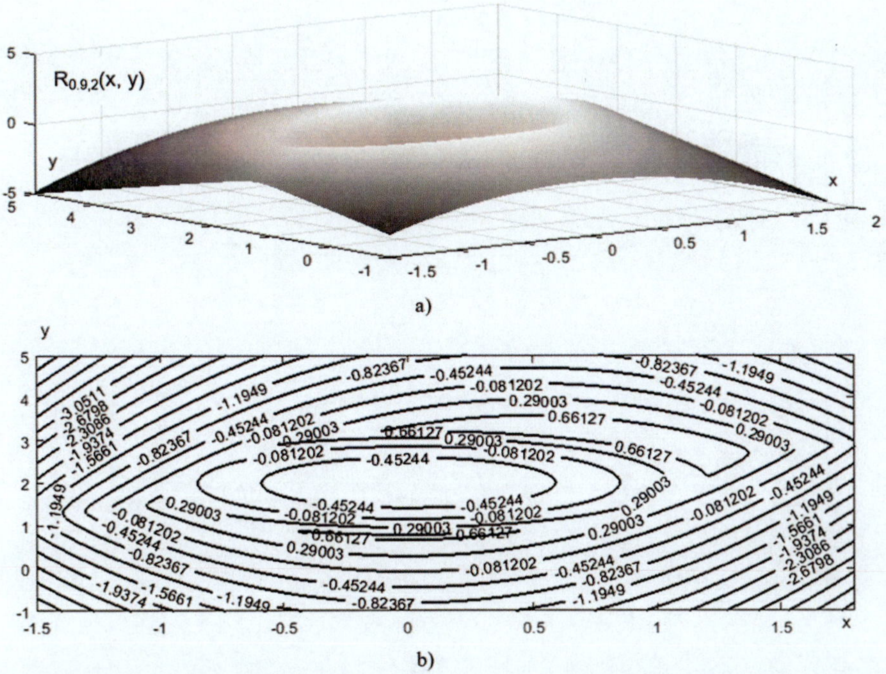

Fig. 10. Painted mesh plot (a) and contour plot (b) of $R_{\alpha 2}$-function (22), $\alpha = 0.9$.

$$w_1(x,t) = (x_1(t) - x), \quad w_2(x,t) = (x - x_2(t)), \quad x_1(t) = P_1^{(1)}(t), \; x_2(t) = P_2^{(1)}(t),$$

$$R_\alpha(w_1, w_2) = R_\alpha(t, x) = (x_1(t) - x) \wedge_\alpha (x - x_2(t)) = \frac{1}{1+\alpha}\Big((x_1(t) - x) + (x - x_2(t)) -$$

$$- \sqrt{(x_1(t) - x)^2 + (x - x_2(t))^2 - 2\alpha(x_1(t) - x)(x - x_2(t))}\Big) \tag{23}$$

In relation (23) the polynomials $x_1(t)$ and $x_2(t)$ are the one-dimension third-order Bezier polynomials, given by the relation (5). The coordinates x for base points of these polynomials $\mathbf{P}_{0,1}$, $\mathbf{P}_{1,1}$, $\mathbf{P}_{2,1}$, $\mathbf{P}_{3,1}$, $\mathbf{P}_{0,2}$, $\mathbf{P}_{1,2}$, $\mathbf{P}_{2,2}$ and $\mathbf{P}_{3,2}$ were given at Example 9.

Painted mesh plot for $R_{\alpha 2}$-function (23) for value $\alpha = 0.9$ is presented at Fig. 11, a, and the corresponding contour plot of this function – at Fig. 11, b.

More sophisticated examples of using R-functions for description of mechanical stresses, temperature, electric and magnetic fields distribution, are given in monographs [15]. In any case, such presentation of scientific graphics is very effective for use in cloud calculations in emerging networking. The presentation of necessary digital data becomes more compact, as well as the time for solving sophisticated scientific problems is also significantly reduced.

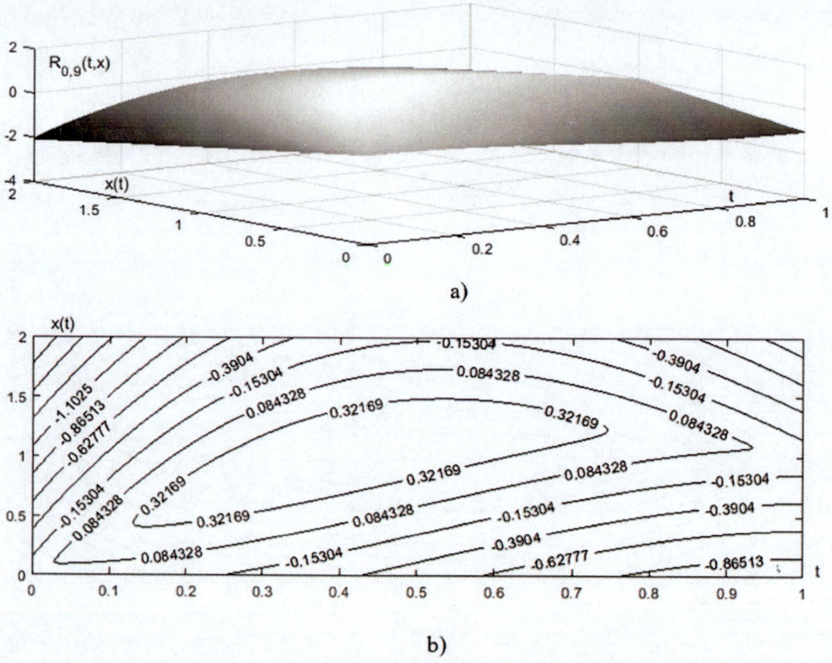

Fig. 11. Painted mesh plot (a) and contour plot (b) of R_α-function (23), $\alpha = 0.9$.

7 Conclusion

In the age of Emerging Networks and Digital Transformations, multiple Smart Applications appear, which provide easy and efficient visualization of complex 2D/3D concepts and figures for their better comprehension for experts, engineers and researchers.

The approaches to graphic data interpolation, examined in the given chapter, provide effective tools for support of Cloud Computing, Smart Analytics in Emerging Networking and, commonly, for Digital Transformation systems.

Bezier curves, considered in Sect. 3, are the suitable methods for describing computer data in modern CAD and CAM systems for automatic designing of complex engineering constructions [5, 6]. Such presentation of complex plane and spatial curves is generally more effective than spline interpolation both by compact graphic data presentation and by small calculation time [1–4, 11, 12].

Levitan polynomials, considered in Sect. 4, are the suitable interpolation functions for describing signals in communication systems and computer networks, especially for extra-high-frequency signals, which are widely used at modern communication equipment for increasing the rate of data transmission. Provided research shows that for high slope signals with quasi-random digital data interpolation of signal function by the Levitan polynomials gives better precision, than interpolation by Nyquist rows.

Rvachov *R*-functions, considered in Sect. 6 of this chapter, is an advanced apparatus for interpolation of solutions for sophisticated scientific problems without solving the complex sets of differential equations. By using such an approach in emerging networking both the necessary amount of graphics data becomes compact and the time of solving sophisticated scientific problems is generally reduced. The theory of *R*-functions is based on computer algebra and arithmetic-logic relations. Basic methods of describing simple and complex computer graphics using arithmetic-logic relations are described and analyzed in examples in Sect. 5.

Generally, the methods of images interpolation in scientific and engineering graphics, described and analyzed at this chapter, is the suitable tools for reducing both the amount of graphic data and the time of solving the sophisticated simulation problems in cloud calculations in emerging networking. The presented results are very interesting and important for experts on creating novel software for calculations in clouds and networks with graphic presentation of simulation results.

References

1. Luntovskyy, A., Guetter, D., Melnik, I.: Planung und Optimierung von Rechnernetzen. Methoden, Modelle, Tools für Entwurf, Diagnose und Management im Lebenszyklus von drahtgebundenen und drahtlosen Rechnernetzen: Planung von Rechnernetzwerke theoretisch anspruchsvoll und praxisnah, Springer/ Vieweg+Teubner Wiesbaden (2012). (in German), 415 Seiten, 245 Abb., ISBN: 9783834814586 (https://www.springer.com/gp/book/9783834814586#otherversion=9783834882424). https://doi.org/10.1007/978-3-8348-8242-4
2. Klymash, M., Beshley, M., Luntovskyy, A. (eds.). Future intent-based networking: on the qos robust and energy efficient heterogeneous software-defined networks. In: LNEE 831, 28 chapters, monograph, on 15.01.2022, Book, Hardcover, XXI, 530 p, 1st ed. January 2022, Springer International Publishing (ISBN 978–3–030–92433–1). https://doi.org/10.1007/978-3-030-92435-5
3. Luntovskyy, A., Guetter, D.: Highly-distributed systems: IoT, robotics, mobile apps, energy efficiency, security, Springer Nature Switzerland, Cham, Monograph, March 2022, ISBN-10: 3030928284, ISBN: 978–3–030–92828–5, 1st ed. 2022, XXXII, 321 p., 189 color figures (Foreword: A.Schill). https://doi.org/10.1007/978-3-030-92829-2
4. Luntovskyy, A., Spillner, J.: Architectural transformations in network services and distributed systems: service vision. Case Studies, Monograph, XXIV, 344 p., 238 pict. (ISBN: 9-783-6581-484-09), Springer, Berlin, April 2017. http://www.springer.com/de/book/9783658148409. https://doi.org/10.1007/978-3-658-14842-3
5. Chang, K.H.: Machining simulation using solidworks CAM. SDC. 290 p. (2021)
6. Schilling, P.J., Shih, R.H.: Parametric Modeling with Solidworks. SDC. 612 p. (2021)
7. Epperson J.F.: An introduction to numerical methods and analysis. Revised Edition. Wiley-Interscience. 590 p. (2007)
8. Mathews, J.H., Fink, K.D.: Numerical Methods. Using MATLAB. Amazon. 696 p. (1998)
9. Lathi, B.P.: Modern digital and analog communication systems. Holt. Rinehart and Whinston Inc. 720 p. (1989)
10. Draper, N.R. Smith H.: Applied Regression Analysis. 3rd edn. 736 p. (1998)
11. Rogers, D.F., Adams, J.A.: Mathematical Elements for Computer Graphics. Second Edition, McCraw-Hill, New York, 512 p. (1990)
12. Varga, R.S.: Functional analysis and approximation theory in numerical analysis, Society of Industrial and Applied Mathematics, Philadelphia, Pennsylvania, 86 p. (1971)
13. de Boor, C., Pinkus, A., Totik, V. (eds.). SAT - surveys in approximation theory (Online). https://www.emis.de/journals/SAT/ (2022)

14. Melnyk, I., Luntovskyy, A.: Estimation of energy efficiency and quality of service in cloud realizations of parallel computing algorithms for IBN. In: Klymash, M., Beshley, M., Luntovskyy, A. (eds.) Future Intent-Based Networking. LNEE, vol. 831, pp. 339–379. Springer, Cham (2022). https://doi.org/10.1007/978-3-030-92435-5_20
15. Rvachov, V.L.: Teoriya R-funktciy i nekotorye ee prilozheniya [Theory of R-functions and some aspects of its application] (published in Russian). Kyiv, Naukova dumka, 552 p. (1982)
16. Bezier's, P.-E.: Biography (online). http://www.edubilla.com/inventor/pierre-etienne-bezier/
17. Levitan's, B.M.: Biography (online). https://www.amazon.com/Boris-Moiseevich-Levitan/e/B001I7GO7W%3Fref=dbs_a_mng_rwt_scns_share/
18. Rvachov's, V.L.: Biography (online). https://prabook.com/web/vladimir_logvinovich.rvachev/163887/
19. Abramovich, M., Stegun, I. (eds.) Handbook on Mathematical Functions with Formulas, Graphs and Mathematical Tables. National Bureau of Standards, Applied Mathematical Series, 55, Washington, 1046 p. (1964)
20. Gubner, J.A.: Probability and random processes for electrical and computer engineers. Cambridge University Press, Cambridge. http://www.cambridge.org/gb/academic/subjects/engineering/communications-and-signal-processing/probability-and-random-processes-electrical-and-computer-engineers (2006)
21. Papoulis, A.: Systems and Transforms with Applications in Optic, McGraw-Hill Book Company, New York, 474 p. (1968)
22. Redwood, B., Schoffer, F., Garret, B.: The 3D Printing Handbook: Technologies, design and applications, 3D Hubs, 304 p. (2017)
23. Gibson, I., Rosen D., Stucke B.: Additive Manufacturing Technologies: 3D Printing, Rapid Prototyping, and Direct Digital Manufacturing. Springer, 519 p. (2015). https://doi.org/10.1007/978-1-4939-2113-3
24. Slepian, D.: Key Papers in the Development of Information Theory, IEEE Press, 420 p. (1988)

Digitalisation in Shopping: An IoT and Smart Applications Perspective

Oliver Cvetkovski[1], Panagiotis Gkikopoulos[2], and Josef Spillner[2(✉)]

[1] University of St. Cyril and Methodius, Skopje, North Macedonia
[2] Zurich University of Applied Sciences, Winterthur, Switzerland
{pang,josef.spillner}@zhaw.ch

Abstract. With online shopping gaining in convenience and coverage, physical shopping is under pressure to innovate. Consequently, smart shopping concepts based on digitalisation are increasingly explored and piloted in urban and rural environments. Beacons lure in customers with push messages and guide their path to desired products in the shops. Doors open automatically for eligible shoppers. Electronic shelf labels are able to adjust prices depending on external factors such as weather and shopping trends. Automated basket scanning and new forms of payment complete the changing experience. Restocking and floor cleaning robots are making inroads. Hence, innovative solutions for smart shopping combine the Internet of Things (IoT) and cyber-physical systems (CPS), and require an appropriate research infrastructure for further exploration. This chapter describes the experience with the IoT/CPS-focused smart shopping lab environment at Zurich University of Applied Sciences and discusses the software and networking design of smart applications in this domain, such as social shopping and product navigation apps, as well as ongoing research challenges.

Keywords: Smart city · Smart shopping · Internet of Things · Cyber-physical systems · Real-time location services

1 Introduction

Smart cities and regions have been a visionary application domain for computer science and engineering for several years. Since the late 2010s and early 2020s, even mid-sized cities have been adopting concepts of technology-supported smartness on a broad and strategic basis. City planners, officials and even private actors in the public space are facing the need to structure their activities and set priorities to cater more smartly to human demands. Figure 1 shows the commonly used smart city wheel that breaks down the activities into more fine-grained topics around smartness. That smartness is often enabled by emerging hardware and software technology and yet would bring real value for citizens, businesses and other city actors. Thus, investigating the practical value of

M. Klymash et al. (Eds.): TCSET 2022, LNEE 965, pp. 197–213, 2023.
https://doi.org/10.1007/978-3-031-24963-1_11

technology becomes an essential task for researchers to contribute to this value creation. Topics in smart cities benefiting from new technological approaches include smart mobility, smart living, smart economy and others, which can then be further broken down to concrete realisations such as smart buildings or smart vehicle sharing.

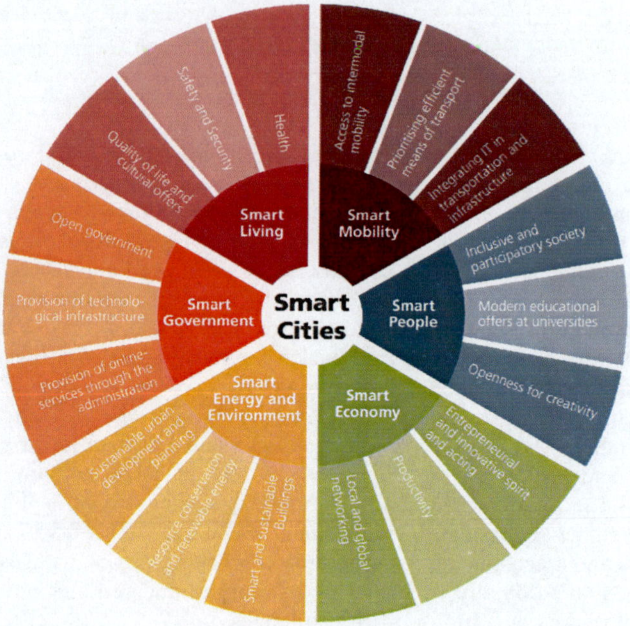

Fig. 1. Smart city wheel reproduced from [1]

Nevertheless, despite trade and commerce being one of the key drivers behind the creation and proliferation of cities ever since the origins of civilisation, smart shopping concepts have not been as prominent as other topics in the context of research on smart cities and regions. At its core, smart shopping is mainly located in the smart economy sector of the wheel, but also touches on smart living and smart mobility and potentially other sectors. It acknowledges the need to innovate for convenience to be able to compete with the rise of online shopping, which in Switzerland alone surpasses 15 billion Swiss francs annually [2], while maintaining the quality of life factors for cities. With emphasis on people-centric smartness, smart shopping specifically refers to the ability of consumers to maximise the convenience, efficiency and safety in obtaining goods or services at any time and at any place. The term place is meant to be cyber-physical, blurring the distinction between the physical world and online services. Hence, smart shopping requires the complementation of online shops and existing physical stores with technology-supported stores that connect to the consumer's devices as clients to on-device and online services. Such technology, if designed and implemented properly, can speed up the shopping process while retaining the advantages of physical stores in local tax generation, anonymity, sustainability and decentralised innovation. Evidently, the supporting technology itself then should be cyber-physical, lending to an advancement of concepts around IoT and CPS as an interface between software and the world, as well as multi-modal human-computer interfaces (HCI).

Industry progress around isolated smart shopping trials within city areas has become visible but is not yet well understood, especially concerning its potential to integrate into a city-wide smart shopping experience that also covers smaller shops, but also concerning technological limitations and further opportunities. Examples from individual chains include cashierless stores as known from Amazon Go or Avec Box [3], and e-commerce integration for Point-Of-Sale (POS) to unify physical and online purchases in so-called cross-channel and omni-channel sales [4]. This cyber-physical combination has seen raising prominence during the COVID-19 pandemic with terms such as click-and-collect or click-and-meet [5]. Another consumer-facing example is the Snäx disposer for convenience food at workplaces that is opened with a mobile application based on a QR code, and billed based on products taken out. In contrast, a smart store management example is the Ubica scan robot to regularly produce digital twins of shops that resulted from the European research project REFILLS [6,7]. A thorough analysis of such industrial trends is therefore mandatory in order to identify knowledge gaps and research questions, and in order to think about the further advancement of such technologies with human needs and preferences across cities and regions in mind. An extended scope that considers also the outdoor experience in the physical world and the associated network communication between devices and humans is therefore necessary.

In this book chapter, the custom-built IoT/CPS research infrastructure and the early-stage research experience on smart shopping [8] at Zurich University of Applied Sciences is summarily described and presented to the reader. The focus is on the underlying devices and network topologies of the smart shopping lab, as well as on smart applications which guide the stakeholders including consumer and store owner, combining three places: online, physical outdoors, and physical indoors. Due to the emergence of the application domain and the associated research field, the ambition of the chapter is not to present well-founded theories. Rather, we hope that the reader will develop a broad understanding of the domain and of applied research methods to advance the state of technology in it, and is subsequently able to study further literature that focuses on narrow problems.

2 Goals, Technology Support and Challenges

2.1 Smart Shopping Goals

As outlined before, smart shopping shall enable consumers to more efficiently and conveniently obtain products and services, in particular when nearby opportunities arise embedded into ongoing activities such as sightseeing, commuting or eating out. The confirmation of an opportunity can be based on various factors, ranging from knowledge of the consumer needs to massive price reduction. To be effective, the process often needs to start at an earlier point, converting potential customers into actual customers through the advertisement of valuable information, offers and opportunities. There are in fact even more stakeholders involved beyond the potential consumers and actual consumers when considering smart shopping processes holistically.

The two main groups in focus for a goal and technology analysis are store owners, selling products in physical stores on behalf of the producers, and consumers, interested in purchasing these products and thus acting as clients and customers in these stores. Both need to be connected by a shared vocabulary and common understanding of the naming, availability, pricing and pickup/delivery modalities of products. Likewise, both represent certain sub-groups including prospective customers, or future store owners looking for a suitable place to open their shop. Further stakeholders involve city planners looking for more bustling downtowns (usually translating into more sales tax income), city traffic planners getting more data points to plan logistics infrastructure, and suppliers wanting to optimise their respective shipments to store owners.

The main goal of a consumer is then to get all desired products in a short amount of time, at a low price and without obstacles such as overrun shops (with long checkout queues) or items being out of stock, including by emergent rebound effects or paradoxons known from other crowd navigation systems [9]. This may lead to a combination of store visits, or even a combination of physical and online processes, rather than just being confined to a single store. The main goal of the store owner is to sell as much as possible, even products that the consumer may not have had on the shopping list beforehand. A goal conflict is imminent. It could be resolved by targeting different groups of consumers who explicitly and dynamically select their preferences in terms of wanting to see additional offers or not. Nevertheless, there may also be shared goals, such as a store owner wanting to empty their stock of a particular kind, and being able to advertise special offers; or a store owner offering regionally produced products that a certain segment of customers is keen on buying.

2.2 Technology Support

Smart shopping is enabled by different technologies and data sources depending on the target group, primarily consumers and store owners. It is inherently data-driven, with part of the data being acquired from the real world by sensing and signalling, and part of the data being input by users from the aforementioned target groups. Figure 2 shows a schematic view of selected smart shopping processes and enabling technologies, thereby relating target groups, data sources and the digital support for smart shopping processes enabled by them.

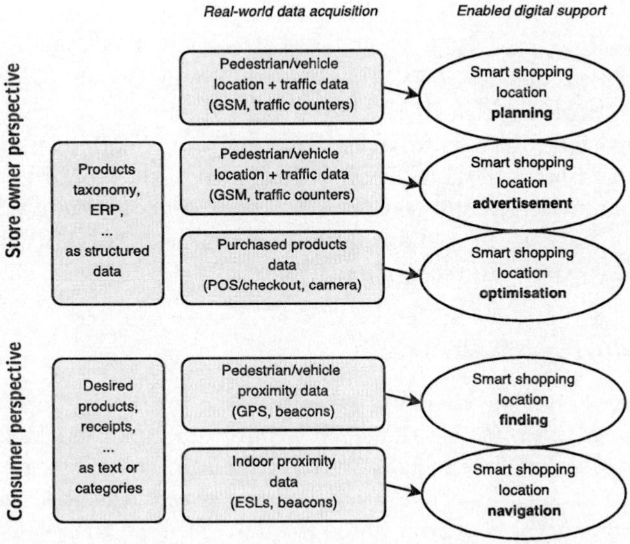

Fig. 2. IoT and real world data acquisition perspective on smart shopping processes; non-exhaustive selection

From a technology and data perspective, smart shopping thus involves the following components and processes:

- Planning: Analysis of pedestrian (prospective customer) frequency in any area, along with existing offerings, to determine optimal placement of new shops with attractive product and service portfolios. This information is generally available from mobile phone operators, given the ubiquity of phones, but also increasingly from urban traffic sensing and public transport operators.
- Generic Advertisement: Information about offered products and services, as well as procurement channels, to interested persons independently from their location. This information is typically reflected in websites, apps and e-commerce shops, based on product management databases and inventory systems, but can also be broadcast to the physical world through beacon signals.
- Specific Advertisement: Attracting pedestrians and other nearby persons to stores in real time, using real-time location services (RTLS). As technological foundation to determine proximity, Bluetooth Low Energy (BLE) beacons and WiFi access points (APs) can be used in addition to Global Navigation Satellite Systems (GNSS) such as GPS or Galileo. For social shopping based on crowd recommendations, the detection of nearby users via their mobile devices based on BLE becomes another option.
- Finding: Matching information about desired products and services. The matching can be performed through various algorithms depending on the data representation, ranging from structured to unstructured (text, voice) input from the human side.

- Navigation: Electronic shelf labels (ESLs) based on LCD or e-ink displays. Inside the store, dozens of ESLs per square meter are able to communicate through protocols such as BLE. Due to their fixed position, they can be used to aid navigation, but also to generate attention through display interaction.
- Optimisation: Smart baskets. When a consumer takes a product from a shelf, either a camera or a weight sensor at checkout time determines the products carried in a basket. This information can be used to verify the shopping process and to generate recommendations before leaving the shop.

2.3 Technological Challenges

There are several challenges associated with these processes, both on the hardware side and on the software side. While pure functional engineering is often possible, the challenges emerge from the desire to gain holistic solutions that take human preferences such as convenience, privacy preservation, social interaction and low environmental footprint into account. Four of these challenges shall be presented briefly, along with corresponding research fields across computer science and engineering:

1. ESL and beacon operation and communication. Low-maintenance and energy-efficient operations are essential to operate smart shopping at scale. New trends in hardware encompass energy harvesting and solar-powered devices such as ESLs and beacons, as well as complete suspend mode apart from scheduled communication slots. How to configure the physical deployment of devices and their configuration is not yet well explored. Innovative ideas encompass the semi-automated registration of ESL positions according to base data from a Building Information Model (BIM) that describes the room properties including floor, wall and ceiling materials, and the resilient transmission of offers through Bluetooth payload without requiring a working Internet connection on consumer devices despite TCP/IP being likely the most widely activated option.
2. Precise determination of proximity. The trade-off between price and precision is well understood and also applies in the improvement of heuristic metrics, such as distance between a device and a signal, by running trilateration, averaging or voting algorithms over a set of signals. The challenge is further complicated by the use of heterogeneous devices such as access points (AP), BLE/ESL devices, and mobile phones or smart watches representing the user position. For instance, sometimes WiFi-based trilateration might be the only option but at reduced precision.
3. Product capturing. Object identification based on training data acquired via Machine Learning (ML) and image segmentation without training are the two primary methods to let store owners take photos of product shelves and rapidly feed them in structured form with price information into an offerings database. If available, offers can be extracted from Enterprise Resource Planning (ERP) databases, in particular when quantity information can be inferred dynamically from modern POS with ERP connectivity.

4. Product matching. Users enter their desired products via voice, in free text form, or by browsing categories. Methods such as Speech-To-Text (STT) and Natural Language Processing (NLP) are required to be able to interpret human input on needed products and to calculate matches against the knowledge of offered products. Users might also have a picture of a favourite product from a trip to another country and look for specialised shops nearby that offer that product, leading to an object detection method. In addition to 1:1 matching, 1:n matching might be performed, for instance based on cooking recipes that require several ingredients from a supermarket, or based on office setup plans that require different devices and cables from a tech shop.

Further challenges such as privacy-preserving recommender systems based on the purchasing behaviour of consumers or crowd campaigns against foodwaste are emerging and should be dealt with by researchers in the coming years.

3 Research Infrastructure

In order to solve the research challenges and drive technological progress, Zurich University of Applied Sciences has set up a smart shopping lab as a research infrastructure and is partnering with stakeholders such as cities, city vendor associations, store operators and technology providers on this basis. This cyber-physical constellation, consisting of both the physical infrastructure and the software prototypes running interactively or as background services, forms one of the world's first 'smart shopping'-focused research and innovation collaborations and contributes to the refinement of technology-based smart city concepts through research work and student projects.

3.1 Physical Infrastructure

Following a real-lab [10] approach, an exemplary smart shopping lab shelf has been set up to host sample products along with technological support to acquire them. Pictured in Fig. 3, it carries several beacons and ESLs along with further communication and edge computation hardware such as a BLE-capable AP. Additionally, tablets and smartphones are available for interactive demonstrations along two paths, one for the store owner (e.g. to capture new products, check the inventory or process customer service requests) and one for the consumer or prospective consumer. To communicate research results, QR codes provide links to further resources such as publications or download links. Researchers with interest in innovative hardware, such as solar-powered ESLs, can use the shelf as a suitable testbed for their ideas.

The connectivity between the individual things and devices is a key concern of the research infrastructure. Apart from a regular WiFi, a specific authentication-less IoT WiFi is available for registered MAC addresses. Several Bluetooth wireless communication versions can be compared: 4.0 (introduction of BLE), 4.1 (introduction of peer-to-peer communication with flexible client/server roles),

Fig. 3. Smart shopping lab shelf

4.2 (compressed transmission), 5.0 (higher distances up to 100 m), and 5.1 (direction finding). Especially the BLE 5.1+ features such as centimetre-level direction detection [11] are promising towards future application designs, but require more time to propagate into affordable devices.

3.2 Software Architecture and Topology

A complex cyber-physical domain like smart shopping requires being modelled and implemented as a capable distributed system to function effectively. The system architecture spans across hardware including a mobile device for signal discovery and user interactions, beacons and ESLs emitting signals, WiFi access points allowing communication and thus also emitting signals, and backend services deployed at the edge (i.e. on a programmable AP) or on dedicated servers. Moreover, it encompasses software for product identification and search, product and shopping list specification, location and proximity determination, and route calculation for complex shopping lists. Hence, it is inherently a distributed system and as such concerned with coverage area, reliability and performance among other relevant metrics. Figure 4 summarises the distributed software architecture and highlights the network topology between the components to permit communication and messaging between the distributed system components.

3.3 Software Applications

Two prototypical software applications for use on mobile devices have been created to support demonstration paths for store owners and consumers, as well as further research. The applications are coupled but could also be deployed standalone in production.

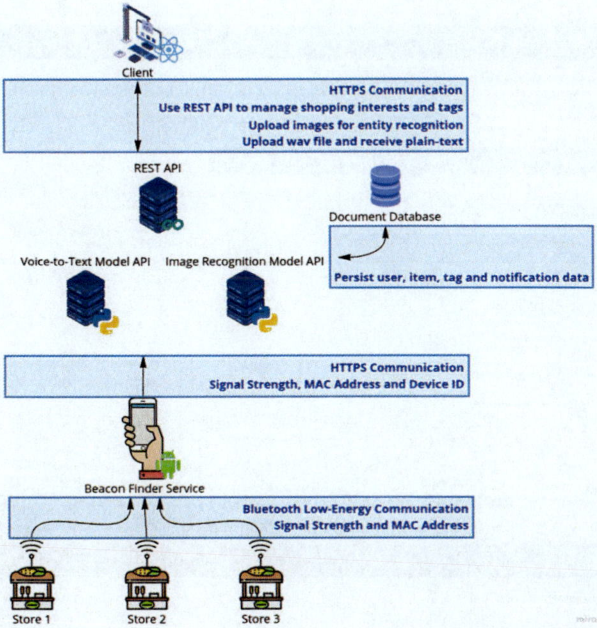

Fig. 4. Smart shopping software architecture and communication topology

Figure 5 shows the prototype of the consumer-facing application. Consumers look for products or define their shopping lists. They can enter full-text information and/or select from a number of pre-defined tags following standard product taxonomies, which are for instance derived from the ERP. The input modality could be extended to capture voice, recipes and so forth in order to extract tags automatically. All inputs are either stored on the consumer's mobile device or, if network connectivity is provided, in a networked database, complementing shop and product position data. At the bottom, a notification area informs the consumers about being in proximity to a store or shelf carrying a matching needed product. This functionality can be triggered based on GNSS, WiFi or BLE proximity. Hence, depending on the needs concerning resilience and self-sufficiency, as well as operational constraints, the application can be implemented to work in centralised or decentralised mode. For the prototype, the centralised implementation with BLE triggering has been chosen, rendering the mobile device into a stateless part of the system.

In contrast, Fig. 6 shows the store owner perspective. Store owners define the tags and create the inventory of products based on photos taken directly from within the application. Each item is curated with at least a name and a price, and potentially a number of tags relating to a taxonomy. Moreover, any deployed beacons can be associated to represent the shop, the shelf or the product itself in the physical world, providing the link to the consumer-facing application by describing the object's physical boundaries as well as further characteristics, such as the product tags, name and price.

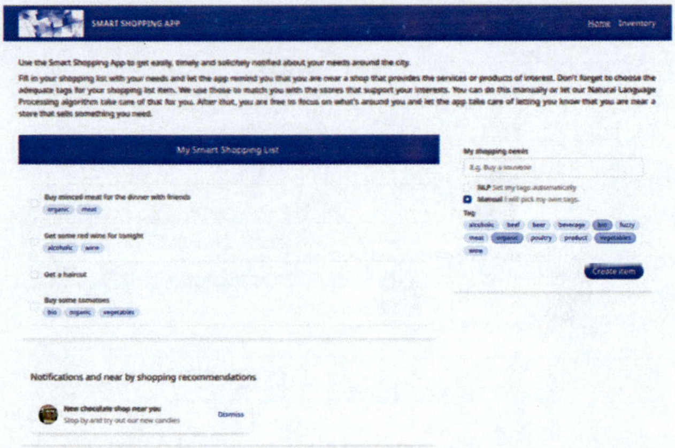

Fig. 5. Smart shopping mobile app prototype, consumer perspective

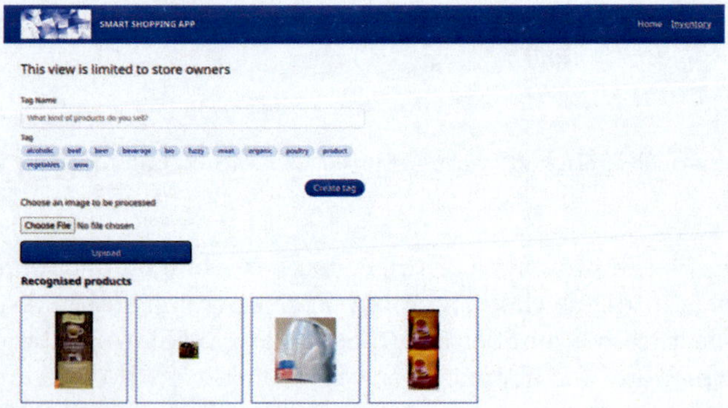

Fig. 6. Smart shopping mobile app prototype, shop owner perspective

Not every potential consumer would however install an application only for shopping in a particular city or region, especially when spending little time there as a tourist or business traveller. With this adoption constraint is mind, further interaction concepts such as integration of the functionality into commonly used tourist applications, or even physical infopoints for basic needs of visitors, are being investigated based on the existing prototype.

4 Proximity Experiment and Findings

As explained, the research infrastructure carries multiple experiments and serves as a real-lab-level validation environment. Due to space constraints, only one of the experiments already conducted will be presented in detail. Its purpose has been the understanding of beacon-based indoor navigation in order to permit more personalised shopping, taking the preferences of customers into account to guide them straight to the desired products.

4.1 Indoor Navigation Experiment Design

Using the research infrastructure for smart shopping, the selected experiment is designed to find answers to the second technological challenge, the precise determination of proximity between a consumer and a product. Specifically, the applied research question guiding the experiment process is: Given the usually unreliable Bluetooth signal strength of beacons, does the multitude and density of ESLs, each of them acting as a beacon, aid in translating signal strength to a reliable distance metric, so that more precise indoor navigation to the target shelf becomes possible? Notably, is the trilateration quality of a line or array of ESLs good enough to be able to avoid a BLE-capable AP as an additional reference point, and thus to reduce the hardware budget? Hence, the experiment investigates a price versus precision question that store owners and their technology partners are confronted with in practice.

Indoor navigation based on wireless communication, sonar or camera feeds or even device sensors has been a research topic for about twenty years [12,13]. They originate from autonomous and robotic navigation but have increasingly been applied to humans, for instance in emergency guidance. This experiment therefore does not reveal fundamentally new techniques, but outlines how to apply specifically recently investigated methods relying on wireless signal strengths [14–16] to the domain of smart shopping where beacon signals are available from ESLs.

The experiment does not consider navigating to the right level of the shelf or the right product itself, which is however also not necessary due to the ability of ESLs to signal the right destination on the display itself or appropriate photos being shown on the mobile device. The signalling of the successful end of the navigation, in particular in crowded environments, is a human-computer interface concern and would warrant separate research. Hence, the research problem can be considered solved if the consumer steps in front of the correct shelf and rapidly gets to select the right product based on that additional signalling. Figure 7 summarises the experiment setup involving the ESLs and the WiFi access point, connecting the ESLs via Bluetooth and the consumer's mobile phone via WiFi signals. Depending on the hardware capabilities, the AP either runs an edge version of the product database or is furthermore connected to a server or cloud service that manages the relevant product data. The AP would not need to have a BLE interface. The hypothesis is that with a spatially distributed grid of ESLs, the consumer's distance as well as the angle from the shelf front can be sufficiently determined by the beacon signal strength in order to give navigation advice.

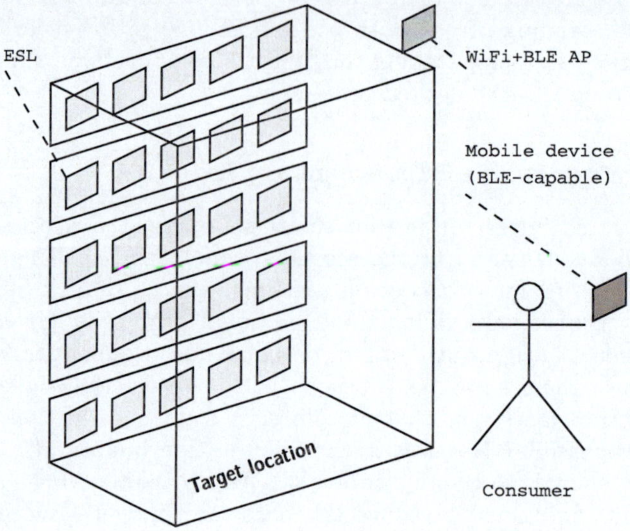

Fig. 7. Distance and angle determination to shelf front array of ESLs

Mathematically, the shelf has a width of 0.8 m and a height of 1.7 m. The two reference positions are 2.5 m in front of the shelf, with a shift of 2 m to the left and to the right of the shelf's centre. Through trilateration, this translates to a detectable shift in distance by approximately $\sqrt{2.5^2 + 2.35^2} - \sqrt{2.5^2 + 1.65^2}$, roughly 43 cm. Measured but not considered further for the analysis is the detectable shift in altitude with small distance variations by roughly 10 cm assuming the mobile device is held at half the shelf height.

4.2 Findings

In an outdoor baseline experiment, in cooperation with a physical store, a Minew BLE 5.0 beacon was placed behind the store's glass front in order to detect its signal strength based on the distance and with the help of GPS as means of verification. Then, due to the signal's heavy deviation, a stack of five beacons was placed just next to the single beacon, and measurements were repeated a few times. Figure 8 summarises the results. It is clear that the signal stability has improved with the stack, covering more waypoints, but it still varies occasionally. Nevertheless, this baseline confirms that using multiple beacons is helpful, at the expense of redundant hardware investment. Moreover, it confirms that even economic beacons are sufficient to achieve the precision needed in outdoor environments, where deviations of one or two metres are tolerable.

The subsequent experiment based on the shelf front array of ESLs is considered next. The consumer needs to pass through a corridor and enter an adjacent room that contains the shelf. The signal diffusion is therefore disturbed by walls and other obstacles. No reference positioning as with GPS exists, but the calculated distance should consistently shrink on the consumer's path towards the

Fig. 8. Signal strength of a beacon in an outdoor scenario: single (left), stack (right)

shelf and, implicitly, towards the desired product on the shelf. In line with this argumentation, the angle of a consumer to the shelf front should be determinable by differences in distance between the two sides of the shelf. All measurements were repeated 100 times with one second inter-measurement interval, resulting in about 10 min continuous observation. While not practical for real navigation scenarios, this setting shall represent the upper limit on obtainable precision.

The results of a static set of measurements from both a left angle and a symmetric right angle, each located about two meters from the shelf, is shown in Fig. 9. All signals strengths from the 100 rounds are aggregated through a mean average function. More sophisticated techniques such as outlier detection and voting (over independent observers) would be possible but are omitted from this exemplary experiment. Still, the ESL grid does not give any information from which distance or angle could be inferred reliably. Evidently, several signals deviate stronger than others. This behaviour confirms the baseline observation. The differences in deviation and signal attenuation could be learned during one or more calibration rounds to influence the interpretation of raw measurements by translating them into device-independent and situation-independent values.

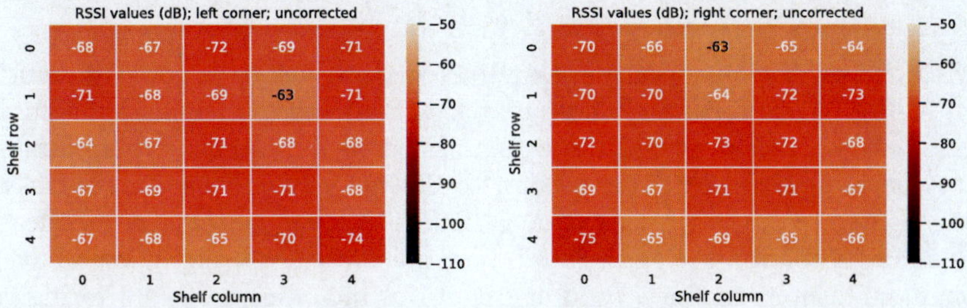

Fig. 9. Signal strength of a grid of beacons/ESLs on a shelf

Three characteristic signal strengths (Received Signal Strength Indication, RSSI) of BLE transmitters are shown for reference in Fig. 10. Apart from being device-specific and placement-specific, including any obstacles to the line of sight (LOS), their behaviour over time depends on the battery charge level among other factors. Given the lifetime of several months, the influence of this factor can be dismissed, and the RSSI spectrum can be reduced to a simplified mean value given a non-disturbed LOS which, while not being realistic in physical scenarios, should be sufficiently valid to find a first solution approach in a lab environment.

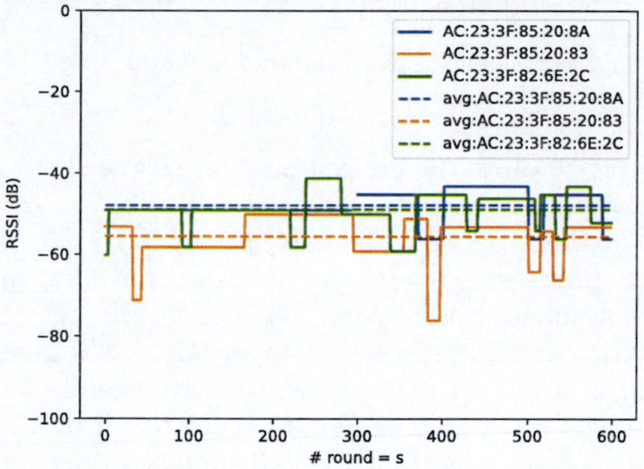

Fig. 10. Calibration function to achieve characteristic RSSI profile per device

Consequently, each measured RSSI of a device i in a fixed set of devices $rssi$ subject to prior calibration can be adjusted by this mean value as explained in Eq. 1. Due to signal propagation behaviour, the calibration needs to be performed at varying distances; in this experiment, at 1.50 m for roughly matching the shelf front layout.

$$rssi\_adj_i = rssi_i + avg(rssi_i) - avg(rssi) \tag{1}$$

With the corrections applied, the adjusted RSSI values appear differently as shown in Fig. 11. Interpreting these matrices is unnecessarily hard for humans, and due to the problem setting can be simplified by a columnar projection, referring to the reduction of 3D navigation to 2D navigation, the usual task when holding a mobile device at the same altitude. In this projection, all measurements across a column are aggregated, here specifically through a mean average although again more sophisticated aggregations including outlier filtering could be possible.

The result of the columnar projection is shown in Fig. 12. For the left corner, the results are satisfying and show a monotonic RSSI difference of approximately 6 dB for the distance difference of 80 cm. For the right corner, the results are

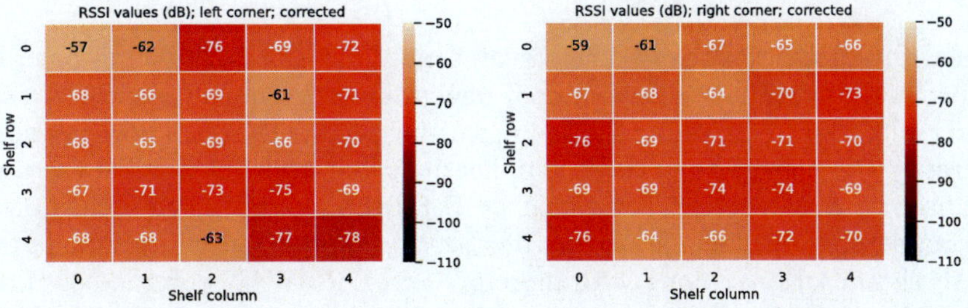

Fig. 11. Signal strength of a grid of beacons/ESLs with correction applied

not satisfying due to a bimodal minimum RSSI distribution. This implies that although the shelf front array technique is promising, it requires further research work to better understand the underlying cause of the result data differences. Apart from using better quality hardware, calibration using the txPower signal flag would be a possible solution that can be attempted within the same shelf lab environment.

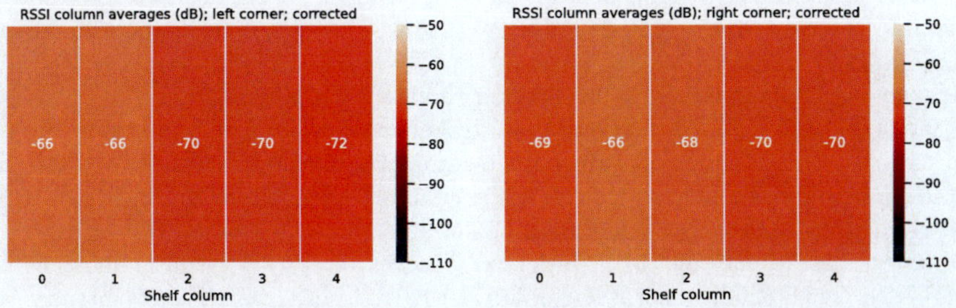

Fig. 12. Columnar signal strength after applying corrections

5 Conclusion and Outlook

This chapter has given a broad overview about the topic of smart shopping and its conceptual evolution along with research in software, hardware, networking and signalling. It has specifically introduced a taxonomy of smart shopping location support enabled by mobile devices and pervasively deployed things, a lab environment centered around a smart shopping shelf, and an exemplary experiment related to real-time location services and navigation.

Much larger economic transitions involving circular economies, autonomous cargo networks and omni-channel distribution are ahead of developed societies. Smart shopping can ensure on an individual level that consumer preferences are taken into account in this evolving macroeconomic landscape, and could be further combined with gamification and social links for instance to foster more regional product circulation. The research needs in smart shopping have been outlined and underlined with an indoor navigation experiment relative to a set of already deployed ESLs which showed promising results that can however not yet be directly transferred to practical applications due to deficiencies in the underlying hardware. New technology such as BLE 5.1+, as well as combined devices (beacon, sensor, display), are being deployed in stores and will gradually mitigate these limitations, along with more 5G and LoRaWAN coverage of outdoor areas which in Switzerland already reaches 98% and 97%, respectively. With a physical lab environment and software research prototypes available, innovative science-based solutions for smart shopping can be transferred to smarter cities and regions in the future.

References

1. SwissEnergy. Smart city guideline report. Smart City Wheel, based on SwissEnergy / SFOE programme (2019)
2. Zumstein, D., Oswald, C., Brauer, C.: Online retailer survey 2021: empirical findings on the e-commerce boom in switzerland and austria. Tech. rep., ZHAW Zürcher Hochschule für Angewandte Wissenschaften, Winterthur (2021). https://doi.org/10.21256/zhaw-2414, https://digitalcollection.zhaw.ch/handle/11475/23116
3. Schögel, M., Lienhard, S.D.: Marketing Review St. Gallen (2020)
4. Kaewprathum, T.P.: Architectural analysis of retail omni-channel and integration of cash it point-of-sale software with e-commerce platform. Master's thesis, Karlstad University, Faculty of Health, Science and Technology (2018). http://urn.kb.se/resolve?urn=urn:nbn:se:kau:diva-68934
5. Zimmer, M., McElrone, M., Anderson Steeves, E.T.: J. Acad. Nutri. Dietetics **121**(12), 2464 (2021). https://doi.org/10.1016/j.jand.2021.05.015, https://www.sciencedirect.com/science/article/pii/S2212267221003397
6. Kümpel, M., de Groot, A., Tiddi, I., Beetz, M.: In: Hammar, K., et al., (eds.): Proceedings of the Joint Ontology Workshops co-located with the Bolzano Summer of Knowledge (BOSK 2020), Virtual & Bozen-Bolzano, Italy, 31 August 7 October 2020, CEUR Workshop Proceedings, vol. 2708, (CEUR-WS.org, 2020) (2020). http://ceur-ws.org/Vol-2708/robontics2.pdf
7. Caccavale, R., et al.: IEEE Robot. Autom. Lett. **5**(3), 4471 (2020). https://doi.org/10.1109/LRA.2020.3000427
8. Cvetkovski, O., Spillner, J.: Privacy-preserving proximity advice for consumers in smart shopping areas. TIn: he 11th International Conference on Cloud Computing in Emerging Markets - Virtual Pre-Conference Workshop (2022). 1st Runner up - The IEEE CCEM Student Project Showcase Proposals
9. Bittihn, S., Schadschneider, A.: J. Stat. Mech. Theory Exper. **2021**(3), 033401 (2021). https://doi.org/10.1088/1742-5468/abdeae
10. Höehner, N., Rodewald, J., Mints, M.O., Kammerlohr, V.: In: The 17th International Conference on Virtual-Reality Continuum and Its Applications in Industry, VRCAI 2019 (Association for Computing Machinery, New York (2019). https://doi.org/10.1145/3359997.3365727
11. Sambu, P., Won, M.: An experimental study on direction finding of bluetooth 5.1: Indoor vs outdoor (2021). https://doi.org/10.48550/ARXIV.2103.04121, https://arxiv.org/abs/2103.04121

12. Golding, A., Lesh, N.: in Digest of Papers. Third International Symposium on Wearable Computers, pp. 29–36 (1999). https://doi.org/10.1109/ISWC.1999.806640
13. Hada, Y., Takase, K.: In: Proceedings 2001 IEEE/RSJ International Conference on Intelligent Robots and Systems. Expanding the Societal Role of Robotics in the the Next Millennium (Cat. No.01CH37180), vol. 2, pp. 1005–1010 (2001). https://doi.org/10.1109/IROS.2001.976300
14. Zhang, D., Portmann, M., Tan, A.-H., Indulska, J. (eds.): UIC 2009. LNCS, vol. 5585. Springer, Heidelberg (2009). https://doi.org/10.1007/978-3-642-02830-4
15. Mackey, A., Spachos, P., Plataniotis, K.N.: In: 2018 IEEE Global Conference on Signal and Information Processing (GlobalSIP), pp. 947–950 (2018). https://doi.org/10.1109/GlobalSIP.2018.8646581
16. Fazio, M., Celesti, A., Villari, M.: In: 2020 IEEE 25th International Workshop on Computer Aided Modeling and Design of Communication Links and Networks (CAMAD), pp. 1–6 (2020). https://doi.org/10.1109/CAMAD50429.2020.9209253

Increasing Functional Stability of Telecommunications Network in the Depressed Zone of HPS Reservoir

Pavlo Anakhov[1](✉), Viktoriia Zhebka[2,3], Lyubov Berkman[2], and Viktoriia Koretska[2]

[1] National Power Company "Ukrenergo", S. Petliuri Street, 25, Kyiv 01032, Ukraine
anakhov@i.ua

[2] State University of Telecommunications, Solomyanska Street, 7, Kyiv 03110, Ukraine
viktoria_zhebka@ukr.net

[3] Borys Grinchenko Kyiv University, Bulvarno-Kudriavska Street, 18/2, Kyiv 04053, Ukraine

Abstract. Sphere of influence of the reservoirs of large HPS involves colossal massifs of rocks. The complex of geophysical fields and processes, mechanical and electrical transformations causes changes in the geophysical situation of the local environment in the depressed zone, which determine the need to make recommendations for the protection of telecommunications. A method for developing measures to protect the telecommunications network from the effects of destructive influences, which includes collecting information on their impact on hardware resources, their analysis and development of appropriate countermeasures. The conditions of functional stability of the telecommunication network are formulated, which are represented by the resistance of the network infrastructure components to the impact of hazards, the ability to reconfigure the operational systems and the transmission network. To verify network protection measures, a matrix of compliance with threats has been developed, the occurrence of which may be due to processes in the depressed zone.

Keywords: Destructive influence · Connectivity · Infrastructure components · Correspondence matrix · Reservation · Reconfiguration

1 Introduction

Among all human engineering activities, large hydropower plants with reservoirs have the greatest impact on the natural environment [1]. They draw into the sphere of their influence colossal massifs of rocks. Complex force fields that arise during that process, extend to considerable depths, cause mechanical and filtration deformations of rocks, their physical and chemical transformation [2].

Superficial analysis of the literature (for example, [3, 4]) has showed that a significant number of large reservoirs are intended for integrated use, in particular for the needs of power stations, primarily hydraulic (HPS). Considering this feature, we agree to understand the term reservoir abbreviated name "HPS reservoir".

M. Klymash et al. (Eds.): TCSET 2022, LNEE 965, pp. 214–230, 2023.
https://doi.org/10.1007/978-3-031-24963-1_12

The study of the effects of reservoirs on the local geological environment has revealed the presence of a depressed zone, the size of which is determined by the range of geophysical fields, and the duration of existence corresponds to the cycles of these fields. Geophysical fields will be understood as physical fields of the Earth, as well as fields represented by the values of their geodynamic parameters given in space [5].

Descriptions of the fields of the depressed zone of the reservoir are presented in Table 1.

Table 1. Geophysical fields of the reservoir depressed zone (modified from [6]).

Field/Process	Limits of action	Field cycle
Deformation field/Lowering of the earth's surface, which can result in landslides or mountain waste debris	The volume of the funnel of the earth's surface lowering is two orders of magnitude greater than the volume of the reservoir, the deflection extends away from the reservoir for tens of kilometers	The relaxation time, during which the initial stress in the cortex is reduced by *e* times, can be from 2,5 to 30 years
Groundwater level/Change in pore pressure of groundwater	The width of the zone of hydrogeological influence of the underground prism of filtration waters usually varies within 0.1–6.0 km away from the shores	Groundwater support zones are characterized by two stages of formation – intensive uplift, when 70–90% of the support is formed (duration 2–10 years); slow rise to the limit position (duration 15–20 years and more)
Microseismic vibration field/Vibration. In addition, in the last century, the effect of reservoir triggered/induced seismicity was discovered [7]	The effective radius of influence depends on the amplitude and frequency of vibrations	It is determined by: - hydrological microseismogenic processes – variations in water level during filling of the reservoir and cycles of replenishment/operation; - meteorological and seasonal, tidal, drift, seismic fluctuations of the level; draft, ripples, wind waves; - microseismic processes – oscillations of the rowing body; vibrations from power plant units and powerful pumps; vibrations from falling water

(*continued*)

Table 1. (*continued*)

Field/Process	Limits of action	Field cycle
Atmospheric processes/evaporation from the water surface with the following: formation of the clouds and breezes; precipitation	The width of the action zone of the air masses on land can be up to 10 km, and on large lakes – up to 30–45 km [8, 9]	Determined by atmospheric processes: - microseismogenic – breeze [4, 10, 11], which initiates vibrations of protruding objects [12]; - formation of clouds [11, 13], which are a factor in the formation of atmospheric electricity [14]; - precipitation [13], which is a factor in the formation of infiltration nutrition [15, 16]

The study of geophysical processes shows that the action and development of some phenomena always creates the preconditions for the emergence and development of others. In the late 1930s, A. Ivanov reported the discovery of a seismic effect of the 2nd kind, the essence of which is that the geological environment under the action of a seismic field generates an electromagnetic field [17]. Analysis of variations in the electromagnetic radiation of the geological environment has showed that they are determined by the mechanisms of energy conversion of these processes into the energy of the electromagnetic field (see Table 2).

Table 2. Mechanisms of electromagnetic field generation due to geophysical processes [18].

Mechanical and electrical converters	Geophysical processes	
	Changing the pore pressure of the groundwater	Deformations due to subsidence of the earth's surface and vibrations
Electrokinetic effect	+	+
Electromagnetic induction	+	+
Piezo effect	–	+

The complex of geophysical fields and processes, mechanical and electrical transformations causes changes in the geophysical situation of the local environment of the reservoir in the depressed zone (see Fig. 1).

The variability of the geophysical environment of the local environment of hydroelectric reservoirs and its changes in comparison with the "local environment without reservoir" determine the need to make recommendations for the protection of one of the

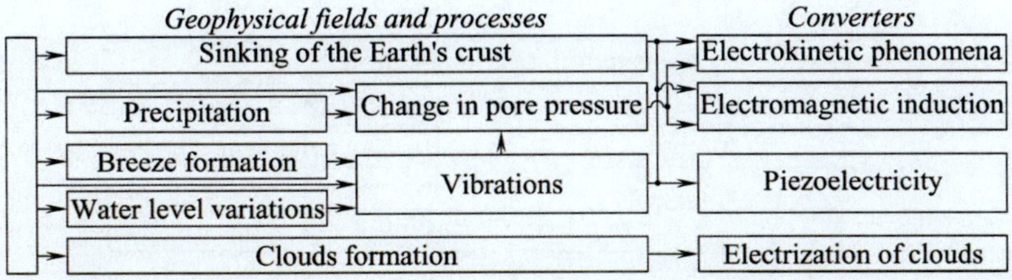

Fig. 1. Changes in the geophysical situation of the local environment of the reservoir due to geophysical fields and processes, mechanical and electrical transformations in the depressed zone.

main industries in electricity generation and transmission – telecommunications (e.g. [19–21]).

The aim of the article is to study the functional stability of telecommunication network equipment in the geophysical situation of the depressed zone of the HPS reservoir.

To achieve this goal it is necessary to solve the following tasks:

– analyze network threats and their consequences;
– develop a scheme for the application of protection measures;
– verify protection measures.

2 Research Method

According to the provisions of the project method, a method for developing measures to protect the telecommunications network from the effects of destructive effects (DE) has been developed, which includes the phased collection of information about the DE effect on the infrastructure components of the telecommunications network, their analysis and development of appropriate countermeasures [22].

At the initial stage, a list of the threats is determined, the occurrence of which may be due to the processes caused by the changes in the geophysical situation of the local environment of the reservoir in the depressed zone. This list has been determined during the analysis of changes in the situation.

At the second stage, the analysis of destructive influences of functioning of a telecommunication network is carried out.

At the final stage, a synthesis of possible measures to reduce the damage caused by the certain threats is performed. The generalized scheme of application of measures of protection against dangerous events is shown in Fig. 2.

The proposed scheme is designed to develop an action plan to prevent hazards. Short-term hazard forecasting is based on current geophysical field measurements and development simulations, and is performed to alert the public to hazards and collect data. Data are used in long-term forecasting. It is used to assess risks and their acceptable levels for declaring the safety of telecommunications, deciding on their location and operation, developing measures to prevent and prepare for accidents. The list of protection measures includes:

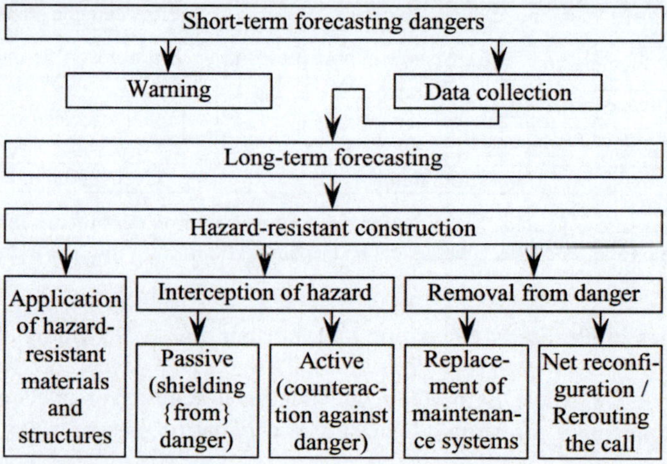

Fig. 2. Scheme of the protection measures application of the telecommunications means from dangers (modified from [23]).

- the use of resistant to certain hazards materials and structures;
- interception of danger, which involves shielding the object or its most vulnerable and responsible elements, from danger, or shielding danger from the object, as well as counteracting the danger;
- replacement of systems for ensuring efficiency (power supply, ventilation and air conditioning, fire alarm, fire extinguishing, warning, etc.);
- network reconfiguration, which is to change the route of transmission of messages or rerouting the call.

Table 3. Matrix of compliance of the protection measures of the information transmission network to threats, the occurrence of which may be due to processes caused by changes in the geophysical situation of the local environment of the reservoir in the depressed zone.

Protection measure	Threats*						
	1	2	3	4	5	6	7
1. Short-term forecasting	a_{11}	a_{12}	a_{13}	a_{14}	a_{15}	a_{16}	a_{17}
2. Notification; data collection for long-term forecasts	a_{21}	a_{22}	a_{23}	a_{24}	a_{25}	a_{26}	a_{27}
3. Application of stable materials and structures	a_{31}	a_{32}	a_{33}	a_{34}	a_{35}	a_{36}	a_{37}
4. Passive interception of a danger	a_{41}	a_{42}	a_{43}	a_{44}	a_{45}	a_{46}	a_{47}
5. Active interception of a danger	a_{51}	a_{52}	a_{53}	a_{54}	a_{55}	a_{56}	a_{57}
6. Replacement of maintenance systems	a_{61}	a_{62}	a_{63}	a_{64}	a_{65}	a_{66}	a_{67}
7. Network reconfiguration / Rerouting the call	a_{71}	a_{72}	a_{73}	a_{74}	a_{75}	a_{76}	a_{77}

*1 – subsidence of the earth's surface, landslides, landslides or debris; 2 – increase in groundwater level (flooding), change in pore pressure; 3 – vibrations, earthquakes; 4 – breeze; 5 – precipitation; 6 – variations in water level, floods; 7 – electromagnetic radiation of the medium.

To verify network protection measures, a matrix of compliance of protection measures with threats, the occurrence of which may be due to processes in the depressed zone, has been developed (see Table 3).

3 Ways to Protect the Telecommunications Network from Hazards

3.1 Conditions of Functional Stability

Let's define functional stability of a telecommunication network, as resistance of the network infrastructure components to influence of dangers, ability to reconfiguration of systems of maintenance of serviceability and a transmission network.

The resilience of network infrastructure components to the impact of hazards is ensured through the use of stable materials and structures, passive and active interception of hazards and is estimated by the formula:

$$\forall i_I \in I, I_{i_I} < S_{i_S}, i_S \in S, i_I = \overline{1, n_I}, i_S = \overline{1, n_S}, \quad (1)$$

where $\forall$ – quantifier of generality; i_I, i_S – identifiers of destructive effects on equipment and indicators of equipment stability, respectively; I, n_I – the set and quantity of the destructive effects on equipment; S, n_S – the set and quantity of the indicators of resistance of the equipment to these influences; I_{i_I}, S_{i_S} – the magnitude of the effects on the equipment and the resistance of the equipment to them, respectively.

Replacement of maintenance systems is provided by their reservation. The probability of operation of the redundant system P_Σ, formed by parallel-connected circuits-systems is estimated by the formula [24]:

$$P_\Sigma = 1 - \prod_{i=1}^{n} (1 - P_i), \ i = \overline{1, n}, \quad (2)$$

where P_Σ – the probability of connectivity of the path formed by parallel connected chains-systems, P_i – the probability of the operability of each of the systems.

Reconfiguration of the transmission network is provided by redundancy of nodes and communication lines, and is estimated by the formula [25]:

$$\chi(G) \geq 2, \ \lambda(G) \geq 2, \ P_{ij}(t) \geq P_{ij}^{normalized}, \ i \neq ji, \ j = \overline{1, n}, \quad (3)$$

where $\chi(G)$ – number of vertex connectivity (the smallest number of vertices (nodes), the extraction of which together with the incident edges (communication lines) leads to a disconnected or single-vertex graph); $\lambda(G)$ – number of edge connectivity (the smallest number of the edges that remove a disconnected graph); $P_{ij}(t)$ – the probability of the connectivity (the probability that the message from node i to node j will be transmitted in a time not exceeding t).

3.2 Analysis of Destructive Influences

Analysis of the destructive effects on the functioning of the telecommunications network is performed in order to identify its most vulnerable elements (see Tables 4, 5).

Table 4. Typical consequences of the DE action [26].

Destructive influences	Consequences of the DE action
Landslides	Destruction of underground gas pipelines keeping cables under excess pressure; damage to line and cable structures
Earthquake	Destruction of external telecommunications facilities; ruptures of gas pipelines holding cables under excess pressure and ruptures of cables
Strong wind (taking into account that the speed of the breeze is 1–5 m/s, we will estimate its influence as insignificant)	Collapse of overhead lines, radio towers; gust, damage to overhead lines
Floods	Flooding of cable sewerage; potential damage to cables; cable locking

Table 5. Matrix of destructive impacts of the depressed zone and their impact on the components of the telecommunications network infrastructure [27].

Infrastructure components	**Destructive influences**			
	Landslides	**Strong wind**	**Rain**	**Floods**
Overheaf line support	M	S	I	S
Antenna	M	S	S	M
Electronics	I	I	M	I
Premises for installation of the devices of the telecommunication network	I	I	S	S
Fiber optic cable	S *	I	I *	I
Symmetrical and coaxial cables	S **	S	M	S
Power supply system	M	I	M	S
Backup power supply	S **	I	M	S
Satellite earth communication stations	M	S	S / M	M
Heating, ventilation and air conditioning systems	I	I	I	I **

I: insignificant; M: moderate; S: significant

* When laying an underground fiber-optic cable, the risk of landslides is taken into account, the most common causes of which are the weakening of the strength of the rocks when they are wet by precipitation.

** The degree of influence of the factor is largely determined by the location of the resource.

3.3 Verification of the Protection Measures

Table 6 shows examples of the protection measures application against natural disasters, in accordance with the elements of the matrix of Table 3.

Table 6. Examples of the protection measures application against the destructive effects of the depressed zone (ME – Matrix element).

ME	Examples of protection measures
a_{11}	- environmental monitoring (organization of early detection systems);
	- monitoring of telecommunication network infrastructure components
a_{21}	- collection of statistical data for long-term forecasts;
	- physical and geographical zoning of the territories to identify climatic conditions of operation;
	- warning of the threat of danger
a_{31}	- increasing the bearing capacity of the structures by increasing the size of the bearing elements and strength of materials, as well as a number of the design measures
a_{41}	- restrictions on a construction in landslide-prone areas; assignment of telecommunications facilities from landslide-prone slopes; increasing the stability of landslide-prone slopes;
	- installation of supporting structures between external objects and steep landslide-prone slopes;
	- increasing the stability of landslide-prone slopes;
	- periodic inspection;
	- see Fig. 3
a_{51}	- "exchange" of a shear mass into a series of smaller collapses, which is performed by dividing the shear mass with the help of explosions in wells into separate blocks. During an earthquake or "cutting" the slope, the blocks of the array will move down the shear surface with a smaller mass compared to the whole massif
a_{61}-a_{67}	- duplication of the power sources and power supply cables;
	- use of reserve generating capacity due to diesel generators, autonomous wind and solar power plants;
	- in some cases, backup equipment is used to increase the reliability of the technological air conditioning system of communication centers;
	- detected defective blocks and elements are subject to immediate replacement (dismantling)
a_{71}-a_{77}	- reconfiguration of the complex of the telecommunication technical means;
	- rerouting the call;
	- compliance with the condition of spatial separation of communication lines;
	- compliance with the condition of separation of communication lines by the nature of the transmitted signal
a_{12}	- installation of the early detection systems of the emergencies;

(continued)

Table 6. (*continued*)

ME	Examples of protection measures
	- installation of the flood sensors and control systems for cable ducts; installation of the early detection systems of the emergencies;
	- environmental monitoring (organization of the early detection systems of the natural disasters);
	monitoring of the telecommunications network infrastructure components
a_{22}	- data collection for long-term forecasts, in particular with the use of machine learning;
	- collection of statistical data for long-term forecasts;
	- physical and geographical zoning of the territories to identify climatic conditions of operation;
	- warning of the threat of danger
a_{32}	- protection of the underground metal structures from corrosion by insulating coatings, tread, cathode or electric drainage systems, linear protective earthing
a_{42}	- restrictions on the location of telecommunications facilities in areas of possible flooding;
	- artificial increase of planning marks of the territory surface; normative soil compaction during backfilling of ditches and trenches; ensuring proper drainage of surface water runoff, construction of protective structures;
	- use of rubber gland seals for penetration into cable ducts, taking measures to ensure tightness of the manhole cover to the neck of the cable well, use of flexible joints for pipelines to keep cables under excess pressure and simulation of seismic actions; installation of vibration sensors or vibration damping systems;
	- laying of communication cables in gas pipelines for keeping cables under excess pressure;
	- installation of waterproof partitions in cable sewers;
	- sealing the ends of plastic pipes with waterproof filler in hatches / mines of underground infrastructure;
	- see Fig. 3
a_{52}	- ensuring drainage of the surface water runoff;
	- installation of drainage pumps in cable sewers
a_{13}	- environmental monitoring (organization of early detection systems);
	- monitoring of the telecommunications (obtaining and processing information about the state of the object, phenomena and processes directly at the installation site of the object);
	- the futility of short-term forecasts is discussed

(*continued*)

Table 6. (*continued*)

ME	Examples of protection measures
a_{23}	- collection of statistical data for long-term forecasts;
	- physical and geographical zoning of the territories to identify climatic conditions of operation;
	- warning of the threat of danger
a_{33}	- using the materials, structures and structural schemes that provide the lowest values of seismic loads (light materials, seismic insulation, other systems of dynamic regulation of seismic load)
a_{43}	- compliance with standards of protection against strong winds;
	- shielding of seismic waves with a layer of loose rock;
	- the use of rubber gland seals for penetration into cable ducts;
	- taking measures to ensure the tightness of the hatch cover to the neck of the cable well;
	- the use of flexible joints in gas pipelines to keep cables under excess pressure
a_{53}	- excitation of weak seismicity by artificial sources in order to relieve tectonic stress;
	- installation of oscillation damping systems;
	- reducing the magnitude of inertial seismic loads on structures by regulating their dynamic characteristics during the oscillating process, and control the mechanism of deformation of structures during earthquakes. Adjustment of dynamic parameters is carried out in order to avoid resonant increase of amplitudes of oscillations or to reduce resonant effects
a_{14}	- environmental monitoring (organization of the early detection systems of the natural disasters), in particular, the speed and direction of the wind are determined;
	- monitoring of the telecommunications network infrastructure components
a_{24}	- collection of statistical data for long-term forecasts, in particular, the wind speed and direction are determined;
	- physical and geographical zoning of the territories to identify climatic conditions of operation;
	- warning of the threat of danger
a_{34}	- increasing the resistance of the materials, used in external telecommunications facilities, to earthquakes
a_{44}	- seismic construction; restrictions on construction within seismically active tectonic faults;
	- installation of additional structures for compensation of wind loads (vertical struts, rope extensions);

(*continued*)

Table 6. (*continued*)

ME	Examples of protection measures
	- strengthening of supports by steel wires (at the expected wind speed more than 40 m / s);
	- use of fastening between supports;
	- increasing the maximum allowable tensile load due to cable clamps;
	- laying thin cables in a bundle wound on an optical cable, which is intended for suspension on the supports of overhead communication lines, catenary and auto-blocking railways, power lines, lighting poles, between buildings or structures or on a dielectric cable
a_{54}	- connection of artificial dampers to protect cables from vibration
a_{15}	- environmental monitoring (organization of the early detection systems of the natural disasters);
	- monitoring of the telecommunications network infrastructure components
a_{25}	- collection of statistical data for long-term forecasts;
	- physical and geographical zoning of territories to identify climatic conditions of operation;
	- warning of the threat of danger
a_{35}	- protection of metal structures from corrosion by insulating coatings
a_{45}	- installation of the waterproof doors and waterproof partitions in cable sewers;
	- sealing the ends of plastic pipes (in hatches / mines of underground infrastructure) with polyfoam filler;
	- see Fig. 3
a_{55}	- installation of drainage pumps in cable sewers
a_{16}	- environmental monitoring (organization of the early detection systems of the natural disasters);
	- monitoring of the telecommunications network infrastructure components
a_{26}	- collection of statistical data for long-term forecasts;
	- physical and geographical zoning of territories to identify climatic conditions of operation;
	- warning of the threat of danger
a_{36}	- protection of metal structures from corrosion by insulating coatings
a_{46}	- restrictions on the location of telecommunications facilities in areas of possible flooding; installation of heavy concrete structures in places of possible post-rain swelling of the soil; installation of supporting structures between external objects and steep landslide-prone slopes

(*continued*)

Table 6. (*continued*)

ME	Examples of protection measures
	To protect against the destructive impact of the wave:
	- construction of coastal protection walls, embankments and dams;
	- the use of forest plantations on the shores, which will dissipate the energy of the waves, reduce the width of the floodplain, detain floating objects (logs, small ships, debris), which can significantly increase the destruction of rattlesnakes;
	- see Fig. 3
a_{56}	- suppression of the long sea waves in the port approach channel, which is to determine the natural frequency of the channel, which depends on its morphometric characteristics, and further change the geometric dimensions of the channel, channel depth, to prevent resonance between incident wave and natural oscillations reservoirs;
	- ensuring drainage of surface water runoff;
	- the use of drainage pumps in cable ducts
a_{17}	- environmental monitoring (organization of the early detection systems of the natural disasters);
	- monitoring of the telecommunications network infrastructure components
a_{27}	- collection of statistical data for long-term forecasts;
	- physical and geographical zoning of territories to identify climatic conditions of operation;
	- warning of the threat of danger
a_{37}	- the use the materials resistant to thermal and dynamic influences, impulse overvoltages;
	- the use of lightning-resistant cables with high conductivity of the sheath (aluminum) and high electrical insulation strength;
	- to protect information from transmission errors, noise-tolerant coding with error detection and error correction is used
a_{47}	- lightning protection system, which consists of a lightning conductor (lightning receiver, current collector), earthing switch, equipotential bonding system (PAS), surge protection devices;
	- protection of cables with insulating coatings, tread, cathode or electric drainage systems, linear protective earthing;
	- protection of cables by calibration, cascade protection and grounding of communication lines, installation of arresters and fuses; installation of lightning rods on overhead communication lines and cables – on cable lines;
	- see Fig. 3

(*continued*)

Table 6. (*continued*)

ME	Examples of protection measures
a_{57}	- for the extreme case of electromagnetic radiation – lightning: the main difference between active lightning protection and passive lightning protection devices is the presence of an active lightning receiver, which response to increasing electromagnetic field strength that occurs when approaching a storm front

To protect line and cable structures from threats, it is proposed to lay them underground. Figure 3 presents the examples of enhanced precautions for the protection of underground cable structures.

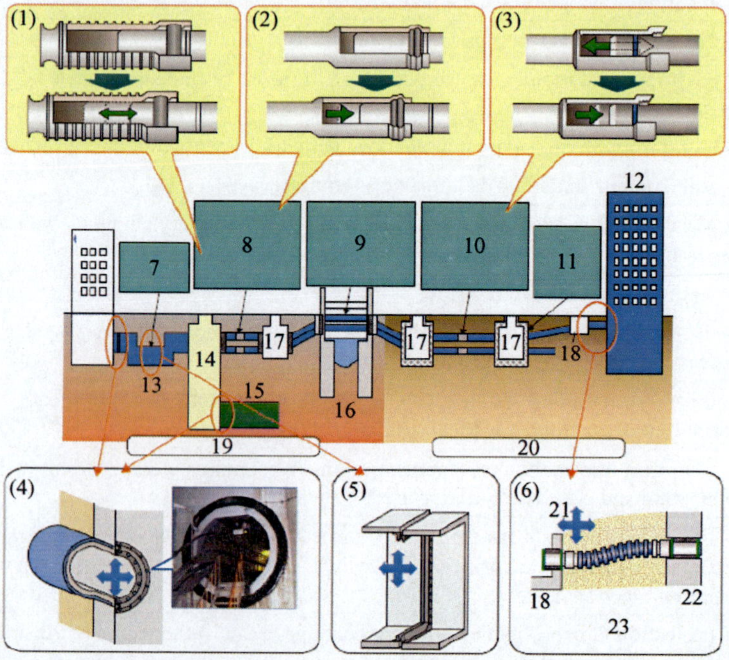

Fig. 3. Examples of precautionary measures for the protection of the underground line and cable structures from the experience of Japanese telecommunications specialists: (1) – sliding connection for inspection wells (connecting pipeline coupling); (2) – sliding connection for gas pipelines; (3) – sliding connection with stopper; (4) – flexible connection for penetrating the wall of the cable shaft; (5); – flexible connection of cable channels; (6) – flexible connection of sections of the gas pipeline for penetration into the building; 7 – flexible connection; 8 – sliding connection + coupling; 9 – sliding connection with stopper + concrete cable tray; 10 – sliding connection with stopper + coupling; 11 – reinforced concrete manhole cover; 12 – building of the user of services; 13 – cable channel; 14 – cable mine; 15 – cable sewerage; 16 – penetration of the bridge crossing; 17 – inspection well; 18 – inspection well; 19 – normal soil; 20 – water-saturated soil; 21 – directions of displacements; 22 – the wall of the building; 23 – flexible corrugated gas pipeline

4 Discussion of the Research Results and Conclusions

The result of the study of the functional stability of the components of the telecommunications network infrastructure in the depressed zone of the HPS reservoir is the realization that the local environment needs physical and geographical zoning of the territory. This is due to the need to determine the climatic conditions of operation of the network, which, in turn, determine the measures to maintain its efficiency.

Threats to the network are geophysical fields of the depressed zone of the HPS reservoir (deformations and microseismic oscillations, pore pressure of groundwater, atmospheric processes), supplemented by an electromagnetic field of geophysical origin. These threats are presented in the form of a scheme of changes in the geophysical situation of the local environment of the reservoir due to geophysical fields and processes, mechanical and electrical transformations in the depressed zone.

To develop an action plan to prevent hazards, a proven scheme of application of measures to protect telecommunications has been used. The conditions of functional stability of the telecommunication network are formulated.

The effectiveness of the protection measures scheme is confirmed by the verification of protection measures against threats, which, in fact, are recommendations for improving the functional stability of the telecommunications network in the depressed zone of the HPS reservoir.

References

1. Gupta, H.K., Rastogi, B.K.: Dams and Earthquakes, p. 229. Elsevier Scientific Publishing Company, Amsterdam (1976)
2. Costain, J.K., Bollinger, G.A.: Review: research results in hydroseismicity from 1987 to 2009. Bull. Seismol. Soc. Am. **100**(5A), 1841–1858 (2010). https://doi.org/10.1785/0120090288
3. ENTSO-E Continental Europe Operation Handbook. Policy 6: Communication Infrastructure
4. Kodjeks systjemy pjerjedachi. [Transmission System Codex] (Resolution of the National Commission of Ukraine for State Regulation of Energy and Utilities of March 14, 2018 №309)
5. Osnovni naprjamky rozvytku (koncjepcija) tjeljekomunikacijnoi mjerjezhi NEK "Ukrjenergo" na 2020–2029 roky. [The main directions of development (concept) of the telecommunication network of National power company of Ukraine "Ukeenergo" for 2020–2029 (Decision of the board of NPC "Ukrenergo" from 13.02.2020)
6. Anakhov, P., Zhebka, V., Grynkevych, G., Makarenko, A.: Protection of telecommunication network from natural hazards of global warming. Eastern-Eur. J. Enterp. Technol. №3/10(105), 26–37 (2020). https://doi.org/10.15587/1729-4061.2020.206692
7. Anakhov, P., Makarenko, A., Zhebka, V., Vasylenko, V., Stepanov, M.: Systematization of measures on lightning protection of the objects of telecommunications network. Int. J. Adv. Trends Comput. Sci. Eng. **9**(5), 7870–7877 (2020). https://doi.org/10.30534/ijatcse/2020/138952020
8. Suo, B., Cheng, Y.-S., Zeng, C., Lia, J.: Calculation of failure probability of series and parallel systems for imprecise probability. Int. J. Eng. Manuf. **2**(2), 79–85 (2012). https://doi.org/10.5815/ijem.2012.02.12
9. Barabash, O.V., Bodrov, S.V., Musienko, A.P.: Analiz pobudovy mjerjezhi videokontrolju punktiv mytnogo spostjerjezhennja na osnovi funkcional'no stijkoji systjemy. [Analysis of building a network video surveillance point of customs supervision on the basis of functional stability system]. Zv'jazok. [Telecommunications]. **2**, 8–11 (2014)

10. Recommendation ITU-T L.92 (10/2012). Series L: construction, installation and protection of cables and other elements of outside plant. Disaster management for outside plant facilities
11. Recommendation ITU-T L.1502 (11/2015). Adapting information and communication technology infrastructure to the effects of climate change
12. ITU-T Recommendation M.34 (11/88). Performance monitoring on international transmission systems and equipment
13. GOST (Interstate standard) 15150–69. Ispolnenie dlja razlichnyh klimaticheskih uslovij [Design for various climatic conditions]
14. Stjebljuk, M.I.: Cyvil'na oborona ta cyvil'nyj zahyst. [Civil Defense and Civil Protection]. Znannia-Press, Kyiv, 487 p. (2007)
15. Ospina, A.V., Faulkner, D., Dickerson, K., Bueti, C.: Resilient Pathways: the Adaptation of the ICT Sector to Climate Change, p. 62. ITU, Geneva (2014)
16. ITU-D Study Group 2. Question 6/2: ICT and climate change. Final Report. Geneva, 64 p. (2017)
17. DSTU (State standard of Ukraine) 3396.1–97. Zahyst informaciji. Tjehnichnyj zahyst informaciji. Porjadok provjedjennja robit. [Information protection. Technical protection of information. The order of works.]
18. Recommendation ITU-T Q.12 (11/1988). General Recommendations on telephone switching and signalling. International automatic an semi-automatic working. Overflow–alternative routing–rerouting–automatic repeat attempt
19. Gao, B., Flemings, P.B.: Pore pressure within dipping reservoirs in overpressured basins. Marine Petroleum Geology **80**, 94–111 (2017). https://doi.org/10.1016/j.marpetgeo.2016.11.014
20. Wei, X., Zhang, L., Yang, H-Q., Zhang, L., Yao, Y.-P.: Machine learning for pore-water pressure time-series prediction: application of recurrent neural networks. Geosci. Front. **12**(1), 453–467 (2021). https://doi.org/10.1016/j.gsf.2020.04.011
21. DBN B.1.1-12:2006. Budivnyctvo u sjejsmichnyh rajonah Ukrajiny. [Construction in seismic areas of Ukraine]. Kyiv, 170 p. (2006)
22. Exline, J.D., Levine A.S., Levine, J.S.: Meteorology: an educator's resource for inquiry-based learning for grades 5-9. NP-2006-08-97-LaRC. 146 p.
23. Chang, F.-J., Hsu, K., Chang, L.-C.: (Eds.). Flood Forecasting Using Machine Learning Methods. MDPI, Switzerland, Basel (2019)
24. Ghimire, S., Yaseen, Z.M., Farooque, A.A., Deo, R.C., Zhang, J., Tao, X.: Streamflow prediction using an integrated methodology based on convolutional neural network and long short-term memory networks. Sci. Rep. **11** (2021). Article number: 17497. https://doi.org/10.1038/s41598-021-96751-4
25. Anakhov, P.V.: Patent of Ukraine №118515. Zastosuvannja sposobu prydushennja dovgyh mors'kih hvyl' u portovomu pidhidnomu kanali dlja prydushennja vysoti hvyl', jaki vynykajut' pry obvalenni u vodojmu zsuvnogo masyvu. [Application of method of suppression of long sea waves in port approach channel for suppression of height of waves which arise at collapse in a reservoir of landslide massif]
26. Space weather prediction center. National oceanic and atmospheric administration. https://www.swpc.noaa.gov
27. Efitorov, A.O., Mjagkova, I.N., Shirokij, V.R., Dolenko, S.A.: Prognozirovanie Dst-indeksa, osnovannoe na metodah mashinnogo obuchenija. [Dst Index Prediction Based on Machine Learning Techniques]. Kosmicheskie issledovanija. [Space exploration]. **56**(6), 420–428 (2018). https://doi.org/10.31857/S002342060002493-0
28. Kataoka, R.: Extreme geomagnetic activities: a statistical study. Earth, Planets and Space **72**(1), 1 (2020). https://doi.org/10.1186/s40623-020-01261-8

29. Kozak, M.M.: Linijni sporudy zv'jazku. [Linear communication structures]. Vinnytsia: Higher Vocational School, 317 p. (2009)
30. Majdanjuk, V.P.: Koduvannja ta zahyst informaciji. [Encryption and protection of information]. National Technical University, Vinnytsia, 164 p. (2009)
31. Analiz suchasnogo zarubizhnogo ta vitchyznjanogo dosvidu vlashtuvannja system blyskavkozahistu ob'ektiv jeljektrychnyh mjerjezh. [Analysis of modern foreign and domestic experience in installing lightning protection systems for electrical networks]. Kyiv: Research and Design Center for the Development of the United Energy System of Ukraine, 74 p. (2018)
32. IEC 62305–1:2010. Protection against lightning – Part 1: General principles
33. Anakhov, P., Zhebka, V., Koretska, V., Sokolov, V., Skladannyi, P.: Increasing the functional network stability in the depression zone of the hydroelectric power station reservoir. In: CEUR Workshop Proceedingsthis link is disabled, vol. 3149, pp. 169–176 (2022)
34. Berkman, L., Kriuchkova, L., Zhebka, V., Strelnikova, S.: Universal method of multidimensional signal formation for any multiplicity of modulation in 5g mobile network. In: Klymash, M., Beshley, M., Luntovskyy, A. (eds.) Future Intent-Based Networking. LNEE, vol. 831, pp. 305–321. Springer, Cham (2022). https://doi.org/10.1007/978-3-030-92435-5_18
35. Anakhov, P., Zhebka, V., Korshun, N., Tushych, A., Dovzhenko, T.: Stability method of connectivity automated calculation for heterogeneous telecommunication network. In: CEUR Workshop Proceedings this link is disabled, vol. 3188, pp. 282–287 (2021)
36. Oleg, K.: the diagnostics methods for modern communication tools in the armed forces of Ukraine based on neural network approach. In: Klymovych, O., Hrabchak, V., Lavrut, O., Lavrut, T., Lytvyn, V., Vysotska, V. (eds.) MOMLET 2020 (Modern Machine Learning Technologies Workshop), pp. 198–208 (2020)
37. Babarczi, P., et al.: A mathematical framework for measuring network flexibility. Comput. Commun. **164**, 13–24 (2020)
38. Rahman, F.: Save the world versus man-made disaster: a cultural perspective. IOP Conf. Ser.: Earth Environ. Sci. **235**, 012071 (2019)
39. Weather and Climate Services in Europe and Central Asia. A Regional Review/World Bank Working Paper No. 151. Washington, D.C., 79 p. (2008)
40. Ospina, A.V., Faulkner, D., Dickerson, K., Bueti, C.: Resilient pathways: the adaptation of the ICT sector to climate change. ITU, Geneva, 62 p. (2014)
41. Khaled, Z.E.L., Mcheick, H.: Case studies of communications systems during harsh environments: a review of approaches, weaknesses, and limitations to improve quality of service. Int. J. Distrib. Sens. Netw. **15**(2), 1550147719829960 (2019)
42. Zhebka, V.V.: Patent of Ukraine №147713. Prystrij dlja bagatokanal'nogo pjerjedavannja informaciji. [Device for multichannel information transmission]
43. Zhebka, V.V., Anakhov, P.V.: Monitoryng stalosti informacijno-tjeljekomunikacijnoji systemy i opracjuvannja zahodiv jiji zahystu vid njebjezpjek. [Monitoring the Sustainability of Information and Telecommunication System and Elaboration of Measures to Protect from Dangers]. Mjetrologija ta prylady. [Metrology and instruments]. **1**(87), 23–29 (2021)
44. Kozak, M.M.: Linijni sporudy zv'jazku. [Linear communication structures]. Higher Vocational School, Vinnytsia, 317 p. (2009)
45. Majdanjuk, P.: Koduvannja ta zahyst informaciji. [Encryption and protection of information]. National Technical University, Vinnytsia, 164 p. (2009)
46. Papyrin, L.P.: Eurasian patent №002565. Sposob ponizhenija vysoty i umen'shenija jenergii voln v glubokom vodoeme pri obrushenii v nego opolznevogo massiva. [Method for lowering of the height and decrease of wave energy in a deep reservoir after collapse of slide rock massif in it]

47. Yazar, A., Tusha, S.D., Arslan, H.: 6G vision: an ultra-flexible perspective. ITU J. Future Evol. Technol. **1**(1), 121–140 (2020)
48. Kellerer, W., et al.: How to measure network flexibility? A proposal for evaluating softwarized networks. IEEE Commun. Mag. **56**(10), 186–192 (2018)
49. ITU-T Recommendation G.602. Transmission media characteristics. Reliability and availability of analogue cable transmission systems and associated equipments

Discrete Vehicle Automation Algorithm Based on the Theory of Finite State Machine

Igor Melnyk(✉), Serhii Tuhai, Mykola Surzhykov, and Iryna Shved

National Technical University of Ukraine "Igor Sikorsky Kyiv Polytechnic Institute", Peremogy Avenue 37, Kyiv 03056, Ukraine
imelnik@phbme.kpi.ua

Abstract. In the chapter the model of vehicle control for the linear law of change of its speed in time is considered. Corresponded mathematical model is created using analytical methods of discrete mathematics, set theory, finite state machines theory and mathematical statistics. Based on the analysis of a simple law of motion of a vehicle between two given points in a straight line with a constant decrease in speed, it is theoretically substantiated that the mathematical model of such a law of motion can be realized in space of finite states over time. Such a simple algorithm is very suitable for smart-driving cars. An advantage of the proposed mathematical model of the vehicle control system is its simplicity, as well as the lack of corrective action at those times when the vehicle movement corresponds to the specified law with a slight error. This avoids unnecessary release of vehicle speed and oscillating processes that can occur in the case of a continuous control action if the control system parameters are selected incorrectly. Mathematical models for cases of quasi-stationary and random perturbing action are considered separately and Student's distribution law is used to model random perturbation. It should be noted that the scheme of the finite state machine, obtained in this work, is universal and corresponds to both quasi-stationary and random perturbing action. Proposed algorithm can be realized on the on-board computer of the vehicle, and defining of best time intervals, as well as changing the velocity of vehicle moving, corresponding to the situation on the road, can be provided by the central server, which controls the situation on the road generally. Communication between central and on-board computers is realized by the wireless channels in emerging networks. Model of data exchanging and solving tasks separation in the local wireless network between on-board computer and local server is also proposed. The task of control the velocity of moving vehicle is solved for quasi-stationary perturbing action with control error range of 2–5% and for random perturbing action with control error range of 5–8%.

Keywords: Smart drive · Self-drive · Transport task · Automatic control · Optimal control · Discrete control law · Arithmetic-logic relation · Finite state machine · Emerging networks · Wireless networks · Statistical analysis

M. Klymash et al. (Eds.): TCSET 2022, LNEE 965, pp. 231–245, 2023.
https://doi.org/10.1007/978-3-031-24963-1_13

1 Introduction and Motivation

Car smart driving is a very important and actual scientific and engineering problem today [1, 2]. Modern cars today usually include such Embedded Systems and IoT electronic programmable devices, as HVAC (Heating, Ventilation, Air-Conditioning, windows regulators), sensors of seating mechanism, lighting, braking system and other embedded electronics [2]. Generally, it allows the use of an on-board computer for smart driving the car by the suitable algorithm, without the presence of humans as driver [2]. Such vehicles in technical literature are also called "unmanned" [1, 2]. Also, different digital electronic systems, that help the self-driving car control the car, are elaborated and used today, such as car navigation system, location system, electronic map and map matching, global path planning system etc. [2].

Most advanced industrial corporations, which elaborated the self-driving cars and electronic components for its realization, are presented in Table 1 [2].

Table 1. Well-known industrial corporations, which elaborated smart driving cars end electronic systems for its realizing

Automotive corporations	Electronics devices corporations
BMW, Ford, General Motors, Mercedes-Benz, Nissan, Toyota	Apple, Bosch, Tesla

The most important assistance and security systems for smart-driving are as follows [2].

1. Anti-lock braking system.
2. Electronic Stability Program.
3. Anti-slip regulation.
4. Brake Assist.
5. Hill Descent Control.
6. Adaptive Cruise Control.
7. Automatic Emergency Braking Systems.

But in any case, automatic driving of vehicles can't be realized independently on the on-board system, because the important information about the situation on the road is absent there. Therefore, smart driving is always realized in emerging networks with wireless communication channels. In such a network, an onboard computer finds the optimal way to draw, by the information from the central server, which is transferred by wireless channels. Basic model of the corresponded communication system is presented at Fig. 1. Nowadays, advanced wireless protocols, such as IEEE 802.11p/pWLAN/ pWi-Fi, ITS-G5/Car2Car, LoRA WAN, LTE Cat-M, NB-IoT are used for data transfer between self-driving cars and the central server. In network theory such connection of server and local host is considered also as M2M (Machine To Machine connection) [3, 4].

2 Basic Statement of Problem

The basic problem of self-driving cars is chosen and the suitable driving algorithm is developed. The main requirements to such algorithms are stability of driving with unexpected random factors on the road and avoiding any emergencies, if possible [1, 2].

Generally, the algorithms of smart driving are usually based on the fundamentals of the theory of automatic regulation [5–9].

Today, on land, water, air and other types of transport, widely used computerized systems of automatic control of moving, including self-driving, which are operated on the base of algorithms for optimal control [10, 11]. These algorithms are well grounded theoretically and based on well-known numerical methods of solving systems of differential equations [12, 13]. The undoubted advantages of such algorithms are the high accuracy of the calculation of the controlled parameters of the moving object with taking into account its' physical particularities, as well as possible harmful influence of distributed random factors.

Fig. 1. Network communication system for realizing smart drive conception by M2M connection.

Taking into account the high order derivation with respect to coordinates and velocity for finding the optimal solutions for further vehicle self-driving in such sophisticated algorithms is also possible [10, 11]. Usually, the task of optimal control for the systems with lumps formulated as follows [10, 11]. For the equation of state:

$$\dot{x} = a[x(t), u(t), t], \tag{1}$$

and boundary conditions:

$$x(t_0) = x_0^*, x(t_1) = x_1^*, \tag{2}$$

finding the minimum of functional:

$$\eta = \int_{t_0}^{t_1} F(x(\tau), \dot{x}(\tau), \tau) d\tau, \tag{3}$$

where $x(t)$ – vector of states, $u(t)$ – control function, t_0 – start time, t_1 – end time.

For solving equation system (1–3) usually, depending on formulated optimization task, Lagrange's calculus of variations, Pontryagin's maximum principle and Bellman's dynamic programming method are used [10, 11].

However, in general, the problem of automatic control the moving objects through numerical solving of systems of functional Eqs. (1–3) also has a lot of disadvantages. On the theoretical point of view, the common reason for these shortcomings is that the mathematical formulation of the problem of solving the system of Eqs. (1–3) on a segment of finite length always requires the analysis of an infinite number of states by dividing it into individual segments of infinitesimal magnitude. The contradiction is, that when using numerical methods for solving equations and computer simulation tools, the number of such segments should be limited, and too many of them leads not only to a disproportionate increase in computational time, but also to increase the computational error [12, 13]. In addition, the speed of solving complex systems of functional equations in real time, even with the use of modern high-speed computers and cloud computing technologies, is also limited [14]. Therefore, in some cases, this standard control method may not correspond to modern requirements for high-speed change of complex technical objects parameters and their reliable operation.

Another approach to solving the problems of automatic control of complex technical objects is associated with the applying of mathematical apparatus of modern discrete mathematics, which is generally based on the theory of finite state machines (FSM) [15–18] and fuzzy logic algorithms [19]. This chapter considers a way to generate a discrete algorithm, based on FSM theory and created for solving the problem of self-driving a vehicle under the condition of a known law of its motion.

3 Simple Discrete Law of Vehicle Moving

Let one assume that the vehicle moves from point A to point B along a straight line of length R, and during its movement the speed $V(r)$ decreases linearly in proportion to the traveled distance r. Therefore, at the starting point A the velocity is maximal and is V_0, and at the end point B it is equal to zero. This law of motion in algebraic form can be written through the following simple analytical expression:

$$V(r) = V_0 \frac{R - r}{R}. \tag{4}$$

To rewrite the analytical expression (4) in the discrete form, it is necessary to divide the considered segment of motion [A, B] into N small identical finite segments, the length of which will be:

$$d = \frac{R}{N}. \tag{5}$$

Taking into account (5), in the case of absence of perturbing action, the law of motion (4) can be rewritten in discrete form as follows:

$$V_{i0} = V_0\left(1 - \frac{i}{N}\right),\ V(i, \Delta r) = V_{i0} - \frac{V_0 \Delta r}{R} = V_0\left(1 - \frac{i}{N} - \frac{\Delta r}{R}\right),$$

$$i = 1, 2, \ldots, N,\ \Delta r < d. \tag{6}$$

where i is the number of reference point P_i on the segment $[A, B]$, Δr is the distance from the point P_i to the current point, where the vehicle is located.

4 Discrete Form of the Law of Vehicle Motion in the Case of Stationary and Quasi-Stationary Perturbing Action

The control law of vehicle motion for the point P_i in the case of stationary perturbing action can be formulated as follows. If the value of speed $V(i - 1, d)$ corresponds to the value of $V(i, 0)$, one can assume that the movement of the vehicle complies with law (4) and leave the value of speed unchanged. Otherwise, it is necessary to form a corrective action ΔV and to calculate the difference between the predicted and the actual value of the speed. Corresponded simple arithmetic relation for ΔV_i is follows:

$$\Delta V_i = V(i,\ 0) - V(i-1, d). \tag{7}$$

Taking into account (7), the discrete law of regulation of the vehicle movement can be written through an arithmetic-logical relation. The basic definition of arithmetic-logical relation was given in the paper [18]. The corresponding analytical expression has the following form:

$$V_i = (V(i, 0) \neq V(i-1, d)) \cdot (V(i, 0) - V(i-1, d)) + (V(i, 0) = V(i-1, d)) \cdot V(i-1, d). \tag{8}$$

The disadvantage of the written arithmetic-logical expression (8) is that to calculate the corrective action it is necessary to know the exact value of the velocity $V(i, 0)$, obtained using relations (6). Another way to calculate the corrective action in the presence of stationary perturbing influence is to analyze the values of the corrective actions at the previous times. To perform such calculations, one can use the FSM model [6]. Will simply assume, that the number of previous states of the machine, used to form the corrective action, is k_s, and the corrective action itself is stationary. In this case, the value of velocity error ΔV_i can be calculated as the average value of its value in previous moments of time, and such an estimate from the point of view of the theory of mathematical statistics can be considered the most plausible [20]. For such conditions the corrective action can be calculated as follows:

$$\Delta V_i = \frac{\sum_{j=1}^{k_s} \Delta V_{i-j}}{k_s} \tag{9}$$

Taking into account the obtained analytical expressions (8, 9), as well as generalized discrete relations (6), one can rewrite the law of motion of the vehicle in the case of stationary perturbing action in the form of the following discrete arithmetic-logical relation:

$$V(i,\Delta r)=(i\le k_s)\cdot\left((V_0-\Delta V_i)\cdot\left(1-\frac{i}{N}-\frac{\Delta r}{R}\right)\right)$$
$$+(i>k_s)\cdot\left(\left(V_0-\frac{\sum_{j=1}^{k_s}\Delta V_{i-j}}{k_s}\right)\cdot\left(1-\frac{i}{N}-\frac{\Delta r}{R}\right)\right). \quad (10)$$

where ΔV_i is the current value of the corrective action for state i, defined as the result of speed measurement.

The state diagram of the Moore FSM [15–17], designed to form the law of motion of the vehicle, which is described by relation (10) under the condition $k_s = 2$, is shown in Fig. 2.

In Fig. 2 to indicate the states of the FSM and transitions between them, which correspond to the nodes and edges of the oriented graph [15–17], the following variables have been introduced:

$$S_i = V_i, W_i = \Delta V_i. \quad (11)$$

Taking into account the variables (11), have been introduced to the control law (10), written for the case of stationary perturbation, one can rewrite the arithmetic-logic relation to determine the current state of FSM S_i as follows:

$$S_i=(i\le k_s)\cdot\left((S_0-W_i)\cdot\left(1-\frac{i}{N}\right)\right)+(i>k_s)\cdot\left(\left(S_0-\frac{\sum_{j=1}^{k_s}W_{i-j}}{k_s}\right)\cdot\left(1-\frac{i}{N}\right)\right) \quad (12)$$

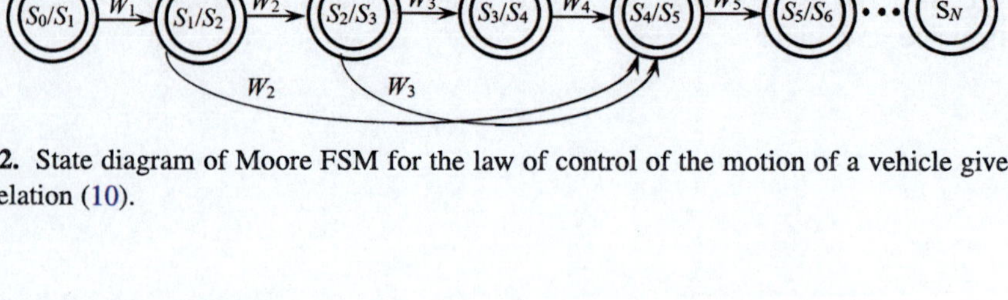

Fig. 2. State diagram of Moore FSM for the law of control of the motion of a vehicle given by the relation (10).

It should be noted that although the law of motion (12) is formulated for the case of a stationary perturbation, the statistical estimation with using relation (9) is based on the maximum likelihood principle [20], which makes it possible to use this relation also for random quasi-stationary perturbations. Typical dependences of the vehicle speed on the distance when using the control law (12) for the cases of stationary and quasi-stationary perturbing action are shown in Fig. 3. From the graphs shown in Fig. 3, it can be seen that in the case of a stationary action, the dependence $V(t)$ fully corresponds to the generalized law of vehicle motion, written as a linear Eq. (4). However, under the condition of a quasi-stationary random impact, the speed corresponds to the predicted one only at discrete reference points P_i, since it is not controlled on the segments $[P_i, P_{i+1}]$. The control error at the points P_i for different values of i in Fig. 3, b, is denoted as ΔV_i, and the predicted speed as V_{i0}. To reduce the average value of this error, one can increase the number of reference points N, in which speed control is implemented. But in some cases, it is namely a large number of discrete reference points leads to additional perturbations [12, 13]. If, in accordance with Fig. 1, the vehicle traffic control algorithm, which is determined by relation (12), is implemented on the on-board computer, then on the central server, according to the general situation on the road, the required number of discrete reference points N and the vehicle predicted speed V_{i0} are determined.

The error ΔV_i, which is calculated by relation (9), is much smaller than in the case of using relation (7). This is due to the fact that in relation (9) is calculated not the current value of the error, but its most plausible error under the condition of quasi-stationary random action.

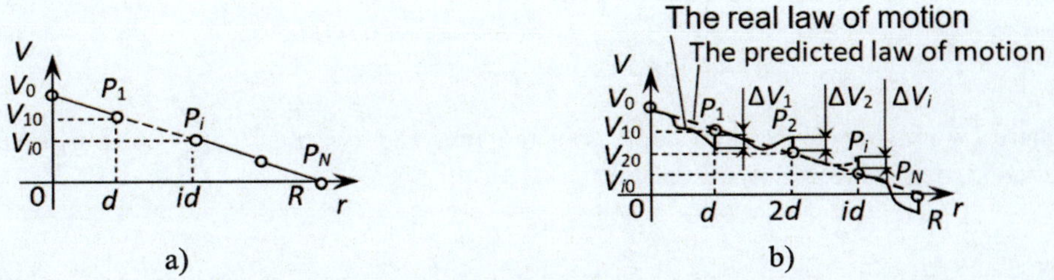

Fig. 3. Dependence of vehicle speed on distance for the discrete law of motion given by relation (12), in the case of stationary (a) and quasi-stationary (b) perturbing action.

It should be noted that an analytical relation similar to ratio (12) can also be updated to the case of random perturbation. Moreover, under this condition, the FSM scheme, which is shown in Fig. 2, also remains unchanged. It is necessary only to correct the analytical relation (9) and the arithmetic-logical relation (12). In the case of random perturbations, it is needed to calculate not the average value of the velocity error ΔV_i, but its most plausible value, which is defined according to the theory of mathematical statistics [20]. The corresponding calculation algorithm will be presented in the next section of this chapter.

5 Discrete Form of the Law of Vehicle Motion in the Case of Random Perturbing Action

Let us rewrite the relation (10) as follows:

$$\begin{aligned} V(i, \Delta r) = (i \le k_s) \cdot \left((V_0 - \Delta V_i) \cdot \left(1 - \tfrac{i}{N} - \tfrac{\Delta r}{R}\right)\right) \\ +(i > k_s) \cdot \left((V_0 - \Delta V_{mp}) \cdot \left(1 - \tfrac{i}{N} - \tfrac{\Delta r}{R}\right)\right) \end{aligned} \tag{13}$$

where ΔV_{mp} is the most plausible value of the error due to the action of a random perturbing factor.

Let us also use the scheme of the FSM, presented in Fig. 2. One can estimate the plausibility of the value ΔV_{mp} according to the chosen distribution law for all previous states of the control system, starting from the current state i. The value of the number of states of the FSM k_{sc}, for which it is necessary to sum the value of ΔV_i in order to find the mathematical expectation of error, will be determined by compliance with the law of distribution of the value ΔV_i to criterion of maximum likelihood. If this criterion is satisfied for the state k_{sc} and it is false for the state $k_{sc} + 1$, the summation should be performed by the number of states k_{sc}. Given that the number of states of the system for which the velocity analysis is performed is small and usually does not exceed 30, according to the theory of mathematical statistics, we can assume that the random variable ΔV_i is corresponded to the well-known Student's distribution law [20]:

$$p(t) = \frac{\Gamma\left(\frac{l+1}{2}\right) \cdot \left(1 + \frac{t^2}{l}\right)^{-\frac{l+1}{2}}}{\sqrt{\pi l} \cdot \Gamma\left(\frac{l}{2}\right)}, -\infty < t < \infty, \tag{14}$$

where l – number of freedom degrees of the analyzed system, $\Gamma(a)$ – Euler gamma-function, defined as follows [20, 21]:

$$\Gamma(a) = \int_0^\infty e^{-t} t^{a-1} dt. \tag{15}$$

If the random variable ΔV is distributed according to Student's law, the interval estimate of its mathematical expectation with the given confidence probability β, is defined as follows [20]:

$$\Delta V^* - t_{l, \frac{1-\beta}{2}} \cdot \frac{s}{\sqrt{k_{sc}}} < \Delta V < \Delta V^* + t_{l, \frac{1-\beta}{2}} \cdot \frac{s}{\sqrt{k_{sc}}}, \; l = k_{sc} - 1 \tag{16}$$

where

$$\Delta V^* = \frac{\sum_{j=0}^{k_s} \Delta V_{i-j}}{k_s}, \; s = \sqrt{\frac{\sum_{j=0}^{k_s} \left(\Delta V_{i-j} - \Delta V^*\right)}{k_s - 1}}. \tag{17}$$

The parameter β, which in mathematical statistics is called the criterion of significance [20], for the control of moving objects can be considered equal to 0.01 [1]. If relations (16, 17) are satisfied under the condition $k_s = j$, but are not fulfilled under the condition $k_s = j + 1$, then the value $k_s = j$ is considered correct and one can assume that $k_{sc} = k_s$. In this case, the corrective action ΔV_i for the current node of the FSM i is calculated as follows:

$$\Delta V_i = \frac{\sum_{j=0}^{k_{so}} \Delta V_{i-j}}{k_{so}} \tag{18}$$

Based on the above considerations and the system of Eqs. (14–18), the arithmetic-logic relation for the required value of the number of states of FSM $k_{sc}(i, j)$ at the current iteration i can be written as follows:

$$\Delta V^*_{i,j} = \frac{\sum_{k=1}^{j} \Delta V_{i-k}}{j}, \ s^*_{i,j} = \sqrt{\frac{j \sum_{k=1}^{j} \left(\Delta V_{i-k} - \frac{\Delta V^*_{i,j}}{j} \right)}{j-1}} \tag{19}$$

$$\begin{gathered} k_{sc}(i,j) = j \cdot \left(\left(|\Delta V_i| \le \Delta V^*_{i,j} - t_{l,\frac{1-\beta}{2}} \cdot s^*_{i,j} \right) | (j < 2) \right) + k_{sc}(i,j-1) \\ \cdot \left(\left(\left(|\Delta V_i| \le \Delta V^*_{i,j} - t_{l,\frac{1-\beta}{2}} \cdot s^*_{i,j} \right) \& \ ((j \ge 2) \, \& (k_{sc}(i,j-1) \ne k_{sc}(i,j-2)) \right) \right) | \\ | ((j \ge 2) \, \& (k_{sc}(i,j-1) = k_{sc}(i,j-2)))), \\ k_{sc}(i,0) = \Delta V_i, j \in 0, \ldots, i. \end{gathered} \tag{20}$$

The block diagram of the algorithm for calculating the value of $k_{sc}(i, j)$, according to which the recurrent arithmetic-logical relation (20) is formed, is shown in Fig. 4.

According to the relations (19, 20), in the case of random perturbing action one can write the following analytical expression for the state of the FSM, shown in Fig. 2:

$$S_i = (i \le k_s) \cdot \left((S_0 - W_i) \left(1 - \frac{i}{N} \right) \right) + (i \le k_s) \cdot \left((S_0 - Q_i) \left(1 - \frac{i}{N} \right) \right) \tag{21}$$

$$Q_i = \frac{\sum_{j=1}^{k_{sc}(i)} W_{i-j}}{k_{sc}(i)} \tag{22}$$

where the current value of k_{sc} is calculated from arithmetic-logic relation (20) with taking into account formulas (19), Q_i – additional variable, defined the corrective action for relation (21).

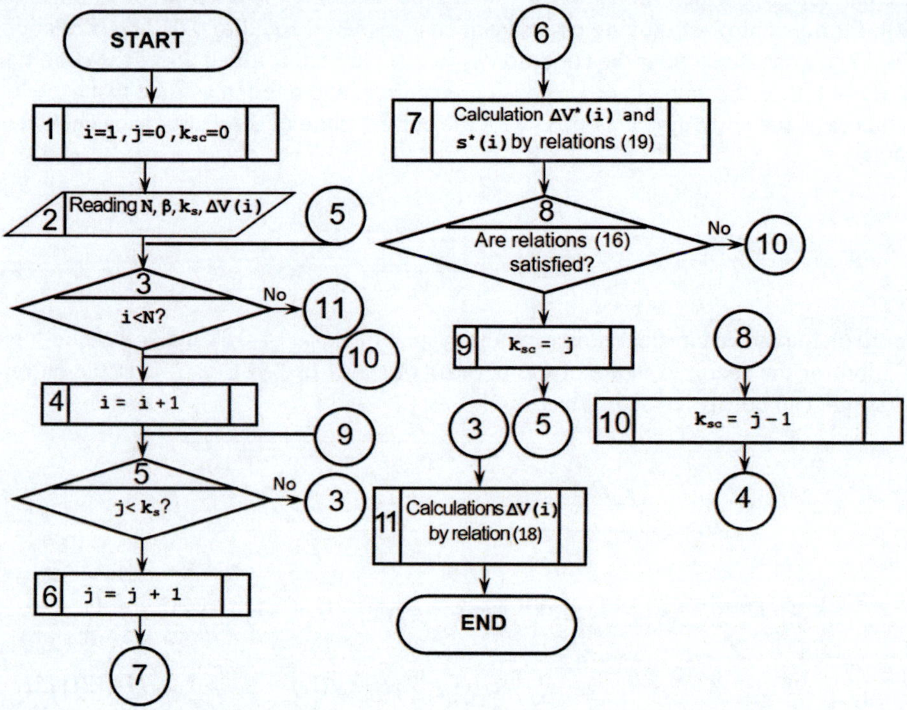

Fig. 4. Block diagram of the algorithm for calculating the corrective action ΔV_i in the case of action the random perturbing factor.

6 Particularities of Realizing of Proposed Discrete Control Algorithm in Considered Network Smart Drive Conception

Firstly, it should be pointed out that relations (12) and (21) for discrete control law for both considered cases of constant and random perturbing actions, are similar and correspond to the scheme of FSM, presented at Fig. 2. Therefore, only the method of calculation of corrected action in relations (12) and (21) is differ.

For the simpler case of quasi-stationary perturbing action, the value of k_s is constant and the proposed algorithm can be fully implemented on the on-board computer. Furthermore, corresponding to relation (12), if perturbing action is absent no control action at this time is necessary.

The considered case of random perturbing action is more sophisticated, but the advantage of the proposed digital control algorithm is that it is simple, similar for both considered cases and implemented on on-board computers. The difference is, that in this case the FSM scheme is dynamic and its operation depends on the random factor. For such a scheme the value of k_s is changeable and defined by sophisticated Eq. (20), which is connected to (22) for defining the control action. Therefore, in case of random perturbing action calculation of corrective action Q_i with taking into account the amount of necessary previous states k_s, defined by arithmetic-logic relation (19), is the task of the central server in correspondence to network model, presented at Fig. 1. But the correct data of previous FSM states are located at local vehicles. Therefore, the scheme of date exchange between central server and on-board computer is as follows.

1. Server sends to the on-board computer the optimal number of basic control points N corresponding to the general road situation. The value of k_s is assumed $k_s = 0$.
2. For the first 2 sages, an on-board computer sends to the server the values of correction actions ΔV.
3. By the values of ΔV_1 and ΔV_2 defined the value ks using relation (20) and sent to the on-board computer.
4. On-board computer keeps in memory the values: $\Delta V_{i-ks}, \Delta V_{i-ks+1}, \ldots, \Delta V_{i-1}$, where i is the number of current state of FSM. At the corresponded time it sent the set of these values to the central server in required format.
5. Corresponding to the road situation the values of k_s and N on on-board computer can be changed by the command from the central server.
6. If $k_s > 2$, the local server calculated the corrective action Q_i using relation (22) and the new value of k_s by relation (19, 20).

Described model of data exchange and separation of calculation tasks is presented in Fig. 5.

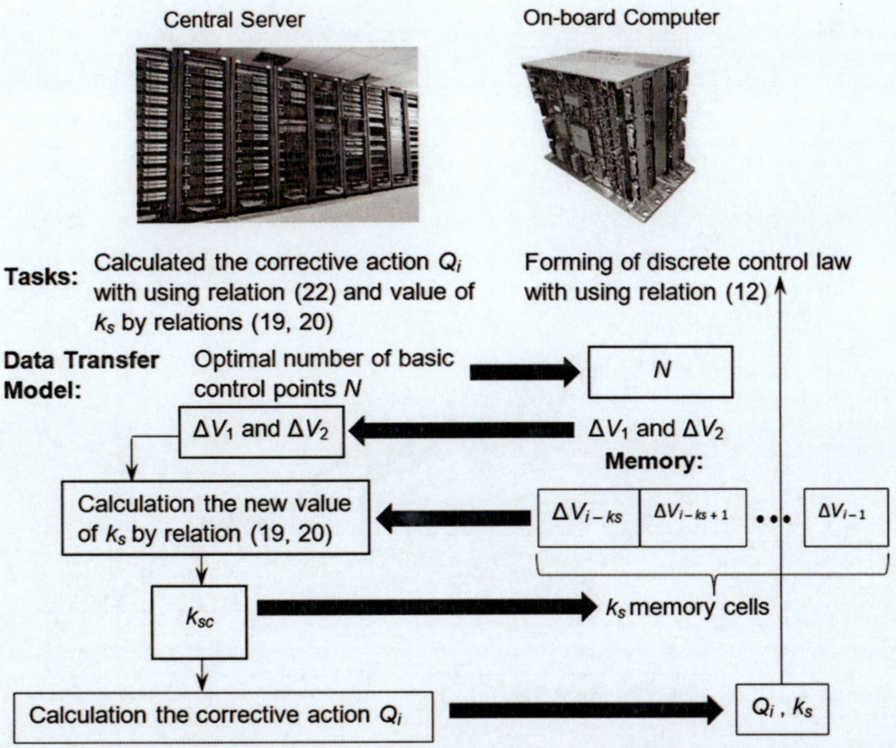

Fig. 5. Model of data exchange and separation of calculation tasks between central server and on-board computer in local network for implementation discrete algorithm of vehicle moving control.

7 The Results of Simulation of Discrete Control System of Vehicle Moving and Its Analyze

The obtained results of calculations for quasi-stationary perturbing action, obtained using relation (12), are shown in Fig. 6, a, and the results of modeling changing of vehicle speed for random perturbing action, obtained with using relation (21) according to the algorithm, the block diagram of which is shown in Fig. 6 – respectively, in Fig. 6, b.

It should be noted that in the case of quasi-stationary perturbing action, the deviation of the velocity from the specified linear law (4) is much smaller, and for a given 10

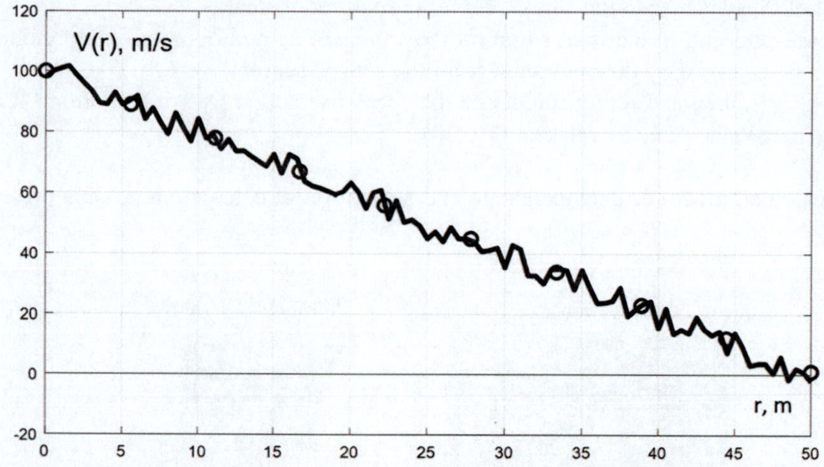

a)

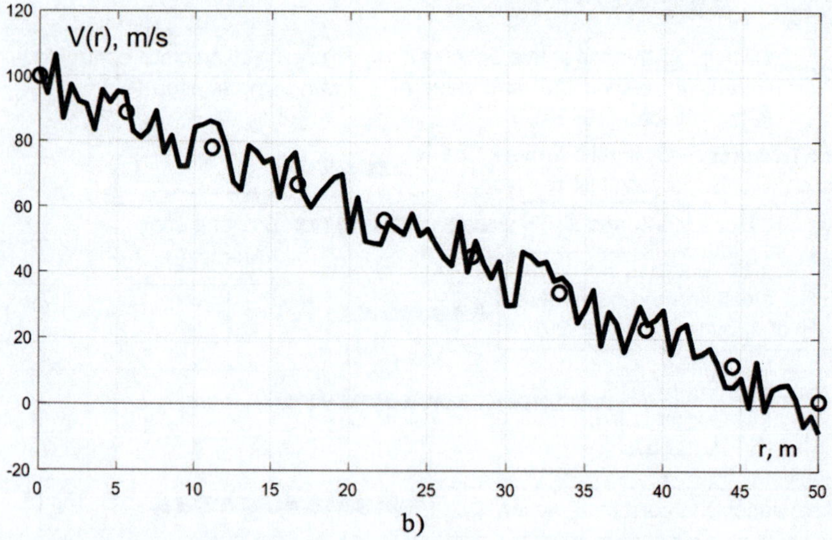

b)

Fig. 6. The results of simulation the process of vehicle moving control in the case of quasi-stationary (a) and random (b) perturbing action.

reference points, the predicted and actual velocity values completely coincide. The error associated with the deviation of the velocity from the linear distribution law (4) does not exceed 2–3%.

Therefore, one can conclude that the formation of statistics on the change in velocity of a moving object in the case of quasi-stationary perturbing action using relation (9) and determining the control law of these statistics is much easier than generating such statistics in the case of presence of random perturbing action with using Student's distribution law and ratios (19–22).

According to the graphical dependence shown in Fig. 6, b, the error of velocity deviation in the presence of random perturbing action is greater and it is approximately 5–8%. Moreover, as can be seen from the graphical dependence, which is shown in Fig. 6, b, even at some reference points the predicted and actual values of the speed may not coincide. The well-known standard mathematical function `random` was used to perform simulation of random action using programming software MATLAB [13].

An important result of the provided research is that the scheme of the FSM, as shown in Fig. 2, is universal and not changed due to the nature of the perturbing function and the selected statistics. One can understand also, that independently of the control law, the control error is dependent on the number of control points N and the number of previous states of the FSM k_s, which have been analyzed. The analysis of the value of such errors may be the subject of further research. In any case, a simple mathematical model has been developed, and it is based on the methods of discrete mathematics [22], on the theory of FSM [15–18] and on the methods of mathematical statistics [20]. Proposed model is significantly simpler, then the sophisticated models, based on the theory of optimal control [5–10]. Therefore, the obtained results, described in this chapter, can be applied not only for the transport problems, but also for modeling other control systems of complex technical objects, as well as for forming and analyzing algorithms of its operation. Discrete control systems are much simpler to implement with different programming components in modern digital electronics, than the similar analog systems [15]. For implementation of proposed digital control method M2M network model of data exchanging, like presented at Fig. 5, can be used.

8 Conclusion

This chapter describes a method of controlling the velocity of a moving vehicle, based on the algorithms of discrete mathematics and FSM theory, which have been described and analyzed. The proposed algorithm is much simpler than the known velocity control methods, based on analyzing the problem of optimal control (1–3) and solving differential equations. The proposed discrete model is presented in the form of a diagram of a Moore FSM. The particularities of implementation of the proposed model in M2M network with local on-board computer and central server, as well as necessary data exchange traffic model also have been presented and analyzed. Test calculations have shown that the error in controlling the velocity of a moving object depends on the chosen control law and the statistical estimation methods, which is used. The case of quasi-stationary and random perturbing action has been considered. In the case of stationary perturbing action this error is absent, in the case of quasi-stationary perturbing action it is approximately

2–5%, and in the case of random action – 5–8%. The theoretical results obtained in this chapter also can be applied for modeling and design of other digital control systems for complex industrial technical objects.

References

1. Assmayr, H., Geyer, D., Schwab, G.: Autonomes Fahren ist der Trend der Zukunft: Synergien zwischen Automotive und Offroad, Agrartechnik, Survey by Company AVL Software and Functions GmbH (2022). https://core.ac.uk/download/pdf/236374514.pdf
2. Luntovskyy, A., Guetter, D.: Self-Driving Cars. In this book
3. Irvine, J., Harle, D.: Data Communication and Networks: An Engineering Approach, p. 300. John Wiley & Sons, LTD., Hoboken (2001)
4. Tanenbaum, A., Wetherall, D.: Computer Networks, 5th edn, p. 960. Pearson Prentice Hall, Upper Saddle River (2010)
5. Kuo, B.C., Golnaraghi, F.: Automatic Control Systems. 8th edn, p. 624. John Wiley and Sons, Hoboken (2003). https://easyengineering.net/automatic-control-systems-by-kuo/
6. Popov, E.P.: The Dynamics of Automatic Control Systems, p. 776. Pergamon (2014) https://mirtitles.org/2022/01/26/the-dynamics-of-automatic-control-systems-popov/
7. Mehta, B.R., Reddy, Y.J.: Industrial Process Automation Systems. Design and Implementation, p. 668. Elsevier, Amsterdam (2015). https://www.sciencedirect.com/book/9780128009390/industrial-process-automation-systems
8. Nagrath, I.J., Gopal, M.: Control Systems Engineering, 5th edn, p. 900. Anshan Publishers, Budapest (2008). https://easyengineering.net/control-systems-engineering-by-nagrath-nw/
9. Ogata, K.: Modern Control Engineering, 5th edn, p. 912. Pearson, Boston (2009). https://www.academia.edu/43692259/Modern_Control_Engineering_Fifth_Edition
10. Ahmed, N.U., Wang, S.: Optimal Control of Dynamic Systems Driven by Vector Measures: Theory and Applications, p. 335. Springer, Heidelberg (2021). https://doi.org/10.1007/978-3-030-82139-5
11. Miroshnik, V., Basu, D.: Dynamic Systems Modelling and Optimal Control: Applications in Management Science, p. 211. Palgrave Macmillan, London (2015)
12. Epperson J.F.: An Introduction to Numerical Methods and Analysis, Revised Edition, p. 590. Wiley-Interscience, Hoboken (2007)
13. Mathews, J.H., Fink, K.D.: Numerical Methods. Using MATLAB. Amazon, p. 696 (1998)
14. Luntovskyy, A., Dietbert, G.: Moderne Rechnernetze: Protokolle, Standards und Apps in kombinierten drahtgebundenen, mobilen und drahtlosen Netzwerken, 481 Seiten, 263 Abb., Springer Nature, Heidelberg (2020). in German, ISBN: 9783658256166. https://www.springer.com/gp/book/9783658256166
15. Lathi, B.P.: Modern Digital and Analog Communication Systems, p. 720. Holt. Rinehart and Whinston Inc., New York (1989)
16. Wang, J.: Formal Methods in Computer Science, p. 294. CRC Press, Boca Raton (2019). https://www.researchgate.net/publication/311302397_Finite_State_Stack_Transducer_for_Functionality_in_a_Multi-Parameter_Environment
17. Wright, D.R.: Finite State Machine. Retrieved from NC State University (2005). https://www4.ncsu.edu/~drwrigh3/docs/courses/csc216/fsm-notes.pdf
18. Melnyk, I., Luntovskyy, A.: Estimation of energy efficiency and quality of service in cloud realizations of parallel computing algorithms for IBN. In: Klymash, M., Beshley, M., Luntovskyy, A. (eds.) Future Intent-Based Networking. LNEE, vol. 831, pp. 339–379. Springer, Cham (2022). https://doi.org/10.1007/978-3-030-92435-5_20

19. Carter, J., Chiclana, F., Khuman, A.S., Chen T.: Fuzzy Logic: Recent Applications and Developments, p. 278. Springer, Heidelberg (2021). https://doi.org/10.1007/978-3-030-66474-9
20. Gubner, J.A.: Probability and Random Processes for Electrical and Computer Engineers. Cambridge University Press, Cambridge (2006). http://www.cambridge.org/gb/academic/subjects/engineering/communications-and-signal-processing/probability-and-random-processes-electrical-and-computer-engineers
21. Abramovich, M., Stegun, I. (eds.): Handbook on Mathematical Functions with Formulas, Graphs and Mathematical Tables. National Bureau of Standards, Applied Mathmatic Series, Washington, vol. 55, p. 1046 (1964)
22. Anderson, J.A.: Discrete Mathematics with Combinatorics, 2nd Edition. University of South Carolina-Spartanburg (2004)

A Modified Federated Singular Value Decomposition Method for Big Data and ML Optimization in IIoT Systems

Olena Hordiichuk-Bublivska(✉), Halyna Beshley, Marian Kyryk, Yuliia Pyrih, Oksana Urikova, and Mykola Beshley

Lviv Polytechnic National University, Lviv 79013, Ukraine
{Olena.v.hordiichuk-bublivska,halyna.v.beshlei,marian.i.kyryk, yuliia.v.klymash,oksana.m.urikova,mykola.i.beshlei}@lpnu.ua

Abstract. Continuous improvement of digital technologies allows for the introduction of new and improvement of existing areas of human life. Medicine, education, industry and many other areas have now, due to the pandemic situation, changed the usual way of functioning. The involvement of modern methods of data processing and management of large-scale systems allows for performing work tasks with less human intervention. The Industrial Internet of Things (IIoT) has become an important stage in the development of industrial systems. The organization of IIoT allows us to flexibly manage the production system, rationally use resources and provide users with a wide range of services. It is necessary to involve a large number of intelligent devices that process data from the environment and transmit it for further processing to control devices. For efficiently and quickly analyzing the information coming from different sensors we should use the methods of working with Big Data. Optimization of information that is not important for further analysis and decision-making can significantly reduce computational time while maintaining high accuracy. Machine learning methods simplify the process of Big Data analysis and create an industrial system based on the information obtained. Because IIoT systems are distributed and consist of many components, federated machine learning is used. The peculiarities of the use of distributed machine learning for Big Data analysis are investigated. Preliminary optimization of data arrays using the Singular Value Decomposition (SVD) method was proposed. The research results proved the proposed SVD method's effectiveness for Big Data pre-optimization in the IIoT systems. Also in the paper, we offered the modified method of SVD, which is adapted to work in the distributed IIoT systems. The research results showed high accuracy and reasonableness of calculations. To improve the efficiency of sparse data analysis in IIoT, it is also suggested to use the Funk-SVD algorithm.

Keywords: Industrial Internet of Things · Big data · Distributed systems · Federated learning · Data optimization · SVD · Funk-SVD

M. Klymash et al. (Eds.): TCSET 2022, LNEE 965, pp. 246–267, 2023.
https://doi.org/10.1007/978-3-031-24963-1_14

1 Introduction

Industrial systems perform many functions that are necessary for a comfortable life. The manufacture of products and the provision of services on a large scale requires a clear organization of the system and the constant processing of large data amounts. Industrial systems have come a long way. From the usual use of devices that perform limited functions and require constant staff supervision, to highly intelligent systems capable of self-service and learning. Centralized networks have been replaced by decentralized ones, which consist of many components. Instead of constant human control, industrial systems perform a lot of tasks independently. Big Data processing is an essential feature of modern industrial systems. As many devices for different purposes are used in production, their condition should be constantly monitored. This approach allows for determining the inefficiency of certain components of the system, and possible safety risks for equipment and personnel. The term Big Data describes the information that is collected from various sources, processed, and presented in a user-friendly form. Big Data in industrial systems is constantly coming from various sensors. Therefore, special information analysis algorithms are used to identify and discard redundant, as well as find the necessary information. Speed of data processing is extremely important because it is necessary to make decisions about system management and troubleshooting very quickly.

The IIoT is a collection of interconnected elements, intelligent devices, data processing algorithms, and control devices. IIoT systems have many advantages over conventional industrial systems. The ability to use many end devices that collect and process data allows responding quickly to various events in the system. Algorithms for data processing and communication between nodes facilitate the rapid exchange of experience and learning of the industrial system. The IIoT is used in many industries: energy, manufacturing, transportation, smart city, etc.

During the pandemic, it is advisable to delegate the transfer of some of the functions previously performed by large numbers of staff to software and hardware. The use of IIoT involves the modernization of the production process and the involvement of a large number of end-users. Production and sales are also integrated. Analysis of the product shows how much it is in demand. In smart manufacture can be decided on the modification of the product or withdrawal from circulation. Continuous monitoring of the efficiency of the enterprise allows determining the suboptimal use of resources and redistributing them automatically. Thus, machine learning and artificial intelligence are important components of the IIoT. Identification of critical situations in the enterprise can be used to avoid their recurrence. Data exchange between different end devices allows learning faster. Because the location of the IoT devices is far apart, the data collection process can cause some difficulties:

- load on communication channels when transmitting large amounts of information constantly from the working nodes to the central device;
- risks of reliability and confidentiality of user data. Special encryption methods should be used to protect the data.

These features have led to the emergence of federated machine learning, in which all nodes are trained on local data and send to the central node only the results, without disseminating personal information [1]. The efficiency of data processing in IIoT systems with the use of federated learning is investigated in the work. It is also proposed to use the Singular Value Decomposition algorithm for preliminary optimization of Big Data processing and disposal of redundant data.

The remainder of the paper is organized as follows. In Sect. 2, we consider the most relevant research on this topic and identify the still existing problem. In Sect. 3, we review the features of smart manufacturing systems. Section 4 discusses the problem of Big Data processing in the IIoT and methods for optimizing it to improve production efficiency. Finally, we conduct an experimental study of the effectiveness of Big Data optimization using the SVD method in Sect. 5 and summarize it in the Conclusion section.

2 Features of Smart Manufacture Systems

Now the world's fourth industrial revolution is happening. Vertical and horizontal integration of industrial systems comes to automate the data processing, creation, and distribution of products. Smart industries can do their job faster and more efficiently, while consuming fewer resources. The reason for this productivity increase is the use of automated control systems and the improved performance of production tasks. Instead of bulky devices that perform a limited range of tasks, smart industries use more autonomous but less powerful ones. The involvement of a large number of sensors that monitor the state of the industrial system, allows for constant monitoring of possible emergencies, to determine the effectiveness of individual components and the system as a whole. The Industry 4.0 program was introduced in Germany to identify the importance of developing smart industrial systems and to explore new ways to improve them. Recognition at the international level of the prospects for improving industrial systems and the impact of this process on humanity indicates a new stage in the development of the manufacturing industry. Based on the development of industry in the country, we can testify to its competitiveness and success in the international arena. Therefore, new standards of digitalisation of production, methods of optimization of Big Data, analysis, and management of industrial systems are constantly being introduced.

The concept of the IIoT involves the use of the latest industry management tools. Distributed sensors constantly collect data on the status of individual system elements. With the help of analysis, algorithms are important information for decision-making on system management. Flexible management methods allow us to respond quickly to events in the workplace and find inventive and effective methods to address them. The IIoT is fully compliant with Industry 4.0's principle of digitizing all stages of production. Important stages are both the process of direct production and sales, i.e. delivery to end-users. If in the past the logistics of goods were done exclusively by people, now it can be integrated into the global industrial system. If the users of certain services have any comments about the manufactured products, or it is not in demand, a smart company can automatically debug production processes to optimize performance. The changes have affected many areas of our lives. Not only traditional production requires modernization.

The energy, health, transportation, and management of entire cities are examples of the IIoT. Due to the pandemic situation, many companies were forced to reduce the number of employees and limit staff gatherings. Therefore, it is very important to ensure the functioning of the industry by providing remote control of the system, and automation of many of the production functions. Given the fact that industrial systems are rapidly evolving and changing, they need to be constantly improved, to introduce new methods of data collection, processing, and storage. Therefore, the management of smart systems must take into account the dynamic changes in software and hardware [2].

Cyber-physical systems (CPS) are used for smart management of industrial systems, which can perform the functions of environmental data collection, processing, and analysis according to certain criteria. Based on the processed information, object can interact with other objects in the system to perform more global tasks. Equipped with state-of-the-art data collection devices, CPS elements can detect and correct the deviation from the established norm in the behavior of local devices with high accuracy. This approach allows us to more carefully determine the features of the industrial system and ensure faster and more efficient execution of tasks. Smart manufacturing opens up prospects for the implementation of functions that were previously unavailable. Flexible management of industrial systems increases the level of quality of services, so it provides an opportunity to upgrade existing industries and create new ones.

Smart manufacture. For transforming classical industrial systems into intelligent production, special algorithms and methods of machine learning should be used. Given the fact that modern large-scale industrial systems consist of a significant number of devices, we can talk about distributed network architecture. Accordingly, data processing algorithms must also be distributed. Adaptation of machine learning methods and deep learning to the features of industrial systems allows achieving better results in productivity. Decentralization of industrial network management processes allows important decisions to be made directly on end devices, which reduces the time of data processing and tasks. Modern intelligent production systems usually use multiple dedicated nodes that process data from the entire network and train the general algorithm of machine learning. Information from different end devices is sent for analysis. However, the functionality of end devices will also allow them to interact independently with users, other components, and the product being manufactured.

The use of smart production can reduce the number of staff who service the process of its work. However, an important feature of the digitalization of enterprises is not the complete abandonment of human participation in management processes, but its replacement where the algorithms of data analysis and artificial intelligence are better able to cope. Operations that require rapid processing of ultra-large data, search for specific information, and rapid response to routine standard tasks are examples of those tasks that people perform slower and less efficiently than machines. Instead, the definition of algorithms for industrial system management, development strategies, and system programming is performed by personnel who also monitor the correctness of production. In determining the priorities in the industrial system, the main advantages and disadvantages of using certain methods are complex decisions that are performed jointly by qualified specialists and data processing algorithms, which allows for the most efficient calculations.

Cyberphysical systems in IIoT. Cyber-physical systems are a combination of the latest effective technologies that should combine hardware and software resources to create a fundamentally new information network. The integration of cyber-physical systems with other technologies, such as the Internet of Things, is expanding opportunities for the development of smart manufacturing. The CPS concept can be seen in personal devices such as a smartphone. While some time ago the phone was just a means of communication, now users can access a huge range of services. Rethinking the role of IIoT devices also involves providing new features to devices that previously performed only certain tasks. Cyber-physical systems blur the distinction between the digital and real worlds. The interaction of users through highly intelligent devices and smart industrial systems provides new opportunities to control the work of production processes [3].

Many industries are now undergoing irreversible change. Modernization of hardware and software requires changes in the production process. The need for sound system management and resource allocation has led to the development of information technology that provides distributed data collection, processing, and storage. Integration of several technologies in one production process allows users to provide a much greater variety of services. Instead of using separate software and hardware to perform specific functions, intelligent production based on cyber-physical systems involves a multifunctional structure. The evolution of the smart industry began with the use of embedded systems (ES) that specialized in performing a narrow range of tasks. Then came the need to integrate embedded systems with network infrastructure. It became clear that it is impossible to process all tasks on one physical device, even if it has powerful computing resources. Over time, the development of ES led to the emergence of cyber-physical systems, which in turn became the basis for the Internet of Things (IoT).

Thanks to the concept of IoT, it has become possible to use things and devices in a new way. Expanding the functionality of data processing devices and their ability to perform certain tasks independently have radically changed the existing info-communication systems. Because the Internet of Things has proven its effectiveness in solving local problems, the IIoT is used for more global problems. Various types of products can be presented as smart, multifunctional, and flexible in the management of the system. Each device can perform its tasks and make local decisions while interacting with other nodes. This approach facilitates the exchange of information between IIoT devices and fast learning. Machine learning models in such systems can be trained faster by combining the experience of different devices.

Thus, the evolution of industrial systems has helped change many areas of human life. Thanks to technical progress, users can now receive ready-made goods and services without worrying about the features of the company that provides them. The IIoT is used in smart industries. Automatic data calculating speeds up the process of making critical decisions. Intelligent algorithms conduct comprehensive monitoring of the system state and identify those production units that need improvement. Changing the parameters of manufacture is now analyzed not manually, but using the methods of machine learning and artificial intelligence. The evolution of digital technologies in industrial systems is shown in Fig. 1.

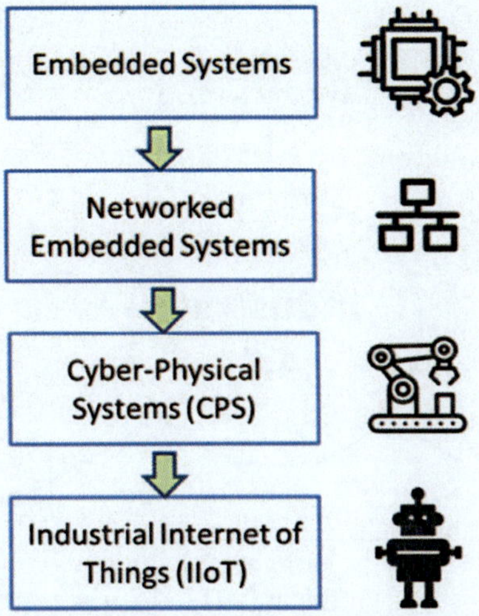

Fig. 1. The evolution of digital technologies in industrial systems.

The use of software to manage industrial systems is much debated. Because many production situations can be handled out of place, centralized system management has not been considered before. Given the development of cyber-physical systems and the Internet of Things, more and more functions of managing complex processes in enterprises are performed by software algorithms. At present, it cannot be said that software plays a key role in the decision-making process in IIoT systems. Instead, the integration of management with the help of special algorithms has become an important part of smart production and allows to correct and eliminate emergencies of various scales. It is important to attract highly qualified specialists for the design and operation of industrial systems that can manage both software and hardware with maximum efficiency [4–6].

The use of the IIoT in the modern world plays a huge role. Many of the functions previously performed by specialists in the workplace have now been forced to be performed online, given the pandemic situation. This approach requires the use of additional software, secure information transmission channels, and fast data processing. The advantages of IIoT are the use of a large number of sensors, controllers, etc. to measure data that replace a significant amount of staff. It is also more efficient to organize the work of staff. Environmental monitoring functions allow you to clean the air whenever the level of pollution exceeds the allowable value. Controlling the movement of people can ensure the preservation of social distance, and optimal attendance of employees in recreation rooms, cafes, etc. Virtual reality features allow managing remotely without physically visiting industrial sites. Also, communication between people can be carried out, avoiding crowds, thus reducing the risk of getting sick. Thus, the IIoT today allows to more efficiently organize the work of production systems, taking care of both their productivity and the health of workers. Industry 4.0 contains many components. Artificial intelligence, the Internet of Things, IIoT, and cloud technologies are promising areas now (Fig. 2).

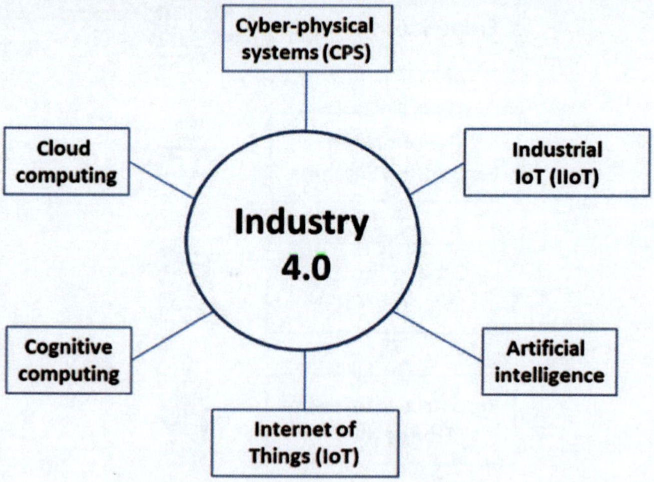

Fig. 2. Industry 4.0 components.

The precondition for the creation of the IIoT was, of course, the Internet of Things. A feature of the Internet of Things is the use of intelligent devices capable of performing certain functions that communicate with each other and control nodes via the Internet. The Internet of Things simplifies a person's everyday life and performs many primitive functions. Collected data used for automating the operation in the IoT system, and are stored on cloud resources [7–9]. The success of the Internet of Things has led to the emergence of its subsectors, able to automate the work of not only individual homes but also large enterprises. The IIoT also contains many end or edge devices. These are sensors, actuators, and PLCs (programmable logic controllers). These devices collect data from different parts of the enterprise on the parameters of operation, performance, and more. Communication via the Internet allows for coordinating the work of many components of the system.

The IIoT, unlike conventional IoT, improves the interoperability between hardware and software. To effectively manage production resources, it is necessary to ensure a high level of performance of individual devices, and use software algorithms that can coordinate them. Continuous data collection, processing, and transmission for storage on cloud resources require the involvement of technologies at all levels of the Open Systems Interconnection (OSI) model.

The main components of the IIoT are:

- smart data collection devices;
- network infrastructure;
- data processing;
- applications.

The use of IIoT allows to organize the work and management of smart production much more efficiently than in conventional centralized systems. The ability to gather and analyze information about abnormal situations not only allows us to quickly eliminate them but also to anticipate and make appropriate decisions in advance.

IIoT systems are flexible to use and scalable. As the need to use resources for production is constantly changing, it is necessary to attract new ones and release surplus ones. Previously, industrial systems consisted of a certain amount of hardware, which was determined by taking into account the projected load. If the volume of production increased sharply, it took time to re-equip the system, and add new devices. In the case of reducing the load on the industrial system, part of the equipment was idle, which is economically unprofitable. In IIoT systems, it is possible to attract cloud resources, which are allocated automatically based on the analysis of the state of the enterprise and the load [10].

The use of IIoT involves a new approach to working with Big Data. Analysis of information from different end devices is used to determine the parameters of the system, which in one way or another affect its performance. Changes in the performance of individual components of the system may indicate problems of local or global nature. The speed of solving such problems is important to ensure the continued functioning of production.

However, the use of the Internet of Things also has certain negative aspects:

- security threats. Because many functions are performed by end devices, there is an increased risk that attackers may disguise themselves as system users and send incorrect data. Processing of user information by central nodes also requires ensuring the confidentiality of data when transmitting through communication channels and direct processing. IIoT involves many users who use their data when interacting with the system. Therefore, ensuring the reliability, and confidentiality of data is a priority;
- compatibility. Because modern industrial systems often use different types of equipment, networking protocols, services, etc., their compatibility is necessary for the proper operation of production.

It should be noted that the advantages of the IIoT far outweigh the disadvantages. Thanks to the use of special communication protocols, data protection methods, and special equipment, it is possible to ensure the reliable and efficient operation of intelligent production. The IIoT system architecture is shown in Fig. 3.

IIoT is an important stage in the development of digital production. Continuous improvement of its components allows to improve the efficiency of industrial systems and eliminate risks and shortcomings. The structure of the IIoT is optimal for the concept of Industry 4.0 and allows the use of various technologies and methods of production organization. The use of machine learning methods, artificial intelligence, and in-depth analysis improves the process of data management and processing [11].

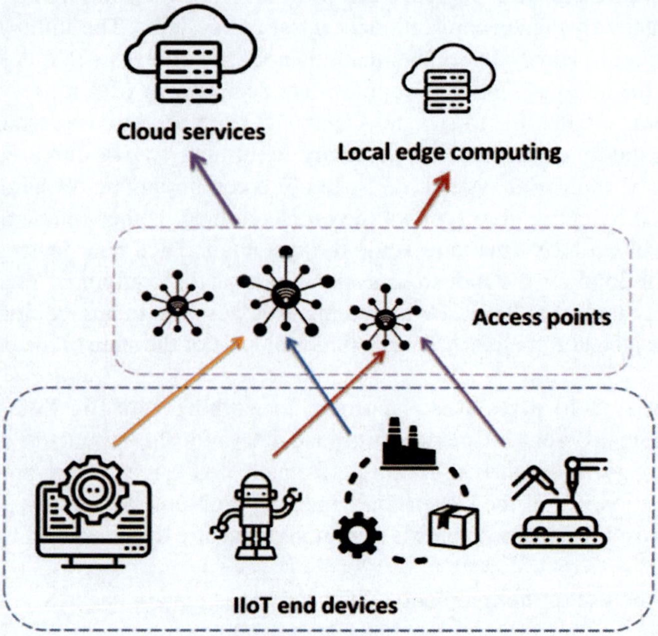

Fig. 3. IIoT system architecture.

3 Big Data and Machine Learning Processing in IIoT

Machine learning in industrial systems. Because conventional computing algorithms are not enough to process large amounts of data from different devices, machine learning techniques help to analyze and optimize information. Thanks to machine learning, IIoT devices can be trained and self-taught, so they can solve problems of different levels of complexity independently and without human assistance [12–15].

Machine learning (ML) is a sub-branch of artificial intelligence and uses statistical data analysis techniques to make decisions that have not been pre-programmed by humans. Machine learning algorithms process the input data and based on them decide on further actions. The effectiveness of ML to solve the problems of classification, clustering, and regression contributes to its implementation in various fields. Industrial intelligent systems also have many challenges that require sophisticated data processing technologies. Determining important values in huge amounts of information, and making decisions based on previous events without the intervention of staff significantly reduces the response time to critical conditions of the industrial system. However, the use of ML in IIoT also has some peculiarities. In particular, the distributed architecture of production systems and the huge number of end devices cause problems in the operation of machine learning algorithms. Also, the constant transfer of Big Data puts a strain on communication channels and computing devices [16].

To solve the above-mentioned problems, a federated machine learning (FL) was proposed. The architecture of industrial systems using federated learning is presented in Fig. 4.

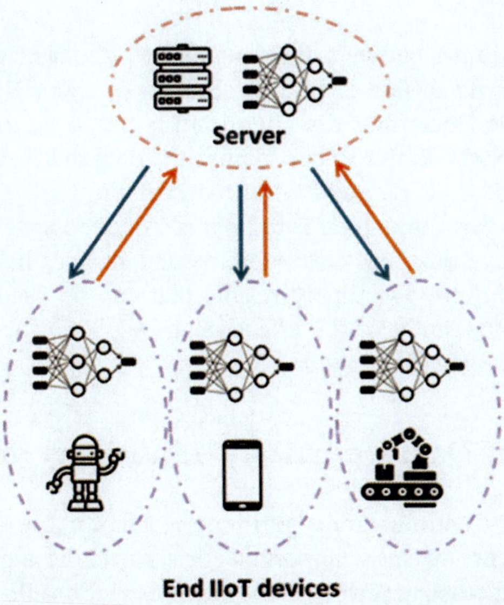

Fig. 4. The architecture of industrial systems using federated learning.

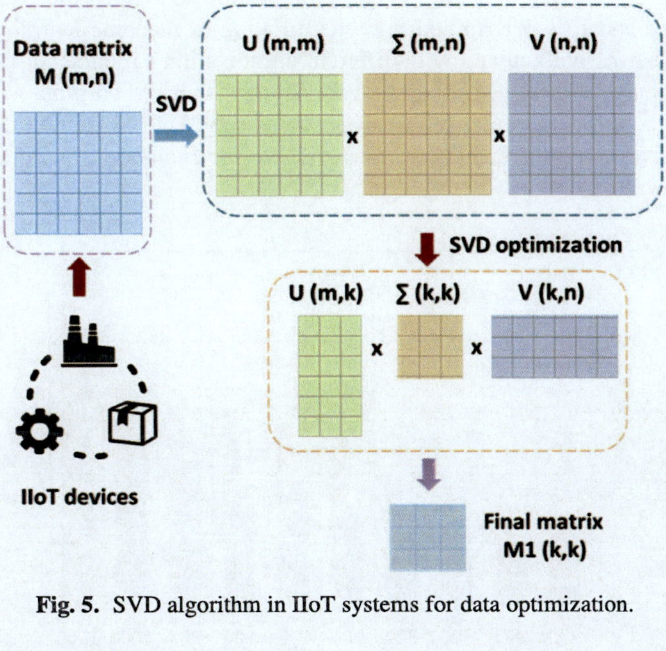

Fig. 5. SVD algorithm in IIoT systems for data optimization.

The peculiarity of FL is that all data is processed directly on end devices using local ML models. Then only the learning results are transmitted to the central device to update the global model, which is then sent to all devices. Thus, it is possible to exchange learning outcomes between different nodes without sending private user data. It also allows for the transmission of smaller amounts of data [17].

The Singular Value Decomposition algorithm is proposed to be used to optimize Big Data in the IIoT. SVD allows to decompose the data matrix into products of three submatrices and reduce their range [18–21]. After multiplying these submatrices, we obtain a matrix of smaller size, which with high accuracy repeats the original (Fig. 5).

Identifying important data and discarding redundant ones helps speed up machine learning algorithms and provide timely results that can be used to manage the IIoT system. An experimental study of the effectiveness of SVD for distributed industrial systems is presented in the next section.

4 Research of Big Data Optimization Efficiency in IIoT Systems

Thanks to the use of data optimization algorithms, it can be chosen from a variety of data sent by different devices, the most important for a particular algorithm. The Singular Value Decomposition method described in the previous section allows for reducing arrays of information while maintaining basic properties. The paper simulates the operation of the machine and federated learning algorithms using the Python programming language [22–26]. The advantages of using the SVD algorithm to reduce the data dimension are also tested and it is determined how the duration of machine learning algorithms processing changes without and with different optimization parameters.

Figure 6 shows the result of the ML and FL algorithms when processing the same amount of data and performing the task of data sorting. Due to the fact that federated learning involves the implementation of training on the end nodes, its calculation takes more time.

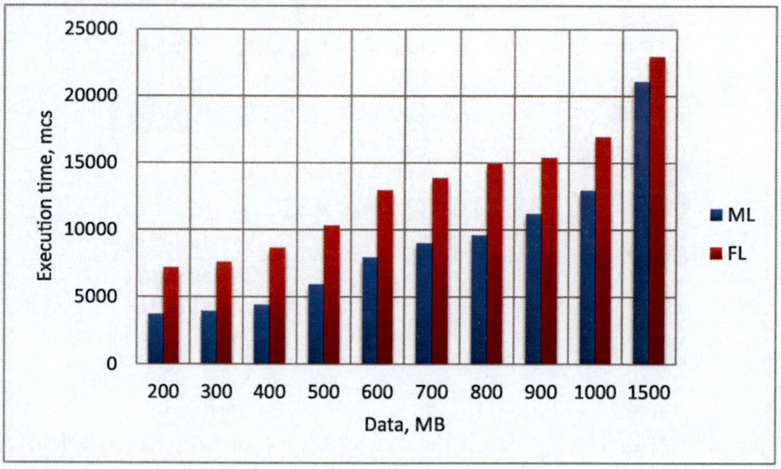

Fig. 6. Dependencies of execution time for machine and federated learning algorithms on the amount of computational data.

We will perform preliminary data optimization using the Singular Value Decomposition algorithm. By changing the number l of rejected singular values in the decomposed matrix, it can reduce the amount of data that is then processed by machine learning algorithms. Figure 7 shows the implementation of the federated learning algorithm using optimization.

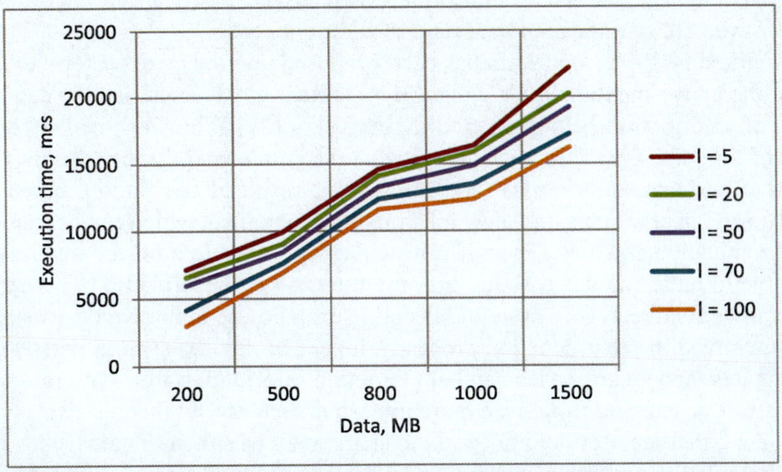

Fig. 7. Dependency of execution time for federated learning algorithms on the number of rejected singular values.

Research on the accuracy of the calculations after the proposed optimization method is shown in Fig. 8.

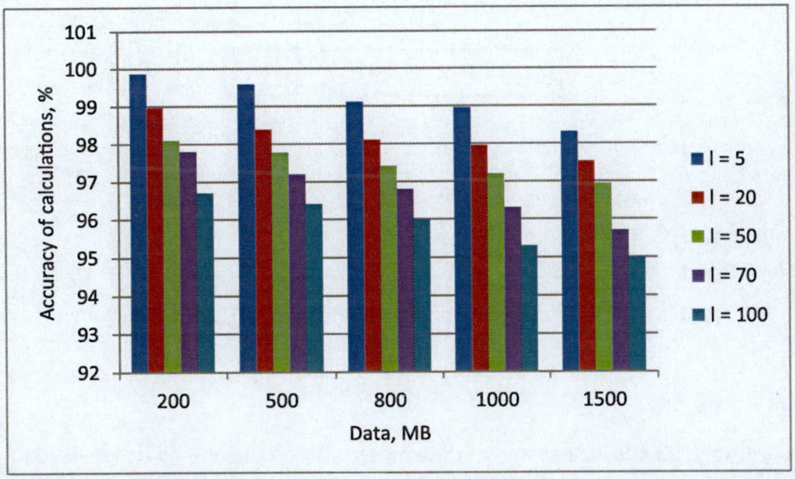

Fig. 8. Dependencies of accuracy of the calculations for federated learning algorithms on the data.

As we can see from the results, the features of the IIoT require the use of special distributed algorithms for data processing. One such method is federated learning, which allows executing machine learning algorithms directly on non-terminal devices. Because federated learning computations take a longer time, the proposed use of the SVD optimization method allows calculations to be performed faster and with higher accuracy. For use in distributed industrial systems, federated learning algorithms are often inconvenient to implement and require additional costs.

A modified FedSVD algorithm has been proposed previously to perform the calculation of the above-mentioned SVD algorithm, taking into account that the data matrix consists of information belonging to different users [27]. The work of the federated method of Singular Data Decomposition is that the data owner can use certain random numbers to perform mathematical operations. As a result of calculations based on the initial data and data for masking, new information is obtained, which can be transmitted via communication channels. Even if private data is leaked to a third party, it will not be possible to understand it without knowing the masking information. This approach is especially effective in IIoT systems, as end-users who use their own devices are now actively involved in the production process. The use of devices such as personal computers, tablets, and smartphones can help to obtain information from different parts of the industrial system and to analyze more important factors. But there is also a growing risk that data, that the user does not want to share, may be stolen or damaged.

The calculation process of the federated Singular Value Decomposition method is shown in Fig. 9.

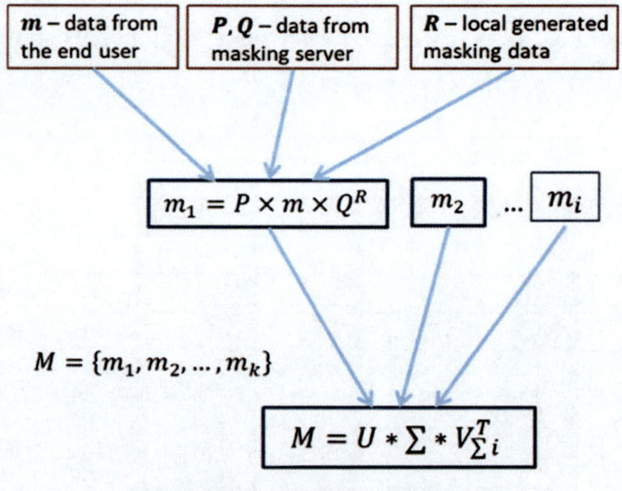

Fig. 9. Federated SVD calculation.

Accordingly, an additional step of data protection on the end device is used to ensure confidentiality, when a special mask is added to the local data, which is known only to the data owner. Therefore, when transmitting data to a central device without knowledge of masking information, private data will be incomprehensible to outsiders.

Figure 9 explains the principle of the federated SVD algorithm work. Local user data is processed before sending it to the central device. Randomly generated numbers P and Q are sent to several selected nodes by a special server. Communication between end-users and the server must be securely protected, as it is a vulnerable place for possible attackers.

The advantages of the federated Singular Value Decomposition algorithm are:

- confidentiality of user data;
- ability to collect data from different devices;
- simplicity and reliability of the algorithm.

The FedSVD algorithm also has certain features. The server that generates random numbers is remote from the end-user. During transmission, masking data can be intercepted. FedSVD also requires additional time for generating random values, and to add and to delete them to user data. Therefore, in this paper, we propose generating all random numbers, used in subsequent calculations, directly at the local node of the data owner.

The proposed scheme is shown in Fig. 10.

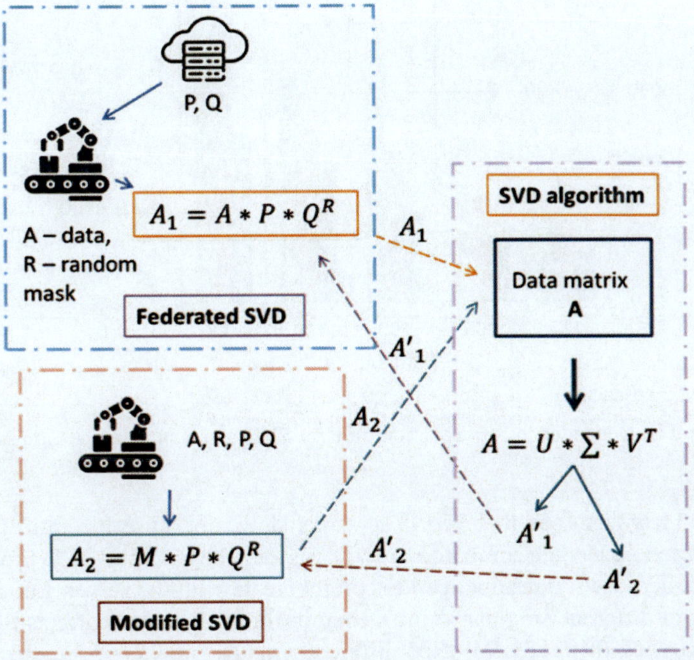

Fig. 10. Schemes of federated and modified federated SVD methods used in the IIoT systems.

As we can see from Fig. 10, the simplification of the pre-processing system allows using only the resources of the local device. In the IIoT systems, this scheme provides to use end or edge devices to work with their data and protect it from damage or interception. The disadvantage of the proposed federated SVD scheme may be the insufficient computing power of local devices for generating random values. Devices of different power are used in IIoT systems. Relatively high device power is required for efficient processing of personal data and their self-protection. So far, the cryptographic stability of random values generated by the end device has not been investigated in the paper. Therefore, the proposed scheme of Singular Value Decomposition is quite convenient for industrial systems, where it is necessary to simplify and speed up the process of generating random data.

Extending the functions of end devices can improve the efficiency of information processing and the confidentiality of personal data. Instead of delegating operations on the user's data to the central nodes, its local processing is possible, which provides when certain conditions are met, less confidentiality risk, and better speed of calculating.

Figure 11 shows the results of data processing using the SVD algorithm and its modifications.

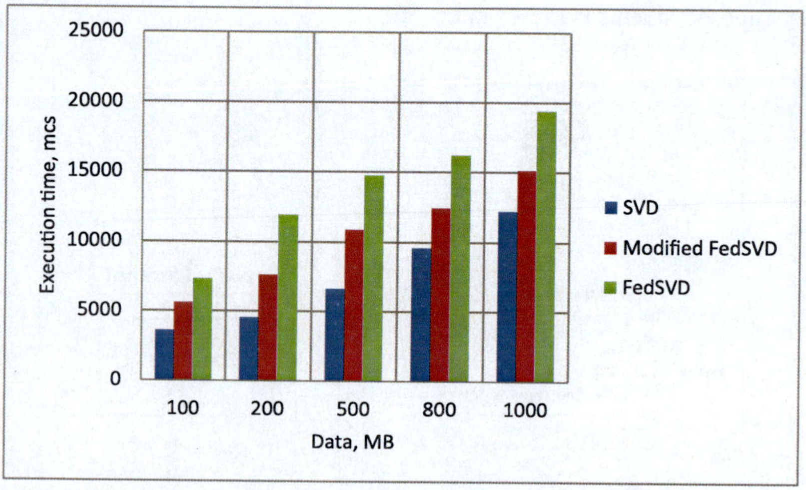

Fig. 11. Dependences of execution time for SVD algorithms on data amount.

At Fig. 11 we can see that FedSVD works slower because it requires additional resources to mask user data compared to SVD. Instead, the modified SVD processes data a little faster because it does not need to communicate with the server. Let's determine how the use of different Singular Value Decomposition algorithms affects the accuracy of data processing (Fig. 12). We used different singular numbers of matrix $\sum$, which are used to establish the accuracy of calculations, according to Fig. 5.

From Fig. 12 we can conclude that the federated method of singular decomposition and its proposed modification work more accurately than conventional SVD. The FedSVD considered in the study does not belong to the federated learning methods, because it collected data from digital devices and calculations on a central computer. Instead, the pre-processing and protection of private information is the responsibility of

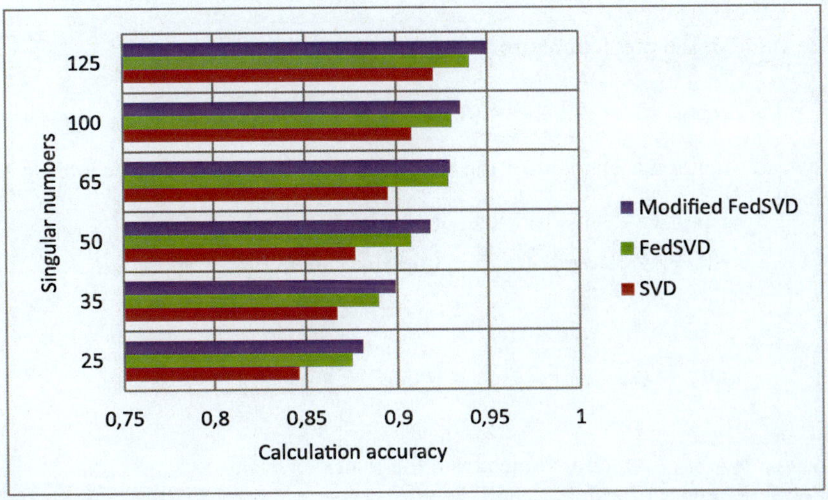

Fig. 12. Dependences of calculation accuracy for SVD algorithms on singular numbers.

the data owner. When using simple data collection, as we can see from Fig. 12, user information can be processed with less accuracy in general. Errors may occur when sending results to the end device. Federated SVD with the use of masking data allows protection and restoration of the result.

Recommendation systems work with data matrices that show existing relationships between products and users, events, etc. Based on them, the most likely future events are determined. The SVD algorithm discussed above, which decomposes the matrix M into three sub-matrices U, $\sum$ and V, determine the unknown element $m_{i,j}$ as follows:

$$M_{i,j} = U_{i,j} \times V_{i,j} + b, \tag{1}$$

where b is the prediction coefficient, selected according to the pre-training parameters of the recommendation system or randomly.

The problem of large-scale recommendation systems is often the sparsity of data matrices. That is, not all cells are filled because there is a lack of information about correspondence between the data. Prediction based on such matrices complicates the work of the SVD algorithm. Also, a lot of information storage resources are allocated to the storage of such tables. For solving these problems, the Funk-SVD algorithm [28] is used. Its feature is that the data M (m, n) is decomposed into the product of two matrices $P(\theta, m)$ and $Q(\theta, n)$, where θ determines the accuracy of the decomposition:

$$M = P^T \times Q. \tag{2}$$

In order to determine the unknown elements in the matrix M the element $m_{i,j} = m_{const}$ is first selected. Then, using the Funk-SVD schedule, can be calculated the predicted value:

$$m_{predicted} = \sum\nolimits_{\theta} \left(p_{i,\theta} T \times q_{j,\theta}\right). \tag{3}$$

The value of the prediction error:

$$N_{error} = \left(m_{const} + m_{predicted}\right)^2. \tag{4}$$

Now any unknown element of the matrix M can be determined according to the equation:

$$m_{i,j} = p_{random} + 2 \times \sigma \times \left(m_{const} - m_{predicted}\right) \times q_{random}, \tag{5}$$

or:

$$m_{i,j} = q_{random} + 2 \times \sigma \times \left(m_{const} - m_{predicted}\right) \times p_{random}, \tag{6}$$

where:

q_{random}, p_{random} – random values from the matrices P^T та Q,

σ – effectiveness of training coefficient.

Since Funk-SVD allows to decompose of the input sparse matrix into two smaller ones, this algorithm works slightly faster than standard SVD.

Figure 13 shows a comparison of the durations of calculation of recommendations for SVD and Funk-SVD algorithms for sparse data matrices of different sizes.

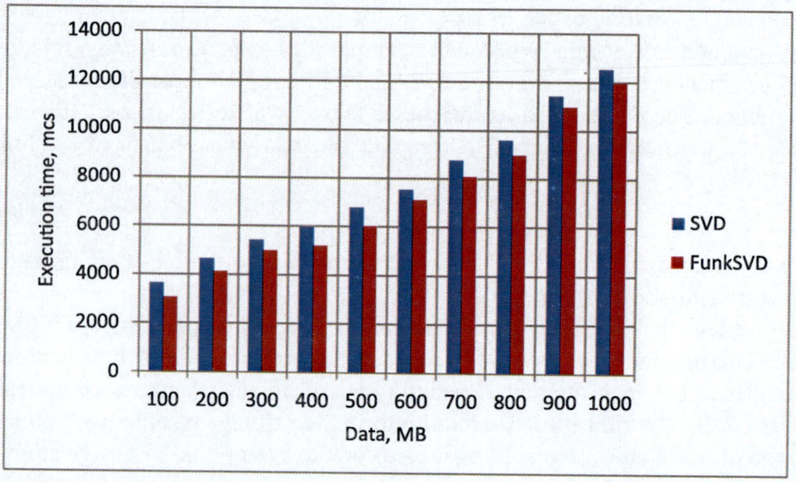

Fig. 13. Dependences of execution time for SVD and Funk-SVD algorithms.

As we can see from Fig. 13, the duration of calculations can be reduced using Funk-SVD. When the enterprise collects data about various processes, phenomena, and relationships between them, there is often a lack of information. Thus, the use of Funk-SVD additionally allows to speed up the determination of the necessary data and the provision of recommendations.

Figure 14 shows the dependencies of the Mean Squared Error (MSE) indicator for two algorithms. MSE is the accuracy of providing recommendations, that is, the square of the mean deviation between the actual value and the prediction:

$$MSE = \frac{(m_{i,jconst} - m_{i,jpredicted})^2}{k}, \quad (7)$$

where k- number of predictions.

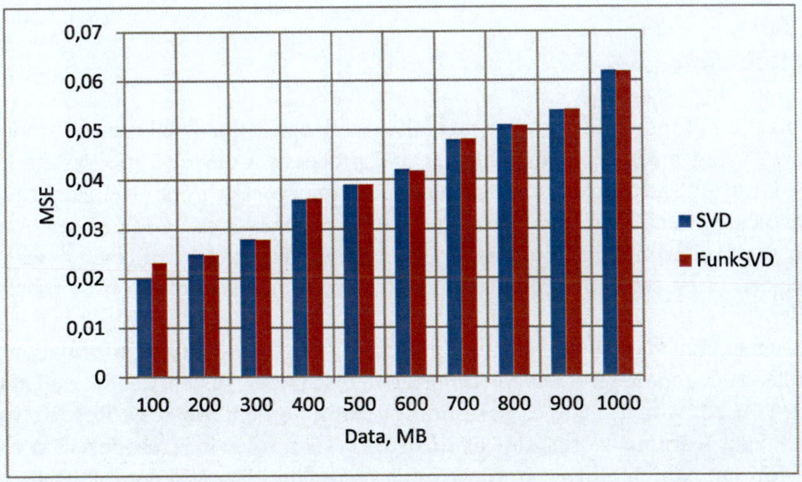

Fig. 14. Dependences of MSE value for SVD and Funk-SVD algorithms.

As we can see from Fig. 14, the SVD and Funk-SVD algorithms work with almost the same MSE value. Therefore, when using Funk-SVD, it is possible to reduce the computation time and the amount of memory needed to store data, while maintaining the accuracy of providing recommendations.

Briefly summarizing, the paper first investigated the effectiveness of using the SVD algorithm to reduce the amount of data coming from the end-users of the federated learning algorithm. The results in Figs. 6, 7 and 8 shows that the SVD algorithm allows for discarding part of the data that is not important for further analysis and reduces the duration of calculations. We can adjust the parameters of the singular decomposition, such as the singular values, to achieve a better balance between accuracy and computational speed. In this part of the study, SVD was performed only on the user side, to reduce the amount of data, then a federated learning algorithm was performed to search for selected values in the information array.

Also, the processing of the SVD algorithm by the central device based on information from different users was investigated (Fig. 11 and Fig. 12). Since it is often difficult to compute complex algorithms such as SVD using federated learning, FedSVD is used, which allows sending data to a central device for processing, previously masked. The study also suggested modifying FedSVD and generating data for masking directly on the local device. This approach is appropriate for cases where the devices have sufficient computing power and the system has high requirements for data processing speed. The accuracy of SVD calculations in this case is the difference between the initial data of the user and data obtained from the central device as a result of singular decomposition. As a result of using personal masking data, the user can remove the mask and restore the data with high accuracy. The use of the Funk-SVD algorithm (Fig. 13 and Fig. 14) allows for improving the efficiency of sparse data processing in IIoT systems. Based on Funk-SVD, a machine learning algorithm can be performed. By passing recommendation accuracy parameters between systems, the training process can be accelerated.

5 Conclusion

Modernization of industrial systems now allows changing the usual type of production. Thanks to using a lot of data collection and processing devices, the system can be managed flexibly and quickly, minimizing human intervention. The IIoT is a large-scale manufacturing system that involves the operation of many smart end devices to collect and analyze data from the environment. However, information is collected in very large volumes and often contains redundancy. Therefore, it is advisable to use optimization methods.

The paper considers the problem of processing large amounts of information in the IIoT systems. For more efficient performance in distributed systems, federated machine learning is used, which involves performing training on end devices. Research shows that federated learning works slower than ML. Therefore, it is proposed to use the SVD algorithm, which allows for identifying redundant data and not taking them into account in further processing. Experimental results indicate a decrease in the duration of calculations of the FL algorithm using SVD. This approach allows organizing the work of machine learning in IIoT and other distributed systems that process Big Data while ensuring the top speed of calculations.

The efficiency of using the federated algorithm of Singular Value Decomposition is also investigated in the work. This algorithm performs data processing and masking before transmission over communication channels. This approach can increase the privacy of users' private information. A slightly improved method of federated Singular Decomposition calculation is also proposed, which involves generating all the parameters for masking directly on local devices. The results of the study showed high accuracy of calculations for the proposed method and ensure the reliability of the data. The results of the research can be used to improve the efficiency of the IIoT and other intelligent systems that process large amounts of data. Thanks to the simulation results, it is clear that the SVD algorithm allows for the optimization of Big Data in industrial systems, which reduces the duration of calculations and allows quickly making important decisions. Also for systems with high requirements for data processing accuracy and reliability, the proposed modified SVD algorithm shows a good result. Funk-SVD algorithm can be effectively used to improve the efficiency of sparse data processing in intelligent industrial systems.

References

1. Klymash, M., Kyryk, M., Demydov, I., Hordiichuk-Bublivska, O., Kopets, H., Pleskanka, N.: Research on distributed machine learning methods in databases. In: 2021 IEEE 4th International Conference on Advanced Information and Communication Technologies (AICT), pp. 128–131 (2021). https://doi.org/10.1109/AICT52120.2021.9628949.
2. Klymash, M., Kyryk, M., Hordiichuk-Bublivska, O., Fabri, L., Kopets, H.: Big data analysis in IIoT systems using the federated machine learning method. In: IEEE 16 International Conference on Advanced Trends in Radioelectronics, Telecommunications and Computer Engineering (TCSET), Lviv-Slavske, Ukraine, 22–26 February 2022 (2022)
3. Przystupa, K., et al.: Distributed singular value decomposition method for fast data processing in recommendation systems. Energies **14**(8), 2284-1–2284-24 (2021)
4. Hassan Reza, M.N., Malarvizhi, C.A.N., Jayashree, S., Mohiuddin, M.: Industry 4.0–technological revolution and sustainable firm performance. In: 2021 Emerging Trends in Industry 4.0 (ETI 4.0), pp. 1–6 (2021). https://doi.org/10.1109/ETI4.051663.2021.9619363
5. Jun, S., et al.: A cost-efficient software based router and traffic generator for simulation and testing of IP network. Electronics **9**(1), 40 (2020). https://doi.org/10.3390/electronics9010040
6. Rezazadegan, R., Sharifzadeh, M.: Applications of artificial intelligence and big data in industry 4.0 technologies. In: Industry 4.0 Vision for the Supply of Energy and Materials: Enabling Technologies and Emerging Applications, pp. 121–158. Wiley (2022). https://doi.org/10.1002/9781119695868.ch5
7. Panchenko, O., Polishuk, A., Seliuchenko, M., Beshley, M.: Method for adaptive client oriented management of quality of service in integrated SDN/CLOUD networks. In: 2017 4th International Scientific-Practical Conference Problems of Infocommunications. Science and Technology (PIC S&T), pp. 452–455 (2017). https://doi.org/10.1109/INFOCOMMST.2017.8246437
8. Klymash, M., Savchuk, R., Pozdnyakov, P., Beshley, M.: The researching and modeling of structures of mobile networks for providing of multiservice radio access. In: Proceedings of International Conference on Modern Problem of Radio Engineering, Telecommunications and Computer Science, pp. 281–282 (2012)
9. Gupta, A., Jain, S.: Optimizing performance of Real-Time Big Data stateful streaming applications on Cloud. In: 2022 IEEE International Conference on Big Data and Smart Computing (BigComp), pp. 1–4 (2022). https://doi.org/10.1109/BigComp54360.2022.00010
10. Kryvinska, N., Bickel, L.: Scenario-based analysis of IT enterprises servitization as a part of digital transformation of modern economy. Appl. Sci. **10**(3), 1076 (2020). https://doi.org/10.3390/app10031076
11. Tang, Z.: A preliminary study on data security technology in big data cloud computing environment. In: 2020 International Conference on Big Data & Artificial Intelligence & Software Engineering (ICBASE), pp. 27–30 (2020). https://doi.org/10.1109/ICBASE51474.2020.00013
12. Klymash, M., Seliuchenko, M., Beshley, M., Redchuk, S.: Increasing wavelengths utilization efficiency in OTNoDWDM network based on local resource distribution method. In: 2015 Second International Scientific-Practical Conference Problems of Infocommunications Science and Technology (PIC S&T), pp. 157–160 (2015). https://doi.org/10.1109/INFOCOMMST.2015.7357300
13. Maharani, M.P., Tobianto Daely, P., Lee, J.M., Kim, D.-S.: Attack detection in fog layer for IIoT based on machine learning approach. In: 2020 International Conference on Information and Communication Technology Convergence (ICTC), pp. 1880–1882 (2020). https://doi.org/10.1109/ICTC49870.2020.9289380

14. Stoshikj, M., Kryvinska, N., Strauss, C.: Service systems and service innovation: two pillars of service science. In: Paper presented at the Procedia Computer Science, vol. 83, pp. 212–220 (2016). https://doi.org/10.1016/j.procs.2016.04.118
15. Abdelmoumin, G., Rawat, D.B., Rahman, A.: On the performance of machine learning models for anomaly-based intelligent intrusion detection systems for the Internet of Things. IEEE Internet Things J. **9**(6), 4280–4290 (2022). https://doi.org/10.1109/JIOT.2021.3103829
16. Romanchuk, V., Beshley, M., Polishuk, A., Seliuchenko, M.: Method for processing multi-service traffic in network node based on adaptive management of buffer resource. In: 2018 14th International Conference on Advanced Trends in Radioelecrtronics, Telecommunications and Computer Engineering (TCSET), pp. 1118–1122 (2018). https://doi.org/10.1109/TCSET.2018.8336390
17. Kim, H., Kim, Y., Park, H.: Reducing model cost based on the weights of each layer for federated learning clustering. In: 2021 Twelfth International Conference on Ubiquitous and Future Networks (ICUFN), pp. 405–408 (2021). https://doi.org/10.1109/ICUFN49451.2021.9528575
18. Uzhga-Rebrov, O., Kuleshova, G.: Using singular value decomposition to reduce dimensionality of initial data set. In: 2020 61st International Scientific Conference on Information Technology and Management Science of Riga Technical University (ITMS), pp. 1–4 (2020). https://doi.org/10.1109/ITMS51158.2020.9259304
19. Hashemipour, S.N., et al.: Big data compression in smart grids via optimal singular value decomposition. In: 2020 IEEE Industry Applications Society Annual Meeting, pp. 1–8 (2020). https://doi.org/10.1109/IAS44978.2020.9334900
20. De, P., Chatterjee, A., Rakshit, A.: Regularized K-SVD-based dictionary learning approaches for pir sensor-based detection of human movement direction. IEEE Sensors J. **21**(5), 6459–6467 (2021). https://doi.org/10.1109/JSEN.2020.3040228
21. Dong, H., Cheng, F., Yang, M., Liu, W., Zhang, X.: On low rank approximation of 3-tensors based on regularized t-SVD. In: 2019 Chinese Automation Congress (CAC), pp. 4827–4830 (2019). https://doi.org/10.1109/CAC48633.2019.8997349
22. Özyurt, F., Kutlu, H., Avci, E., Avci, D.: A new method for classification of images using convolutional neural network based on dwt-svd perceptual hash function. In: 2018 3rd International Conference on Computer Science and Engineering (UBMK), pp. 410–413 (2018). https://doi.org/10.1109/UBMK.2018.8566537
23. Kryvinska, N., Strauss, C.: Conceptual model of business services availability vs. interoperability on collaborative iot-enabled eBusiness platforms. In: Bessis, N., et al. (eds.) Internet of Things and Inter-cooperative Computational Technologies for Collective Intelligence, pp. 167–187. Springer Berlin Heidelberg, Berlin, Heidelberg (2013). https://doi.org/10.1007/978-3-642-34952-2_7
24. Bentbib, A.H., Kreit, K., Labaali, I.: Randomized tensor singular value decomposition for multidimensional data compression. In: 2022 11th International Symposium on Signal, Image, Video and Communications (ISIVC), pp. 1–6 (2022). https://doi.org/10.1109/ISIVC54825.2022.9800729
25. Masiuk, A., Beshley, M., Lavriv, O., Deschynskiy, Y.: Common radio resource management model for heterogeneous cellular networks. In: 2016 13th International Conference on Modern Problems of Radio Engineering, Telecommunications and Computer Science (TCSET), pp. 661–663 (2016). https://doi.org/10.1109/TCSET.2016.7452145
26. Xu, G., et al.: TT-SVD: an efficient sparse decision-making model with two-way trust recommendation in the AI-enabled IoT systems. IEEE Internet Things J. **8**(12), 9559–9567 (2021). https://doi.org/10.1109/JIOT.2020.3006066

27. Chai, D., Wang, L., Fu, L., Zhang, J., Chen, K., Yang, Q.: Federated Singular Vector Decomposition. arXiv:2105.08925v1 [cs.DC]. Accessed 19 May 2021
28. Xiaochen, Y., Qicheng, L.: Parallel algorithm of improved FunkSVD based on GPU. IEEE Access **10**, 26002–26010 (2022). https://doi.org/10.1109/ACCESS.2022.3156969

Check for updates

Smart House Management System

Bohdan Zhurakovskyi[1], Oleksiy Nedashkivskiy[1], Mikhailo Klymash[2], Oleksandr Pliushch[3](✉), and Mykyta Moshenchenko[1]

[1] National Technical University of Ukraine "Igor Sikorsky Kyiv Polytechnic Institute", Kyiv, Ukraine
zhurakovskiybyu@tk.kpi.ua, al_1@ua.fm, nrodan@icloud.com
[2] Lviv Polytechnic National University, Lviv, Ukraine
mykhailo.m.klymash@lpnu.ua
[3] Taras Shevchenko National University of Kyiv, Kviv, Ukraine
opliusch@yahoo.com

Abstract. The purpose of this chapter is to create a system for managing a Smart Home, using automation devices from different developers. A system has been developed that allows you to control home automation devices. Telegram Bot has also been created, which will allow the user to communicate with the IFTTT automation service, which communicates with the services of developers of the home automation devices. To implement the system, the existing systems of the Smart Home were analyzed, and an algorithm was developed that will allow to combinsng of these systems into one interface. The system is created using the Telegram messenger, which acts as an interface and allows you to protect the system using TelegramID. The system includes ChatBot, database, IFTTT system and Web-services of developers.

Keywords: Smart home · IFTTT · Automation · Control system · Service · ChatBot · Telegram · Database · Automation device · Sensor

1 Introduction

Today, the global Internet is one of the most important inventions in human history. Systems for automating the management of electronic devices are being developed, both in large enterprises and in everyday life. Many people dream of connecting their home devices to the Internet and controlling them remotely from a personal mobile device. There are active discussions of "Smart Home" projects, common information space [1] and other options for combining disparate electronic components into a single system [2].

When equipping your home with modern smart devices, there may be a problem of the coordinated operation of devices from different manufacturers. Today, almost every major manufacturer of smart devices has its own so-called ecosystem, element management and transmission of information that occurs within the system [3]. For a person who is not an expert in electronics, programming and microcomputers, there are no opportunities to merge different ecosystems into one, of their own. Therefore, the

M. Klymash et al. (Eds.): TCSET 2022, LNEE 965, pp. 268–283, 2023.
https://doi.org/10.1007/978-3-031-24963-1_15

idea of creating a software and hardware system to control Smart Home devices from different manufacturers is quite relevant.

So the subject of development is a system for controlling Smart Home devices.

The goal is to develop a system for managing Smart Home devices that combines smart devices from different manufacturers in one ecosystem. To achieve this goal, it is necessary to solve the following tasks:

1. Analyze the main ways to build and use software and hardware systems for home automation.
2. Analyze smart devices, and consider their capabilities and manufacturers' proposed control methods.
3. Develop a system for managing Smart Home devices from different manufacturers in one ecosystem.

2 Statement of Research Problem

2.1 Theoretical Foundations of Smart Home Systems Development

We will analyze the components of all typical and non-typical Smart Homes. All systems are linked by a goal based on the desire to automate the home space for a more comfortable living. But there is no system that would not have disadvantages: whether high cost or complexity, or the inability to use the desired user technology [3].

We will consider the options for the building systems, management methods and forms of management of these systems. Then we will analyze the structure of a Smart Home and identify the components that are typically used in most systems.

Also, we will analyze the automation of management through the use of cloud technologies, to control the system remotely, through the use of mobile devices with Internet access. Consider the software component of the system and its levels, functionality and HTTP requests used in user communication with server systems [4].

We will analyze a large number of ready-made solutions for Smart Home systems that exist in the market. Among them: LogicMachine, Z–Wave.Me [2, 11], Wink, SmartThings from Samsung, Wiren Board, MajorDoMo. Most solutions are based on the Hub controllers that connect to sensors and various actuators. The last one is a simple software package that allows you to combine everything in a regular PC without using a separate controller.

2.2 Approaches of Building a Smart House System and Managing Automated Systems Methods

Up to date, there are three approaches to building a Smart Home system [1, 2]:

– centralized systems;
– decentralized systems;
– mixed.

Centralized systems are built based on a central controller, which, in fact, is the only powerful processor. Including executive units with the addition of various control panels. Such systems are common mainly in the private sector. All devices are connected to the controller, which is the center of the system [5]. Individual components in the system may have their own microcontrollers, but only the central controller controls and is responsible for the interaction of the entire system. Central controllers can receive and send control signals through different channels. The main advantage of such systems is not the high cost of the whole system. But the disadvantage is the high dependence of most devices on the central unit, which can fail.

Decentralized systems differ from centralized ones in the absence of a central controller. All devices are connected directly to each other via a bus and are equipped with their autonomous controller [6]. The disadvantage of such systems is the relative complexity and cost of each individual device, but it pays off in reliability, as well as fewer or almost no need for wired networks.

Mixed systems include elements and principles of both of the above systems. As an example of a mixed system, we can mention X10. X10 - widely used in the field of home automation standard that defines protocol and method of transmitting signals over wireless channels or ordinary power lines to electronic modules to which controlled household appliances are connected.

Despite the many advantages, the Smart Home system also has a number of disadvantages. The main problem of the modern Smart Home is the fragmentation of the market [7]. This means that many start-up devices may not work with each other at all. And in the best case, you need to install many applications on your smartphone for each of them. This is not the most convenient option. Some companies offer "turnkey" systems where everything works perfectly, but in this case, it will not be possible to combine your favorite devices: for example, a smart lamp from Xiaomi and a humidifier from Apple.

To solve this problem, it is necessary to understand the modern way of communication with the Smart Home through the Internet. From any computer connected to the global network, you can learn about the state of the house, see CCTV footage and even the current video of them [8]. If necessary, the system can regularly send to its owner by e-mail or on his phone, a report on the condition of the house, the presence of people, the weather and much more. Of course, this information will be available only to homeowners. And also, using the technical capabilities of modern mobile devices, you can create different scenarios of home devices, based on user interaction with the environment. Thus, the development of software and hardware system for controlling Smart Home devices from different manufacturers is relevant.

2.3 Analysis of Methods of Automation of Smart Home Management Systems

When designing a Smart Home, the concept of the Internet of Things can be implemented by transferring data to the cloud from a variety of devices used in the system, where this data will be processed and stored, and by transmitting commands from the user to end-to-end devices. The user will have access to an interface (via a browser or mobile application), designed to monitor data from the sensor, as well as to control devices. In this way, users will get a tool for remote management and monitoring of the entire

Smart Home system. The main advantages of using cloud technologies in a Smart Home system are as follows:

- Through the cloud server it is possible to provide a single interface for Smart Home systems in the case of using devices from different manufacturers with different data protocols [2, 9];
- At any time and in any place the user will have access to all information about what is happening in the house;
- The Smart Home system itself is becoming more flexible, it will be much easier to integrate new devices just by providing them with Internet access without rebuilding the whole system.

In order for the Smart Home management system [10, 11] to work effectively, you need to ensure reliable interaction between the cloud and the home server [12], as well as between remote management devices and the cloud [13, 14]. There is a technology of web services to ensure the interaction of various applications and sites on the Internet. There are mainly two ways to implement web services: SOAP [2, 15] and REST architectural style.

SOAP is a structured messaging protocol in XML format in a distributed computing environment.

REST is an architectural style in which the components of a distributed application interact on the Internet using HTTP requests (mainly GET or POST) [16]. The term "RESTful web service" is used to refer to web services implemented using HTTP and REST principles [17]. The REST style has a number of advantages over SOAP.

2.4 Analysis of Ready-Made Solutions for Smart Home System

A Smart Home is more than a group of IoT things (smart light bulbs, thermostats, security sensors) or so-called *connected devices*. The difference between smart appliances and a full-fledged engineering system "Smart Home" is that the second can be supplemented by the first, but only on smart devices to build a full-fledged Smart Home is impossible. Anyway, for now. A high degree of automation of connected devices and systems, as well as higher efficiency of their work, is achieved only with the help of engineering solutions such as Loxone, KNX, etc., but certainly not related to the Internet of Things. Some smart devices are good in themselves, but they do not create a reliable system. In addition, to integrate them with each other, you need a Hub. When the router is turned off, nothing will work. There are also many ready-made solutions for creating a Smart Home. Each of these solutions have advantages and disadvantages, which allows you to create the most accurate opinion about them.

LogicMachine, which is a real combination for a Smart Home. Different versions of the controller differ from each other in a set of interfaces. The latter has the widest capabilities among the entire LogicMachine family [18].

LogicMachine does not have fixed built-in algorithms but can be programmed according to the specific logic of your project [19]. So you can't do without programming knowledge. In general, on LogicMachine you can build any scenario of a Smart Home by writing your script in LUA.

Archos is an inexpensive system that is an Android-based tablet with a set of additional devices: a mini-camera, motion sensor, temperature and humidity sensor, and smart sockets [20]. The system is wireless, which greatly facilitates installation, but the coverage area is small. If you need to cover a large two-floor cottage, there will be difficulties. There are built-in behavior scripts, as well as a simple and skill-free programming editor. Scenarios are simple, for example, to receive an e-mail and a picture from the surveillance camera when someone approaches the front door.

Z–Wave. Me. Z-wave is a wireless communication protocol for creating Smart Homes [21]. Based on it, there are many solutions: from the usual central controller MiCasaVerdeVera 3 to the board, which turns the RaspberryPi into a Smart Home controller.

For those who prefer to assemble "something of their own" based on the Arduino, you should pay attention to the controller ***Fibaro Home Center*** [22]. It is at the junction between DIY- geek solutions and ready-made systems for ordinary users. It has both built-in automatic scripts that can be activated in a few clicks, and an editor that allows you to write your script in LUA.

Prestigio. An application for Android that allows you to control the brightness and glow color of light bulbs from the same manufacturer [23]. Prestigio has several lines of smart lamps: some can change the brightness and color, and others - only the brightness.

Wink. The simplest solution is if the Smart Home is "too complicated", expensive or pointless. The motto of this device: "We control all gadgets that can only be managed" [24]. This hub is compatible with hundreds of devices from dozens of different manufacturers, and supports Bluetooth, Wi-Fi, Z-Wave, ZigBee [2], Lutron [25] and Kidde [26]. Unfortunately, not all solutions for a Smart Home are good. Wink users have received a lot of negative comments, and Amazon scores only 2.6 points.

Smart Things from ***Samsung***. The system consists of a neat white box hub, it is compatible with several hundred devices and is equipped with a battery that allows you to work up to 10 h. The hub allows you to build the behavior of security cameras, thermostats, smart outlets and other gadgets.

SmartThings support devices for home LANs, connecting to the router via Wi-Fi or cable connection, as well as ZigBee and Z-Wave wireless standards [27]. Thus, the hub works with many devices on the market, but there are still many that it does not support.

Iris. Home automation system with support for many different devices: CCTV cameras, smoke detectors, leak detectors, locks and more. Iris includes a central hub, several separate sensors and modules, as well as a number of kits for automation and security [28].

Wiren Board. Another representative of the budget home automation class. The RS-485 interface supports water and energy meters, power supply control devices (relay modules), digital and analog sensors, and lighting control devices. Measures parameters such as temperature, humidity, light, pressure, and the presence of gases [29]. Thanks to the cloud web interface, the system can be controlled remotely even via GPRS.

Velocio Ace. Miniature (only 2.5 x 2.5 cm) controller with minimum consumption of 5 V. Advanced versions with MiniUSB and RS-232 which costs $ 170. For this money, you get 31 inputs and outputs to connect various sensors and the ability to load your controller programs [30].

Loxone Miniserver. This controller does not have enough batteries, so you need a power supply. And not surprisingly, it is equipped with an EIB / KNX connector, LAN port, microSD slot, integrated WEB-server and much more [31]. You can connect, for example, 8 light groups and 4 radiators to the Miniserver connectors, and then everything depends on the availability of extensions. Miniserver has both built-in behavior scripts and the ability to write your scripts.

MajorDoMo is a free open source software platform for integrated home automation management. The hub is a personal computer (on Windows or Linux), to which you need to connect smart sensors: factory or homemade. But even without the connection of equipment, the system can act as an organizer of a personal information environment [32].

Thus, modern smart devices have a variety of software and web services that allow you to control them remotely. After analyzing the existing technologies for building a Smart Home system and ready-made solutions for home automation management, it becomes obvious the urgency of developing a software and hardware system that combines smart devices in one ecosystem.

2.5 System Development

The system being developed consists of several modules that interact with each other using HTTP queries or webhooks. The following diagram (Fig. 1) demonstrates these modules and gives them a brief description.

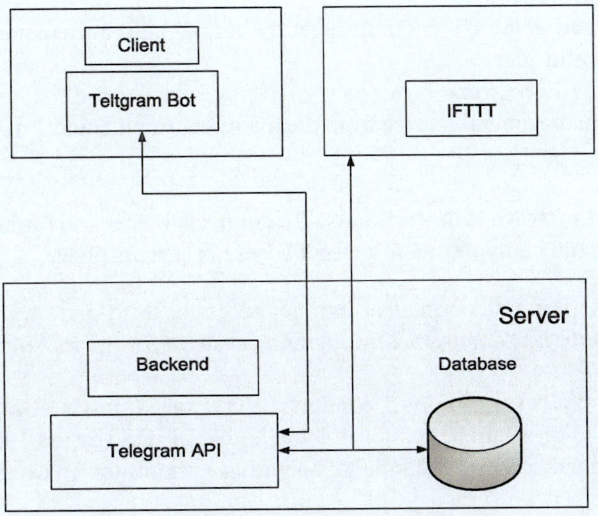

Fig. 1. Architectural scheme of the system.

The system is designed to control home automation components from different manufacturers through one common interface. The system must be able to add new devices and create different scenarios for their interaction with the user.

System Structure. SoftwarE - hardware solution should be a system consisting of a ChatBot written in the Python programming language, which connects the management interface with a Web service, which, in turn, manages home devices by connecting to servers manufacturers of devices. Program code in the Python programming language provides the following functions:

- processing of JSON requests of the management interface and sending of results of execution of requests back;
- organization of interaction of all elements of the system;
- sending management commands in the form of HTTP requests to the IFTTT server;
- implementation of the logic of user interaction with the web interface.

The management interface must provide:

- clear and user-friendly control of devices added to the system;
- device status notification.

The IFTTT automation service integrated into the system allows:

- add new smart devices to the system by connecting to the manufacturer's servers;
- create new scenarios of user interaction with devices, such as trigger of certain interface functions (such as push-up messages) or home automation elements, depending on the user's geolocation.
- Devices attached to the system must meet the following requirements:
- have its own controller;
- connect to a Wi-Fi network;
- have a native management service from the manufacturer included in the list of IFTTT services.

Hardware and Software Requirements. To connect devices and further manage them You need a personal computer with Internet access or a smartphone.

Home Web server with permanent Internet access or third-party web hosting.

System and software requirements (operating systems, browsers, software platforms, etc.).

Personal computers running Windows, browser (mostly Google Chrome). Smartphone-based on Android version 4.4 and higher or iOS version 10.3.3 and higher.

For implementation of the Windows operating system, the PyCharm source code editor is used.

Analysis of Automators. Automators will help us bring together different manufacturers of components for a Smart Home. Such services operate on one simple principle, they create causal links between dozens or hundreds of different Web services used in different fields. Automators are used not only in Smart Homes but also in everyday life, here are some examples of their use:

– turn off Wi-Fi on the phone when leaving home;
– receiving a notification in the messenger when receiving a message in the mail;
– when you add a photo to one social network, it is automatically added to another;
– turn on the phone's silent mode after 23:00.

All Automators are built on causation, i.e. after event A (leaving the house, getting dark, receiving notification, etc.), the next event you programmed is B. From the large number of existing Automators, IFTTT was selected.

IFTTT Automator. IFTTT got its name from the conditional programming expression "if this, then that". The company provides a software platform that connects applications, devices, and services from different developers to run one or more automations that include those applications, devices, and services.

Initially, the sequences were called recipes, but after a recent rebranding, they received a new name - applets. Although the meaning has not changed: you choose a trigger, which then automatically starts the action.

It is always possible to simplify the task and use ready-made applets prepared by IFTTT and other users. You will be offered thousands of ways to automate social networks, e-mail and music services, online publications, and smart gadgets.

IFTTT supports about 320 services. These include Facebook, Evernote, YouTube, Spotify, Giphy, Wikipedia, and Android and iOS. We can say that the specifics of IFTTT – are casual services for a comfortable life. Get at least the support of intelligent Google Assistant and WeMo appliances.

After analyzing this service, the advantages and disadvantages were highlighted:

Advantages:

– complete list of common applications that we use every day;
– support for smart light bulbs and remote control of garage doors;
– simple clean design of Web interface and mobile clients;
– free.

Disadvantages:

– lack of support for the Ukrainian language;
– lack of additional actions and conditions.

So, to sum up, Zapier covers the maximum number of services, which is even too many. IFTTT is easily enough for most common cases, and Flow [33, 34] is far behind competitors and is too focused on the Microsoft ecosystem, but is rapidly increasing its power. Zapier [35–37] and Flow dominate IFTTT thanks to filters and the ability to do multiple actions at once. Zapier is too minimalist, moreover, there is no information about when the service will get mobile versions.

It is better not to find free IFTTT for home purposes. So we chose IFTTT to develop our system.

Configuring the Selected Service. Automation is performed using macro-like applets that connect multiple applications to run automated tasks. You can enable or disable the applet using the IFTTT Website [38] or mobile applications (and/or IFTTT widgets in mobile applications). You can also create your applets or make changes to existing ones through the user-friendly IFTTT interface.

Typically, developers run their "typical" applets, and then the user community "builds what developers never expected". IFTTT applets can use JavaScript, HTTP requests, advanced filtering, and other tools to create new interactions. JavaScript support helps IFTTT partners create robust applets over the more limited "recipes" of years past. It is possible to create a certain custom JavaScript that will filter events automatically, so that, for example, the applet will include multiple light sources in the house, on arrival after 6 pm or only the light on the porch, on arrival home before 6 pm.

Getting Started with IFTTT. IFTTT service is easy to use. Management is made through the application (for Android or Apple iOS), creating a free account you can immediately start working with automation. IFTTT offers a number of the most popular standard and custom applets. The collections combine applets for different platforms - such as iOS, Android and voice assistants, and showcase everything from applets for news services and from weather forecasts to home automation. The My Applets screen allows users to manage the currently included applets and displays the history of those previously used. Users can create their own applets by combining different "services" of the application and setting trigger parameters.

To integrate this service into the system under development, applets based on Webhooks technology, geolocation of Android devices and system actions on Android devices, such as connecting/disconnecting the home Wi-Fi network, were created.

The Webhook on the IFTTT [39] server looks like this: https://ifttt.com/maker_webhooks.

The link consists of three main parts:

- https://maker.ifttt.com/trigger/ - this is the server address;
- trigger {event} - the name of the event that the user comes up with when creating an applet;
- with/key/ gHLhMYNwZjmb7c1NYPGAz - unique access key generated by the service.

Telegram ChatBot. The ChatBot in the popular Telegram messenger was chosen as the interface for managing Smart Home devices.

Currently, ChatBots as modern communication tools have become widely used in many areas of human life to establish contact with Internet users. ChatBots became most popular when they began to be used in messengers and social networks (for example, in Telegram, Viber and Facebook). Thanks to the new virtual communication tool, it is possible to learn about special offers online, receive newsletters and special offers in the field of goods and services, and perform more complex operations. Currently, ChatBots are of the greatest importance in the formation of such a model of behavior that will be as close as possible to humans.

The interface capabilities of the program are quite simple and clear. ChatBot algorithms are based on pre-loaded templates, commands and subsequent execution of any program actions depending on the receipt of the message (Fig. 2).

To create this mechanism and implement it in the system under development, it is necessary to analyze the capabilities of the free Telegram Bot API.

API Telegram is an interface that allows you to obtain information from the Telegram database using HTTP-queries to a special server [40]. The convenience is that you do not need to know in detail how the database is arranged, and what tables and fields of what types it consists of. It is enough that the API query "knows". The syntax of requests and the type of data returned are strictly defined on the side of the service.

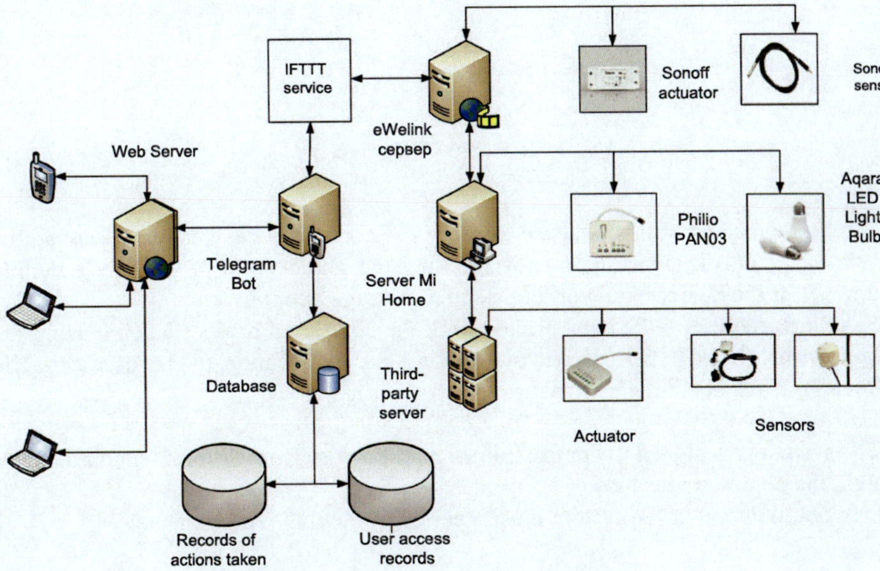

Fig. 2. General network architecture.

Description of Development Results and Model Representation of Processes. As a result of the development, an open control system was created that allows you to add smart devices from different manufacturers that have a built-in controller that is autonomously connected to the servers of manufacturers [41]. The user has the ability to create several scenarios for the interaction of devices with each other, with the environment, the user and the software they use [42].

The following "Use Case" UML diagram depicts possible options for user interaction with the system (Fig. 3).

List of tested scenarios with a model description in the form of UML activity diagrams:

1. Turning on/off the light by setting the chatbot command "Turn on the light".

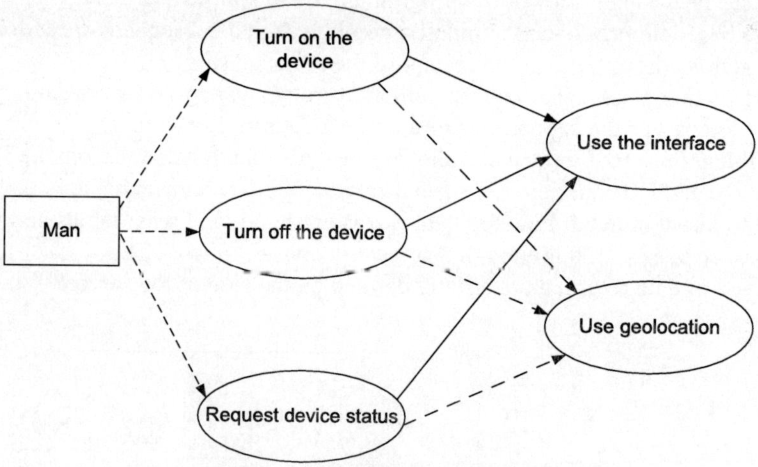

Fig. 3. Use Case UML diagram.

2. Sending a request to the database to check the connected devices when leaving the house (the check is carried out by tracking the geolocation of the user's mobile device). If the devices are working, send a message about it.
3. Sending a request to the database to check the connected devices when leaving the house (the check is carried out by tracking the geolocation of the user's mobile device). If the devices are working, send a message about it.

The system is designed to control home smart devices from different manufacturers through one common interface.

Devices included in the system must meet the following requirements:

- Have your own controller;
- Connect to the WiFi network;
- Have a native control service from the manufacturer, included in the list of IFTTT services.

Any PC running Windows OS, with an installed browser (Google Chrome recommended) and Internet access, or a mobile device with the operating system Android version 4.4 and higher or IOS version 10.3.3 and higher can act as a control device.

The system does not require installation, all control elements are located on the Internet and provide access both from a browser and from applications on a mobile device (Fig. 4, Fig. 5).

To start the system:

1) Connect the smart device to your smartphone through the manufacturer's application;
2) Install the IFTTT application and add the connected device to the list of your devices;
3) Install the Telegram application on your phone or PC.

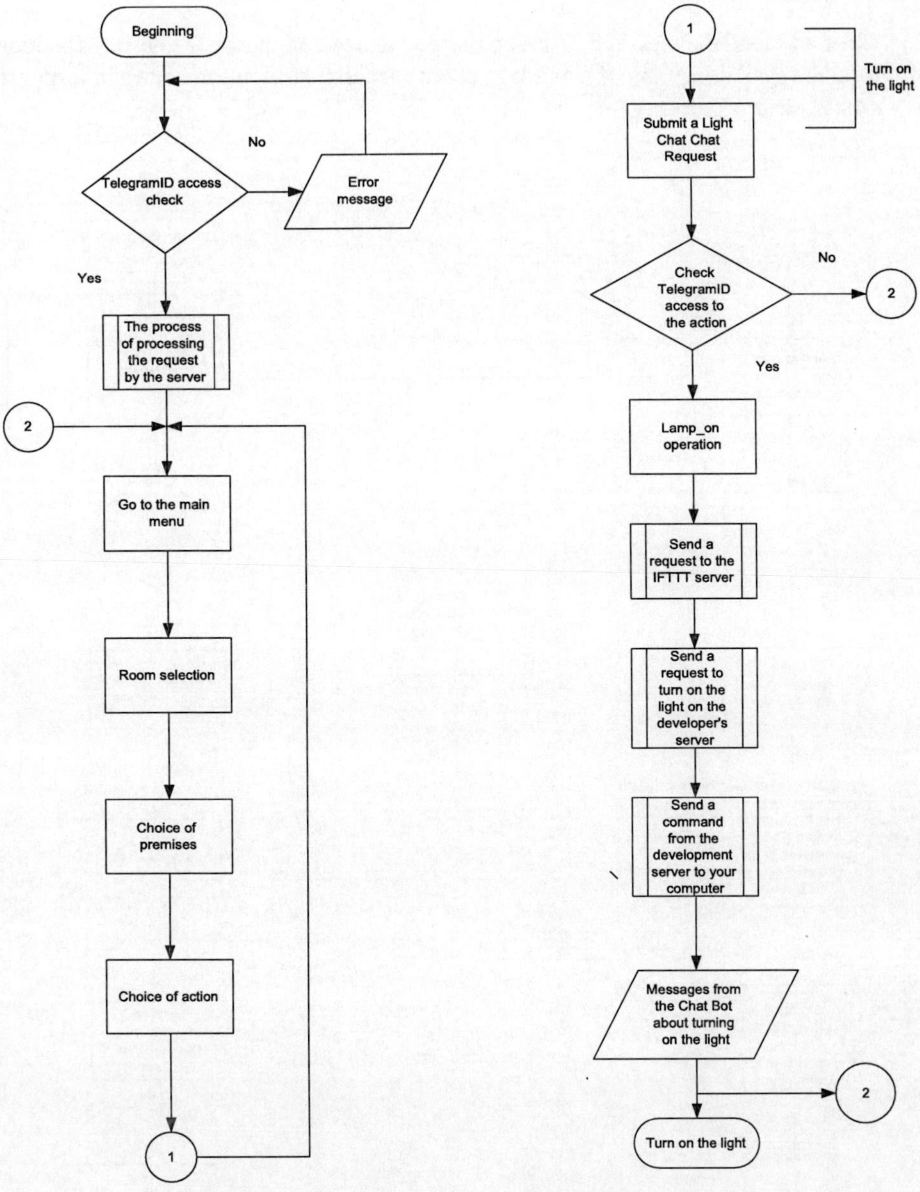

Fig. 4. General algorithm of system operation (Part 1).

Procedure for checking performance:

1) Open a dialog box with a chatbot;
2) Write and send the "/start" command, if the message "Hello, this is your smart home management system!" – you are connected to the system;

3) Write and send the message "Turn on (device name)", or choose the desired function from the available menu, if a message about successful execution comes in response - the system is working.

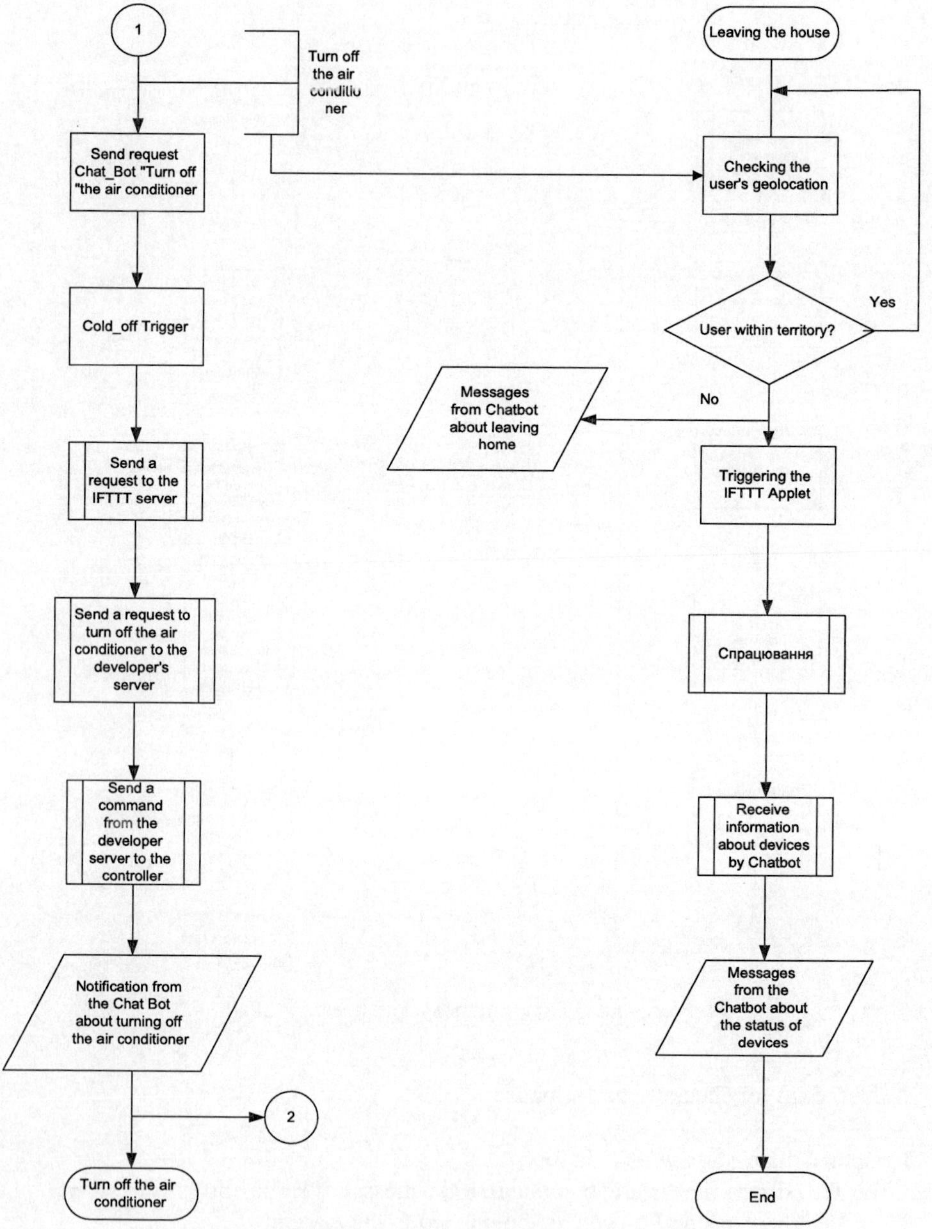

Fig. 5. General algorithm of system operation (Part 2).

3 Conclusions

As a result of work, analyzing the main methods of home automation, types of systems, their features and shortcomings, and the problem of coordinated operation of smart devices from different manufacturers were identified, primarily due to market fragmentation of these same devices.

After analyzing smart devices and considering the methods of control offered by the manufacturers, a software and hardware solution was designed and developed, which is based on several technologies: Python programming technologies; technology of creating NoSQL database MongoDB and working with it; modern Web technologies; SaaS (Software as a Service); IFTTT, which allows you to control smart devices through its interface; the main interface is implemented in the form of a ChatBot based on Telegram API technology.

Thus, the problem is solved in full, the goal is achieved and developed a software and hardware solution for Smart Home smart devices management, which combines smart devices from different manufacturers in one ecosystem.

In the future, it would be interesting to expand the system by integrating many different devices, as well as: implement an improved protection system for safer use; manage smart devices directly through the "native" API, i.e., through the manufacturers' servers, without the use of ancillary services such as IFTTT.

Integrate the system into other, professional home automation systems. For example, the MajorDoMo platform. In this case, the developed system would be an excellent extension in the form of a new interface and several additional automation scenarios, by adding triggers based on user interaction with the mobile device and the environment.

References

1. What is a Smart Home? (2021). https://www.smarthouse.ua/umnyj_dom.html
2. Zhurakovskyi, B., Zeniv, I.: Internet of Things technologies, Tutorial, Igor Sikorsky, p. 503 Kyiv Polytechnic Institute, Kyiv (2021). https://ela.kpi.ua/handle/123456789/42078
3. Shevchenko, O., Bondarchuk, A., Polonevych, O., Zhurakovskyi, B., Korshun, N.: Methods of the objects identification and recognition research in the networks with the iot concept support. In: Paper Proceedings of the Selected Papers on Publishing Papers with CEUR-WS co-located with Workshop on Cybersecurity Providing in Information and Telecommunication Systems (CPITS 2021), vol. 2923, pp. 277–282 (2021). http://ceur-ws.org/Vol-2923/
4. Bots: information for developers (2022).https://tlgrm.ru/docs/bots
5. Zhurakovskyi, B., Boiko, J., Druzhynin, V., Zeniv, I., Eromenko, O.: Increasing the efficiency of information transmission in communication channels. Indon. J. Electr. Eng. Comput. Sci. **19**(3), 1306 (2020). https://doi.org/10.11591/ijeecs.v19.i3.pp1306-1315
6. Zhurakovskiy, B., Tsopa, N.: Assessment technique and selection of interconnecting line of information networks. In: 3rd International Conference on Advanced Information and Communications Technologies (AICT), pp. 71–75 (2019). https://doi.org/10.1109/AIACT.2019.8847726
7. Zhurakovskyi, B., Parkhomey, I., Druzhini, V.: Processing of information in sensory framing. In: Adaptive Systems and Automatic Control, no.1, pp. 42–57 (2018). http://nbuv.gov.ua/UJRN/asau_2018_1_7. (in Ukrainian)

8. Zhurakovskyi, B., Toliupa, S., Otrokh, S., Kuzminykh, V., Dudarieva, H., Zhurakovskyi, V.: Coding for information systems security and viability. In: CEUR Workshop Proceedings, vol. 2859, pp. 71–84 (2021)
9. Druzhynin, V., Toliupa, S., Pliushch, O., Stepanov, M., Zhurakovskyi, B.: Features of processing signals from stationary radiation sources in multi-position radio monitoring systems. In: CEUR Workshop Proceedings, vol. 2746, pp. 46–65 (2020). http://ceur-ws.org/Vol-2746/
10. Home automation controller Fibaro Home Center (2022). https://www.fibaro.com/en/products/home-center-2
11. Zhurakovskyi, B., Fedorova, N., Gavrilko, E., Zeniv, I.: Technologies for the creation of the Internet of Things. Computer workshop, Tutorial, Igor Sikorsky Kyiv Polytechnic Institute, Kyiv, p. 128 (2021). https://ela.kpi.ua/handle/123456789/46169
12. Zhurakovskyi, B., Tsopa, N., Batrak, Y., et al.: Comparative analysis of modern formats of lossyaudiocompression. In: CEUR Workshop Proceedings (2020). http://ceur-ws.org/Vol-2654/paper25.pdf
13. Nedashkivskiy, O.: Precise method of balancing passive optical networks with irregular splitter with two or more outputs. In: IEEE Xplore, pp. 228–231 (2017). https://ieeexplore.ieee.org/document/8020107, https://doi.org/10.1109/AIACT.2017.8020107
14. Nedashkivskiy, O., Havrylko, Y., Zhurakovskyi, B., Boiko, J.: Mathematical support for automated design systems for passive optical networks based on the β-parametric approximation formula. Int. J. Adv. Trends Comput. Sci. Eng. **9**(5), 8207–8212 (2020). https://doi.org/10.30534/ijatcse/2020/186952020
15. Fedorova, N., Havrylko, Y., Kovalchuk, A., Husyeva, I., Zhurakovskiy, B., Zeniv, I.: Software system for processing and visualization of big data arrays. In: ICCSEEA 2022: Advances in Computer Science for Engineering and Education. Part of the Lecture Notes on Data Engineering and Communications Technologies book series, LNDECT, vol. 134, pp 324–336(2022). https://doi.org/10.1007/978-3-031-04812-8_28
16. Nedashkivskiy, O., Semenko, A.I., Kozlov, A.S., Bokla, N.I.: Internet: models, technologies, access speed, prospects. In: Information and TelecommunicationSciences, NationalTechnicalUniversity of Ukraine “Kiev Polytechnic Institute”, no. 1, pp.19–25 (2018). https://doi.org/10.20535/2411-2976.12018.19-25
17. Luntovskyy, A., Nedashkivskiy, O.: Intelligent networking and bio-inspired engineering. In: The Second International Conference on Information and Telecommunication Technologies and RadioElectronics (UkrMiCo’2017), Odessa, Ukraine, Collection of Proceedings of the Scientific and Technical Conference, 11–15 September, pp. 501–504 (2017). https://doi.org/10.1109/UkrMiCo.2017.8095421
18. LogicMachine (2021). https://openrb.com/
19. Automation controllers EvikaLogicMachine (2022). https://www.ixbt.com/home/evika-logicmachine.shtml
20. ARCHOS (2022). https://www.archos.com/us/products/objects/chome/ash/index.html
21. Z–Wave.Me (2022). https://z-wave.me/
22. FibaroHomeCenter (2022). https://www.fibaro.com/en/products/home-center-2
23. Prestigio (2022). https://prestigio.com/
24. Wink (2017). http://wink.in.ua/help/faq
25. Lutron (2022). https://store.smart-maic.com/?gclid=CjwKCAjwyryUBhBSEiwAGN5OCPw24PzQobTzlSMYY98yYL1yzzMAGdXyIdWagmFc_fwBwwwGW5fXwxoCJsEQAvD_BwE
26. Kidde (2022). https://www.kidde.com/home-safety/en/us/
27. SmartThings. One smart app for all smart devices (2021). https://www.samsung.com/ru/apps/smartthings
28. Iris (2022). https://www.ferra.ru/news/smarthome/SmartHome-Iris-07-08-2014.htm

29. Wirenboard.Documentation Center (2021). https://wirenboard.com/wiki/index.php/
30. VelocioAce (2022). https://velocio.net/
31. LoxoneMiniserver (2021). https://domotix.pro/loxone/controller/miniserver
32. Smart Home System MajorDoMo (2021). https://habr.com/ru/post/538896/
33. MICROSOFT FLOW (2018). https://habr.com/ru/post/425847/
34. FLOW (2022). https://powerautomate.microsoft.com/
35. What is Zapier and how can it be useful to you? (2022). https://onlinehelp.mindjet.com/help/MindManager/20/RU/what_is_zapier_.htm
36. Zapier Intro (2022). https://platform.zapier.com/docs/zapier-intro
37. Zapier (2022). https://zapier.com/
38. IFTTT (2022). https://en.wikipedia.org/wiki/IFTTT
39. IFTTT is a free service for automating routine tasks (2022). https://tproger.ru/articles/ifttt/
40. API Telegram (2022). https://core.telegram.org/bots/api
41. Zhurakovskyi, B., Toliupa, S., Druzhynin, V., Bondarchuk, A., Stepanov, M.: Calculation of quality indicators of the future multiservice network. In: Klymash, M., Beshley, M., Luntovskyy, A. (eds.) Future Intent-Based Networking. LNEE, vol. 831, pp. 197–209. Springer, Cham (2022). https://doi.org/10.1007/978-3-030-92435-5_11
42. Zhurakovskyi, B., Moshenchenko, B.M., Zhurakovskyi, V.: Algorithm for detecting and troubleshooting multiservice networks. Aktualnye nauchnye issledovania v sovremennom mire (5), 94–101 (2020). (in Ukrainian)

Searching Extreme Paths Based on Travelling Salesman's Problem for Wireless Emerging Networking

Oleh Pavlenko[1], Anatolii Tymoshenko[1], Oksana Tymoshenko[1], Andriy Luntovskyy[2], Yaroslav Pyrih[3], and Igor Melnyk[4](✉)

[1] Open International University of Human Development "Ukraine", Lvivska Street, 23, Kyiv 03115, Ukraine
TimoshAG@i.ua
[2] BA Dresden University of Cooperative Education, Saxon Academy of Studies, Hans-Grundig-Street 25, 01307 Dresden, Germany
Andriy.Luntovskyy@ba-dresden.de
[3] Lviv Polytechnic National University, S. Bandery Street, 12, Lviv 79013, Ukraine
[4] National Technical University of Ukraine "Igor Sikorsky Kyiv Polytechnic Institute", Peremogy Avenue 37, Kyiv 03056, Ukraine
imelnik@phbme.kpi.ua

Abstract. Nowadays, the world is characterized by Emerging Networking in various areas of human activity and industries. Among the combined networks of today (mobile, wireless, fixed), the networks with a fully connected topology occupy a special place. They play an essential role in the software agents interacting with each other according to the point-to-point (P2P) model, the machine-to-machine (M2M) communication model for robotics and IoT with a specified gateway, energy-efficient clusters, or data centers with a large number of central processing units and radio channels between them, as well as Multiple Input Multiple Output (MIMO) antennas for the 5G and Beyond standards. Typically, the analysis of network systems with a fully connected topology leads to solving the Traveling Salesman's Problem (TSP). A solution, however, requires combinatorial enumeration and needs a massive amount of computing time, even on powerful modern computers (NP-complexity). The given work provides a new TSP algorithm, which is based on the analysis of the digital series by unique sums. That allows one to find the best route between network blocks with a fully meshed topology. The obtained simulation results and computational experiments have shown the very high efficiency of the proposed algorithm. Furthermore, an attempt for a specific TSP-based MAC protocol is herewith offered (OSI layer 2), which distinguishes from existing routing protocols (layer 3) such as OSPF + BGP or RIP + BGP and above (with QoS considering). The considered TSP-based MAC protocol for WSN (Wireless Sensor Network) provides access from a start node (cluster head, gateway) to all other nodes within the limited small net (cluster). The start node can be selected dynamically for future reconfiguration for energy efficiency and long life without frequent battery exhaustion. The start node provides TSP-based polling to all other networking energy-autarky nodes within the cluster. We are confident that the obtained results have more theoretical than applied meaning. Therefore, we offer this work as a WIP (Work-in-Process).

M. Klymash et al. (Eds.): TCSET 2022, LNEE 965, pp. 284–304, 2023.
https://doi.org/10.1007/978-3-031-24963-1_16

Keywords: Emerging Networking · Intra-domain routing protocols · MAC protocols · Mesh networks · Fully connected topology · Traveling salesman's problem · Series of numbers · Graph theory · Extreme paths · Clustering · WSN · Dynamic reconfiguration · Operation research

1 Introduction and Motivation

The classical Traveling Salesman's Problem (TSP) from OR (Operation Research, 1945) theory is one of the most widespread combinatorial optimization problems, which consists in finding the most profitable route on any oriented graph that passes through the specified points or nodes at least once. After that, in the next steps, a return to the previous nodes is possible. In the conditions of the problem, the route profitability criterion, like the shortest common way, the cheapest summarized route, less fuel to reach it, a cumulative criterion, etc., and the corresponding distance and cost matrices are indicated herewith. Sometimes this algorithm possesses a nondeterministic polynomial (NP) complexity via the growth of the considered combinations. Really, the matrix of transitions has an asymmetric character. The TSP is represented by a graph in which the nodes are various operations, and the branches fix the terms of the transition from one operation to another. For N nodes with complete connections, the number of possible paths is a factorial number of (N – 1)!. With N = 20, the value (N –1)! ≈ 122*10**15 have been obtained.

On other hand, network structures of various spheres have common patterns [1, 2], and sometimes complement each other. Thus, the problem of the maximum flow of the network of oil pipelines can be represented by a double graph, in which the bandwidth of the network is equivalent to the shortest path [3]. Therefore, the problem of determining the extreme paths of the network has found wide applications. The search for the shortest network path is implemented by simple algorithms, if the values of each branch are known. The task of determining the shortest path in a computer network is more complex because the time of information transmission from one node to another is probabilistic [4].

The work solves the classical TSP for the mesh networking systems, which are used nowadays in voluminous Emerging Networking (EmN) under decomposition and reconfiguration concepts both. The real dimensions can be divided, therefore, into the smaller dimensions (so-called clusters, autonomous areas), e.g., for N = 20 and, then, for such dimensions, TSP can be deployed and solved.

The following practical scenarios of EmN can be considered in this case for the presented in the work TSP solution:

1. N:N software agents communicate one each other in a P2P network on the run-time platform and then rewind to a gateway (as the start node for TSP considered)
2. M2M communication model for Robotics and IoT with a gateway, as the start node for TSP.
3. Wireless Sensor Networks (WSN) can benefit on TSP solving too because of the gateway polling protocol on OSI data link layer 2.

4. An energy-efficient cluster or data center with $N = 20$ CPUs with direct fiber and radio-links in 100 GHz bandwidth on a short-range $d = 0{,}3$ m one each other. A corresponding network structure with software parallelization is presented in Fig. 1. An energy-efficient powerful cluster is deployed
5. with multiple TFLOPS (~1000 milliards floating operations per second, or 10^{12} -10^{14} FLOPS).
6. MIMO antennas for 5G Beyond, which nowadays reach multi-dimensional features up to 64x64x64.

In all the above-mentioned scenarios a specific TSP-based MAC protocol can be used (OSI layer 2), which distinguishes it from the routing protocols (layer 3) such as OSPF + BGP or RIP + BGP and above (with QoS considering) [14–20].

The considered TSP-based MAC protocol provides access from a start node (cluster head, gateway) to all other nodes with a limited small net (cluster). The start node can be selected dynamically for future reconfiguration for energy efficiency and long life without frequent battery exhaustion (for WSN). The start node provides TSP-based polling to all other networking energy-autarky nodes within the cluster [21, 22].

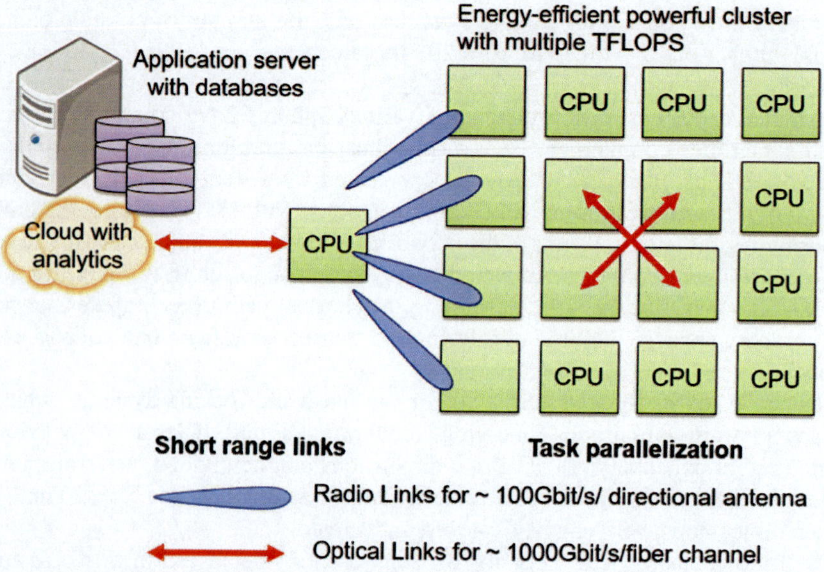

Fig. 1. Energy-efficient computing cluster within a data center with a TSP-based polling protocol.

2 Statement of the Scientific Problem

The general structure of a fully connected topology graph is presented in Fig. 2.

In graph theory, the PSM is defined on a graph $G = (V, A)$, where $V = \{v_1, \ldots, v_n\}$ is a set of n nodes, $n = l{\cdot}k$ and $A = \{(v_i, v_j)|v_i, v_j \in V, i \neq j\}$ is a set of arcs, together with

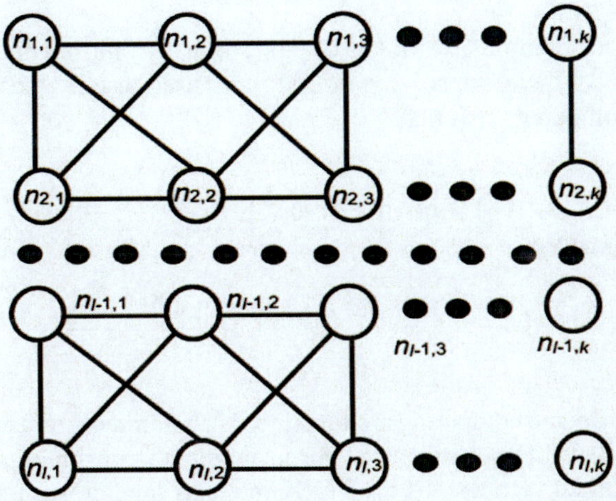

Fig. 2. The structure of a system with a fully connected topology.

a matrix non-negative cost or distance matrix $C = (c_{ij})$, which are generally associated with A. The problem is considered to be symmetric if $c_{ij} = c_{ji}$ for all $(v_i, v_j) \in A$, and asymmetric otherwise. Elements of A are often called edges in the symmetric case. The version of symmetric TSP in which distances satisfy the triangle inequality ($c_{ij} + c_{jk} \geqslant c_{ik}$) is mostly studied as a special case of the problem. The TSP consists in determining the Hamiltonian cycle, often simply called a *tour of minimum cost* [5].

One of the approaches to solving the traveling salesman problem is to formulate its solution in the form of a discrete optimization problem, in which the solutions are represented as variables, and the connections as inequality relations between them. Thus, several options are possible. For example, a symmetric TSP can be represented as a set of edges. Each edge V is associated with a corresponded binary variable $x_{i,j} \in \{0, 1\}$ [5]. Each vertex must communicate through a pair of edges with the rest of the vertices, that is, through an input and output edge [5]:

$$\forall_i \in V, \sum_{j \in V/\{i\}} x_{ij} = 2. \quad (1)$$

Today most known solution of TSP are the follows [5]:

1. Jonson, 1954:

$$\sum_{j \in S, j \notin S} x_{ij} \geq 2. \quad (2)$$

2. Miller, Tucker and Zemlin, 1960:

$$\min\left\{ \sum_{i \in V} \sum_{j \in V/\{i\}} c_{ij}x_{ij} \middle| x\ valid\,(1)(2),\ x_{ij} \in \{0, 1\} \right\}. \quad (3)$$

But pointed out NP complexity, which is proportional to n!, is corresponded also to relations (2) and (3), therefore solving of TSP isn't used usually today for finding the optimal path in computer networks.

3 Demarcation and Assignment of TSP for Simulation in the Networks

In any case, main advantages for using TSP for simulation EmN with full connection topology are follows.

1. The use of distributed computing architectures is undoubtedly essential when solving real-time problems. However, since TSP is quite time-consuming, for example, for a structure with 20 vertices, the time is 30 min and smaller, such a solution should be considered as very effective [14].
2. Since the TSP is sufficiently well known, a comparison with existing implementations of solving this problem is always necessary. So, for example, implementing the TSP solution in the MATLAB environment by reducing it to an optimization problem of integer linear programming for the number of nodes even more than 100 takes seconds. But, on other hand, for full connection network topology using TSP without its modification may be the only suitable solution [14].
3. Surely, modern routing protocols for voluminous irregular TCP/IP networks go beyond TSP and use so-called greedy algorithms. Effective solutions should consider both the network's structural and functional characteristics, QoS, via advantage routes like OSPF of Dijkstra during route calculation [14–22]. However, after the decomposition of big mesh networks into autonomous systems or small clusters with a limited quantity of nodes the TSP-based solutions for node polling can be used for the optimization of the communication between the start node and all further nodes within the mesh. Corresponded model of network decomposition is presented in Fig. 3.

Therefore, in this chapter the original fast algorithm for solving the TSP, based on the rows with unique sums, is proposed and analyzed. The possible network structures, where the use of TSP can be not only effective, but really necessary, will be considered in the next part of this chapter.

4 An Appropriate Case Study

A Wireless Sensor Networks (WSN) is considered below with the sensors for measurement of the contaminants, humidity, pressure, temperature, lighting, radiation, or further useful parameters.

A modern WSN can possess up to $2^{16} = 65.536$ sensors (sensing nodes), like a ZigBee net nowadays. Corresponded information is given at Table 1 [6–8].

A general WSN with a GW (gateway) and multiple sensing nodes with energy harvesting (sunlight, piezo, thermogradient) is depicted in Fig. 4. Via the GW (i.e., a starting node for the considered graph), the data can be transferred to the cloud with the analytical functionalities for the measured data from the sensors of the WSN. And vice versa, via the given GW, the commands aimed at the remote control can be transferred to the sensors, and the clusters and reconfigure them.

We divide such a massive network into multiple small dynamic clusters (segments) with max. N = 20 sensors and use e.g. LEACH approach [6–8, 22] with the dynamic assignment of the sensor for cluster head (CH), or gateway (GW).

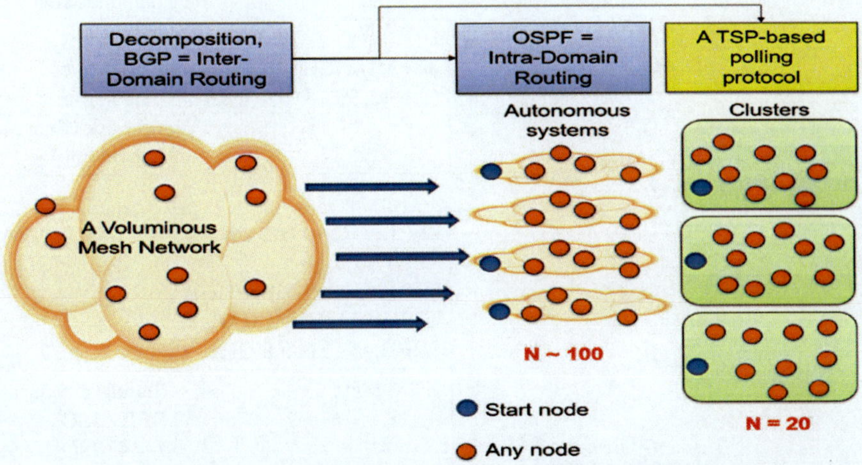

Fig. 3. Decomposition and use of intra-domain routing and TSP-based polling.

Table 1. IoT sensing technologies, radio and contactless.

Type	Title	Frequency, MHz	Purpose
Radio Technologies			
WAN	SigFox	868 MHz (EU), 902 MHz (USA)	Low-power wide-area network
	LoRa	2,4 GHz, 868/915 MHz, 433 MHz, 169 MHz	Low-power wide-area network
	NarrowBand-IoT	700–2100 MHz	4G mobile network
	LTE-CatM	700–2100 MHz	4G mobile network
LAN	WiFi5 IEEE 802.11ac	ISM 2400/5000 MHz	Wireless LAN
	WiFi6 IEEE 802.11ax	ISM 2400/5000 MHz	Wireless LAN
	ZigBee IEEE 802.15.4	900/2400 MHz	Low-power personal-area network

(continued)

Table 1. (*continued*)

Type	Title	Frequency, MHz	Purpose
PAN	6LoWPAN	ISM 2400/5000 MHz	Low-power personal-area network
	Bluetooth IEEE 802.15.1	ISM 2400/5000 MHz	Low-power personal-area network
	EnOcean	868 MHz (EU), 928 MHz (Japan), 902 MHz (USA)	Low-power personal-area network (Energy Harvesting)
Contactless Technologies			
Tags, labels, chipcards	RFID	LW 125–134 kHz, SW 13,56 MHz; UHF 865–869 MHz (EU), UHF 950 MHz (US/Asia), SHF 2,45 GHz/5,8 GHz	Multiple standards
	NFC	13,56 MHz	Standardized RFID, ISO 14443/ISO 15693

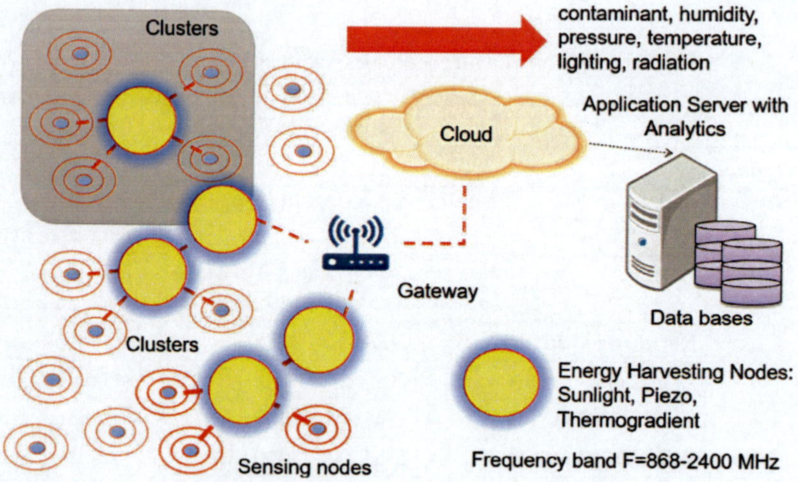

Fig. 4. A general WSN with a GW and multiple sensing nodes with energy harvesting [6–8, 22].

LEACH Clustering approach provides topology optimization for a WSN with more than N = 20 nodes. Efficient dynamic cluster formation with a dynamic selection of nodes and CH (cluster heads) is known as the LEACH method. The basic LEACH Principle is described in Fig. 5.

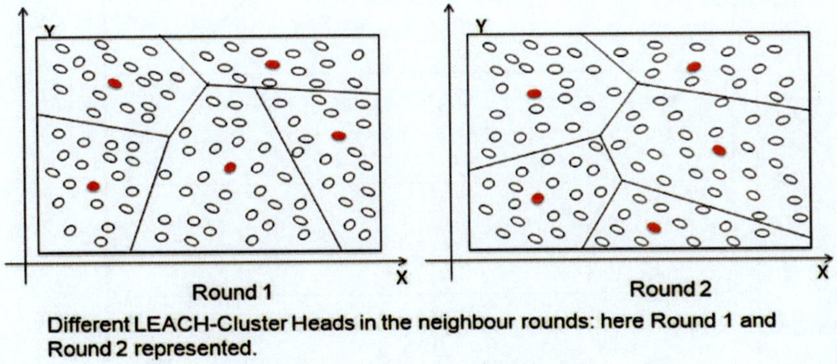

Fig. 5. LEACH-clustering for energy efficiency [6–8, 22].

Efficient dynamic cluster formation with dynamic selection of nodes and CH (cluster heads) is known as the LEACH method. The LEACH Principle can be described as follows see Fig. 4 and Fig. 5. The nodes, which are already CHs, cannot operate again as CHs for 1/P rounds, where P is a wished part of Cluster Heads within the network is. Furthermore, each node possesses the average probability $Z < T(n)$ to become to Cluster Head in each new round. Finally, at the end of the round each node, which did not become CH connects to the next Cluster Head and became Cluster Member (Join Cluster). Each of the Cluster Heads schedules a slotted communication plan (Cluster Schedule) for each node from the own Cluster aimed at the successful data transfer (Slot 1 - Slot m). The procedure extends up to 4 times the average life expectancy of the WSN compared to static solutions:

- directly connected sensor nodes,
- static-coupled nodes and CH in the clusters.

Corresponded graphic dependences are presented in Fig. 6.

Such sensor has more Time To Live (TTL) and need further advanced solutions. Further optimization is possible via TSP solving.

Therefore, we use the TSP algorithm to optimize the standard communication within the selected cluster with $N = 20$ nodes and the GW as a starting node. Finding the most profitable route via the given dynamic WSN cluster with $N = 20$ and CH (a GW) is considered below. A TSP-based polling protocol is offered below.

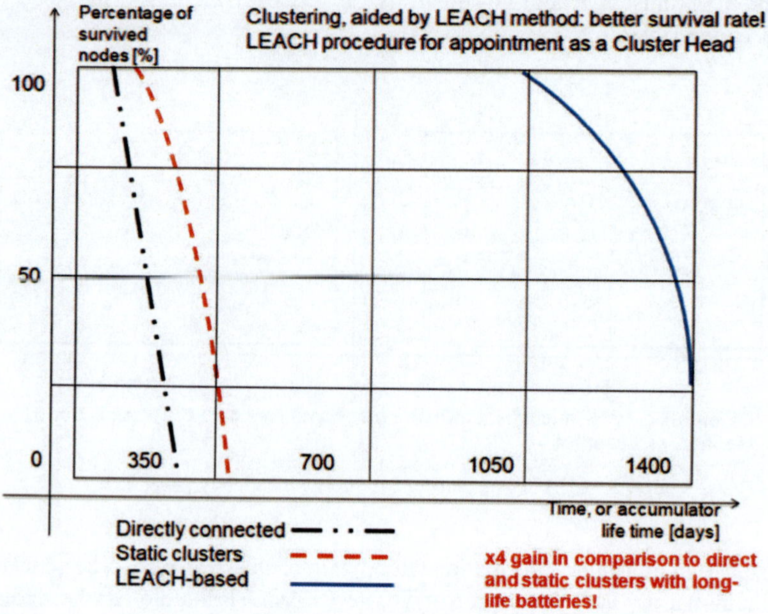

Fig. 6. LEACH: more survivability [6–8, 22].

5 Rows with Unique Sums and Testing of Proposed Algorithm

Corresponding to Fig. 1, modern computer systems use parallelism with a large number of processors and intensive interaction by high-frequency channels between them. There is a need for communication between processors at different points in time. Corresponding to considered in Sect. 2 methods of TSP solving it is necessary to create a sequence of n numbers that have unique sums for any set of elements. It is clear that series of the type r^n, where r and n are natural numbers greater than 1, have the same property. The smallest such series 2^n has unique sums for any set of numbers. The problem arose by creating a series of numbers in which the maximum element is smaller than the series 2^n. For the first time, such a series was proposed in [9]. Let's call this additive set of numbers the series ***A***. The properties of this series are considered in [10]. Unlike the 2^n series, this series has a much smaller maximum element, but the sum of the series is greater.

This series is built for a new set based on the previous one by adding the first element, which is determined according to certain rules. Thus, the previous series is a right difference operator. Later it turned out that for n greater than 6, it is possible to create a better sequence, that is, with a smaller value of the maximum element. Such sequence was also created according to the same principle, but the selection of a new first element for a new sequence is based on a different principle [9]. Checking the possibility of improving the proposed algorithm showed that this series is optimal from the point of view of the minimum of the maximum element [11, 12]. For a small number of elements n less than 9, the check is made by exhaustively going through all possible options. For n greater than 8, a more complex algorithm is used. The search for the best route in a network using graph theory is well known and discussed in books [13, 21]. In the papers [14–19] the problems of routing in the Emerging Networks also have been discussed. But the algorithm proposed in this chapter is based on number theory and is generally new.

We denote this perfect set of additively distinct numbers by ***D***.

Examples of additive series of type ***A*** and type ***D*** at values of n = 10 and n = 20 are given in Table 2.

Table 2. Examples of additive series of type ***A*** and type ***D***.

$n = 1, 2\ldots N$	$2n$	Series ***A, D*** $N = 5$	Series ***A*** $N = 10$	Row ***D*** $N = 10$	Series ***A*** $N = 20$	Row ***D*** $N = 20$
1	1	6	165	148	166116	132568
2	2	9	249	225	249216	199412
3	4	11	291	265	290766	233119
4	8	12	313	285	311562	250115
5	16	13	324	296	321960	258613
6	32		330	302	327170	262936
7	64		333	305	329775	265136
8	128		335	307	331083	266256
9	256		336	308	331737	266826
10	512		337	309	332067	267111
11	1024				332232	267259
12	2048				332316	267336
13	4096				332358	267376
14	8192				332380	267396
15	16384				332391	267407
16	32768				332397	267413
17	65536				332400	267416
18	131072				332402	267418
19	262144				332403	267419
20	524288				332404	267420

In Fig. 7, the first column n is the number of graph nodes, the second column is row 2^n of the third and fourth sequences ***A*** and ***D*** at $n = 10$, the fifth and sixth sequences ***A*** and ***D*** at n = 20. For n less than 7, these sequences coincide.

k	Processor 1	Σ 1	Σ 2	Processor 2	Σ 1	Σ 2	Processor 3	Σ 1	Σ 2
1	1			2			3		
2	2 1	9		1 2	6		1 3	6	
	3 1	11		3 2	11		2 3	9	
	4 1	12		4 2	12		4 3	12	
	5 1	13		5 2	13		5 3	13	
3	2 3 1	20	323	1 3 2	17	195	1 2 3	15	312
	2 4 1	21	365	1 4 2	18	237	1 4 3	18	91
	2 5 1	22	510	1 5 2	19	382	1 5 3	19	445
	3 2 1	20	326	3 1 2	17	199	2 1 3	15	313
	3 4 1	23	105	3 4 2	23	240	2 4 3	21	354
	3 5 1	24	459	3 5 2	24	391	2 5 3	22	505
	4 2 1	21	350	4 1 2	18	215	4 1 3	18	98
	4 3 1	23	87	4 3 2	23	214	4 2 3	21	360
	4 5 1	25	481	4 5 2	25	413	4 5 3	25	476
	5 2 1	22	519	5 1 2	19	587	5 1 3	19	470
	5 3 1	24	465	5 3 2	24	592	5 2 3	22	529
	5 4 1	25	505	5 4 2	25	640	5 4 3	25	494
4	3 2 5 1	33	543	1 3 4 2	29	246	1 2 5 3	28	514
	3 5 2 1	33	528	1 5 3 2	30	592	2 5 1 3	28	530
	2 4 3 1	32	374	1 5 4 2	31	640	1 5 2 3	28	529
	2 5 3 1	33	525	3 5 1 2	30	596	1 5 4 3	31	494
	2 3 4 1	32	366	4 1 5 2	31	432	2 1 4 3	27	356
	4 2 5 1	34	567	4 3 1 2	29	224	4 1 5 3	31	495
	4 3 2 1	32	351	4 5 3 2	36	623	4 2 1 3	27	370
	4 5 3 1	36	496	4 3 5 2	36	416	4 1 2 3	27	362
	2 5 4 1	34	565	4 5 1 2	31	618	4 2 5 3	34	562
	4 3 5 1	36	484	5 3 4 2	36	643	2 5 4 3	34	554
	4 5 2 1	34	550	1 3 5 2	30	397	4 5 1 3	31	501
	5 3 4 1	36	508	1 4 3 2	29	238	4 5 2 3	34	560
5	4 3 5 2 1	45	553	1 5 4 3 2	42	641	1 2 5 4 3	40	563
	2 5 4 3 1	45	574	1 5 3 4 2	42	643	4 1 5 2 3	40	579
	2 5 3 4 1	45	568	4 3 5 1 2	42	621	4 5 2 1 3	40	570
	4 3 2 5 1	45	568	1 4 3 5 2	42	440	4 1 2 5 3	40	564
0	4 3 5 2 1	51	581	2			3		
0	1 4 3 5 2	51	596						
0	1 2 5 4 3	51	605						
0	1 2 5 3 4	51	581						
0	4 3 1 2 5	51	606						

Fig. 7. The digital rows for Processor 1, Processor 2 and Processor 3.

6 Finding Extreme Paths on a Graph with Fully Connected Topology

According to the definition, the additive sequence contains integers, the sum of any elements has a unique value, that is, it does not coincide with the sum of other elements.

The method proposed in the work [9] involves splitting the starting point into initial and final points. Splitting the starting point consists in replacing it with two new ones, in one of which branches only enter, and from the other only branches leave. Thus, one can get the initial and final points. A signal is sent from the starting point to all nodes (processors) with information about the length of the path from the starting point to all other nodes [3].

At each point, after receiving the information, a set is prepared data, which contains:

1) Sequences of traveled routes with the addition of the number of this node.
2) $\sum_1$ - sums of additive numbers of the traveled path.
3) $\sum_2$ - the sum of the lengths of the branches of the traveled path.

Information is transmitted to all nodes that are not contained in the sequences of the corresponding processors.

Figure 7 and Fig. 8 show the scheme of work.

Given a row with 6 vertices P0p P1 P2 P3 P4 P5 P0k, where P0p is the initial (output signals) and P0k is the final peak (input signals).

Each processor is assigned an additively different number: P1 = 6; P2 = 9; P3 = 11; P4 = 12; P5 = 13.

A signal is sent to all 5 processors from the top node. From each processor, information is provided with a set of received paths to all other processors, except for those vertices that have already been bypassed. At the same time, the length of the traveled path is added. An additively different number is also written as the sum of the numbers from processors that are included in the route, except for the initial P0n and the final processor.

Example:
Route 0—2—3—4—1
The sum of the additively different numbers $9 + 11 + 12 = 32$

If the sum of the additive-distinct numbers is the same, that is, sequences arrive that have a different order of route passage, but have the same route points (0**243**1, 0**342**1), then there remains a route with the minimum path length among them.

At each stage, one vertex is added to each sequence and redundant options are discarded. This value is shown in green in the table, as well as the number of routes left and the option reduction factor *R*, in Fig. 7 and Fig. 9 – last columns.

The unique sum of additive-difference numbers allows one to search quickly and compare series that have the same set of vertices but in a different sequence.

So, for 6 processors, an additive series of 5 numbers is taken: 6 9 11 12 13. This data corresponds to Table 2, series ***A, D*** with $N = 5$. In Table 4 such notes are used: N_{rp} – the number of routes on the processor, N_r – the number of all routes, N_{rl} – the number of routes in the left direction, and R – reduction factor.

In Table 2 and Fig. 7, Fig. 8 and Fig. 9 **k** is the number of completed stages. Depends on the number of nodes, namely $n - 1$.

Processors from 1 to 5 - contain the possible routes of the nodes indicated in the columns.

$\sum_1$- sums of additive numbers of the traveled path.

$\sum_2$- the sum of the lengths of the branches of the traveled path.

k	Processor 4	$\sum_1$	$\sum_2$	Processor 5	$\sum_1$	$\sum_2$	N_{rp}	N_r	N_{rl}	R
1	4			5			1	5		
	1 4	6		1 5	6					
2	2 4	9		2 5	9		4	20		
	3 4	11		3 5	11					
	5 4	13		4 5	12					
	1 2 4	15	337	1 2 5	15	308				
	1 3 4	17	74	1 3 5	17	254				
	1 5 4	19	468	1 4 5	18	294				
3	2 1 4	15	330	2 1 5	15	504				
	2 3 4	20	329	2 3 5	20	509				
	2 5 4	22	528	2 4 5	21	557	12	60	30	2
	3 1 4	17	99	3 1 5	17	273				
	3 2 4	20	361	3 2 5	20	332				
	3 5 4	24	477	3 4 5	23	297				
	5 1 4	19	487	4 1 5	18	289				
	5 2 4	22	554	4 2 5	21	356				
	5 3 4	24	471	4 3 5	23	273				
	2 3 1 4	26	360	2 3 1 5	26	534				
	1 3 2 4	26	367	1 3 2 5	26	338				
	1 2 3 4	26	338	1 2 3 5	26	518				
	1 2 5 4	28	537	1 3 4 5	29	303				
4	2 5 1 4	28	547	2 4 3 5	32	560				
	1 5 2 4	28	554	2 1 4 5	27	559				
	1 5 3 4	30	471	4 2 1 5	27	561	12	60	20	3
	3 2 5 4	33	561	4 1 2 5	27	358				
	3 5 1 4	30	496	2 3 4 5	32	558				
	3 5 2 4	33	563	4 3 1 5	29	298				
	2 5 3 4	33	531	4 3 2 5	32	357				
	1 3 5 4	30	483	1 4 3 5	29	297				
	1 2 5 3 4	39	540	1 2 3 4 5	38	567				
	1 5 3 2 4	39	764	2 1 4 3 5	38	562	4	20	5	4
5	1 2 3 5 4	39	747	4 3 1 2 5	38	367				
	2 5 3 1 4	39	562	4 3 2 1 5	38	562				
	4		0	4 3 5 2 1	0	581				

Fig. 8. Digital rows for Processor 4 and Processor 5.

k	Processor 1	Σ1	Σ2	Processor 2	Σ1	Σ2	Processor 3	Σ1	Σ2
1	1			2			3		
2	2 1	9		1 2	6		1 3	6	
	3 1	11		3 2	11		2 3	9	
	4 1	12		4 2	12		4 3	12	
	5 1	13		5 2	13		5 3	13	
3	**2 3** 1	20	323	**1 3** 2	17	195	**1 2** 3	15	312
	2 5 1	22	510	**1 5** 2	19	382	**1 4** 3	18	91
	3 5 1	24	459	**3 5** 2	24	391	**1 5** 3	19	445
	4 2 1	21	350	**4 1** 2	18	215	**2 4** 3	21	354
	4 3 1	23	87	**4 3** 2	23	214	**2 5** 3	22	505
	4 5 1	25	481	**4 5** 2	25	413	**4 5** 3	25	476
4	**2 5 3** 1	33	525	**1 5 3** 2	30	592	**1 2 5** 3	28	514
	4 3 2 1	32	351	**4 1 5** 2	31	432	**1 5 4** 3	31	494
	4 3 5 1	36	484	**4 3 1** 2	29	224	**2 1 4** 3	27	356
	4 5 2 1	34	550	**4 3 5** 2	36	416	**2 5 4** 3	34	554
5	**4 3 5 2** 1	45	553	**1 4 3 5** 2	42	440	**1 2 5 4** 3	40	563
	4 3 5 2 1	51	581	1 4 3 5 2	51	596	1 2 5 4 3	51	605
0	1 4 3 5 2	51	596						
	1 2 5 4 3	51	605						
	1 2 5 3 4	51	581						
	4 3 1 2 5	51	606						

Fig. 9. Digital rows for Processor 1, Processor 2 and Processor 3.

Figure 9 and Table 3 present the data of Fig. 7 and Fig. 8 after discarding paths that are longer than found. This allows to significantly reduce the number of options corresponding to Table 5 at the end of this section.

Figure 10 presents the adjacency matrix of the graph, 6 nodes are given. The intersection is marked in green. The minimum path length 581 was found, and the same path 0 4 3 5 2 1 0 in Fig. 9 is marked in yellow.

Figure 11 presents the adjacency matrix of the graph, 20 nodes are given. This is input to the program. And also, the intersection is marked in green. A minimum path length of 1032 was found, and the same path is shown in Table 4 for 20 nodes.

Table 5 shows the results of the program on different computers with the following parameters:

PC1: Intel ® Core ™ i-3 4130 CPU @ 3.40GHz, RAM 8 GB, Windows 7;

PC2: Intel ® Core ™ 2 Duo CPU E8400 3.40GHz, RAM 3 GB, Windows XP.

Table 3. Data after discarding paths.

k	Processor 4	Σ1	Σ2	Processor 5	Σ1	Σ2	N_{rp}	N_r	N_{rl}	R
1	4			5			1	5		
2	1 4	6		1 5	6					
	2 4	9		2 5	9		4	20		
	3 4	11		3 5	11					
	5 4	13		4 5	12					
3	**1 3** 4	17	74	**1 2** 5	15	308				
	1 5 4	19	468	**1 3** 5	17	254				
	2 1 4	15	330	**3 2** 5	20	332	12	60	30	2
	2 3 4	20	329	**4 1** 5	18	289				
	2 5 4	22	528	**4 2** 5	21	356				
	5 3 4	24	471	**4 3** 5	23	273				
4	**1 2 3** 4	26	338	**1 2 3** 5	26	518				
	1 2 5 4	28	537	**4 1 2** 5	27	358	12	60	20	3
	1 5 3 4	30	471	**4 3 2** 5	32	357				
	2 5 3 4	33	531	**1 4 3** 5	29	297				
5	**1 2 5 3** 4	39	540	**4 3 1 2** 5	38	367	4	20	5	4
	1 2 5 3 4	51	581	4 3 1 2 5	51	606				

–	0	1	2	3	4	5
0	M	28	156	42	41	239
1	28	M	137	20	37	211
2	156	137	M	147	172	143
3	42	20	147	M	26	206
4	41	37	172	26	M	229
5	239	211	143	206	229	M

Fig. 10. The adjacency matrix of the graph for 6 nodes.

The program execution time is specified in seconds. At N nodes the minimum length of the path and its sequence of the passage of the row are also found.

In Table 6 a comparison of the algorithm for the complete enumeration of factorial numbers is presented. There given (N – 1)! options, and the number of options for the proposed algorithm is 2(N-2). Here N is the number of nodes.

N	0	1	2	3	4	5	6	7	8	9	10	11	12	13	14	15	16	17	18	19
0	M	194	112	98	104	75	171	130	51	60	101	112	61	165	145	143	101	85	52	79
1	194	M	198	219	256	249	73	184	161	138	266	236	209	41	176	57	138	275	221	125
2	112	198	M	208	215	183	138	28	74	127	209	42	172	158	239	172	61	176	163	144
3	98	219	208	M	42	61	227	223	137	101	60	209	37	206	79	162	183	102	54	101
4	104	256	215	42	M	37	256	235	152	128	20	208	51	239	119	199	201	73	52	134
5	75	249	183	61	37	M	239	204	126	114	26	172	42	227	138	193	176	41	28	125
6	171	73	138	227	256	239	M	118	125	128	261	179	206	37	209	86	84	256	211	126
7	130	184	28	223	235	204	118	M	86	134	230	69	189	143	246	165	54	201	183	149
8	51	161	74	137	152	126	125	86	M	54	151	92	104	126	164	118	51	134	101	73
9	60	138	127	101	128	114	128	134	54	M	134	145	78	114	112	84	86	138	85	20
10	101	266	209	60	20	26	261	230	151	134	M	198	57	247	138	209	201	54	51	142
11	112	236	42	209	208	172	179	69	92	145	198	M	172	196	252	204	98	156	158	165
12	61	209	172	37	51	42	206	189	104	78	57	172	M	190	102	153	152	79	20	85
13	165	41	158	206	239	227	37	143	126	114	247	196	190	M	178	52	98	249	199	106
14	145	176	239	79	119	138	209	246	164	112	138	252	102	178	M	126	195	178	122	98
15	143	57	172	162	199	193	86	165	118	84	209	204	153	52	126	M	112	221	165	69
16	101	138	61	183	201	176	84	54	51	86	201	98	152	98	195	112	M	183	151	98
17	85	275	176	102	73	41	256	201	134	138	54	156	79	249	178	221	183	M	60	152
18	52	221	163	54	52	28	211	183	101	85	51	158	20	199	122	165	151	60	M	97
19	79	125	144	101	134	125	126	149	73	20	142	165	85	106	98	69	98	152	97	M

Fig. 11. The adjacency matrix of the graph for 20 nodes.

Table 4. Path for 20 nodes.

0	8	11	2	7	16	6	13	1	15	19	9	14	3	4	10	17	5	18	12	0
–	81	92	42	28	54	84	37	41	57	69	20	112	79	42	20	54	41	28	20	61

7 Formalization of Proposed Algorithm

The proposed algorithm of TSP solving is divided into N stages, where N is the number of nodes in the fully complete topology graph.

In the sentence formulation it can be described as follows.

The Preliminary Step. Let's choose the main processor, or the main node of the network, and mark it as P0. This node has to start the work, synchronized the operation of all other nodes for further providing of common work and receiving the correct result. Let's mark the other N – 1 nodes as P1, P2, …, P(N – 1). At all next steps 1 node to the end of each route is added.

The First Step. The main node P0 sent to other nodes P1, P2, …, P(N – 1) the information about the route map, including node numbers and paths between them.

The Second Step. Each node of the network, except of the main, at the correct time, which is corresponded to the given number, sends the necessary information to each other node. This information includes: current route, the sum of the additive number of this route, and the length of the path to receiving node. After finishing the second step each node has N – 2 routes to other nodes, total values of additive numbers and length of routes.

Table 5. Results of the program on different computers.

N	PC1 time in seconds	PC2 time in seconds	Path length	Number
4	0.0007	0.0002	627	0 2 1 3 0
5	0.0008	0.0003	675	0 2 1 3 4 0
6	0.0009	0.0003	683	0 2 1 3 4 5 0
7	0.0012	0.0005	698	0 2 6 1 3 4 5 0
8	0.0013	0.0008	704	0 2 7 6 1 3 4 5 0
9	0.0021	0.0015	717	0 5 4 3 1 6 7 2 8 0
10	0.0105	0.0038	737	0 5 4 3 9 1 6 7 2 8 0
11	0.0289	0.0101	746	0 5 10 4 3 9 1 6 7 2 8 0
12	0.0758	0.0297	806	0 5 10 4 3 9 1 6 7 2 11 8 0
13	0.1197	0.0951	820	0 5 10 4 3 12 9 1 6 7 2 11 8 0
14	0.2983	0.3306	825	0 5 10 4 3 12 9 1 13 6 7 2 11 8 0
15	1,074	1.2506	918	0 9 8 11 2 7 6 13 1 14 3 4 10 5 12 0
16	4.2615	4.9701	925	0 9 8 11 2 7 6 13 1 15 14 3 4 10 5 12 0
17	17.7792	20.5296	945	0 9 8 11 2 7 16 6 13 1 15 14 3 4 10 5 12 0
18	75.9745	86.8515	1014	0 9 8 11 2 7 16 6 13 1 15 14 3 12 4 10 5 17 0
19	327.1397	370.7264	1020	0 9 8 11 2 7 16 6 13 1 15 14 3 4 10 17 5 18 12 0
20	1412,084		1032	0 8 11 2 7 16 6 13 1 15 19 9 14 3 4 10 17 5 18 12 0

Next Steps. Each node forms and sends the corresponded information to other nodes. But such block of information is formed only in the case, when the corresponded node is absent in the current route. In the specific case, when the node receives the new route with the same additive sum, from 2 routes the route with a smaller length is chosen.

Table 6. Comparison of the algorithm for the complete enumeration of factorial numbers.

N	$(N-1)!$	$2^{(N-2)}$
1	1	1
2	1	1
3	2	2
4	6	4
5	24	8
6	120	16
7	720	32
8	5,040	64
9	40,320	128
10	362,880	256
11	3,628,800	512
12	39,916,800	1 024
13	479 001 600	2 048
14	6,227,020,800	4 096
15	87 178 291 200	8 192
16	1,307,674,368,000	16 384
17	20,922,789,888,000	32 768
18	355,687,428,096,000	65 536
19	6,402,373,705,728,000	131 072
20	121,645,100,408,832,000	262 144
21	2,432,902,008,176,640,000	524,288

The sum of additive numbers is formed only in the case, when it corresponded to the same sequence of nodes, but in another order. Therefore, the number of routes is significantly reduced.

The additive numbers of the route are formed using following relation:

$$S_{i,t}\sum_{j=1}^{k} L_j. \tag{4}$$

where $S_{i,t}$ – additive number of node t relative to node i, k – number of transit nodes between i and t, j – counter, L_j – assigned value of node j corresponding to the basic digital row A.

The flowchart of proposed algorithm is presented at Fig. 12.

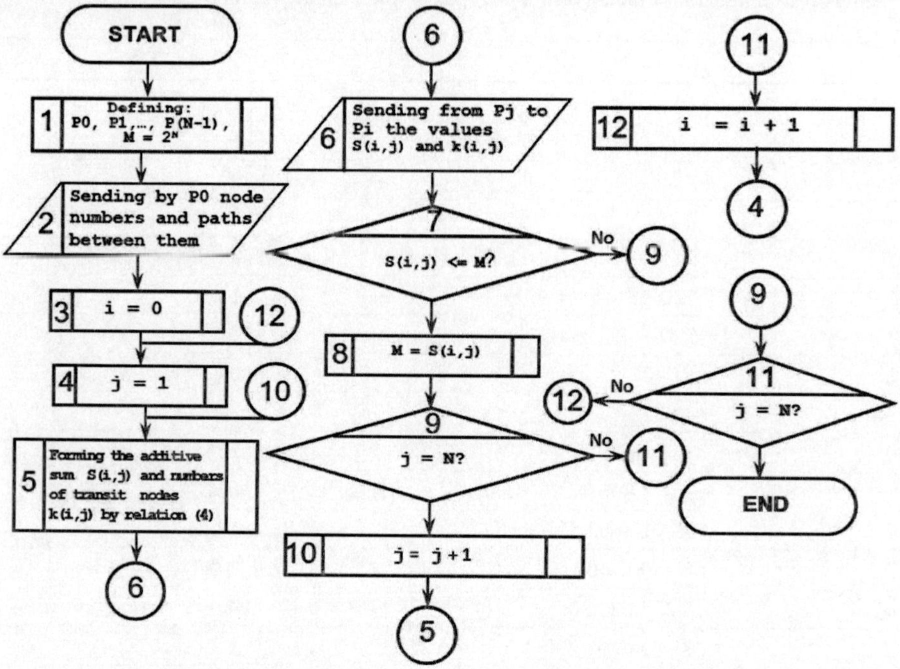

Fig. 12. Flowchart of the proposed TSP algorithm, based on digital series.

8 Conclusions and Outlook

A proposed algorithm for solving the TSP using additively different numbers has been developed. The unique sum of numbers allows you to quickly search and compare current routes that have the same set of vertices but in a different sequence and choose a shorter route path.

The algorithm reduces the number of options by discarding redundant routes.

The speed of work at 20 peaks is 23.5 min.

The number of options is $122 \cdot 10^{15}$ (complete enumeration), and after discarding extra routes, 262 144 options remain.

The project of using additive numbers to solve the TSP was tested on low-capacity personal computers.

These results demonstrate the effectiveness of using additive numbers to operate parallelism in a multicore machine (computing cluster) and the reduction of routes with the same set of nodes.

The results of the experiments in the paper are given for various graphs with the number of nodes from 4 to 20. The time to complete the task increases approximately four times at each step. When performing such a task with parallel computers, there are prospects for increasing the number of nodes.

Such studies can be applied to organizing the operation of a computer network with N devices or nodes that have to work simultaneously in a fully connected topology.

For example, in WSN, energy-efficient computing clusters or data centers with interconnected processor units, M2M communication model for Robotics and IoT with a gateway, as well as MIMO antennas for 5G and Beyond.

Really, the proposed in this chapter algorithm for solving TSP can be used today for forming a new generation of effective data exchange protocols in emerging networks with fully connected topology.

Using the proposed algorithm for defining the route in networks with defined connections between nodes, for example, with a cell topology, star topology, or complex mixed topology, is the subject of further investigation.

We are confident that the obtained results have more theoretical than applied meaning. Therefore, we offer this work as a WIP (Work-in-Process).

An attempt for a specific TSP-based MAC protocol is offered (OSI layer 2) herewith too, which distinguishes from existing routing protocols (layer 3) such as OSPF + BGP or RIP + BGP and above (with QoS considering).

The considered TSP-based MAC protocol for WSN provides access from a start node (cluster head, gateway) to all other nodes with a limited small net (cluster). The start node can be selected dynamically for future reconfiguration for energy efficiency and long life without frequent battery exhaustion. The start node provides TSP-based polling to all other networking energy-autarky nodes within the cluster.

Acknowledgment. The authors' acknowledgment belongs to the Open International University of Human Development "Ukraine" in Kyiv, Ukraine, for supporting this research and gaining inclusive development possibilities and human equality.

References

1. Deo, N.: Graph Theory with Applications to Engineering and Computer Science (Dover Books on Mathematics), 1st edn. Dover Publications (2016)
2. Benjamin, A., Chartrand, G., Zhang, P.: The Fascinating World of Graph Theory by Arthur Benjamin. Princeton University Press (2017)
3. Kostenko, L.I., Tymoshenko, A.G., Traynin, E.Z.: Electronic modeling of operations research tasks. Academy of Sciences of Ukraine, Institute of Cybernetics. Sciences, p. 163 (1973)
4. Dibrova, M.O: The method of forming non-intersecting paths in software-configured networks. Bull. Univ. "Ukraine" Inf. Comput. Cybern **1**(18), 131–136 (2017)
5. Rego, C., Gamboa, D., Glover, F., Osterman, C.: Traveling salesman problem heuristics: Leading methods, implementations and latest advances. European Journal of Operational Research **211**(3), 427–441 (2011)
6. Luntovskyy, A., Beshley, M., Klymash M. (eds.): Future Intent-Based Networking: on the QoS Robust and Energy Efficient Heterogeneous Software-Defined Networks, LNEE 831 Springer, 28 chapters, vol. XXI, p. 530 (2022). ISBN 978–3–030–92433–1
7. Luntovskyy, A., Guetter, D.: Highly-Distributed Systems: IoT, Robotics, Mobile Apps, Energy Efficiency, Security, p. 350. Springer Nature, Heidelberg (2022). ISBN: 978-3-030-92828-5
8. Zaigham, M. (ed.): Fog Computing: Concepts, Frameworks and Technologies. Springer, London (2017). ISBN 978–3–319–94890–4
9. Aloyan, G.S., Timoshenko, A.G.: Method of constructing a series of additively different numbers. Kibernetika, no. 6 (1969). https://ur.booksc.eu/book/6416122/22173e

10. Tymoshenko, A.G., Luzan, K.O., Rohacheva, T.V.: Creation of a series of integers with unique sums of elements. Sci. Notes Natl. Acad. Sci. **20**(Special issue: in two parts. Part 2), 516 (2002)
11. Pavlenko, O.Yu.: Determination of the optimal sequence of additively distinct numbers. In: European Scientific Discussions. Proceedings of X International Scientific and Practical Conference Rome, Italy, 15–17 August 2021, pp. 75–79 (2021)
12. Anderson, J.A.: Discrete Mathematics with Combinatorics. Second Edition. University of South Carolina-Spartanburg (2004)
13. Irvine, J., Harle, D.: Data Communication and Networks: An Engineering Approach. John Wiley & Sons LTD., Hoboken (2001)
14. Yeremenko, O.: Expert Meaning Interview, p. 2 (2022)
15. Yeremenko,O., Lemeshko, O., Yevdokymenko, M., Hailan, A., et al.: Fault-tolerant multicast routing with path and bandwidth protection. In: Conference Paper (2022)
16. Lemeshko, O., Papan, J., Yevdokymenko, M., Yeremenko, O.: Advanced tensor solution to the problem of inter-domain routing with normalized quality of service (2022)
17. Yeremenko, O., Lemeshko, P., Yevdokymenko, M., Baranovskyi, O., et al.: Complex investigation of the compromise probability behavior in traffic engineering oriented secure routing model in software-defined networks, Chapter (2022)
18. Yeremenko, O., et al.: Models and methods of secure routing and load balancing in infocommunication networks, Chapter (2022)
19. Yeremenko, O., Lemeshko, O., Yevdokymenko, M.: New Trends in Fast Reroute, Chapter (2022)
20. Papan, J., Filipko, A., Chovanec, T., Yeremenko, O.: Fault-tolerant multicast routing in infocommunication network with path and bandwidth protection, Article (2021)
21. Tanenbaum A., Wetherall D.: Computer Networks. Fifth edition. Pearson Prentice Hall. (2011)
22. Luntovskyy, A., Gütter, D.: Moderne Rechnernetze: Protokolle, Standards und Apps in kombinierten drahtgebundenen, mobilen und drahtlosen Netzwerken, 481 Seiten, 263 Abb., Springer, Heidelberg (2020). in German, ISBN: 9783658256166. https://www.springer.com/gp/book/9783658256166

A Survey of Intrusion Detection Methods in Wireless Networks

Anatolii Babaryka, Ivan Katerynchuk(✉), and Ivan Chesanovskyi

Bohdan Khmelnytskyi National Academy of the State Border Guard Service of Ukraine, Khmelnytskyi, Ukraine
zxk.ivan@gmail.com

Abstract. With the growing popularity of wireless networks, there is an urgent task of ensuring a high level of their security. The broadband nature of radio waves determines the specifics of possible attacks on wireless networks and methods for countering such threats. The paper describes the characteristic vulnerabilities of wireless networks and, based on these properties, identifies a number of possible types of attacks on wireless networks. Based on the analysis of modern research in the field of Information Systems Security, the classification of intrusion detection systems is proposed by monitoring methods, by the nature of responding to threats, by the nature of the data processing process, by security levels, by time characteristics, by the place of data processing and by the type of anomaly recognition.

Keywords: Analysis · Intrusion detection · Taxonomy of IDS · Wireless network

1 Introduction

The current level of evolution of computer technologies has led to a huge scale of functioning of computer networks. The world is becoming increasingly dependent on devices and sensors connected to networks. At the same time, the number of threats to the security and confidentiality of information in computer systems is growing. Therefore, specialized security tools are needed to ensure the normal functioning of such systems and prevent intrusions.

Intrusions into distributed computer networks can be organized either directly from the systems themselves (internal attacks) or from outside (external attacks). Internal attacks are carried out directly from the system using the rights of authorized users or hosts, or users trying to get extended access rights to information resources. External attacks are usually carried out by interfering with computer networks by unauthorized users-intruders.

Distributed computer systems have two components: hardware and software. Both components have different vulnerabilities. Hardware threats cause damage to the hardware of systems. Software threats target the data processed in such systems.

Features of the architecture of computer networks and the environment for distributing information signals also affect the choice of attackers to carry out intrusions into

M. Klymash et al. (Eds.): TCSET 2022, LNEE 965, pp. 305–316, 2023.
https://doi.org/10.1007/978-3-031-24963-1_17

such systems. According to ETSI 28 ETR 332 [1], the threats can be grouped into the following categories: integrity breach threats, privacy breach threats, accessibility breach threats, and specific threats.

Nowadays, wireless networks are widely used. along with such advantages of these systems as mobility, bandwidth, and relative cheapness of network infrastructure elements, wireless networks are characterized by a number of vulnerabilities (see Fig. 1).

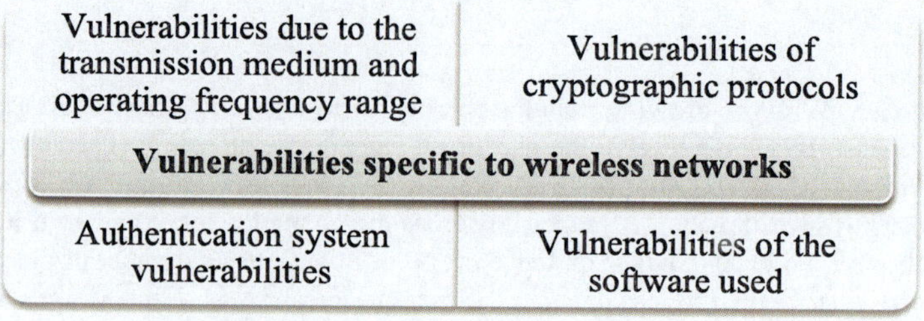

Fig. 1. Typical vulnerabilities in wireless networks.

In accordance with shown in Fig. 1 characteristic vulnerabilities, we can distinguish the following types of attacks inherent in wireless networks, which are shown in Fig. 2.

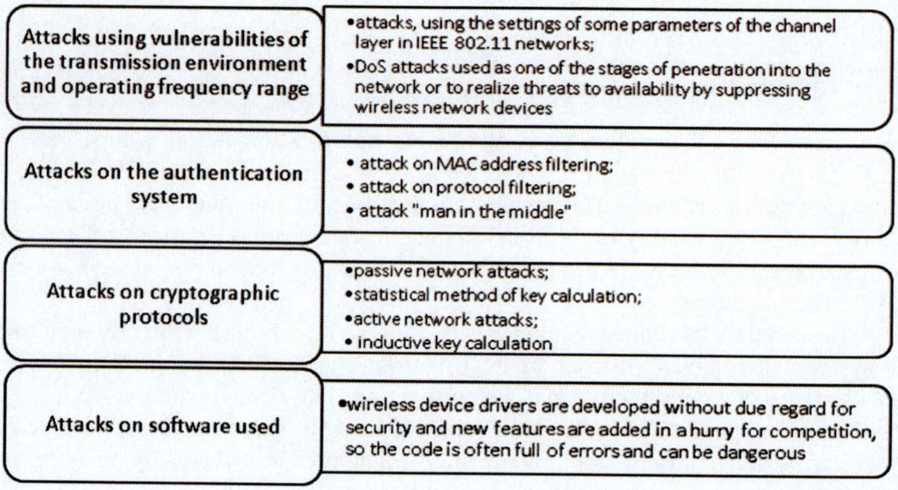

Fig. 2. Types of attacks on wireless networks.

Research conducted in the field of wireless network security is carried out in the following areas [2]:

– encryption and authentication technologies;

- research of intrusion detection models;
- research of security and network security strategy models;
- research of routing security models;
- research of attack models;
- research of key management models, etc.

In order to detect the facts of unauthorized access to computer systems, the Intrusion Detection System (IDS) has being developed. Monitoring and analysis of network communications is the basis for detecting anomalies and intrusions by IDS [3].

2 Taxonomy of IDS

Intrusion Detection System consider as probable attacks the behavior that can cause threats to the privacy, integrity, reliability, and availability of data [4].

Depending on the monitoring method, IDS can be classified as follows [5–8]:

- Host-based intrusion detection systems (HBIDS);
- Network-based intrusion detection systems (NBIDS);
- Hybrid Intrusion Detection Systems (HIDS).

HBIDS monitor and analyze internal elements of the computing system (host), track changes in important files on the host, and unusual CPU activity [9].

NBIDS monitor and analyze packet traffic on network interfaces, check entire IP packets, packet structure, multiple attempts to disrupt host access, etc. [10].

HIDS combine the advantages of HBIDS and NBIDS, while forming a well-organized system through the use of mobile agents.

By the nature of threat response, Intrusion Detection Systems can be classified as active, passive, and hybrid systems.

Passive systems only notify the user or administrator of a detected threat. Active systems resist intrusions by changing network access rules, restricting suspicious traffic, or blocking it. Active systems, in turn, can be "attacking" or "attacked". "Attacked" IDS are able to change the state of the attacked system by blocking network connections, destroying suspicious processes, and making changes to security rules. "Attacking" systems try to attack the attacker's platform themselves. Hybrid systems detect and counter threats automatically.

By the nature of the data processing process, Intrusion Detection Systems are divided into systems that process data continuously and systems that process data at regular intervals.

According to security levels, Intrusion Detection Systems are classified into high and low scale.

According to time characteristics, Intrusion Detection Systems can be divided into two groups: systems that attempt to detect intrusions in real time, and systems that process audit data with some delay.

According to the data processing location, Intrusion Detection Systems are classified as centralized and distributed depending on where the audit data is processed.

Modern Intrusion Detection systems can be classified according to the type of anomaly recognition [3]:

- Signature-based Intrusion Detection Systems (SIDS);
- Anomaly-based Intrusion Detection Systems (AIDS).

SIDS analyze systems by monitoring events according to a pre-defined pattern (behavior model) [11]. This behavior pattern is called a signature. Signatures are defined not only for individual methods of conducting attacks, but also for groups of attacks.

AIDS monitors systems to detect unusual (abnormal) behavior on hosts and on the network [12]. Anomaly detectors generate profiles of normal behavior of hosts, network connections, and users.

Various metrics are used to determine whether the analyzed activity on the network or on the host belongs to the formed profile of normal behavior. Based on the features of the above metrics, AIDS can be divided into categories (see below) Fig. 3):

- Methods based on statistical approaches [13];
- Methods based on nonparametric approaches [14];
- Methods based on neural network approaches [15, 16];
- Methods based on genetic algorithms, etc. [17–20].

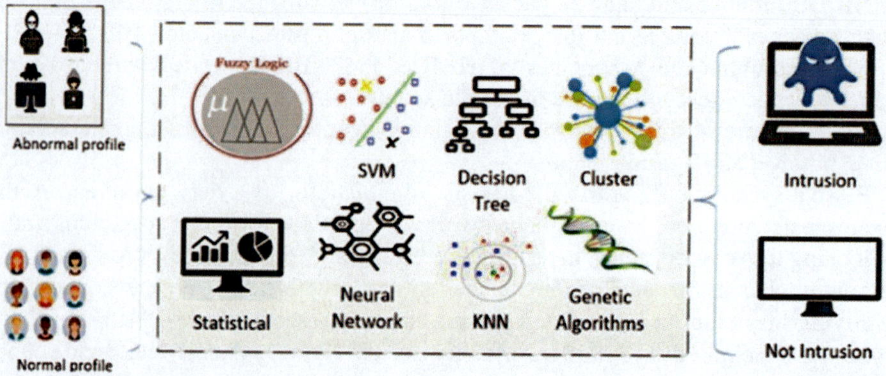

Fig. 3. Conceptual approaches to the functioning of anomaly-based intrusion detection systems [3].

The taxonomy of Intrusion Detection Systems is shown in Fig. 4.

Therefore, an important aspect of building IDS is the formation of rules that distinguish standard behavior (no attack) from non-standard behavior (state of attack). The main formation of these rules is algorithms for detecting intrusion. Patterns of known intrusions or system vulnerabilities can be used to identify intrusions.

Also, the identification of attacks can be carried out based on the detection of anomalies, which manifests itself as a deviation from the standard behavior.

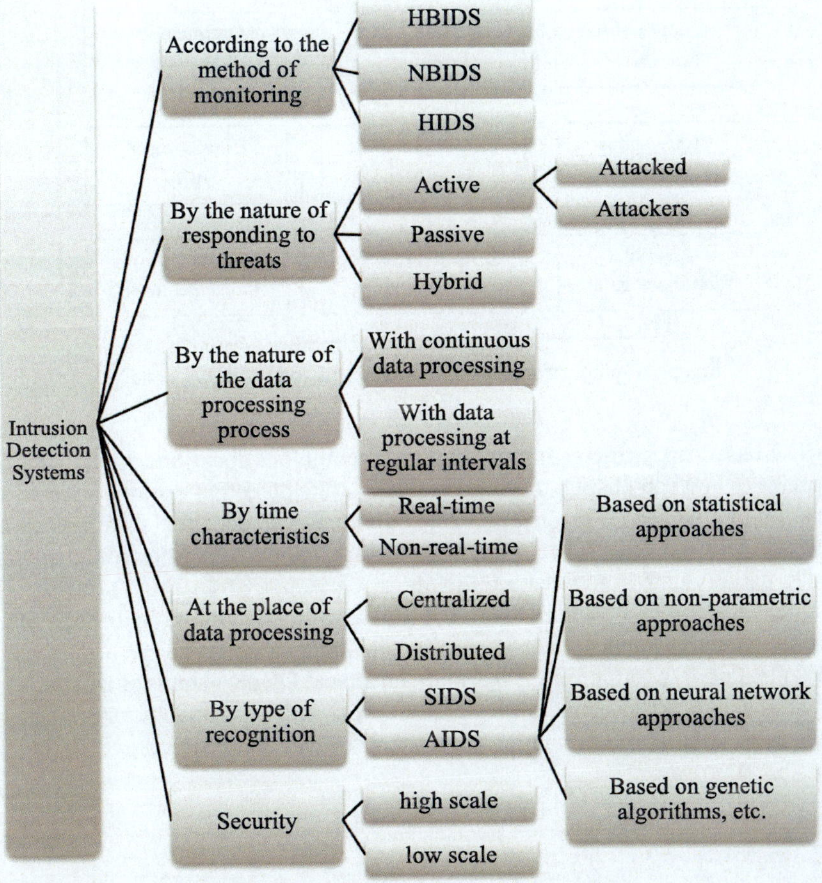

Fig. 4. Taxonomy of intrusion detection systems.

One of the promising areas of development of IDS is models based on artificial immune systems [26]. This technology stands out among others for its high flexibility, sensitivity to various types of attacks and the ability to adapt to different operating conditions.

Figure 5 shows the basic structure of adaptive IDS based on the principle of functioning of the artificial immune system.

A special feature of these systems is the presence of an adaptation module and a number of correction algorithms. They provide additional system configuration when searching for anomalies in the state of the information system or traffic.

These systems occupy an intermediate place between systems based on statistical approaches (formation of the state vector and application of criteria for assessing anomalies) and systems based on artificial intelligence technologies (availability of an adaptation mechanism).

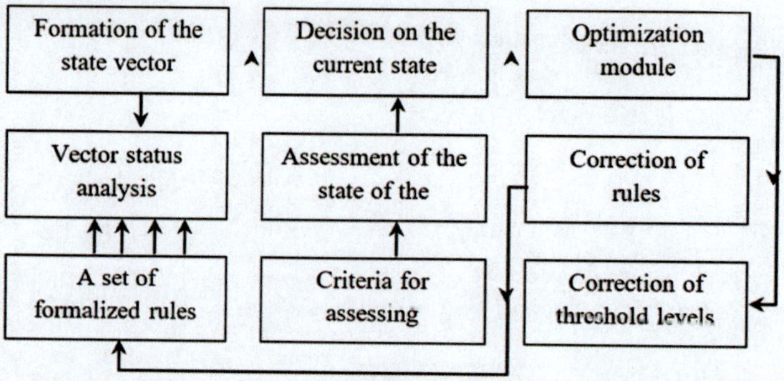

Fig. 5. Typical structure of adaptive intrusion detection systems.

AIDS based on artificial immune systems combines local and global search for abnormalities and can change the rules and adjust the detection criteria. This allows to find the optimal solution for different rule structures and specified criteria and makes them extremely effective when searching for anomalies with known signatures and allows to quickly tune in to potential threats.

Figure 6 and Fig. 7 show the effectiveness of adapting IDS based on artificial immune systems to potential attacks. The plausibility indicator was used as a performance indicator *TPR* (1). Possible traffic states: Normal - no attack, DDoS – status of an attack aimed at denial of access, R2L – unauthorized remote access, U2R – unauthorized access with administrator rights, probe-scanning the system for vulnerabilities.

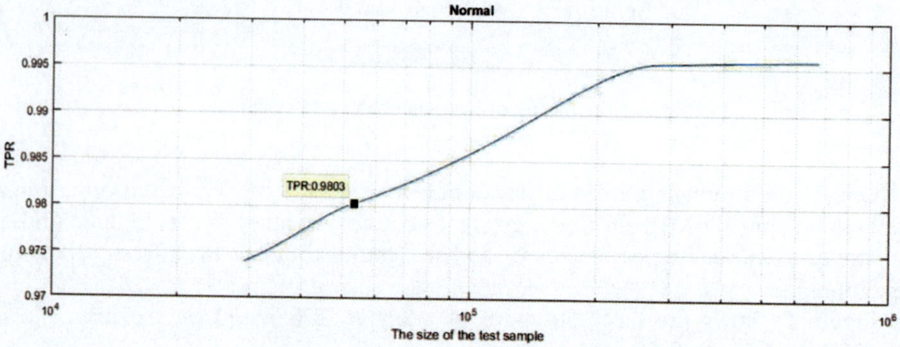

Fig. 6. TPR of the "Normal" state.

The simulation results show a steady trend of increasing the reliability of traffic state detection with increasing sample size. The dot in Fig. 6 indicates the TPR value achieved without system adaptation.

Shown in Fig. 7 dependence of TPR on the sample size shows that the use of an adaptation mode in AIDS based on models of artificial immune systems can significantly increase the level of likelihood for all types of attacks. It is worth noting that as the volume of the test sample increases, the reliability of detecting anomalies of all types increases, regardless of how fully and accurately the signatures are formed.

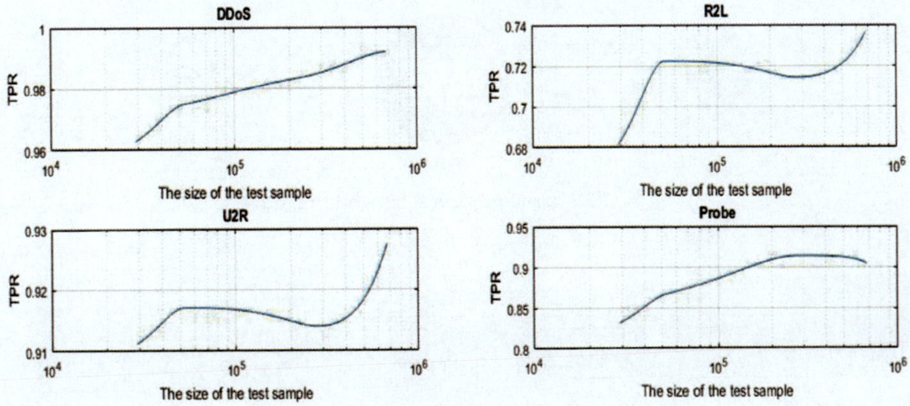

Fig. 7. TPR of attack states.

Unlike models based on neural networks, AIDS based on models of artificial immune systems undergo replication of the internal state at different stages of adaptation, which makes it possible to more accurately study the properties of abnormal states of the system and form reliable signatures of known attacks.

Due to the fact that there are more and more attack options every day, even with the constant updating of databases, none of the attack detection technologies can provide accurate identification of each intrusion. One way to overcome this problem is to use methods based on neural network approaches [21–24, 27, 28].

The neural network is able to analyze information and assess the consistency of data with characteristics recognized like expert systems. This puts this type of AIDS out of competition when detecting attacks of an unknown or insufficiently described type. However, the use of neural networks is preceded by training on a sample of examples, the volume and quality of which are crucial.

One of the important advantages of neural networks is their ability to take into account the characteristics of attacks, identifying elements that are not similar to those studied.

The IDS model based on the neural network is shown in Fig. 8. Each neural network "learns" to detect a separate attack class and can have a different structure. A typical structure of a separate neural network is shown in Fig. 9.

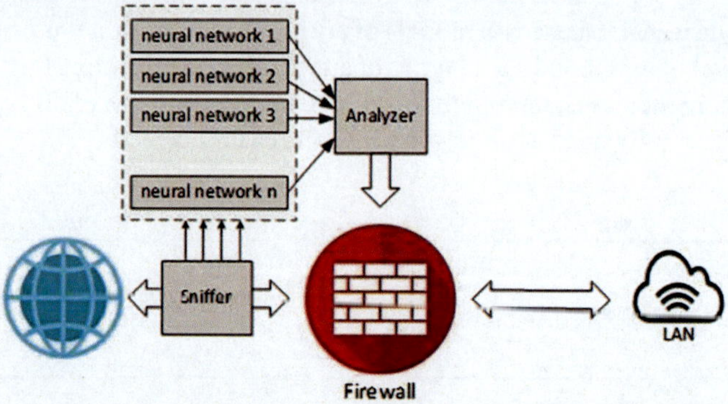

Fig. 8. IDS based on neural network methods.

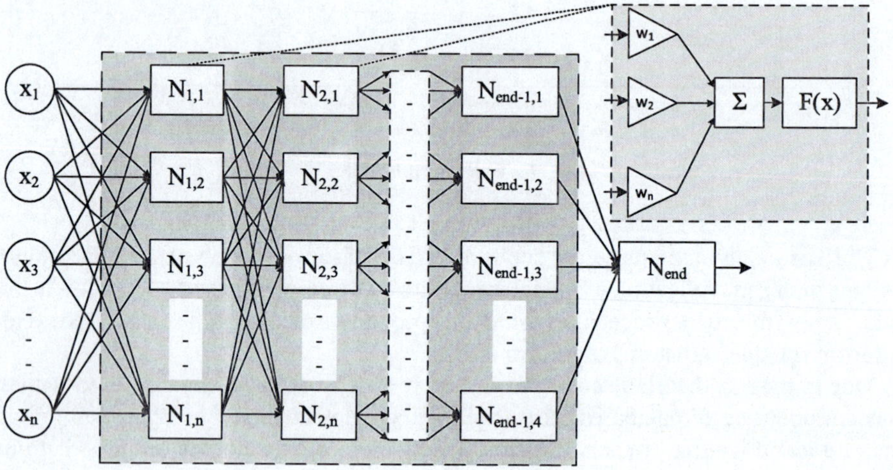

Fig. 9. Neural network.

Depending on the number of features, completeness of characteristics, and their entropy, the neural network of each type of attack contains a different number of neurons in the inner layer.

Figure 10 presents the results of a study of the effectiveness of detecting the state of AIDS traffic based on neural networks (number of neurons: 9 – 21; number of features: 10; states: no attack (Normal), Attack state (Attack), detection of an unknown attack (Unknown)). The test base from the previous example was used for modeling. The maximum TPR value is achieved with a certain number of neurons in the hidden layer. This number depends on the set of features that are used for recognition. The graphs show the ability of IDS to detect unknown attacks when configuring a neural network for a specific type of attack.

AIDS neural network methods include support vector machine (SVM), K-nearest neighbor (KNN), naive Bays, logistic regression (LR), decision tree, and artificial neural

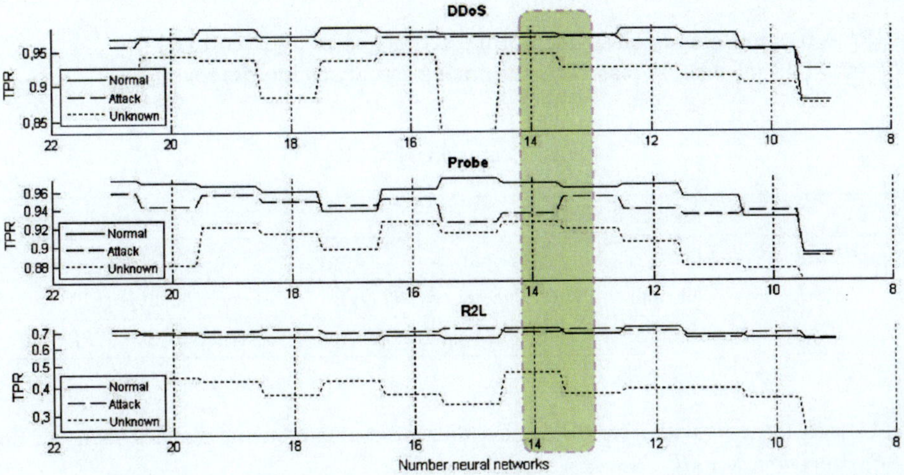

Fig. 10. Efficiency of detecting the state of AIDS traffic based on neural networks.

network (ANN). The advantages and disadvantages of these methods are shown in Fig. 11 [25].

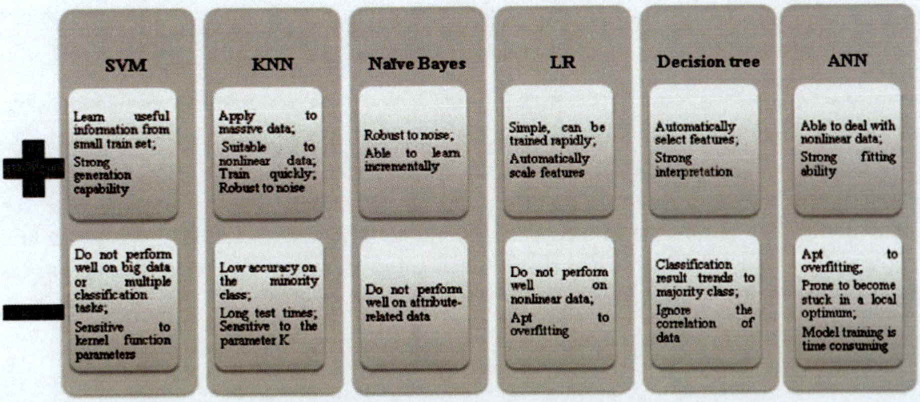

Fig. 11. Advantages and disadvantages of neural network methods.

3 Criteria for Evaluating the Effectiveness of IDS

To evaluate the effectiveness of using Intrusion Detection Systems, criteria based on a two-class classifier are often used. At the same time, as a result of the classifier's operation, there are four possible answers to the classifier (see Table 1):

TP – IDS correctly classified suspicious activity as interference;

TN – IDS correctly classified suspicious activity as normal activity (not a threat);

FP – IDS mistakenly classified normal activity as an intervention;
FN – IDS mistakenly classified suspicious activity as interference.

Table 1. Confusion matrix for IDS

	Predicted class		
Actual class	Class	Normal	Attack
	Normal	True negative (TN)	False positive (FP)
	Attack	False Negative (FN)	True positive (TP)

Usually systems with a two class classifier are evaluated using metrics such as *TPR, FPR, Precision, Recall*:

True Positive Rate (*TPR*), which reflects the ratio between the number of correctly predicted attacks and the total number of attacks on the system

$$\text{TPR} = \frac{TP}{TP + FN}. \tag{1}$$

False Positive Rate (*FPR*), which reflects the ratio between the number of cases that were mistakenly classified as an attack and the total number of cases where no attacks occurred

$$\text{FPR} = \frac{TP}{FP + TN}. \tag{2}$$

Precision characterizes the fate of attacks that were correctly classified, and the completeness indicator – what fate of objects of the true class out of all objects of the positive class the algorithm found. *Recall* demonstrates the algorithm's ability to find a given class, and *Precision* - the ability to distinguish this class from other classes is positive, how many of them are actually positive

$$Precision = \frac{TP}{TP + FP}, \tag{3}$$

$$\text{Recall} = \frac{TP}{TP + FN}. \tag{4}$$

Metrica F-score represents a joint assessment *Precision* and *Recall:*

$$F - score = \frac{2 \times \text{Recall} \times Precision}{\text{Recall} + \text{Precision}}. \tag{5}$$

4 Conclusions

The paper analyzes the methods of functioning of Intrusion Detection Systems. Based on the analysis of the literature and methods of constructing these systems, the classification of IDS is proposed. The main approaches to evaluating the effectiveness of IDS are described. Given that wireless networks use radio waves to transmit information and the broadband nature of this carrier, such systems are very vulnerable to possible attacks and unauthorized access.

The direction of further scientific research is the study of Intrusion Detection Systems based on sufficiently new techniques, so-called "honey pots", and the peculiarities of the application of IDS in wireless networks built using various technologies (Wi-Fi, LoRaWAN, NB-IoT, Sigfox, etc.).

Acknowledgements. This work was carried out within the framework of the research work 0121U11425 "Features of using video surveillance systems with automated video information processing functions and performance criteria" of the National Academy of the State Border Service of Ukraine.

References

1. ETSI ETR 332: Security techniques advisory group (STAG). Security requirements capture (1996)
2. Wen, W., Shang, C., Dong, Z., Keh, H.C., Roy, D.S.: An intrusion detection model using improved convolutional deep belief networks for wireless sensor networks. Int. J. Ad Hoc Ubiq. Comput. **36**(1), 20–31 (2021)
3. Khraisat, A., Gondal, I., Vamplew, P., Kamruzzaman, J.: Survey of intrusion detection systems: techniques, datasets and challenges. Cybersecurity **2**(1), 1–22 (2019). https://doi.org/10.1186/s42400-019-0038-7
4. Wanda, P.: A survey of intrusion detection system. Int. J. Inf. Comput. **1**(1), 1 (2020)
5. Sobh, T.S.: Wired and wireless intrusion detection system: classifications, good characteristics and state-ofthe-art. Comput. Stand. Interfaces **28**(6), 670–694 (2006)
6. Song, W., et al.: A software deep packet inspection system for network traffic analysis and anomaly detection. Sensors **20**(6), 1637 (2020)
7. Albers P., Camp O., Percher J., Jouga B., Puttini R.: Security in ad hoc networks: a general intrusion detection architecture enhancing trust based approaches. In: Proceedings 1st International Workshop on Wireless Information Systems (WIS-2002), pp. 1–12 (2002)
8. Sharma, R., Athavale, V.A.: Survey of intrusion detection techniques and architectures in wireless sensor networks. Int. J. Adv. Netw. Appl. **10**(4), 3925–3937 (2019)
9. Creech, G., Hu, J.: A semantic approach to host-based intrusion detection systems using contiguous and discontiguous system call patterns. IEEE Trans Comput. **63**(4), 807–819 (2014)
10. Bhuyan, M., Bhattacharyya, D., Kalita, J.: Network anomaly detection: methods, systems and tools. IEEE Commun. Surv. Tutor. **16**(1), 303–336 (2014)
11. Khraisat, A., Gondal, I., Vamplew, P.: An anomaly intrusion detection system using C5 decision tree classifier. In: Ganji, Mohadeseh, Rashidi, Lida, Fung, Benjamin C M., Wang, Can (eds.) PAKDD 2018. LNCS (LNAI), vol. 11154, pp. 149–155. Springer, Cham (2018). https://doi.org/10.1007/978-3-030-04503-6_14
12. Hall, J., Barbeau, M., Kranakis, E.: Anomaly-based intrusion detection using mobility profiles of public transportation users. In: IEEE International Conference on Wireless and Mobile Computing, Networking and Communications, pp. 17–24 (2005)
13. Chao, L., Wen, S., Fong, C.: CANN: an intrusion detection system based on combining cluster centers and nearest neighbors. Knowl.-Based Syst. **78**, 13–21 (2015)

14. Elhag, S., Fernández, A., Bawakid, A., Alshomrani, S., Herrera, F.: On the combination of genetic fuzzy systems and pairwise learning for improving detection rates on intrusion detection systems. Expert Syst. Appl. **42**(1), 193–202 (2015)
15. Kaissar, A., Nassif, A.B., Injadat, M.: A survey on network intrusion detection using convolutional neural network. In.: ITM Web of Conferences, vol. 43, p. 01015 (2022)
16. Sajja, G.S., Mustafa, M., Ponnusamy, R., Abdufattokhov, S.: Machine learning algorithms in intrusion detection and classification. Ann. Romanian Soc. Cell Biol. **25**(6), 12211–12219 (2021)
17. Aslahi-Shahri, B.M., et al.: A hybrid method consisting of GA and SVM for intrusion detection system. Neural Comput. Appl. **27**(6), 1669–1676 (2015). https://doi.org/10.1007/s00521-015-1964-2
18. Almomani, O.: A feature selection model for network intrusion detection system based on PSO, GWO FFA and GA algorithms. Symmetry **12**(6), 1046 (2020)
19. Cho H., Andreev A., Kalinin M. Moskvin D., Zegzhda D.: Mismatch-Resistant Intrusion Detection with Bioinspired Suffix Tree Algorithm. In: Jahn, C., Ungvári, L., Ilin, I. (eds) Algorithms and Solutions Based on Computer Technology. Lecture Notes in Networks and Systems, vol 387. Springer, Cham (2022). https://doi.org/10.1007/978-3-030-93872-7_1
20. Reddy, A.V.S., Reddy, B.P., Sujihelen, L., Mary, A.V.A., Jesudoss, A., Jeyanthi P.: Intrusion detection system in network using decision tree. In: 2022 International Conference on Sustainable Computing and Data Communication Systems (ICSCDS), pp. 1186–1190 (2022)
21. Rajendra, P., Shivashankar, P.: Multitier energy system review on secure intrusion detection system in MANETs. In: 2017 2nd IEEE International Conference on Recent Trends in Electronics, Information & Communication Technology (RTEICT), pp. 1722–1726 (2017)
22. Chua, T.H., Salam. I.: Evaluation of Machine Learning Algorithms in Network-Based Intrusion Detection System. https://arxiv.org/abs/2203.05232 (2022)
23. Srivastava, N., Chaudhari, A., Joraviya, N., Gohil, B.N., Ray, S., Rao, U.P.: A review of machine learning-based intrusion detection systems on the cloud. In: Rao, U.P., Patel, S.J., Raj, P., Visconti, A. (eds.) Security, Privacy and Data Analytics. LNEE, vol. 848, pp. 303–317. Springer, Singapore (2022). https://doi.org/10.1007/978-981-16-9089-1_25
24. Unnisa, A.N., Yerva, M., Kurian, M.Z.: Review on intrusion detection system (ids) for network security using machine learning algorithms. Int. Res. J. Adv. Sci. Hub **4**(3), 67–74 (2022)
25. Liu, H., Lang, B.: Machine learning and deep learning methods for intrusion detection systems: a survey. Appl. Sci. **9**(20), 4396 (2019)
26. Skatkov, A.V., Briukhovetskyi, A.A.: An adaptive model of intrusion detection in computer networks based on artificial immune system. Electr. Comput. Syst. **12**(88), 102–111 (2013)
27. Katerynchuk, I., Balendr, A., Komarnytska, O., Islamova, O., Ordynska, I., Chubina, T.: Training of remotely piloted aircraft operators for border surveillance in Ukraine based on the European Union standards. Revista Romaneasca pentru Educatie Multidimensionala **13**(3), 134–150 (2021)
28. Katerynchuk, I., Rachok, R., Mul, D., Balender, A.: Modelling of radio waves propagation and creation of radio networks using geoinformation systems. In: 2016 13th International Conference on Modern Problems of Radio Engineering, Telecommunications and Computer Science (TCSET), pp. 677–681. IEEE (2016)

Applied Steganographic System for Hiding Textual Information on Audio Files

Serhii Buchyk, Serhii Toliupa(✉), Nataliia Lukova-Chuiko, Oksana Khomenko, and Yurii Serpinskyi

Taras Shevchenko National University of Kyiv, Bohdan Hawrylyshyn Street 24, Kyiv UA-04116, Ukraine
tolupa@i.ua

Abstract. The article presents an applied steganographic system of hiding textual information in audio files on the basic of an advanced approach that combines the methods of steganography and cryptography. The method of the least significant bit (LSB) was used as a steganographic transformation algorithm. The low level of reliability of the method can be improved by introducing a cryptographic layer, the feasibility of which is justified in the article. The improvement is to add one of the modern symmetric encryption algorithms – the AES algorithm in CBC mode. The applied steganographic system for hiding textual information in audio files, taking into account the improvement of the LSB method, is implemented in the Microsoft Visual Studio 2019 environment using the cryptographic libraries that are present in it. The application is developed in the C++ programming language. NIST tests were used to assess the stability of the implemented system to steganalysis. The proposed approach is considered in order to anticipate its role in emerging networks.

Keywords: Steganography · Steganographic system · Least significant bit · AES-CBC · An audio file · NIST-tests

1 Introduction

To ensure confidentiality, not only cryptography methods are used, but also steganography methods. Although each of these sciences has its areas of application, they can be combined to increase the efficiency of information protection (for example, for the transmission of cryptographic keys) or to form a reliable secure communication system [1].

Steganography is the science of hiding messages, not by turning them into something unrecognizable, as it happens in cryptography, but by going unnoticed [2]. In general, steganography can use in the form of containers anything from digital media such as images, audio, and video to network protocols of the OSI reference model or protocol for emerging networks.

In this study, attention is paid only to steganography in the audio environment. However, the presented method can be adapted for hiding not only in audio files. It is

M. Klymash et al. (Eds.): TCSET 2022, LNEE 965, pp. 317–334, 2023.
https://doi.org/10.1007/978-3-031-24963-1_18

believed that the use of steganography in audio is difficult because the human auditory system (HAS) is more sensitive to small changes in audio data than the human visual system (HVS) [3]. Although audio files are potentially more profitable for steganographic embedding since larger amounts of textual information can be hidden without noticeable degradation.

Some of the studies published in the field of audio steganography look at a set of common approaches such as Least Significant Bit (LSB), Echo Hiding, Spread Spectrum (SS), and Phase Coding, discussed in [4].

The scheme of hiding data in video and audio files, proposed in [5], uses 4LSB and a phase coding algorithm to preserve the quality of the video file even after embedding secret data.

The use of audio steganography with RC4 streaming cipher encryption text is developed in [6].

The proposed approach [7] to audio steganography used a combination of DCT coefficient calculation and AES encryption scheme to improve module security. [8] proposes a method of rounding the values of image elements to modify the smallest bit in steganographic protection problems.

The developed existing steganographic methods can be used both for the hidden storage of information and for the protection of copyright to various intellectual property objects by attaching digital identification labels and "watermarks".

Most software solutions, where a sound signal is used as a container, offer only the LSB method to embed information, which can be tried to explain the complexity of implementing alternative methods. However, the use of this simple method in combination with other methods no longer makes it so at all and allows you to implement new approaches that improve the basic characteristics of the steganosystem based on them.

The proposed approach in this article is also implemented by a combination of two steganographic and cryptographic techniques.

The described improved algorithm can be used at the network systems for providing information security and also to enhance the security of next-generation networks, for the secure exchange of sensitive data over emerging networks. Given that the information steganography methods can be used for hiding secret messages in the temporal behavior of the traffic or the storage fields in the network protocol, which composes CTCs (covert timing channels) and CSCs (covert storage channels) accordingly [9], it makes sense to use the method presented below to build network covert channels in the new network environment, for example, covert channels based on streaming media networks that hide secret messages in audio and video traffic and use popular streaming media applications as the carrier, and covert channels based on IPv6 network where IPv6 header and its extensions have many reserved fields or other fields which can embed information, thus leading to many possible covert channels.

2 Task Formulation

The study aims to modify the LSB algorithm for the hidden storage and transmission of text information in sound files through the additional use of the cryptographic method and pseudo-random numbers to create a stable crypto key.

The object of research is the process of encoding and decoding textual information in audio files, using the previous cryptographic transformation of the message.

The subject of the study is LSB's steganographic method for hiding textual information in audio files and the AES cryptographic algorithm.

Thus, the task is to improve the LSB algorithm to hide text information in sound files by additionally using the cryptographic method and pseudo-random numbers to create a stable crypto key, as well as to assess how improving the approach affects the main criteria of the steganographic system.

3 Solving the Task

We will define some basic terms for a correct understanding of the content of the article.

The *steganographic system (steganosystem)* is a set of $\Sigma = (X, M, K, Y, E, D)$, empty containers X, messages M, keys K, filled containers Y and transformations E and D that bind them (embedding and extraction algorithms).

A container (media, cover) is secret information in which confidential data (messages) will be hidden. In computer steganography, any file or data stream can serve as a container. Due to its redundancy, most often digital containers are images, audio, or video signals.

A stegoconteiner (stegosignal, stegofile, stegobject) is a modified version of a container containing a hidden message as a payload.

A message (payload) is information whose privacy the sender wants to keep. With the help of knowledge of passwords, the recipient can obtain secret information from a regular audio file that contains an embedding, that is, it is a stegofile.

An initialization vector (IV) is a randomly generated string, always 16 bytes long, used in the AES algorithm to create encrypted text. The initialization vector is saved so that it can be used to extract the message later.

A sample is a relatively small digitized sound fragment.

In the developed software implementation, one of the basic methods of steganography was improved – the Least Significant Bits (LSB) technique. The classical interpretation of the method of replacing the least significant bit is possible by replacing the least significant bits in the container with bits of the hidden message.

Below are the general basic steps of the LSB technique [10]*:*

- audio file bytes are converted to a bit sequence;
- all message symbols are converted to an equivalent binary system;
- the least significant bit of audio is replaced by a bit of the message.

To improve steganographic technology, additional cryptographic protection has been introduced in the form of the AES-CBC algorithm, according to which a secret message is encrypted using a key-password before embedding.

The AES algorithm is implemented in CBC (Cipher Block Chaining) mode, which allows encrypting messages of greater length by segmenting information with fixed-length blocks and applying it to the first segment of the XOR operation with a random value of the initialization vector. This initialization vector can be zero, which is not recommended from the point of view of the security of the algorithm. The result of this operation is encrypted with a key, and thus we have a new encrypted block. Next, XOR is applied to the next message block along with the encrypted text received in the previous step, which is also encrypted for subsequent blocks.

Alternatively, the value of the initialization vector could be made programmatically fixed, but in this software solution, its value is generated during each attempt and used to receive a hidden message from the audio file in the future. Thus, if the initialization vector is a secret, then the algorithm uses two passphrases.

Thus, according to Fig. 1, an audio file container, messages, and a crypto key are submitted to the input of the program.

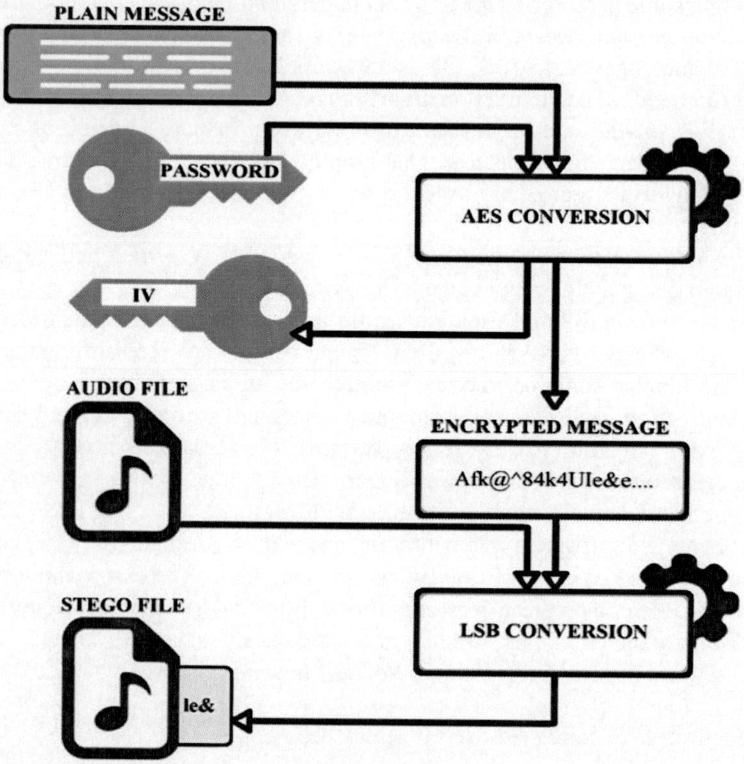

Fig. 1. Scheme of the process of hiding the message according to the LSB-AES algorithm.

The input audio file and the messages encrypted by the cryptographic algorithm are programmatically converted to a binary representation.

Then, according to the steganographic algorithm, the message bits replace the defined bits of the input audio file that form the output audio file with the embedding.

Given that each sample of the audio file contains 1 bit of the message, the sequence after the attachment does not differ from the original one, provided that the least significant bit of the original sequence coincides with the nested bit of information. Otherwise, the value of the last bit will be with a probability of equal to 50% increased, or reduced, by 1.

The process of finding a hidden message that is schematically depicted in Fig. 2 is the opposite.

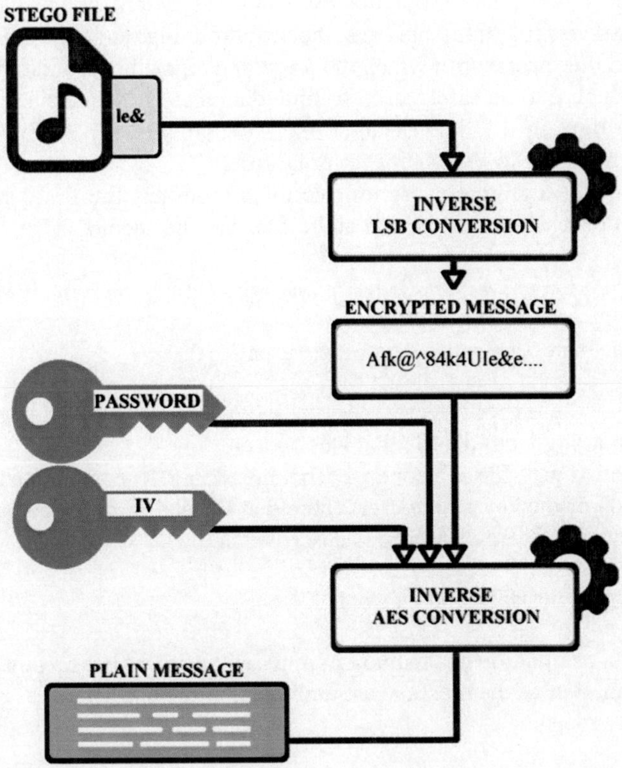

Fig. 2. Scheme of the process of retrieving the message according to the LSB-AES algorithm.

The user needs to know the password, and initialization vector (IV). Using the correct password and the value of the IV bits, the message will be restored, decrypted, and converted to text form.

For added security, a less secure message was embedded that looks like a secret message – a password randomly generated in compliance with security practices.

This is ensured by embedding the honeypot after encryption, making it more visible. Thus, when extracting data using steganalysis methods, the attacker will receive a message that will be perceived as a secret. If the attacker continues his steganalysis, he will have encrypted information that looks like the usual load of an audio file, and with a high probability, it will not be paid attention to. Thus, given the manipulations described above, the message will be almost "invisible".

3.1 Description of the Software Implementation

For a practical demonstration of the work of an improved algorithm of work, a software application was developed that implements the algorithm presented in the study. The application system of steganographic protection of information in audio files using a cryptographic algorithm is implemented using the Microsoft Visual Studio 2022 environment and cryptographic libraries, the programming language is C++.

As an audio file format with which the presented algorithm works, the WAV format was chosen, which can be safely used to hide data and is ideal for implementing the proposed algorithm: in the field of audio data, uncompressed and raw data obtained directly from an analog-to-digital converter is stored.

The developed program consists of two main modules: the first part is the actual concealment of text information in the audio file, and the second is the reverse process of finding it.

The procedure for hiding a message in an audio file can be described as follows:

1. Filling in the necessary input data by the user.

   - Determine the text of a secret message in a way to choose from: entering by the user or reading from the file that was chosen.
   - Definition of wav-file as a container for embedding, its conversion to binary form.
   - Definition of the key by the user: entered at the choice or based on the results of the password generator, which is described below.

2. Preprocessing a message and a signal.

   - There is a calculation of the maximum possible size of the text that is planned to be hidden. Bits of the message are embedded if the text size does not exceed this size.

3. The module for hiding a message using an advanced steganographic algorithm.

   - Generate a password as a honeypot and add it to the beginning of the message.
   - Message bits are encrypted according to the AES-CBC algorithm.
   - Hiding information in an audio cover using LSB encoding: the resulting values gradually replace the least significant bit of each audio. The beginning of the replacement bits starts after the header in the audio data area after the built-in signature-honeypot is applied as an additional security measure that was previously described.
   - Modified audio samples are recorded to the file forming the output audio signal as a stegoobject.

4. Sending a stegofile - a modified audio file - to the recipient at the destination through a communication channel, keys through separate ones.

The procedure for finding a message from an audio file can be described by the following sequence:

1. Filling in the necessary input data by the user.

   - Selecting an audio file in which plan to find hidden text information.
   - Entering the password and initialization vector, in theory, obtained from the sender, their verification.
   - Assumptions about the number of characters in a message.

2. Preprocessing the audio file and checking the keys.
3. Module of recovery hidden message by the advanced steganographic algorithm.

   - Recovering an encrypted hidden message using LSB decoding: gradually extracting the LSB of each byte from the stegofile audio area skipping the honeypot hide area and finding the message that is encrypted. The duration of the process is determined by the number of characters of the hidden message assumed by the user
   - Recovery of a hidden message, presentation in ASCII format, the ability to save it to a file: successful application to an encrypted message of the decryption process following the AES-CBC algorithm and its demonstration to the user, provided that the keys are entered correctly.

4. After entering the correct password and initialization vector, the message will be successfully displayed.

The general interface of the program is presented in Figs. 3, 4, 5, 6, 7 *and* 8, *which was presented in* [11], *and consists of 3 interface components that give the user the ability to interact to hide textual information:*

- Form 1.

It is the main form in which everything happens, for which two modules are introduced.

The first (see Fig. 3) is embedding, for hiding encrypted data in audio, and the second (see Fig. 4) – is for deletion, to obtain data from this audio. An audio file container, a message entered by the user or from a text file, and a password are served at the entrance, in case the appropriate option is selected to implement additional cryptographic protection of the message.

When the form is filled out correctly, when you click on the "Hide" button, after execution, the program will report the success of the operation and how many of the least significant bits possible were involved, as shown in Fig. 5 or report the error and its cause.

The password in the software model is formed to choose from: it is set by the user or, if desired, generated using pseudo-random numbers, for which a password generator based on PRNG is developed as additional functionality for the user.

- Form 2.

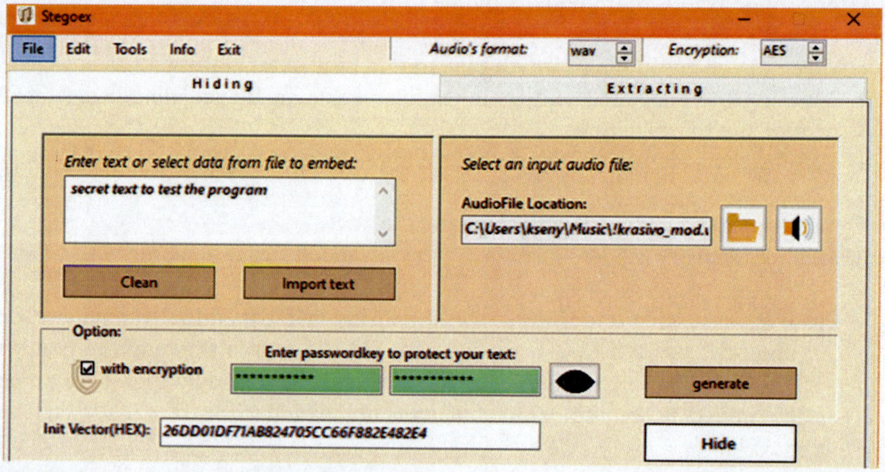

Fig. 3. General interface of the application program. Hiding module with an example of filling in data.

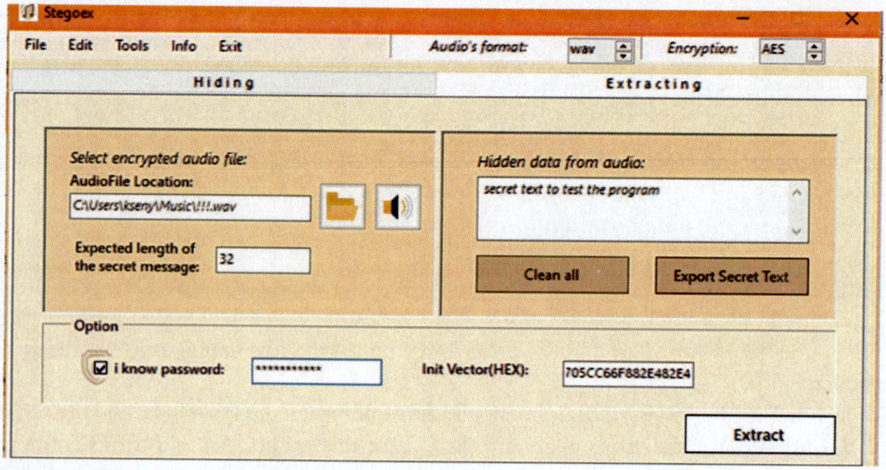

Fig. 4. General interface of the application program. Extraction module with an example of filling in data.

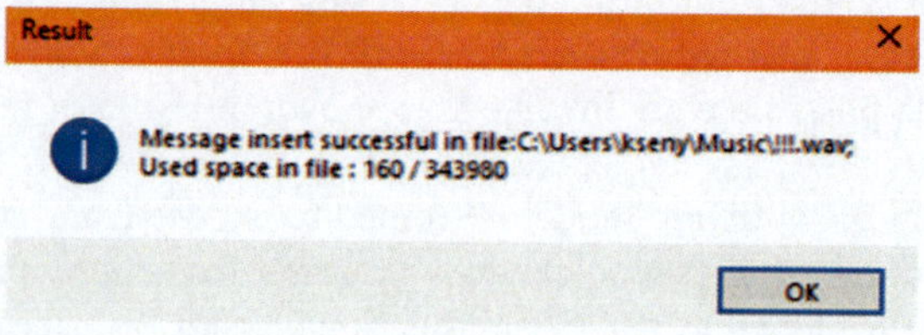

Fig. 5. Notification of the success of the embedding.

Designed to generate a password following the parameters selected by the user, is called in modal mode when clicking on the "generate" button or through the tool- > Password Generator context menu item. As depicted in Fig. 6. The minimum allowed length is 4 characters, in case of non-fulfillment of the condition, a corresponding warning appears. It should be noted that the password is generated using PRNG and checked for reliability. In particular, the capabilities of the generator are directly used to create a honeypot. The application can set a custom password length of 4 to 19 characters, as well as checks for reliability.

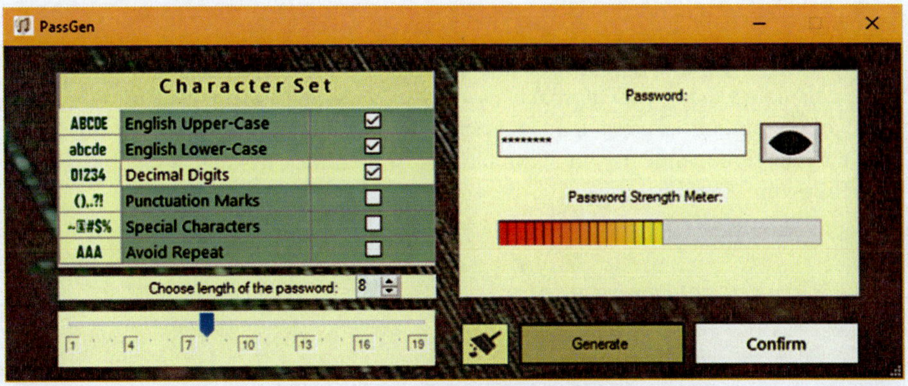

Fig. 6. Password generation form.

The password is formed to choose from: set by the user or, if desired, generated using pseudo-random numbers.

To obtain a pseudo-random number, two generators are combined:

- Mersenne twister – or rather mt19937, is based on the properties of prime numbers;
- std::random_device – it is an evenly distributed random number generator for generating truly random numbers that produce an undetermined random number.

In the program for the formation of pseudo-random sequences, one PRNG is initialized as a result of the action of another PRNG (see Fig. 7), then the generated number is used to generate a password.

The generated number is directly used to generate a password when choosing from 6 arrays of digits, Upper and lowercase English letters, some punctuation marks, or special characters with a repetition of the operation until the selected password length is reached.

Fig. 7. Initialization of generators.

The stability of the generated password against a particular attack (brute force attack) can be calculated using the approach used in [12], which calculates the information entropy of the random process that generated the password. If each character in a password is created independently and with uniform probability, the entropy in bits is given by the following Eq. (1):

$$H = L\log_2 N = \frac{logN}{log2}, \tag{1}$$

where N – number of possible characters, L – number of characters in passwords.

According to RFC 4086 [13], it is recommended to use keys with an entropy value of 29 bits for online attacks up to 96 bits for cryptographic keys. Currently, it is recommended to use passwords with an entropy of at least 80 bits.

The power of the key space of possible generated passwords according to the algorithm used in the program makes it impossible to brute force attack complete.

- Form 3.

To play media files, including audio files (see Fig. 8), use a special added element of the item panel – COM component – Windows Media Player.

3.2 Evaluation of Efficiency

When using steganography, the following characteristics are important [14]:

- capacity – characterizes the amount of secret data relative to the file size, where you can embed in a container using a certain method;
- robustness – indicates the resistance of stegofiles to deliberate and unintentional attacks;
- invisibility – describes the similarity of the filled container and empty container degradation level after hiding the secret message.

Maintaining the balance of all characteristics is important and somewhat difficult to implement. In general, LSB is a very simple method and has an advantage in the criteria of invisibility, but this method becomes weak and predictable if you perform it traditionally.

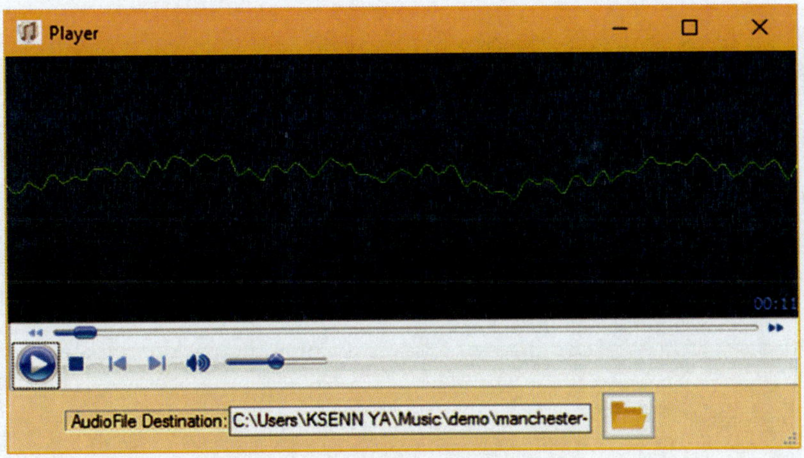

Fig. 8. Built-in audio playback form.

The bandwidth of the stegochannel can be found using the following equation:

$$size = \frac{F}{BPS} - 132, \tag{2}$$

where F – container file size, BPS – the number of bytes per sample (this program is designed for BPS, which is equal to 16 bytes).

Since there is simply a replacement of the least significant bit, it is logical to say that the addition of a cryptographic component to the method does not affect the throughput of the secure communication channel created by the system, which can be increased only in cases of increasing the power of filling the container.

The developed steganographic tool is built in such a way that it does not change the properties of audio files for the sake of embedding bits of the message and does not leave any suspicious signatures, therefore, when analyzing the stegofile in this way, it is impossible to detect the presence of hidden messages, and even more so to identify the use of this particular utility, which is an advantage.

As you know, the least discharges contain very little useful information. Their filling with additional information practically does not affect the quality of perception. Therefore, in general, the difference between empty and filled containers should not be felt by human perception organs. The signal quality increases with increasing signal size, as the ratio of signal bits to the number of bits of the message increases, thus distortion decreases.

Distortions made by replacing the least significant bit in a real beep can be detected by visual analysis of the signal spectrum. Verification of the audibility of distortions (in other words, the possibility of detecting attachments) is carried out visually using the Audacity 1.3 sound editor (see Fig. 9).

Based on the comparison of oscillograms, it can be argued that there are no hidden messages.

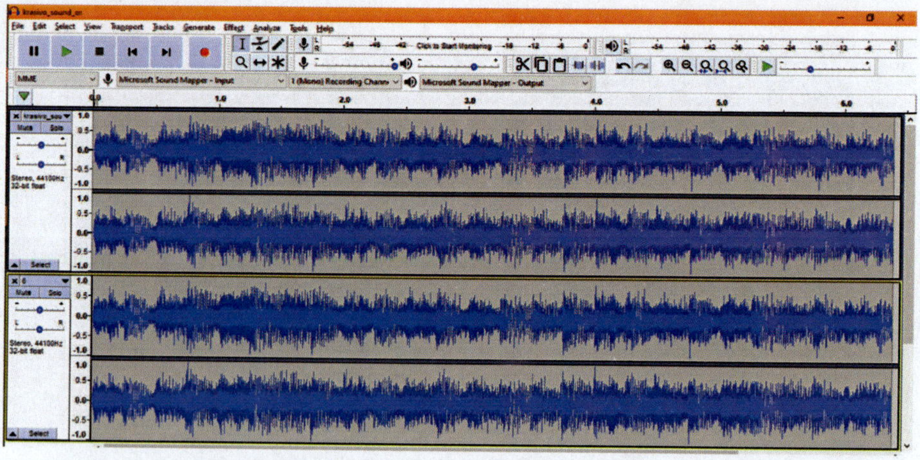

Fig. 9. Oscillogram of "blank" audio file (top) and with a hidden message (bottom).

Table 1. Characteristics of standards NIST-tests.

Test's №	Name/Description
1	Frequency (Monobits) Test
	The focus of the test is the proportion of zeroes and ones for the entire sequence. The purpose of this test is to determine whether that number of ones and zeros in a sequence are approximately the same as would be expected for a truly random sequence
2	Test For Frequency Within A Block
	The focus of the test is the proportion of zeroes and ones within M-bit blocks. The purpose of this test is to determine whether the frequency of ones is an M-bit block is approximately M/2
3	Runs Test
	The focus of this test is the total number of zero and one runs in the entire sequence, where a run is an uninterrupted sequence of identical bits. The purpose of the runs test is to determine whether the number of runs of ones and zeros of various lengths is as expected for a random sequence
4	Test For The Longest Run Of Ones In A Block
	The focus of the test is the longest run of ones within M-bit blocks. The purpose of this test is to determine whether the length of the longest run of ones within the tested sequence is consistent with the length of the longest run of ones that would be expected in a random sequence
5	Random Binary Matrix Rank Test
	The focus of the test is the rank of disjoint sub-matrices of the entire sequence. The purpose of this test is to check for linear dependence among fixed length substrings of the original sequence
6	Discrete Fourier Transform (Spectral) Test

(*continued*)

Table 1. (*continued*)

Test's №	Name/Description
	The focus of this test is the peak heights in the discrete Fast Fourier Transform. The purpose of this test is to detect periodic features (i.e., repetitive patterns that are near each other) in the tested sequence that would indicate a deviation from the assumption of randomness
7	Non-Overlapping (Aperiodic) Template Matching Test
	The focus of this test is the number of occurrences of pre-defined target substrings. The purpose of this test is to reject sequences that exhibit too many occurrences of a given non-periodic (aperiodic) pattern
8	Overlapping (Periodic) Template Matching Test
	The focus of the test is the longest run of ones within M-bit blocks. The purpose of this test is to determine whether the length of the longest run of ones within the tested sequence is consistent with the length of the longest run of ones that would be expected in a random sequence
9	Maurer's Universal Statistical Test
	The focus of this test is the number of bits between matching patterns. The purpose of the test is to detect whether or not the sequence can be significantly compressed without loss of information
10	Linear Complexity Test
	The focus of this test is the length of a generating feedback register. The purpose of this test is to determine whether or not the sequence is complex enough to be considered random
11	Serial Test
	The focus of this test is the frequency of each and every overlapping m-bit pattern across the entire sequence. The purpose of this test is to determine whether the number of occurrences of the 2m m-bit overlapping patterns is approximately the same as would be expected for a random sequence. The pattern can overlap
12	Approximate Entropy Test
	The focus of this test is the frequency of each and every overlapping m-bit pattern. The purpose of the test is to compare the frequency of overlapping blocks of two consecutive/adjacent lengths (m and m + 1) against the expected result for a random sequence
13	Cumulative Sum (Cusum) Test
	The focus of this test is the maximal excursion (from zero) of the random walk defined by the cumulative sum of adjusted (-1, + 1) digits in the sequence. The purpose of the test is to determine whether the cumulative sum of the partial sequences occurring in the tested sequence is too large or too small relative to the expected behavior of that cumulative sum for random sequences

(*continued*)

Table 1. *(continued)*

Test's №	Name/Description
14	Random Excursions Test
	The focus of this test is the number of cycles having exactly K visits in a cumulative sum random walk. The purpose of this test is to determine if the number of visits to a state within a random walk exceeds what one would expect for a random sequence
15	Random Excursions Variant Test
	The focus of this test is the number of times that a particular state occurs in a cumulative sum random walk. The purpose of this test is to detect deviations from the expected number of occurrences of various states in the random walk

The basic idea of using a method with cryptography instead of the usual in improving the statistical properties of an object, thus this method of attachment becomes more stable, in particular to statistical methods of steganalysis. Even more, stability is provided by the use of the AES algorithm, which also involves the initialization vector, which is used with a secret key to avoid repetitions in the process of data encryption, which makes it impossible for hackers using a dictionary attack to decrypt an encrypted message by detecting a template.

It is known that there are various steganalysis methods, including those based on the study of the statistical properties of extracted messages [15, 16]. Therefore, to assess the stability of the investigated method, the embedded message is subject to testing for pseudo-randomness using the so-called NIST tests [17], based on which there is a developed finished solution [18] used in the work.

These tests focus on different types of non-randomness that can exist in a sequence and their goal is to determine the degree of randomness of the resulting binary sequences.

Each test calculates the value of the test statistics using the calculated values of the P-value, as a result of which the conclusion is made about the acceptance or rejection of the null hypothesis about the randomness of the binary sequence.

The set of NIST tests includes 15 statistical tests [19], the purpose of which is to determine the degree of randomness of binary sequences formed by hardware or software generators of pseudo-random numbers. A brief description of each of the tests is given in Table 1.

A cryptographically converted message (see Fig. 10) and ordinary (see Fig. 11) with a length of 40 KB were tested for pseudo-randomness.

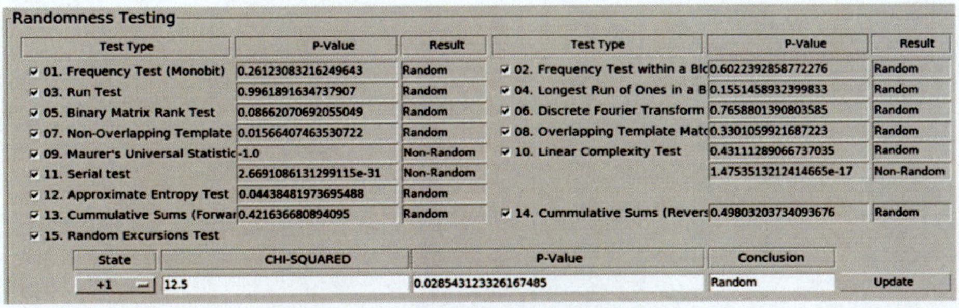

Fig. 10. Encrypted message test results.

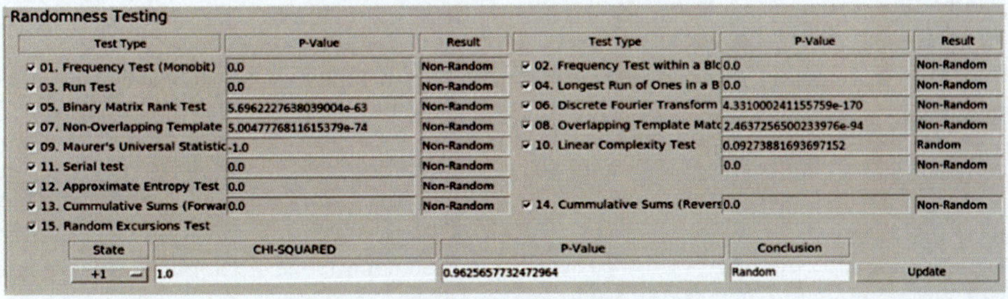

Fig. 11. Simple message tests results.

The relationship between the results obtained can be presented graphically (see Fig. 12, 13):

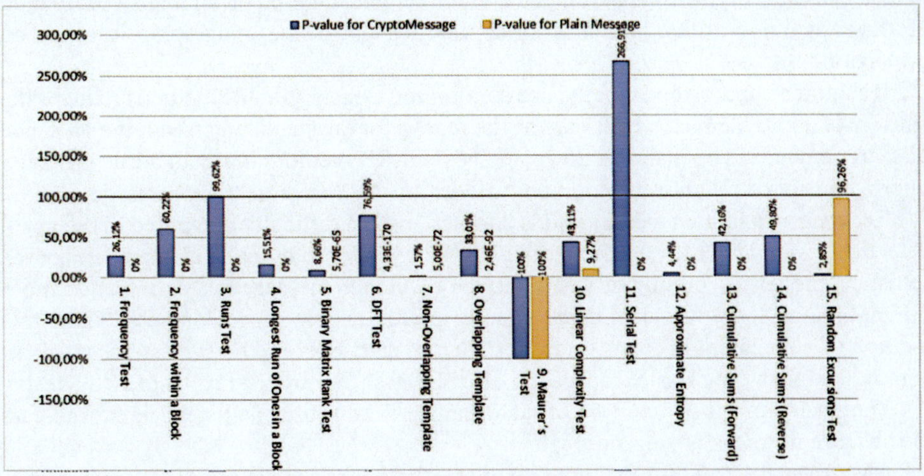

Fig. 12. Results of testing an encrypted and open message using the NIST test package.

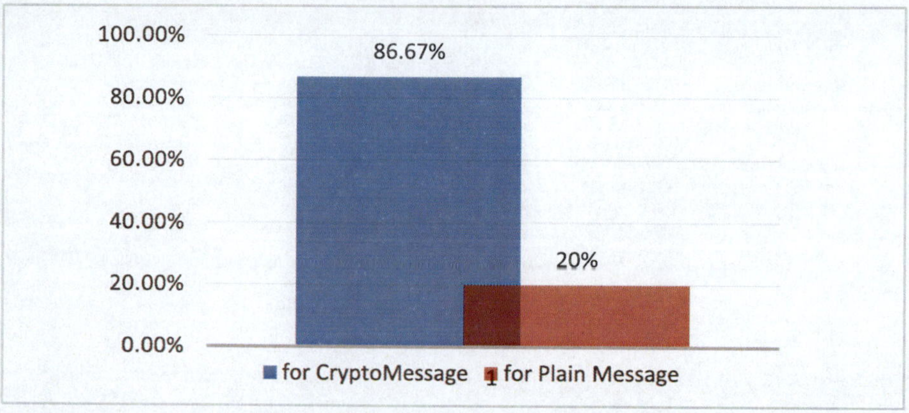

Fig. 13. The ratio of the results of successful completion of 15 tests of an encrypted and simple message.

Based on the results obtained, it can be argued that there are no steganographic investments in the embedding of an encrypted message since the number of successfully passed tests is the majority, or rather 86.67%, which cannot be said about the test result for a simple message, where this figure is only 20%.

4 Conclusions

The LSB steganographic algorithm for hiding text information in sound files was improved by the additional use of the cryptographic method and pseudo-random numbers to create a stable crypto key, developed and implemented based on an improved method of steganographic utility, which is an applied system of steganographic information protection.

It should be understood that the use of common steganographic methods reduces the entropy of audio file bits, which may be the reason for the statements about the presence of secret embeds. The main disadvantage of the LSB method, low reliability, has been improved using a reliable cryptographic algorithm AES to protect the message.

Recovering a hidden message adds another level of difficulty compared to the usual detection of a hidden message. In particular, the use of cryptography elements improves statistical properties, preliminary encryption of a hidden message makes detection more difficult, since encrypted data usually has a greater degree of randomness. Thus, this method of attachment becomes more resistant to statistical methods of steganalysis in comparison with the classical approach of LSB, as shown by the results of NIST tests.

The results of visual analysis of histograms of the source and stegosignal indicate the absence of traces of steganography.

Also, the developed utility based on the method does not leave any signatures from the point of view of the structural method of steganalysis.

Based on the results obtained, it can be argued that the proposed approach is reliable and effective, so the use of LSB-AES techniques can be proposed to ensure safe data transmission. The proposed improved algorithm can be used in the next-generation networks, it makes sense to use the method to build network covert channels in the new network environment, for example, covert channels based on streaming media networks or covert channels based on IPv6 network, etc.

References

1. Hazra, T.K., Samanta, R., Mukherjee, N., Chakraborty A.K.: Hybrid image encryption and steganography using SCAN pattern for secure communication. In: 8th Annual Industrial Automation and Electromechanical Engineering Conference (IEMECON) , Bangkok, pp. 370–380 (2017)
2. Katzenbeisser, S., Petitcolas, F.A.P.: Information Hiding Techniques for Steganography and Digital Watermarking, pp. 121–148. Artech, Boston (2000)
3. Basu, P.N., Bhowmik, T.: On embedding of text in audio-a case of steganography. In: Proceedings of International Conference on Recent Trends in Information, Telecommunication and Computing, pp. 203–206 (2010)
4. Kaur, N., Behal, S.: A survey on various types of steganography and analysis of hiding techniques. Int. J. Eng. Trends Technol. (IJETT) **V11**(8), 388–392 (2014). ISSN: 2231–5381
5. Bhagat, V.B., Kulurkar, P.N.: Audio and video steganography: using lsb and phase encoding algorithm. IJPRET **3**(9), 1640–1648 (2015)
6. Jian, C.T., Wen, C.C., Binti Ab Rahman, N.H., Hamid, I.R.B.A.: Audio steganography with embedded text. In: IOP Conference, vol. 226, no. 1, pp. 1–10 (2017)
7. Bandi, S, Rddy, H.S.M.: Combined audio steganography and AES encryption to hide the text and image into audio using DCT. J. Recent Technol. Eng. **8**(3), 1732–1738 (2019)
8. Buchyk, S., Tolyupa, S., Symonychenko, Y., Symonychenko, A., Platonenko, A.: Improvement of steganographic methods based on the analysis of image color models. In: Cybersecurity Providing in Information and Telecommunication Systems, Kyiv, Ukraine, pp. 117–124. (2021)
9. Tian, J., Xiong, G., Li, Z., Gou, Z.G.: A survey of key technologies for constructing network covert channel. Secur. Commun. Netw. **2020**, Article ID 8892896, 20 (2020). https://doi.org/10.1155/2020/8892896
10. Pooyan, M., Delforouzi, A.: LSB-based audio steganography method based on lifting wavelet transform. In: ISSPIT 2007, Egypt (2007)
11. Buchyk, S.S., Khomenko, O.V.: Applied program for hiding textual information in audio files. In: Applied Systems and Technologies in the Information Society: V International Scientific and Practical Conference, Kyiv, pp. 27–32 (2021)
12. Buchyk, S., Piatyhor, V.: Diceware password generation algorithm modification based on pseudo-random sequences, Kyiv (2021)
13. Eastlake, D., Schiller, J., Crocker, S.: Randomness requirements for security (2005)
14. Al-Ani, Z.K., Zaidan, A.A., Zaidan, B.B., Alanazi, H.O.: Overview: main fundamentals for steganography. J. Comp. **2**(3), 158–165 (2010)
15. Korzhik, V., Fedyanin, I., Godlewski, A., Morales-Luna, G.: Steganalysis based on statistical properties of the encrypted messages. In: Rak, J., Bay, J., Kotenko, I., Popyack, L., Skormin, V., Szczypiorski, K. (eds.) MMM-ACNS 2017. LNCS, vol. 10446, pp. 288–298. Springer, Cham (2017). https://doi.org/10.1007/978-3-319-65127-9_23
16. Akhrameeva, K., Korzhik, V., Nguyen, C.: Detection of video steganography with the use of universal method based on NIST-tests. Proc. Telecom. Univ. **6**(1), 72–78 (2020)
17. National Institute of Standards and Technology, p. 131 (2010). https://www.nist.gov/publications/statistical-test-suite-random-and-pseudorandom-numbergenerator-cryptographic. Accessed 20 Nov 2021

18. Python implementation of NIST's A Statistical Test Suite for Random and Pseudorandom Number Generators for Cryptographic Applications. https://github.com/stevenang/randomness_testsuite. Accessed 20 Nov 2021
19. A Statistical Test Suite for Random and Pseudorandom Number Generators for Cryptographic Applications. https://nvlpubs.nist.gov/nistpubs/Legacy/SP/nistspecialpublication800-22r1a.pdf. Accessed 20 Jan 2022
20. Yudin, O., Symonychenko, Y., Symonychenko, A.: The method of detection of hidden information in a digital image using steganographic methods of analysis. In: IEEE International Conference 2019 on Advanced Trends in Information Theory (ATIT), Kyiv, Ukraine, pp. 262–266(2019)
21. Yudin, O., Ziubina, R., Buchyk, S., Frolov, O., Suprun O., Barannik, N.: Efficiency assessment of the steganographic coding method with indirect integration of critical information. In: IEEE International Conference 2019 on Advanced Trends in Information Theory (ATIT), Kyiv, Ukraine, pp. 36–40 (2019)

Reliability Assessment of UAV Fleets

Elena Zaitseva[1(✉)], Vitaly Levashenko[1], Nicolae Brinzei[2], Andriy Kovalenko[3], Marina Yelis[4,5], Viktors Gopejenko[6,7], and Ravil Mukhamediev[4,5]

[1] University of Zilina, 01026 Žilina, Slovakia
elena.zaitseva@fri.uniza.sk
[2] Centre de Recherche en Automatique de Nancy, University of Lorraine, Nancy, France
[3] Kharkiv National University of Radio Electronics, Nauky Avenue 14, Kharkiv 61166, Ukraine
[4] Institute of Information and Computational Technologies, Almaty 050010, Kazakhstan
k.marina92@gmail.com
[5] Institute of Automation and Information Technology, Satbayev University (KazNRTU), Almaty 050013, Kazakhstan
[6] Ventspils University of Applied Sciences, International Radio Astronomy Centre, LV-3601 Inzhenieru Street 101, Ventspils, Latvia
[7] ISMA University of Applied Sciences, Lomonosov Street, 1, Riga 1019, Latvia
viktors.gopejenko@isma.lv

Abstract. UAVs have a great potential of application for monitoring, search, detection, communication, delivery and transportation of cargo in various sectors of economy. In spite of this, the existing software and hardware, as well as legal limitations, prevent the wide application of UAVs. There is intensive research related to automation and optimization of missions of one or more UAVs in various application areas. However, the execution of missions of both individual vehicles and their homogeneous or heterogeneous groups depends on the reliability issues of these technical devices, UAVs fleets, and control systems. In this paper, we present models for assessing fleet reliability of UAVs that are managed centralized or decentralized. The method is based on the representation of the fleet as a Binary-State System. The following topologies are considered: (a) a homogenous irredundant drone fleet, (b) a homogenous hot stable redundant drone fleet, (c) a heterogeneous irredundant drone fleet, and (d) a heterogeneous hot stable redundant drone fleet. For the listed topologies, reliability estimates were obtained as a function of the number of primary and redundant UAVs.

Keywords: UAVs · Reliability · Binary-State System · Importance analysis · Structure function · Availability

1 Introduction

A UAV is a relatively new technology used for monitoring, search, communication, cargo delivery, etc. [1] The use of UAVs in practice is justified by their relatively low cost [2], the availability of devices on the market [3], the possibility of use in hard-to-reach places [4], and the possibility of adaptation to special tasks [5] and relatively simple control methods [6].

M. Klymash et al. (Eds.): TCSET 2022, LNEE 965, pp. 335–357, 2023.
https://doi.org/10.1007/978-3-031-24963-1_19

These advantages provide significant potential for UAV applications ranging from the entertainment industry to geological exploration [7–10]. According to estimates, the UAV market will reach $127 billion, of which 36% will be used for infrastructure, 26% for agriculture, 10% for transportation [11] (Fig. 1).

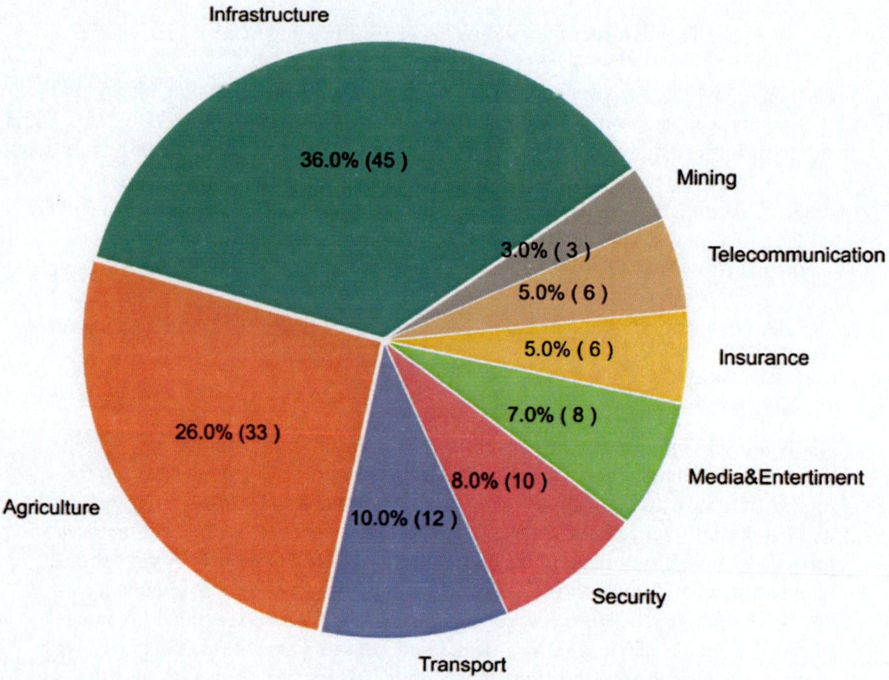

Fig. 1. Future market of UAVs.

Practical applications of UAVs are considered in close connection with artificial intelligence technologies. In essence, there is a fusion of intelligent data processing and UAVs. A set of methods and means of Intelligent Unmanned Aerial Vehicle Technology (IUAVT) is being formed) [12]. There are three main categories of limitations that limit the widespread use of technology in various sectors of the economy [10]:

1. Technical. Limitation of battery capacity, flight time, payload, sensor sensitivity, dependence on weather conditions, limitations in computing power on board, etc.;
2. Legal. The impossibility of some use options within the city limits and restrictions on the use of UAVs weighing more than 250 g [13];
3. Software and algorithmic.

The third group of limitations is caused by the relatively low computing power of the onboard computer, and the insufficient development of algorithms related to the main applications of UAVs, including flight planning. Flight planning is defined by the mission performed by one or more UAVs. Among such tasks performed during the flight, the literature mentions: search operations, routing for a set of locations, area coverage, data collection and recharging in a wireless sensor network (WSN), allocation of communication channels and computing power for mobile devices, and the operational aspects of a self-organizing drone network [5]. However, UAVs and ground-based communication and control systems as technical devices are not reliable. The limited reliability of technical means should be taken into account in the planning process, especially in those tasks that are critically time-dependent (search for victims, application of fertilizers and herbicides, organization of communication in emergency areas, etc.).

In this paper, we consider approaches to estimating the reliability of drone fleets using the Binary-State System representation of the fleet.

The paper consists of the following sections:

Related works section, analyzes the current state of research on UAV applications, reliability assessment methods.

The third section discusses the methods for analyzing drone fleet reliability and the results.

The fourth section describe the fleet structure and functions of unmanned aerial vehicles.

Fifth section consist on analysis of the availability of UAV fleets based on a structure function.

In the sixth part analyzing and evaluating the impact of components' failures on a system.

In the last section, conclusion, we summarize the need to calculate UAV fleet reliability when executing missions of different types and particular responsibility. In this paper, this gap is partially filled. Methods of evaluation are described in the paper.

2 Related Works

The IUAVT is used for a variety of tasks, including:

1. Monitoring:
- For mapping, which usually requires overlapping images to produce a quality map [14, 15];
- Monitoring along a predetermined route (structures [16], technical facilities [17] pipelines [18, 19];
- Monitoring using special equipment, such as thermal and multispectral cameras, gamma spectrometers, aeromagnetic sensors etc., when it is necessary to take into account the peculiarities of the equipment during missions [20, 21];
- Monitoring-inspection of casualty areas during emergencies [22–24];
- Traffic monitoring [25];
- Wildlife monitoring [15, 26].

2. Detection and identification [27, 28]:
- conditions of structures, harvested crops, animals [29];
- violations of laws, such as hunting [30];
- air or aquatic environment, periodically at predetermined points in time [28, 31, 32].

3. Search:
- Protecting and Searching for Animals [33, 34];
- Injured people [35, 36];
- Minerals [37, 38];
- etc.

4. Delivery [39] and transportation of cargo, including oversized cargo on suspension [40].

5. Organization of communications:
- Collection of data from installed sensors: in precision farming [41], wireless sensor networks [42]:
- In natural disaster areas [43–45].

Figure 2 shows the distribution of publications on UAV applications.

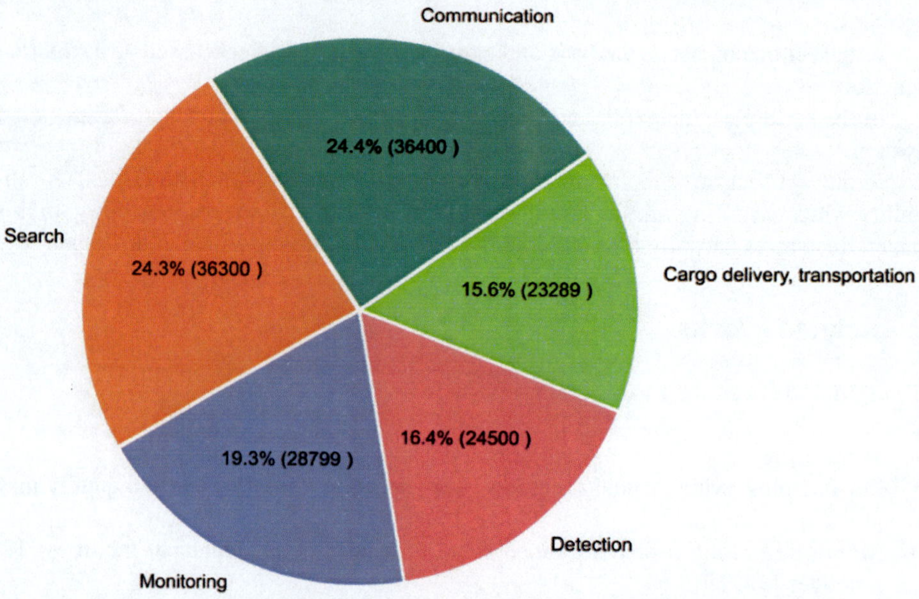

Fig. 2. Number of publications according to Google Scholar since 2018.

The effective applications of UAVs in the implementation of the discussed problems are caused by some characteristics and properties of itself UAVs. One of the important properties of UAVs application is the definition of the optimal coverage path planning (CPP) [46–48]. There are two modifications of this problem: CPP1 - without influence on the surface analyzed; CPP2 - with influence on the surface coated.

Another important property of UAVs is reliability, which allows determining the conditions of UAVs' functioning performance at the specified level [49]. The reliability analysis or reliability engineering is a knowledge domain, which includes many approaches, methods, and algorithms for reliability evaluation and risk assessment of complex systems. The choice and application of methods for reliability analysis in every specified case depend on a system structure, its type, and area of applications. UAVs are complex systems from point of view of reliability analysis. In the study [50] the classification of UAV reliability analysis depending on different parameters has been introduced. This classification is presented in form of a parameters' matrix, which takes into consideration such aspects as the number of drones in exploitations, homogenous or heterogeneous structure of UAVs' fleet, and type redundancy in drone fleets. This classification matrix of UAV's fleet reliability is shown in Table 1. In this matrix types of UAVs' structures as a single drone, UAVs' fleet, and UAVs' multi-fleet are considered. These structures can implement one function (it is indicated by "1") or can be multi-functions (this parameter is indicated as MF). The UAVs' fleet and UAVs' multi-fleet can be formed by homogenous drones or can be heterogeneous, which is indicated by "Y" in the column "Heterogeneous" and the symbol "N" in this column specifies a homogenous type of drone fleet. A single UAV or UAV's fleet can be represented by two mathematical models: *Binary-State System* (BSS) or *Multi-State System* (MSS). A BSS is a mathematical model which has two states in the system functioning and two states of components in the mathematical representation. A MSS is a mathematical model, which allows consideration in the mathematical representation of more than two states for a system functioning and the system's components states. MSS permits to describe and analyze a system in more detail: it is possible to analyze not the system failure only but the system degradation too. However, the methods for MSS reliability analysis request higher computational resources [51]. There are some investigations of a MSS used for reliability analysis of UAVs [50, 52, 53]. But most of the studies in UAVs' reliability use a BSS for the mathematical representation of the investigated system [50, 54]. Besides the specified parameters of UAV or UAV's fleet, the specifics of reliability should be taken into consideration too. For UAVs, the first should be to consider the redundancy of the system. Typically, two types of redundancy are investigated: active redundancy and standby redundancy [55]. A standby redundancy can be hot, cold, and warm standby sparing [56]. The use of every one of these types of redundancy is caused by the request for the time or energy resources of the specified application of UAVs. The cold redundancy is typically used in the system if it is critical for energy consumption: the spare component is started functioning only when the worked component fails. Hot standby sparing is used as a failover mechanism to provide the reliability of a system, where the recovery time is critical. The mathematical model for this type of redundancy is equal to active redundancy if the switching delays and failures are not taken into consideration. Warm standby sparing compromises the energy consumption and the recovery time. The spare components are partially powered up when the primary component is operational and it is fully powered up only after the primary component fails.

The methods of reliability analysis depend on the structure of the system. The methods for reliability evaluation of a single drone differ from the methods of UAVs' fleets. According to a study [50, 53] most often used structure for UAVs is the UAV's fleet. A UAVs' multi-fleet is not often used because needs many resources (energy, financial, etc.). A UAV's fleet allows archive good results in such problems as monitoring [57], transportation [58], and agriculture [59]. Therefore, this structure is considered in the paper. The parameter of fleet functions has some correlation with the homogenous and

Table 1. Matrix of UAVs' fleet reliability assessment parameters from [50].

UAV/UAVs' fleets					Reliability		
Type	Parameters			Irredundant	Redundant		
	Functions	Mat. Model	Heterogeneous		Hot standby	Cold standby	Warm standby
Single UAV	1	BSS	–				
		MSS	–				
	MF	BSS	–				
		MSS	–				
UAVs' fleet	1	BSS	Y	x	x		
			N	x	x		
		MSS	Y				
			N				
	MF	BSS	Y	x	x		
			N	x	x		
		MSS	Y				
			N				
UAVs' Multi-fleet	MF	BSS	Y				
			N				
		MSS	Y				
			N				

heterogeneous types of UAVs' fleet. There are problems in which these parameters cannot be joined and should be considered as two independent parameters. In this study, we assume the correlation of these parameters: the homogenous fleet implements one function and the heterogeneous fleet can have some functions for the implementation and every fleet drone is with one function. Typically, the methods for the reliability analysis of homogenous [60] and heterogeneous [61] fleets are different. In this paper, we propose a new method of the UAVs' fleet reliability evaluation, which can be used, for both homogenous and heterogeneous fleets. From point of view of redundancy, this method can be applied for irredundant drone fleets or fleets with hot redundancy. The hot redundancy is considered because in most applications of UAVs' fleet the time is critical for the spare drone start.

3 Method and Results

The proposed method is based on BSS in mathematical representation of the UAVs' fleet. In particular, the structure function-based method is developed. The structure function as mathematical model in reliability analysis is one of simplest mathematical representation

[62, 63]. This function maps the set of system components' states to one of the system states. In case of the BSS this function agrees with a Boolean function. Let us consider a system of n components (drones). The i-th drone functioning is denoted by variable $x_i (i = 1, \ldots, n)$ where $x_i = 1$ if the drone is functioning state and $x_i = 0$ if it fails. The system state (its reliability) depends on the states of the components (drones). This depending is defined by the structure function [62, 63]:

$$\phi(x_1, \ldots, x_n) = \phi(\boldsymbol{x}) : \{0, 1\}^n \to \{0, 1\}, \tag{1}$$

where $\boldsymbol{x} = (x_1, \ldots, x_n)$ is a vector of the system components states (state vector); $\phi(\boldsymbol{x}) = 1$ if the system is functioning and $\phi(\boldsymbol{x}) = 0$ if the system is failure.

A UAVs fleet is a coherent system. It means that all component are (a) relevant for the system ($\phi(1_i, \boldsymbol{x}) \neq \phi(0_i, \boldsymbol{x})$ for some state vectors) and (b) failure of any component cannot result the improvement in the system functioning ($\phi(1_i, \boldsymbol{x}) \geq \phi(0_i, \boldsymbol{x})$ for any component), where $\phi(1_i, \boldsymbol{x}) = \phi(x_1, \ldots, x_{i-1}, 1, x_{i+1}, \ldots, x_n)$ and $\phi(0_i, \boldsymbol{x}) = \phi(x_1, \ldots, x_{i-1}, 0, x_{i+1}, \ldots, x_n)$. The structure function (1) for a coherent system is monotonically non-decreasing [62].

The structure function advantages are possibility of the representation of a system of any structural complexity and simplify methods for reliability evaluation. The disadvantage of this mathematical model is large dimension of system of many components: the structure function dimension is exponential increase depending on the number of the system components [62]. For the decision of this difficulty in reliability analysis, some approaches have been proposed. One of them is use of Binary-Decision Diagram (BDD), which has been developed in Boolean Algebra [64] and are effective for the processing of the function of large dimensional [63]. The alternative approach is based on the structure function representation in form of survival signature [65].

4 Structure Function of UAV'S Fleet

Need to note that the reliability of the system is a complex characteristic, which includes some indices and measures. The reliability analysis of a UAV's fleet drone supposes the calculation of some indices and measures. The drone fleet reliability in the stationary state is studied in this paper. The assumption about equivalent properties of multi-functions and heterogeneous in this study takes into consideration. It allows us to consider the mathematical model in form of the structure function (1). The structure function should be defined for the fleet:

- Homogenous irredundant drone fleet (see Fig. 3 (a));
- Homogenous hot stable redundant drone fleet (see Fig. 3 (b));
- Heterogeneous irredundant drone fleet (see Fig. 3(c));
- Heterogeneous hot stable redundant drone fleet (see Fig. 3 (d)).

These fleets can include a Main Drone Fleet (MDF) for the implementation of the objective action, a Reverse Drone Fleet (RDF) for the support of the functioning state of a MDF and Control Unit (CU) for the coordination of drones actions.

One of the possible ways to construct structure function is based on the application of typical structures of systems, which are series, parallel and k-out-of-n systems. There are two definitions of k-out-of-n system. A k-out-of-n:G is functioning if k or more of n components are in the working states. A k-out-of-n:F is a system that fails if at least k components are failed. In this study k-out-of-n:G type is used only, therefore the nomination k-out-of-n will be used for this system. It is well known that the system k-out-of-n is a generalization of two other typical structures: 1-out-of-n corresponds to a parallel system and n-out-of-n agrees with a series system. Therefore, the drone fleet functioning is presented and studied as k-out-of-n system in this paper.

A homogenous irredundant drone fleet with decentral control (see Fig. 3 (a)) can be presented by two topologies. One of them has the distribution control and the other topology includes the central CU. A decentral control topology is interpreted as the n-out-of-n system or the series system with the structure function:

$$\phi(x) = \bigwedge\nolimits_{w=1}^{n} x_w, \tag{2}$$

where $\bigwedge_{w=1}^{s}$ is the symbol of Boolean operation AND for w variables.

The structure function of a homogenous irredundant drone fleet with the central CU (see Fig. 3 (a)) can be defined as the series system of the MDF and CU, where drone fleet is the series system:

$$\phi(x) = \left(\bigwedge\nolimits_{w=1}^{n-1} x_w\right) \bigwedge x_n = \bigwedge\nolimits_{w=1}^{n} x_w, \tag{3}$$

where the variables from x_1 to x_{n-1} are represent the states of the drones of the MDF and variable x_n represents the states of the CU.

A homogenous hot stable redundant drone fleet can be presented by two topologies too (see Fig. 3 (b)). Similar to the previous structure of UAV's fleet, this structure can have distribution control and the central CU for fleet control. The hot stable redundant can be presented as the k-out-of-n. According to [66] the structure function of the k-out-of-n system and analysis of its reliability can be based on an analysis of minimal paths set. The structure function of such a system is defined as the unit of minimal paths:

$$\phi(x) = \bigvee\nolimits_{Q_k} \bigwedge\nolimits_{w=1}^{k} x_{i_w}, \tag{4}$$

where $\bigvee_{w=1}^{k}$ is the symbol of Boolean operation OR for k variables; the k variables x_{i_s} (s = 1, ..., k) represent the states of the drones which are formed by the minimal paths in the topology of k-out-of-n system and the number of the paths Q_k is defined as:

$$Q_k = \binom{n}{k} = \frac{n!}{(n-k)!k!}. \tag{5}$$

The structure function of k-out-of-n system (3) is formed by the Q_k implicants and each implicant has k literals.

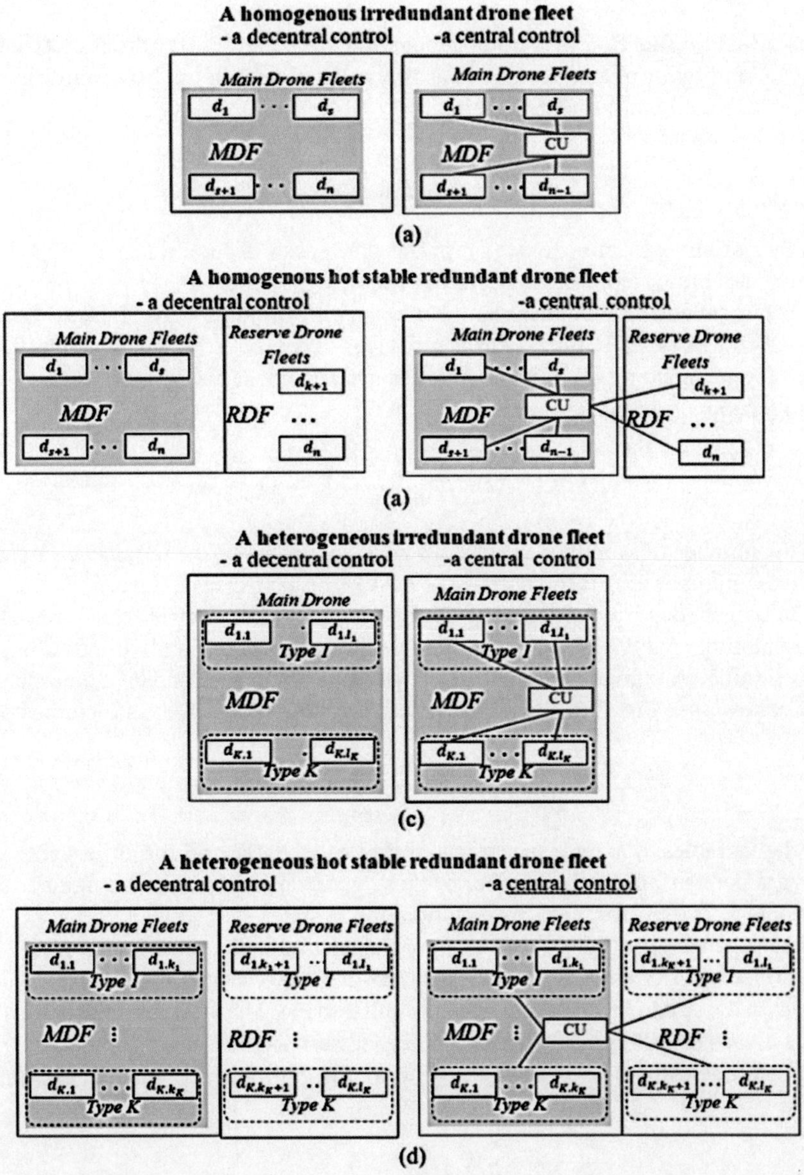

Fig. 3. The topologies of (a) a homogenous irredundant drone fleet, (b) a homogenous hot stable redundant drone fleet, (c) a heterogeneous irredundant drone fleet, and (d) a heterogeneous hot stable redundant drone fleet.

The structure function of a homogenous hot stable redundant drone fleet with the central CU in Fig. 3 (b) can be defined as the series system of the MDF and CU, where the drone fleet is the k-out-of-n system:

$$\phi(\boldsymbol{x}) = \left(\bigvee_{Q_k} \bigwedge_{w=1}^{k} x_{i_w}\right) \bigwedge x_n, \tag{6}$$

where the variables from x_1 to x_{n-1} represent the states of the drones of the MDF and RDF and variable x_n represents the states of the CU.

A heterogeneous irredundant drone fleet with decentral control (see Fig. 3 (c)) is formed of UAVs of K types. Drones of every type of irredundant drone fleet are formed a series system. All of these K series systems must be working for the work of a heterogeneous irredundant drone fleet:

$$\phi(\boldsymbol{x}) = \left(\bigwedge_{r=1}^{K} \bigwedge_{w=1}^{l_K} x_{r,w}\right) = \bigwedge_{w=1}^{n} x_w, \tag{7}$$

where the number of variables $x_{r,w}$ for all possible values of the parameters r and w is equal to the number n; l_r is the number of UAVs of type r (r = 1, …, K).

A heterogeneous irredundant drone fleet with decentral control from point of view of the reliability analysis is a series system that can be worked if all drones of all types are functioning. A similar topology has a heterogeneous irredundant drone fleet with central control (see Fig. 3 (c)), which is a series system too with the structure function:

$$\phi(\boldsymbol{x}) = \left(\bigwedge_{r=1}^{K} \bigwedge_{w=1}^{l_K} x_{r,w}\right) \bigwedge x_n = \bigwedge_{w=1}^{n} x_w, \tag{8}$$

where the variables $x_{r,w}$ represent states of drones of MDF and for all possible values of the parameters r and w the number of these variables is equal to the number n − 1; the variable x_n is used for the representation of CU states.

A heterogeneous hot stable redundant drone fleet with distributed control (see Fig. 3 (d)) is formed by UAVs of K types and every one of these types of drones is the k-out-of-n system with the structure function which is similar to (6). The working state of the fleet is possible if each drone' type implements specified activity. The structure function of this type of heterogeneous drone is a series system of K components which are kr-out-of-lr systems:

$$\phi(\boldsymbol{x}) = \bigwedge_{r=1}^{K} \left(\bigvee_{Q_r} \bigwedge_{w=1}^{k_r} x_{r,i_w}\right), \tag{9}$$

where Q_r is the number of minimal paths for each type of UAVs which is computed according to (5) and k_r is the number of minimum required working drones for type r (r = 1, …, K).

A heterogeneous hot stable redundant drone fleet with the central CU (see Fig. 3 (d)) in reliability analysis point of view is formed by K kr-out-of-lr systems which are interconnected in series and connected in series with the control element:

$$(\boldsymbol{x}) = \left(\bigwedge_{r=1}^{K} \left(\bigvee_{Q_r} \bigwedge_{w=1}^{k_r} x_{r,i_w}\right)\right) \bigwedge x_n, \tag{10}$$

The structure functions (2)–(4) and (6)–(10) are defined for homogeneous and heterogeneous UAVs' fleets that can be irredundant or hot stable redundant. The introduced structure function for all types of irredundant fleets (2), (3), (7), and (8) are equal and are presented as series topology. Therefore, the reliability analysis for these fleets can be implemented as for one series system. Hot stable redundant drone fleets have different structure functions (4), (6), (9), and (10) and for every one of them should be proposed reliability evaluation.

5 Availability of UAV's Fleet Defined Based on a Structure Function

The reliability analysis of the system based on the structure function is studied well [62, 66]. The system availability is defined depending on the probabilities of the system components functioning state, which is defined as:

$$p_i = \Pr\{x_i = 1\}, \tag{11}$$

and the probabilities of the system components failure, which are:

$$q_i = 1 - p_i = \Pr\{x_i = 0\}, \tag{12}$$

The system probability to be in the functioning state is system availability [62, 66]. Because the UAV's fleet is a coherent system than the availability can be defined as:

$$A = \Pr\{\phi(\boldsymbol{x}) = 1\}, \tag{13}$$

and the system unavailability is defined as:

$$U = 1 - A = \Pr\{\phi(\boldsymbol{x}) = 0\}. \tag{14}$$

As the first, the series structures of the irredundant fleets are considered. A homogenous irredundant drone fleet with decentral control (Fig. 3 (a)) is formed for drones that have equal characteristics, in particular, the probability of the drone to be in a working state: $p_i = p_j = p$ for $i \neq j$. It allows us to define this fleet availability A_{HoID} according to (2) and (13) as:

$$A_{HoID} = \Pr\left\{\bigwedge_{w=1}^{n} x_w\right\} = \prod_{w=1}^{n} p_w = p^n, \tag{15}$$

The availability of a homogenous irredundant drone fleet with central CU (see Fig. 3 (a)) A_{HoIC} is similar to the availability of a homogenous irredundant drone fleet with decentral control. But the definition of this availability should take into consideration of the probability of functioning of the CU p_n:

$$A_{HoIC} = \Pr\left\{\left(\bigwedge\nolimits_{w=1}^{n-1} x_w\right) \bigwedge x_n\right\} = \left(\prod\nolimits_{w=1}^{n-1} p_w\right) \cdot p_n = p^{n-1} \cdot p_n, \tag{16}$$

where p is the probability of the working state of UAVs of MDF and p_n is the probability of CU being in the working state.

The definitions of availability (15) and (16) for irredundant homogenous UAV fleets allow us to provide the analysis and comparison of the availability of these homogenous fleets (see Fig. 4). The curves of availabilities in Fig. 4 reflect the average values of availabilities of these two fleets. They have been computed for the probabilities of UAV and CU, which change from 0.600 to 0.999 accordingly and the probability of CU functioning is more than the probability of UAV working state. According to this study, a homogenous irredundant drone fleet with central CU has the best availability for the fleet which consists of less than 12 UAVs.

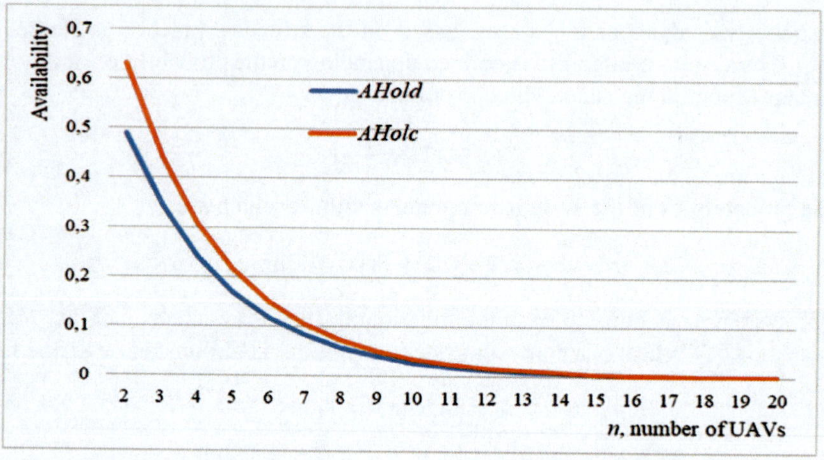

Fig. 4. The availabilities of homogenous irredundant UAV fleets.

The availability of heterogeneous irredundant fleets should take account of some types of UAVs. The probability of UAV of type r functioning is denoted as p_r, where r = 1, …, K, and lr is the number of UAVs of type r. The availability of a heterogeneous irredundant drone fleet with decentral control (see Fig. 3 (c)) A_{HeID} according to (7) and (13) is:

$$A_{HeID} = \Pr\left\{\left(\bigwedge\nolimits_{r=1}^{K} \bigwedge\nolimits_{w=1}^{l_K} x_{r,w}\right)\right\} = \prod\nolimits_{r=1}^{K} p_r^{l_r}, \tag{17}$$

The availability of a heterogeneous irredundant drone fleet with central CU (see Fig. 3 (c)) A_{HeIC} according to (8) and (13) and the assumption of the probability p_n of the working state of CU is:

$$A_{HeIC} = \Pr\left\{\left(\bigwedge\nolimits_{r=1}^{K} \bigwedge\nolimits_{w=1}^{l_K} x_{r,w}\right) \bigwedge x_n\right\} = \left(\prod\nolimits_{r=1}^{K} p_r^{l_r}\right) \cdot p_n, \tag{18}$$

The analysis of the availabilities of heterogeneous irredundant drone fleets is shown in Fig. 5. The curves of availability have been computed for the different number of drone types and the number of drones in the fleet. The number of drone types is taken into account too: from 2 to 5 types have been evaluated for these fleets of 20 drones.

This experimental investigation shows that the increase in the number of UAV types has a threshold and the larger number of drone types has no positive impact on fleets' availability. The best solution for a fleet that consists of less than 7 drones is two types.

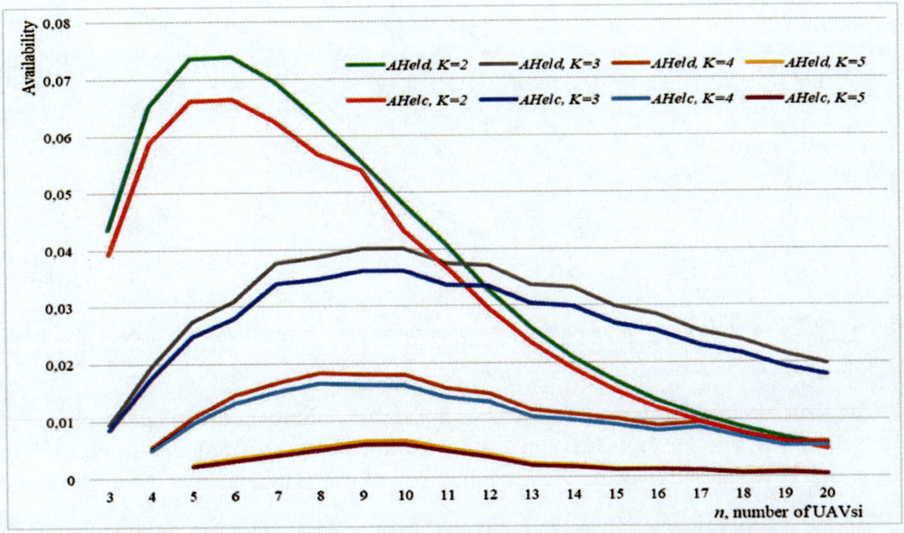

Fig. 5. The availabilities of heterogeneous irredundant UAV fleets.

The background topology of hot stable redundant drone fleets is the k-out-of-n system. A homogenous hot stable redundant drone fleet with decentral control (see Fig. 3 (b)) is k-out-of-n system with identical components. The availability of this system, for example, is defined in [66, 67] and for this UAV's fleet is:

$$A_{HoRD} = \Pr\left\{ \bigvee_{Q_k} \bigwedge_{w=1}^{k} x_{i_w} \right\} = \sum_{s=k}^{n} \binom{n}{s} \cdot p^s \cdot q^{n-s}, \tag{19}$$

The availability of a homogenous hot stable redundant drone fleet according to (19) depends on the number of part of working drones' k (see Fig. 6). Therefore, the analysis of this fleet availability is provided depending on the number of drones in the fleet and the specified number of working drones. This analysis shows that the increase in the number of specified working drones for the fixed number of drones in the fleet results in the deterioration in reliability. Similar to the previous experiments, Fig. 4 shows the average dependences, and the studies were performed for a set of probabilities of the operational state of drones from 0.6 to 0.0999.

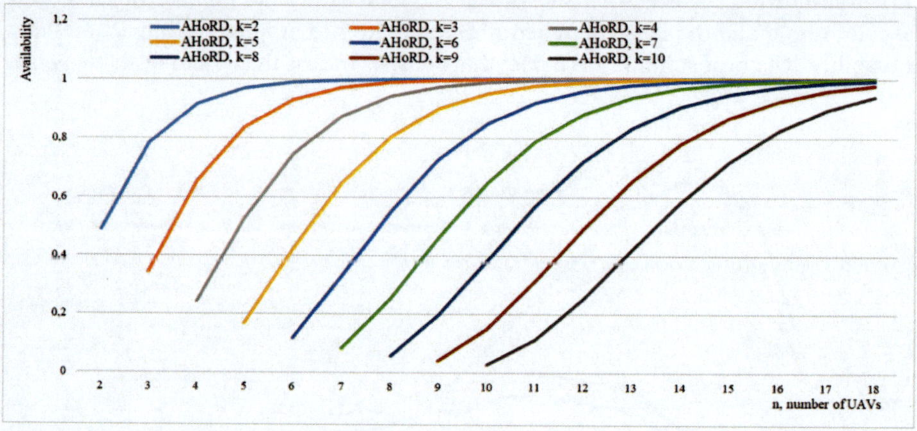

Fig. 6. The availability of homogenous hot stable redundant drone fleet with decentral control.

The structure function of a homogenous hot stable redundant drone fleet with central CU in Fig. 3 (b) is defined by (1.5) and besides k-out-of-n system includes the component CU, which is in series. Therefore, the availability of this fleet is:

$$A_{HoRC} = \Pr\left\{\left(\bigvee_{Q_k}\bigwedge_{w=1}^{k} x_{i_w}\right)\bigwedge x_{n},\right\} = \left(\sum_{s=k}^{n-1}\binom{n-1}{s}\cdot p^{s}\cdot q^{n-s}\right)\cdot p_n, \qquad (20)$$

The availability of a homogenous hot stable redundant drone fleet with the central CU is similar to the availability of a homogenous hot stable redundant drone fleet with decentral control. The change of the fleet control from decentralized to centralized leads to a slight decrease in the reliability of this fleet with central CU (see Fig. 7).

A heterogeneous hot stable redundant drone fleet with distributed control (see Fig. 3 (d)) consists of drones of types and has structure function (9). The availability of this fleet according to (13) and the definition of the availability of the k-out-of-n system in [66, 67] is computed as:

$$A_{HeRD} = \Pr\left\{\bigwedge_{r=1}^{K}\left(\bigvee_{Q_r}\bigwedge_{w=1}^{k_r} x_{r,i_w}\right)\right\} = \prod_{r}^{K}\sum_{s=k_r}^{l_r}\binom{l_r}{s}\cdot p_r^{s}\cdot q_r^{l_r-s}. \qquad (21)$$

The evaluation of the average availability of a heterogeneous hot stable redundant drone fleet with distributed control is in Fig. 8, which shows that the increase in the number of UAVs in the fleet leads to a decrease in its availability. Therefore, the heterogeneity of the fleet increasing without a due need is unjustified from the point of view of availability.

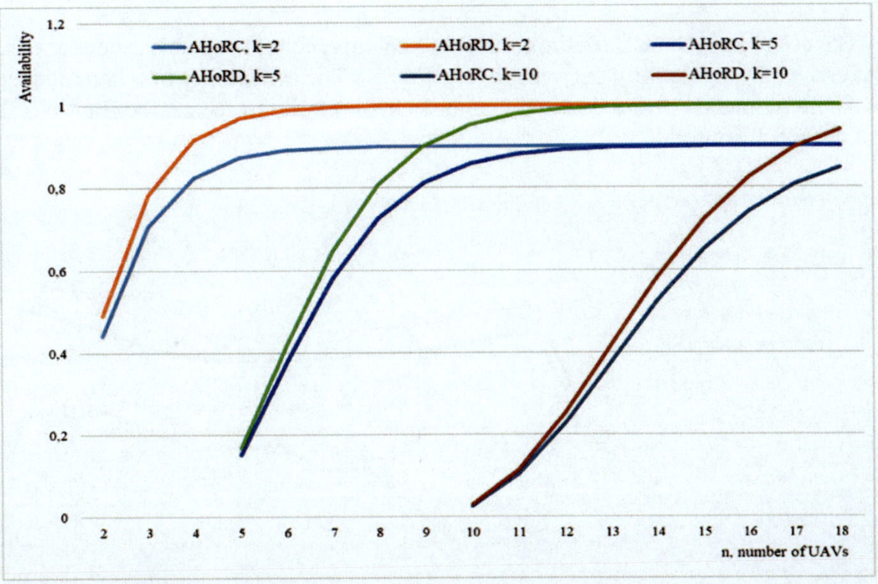

Fig. 7. The comparison of availability of a homogenous hot stable redundant drone fleets with decentral control and central CU.

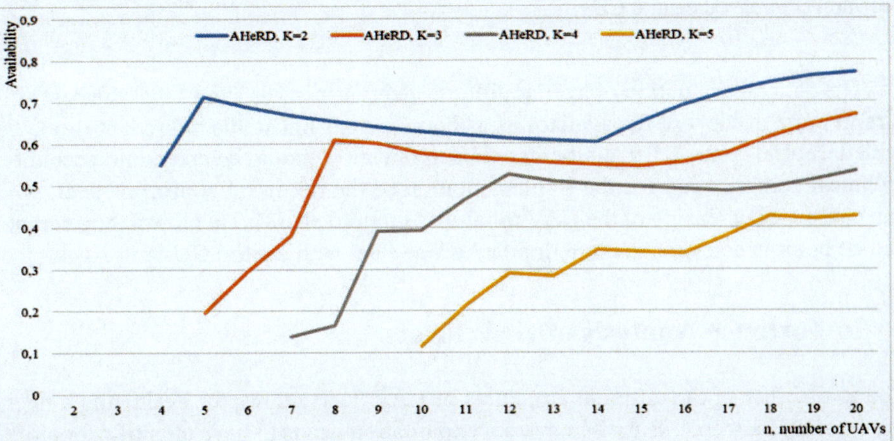

Fig. 8. The availability of a heterogeneous hot stable redundant drone fleet with distributed control.

The availability of a heterogeneous hot stable redundant drone fleet with central CU (see Fig. 3 (d)) A_{HeRC} is defined in a similar way:

$$A_{HeRD} = \Pr\left\{\left(\bigwedge_{r=1}^{K}\left(\bigvee_{Q_r}\bigwedge_{w=1}^{k_r} x_{r,i_w}\right)\right)\bigwedge x_n\right\} = \left(\prod_{r}^{K}\sum_{s=k_r}^{l_r}\binom{l_r}{s}\cdot p_r^s\cdot q_r^{l_r-s}\right)\cdot p_n, \tag{22}$$

The use of a central CU for the control in a heterogeneous hot stable redundant drone fleet causes the availability decreasing (see Fig. 9). The availability of a heterogeneous hot stable redundant drone fleet with central CU is similar to the availability of a fleet with decentral control.

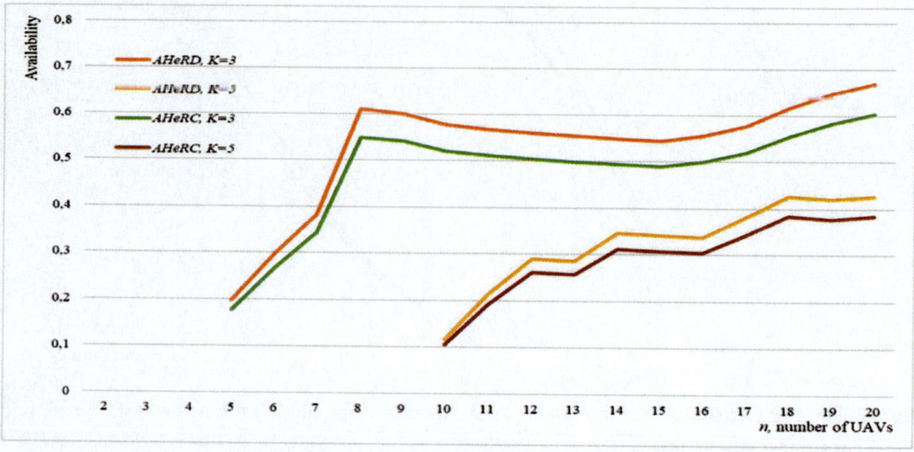

Fig. 9. The comparison of availability of a heterogeneous hot stable redundant drone fleets with decentral control and central CU.

The general evaluation of availabilities of fleets structures shows that the best structure in point of view of the availability is homogenous hot stable redundant drone fleet with decentral control. But un the practical application should be taken into account the technical complication for the implementation of the decentral control of fleet, which can cause the decreasing of the UAV reliability properties [68]. Therefore, the alternative can be homogenous hot stable redundant drone fleet with central CU.

6 Importance Analysis of UAV Fleets

The availability of the different structures of UAV fleets allows for evaluating a fleet as the system as a whole. But often system exploitation needs to have quantification of the influence of some components' failure on system reliability or availability. For example, quantification of the influence of failure of different drone types on a fleet functioning can be useful for the maintenance of a UAV's fleet. The analysis and evaluation of the influence of specified components functioning or their failure on a system is part of the reliability analysis which is known as importance analysis [69]. Methods of importance analysis allow computing of Importance Measures (IM), which quantify the impact of specified system components on a system failure or performance. There are different types of IMs that quantify various aspects of a system failure or operation in case of failure or recovery of a failed component. Most of them depend on the probability of the component being in working or failed states and they can be computed based on data, which is collected in system exploitation. Structural Importance (SI) is one of IMs which is based on the topological properties of the system and can be used in the step of the system development. Where the information about probabilities of components failure is absent. The SI evaluates the probability of the system failure depending on specified component fails and it is computed as a relative number of state vectors at which a failure of the i-th system component causes a system failure [69]. This definition supposes the identification of situations in which a failure of a fixed component causes the system failure. In the case of a drone fleet analysis, this measure allows us to evaluate the probability of a fleet failure if a specified drone or CU fail.

There are different algorithms, which allow the computation of the SI [67, 69]. One of the known algorithms is based on the application of Direct Partial Boolean Derivatives (DPBD) [63]. According to the definition of this derivative in Boolean algebra, DPLD with respect of the i-th variable allows defining the sets of a Boolean function variables values for which the change of the i-th variable value results in the change of the Boolean function value [70, 71]. In reliability analysis DPLD of the structure function (1) allows the definition of the system component states (state vectors) for which the change of the i-th component state results in the change of the system state [63]:

$$\frac{\partial\phi\left(j \to \tilde{j}\right)}{\partial x_i(s \to \tilde{s})} = \begin{cases} 1, & if\ \phi\left(s, \boldsymbol{x}\right) = j\ and\ \phi(\tilde{s}, \boldsymbol{x}) = \tilde{j}\right) \\ 0, & other \end{cases} \tag{23}$$

where s, $\tilde{s}$, j, $\tilde{j}$ $\in$\{0,1\} and $s \neq \tilde{s}, j \neq \tilde{j}$; the change of the variable value from s to $\tilde{s}$ and the function value from j to $\tilde{j}$ is defined by the symbol $\to$.

In this paper, an unrecoverable fleet is considered, therefore the SI for the system failure depending on its components failure is used in the analysis. In this case, the DPLD is defined as:

$$\frac{\partial\phi(1 \to 0)}{\partial x_i(1 \to 0)} = \begin{cases} 1, & if\ \phi(1, \boldsymbol{x}) = 1\ and\ \phi(0, \boldsymbol{x}) = 0) \\ 0, & other \end{cases} = \overline{\phi(0_i, \boldsymbol{x})} \wedge \phi(1_i, \boldsymbol{x}), \tag{24}$$

The definitions of the SI and DPLD (1.23) have an obvious correlation, which can be used for the SI computation based on DPLD [63]:

$$\mathrm{SI}_i = \mathrm{TD}\left(\frac{\partial\phi(1 \to 0)}{\partial x_i(1 \to 0)}\right) = \frac{\rho}{2^{n-1}}, \tag{25}$$

where TD(.) is the truth density of the argument and this value agrees with the relative number of vectors for which the argument takes a nonzero value; ρ is the number of the system states for which the i-th system component failure causes the system failure or it is non-zero values of the DPBD (24).

The SIs for considered structures of drone fleets are computed based on their structure function definitions (2)–(4), (6)–(10) according to (25). The computation of SIs of a drone of homogenous and heterogeneous irredundant drone fleets (Table 2) are computed as SI of a series system. The SIs of CU in irredundant drone fleets with central CU are computed based on a series system too. The definition of SI of UAV in a homogenous hot stable redundant drone fleet with decentral control is SI of k-out-of-n system defined in [72]. In the study [73] the chain rule for the calculation of DPBD of complex structure function represents a system has been introduced. In the study [73], a complex system is interpreted as a system that can be decomposed to a set of based structures as series, parallel, and k-out-of-n. The chain rule is used for the definition of SIs of a drone and CU for a homogenous hot stable redundant drone fleet with central CU in Table 2. Based on the chain rule the SIs of a drone in heterogeneous hot stable redundant drone fleets with decentral control and central CU have been defined too.

Table 2. The SIs of the components of drone fleets.

A fleet structure	SI	
	The *i*-th drone	CU
Homogenous irredundant drone fleet with distributed control	$\frac{1}{2^{n-1}}$	
Homogenous irredundant drone fleet with central CU	$\frac{1}{2^{n-1}}$	$\frac{1}{2^{n-1}}$
Homogenous hot stable redundant drone fleet with decentral control	$\frac{(n-1)!}{2^{n-1}\cdot(n-k)!\cdot(k-1)!}$	
Homogenous hot stable redundant drone fleet with central CU	$\frac{(n-2)!}{2^{n-1}\cdot(n-1-k)!\cdot(k-1)!}$	$\frac{(n-2)!}{2^{n-2}\cdot(n-1-k)!\cdot(k-1)!}$
Heterogeneous irredundant drone fleet with decentral control	$\frac{1}{2^{n-1}}$	
Heterogeneous irredundant drone fleet with central CU	$\frac{1}{2^{n-1}}$	$\frac{1}{2^{n-1}}$
Heterogeneous hot stable redundant drone fleet with decentral control	$\left(\prod_{r}^{K}\sum_{s=k_r}^{l_r}\frac{(l_r-1)!}{2^{l_r-1}\cdot(l_r-k_r)!\cdot(k_r-1)!}\right)$	
Heterogeneous hot stable redundant drone fleet with central CU	$\left(\prod_{r}^{K}\sum_{s=k_r}^{l_r}\frac{(l_r-1)!}{2^{l_r-1}\cdot(l_r-k_r)!\cdot(k_r-1)!}\right)$	$\left(\prod_{r}^{K}\sum_{s=k_r}^{l_{r-1}}\frac{(l_r-2)!}{2^{l_r-2}\cdot(l_r-k_r-1)!\cdot(k_r-1)!}\right)$

7 Conclusion

UAVs are widely used for monitoring, search, communications, etc. A significant number of scientific works are devoted to solving the problems of mission optimization for solving the problems of territory coverage, selection of the optimal route under various conditions of application of individual UAVs and groups of vehicles. Much less attention is paid to the reliability analysis of missions by groups (fleets) of UAVs.

Calculation of UAV fleet reliability is necessary to assess the required degree of fleet redundancy in the execution of missions of different types and especially high-responsibility ones. This paper partially closes this gap. The paper describes evaluation methods.

Availability of UAV's fleet based on a structure function. In this paper, a new method for the reliability estimation of drone fleet is proposed. This method is based on the fleet representation by the structure function (1). The structure function is one of the mathematical models in reliability engineering which allows the representation of a system of any structural complexity. It is confirmed by a drone fleet analysis. The proposed method can be used for the analysis of homogenous and heterogeneous fleets which have different types of control. The formulas for the availability calculation for 8 types of drone fleets are defined (15)–(22). These definitions of availability can be used for the simplified evaluation of the fleets of different types (structures). The proposed definitions of availability for different types of drone fleets allow for the providing of fleets' reliability analysis with reduced computational complexity. In addition, these definitions allow us to evaluate the availability behavior of the considered types of UAV's fleet and propose recommendations for their structure (numbers of drones, number of reserved drones, number of drones' types). These recommendations can be formed according to the curves of availabilities changes depending on the structural parameters of fleets that are shown in the diagrams in Figs. 4–9. The influence of a drone breaks down to the fleet failure is studied through the importance analysis approach. The measures of SIs are defined in Table 2 for considered structures of fleets. In the future investigation the analysis of these measures will be implemented in more detail and other IMs, for example, Birnbaum Importance will be studied too.

Acknowledgments. This research was funded by the Slovak Research and Development Agency (reg.no. SK-FR-2019- 0003) and the French Ministry for European and International Affairs (Hubert Curien Partnership Stefanik grant no. 45127XL) under the project "Mathematical Models based on Boolean and Multiple-Valued Logics in Risk and Safety Analysis" and jointly with the support of Operational Program Integrated Infrastructure 2014 - 2020 of the project: Intelligent operating and processing systems for UAVs, code ITMS 313011V422, co-financed by the European Regional Development Fund".

This research also has been funded too by the Science Committee of the Ministry of Education and Science of the Republic of Kazakhstan, Grant No. AP08856412, "Development of intelligent data processing and flight planning models for precision farming tasks using UAVs", Grant No. BR10965172, "Space monitoring and GIS for quantitative assessment of soil salinity and degradation of agricultural land in the South of Kazakhstan".

References

1. Shakhatreh, H., et al.: Unmanned Aerial Vehicles (uavs): a survey on civil applications and key research challenges. IEEE Access **7**, 48572–48634 (2019)

2. Denora, M., et al.: Validation of rapid and low-cost approach for the delineation of zone management based on machine learning algorithms. Agronomy **12**, 183 (2022)
3. Tsolakis, N., Harrington, T.S., Srai, J.S.: Digital supply network design: a circular economy 4.0 decision-making system for real-world challenges. Prod. Plan. Control. 1–26 (2021)
4. Giordan, D., et al.: The use of unmanned aerial vehicles (UAVs) for engineering geology applications. Bull. Eng. Geol. Env. **79**, 3437–3481 (2020)
5. Otto, A., Agatz, N., Campbell, J., Golden, B., Pesch, E.: Optimization approaches for civil applications of unmanned aerial vehicles (uavs) or aerial drones: a survey. Networks **72**, 411–458 (2018)
6. Chittoor, P.K., Chokkalingam, B., Mihet-Popa, L.: A review on UAV wireless charging: fundamentals, applications, charging techniques and standards. IEEE Access **9**, 69235–69266 (2021)
7. Tsouros, D.C., Bibi, S., Sarigiannidis, P.G.: A review on UAV-based applications for precision agriculture. Information **10**, 349 (2019)
8. Mohan, M., et al.: UAV-supported forest regeneration: current trends, challenges and implications. Remote Sens. **13**, 2596 (2021)
9. Martinez, J.L., et al.: Comparison of satellite and drone-based images at two spatial scales to evaluate vegetation regeneration after post-fire treatments in a mediterranean forest. Appl. Sci. **11**, 5423 (2021)
10. Mukhamediev, R.I., et al.: Review of some applications of unmanned aerial vehicles technology in the resource-rich country. Appl. Sci. **11**, 10171 (2021)
11. Global market for commercial applications of drone technology valued at over $127BN. https://pwc.blogs.com/press_room/2016/05/global-market-for-commercial-applications-of-drone-technology-valued-at-over-127bn.html. Accessed 16 Sep 2021
12. Mukhamediev, R., et al.: Intelligent unmanned aerial vehicle technology in urban environments. In: Alexandrov, D.A., Boukhanovsky, A.V., Chugunov, A.V., Kabanov, Y., Koltsova, O., Musabirov, I. (eds.) Digital Transformation and Global Society. DTGS 2020. Communications in Computer and Information Science, vol. 1242, pp. 345–359. Springer, Cham (2020). https://doi.org/10.1007/978-3-030-65218-0_26
13. (https://www.gov.kz/), Министерства РК: 277 unmanned aerial vehicles are registered in Kazakhstan, https://online.zakon.kz/Document/?doc_id=37316914&pos=3%3B-82#pos=3;-82. Accessed 20 Jun 2022
14. Drone mapping software. https://www.opendronemap.org/. Accessed 20 Jun 2022
15. Han, Y.-G., Yoo, S.H., Kwon, O.: Possibility of applying unmanned aerial vehicle (UAV) and mapping software for the monitoring of waterbirds and their habitats. J. Ecol. Environ. **41**, 21 (2017). https://doi.org/10.1186/s41610-017-0040-5
16. Chen, Q., Wen, X., Wu, F., Yang, Y.: Defect detection and health monitoring of steel structure based on UAV integrated with image processing system. J. Phys: Conf. Ser. **1176**, 052074 (2019)
17. Montoya, J.C., Muñoz, C.Q., Márquez, F.P.: Remote condition monitoring for photovoltaic systems. In: Non-Destructive Testing and Condition Monitoring Techniques for Renewable Energy Industrial Assets, pp. 133–142 (2020)
18. UAVs for monitoring pipelines and oil and gas facilities, https://brlab.ru/scopes/monitoring/. Accessed 16 Sep 2021
19. Ribeiro, R.G., Junior, J.R., Cota, L.P., Euzebio, T.A., Guimaraes, F.G.: Unmanned aerial vehicle location routing problem with charging stations for belt conveyor inspection system in the mining industry. IEEE Trans. Intell. Transp. Syst. **21**, 4186–4195 (2020)
20. Aasen, H., Honkavaara, E., Lucieer, A., Zarco-Tejada, P.: Quantitative remote sensing at ultra-high resolution with UAV spectroscopy: a review of sensor technology, measurement procedures, and data correction workflows. Remote Sens. **10**, 1091 (2018)

21. Jakob, S., Zimmermann, R., Gloaguen, R.: The need for accurate geometric and radiometric corrections of drone-borne hyperspectral data for mineral exploration: MEPHySTo—a toolbox for pre-processing drone-borne hyperspectral data. Remote Sensing **9**, 88 (2017)
22. Luo, C., Miao, W., Ullah, H., McClean, S., Parr, G., Min, G.: Unmanned aerial vehicles for disaster management. In: Durrani, T.S., Wang, W., Forbes, S.M. (eds.) Geological Disaster Monitoring Based on Sensor Networks. SNH, pp. 83–107. Springer, Singapore (2019). https://doi.org/10.1007/978-981-13-0992-2_7
23. Martínez-Vargas, A., Rodríguez-Cortés, G.L., Montiel-Ross, O.: Comparative representations of a genetic algorithm to locate unmanned aerial vehicles in disaster zones. http://www.engineeringletters.com/issues_v27/issue_2/EL_27_2_14.pdf. (2019)
24. Kamilaris, A.X., Prenafeta-Boldú, F.: Disaster monitoring using unmanned aerial vehicles and deep learning. In: Disaster Management for Resilience and Public Safety Workshop (2018)
25. Ho, F., Salta, A., Geraldes, R., Goncalves, A., Cavazza, M., Prendinger, H.: In: 18th International Conference on Autonomous Agents and multiagent systems (Aamas 2019), Montreal, Canada 13–17 May 2019
26. Medvedev, A.A.; Alekseyenko, N.Y.A.Y., Karpenko, I.O.: Monitoring of wildlife in specially protected natural areas using unmanned aerial vehicles. Bull. Samara Sci. Cent. Russian Acad. Sci. **17**, 304–309 (2015)
27. Mohd Noor, N., Abdullah, A., Hashim, M.: Remote Sensing UAV/drones and its applications for urban areas: a review. IOP Conf. Ser. Earth Environ. Sci. **169**, 012003 (2018)
28. Report, S.: Drone inspections help cut pollution by half. https://gulfnews.com/uae/environment/drone-inspections-help-cut-pollution-by-half-1.2263928. Accessed 16 Sep 2021
29. Tatum, M.C., Liu, J.: Unmanned aircraft system applications in construction. Procedia Eng. **196**, 167–175 (2017)
30. Bondi, E., et al.: Automatic detection of poachers and wildlife with UAVs. In: Fang, F., Tambe, M., Dilkina, B., Plumptre, A. (eds.) Artificial Intelligence and Conservation. Artificial Intelligence for Social Good, pp. 77–100. Cambridge University Press, Cambridge (2019). https://doi.org/10.1017/9781108587792.005
31. UAV radiation mapping drone and AARM system. https://www.kromek.com/product/aerial-radiation-mapping-drone. Accessed 16 Sep 2021
32. Automation, industrial: air pollution can now be detected using UAVs with gas sensors. https://vk.com/@industriation-uroven-zagryazneniya-vozduha-teper-mozhno-opredelit-s-pomosc
33. Corcoran, E., Winsen, M., Sudholz, A., Hamilton, G.: Automated detection of wildlife using drones: synthesis, opportunities and constraints. Methods Ecol. Evol. **12**, 1103–1114 (2021)
34. Liang, X., Zhao, S., Chen, G., Meng, G., Wang, Y.: Design and development of ground station for UAV/UGV heterogeneous collaborative system. Ain Shams Eng. J. **12**, 3879–3889 (2021)
35. Alhaqbani, A., Kurdi, H., Youcef-Toumi, K.: Fish-inspired task allocation algorithm for multiple unmanned aerial vehicles in search and rescue missions. Remote Sens. **13**, 27 (2020)
36. Alotaibi, E.T., Alqefari, S.S., Koubaa, A.: LSAR: multi-UAV collaboration for search and rescue missions. IEEE Access **7**, 55817–55832 (2019)
37. King, E.: Aerial drones in mineral exploration. https://www.geologyforinvestors.com/aerial-drones-in-mineral-exploration. Accessed 16 Sep 2021
38. Jackisch, R., Zimmermann, R., Lorenz, S., Saartenoja, A., Pirttijärvi, M., Gloaguen, R.: Integration of drone-borne hyperspectral and geomagnetic data. A combined approach in geologic remote sensing. In: 20th EGU General Assembly (2018)
39. Murray, C.C., Raj, R.: The multiple flying sidekicks traveling salesman problem: parcel delivery with multiple drones. Transp. Res. Part C: Emerg. Technol. **110**, 368–398 (2020)
40. Villa, D.K., Brandão, A.S., Sarcinelli-Filho, M.: A survey on load transportation using multirotor Uavs. J. Intell. Rob. Syst. **98**, 267–296 (2019)
41. Bacco, M., Berton, A., Gotta, A., Caviglione, L.: IEEE 802.15.4 Air-ground UAV communications in smart farming scenarios. IEEE Commu. Lett. **22**, 1910–1913 (2018)

42. Bouhamed, O., Ghazzai, H., Besbes, H., Massoud, Y.: A UAV-assisted data collection for wireless sensor networks: autonomous navigation and scheduling. IEEE Access **8**, 110446–110460 (2020)
43. Busnel, Y., Caillouet, C., Coudert, D.: Self-organized disaster management system by distributed deployment of connected Uavs. In: 2019 International Conference on Information and Communication Technologies for Disaster Management (ICT-DM) (2019)
44. Hasan, N.U., Valsalan, P., Farooq, U., Baig, I.: On the recovery of terrestrial wireless network using cognitive Uavs in the disaster area. Int. J. Adv. Comput. Sci. Appl. **11**, 819–822 (2020)
45. Aljehani, M., Inoue, M.: Performance evaluation of multi-UAV system in post-disaster application: validated by Hitl simulator. IEEE Access **7**, 64386–64400 (2019)
46. Vazquez-Carmona, E.V., Vasquez-Gomez, J.I., Herrera-Lozada, J.C., Antonio-Cruz, M.: Coverage path planning for spraying drones. Comput. Ind. Eng. **168**, 108125 (2022)
47. Araujo, J.F., Sujit, P.B., Sousa, J.B.: Multiple UAV area decomposition and coverage. In: 2013 IEEE Symposium on Computational Intelligence for Security and Defense Applications (CISDA) (2013)
48. Hong, Y., Jung, S., Kim, S., Cha, J.: Autonomous mission of multi-UAV for optimal area coverage. Sensors **21**, 2482 (2021)
49. Petritoli, E., Leccese, F., Ciani, L.: Reliability and maintenance analysis of unmanned aerial vehicles. Sensors **18**, 3171 (2018)
50. Fesenko, H., Kharchenko, V., Bardis, N., Kor, A.-L., Brezhniev, Y.: Drone fleet survivability evaluation based on Lanchester's modified deterministic model. Int. J. Circ. Syst. Sig. Process. **14**, 775–781 (2020)
51. Natvig, B.: Multistate system reliability. Wiley Encyclopedia of Operations Research and Management Science. (2011)
52. Kliushnikov, I., Kharchenko, V., Fesenko, H.: An unmanned aerial vehicle as a multi-state system. In: Proceedings - 16th International Conference on Advanced Trends in Radioelectronics, Telecommunications and Computer Engineering, TCSET 2022, pp. 291–296 (2022)
53. Dui, H., Zhang, C., Bai, G., Chen, L.: Mission reliability modeling of UAV SWARM and its structure optimization based on importance measure. Reliab. Eng. Syst. Saf. **215**, 107879 (2021)
54. Rusnak, P., Kvassay, M., Zaitseva, E., Kharchenko, V., Fesenko, H.: Reliability assessment of heterogeneous drone fleet with sliding redundancy. In: 2019 10th International Conference on Dependable Systems, Services and Technologies (DESSERT) (2019)
55. Amari, S.V., Dill, G.: A new method for reliability analysis of standby systems. In: 2009 Annual Reliability and Maintainability Symposium (2009)
56. Tannous, O., Xing, L., Rui, P., Xie, M., Ng, S.H.: Redundancy allocation for series-parallel warm-standby systems. In: 2011 IEEE International Conference on Industrial Engineering and Engineering Management (2011)
57. Kumar, A., Yadav, A.S., Gill, S.S., Pervaiz, H., Ni, Q., Buyya, R.: A secure drone-to-drone communication and software defined drone network-enabled traffic monitoring system. Simul. Model. Pract. Theory **120**, 102621 (2022)
58. Yu, Y., Shi, C., Shan, D., Lippiello, V., Yang, Y.: A hierarchical control scheme for multiple aerial vehicle transportation systems with uncertainties and state/input constraints. Appl. Math. Model. **109**, 651–678 (2022)
59. Behera, T.K., Bakshi, S., Sa, P.K.: Vegetation extraction from UAV-based aerial images through deep learning. Comput. Electron. Agric. **198**, 107094 (2022)
60. Maurya, H.L., Singh, P., Behera, L., Verma, N.K.: Control design for unmanned aerial vehicle using fractional order backstepping control and extended homogenous control. In: 2021 International Symposium of Asian Control Association on Intelligent Robotics and Industrial Automation (IRIA). (2021)

61. Levitin, G., Xing, L., Dai, Y.: Heterogeneous 1-out-of-N standby systems with limited unit operation time. Reliab. Eng. Syst. Saf. **224**, 108532 (2022)
62. Schneeweiss, W.G.: A short boolean derivation of mean failure frequency for any (also non-coherent) system. Reliab. Eng. Syst. Saf. **94**, 1363–1367 (2009)
63. Zaitseva, E., Levashenko, V., Kostolny, J.: Importance analysis based on logical differential calculus and binary decision diagram. Reliab. Eng. Syst. Saf. **138**, 135–144 (2015)
64. Akers: Binary decision diagrams. IEEE Trans. Comput. **C-27**, 509–516 (1978)
65. Coolen, P.A., Coolen-Maturi, T.F.: Generalizing the signature to systems with multiple types of components. Adv. Intell. Soft Comput. **170**, 115–130 (2012)
66. Heidtmann, K.D.: Improved method of inclusion-exclusion applied to K-out-of-N systems. IEEE Trans. Reliab. **R-31**, 36–40 (1982)
67. Rausand, M., Hoyland, A.: System Reliability Theory: Models, Statistical Methods, and Applications, https://www.scirp.org/(S(lz5mqp453edsnp55rrgjct55))/reference/ReferencesPapers.aspx?ReferenceID=1821349
68. Vachtsevanos, G., Ludington, B.: Unmanned aerial vehicles: challenges and technologies for improved autonomy. WSEAS Trans. Syst. **5**(9), 2164–2171 (2006)
69. Kuo, W., Zhu, X.: Importance Measures in Reliability, Risk, and Optimization: Principles and Applications. Chichester, Wiley, UK (2012)
70. Talantsev, A.D.: Analysis and synthesis of certain electric circuits by means of special logical operators. Autom. Remote Control **20**, 874–883 (1959)
71. Steinbach, B., Posthoff, C.: Boolean Differential Equations, Morgan & Claypool Publishers, San Rafael (2013)
72. Kvassay, M., Zaitseva,E., Levashenko, V., Kostolny, J.: Binary decision diagrams in reliability analysis of standard system structures. In: Proceedings of 2016 International Conference on Information and Digital Technologies (IDT), pp. 164–172 (2016)
73. Kvassay, M., Zaitseva, E.: Topological analysis of multi-state systems based on direct partial logic derivatives. In: Lisnianski, A., Frenkel, I., Karagrigoriou, A. (eds.) Recent Advances in Multi-state Systems Reliability. SSRE, pp. 265–281. Springer, Cham (2018). https://doi.org/10.1007/978-3-319-63423-4_14

Method of Cyber-Resilience Information-Control System Synthesis of Mosaic Structure

Yuriy Danyk[1] (✉) and Valery Shestakov[2]

[1] National Technical University of Ukraine "Igor Sikorsky Kyiv Polytechnic Institute", Kyiv, Ukraine

[2] Zhytomyr Polytechnic State University, Zhytomyr, Ukraine

Abstract. The article presents the results of the development of a model and method of synthesis of cyber-resilience information-control system of mosaic structure. The mathematical apparatus of multicriteria optimization according to a non-linear trade-off scheme was used in the methods. The model and method provide an opportunity to build of a mosaic structure cyber resilience ergatic information and control system. The system has to be with an open architecture that is rational in terms of structure and characteristics of automated workplaces.

Keywords: Cyber-resilience · Information-control system · Mosaic structure · Synthesis

1 Introduction

Based on the experience of actions in crisis situations of different nature and origin a timely and adequate response to changes in the situation can be achieved by the complex use of means and systems: data acquisition, automation of calculation and forecasting processes, decision-making and operational management, implementation of protection, proactive and active actions (influences) [1]. The most rational way of complex and effective application of disparate, various means and systems is integration based on the use of advanced cyber information technologies and modern concepts of organization and management of actions..

The main trends in the field of information technology development are [2]: data fabric, artificial intelligence, augmented reality, digital platforms, cyber-physical systems, geo-information systems.

Numerous studies in the field of information technologies prove that with their development:

- the number and complexity of cyber attacks on information systems is growing exponentially;
- the spectrum of cyber incidents is expanding exponentially;
- the means of ensuring information and cyber security in most cases lag behind in the ability to effectively resist innovative destructive cyber influences;

M. Klymash et al. (Eds.): TCSET 2022, LNEE 965, pp. 358–366, 2023.
https://doi.org/10.1007/978-3-031-24963-1_20

– the weight of cyber resilience in the system of indicators of the quality and efficiency of information systems increases, which, according to the terminology of NIST [3], is the ability of the information system to resist cyber threats and resume its work in the event of cyber incidents.

The concept of organizing and conducting "mosaic actions" is widely used today [4]. The distinguishing feature of concept is the conduct of actions by a mosaic structure (MS) of situational organizational and technical complexes (SC) in terms of composition, organization and application conditions [5].

The integration of individual forces and means of the MS into the SC, monitoring their condition and forecasting the ability to perform tasks as part of the SC, control of the configuration and application of the SC is carried out by a specialized information-control system (ICS), which in itself is a complex energetic system with an open architecture.

Since the ICS is formed from a set of different types of computerized means with specialized software, cyber resistance is one of the main indicators of the effectiveness of its application.

In particular the questions of the theory of ICS construction development in conditions of crisis situations was reflected in [6, 7] and others. The results of the analysis of the publications prove that:

– in most of the works, the general principles of building complex ergatic systems are presented, or individual issues of automating the tasks of information-cybernetic integration, rational choice and situational control of the configuration of a complex organizational and technical complex are revealed;
– the various simplifications are used that reduce the efficiency and effectiveness of decisions;
– the theoretical principles that would allow obtaining a comprehensive solution for building a cyber-resilience ICS is currently at the stage of research.

Taking into account the number of approaches given above, based on [3, 8, 9] and the materials of numerous publications in professional publications, we can come to the conclusion that a single established methods for ensuring cyber resilience has not been formed, methods the cyber resilience of ICS is developing.

In the course of the conducted research, the individual results of which are presented in [10, 11], it is found that each SC of the mosaic structure operates according to its own script, which is produced using ICS. It consists from of means: communication; automation on the basis of computer equipment with appropriate software, including that which implements artificial intelligence, which are combined into a computer network with an open architecture; information protection.

So, there is a contradiction between the need for cyber resilience operational command and control of a situational complex of a mosaic structure and the lack of theoretical foundations for building a cyber resilience ICS of such a complex.

These contradictions can be resolved by development and amplification of the theoretical foundations of building a cyber resilience ICS as an integrating component of the situational complex of the mosaic structure (with elements of intellectualization and automation of control tasks).

The purpose of the article is the present of the development results of a model and synthesis method of cyber-resilience information-control system as an integrating component of the situational complex of the mosaic structure.

2 Model of Cyber-Resilience Information-Control System

The functional purpose of the ICS is the timely provision of information on the rational distribution of available disparate and diverse forces and means to the IC control body, the formation of SC configurations and their adjustment in accordance with the existing situation, control of the use of SC in accordance with the approved action plan in conditions of complex destructive cyber influences R.

It is predicted that the software-hardware core of the ICS from the set of computerized means and specialized software available in the mosaic structure can synthesize its heterogeneous (independent of the operating system), cyber-resilience $k*$ configuration from the set K with the corresponding set of logs E^L and metadata E^M characteristic of the synthesized situational system, which will allow to ensure the performance of a certain set of functions F, namely:

- the integration of ICS into a single information and combat space;
- the collection, processing of data from means (systems) of obtaining information; organization of information storage; control of the ability of the members of the MC to perform the assigned tasks; implementation of operational (operational-tactical) calculations;
- providing decision-makers with relevant, reliable and timely information; information and analytical support for control decision-making based on modeling technologies, analysis of situations, forecasting of their development scenario, collective work with information, etc.;
- the formation of signals, switching commands of nodes of the communication matrix;
- the prediction of cyber threats, control of sets of logs and metadata (in terms of [12] – means of classification, arrangement and characterization of data or content), timely detection of anomalies in them, detection of cyber impacts and recovery in the event of cyber incidents;
- the organization of distributed processing of large data sets, machine learning;
- the generation of operational documents;
- the control and display of the current (operational) situation and its transfer to consumers (event planning, modeling and forecasting) in a time regime close to the real one.

The implementation of each function of the ICS at a certain time interval is performed with a certain level of quality q depending on the characteristics of the selected hardware and software tools (HST) from the set Z for the implementation of the corresponding function.

Then ICS can be filed through the specified components as follows:

$$S = (F, E, Z), \tag{1}$$

where $F = \{f_1, f_2, ..., f_N\}$ is an ordered set of ICS functions; $E = E^L \times E^M$ is an ordered set of characteristic logs $E^L = \{e_1^L, e_2^L, ..., e_P^L\}$ and metadata $E^M = \{e_1^M, e_2^M, ..., e_H^M\}$; $Z = \{z_1, z_2, ..., z_W\}$ is an ordered set of HST.

For a set F of functions N, it is possible to form a set Q of levels of the quality of functioning of the ICS, consisting of partially ordered $\{Q^i\}$, non -intersecting subsets with elements $q_k^i \in Q^i$, $i = \overline{1, N}$. In turn, on the set Q, you can build a set of vectors $A = \{A_j\}$, $A_j = \{q_i^j\}$, where each vector A_j determines a certain level of quality of implementation of a set of ICS functions F.

At the same time, a partial order must be introduced on the set of vectors A, such that for any two vectors A_i and A_j. The vector A_i is better A_j, if for each function f_m from the set F the quality of its implementation $q_m^i \in_i$ is not worse than the quality of function implementation $q_m^j \in a_j$.

Then, based on [13], the cyber resilience of the ICS can be defined as follows: the system is cyber-resilience in a certain time interval Δt, if for any set of destructive influences $R\Delta t$ during this interval Δt there is at least one workable distribution of available HST $Z\Delta t$, which ensures, taking into account the compliance of the general scheme of the new system of collection and delivery of logs and metadata, the implementation of a set of functions F with a quality level not lower than A_{lim}:

$$\forall\left(R_{\Delta t} \in R, e_m^L \in E_{\Delta t}^L, e_l^M \in E_{\Delta t}^M\right), \exists k^*(Z\Delta t) \in K, A\left(k^*, e_m^L, e_l^M\right) \geq A_{\text{lim}} \quad (2)$$

Thus, to assess the cyber resilience of an ITS, it is necessary to have models that identify such mappings:

1. Threat model:

$$\rho : R \times T \to R_{\Delta t}. \quad (3)$$

2. Log degradation model;

$$\lambda : E^L \times R_{\Delta t} \to E_{\Delta t}^L. \quad (4)$$

3. Metadata degradation model:

$$\mu : E^M \times R_{\Delta t} \to E_{\Delta t}^M. \quad (5)$$

4. Model of failure (degradation) of the HST:

$$\xi : Z \times R_{\Delta t} \to Z_{\Delta t}. \quad (6)$$

5. Model of the quality of the functioning of the ICS:

$$\alpha : K(Z_{\Delta t}) \times E_{\Delta t}^L \times E_{\Delta t}^M \to A. \quad (7)$$

3 Method of Cyber-Resilience Information-Control System Synthesis

The synthesis of the ICS configuration for a specific MS to solve the relevant tasks in the existing conditions is carried out by means of the software and hardware core of the ICS. The software part of the ICS core is a specialized software that implements models and variants of MS configuration, methods of ICS and SC synthesis.

In the event that the capabilities of the ICS are insufficient to fulfill the target task, control teams are created to adapt it to existing conditions or reconfiguration in accordance with new tasks.

The synthesis of ICS is proposed to be carried out according to the classical scheme:

- the determination of factors that influence the performance of ICS tasks, quality indicators and criteria for the effectiveness of its application;
- The functional synthesis;
- the implementation of structural and parametric synthesis.

Determination of quality indicators and efficiency criteria of the Cyber-resilience Information-control System. The selection and determination of indicators of the quality and efficiency of the ICS is a rather complex theoretical and practical task. Since the ICS is a component of the SC, in accordance with the principle of the priority of the final goal, in the course of evaluating the effectiveness of the functioning of the subsystems of the SC, one should proceed from how the ICS affects the effectiveness of the targeted application of the SC.

Therefore, the effectiveness of the ICS is understood as the success of the task of managing the formation and application of the SC.

It is proposed to evaluate the effectiveness of the ICS by the indicator an effectiveness of control. This indicator characterizes the ability of the ICS to ensure the achievement of the results of the targeted application of the CS. It is an integral indicator of the system of three partial groups.

The first group provides an assessment of the fulfillment of the purpose of the ICS, namely: readiness to perform tasks as intended, reliability of decisions, throughput, system capacity, compatibility, control efficiency.

The second group determines the level of ensuring the functional stability of the ICS (the ability of the system to perform the tasks defined for it under the influence of unintentional and intentional (organized) influences). The quality indicators of this group include: survivability, reliability, security, cyber resistance.

The third group of criteria makes it possible to determine the extent to which the requirements of specific control laws are met in this system. These criteria are mostly qualitative in nature.

If the effectiveness of the ICS is reduced to indicators from the first two groups - control efficiency, which is numerically expedient to represent through the time of the control cycle T_{CC}, and the indicator of cyber resistance of the ICS C_{CR} with a fixed number of communication channels N_{Ch}, then the following criteria can be presented:

$$e_h = \min T_{CC}^{(i)},\ h \in H,\ C_{CR}^{(i)} \le C_{CR}^{зад},\ i = 1,\ldots,N. \quad (8)$$

$$e_{h+j} = \max C_{KC}^{(i)},\ h + j \in H,\ T_{CC}^{(i)} \leq T_{CC}^{зад}. \tag{9}$$

$$E \rightarrow \min \max \left\{ T_{CC}^{(i)}, C_{CR}^{(i)} | N_{Ch} = const \right\},\ i = 1, \ldots, N; \tag{10}$$

where $T_{CC}^{(i)}$ is the estimate of the average time of the control cycle in the ICS; $C_{CR}^{(i)}$ is the evaluation of the cyber resistance indicator; i is the serial number of the alternative version of the ICS.

3.1 A Functional Synthesis

The task of functional synthesis will be reduced to finding the best mapping on the set Z of ICS functions F:

$$f_i : F \times E \times Z \rightarrow extrQ^i, \tag{11}$$

where $Q^i = \left\{ q_1^i, q_2^i, \ldots, q_{K_i}^i \right\}$ is the ordered finite set of quality levels of the implementation of the i function.

In accordance with the above, for a set of functions F there is a set Q of levels of the quality of the functioning of the ICS, consisting of non-intersecting subsets $\{Q^1, Q^2, \ldots, Q^N\}$ with elements $q_j^i \in Q^i$, $i = \overline{1, N}$, ordered according to the condition $q_1^i \leq q_2^i \leq \ldots, \leq q_{K_i}^i$, $i = \overline{1, N}$.

$$Q = \begin{Vmatrix} q_1^1 & \cdots & q_K^1 \\ \vdots & \ddots & \vdots \\ q_1^M & \cdots & q_{K_N}^M \end{Vmatrix} = q_j^i,\ i = \overline{1, M},\ j = \overline{1, K_i}. \tag{12}$$

3.2 A Structural-Parametric Synthesis

The structural-parametric synthesis of ICS was carried out according to the following method:

1. The tuples of the *i*-th control situation, a typical HST, a source of information of the ICS according to the data of previously (a priori) formed databases and knowledge are determined:

$$U_i = \left\{ P_i^U, F_j^U, I_f^U \right\},\ i = 1 \ldots I,\ j = 1 \ldots J,\ f = 1 \ldots F; \tag{13}$$

$$z_j = \left\{ F_j^z, I_j^z(I_f^{ID}), TX_j^z \right\}; \tag{14}$$

$$ID_f = \left\{ I_f^{ID}, TX_f^{ID} \right\}; \tag{15}$$

where P^U – a unique feature of the control situation; F^U – a set of tasks for solving the control situation; I^U – a set of information needs for solving the control situation; F^z – a set of tasks that provides a solution (performs) of the HST; I^z – a set of information needs of the HST; TX^z – a set of technical characteristics of the HST; I^{ID} – set of information possibilities of the information source; TX^{ID} – set of characteristics of the information source.

2. The rational number of HPS N_z was determined, which will ensure the minimum time of execution of the control cycle T_S, a high level of reliability of decisions D_S, better information redundancy IN_S to ensure the process of formation and control of SC:

$$\begin{cases} T_S(N_z) \to \min, & npu\ T_S(N_z) \le T_S(N_z), \\ D_S(N_z) \to \max, & npu\ D_S(N_z) \ge D_{S\,zp}(N_z), \\ IN_S(N_z) \to \max, & \\ npu\ IN_S(\min N_z) \le IN_S(N_z) \le IN_S(\max N_z). & \end{cases} \quad (16)$$

The solution of a multi-criteria problem is possible on the basis of a non-linear scheme of compromises according to [14]. The determination of normalized partial criteria and weighting factors was carried out on the basis of the algorithm developed by the authors and given in [15].

Then, the optimization model looks like:

$$\delta_z = \gamma_{TO}(1 - T_{SO})^{-1} + \gamma_{DO}(1 - D_{SO})^{-1} + \gamma_{INO}(1 - IN_{SO})^{-1} \to \min; \quad (17)$$

$$N_{ARM}^{rational} = N_{ARM} \quad if \quad \delta_{ARM} = \min. \quad (18)$$

where $\gamma_{TO}, \gamma_{DO}, \gamma_{INO}$ are the normalized weighting coefficients of the corresponding indicators.

Performing a sequence of mutual mappings (multiplication of corresponding elements) tuples of the control situation, HST and information

$$S = \left\{ P^U, F_j^{\bmod\ U}, I_j^{\bmod\ ID,U}, TX_j^{\bmod\ ID,z} \right\}, \quad (19)$$

which, in accordance with the control situation, contains a description of the HST of system, the information needs of the system $I_j^{\bmod\ ID,U}$, the quality of the initial data (proposals for a decision, or a decision) $TX_j^{\bmod\ ID,z}$.

The parameters that make up the ICS tuple were used to calculate their weighting coefficients for each HST. Next, finding the sums of significant items from the list of tuple parameters and weighting coefficients for each HST was realized.

Thus, the structure of the ICS will be rational, the HST of which ensure the high-quality performance of the largest number of tasks of the ICS $F_j \to \max$; the largest (necessary) number of sources of information $I_j \to \max$, with the best technical characteristics of the APR, are involved. The previously obtained rational number of HST $N_z^{rational}$ was a limitation for the defined system of criteria requirements.

The solution of the multi-criteria problem of structural-parametric synthesis was carried out using a nonlinear scheme of compromises in the form of a discrete convolution:

$$\Psi_j = GF_{j0}(1 - F_{j0})^{-1} + GI_{j0}(1 - I_{j0})^{-1} + GTX_{j0}(1 - TX_{j0})^{-1} \rightarrow \min, \tag{20}$$

where $F_{j0}, GI_{j0}, GTX_{j0}$ are normalized criteria values; $GF_{j0}, GI_{j0}, GTX_{j0}$ are weight coefficients of the relevant criteria.

Evaluating the effectiveness of system configuration results with condition control:

$$\Omega_{POR}^{\min} \leq \Omega < \Omega_{POR}^{\max},\ \Omega = \sqrt{(F_Z - F_S)^2 + (I_Z - I_S)^2 + (TX_Z - TX_S)^2}. \tag{21}$$

To ensure security from destructive cyber influences, taking into account that the system must process large volumes of traffic in real time, it is necessary to ensure placement on the border of the perimeter of network infrastructures or on the ICS and include the following modules in its composition:

- the collection and analysis of network traffic from any active network equipment;
- the detection of external and internal network threats with a quick search based on specified criteria for violations and anomalies, network traffic filtering;
- the analysis and detection of abnormal components of traffic (including encrypted);
- the network segmentation and construction of topology graphs (visualization);
- the detection of unauthorized network activity;
- the integration with external solutions and signature analysis channels;
- the automated detection of the distribution of malicious code.

When the control situation changes, the cycle is repeated according to the method.

4 Conclusions

To ensure the rational distribution by task, place and time of spatially dispersed, heterogeneous and diverse forces and means of the MS, to implement their effective control in accordance with the crisis situation in the conditions of complex destructive cyber influences, a model and method for the synthesis of a cyber-resistant information control system of the MS have been developed.

The model and method of synthesis of ICS allow to automate the process of determining the rational in terms of the structure and characteristics of automated workplaces, cyber-resistant ICS. At the stage of structural-parametric synthesis of the method, the mathematical apparatus of multi-criteria optimization according to the nonlinear scheme of compromises was used.

According to the results of the research carried out by the authors during the deployment of the ICS of the specialized center of operational control of forces and means, with the use of the developed method, an increase in its cyber resistance was obtained in the range of up to 95–98%.

References

1. Danyk, Y., Shestakov, V.: Ways of reducing civilian casualties during wars and armed conflicts of modern times. High-tech aspects. The actual problems of the world today. Vol. 2, p. 15 – 30. London (2019)
2. Gartner top strategic technology trends for 2021. https://www.gartner.com/marterwithgartner/gartner-top-strategic-technology-trends-for-2021. Accessed 21 Jan 2022
3. . Ross, R., Pillitteri, V., Graubart, R., Bodeau, D., Mcquaid, R.: Developing cyber-resilient systems: a systems security engineering approach. In: NIST Special Publication vol. 2, pp. 800–160 Rev. 1. (2021). https://doi.org/10.6028/NIST.SP.800-160v2r1
4. Mosaic warfare and human–machine symbiosis. https://doi.org/10.32386/scivpro.000024
5. Creating cross-domain kill webs in real time. https://www.darpa.mil/news-events/2020-09-18a
6. Situational centers. Theory and practice. A collection of articles. Science ed. Morozov A.O., Kuzmenko G.E., Litvinov V.A. Inst. Problems of Mathematical Machines and Systems of the National Academy of Sciences of Ukraine, Kyiv (2009)
7. Zagorka, O.M., Pavlikovskyi, A.K., Zagorka, I.O., et al.: Synthesis of the organizational structure of the control of a group of troops (forces): methodical aspect. Mod. Inf. Technol. Sphere Secur. Defen. **1**(25), 160–165 (2016)
8. T. H. Llansó, D. A. Hedgecock, and J. A. Pendergrass. The State of Cyber Resilience: Now and in the Future / Johns Hopkins APL Technical Digest, Vol. 35, Num. 4 (2021)
9. Cyber Resilience Review Method Description and Self-Assessment User Guide. https://www.us-cert.gov/ccubedvp/self-service-crr, last accessed 2022/04/21
10. Danyk Yu.G., Shestakov V.I. Methodology of the synthesis of the automated information-control system of the situational intelligence-strike complex. Collection of scientific works of the National Academy of the State Border Service of Ukraine. Series: military and technical sciences. Khmelnytskyi: NADPSU, 3 (81), p. 316 - 340. (2019)
11. Danyk Yu., Shestakov V. Methodology of Situational Control System Synthesis. 2019 3rd International Conference on Advanced Information and Communications Technologies, AICT 2019. https://ieeexplore.ieee.org/document/8847914
12. ANSI/NISO Z39.85–2012 The Dublin Core Metadata Element Set (2012)
13. Korolev, A.N., Kem, O.V., Mironychev, V.A.: Modeling of functionally stable navigation and information systems. Izv. Universities. Instrumentation. **61**(8), 652–659 (2018)
14. A. N.Voronyn, Yu. K. Ziatdinov, M. V. Kukpinskyi. Multi-criteria solutions: models and methods: monograph. NAU, Kyiv (2011)
15. Yu. G. Danyk, O. V. Lagodny, O. O. Pisarchuk, K. O. Sokolov, S. V. Chernyshuk, S. V. Tymchuk, V. I. Shestakov. Multicriteria mathematical models of situational control and self-organization in complex information systems: monograph. PP "Ruta", Zhytomyr (2016)

Neural Network Architecture with Oscillatory Synaptic Blocks for Signal Processing of Spatially Spaced Multiband Radar Complex

Roman Peleshchak, Vasyl Lytvyn, Ivan Peleshchak(✉), and Oleh Veres

Lviv Polytechnic National University, Lviv 79000, Ukraine
vasyl17.lytvyn@gmail.com

Abstract. One of the promising areas of radar development is the development of methods for target recognition using spatially spaced multi-band radar complex, which are based on neural networks. Spatially spaced multi-band radar complex consists of spatially spaced radar stations, which jointly conduct radar surveillance of targets. Signal scattering diagrams are obtained from the radar complex, which are vectors of features supplied to the neural network input. This paper proposes a neural network with parallel blocks of oscillatory synaptic connections, which allows real-time (information for processing comes with a fairly high frequency) recognition of noise signals reflected from radar targets based on information resonance.

Keywords: Neural Network · Oscillatory Synaptic Blocks · Multiband Radar Complex · Signal Processing

1 Introduction

Research in the field of pattern recognition and the development of technology contributes not only to the detection, but also to the classification of types of aerial objects. Non-parametric recognition methods (minimum average distance, k-nearest neighbors and neural networks) are used to create a classifier when the probability density distribution function of distant portraits is unknown. Therefore, a promising direction of information technologies is the development of neural network methods for recognizing radar targets based on a spatially distributed multi-band radar complex. Such a complex consists of radar-spaced radar stations that conduct joint radar target detection. The basic idea of a multi-band radar system is that radar systems operate at different frequencies with different polarizations and are located in different places. This allows you to simultaneously detect dozens of radar targets and recognize them in real time using a neural network with parallel blocks of oscillatory synaptic connections. The input features for recognition are target scattering diagrams, which are integral characteristics of the effective target scattering areas.

Hybrid neural networks (neural fuzzy networks, wavelet neural fuzzy networks, neural networks with a sensory layer in the form of Fourier series) are used to solve a wide

M. Klymash et al. (Eds.): TCSET 2022, LNEE 965, pp. 367–374, 2023.
https://doi.org/10.1007/978-3-031-24963-1_21

range of Data Mining problems (prediction, emulation, identification of non-stationary chaotic processes, clustering, classification, intelligent control and diagnostics of states of complex objects) under conditions of uncertainty, nonlinearity, stochasticity, various kinds of interference and interference. These hybrid neural networks are capable of functioning in real time. Such systems include Wang-Mendel architecture [1], adaptive neuro-fuzzy Takagi-Sugeno-Kang systems [2], wavelet-fuzzy networks [3], adaptive wavelet-fuzzy systems with W-neurons [4] and Fourier neural systems [5]. However, these systems have a number of limitations: a clearly defined form of the type-2 membership function; real-time operation is associated with the model reduction procedure; do not take into account the neuron's own dynamics; do not take into account the accumulation of impulses in neurons; do not take into account non-locality over time. These models of neural systems did not consider an approach to the recognition of multispectral images (multispectral electromagnetic signals) emitting external oscillatory objects (reflected radar signals from different types of air targets, non-stationary signals emitted by charged nanoparticles of heterophase plasma, formed by combustion of solid fuel during rocket flight, high-temperature emissions of industrial aerosols, ionized dust and gas formations in the atmosphere their interaction with the Earth's atmosphere, the accumulation of cosmic dust under conditions of ionizing radiation) and did not take into account the number of signal pulses fed to the input of neurons and the relaxation time of the neuron after its action potential.

Research in the field of spectral image recognition and the development of information technologies contribute to the creation of modern radar stations that are capable not only of detecting, but also automatically classifying types of aerial objects.

Modeling of the characteristics of secondary radiation as features of radar recognition and indicators of the quality of radar recognition of aerial objects is proposed in [6]. The author [7] compares the effectiveness of aerial object recognition algorithms when used as features of scatter diagrams or distant portraits.

In works [8], a type-2 wavelet-neuro-fuzzy system and a high-speed algorithm for its learning were proposed, in which the problem of model reduction was solved on the basis of parallel work of ensembles of wavelet-neuro-fuzzy systems of type-1. The problem of model reduction was also solved on the basis of ensembles of Fourier neurons [9, 10] and their subsequent optimal combination.

In this work, a neural network with parallel blocks of oscillatory synaptic connections in the hidden layer is proposed, which enables the recognition of radar targets in real time (information for processing arrives with a sufficiently high frequency) based on informative features (effective areas of target scattering) and information resonance.

2 Mathematical Model of a Neural Network with Parallel Blocks of Oscillatory Synaptic Connections

A neural network detector with parallel blocks of oscillatory synaptic connections (Fig. 1), whose basic elements are oscillatory neurons with their own dynamics, is proposed to solve the problem of spectral signal recognition in real time.

Effective target scattering areas, polarization features, and modulation features can be used as features of target recognition in narrowband sounding. In particular, in this work,

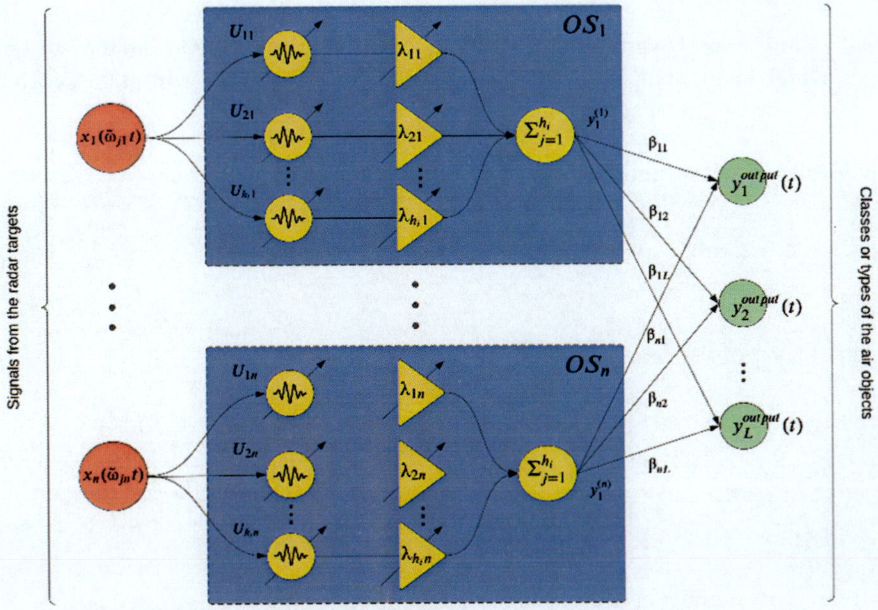

Fig. 1. A hybrid neural network with parallel blocks of oscillatory synaptic connections in a hidden layer.

the features of recognition of radar aerial targets (cruise missile, 122-caliber missile, 227-caliber missile) are their effective target scattering areas (Table 1), which are proportional to the scattering intensity of reflected electromagnetic waves represented by the scattering diagram.

An oscillatory neuron is quite close in architecture to a classical formal neuron with n inputs, but instead of the usual tunable synaptic weights, it has parallel blocks of oscillatory synaptic connections $OS_j (j = 1, 2, ..., n)$, in which the tunable parameters are not only the synaptic weights λ_{ji}, but also parameter ω_{0ji} (adjustable parameter when training an oscillatory neural network), which describes the natural frequency of the j-th neuron to which the i-th external signal arrives.

The structure of a three-layer neural network with a hidden layer in the form of parallel blocks of synaptic connections was used in the simulation. A reflected electromagnetic signal from one air target is applied to the input of each unit. The number of inputs of the first sensor layer is equal to the number of effective target scattering areas. The number of neurons of the output layer is equal to the number of aerial objects, that is, the number of classes.

The functioning of a neural network with parallel blocks of oscillatory synaptic connections is described by the following system of equations:

- input layer of the neural network

$$\overline{x}(\tilde{\omega}_{ji}t) = \left(x_1\left(\tilde{\omega}_{ji}t\right), x_2\left(\tilde{\omega}_{ji}t\right), ..., x_i\left(\tilde{\omega}_{ji}t\right), ..., x_n\left(\tilde{\omega}_{ji}t\right)\right), \quad (1)$$

$x_i(\tilde{\omega}_{ji}t) = \overline{\sigma}_i^{targ} \cdot e^{-\beta t}\sin(\tilde{\omega}_{ji}t),\ i = 1, 2, ..., n$ - the signal reflected from the i-th aerial target, which enters the j-th oscillatory neuron of the hidden layer from the receiving antenna of the i-th radar station.

- hidden layer with parallel blocks of oscillatory synaptic connections

$$y_i^{(1)} = \sum_{j=1}^{h_i} \lambda_{ji}(t) U_{ji}(x_i(\tilde{\omega}_{ji}t)) + b_i^{(1)}, \tag{2}$$

- output layer of the neural network

$$y_l^{output}(t) = f_l\left(\sum_{i=1}^{n} \beta_{il} y_i^{(1)} + b_l^{output}\right) = \frac{e^{\left(\sum_{i=1}^{n} \beta_{il} y_i^{(1)} + b_l^{output}\right)}}{\sum_{l=1}^{n} e^{\left(\sum_{i=1}^{n} \beta_{il} y_i^{(1)} + b_l^{output}\right)}}, \tag{3}$$

$l = 1, 2, ..., n$ - number of neurons in the output layer (number of classes); $i = 1, 2, ..., n$ - the number of neurons in the hidden layer with parallel blocks of oscillatory synaptic connections; $b_i^{(1)}$ - displacement vector for neurons in the hidden layer; $t = 0, 1, 2, ...$ - current discrete time; n - the number of recognizable aerial objects; $\overline{\sigma}_i^{targ}$ - the average value of the effective area of the i-th target [6]; β_{il} - synaptic connections between neurons of the hidden layer with blocks of oscillatory neurons and neurons of the output layer; b_l^{output} - displacement vector for neurons of the output layer; $y_i^{(1)}$ - value of oscillator neurons at the output in the hidden layer, which implement a complex response to the input signal similar to a field of fuzzy sensory neurons [10].

The outputs of oscillatory neurons are determined both by adjustable synaptic weights $\lambda_{ji}(t)$ and by linear oscillatory functions $U_{ji}(x_i(\tilde{\omega}_{ji}t))$ obtained from olutions of the oscillatory differential equation

$$\frac{d^2 U_{ji}}{dt^2} + 2\delta \frac{dU_{ji}}{dt} + \omega_{0ji}^2 U_{ji} = \overline{\sigma}_i^{targ} \cdot e^{-\beta t}\sin(\tilde{\omega}_{ji}t)\omega_{0ji}, \tag{4}$$

$2\delta(N_j, N_{cj}) = \tilde{\delta}\tanh\left(\frac{N_j - N_{cj}}{\sigma^2}\right)$ - attenuation amplitude parameter of the j-th neuron; N_k; N_{ck}; σ^2 - the number of impulses arriving at the j-th neuron, the threshold value of impulses of the j-th neuron, and the variance, respectively; β - a parameter that describes the attenuation of reflected electromagnetic waves from a radar target.

The solution of the linear oscillator Eq. (4) has the form

$$U_{ji}(x_i(\tilde{\omega}_{ji}t)) = \frac{\overline{\sigma}_i^{targ} e^{-\beta t}\omega_{0ji}^2\left(-2(\beta-\delta)\tilde{\omega}_{ji}\cos[\tilde{\omega}_{ji}t] - \left(\beta^2 - 2\beta\delta + \delta^2 + \omega_{ji}^2 - \tilde{\omega}_{ji}^2\right)\sin[\tilde{\omega}_{ji}t]\right)}{\beta^4 - 4\beta^3\delta + \delta^4 + \left(\omega_{ji}^2 - \tilde{\omega}_{ji}^2\right)^2 + 2\delta^2\left(\omega_{ji}^2 + \tilde{\omega}_{ji}^2\right) - 4\beta\delta\left(\delta^2 + \omega_{ji}^2 + \tilde{\omega}_{ji}^2\right) + 2\beta^2\left(3\delta^2 + \omega_{ji}^2 + \tilde{\omega}_{ji}^2\right)} \tag{5}$$

$$\omega_{ji} = \sqrt{\omega_{0ji}^2 - \delta^2}.$$

In particular, in the case of a nonlinear oscillatory neuron with its own dynamics ω_{0ji}, the oscillatory neuron is able to generate pulses in the absence of external non-stationary signals $x_i(\tilde{\omega}_{ji}t) = 0$, when the number of pulses entering the neuron N_j reaches a threshold value of N_{cj}, $N_j \geq N_{cj}$ [11, 12]. Such a non-linear neuron can be considered as a threshold device that converts an input non-stationary signal into a sequence of output pulses due to the "imposition" of the dynamics of the input non-stationary signal on the own dynamics of the non-linear neuron. Then the task of the dynamics of a nonlinear oscillatory neuron in the presence of external nonstationary signals of different shape, frequency, and amplitude is to determine the frequency-time and time dependence of the morphology of the signal at the output of the nonlinear oscillatory neuron and to establish the criterion for the occurrence of resonance effects in the nonlinear oscillatory neuron.

3 An Algorithm for Learning the Parameters of a Neural Network with Parallel Blocks of Oscillatory Synaptic Connections

The learning algorithm consists of the following steps:

1. Initialization of the network: weights and offsets of the network take small random values. We set the learning rate coefficient $\eta (0 < \eta < 1)$ and the maximum value of the root mean square error of learning $E_{\max}$.
2. We determine the reference value $y_l^{(e)}$ for each element of the training sample from the condition of information resonance

$$y_l^{(e)} = \begin{cases} A_i^{res}(t) = \sum\limits_{i=1}^{n}\sum\limits_{j=1}^{h_i} \dfrac{\overline{\sigma}_i^{targ}\beta_{il}(t)\omega_{0ji}^2(t)}{2\sqrt{(\beta-\delta)^2\left(\omega_{0ji}^2(t)-\delta^2\right)}} \; if \left|\tilde{\omega}_i - \sqrt{\left|\omega_{0ji}^2(t) - \delta^2 - (\delta-\beta)^2\right|}\right| \leq \varepsilon \\ 0, \; if \left|\tilde{\omega}_i - \sqrt{\left|\omega_{0ji}^2(t) - \delta^2 - (\delta-\beta)^2\right|}\right| > \varepsilon \end{cases}, \quad (6)$$

$0 < \varepsilon < 1$ predetermined.

3. Pairs from the training sample are fed to the neural network $\left\{x_i(\tilde{\omega}_{ji}t), y_l^{(e)}\right\}_{i,l=1}^{n}$.
4. We calculate
$$E_l(t) = \frac{1}{2}\left(y_l^{(e)}(t) - y_l^{output}(t)\right)^2 = \frac{1}{2}\left(y_l^{(e)}(t) - f_l\left(\sum_{i=1}^{n}\beta_{il}y_i^{(1)} + b_l^{output}\right)\right)^2.$$
5. We sum $E = \frac{1}{n}\sum\limits_{l=1}^{n} E_l$.
6. If $E < E_{\max}$, then we complete the learning algorithm. If $E \geq E_{\max}$, then a new learning cycle is started and we proceed to step 7.
7. We calculate derivatives $E_l'(\omega_{0ji})$, $E_l'(\lambda_{ij})$, $E_l'(\beta_{il})$.
8. After that, the weights of the network are updated $\omega_{0ji}(t+1) := \omega_{0ji}(t) - \eta E_l'(\omega_{0ji})$,, $\lambda_{ji}(t+1) = \lambda_{ji}(t) - \eta E_l'(\lambda_{ij})$, $\beta_{il}(t+1) = \beta_{il}(t) - \eta E_l'(\beta_{il})$ and we proceed to step 2.

4 Computer Experiment and Analysis of Results

The proposed model of a neural network with parallel blocks of synaptic connections (Fig. 1) is programmed in Python to study its functional properties.

For the computer experiment, features $x_i(\tilde{\omega}_{ji}t)$ (effective scattering areas of targets, which are displayed by scattering diagrams, which are integral characteristics of these areas) were selected during incoherent multi-frequency sounding by a radar signal. Sounding is considered incoherent if the phase-time relations of the signal at different frequencies are random. In this case, only amplitude information is used, which depends on the interference effects of secondary radiation at sufficiently high frequencies. Amplitude information is determined in the high-frequency approximation by the expression of the effective area of the target (Table 1) [6, 13] from which the signal of the group secondary emitter with inseparable and unobscured shiny elements is scattered.

In general, the signs of radar target recognition are:

- frequency-averaged values of the effective area of the target $\overline{\sigma}_i^{targ}$; values of normalized correlation functions of signal amplitude fluctuations at different frequencies;
- incoherent distant portraits.

In order to analyze the effectiveness of a neural network with parallel blocks of oscillatory synaptic connections, the results of the simulation of the recognition procedure are presented on the example of 3 aerial objects (Table 1).

Table 1. The value of the effective areas of target scattering

№	Radar target	$\overline{\sigma}_i^{targ}$, m2	$\overline{\sigma}_i^{targ}$, dB
1	The cruise missile	1	0
2	Missile, caliber 122 mm	0,009	−20
3	Missile, caliber 227 mm	0,018	−17

For training and testing the neural network, experimental data on the effective scattering areas of electromagnetic waves from air objects, given in Table 1, are used. Based on these data, 1000 values of the effective scattering areas were generated using a uniform distribution.

The evaluation characteristic of the recognition efficiency is the full probability of a recognition error when objects of different L classes are equally likely to appear

$$P_{err} = \frac{1}{L}\sum_{l=1}^{L}(1 - P_{ll}), \tag{7}$$

P_{ll} - the conditional probability of making a decision that the target belongs to class l under the condition that it belongs to class l.

Table 2. Accuracy of recognition of aerial targets at different signal/noise ratios.

№	SNR, dB	Feature: effective areas of target dispersion, $\overline{\sigma}_i^{targ}$, m^2		
		Missile, caliber 122 mm	Missile, caliber 227 mm	The cruise missile
		0,009	0,018	1
1	5	0,94	0,96	0,97
2	10	0,965	0,972	0,985
3	20	0,967	0,976	0,987
4	30	0,968	0,977	0,988

Table 2 shows the results of recognition based on information resonance of radar air targets using a hybrid neural network with hidden parallel oscillator blocks. When using the feature of the effective scattering areas of electromagnetic signals as targets, aerial objects with larger $\overline{\sigma}_i^{targ}$ have the highest probability of their direct recognition. As the signal-to-noise ratio decreases, the probability of correctly recognizing targets decreases. This is because the hybrid neural network is properly trained and these types of aerial targets are correctly separated by reflected electromagnetic portraits.

5 Conclusions

A hybrid neural network with parallel blocks of oscillatory synaptic connections in the hidden layer has been developed, which enables real-time recognition of noisy signals reflected from radar targets based on information resonance.

A criterion for the occurrence of information resonance is proposed, in which there is a sharp increase in the amplitude of the useful signal at a low signal-to-noise ratio.

An algorithm for learning a hybrid neural network with parallel blocks of oscillatory synaptic connections based on gradient descent has been developed.

A computer experiment showed that aerial objects with the largest effective target scattering area have the highest probability of correct target recognition. As the signal-to-noise ratio decreases, the probability of correctly recognizing targets decreases.

References

1. Wang, Y., Liu, H., Jia, W., et al.: Deep fuzzy rule-based classification system with improved wang-mendel method. IEEE Trans. Fuzzy Syst. (2021). https://doi.org/10.1109/TFUZZ.2021.3098339
2. Li, J., Yang, L., Qu, Y., Sexton, G.: An extended Takagi–Sugeno–Kang inference system (TSK+) with fuzzy interpolation and its rule base generation. Soft. Comput. **22**(10), 3155–3170 (2017). https://doi.org/10.1007/s00500-017-2925-8
3. Yen, V.T., Nan, W.Y., Van Cuong, P.: Recurrent fuzzy wavelet neural networks based on robust adaptive sliding mode control for industrial robot manipulators. Neural Comput. Appl. **31**(11), 6945–6958 (2018). https://doi.org/10.1007/s00521-018-3520-3

4. Bodyanskiy, Ye., Pliss, I., Vynokurova, O.: Hybrid waveletneuro-fuzzy system using adaptive W-neurons. Wissenschaftliche Berichte, FH Zittau/Goerlitz 106, 301–308 (2010)
5. Gashler, M.S., Ashmore, S.C.: Training Deep Fourier Neural Networks to Fit Time-Series Data. In: Huang, D.-S., Han, K., Gromiha, M. (eds.) ICIC 2014. LNCS, vol. 8590, pp. 48–55. Springer, Cham (2014). https://doi.org/10.1007/978-3-319-09330-7_7
6. Shirman, Y.D.: Computer Simulation of Aerial Target Radar Scattering, Recognition, Detection, and Tracking. Artech House, Norwood (2002)
7. Shin, C., Veremyev, V.I.: Comparative analysis of the effectiveness of radar target recognition by scatterplots and range portraits using neural network algorithms. Nat. Tech. Sci. **6**, 219–222 (2006)
8. Bodyansky, E.V., Vinokurova, E.A.: Type-2 wavelet-neuro-fuzzy system and its learning algorithm in problems of intelligent information processing. Adapt. Autom. Contr. Syst. **17**(37), 139–148 (2011)
9. Andreev, S.N., Firsov, V.G., Pavlovich, A.V.: A neurological method for the analysis of multispectral aerial photographs.In: XXIII international scientific and technical conference "Radar, navigation, communications", pp. 405–413, (2017)
10. Andreev, S.N., Firsov, V.G.: Synthesis of neural network and holographic methods. Heliogeophys. Res. **9**, 23–29 (2014)
11. Peleshchak, R., Lytvyn, V., Bihun, O., Peleshchak, I.: Structural transformations of incoming signal by a single nonlinear oscillatory neuron or by an artificial nonlinear neural network. Int. J. Intell. Syst. Appl. (IJISA)) **11**(8), 1–10 (2019)
12. Lytvyn, V., Vysotska, V., Peleshchak, I., Rishnyak, I., Peleshchak, R.:Time dependence of the output signal morphology for nonlinear oscillator neuron based on van der pol model. International J. Intell. Syst. Appl. (IJISA)) **10**(4), 8–17 (2018)
13. Opanasyuk, I.: Artillery monitoring battlefield radars and prospects of their development. Military-technical collection 12 (2015)

Modeling, Simulation and Visualization of Acoustic Coherent Images

Iryna Hvozdeva(✉) and Volodymyr Myrhorod

National University "Odessa Maritime Academy", 8, Didrikhsona Street, Odessa 65029, Ukraine
onopchenko.im@gmail.com

Abstract. This study is devoted to the consideration of methods for creating mathematical models of acoustic coherent images in relation to the tasks of monitoring the underwater environment of the World Ocean using modern distributed information and telecommunication networks. The combination of the theoretical approach to constructing an acoustic image as a section of the secondary field at the aperture of the receiving antenna, and the experimental method for obtaining such images with instrumental measuring instruments of a hydroacoustic tank was chosen as the main research method. Comparison of theoretical and experimental results makes it possible to verify the proposed mathematical models of acoustic coherent images. The proposed models, confirmed experimentally, make it possible to perform computer simulation of the processes of formation and processing of acoustic coherent images (ACI) to solve the problems of creating and improving methods for processing such images for their recognition and classification, as well as for testing the technical means of underwater monitoring networks during their tests.

Keywords: Networked simulation and visualization · Underwater environment · Secondary field · Acoustic coherent image · Mathematical model · Hydroacoustic tank

1 Motivation and Demarcation

The need for research in the chosen direction is due to the importance of the problems of studying the underwater environment of the World Ocean and the problems of involving this environment in economic activity. A number of significant activities and businesses are associated with the use of the underwater environment of the World Ocean and its bottom, since they provide an infrastructural connection between the continents. Numerous technical facilities are currently located directly in the ocean environment. Accordingly, such objects require the creation and safe operation of navigation, communication and control facilities, which together form a complex distributed information and telecommunications network.

Due to the nature of the environment, the main means of studying the underwater environment and objects located in it and at the bottom is the use of acoustic waves. Only at relatively short distances, corresponding to tens of meters, optical means can be

M. Klymash et al. (Eds.): TCSET 2022, LNEE 965, pp. 375–395, 2023.
https://doi.org/10.1007/978-3-031-24963-1_22

used. Therefore, studies of the problems of propagation of acoustic waves in a layered inhomogeneous medium, refraction of acoustic waves and diffraction on the surfaces of objects are of particular importance. If there are proven mathematical models for solving the problems of acoustic waves propagation in a layered inhomogeneous medium, then for solving diffraction problems such models have not yet been sufficiently developed. Therefore, studies devoted to the construction of adequate models of the diffraction of acoustic waves on the surfaces of objects with a complex structure, which are confirmed by experimental data, are relevant.

The secondary acoustic field recorded on the sensor system of such a field is a multidimensional data array; therefore it is represented as a spatial section of the acoustic image of the studied object.

The processes of propagation of a probing acoustic field in an inhomogeneous medium and their interaction with the surface of investigated objects are of a complex and still insufficiently studied nature under the influence of various factors. Therefore, to solve subsequent problems of recognition and classification, it is necessary to single out only such a set of feature-forming factors that create a recognizable image of the studied object. Therefore, the Kirchhoff approximation, which was subsequently confirmed in a full-scale experiment, is taken as the limitations adopted in the present study. Such a limitation corresponds to the usually existing difference in the wave sizes of the investigated objects or their parts in comparison with the wavelength of the probing field or the length of the wave packet. This limitation is due to the practical need to ensure a high spatial-angular resolution of sounding tools in the conditions of limited dimensions of autonomous underwater vehicles and, accordingly, the wave sizes of receiving-emitting antennas.

The General Aim of This Work is as Follows: Computer modeling of the processes of formation of secondary acoustic fields of complex reflection surfaces using the results of semi-nature experiments for subsequent processing of the obtained images and their transmission over communication networks.

2 Introduction and Foundations

Known methods and means of mathematical modeling allow solving a wide range of theoretical and practical problems related to the formation and processing of coherent images in acoustic sounding systems. These systems are used to obtain information about objects in such applied areas as underwater geological exploration, seismic acoustics, medical and technical diagnostics. Acoustic images in sounding systems are formed on the basis of secondary acoustic wave fields of the studied objects. The main feature of these fields is a high degree of coherence, which leads to the preservation of the phase distribution of the spatial components of the wave field and their interference in the receiving plane.

The complexity of the problems of mathematical modeling of the processes of acoustic images formation is due to the significant difficulties in the mathematical description of interference phenomena in secondary acoustic fields. In contrast to optical images, the essence of the problem of formation and processing of acoustic images is to establish the relative spatial position of the areas of local reflection of acoustic waves, which is determined by the features of the studied surfaces of objects and the spatial resolution of acoustic sounding systems, both in the longitudinal and transverse directions.

Underwater acoustic networks are considered in detail in [1]. Following [1], an underwater acoustic network is a technology that makes it possible to implement environmental monitoring tasks. Various architectures for 2D and 3D subsea sensor networks are discussed, and the characteristics of the subsea channel are described in detail. The main problems associated with the development of effective network solutions related to the underwater environment are detailed, a cross-level approach to the integration of all communication functions is proposed.

Underwater sounding and imaging is a vast field with applications including biological research [2], bathymetry [3], debris search [4], defense surveillance [5], and others. To date, sounding of sea water is carried out using hydroacoustic systems that use acoustic waves to obtain high-resolution images of objects.

"Underwater wireless sensor networks (UWSNs) have been proposed for ocean monitoring. In comparison with ocean monitoring technologies currently in use, UWSNs have the potential to revolutionize ocean monitoring applications by enabling (quasi) real-time data acquisition" [6]. The problems of improving the UWSN are also devoted to the works [7, 8, 9].

Hydroacoustic robotic monitoring in the depths of the Black Sea is an urgent issue due to its importance in the water area for ensuring both economic activity and solving ecological and environmental problems [10, 11], as well as archaeological research [12]. Theoretical foundations of the processes of acoustic coherent images formation are considered in the works of the authors [13, 14, 15].

The main idea (hypothesis) of the authors' research is that the coherent probing field forms a secondary reflection field in the form of a superposition of wave fields of local reflection regions. Such reflection regions mainly consist of central Fresnel zones in the vicinity of points where the normal to the reflecting surface is directed to the signal reception point. The level of the reflected signal corresponds to the square root of the product of the main radii of curvature at the reflection point. This is true, according to the Kirchhoff theorem, if such radii are many times greater than the wavelength of the probing field, which is almost always observed in practice. Then the secondary field is formed in the form of interference of secondary fields of local reflection regions, if the wave length of the probing packet exceeds the longitudinal distance between such regions. If such a condition can be eliminated, the local reflection areas are completely separated and unambiguously form a recognizable image of the sounding object. This appearance differs significantly from the optical one, which is mainly associated with the contour of the object, as a recognizable signature.

We believe that information about the relative position and intensity of local reflection areas is an important feature-forming factor sufficient to solve subsequent problems of recognition and classification of the investigated objects. Of course, acoustic coherent images have little in common with optical ones. The closest analogue may be images obtained using lasers in non-destructive testing. An exception is the corner structures on the surface of the probing object, for which such a hypothesis is not confirmed.

Studies in the hydroacoustic tank (because they are free from the influence of the propagation medium) make it possible to establish the true signature of sounding objects and, thereby, ensure the adequacy of the subsequent stages of modeling the interaction of sounding signals with real investigated objects.

3 Mathematical and Computer Modeling

The complexity of the problems of mathematical and computer modeling of ACI formation of is due to significant difficulties in the mathematical description of interference phenomena in secondary acoustic fields reflected from complex surfaces of objects. The essence of the problem of ACI formation and processing is to establish the relative spatial location of local reflection areas of acoustic waves, which is determined by the characteristics of the studied surfaces and the spatial resolution of acoustic sounding systems installed on underwater vehicles.

Using the Symbolic Math Toolbox MATLAB software, the authors have developed programs that allow carry out the computer simulation of secondary acoustic fields that form images of complex reflection surfaces of objects. The chosen method of reproducing secondary acoustic fields in the form of angular spectra in the language of symbolic mathematics MAPLE is due to the rigor and correctness of the results obtained in an analytical form, its limitations follow only from theoretical assumptions of the possibility of describing secondary acoustic fields by angular spectra.

The simulated angular spectra are presented in the form of the following dependencies:

- in Cartesian coordinate system:

$$\mathrm{F}(k_x/z_0, k_y/z_0) = \frac{1}{(2\pi)^2} \int_{-\infty}^{\infty}\int_{-\infty}^{\infty} \dot{u}(\zeta, \xi, 0) \cdot \exp[-j(k_x\zeta + k_y\xi)]d\zeta d\xi, \quad (1)$$

- in a cylindrical coordinate system:

$$\mathrm{F}\left(k\frac{r}{z_0}, \theta\right) = \frac{1}{(2\pi)^2} \int_{0}^{\infty}\int_{0}^{2\pi} \dot{u}(\rho, \varphi) \cdot \exp\left[-jk\frac{r}{z_0}\rho \cdot \cos(\varphi - \theta)\right]\rho d\rho d\varphi. \quad (2)$$

The developed software make it possible to directly implement dependences (1) and (2) for a given configuration of a complex reflecting surface, as well as for the distribution of the acoustic field on this surface. When performing computer simulation of angular spectra in numerical form, specific numerical values of input variables are substituted into the obtained analytical expressions, on the basis of which data arrays of the spatial distribution of secondary acoustic fields are formed. At each modeling step, the correctness of the proposed approach is verified by analyzing known analytical expressions for the angular spectra of secondary acoustic fields of reflecting surfaces of a simple configuration (limited plate, disk, etc.). Complex surfaces are modeled by a

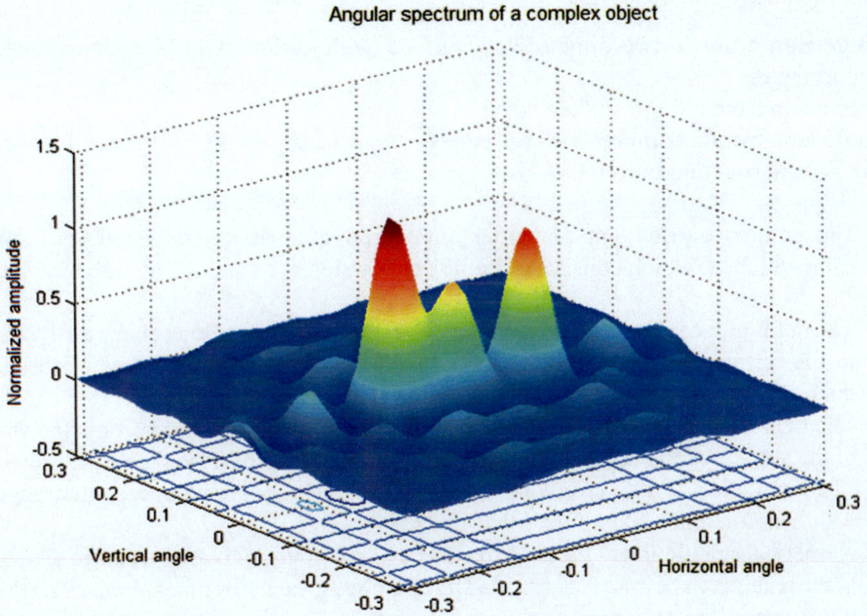

Fig. 1. ACI angular spectrum of a complex source consisting of three objects of various configurations.

combination of simple structures, taking into account the phase relationships between them: in Fig. 1 is a complex source consisting of three objects of different configurations.

Taking into account the specific features of acoustic images, the approach, proposed by the authors to their computer simulation, assumes that the formation of angular spectra in numerical form fully corresponds to analytical expressions, and distortions that are possible when using the discrete Fourier transform are excluded. Therefore, the angular spectra are formed in an analytical form in the language of symbolic mathematics, and their verification by test examples guarantees the correctness of the simulation results.

3.1 Computer Simulation of Spatially Distributed Multidimensional Signals, that Forming Acoustic Images

Numerical implementation of ACI spatial models is based on the theoretically substantiated and experimentally confirmed principle of formation of secondary acoustic fields of complex reflection surfaces using the Kirchhoff approximation. According to the Kirchhoff approximation, the secondary field of an object is a superposition of the secondary fields of a finite number of secondary (virtual) sources located on the reflecting surface and representing the neighborhood of the first Fresnel zone of the stationary phase points. The unification of the fields of local sources is carried out taking into account the phase relationships between them, and the partial components themselves are determined by the shape of the object's surface in the vicinity of the points of the stationary phase.

Fields of local sources can be divided into several types depending on the conditions of acoustic waves reflection from the object surface:

- reflection from a two-dimensional surface with different radii of curvature in coordinates;
- reflection from a finite surface (such as a limited cylinder);
- reflection from a complex surface (edges, corners, etc.) with a multi-leaf fluctuating backscattering diagram (BD).

The authors carried out computer simulation of spatial models of ACI using experimental data, which consists of the following steps:

1) The fields of local sources on a complex reflection surface are determined analytically in the form of angular spectra - two-dimensional Fourier transforms of the field distribution in the vicinity of the stationary phase points (SPP).
2) The fields of local sources are simulated numerically, taking into account the shape of the surface of the object and the distribution of the field in the vicinity of the SPP. The simulation results are two-dimensional discrete arrays of observation angles, which are the initial data for further image processing.
3) Numerical calculation of the fields on the basis of which ACIs are formed in the plane of the receiving aperture is performed by combining the fields of local sources, taking into account the phase relationships between them, as well as taking into account the observation angles.
4) Models of angular sections using experimental data are constructed taking into account the distributions of amplitudes, time shifts, as well as phase distributions of individual highlights reflections for different viewing angles of the object.
5) The noise field in the object plane is formed in the form of spatial-temporal angular noise of different intensity, which distorts not only the amplitude, but also the phase relationships between highlights reflections.

Computer simulation of the spatial distribution of ACI has been performed for a number of possible typical models of reflection surfaces. The simulation results are verified by experimental data.

Figure 2 and Fig. 3 show the angular spectra of the most important types of surfaces, for example, angular spectrum of an object with spatial noise fields – Fig. 2; the angular spectrum of the superposition of interfering objects in the noise field – Fig. 3.

The results of computer simulation shown in Fig. 5 illustrate the most important specific features of ACI formation, namely: the interference of the spatial components of the highlight secondary field of an object in the plane of the receiving aperture in the presence of several local sources, as well as the speckle structure of the secondary field in the presence of noises in the object plane, caused by their interference.

3.2 Spatial-Temporal Models of Angular Sections of Acoustic Images

General Problem of Computer Simulation Based on Experimental Data. Computer modeling using the data of a full-scale large-scale experiment in a hydroacoustic tank (hat) was carried out in order to confirm and refine the theoretically substantiated model for the formation of secondary acoustic fields of complex surfaces of objects in the

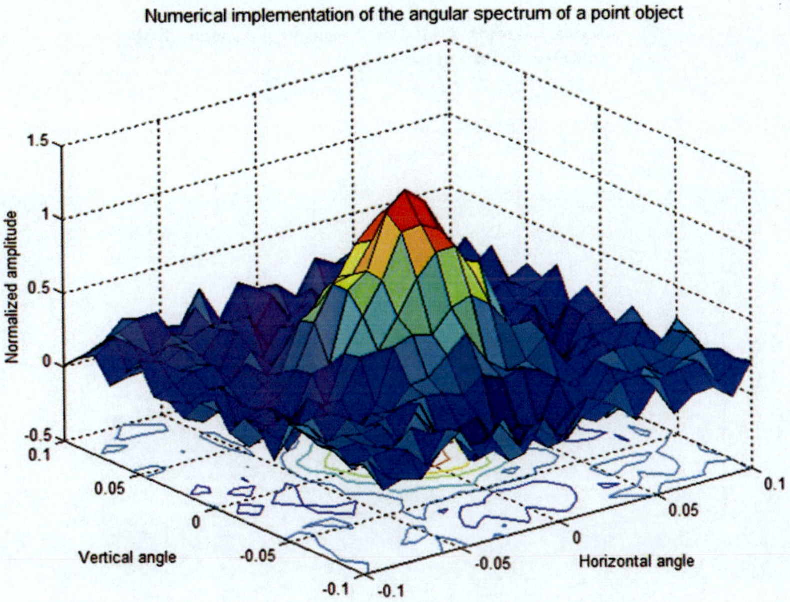

Fig. 2. Angular spectrum of a point object in the field of spatial noise.

form of a superposition of highlights reflections structurally related to the features of the investigated surface.

The data obtained as a result of full-scale simulation in the HAT are practically free from the distorting influence of the inhomogeneous propagation medium. This allowed the authors to experimentally investigate the physical nature of the formation of secondary acoustic fields of objects while observing the scale ratios of the space-wave dimensions of the investigated model, the carrier frequency of the probing signal, and its coherence intervals.

When conducting computer simulation using the data of a full-scale experiment, the following tasks were set and solved:

- obtain and generate in numerical form databases of angular sections of the ACI of the scale model studied in the HAT with a different level of spatial resolution;
- select in the databases the highlight component for different angular sections of ACI (configurations) in the form of amplitude and spatial-temporal distribution of the scattering function (SF);
- establish the connection between the spatial distribution of highlight reflections and the structural features of the model surface;
- to solve the problem of reconstructing the phase distribution of the SF for various configurations;
- to carry out the synthesis of ACI angular sections in numerical form for various configurations as a spatial-temporal convolution of probing signals with the reconstructed SF;

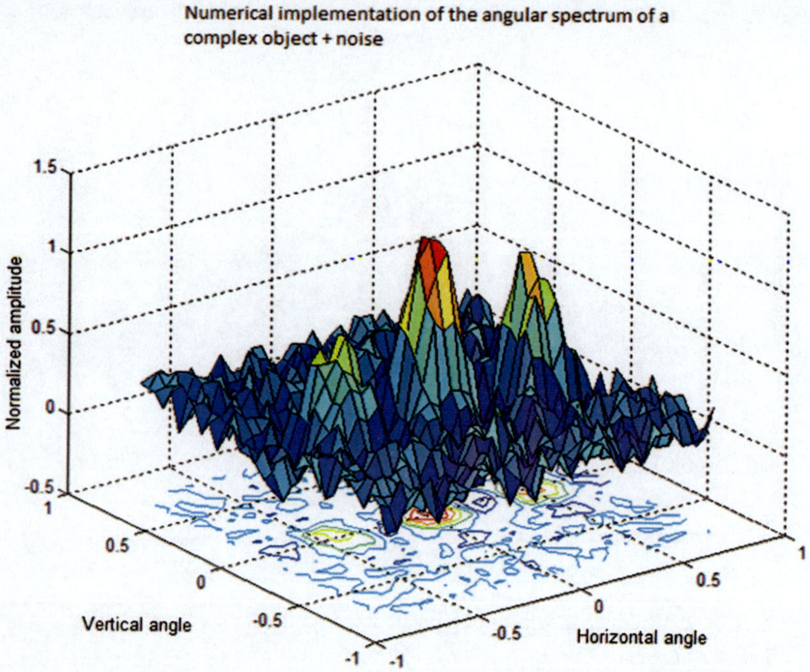

Fig. 3. Angular spectrum of superposition of interfering objects in the noise field.

- to perform a numerical comparison of real measured angular sections of ACI with their synthesized model counterparts according to two criteria: correlation and root-mean-square;
- to perform a numerical assessment of the stability of the ACI highlight structure to variations in the viewing angles of the model and the distance to it;
- to develop software tools that allow computer simulation of the formation of ACI angular sections and their visualization.

The structure of the software package for computer simulation of ACI angular sections and their comparison with experimental data is shown in Fig. 4.

Computer Simulation of Coherent Union of Partial Components of the Secondary Field. To improve the accuracy of the numerical implementation of multidimensional signals that form ACI, the authors refined the shape of the envelope of the emitted signal. For this purpose, in the process of full-scale modeling in the hydroacoustic tank, measurements were made of the shape of the front of the probing field incident on the surface of the scale model. To do this, at the location of the model in the HAT at a given distance, a measuring hydrophone was located, the signal of which was recorded by the receiving aperture during several sounding cycles.

When restoring the angular sections we performed the approximation of the real front of the probing signal of the Gaussian curve $s(t) = s_0 \cdot \exp(-t^2/2\sigma_{st}^2)$, where s_0 – the probing signal amplitude. It should be noted that insignificant differences between the

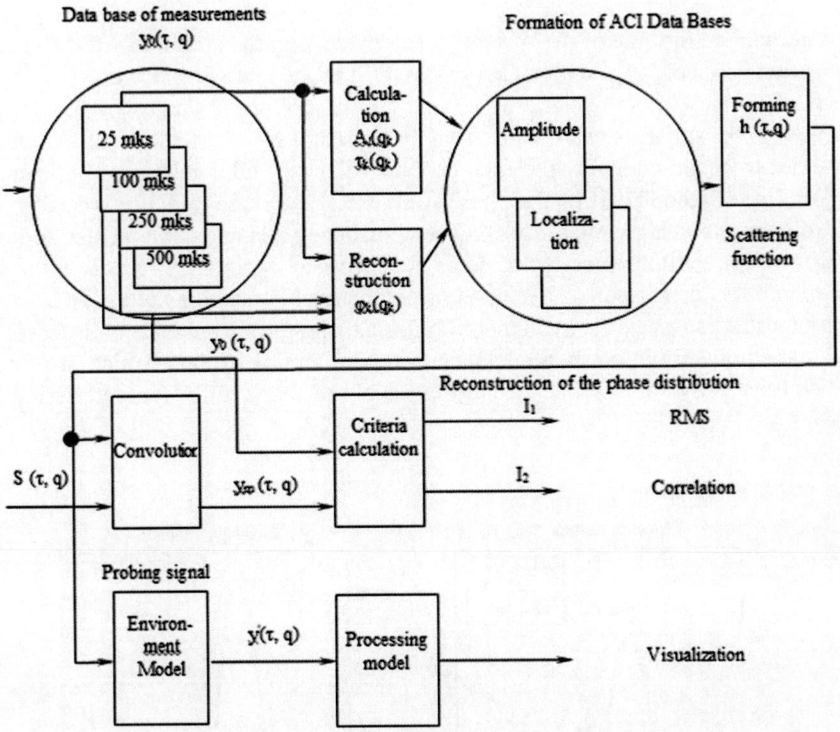

Fig. 4. The structure of software for computer modeling of angular sections of ACI.

approximated amplitude value and the value measured using the calibration equivalent sphere do not exceed the level of the noise component in the HAT.

Experimental refinement of the shape of the envelope based on the data of direct measurement of the deformation of the probing field front made it possible, firstly, to significantly improve the accuracy of computer modeling of angular sections, and secondly, to improve the accuracy of the spatial localization of highlights components by the maximum signal level recorded by the receiving path.

An essential difference between modeling coherent fields and signals is the need to take into account the phase component when combining partial components. In computer simulation using the results of measurements in the HAT, this possibility is difficult to implement due to the high frequency of the probing signal. Therefore, the phase structure of the partial components of the field is taken into account by forming the input data in the form of arrays of complex numbers in the following sequence:

- combined array of ACI angular section for a given configuration is formed as a union $N(q_i)$ of arrays, where $N(q_i)$ is the number of recorded highlights reflections in a given configuration;
- each of the sub-arrays corresponding to individual specular reflections occupies a window in the combined array, into which the time delay of this specular reflection

is entered, and the size of the window also depends on the amplitude of the specular reflection in accordance with its approximation of the Gaussian curve;

- each of the subarrays consists of complex numbers, the modulus of the complex number of the subarray cell is calculated by Gaussian approximation, and the phase is assumed to be the same for all elements of the subarray and is calculated by a separate algorithm for reconstructing the phase structure of the scattering function (SF);
- the efficiency and high reliability of the method developed and implemented in numerical form for combining coherent fields is confirmed by the illustrations of the measured and reconstructed implementations of the angular sections of the ACI, as well as numerical estimates for a number of criteria using probing signals of different spatial resolution. Attention should be paid to the high value of the correlation criterion, which most fully characterizes the closeness of the shape of the real interfering field and its model.

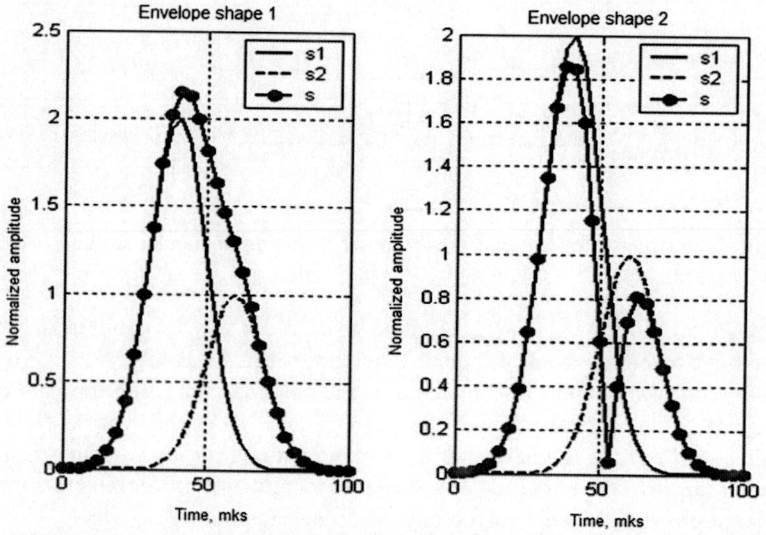

Fig. 5. Examples of coherent combination of specular reflections in in-phase and anti-phase combinations: s1 – source field 1, s2 – source field 2, s – result of coherent fusion of highlights reflections s1 and s2.

Figure 5 shows examples of coherent combination of specular reflections, taking into account the phase relationships and the shape of the envelope with in-phase and anti-phase combinations of partial components, demonstrating interference effects during the formation of ACI.

Reconstruction of the Amplitude and Space-Time Distributions of the Scattering Function. For the reconstruction of a multidimensional SF, the initial data are the elements of its angular sections measured in the HAT using a high-resolution probing signal in a given range of angles with acceptable discreteness. In these ranges, the angles between the sound wave propagation direction and the tangent to the model surface elements are such that the glare component of the secondary sound field, which forms the ACI, significantly prevails over the diffuse component. The indicated angles are the angles of the model view in the horizontal plane, since in the experiment the receiving-emitting hydroacoustic system is stationary, and the model is rotated by the HAT equipment in the horizontal plane relative to its geometric center, the position of which is fixed in space and is the reference point.

For each angular position of the model, the following data were experimentally obtained:

1) $N(q)$ is the number of highlights reflections, the level of which exceeds the noise level of the pool;
2) $A(q)$ are the amplitudes of highlights reflections reduced to an equivalent sphere with a radius of 1 m;
3) $T(q)$ are the relative delays in microseconds, reduced to the model rotation axis, and measured by the maximum of each specular reflection.

The results of applying the developed algorithm are represented in Fig. 6 and in Fig. 7, which show:

- reconstructed angular section of the SF, taking into account the phase distribution: fragment 1;
- distribution of glare reflections in comparison with the angular sections of ACI of various degrees of detail when using probing signals of various resolutions: fragment 2.

The intercorrelation coefficients of the reconstructed angular sections according to the developed algorithm with the actually measured angular sections using a high-resolution probing signal amounted to 0.9…0.95. The need to take into account phase information is essential, since a similar comparison without taking into account the phase distribution leads to a drop in the value of the cross-correlation coefficient to 0.5… 0.7, which is not statistically significant.

The multidimensional nature of the SF required the combination of the ACI angular sections in the studied ranges of angular configurations. Such a combination is made taking into account the deformation of time delays caused by the angular rotation of the model in the horizontal plane. Since the studied model is symmetrical with respect to the vertical plane, it was assumed that the SF has the same symmetrical structure. The angular spatial dependence of the SF is illustrated in Fig. 8.

The Method of Phase Distribution Reconstruction. The method of phase distribution reconstruction (MPDR) of the scattering function proposed by the authors is based on the following assumption: the scattering surface described by a deterministic function has a deterministic sf (including the phase distribution) determined only by the surface shape and observation angles. Accordingly, a stochastic uneven surface has a random sf, in which the phase distribution is to be determined depending on the type of surface irregularity distribution density.

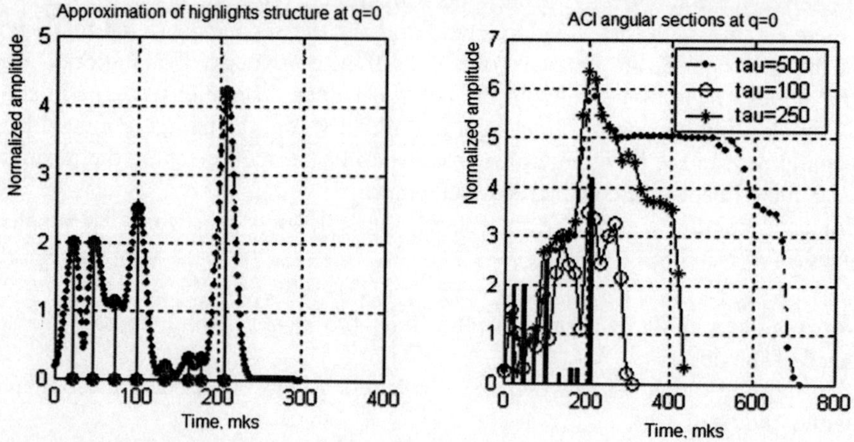

Fig. 6. The results of the numerical implementation of the algorithm for reconstructing the angular sections of the SF at. $q = 0°$

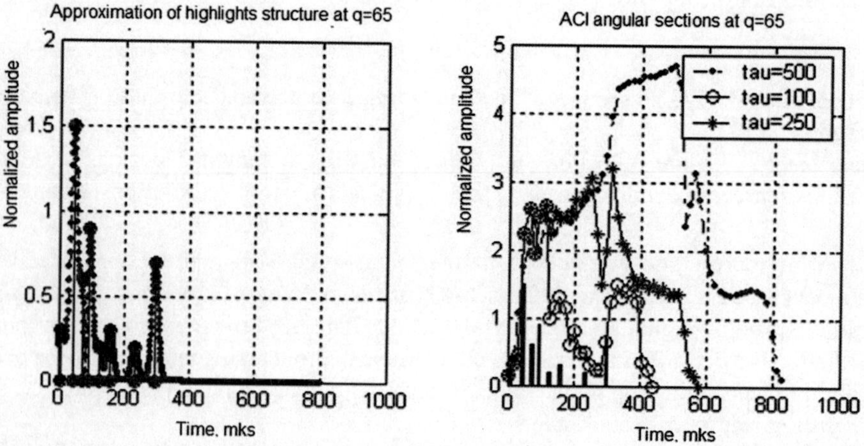

Fig. 7. The results of the numerical implementation of the algorithm for reconstructing the angular sections of the SF at $q = 65°$

Based on the accepted hypothesis, the method for reconstructing the phase distribution of the SF of the object model is based on the following data:

1) the amplitude and spatial-temporal distribution of the SF of the object model are experimentally determined;
2) as initial data, there are a number of implementations of the angular sections of the SF for signals of different spatial resolution;
3) the surface of the model is deterministic and its SF depends only on the observation angles.

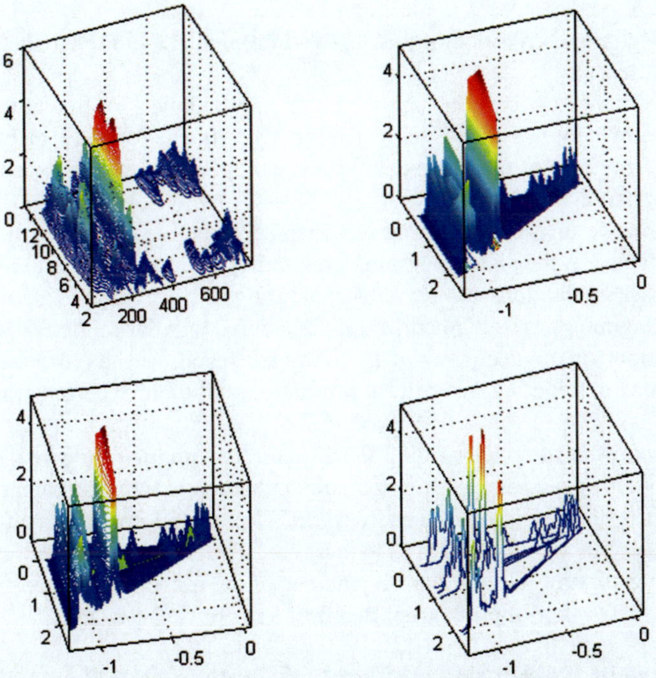

Fig. 8. Angular spatial dependence of the SF.

The basis of the phase distribution reconstruction method, confirmed by experimental measurements, is the sequential combination or elimination of the contributions of partial highlights reflections with their inherent amplitude, spatial-temporal and phase distributions.

The use of MPDR made it possible to determine the phase relationships between individual highlights reflections, since there is a one-to-one correspondence between the number of points where the phase front meets a given highlight reflection and the number of phase front convergence points relative to a given highlight reflection.

The numerical implementation of the proposed MPDR consisted in the development of an algorithm for sequential combining - elimination. The proposed algorithm is based on the mechanism of formation of an interference field. The advantage of this algorithm lies in the simplicity of implementation, and the disadvantage is in the sensitivity of the solution to the amplitude of the first flare. With its relatively small value, significant errors in the result are possible.

When solving the considered applied problem of processing experimental measurement data in the HAT, we fulfilled the reconstruction of the phase distribution of all angular sections of the ACI in the range of angles (0°… ± 30°), (180°… ± 20°), as well as for the configuration $q = 65°$. In this case, four types of signals were used with duration corresponding to the length (0.065; 0.272; 0.682; 1.364)Y_{max}, where Y_{max} is the maximum length of the object. The first of these signals was used to determine the amplitude and spatial-temporal distribution of the SF, the last (maximum length) – to determine the phase distribution of the SF, two intermediate – to refine the solutions and verify the results.

Since a reference calibration sphere of known radius was located at the location of the object, this made it possible to select the received signals for glare, diffuse, and noise components. The level of the diffuse and noise components did not exceed 10% of the calibration signal level.

Comprehensive information about the phase distribution of the angular sections of the SF is contained in the phase diagrams illustrating the amplitudes and relative phases of both highlights reflections and partial sums during the formation of the reflected field. Such phase diagrams were obtained for all considered angular sections of the SF.

The criterion for the adequacy of the obtained results is the closeness of the ACI angular sections obtained as a result of simulation and their experimentally measured realizations.

An indirect result of applying the MPDR is the restoration of the phase distribution of the backscattering diagram (BSD) of an object, which is an independent task of great importance. It is not possible to measure the BSD phase component with instrumental tools even under practically ideal HAT conditions. The phase component of the BSD is necessary for estimating the angular dimensions of the reflection surfaces of objects, as well as for solving other important applied problems of ACI research.

Reconstruction of the Angular Sections of Acoustic Coherent Images and Their Comparison with Experimental Data. Reconstruction of the ACI angular sections for each of the studied configurations of the model location was performed by spatial temporal convolution of the complex SF with the measured amplitude, spatial-temporal, reconstructed phase distributions and a given probing signal. Numerical reconstruction of the ACI angular sections was carried out by forming data arrays with a given sampling step, both for the SF and probing signals, obtaining arrays of model angular sections, forming their envelopes, and comparing them with experimentally measured angular sections according to two criteria: root-mean-square (d) and correlation (r).

The scheme of numerical reconstruction of the ACI angular sections and their comparison with the experimental data is shown in Fig. 9. The experimentally measured implementations of the angular sections were previously subjected to spline approximation with a grid step corresponding to the dimension of the model arrays. The dimensions of the compared arrays, depending on the type of used signals, were 400…1500, which ensured the reliability of the comparison results.

The choice of two comparison criteria is determined by the task: the correlation criterion characterizes the degree of correspondence between the shape of the processes, which is the most significant in the study of their interference, and the root-mean-square criterion characterizes the correspondence of energy relations. The numerical values of the matching criteria are statistically significant with a confidence probability of at least 0.95…0.99, which is ensured by sufficiently large dimensions of the compared arrays.

Since the correlation criterion characterizes the conformity of the shape of the reconstructed ACI angular section with the shape measured in the experiment, the numerical value of this criterion is a numerical measure of the correspondence of the reconstructed phase distribution to its actual character.

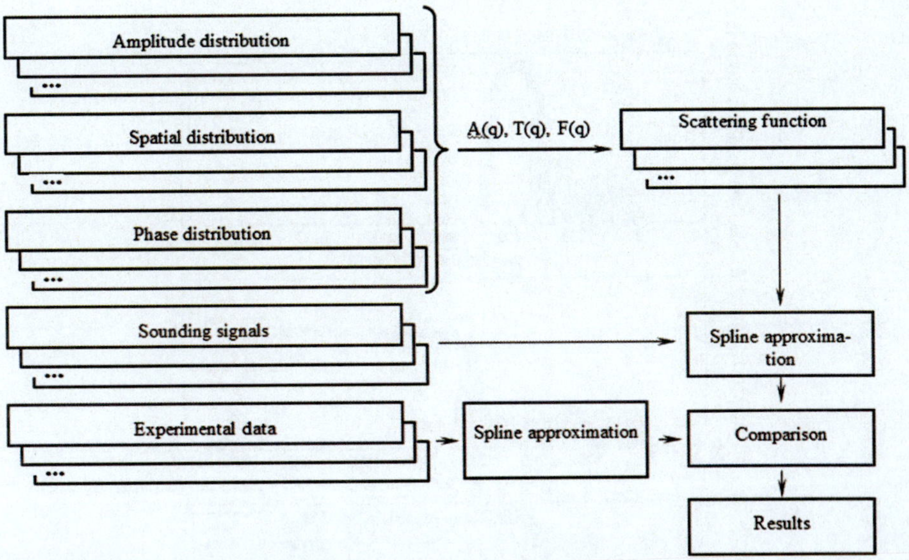

Fig. 9. Reconstruction of ACI angular sections.

Therefore, if as a result of comparison for any configuration, two of the three values of this criterion turned out to be lower than 0.9 (for three analyzed probing signals), then the phase distribution of this configuration was subjected to additional consideration and refinement until the established threshold value of the criterion was reached. The root-mean-square criterion was used to control the correctness of calibration, the correspondence of scales, the correctness of normalization, and also to control the results of determining the effective scattering surface (ESS).

The results of reconstruction and comparison of ACI angular sections for various configurations of the model location and types of probing signals are illustrated in Fig. 10 – Fig. 12.

As was established as a result of processing the data of computer and natural experiments, high values of the correlation criterion primarily indicate the reliability of the results obtained using the method of reconstructing the phase distribution of the SF, as well as the possibility of modeling the real picture of the formation of the secondary acoustic field and the corresponding ACI based on the proposed approach.

The most important achieved result is the confirmation of the adequacy of the reconstructed SF of surface, which makes it possible (by mathematical modeling) to study the methods for the ACI formation, that are not reproducible in HAT and natural conditions. Indeed, by setting the probing field in an arbitrary form, the corresponding ACI can be obtained on the basis of the spatial-temporal convolution of the reconstructed SF, which makes it possible to solve a wide range of problems in testing methods and algorithms for the formation of ACI at solving the various applied problems.

Assessment of the Stability of Acoustic Coherent Images When Changing the Distance and Viewing Angles. Viewing distance r is the distance between the plane of the receiving aperture and the object plane.

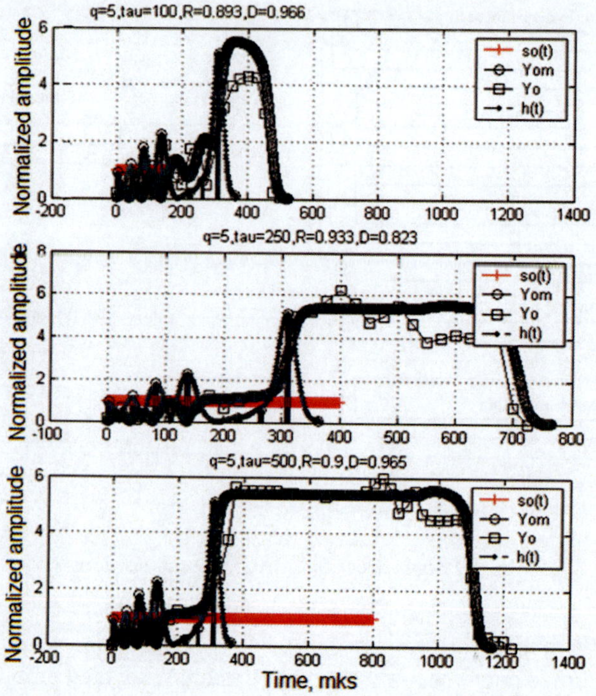

Fig. 10. The results of synthesis and comparison for various types of probing signals at $q = 0°$

ACI in the idealized version is completely determined by the SF of the surface, which, in the Kirchhoff approximation, coincides with the angular spectrum – the Fourier transform of the field distribution on the surface. Under real conditions, the distribution of the field in the plane of the aperture uniquely depends on the observation distance and the conditions that determine the boundaries of the Fraunhofer zone. When waves are diffracted by some object whose characteristic linear size is equal to z, the Fraunhofer zone begins at a distance R that satisfies the well-known relation from the classical theory of diffraction $R \gg z^2 \cdot \cos^2 q/\lambda$, where λ – is the wavelength, q is the bearing. In terms of angular relations, this condition is represented as

$$\Delta\theta_c/\Delta\theta_p \ll \pi/4, \tag{4}$$

where $\Delta\theta_c$ – is the angle at which the scattering surface area is visible from the observation point, $\Delta\theta_p$– the BSD angular width.

In other words, the plane of the receiving aperture will be in the Fraunhofer zone if the angle at which the scattering area or the entire reflection surface is visible is much less than the angular width of the BSD or the angular spectrum, the latter value being estimated by the Rayleigh criterion $\Delta\theta_p \approx \lambda/P_r(\vec{z})$, where $P_r(\vec{z}) = z \cdot \cos q$. Thus, the Fraunhofer zone is defined by the boundary $R >> \pi\left[P_r(\vec{z})\right]^2 \Big/ 4\lambda$.

Starting from the boundary of the Fraunhofer zone, the change in phase relationships depending on the distance can be neglected, the object can be considered as a point

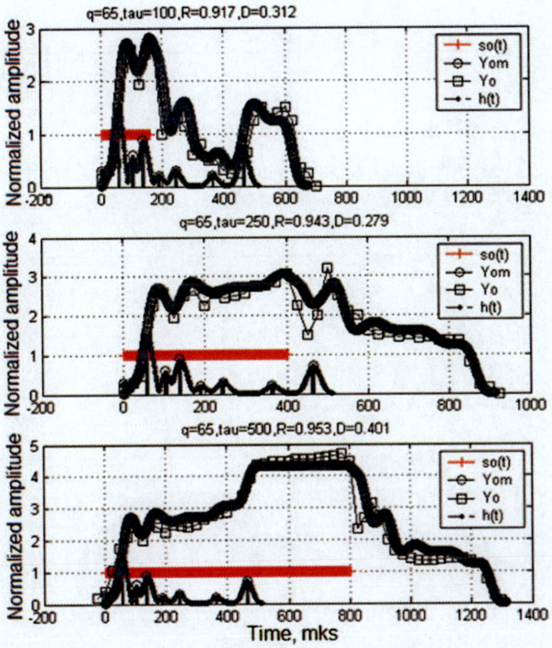

Fig. 11. The results of synthesis and comparison for various types of probing signals at $q = 65°$

object. In this case, its angular spectrum completely determines the ACI in the plane of the receiving aperture. If this condition is not met, then the dependence of the phase front on the observation distance cannot be neglected, and the ACI obtained in the plane of the receiving aperture may be deformed. The physical reasons that determine the degree of such deformation are the need to take into account the spherical phase front in the form of factors with a quadratic phase, as well as the possible violation of the phase relationships between individual highlight on a complex surface of an object.

Checking the given ratio for the measurement conditions in the HAT shows that the plane of the receiving aperture can be considered to be in the Fraunhofer zone only for narrow ranges of observation angles in the horizontal plane q = (−6... + 6) and (174...186). For all other configurations for which measurements were made, the conditions of the far zone (Fraunhofer zone) are not met. In connection with this circumstance, caused by the limitations of the measurement conditions in the HAT, additional experimental studies of the influence of the distance on the nature of the highlights distribution were carried out. Such measurements and their results are described in [13]. As established by the results of measurements in the HAT and their processing, a change in the distance by 30% under the experimental conditions is not critical and does not affect the nature of the ACI in the plane of the receiving aperture. When using signals of high spatial resolution, a high correlation between the angular sections of ACI is observed. The highest level of such a correlation is observed for the space-time position of highlights associated with the characteristic features of the surface. Somewhat lower is the level of correlation between the amplitude ratios of highlights reflections. In general, it can be

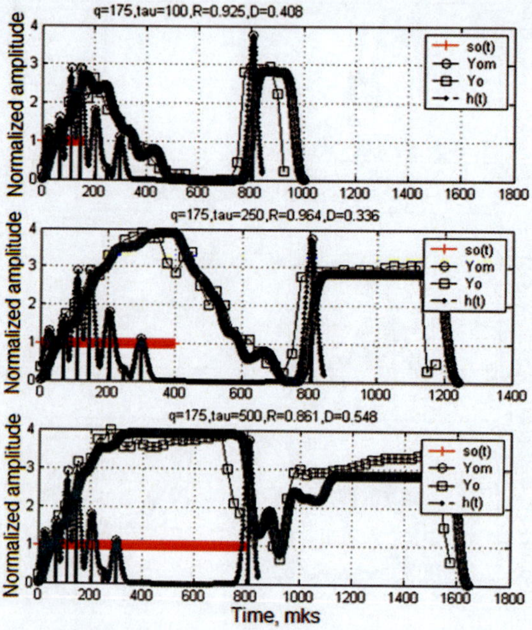

Fig. 12. The results of synthesis and comparison for various types of probing signals at $q = 175°$

concluded that the main sign-forming factors of ACI are resistant to changes in distance under the conditions of experiments in the HAT.

The investigated scale model is symmetrical with respect to its central vertical section, but it is not an axisymmetric body, therefore the nature of the SF of its surface also depends on the viewing angles in the vertical plane, namely, on the roll angle. Strictly speaking, since the SF is multidimensional, it is also necessary to include the dependence on this argument in its description. This dependence has been experimentally established for all configurations and stated roll angles. For the specified roll angles, a numerical assessment was made the evaluation of the correlation between the SF angular sections and the nature of the change in this relationship [13]. The data obtained are of a general nature. Adjacent angular sections of the SF with a change in the roll angle are characterized by a high level of correlation of the envelopes, in particular: $r_{0\text{–}3} = 0{,}810$; $r_{3\text{–}6} = 0{,}897$; $r_{6\text{–}9} = 0{,}806$. However, as the angle increases, the values of the correlation coefficients noticeably decrease, in particular, for the illustrated configuration q = 65°, we obtain $r_{0\text{–}6} = 0{,}597$. In this case, the number of analyzed readings is (40…160), which guarantees the statistical significance of the obtained estimates.

The general nature of the obtained results is that small variations in the viewing angle in the vertical plane do not significantly influence on the amplitude and space-time distributions of the SF. For large deviations, such stability is not observed, and changes in the SF with a change in the angle of roll occur the sooner, the greater the angle of the object's view. Thus, for a complete characterization of the ACI, it is necessary to take into account in the SF the dependence on both observation angles, which corresponds to the two-dimensional nature of the scattering surface [13].

The dependence of the SF from the angle in the horizontal plane is decisive for the investigated scale model. Nevertheless, it is of considerable interest to study the variability of the SF elements with a change of the angle of view, both to clarify the required discretization step for this angle, and to evaluate the angular dependence of the BSD of individual local reflection areas. Such a numerical assessment was made for all experiments performed at the HAT [13].

In general, the level of correlation in the studied range of angles is quite high and amounts to $r_{0-5} = 0.924$; $r_{0-10} = 0.806$; $r_{0-15} = 0.803$; $r_{5-10} = 0.815$, that is, adjacent angular sections of the SF of the studied model have a high level of correlation when using the maximum spatial resolution signal. Increasing the argument of the cross-correlation function, i.e. difference angle, leads to a decrease in its level, and the angular correlation interval decreases with increasing observation angle. For observation angles in the range (150…180) the interval of angular correlation is significantly lower, which is explained by the presence of developed corner structures in this part of the model and the presence of possible re-reflections on edges and corners that distort the shape of the envelope. The main result of the performed numerical analysis is the validity of the chosen discretization step in the angle in the horizontal plane. Discretization of the observation angle limits the upper frequency of the angular spatial harmonic, which can be determined from the measurement data, however, such frequencies, which significantly affect the character of the SF of the model under study and exceed the level of 0.1 m of HAT noise, have not been established in the experiment [13].

To estimate fluctuations in the position of the stationary phase points (SPP) (phase centers of reflecting surfaces), the angular sections of the SF are reduced to $q = 0°$ when the model is rotated relative to its geometric center by taking into account the phase factor in the angle of rotation, and then a repeated correlation analysis of the reduced angular sections is carried out [13].

The result of the correlation analysis of the given implementations of the SF angular sections is essential and consists in the fact that the level of correlation in the range (0…20) along the envelope of the SF sections rises to 0.9. In this case, the spatial-temporal distribution of the SF almost unambiguously indicates the position of local reflection regions associated with the features of the scattering surface of the model. Thus, the relative position of local reflection regions on the model surface is its stable characteristic, practically unchanged under the conditions of the experiment. Variation in the amplitude distribution indicates a change in the size of the first Fresnel zone associated with a variation in the observation angle. This leads to a shift of the phase centers of the local reflection regions over the surface of two-dimensional curvature, and the phase distribution is significantly deformed. However, such a shift for a deterministic surface can be uniquely determined, which makes it possible to establish a deterministic relation for the phase distribution.

More sophisticated examples of using *R*-functions for describing of mechanical stresses, temperature, electric and magnetic fields distribution, are given in monographs [14]. In any case, such presentation of scientific graphics is very effective to using at clouds calculations in emerging networking. The presentation of necessary digital data becomes more compact, as well as the time for solving sophisticated scientific problems is also significantly reduced.

4 Conclusion

1. A complex of interacting software tools for the numerical implementation of the created models for the formation of ACI with using the experimental data of modeling in the HAT has been developed.
2. Exact analytical solutions have been obtained for a number of problems of forming ACI angular spectra for various types of reflective surfaces based on the methods of symbolic mathematics of the MAPLE package.
3. Numerical implementation of spatial models of ACI objects with complex reflection surfaces was carried out, taking into account the phase relationships between the acoustic fields of local sources for the following types of surfaces: two-dimensional, three-dimensional, as well as reflections from a complex surface with a multilobe backscattering diagram.
4. A computer implementation of spatial-temporal models of ACI angular sections was carried out using experimental data obtained as a result of semi-natural scale modeling in the HAT. This made it possible to confirm and refine the theoretically substantiated models for the formation of ACI for objects with complex surfaces. Including the following tasks were solved:

4.1. The databases of angular sections of the ACI of the object studied in the HAT with different levels of spatial resolution were obtained and formed in numerical form. In the databases, the highlight component for various ACI angular sections (configurations) is identified in the form of the amplitude and spatial-temporal distribution of the scattering function.

4.2. The connection of highlight reflections with the structural features of the reflection surface of the investigated object has been established.

4.3. The problem of reconstructing the phase distribution of the scattering function for various configurations is numerically solved on the basis of the proposed method for reconstructing the phase distribution, which makes it possible to restore the phase distribution of the backscattering diagram of an object and solve the problems of reconstruction of ACI.

4.4. Models of ACI angular sections for various combinations of the relative position of the object and the receiving aperture are numerically implemented in the form of spatial-temporal convolutions of probing signals with a complex-valued scattering function with measured amplitude, spatial-temporal, and reconstructed phase distributions. The obtained results are numerically compared with the experimentally measured ACI angular cross sections according to two criteria (root mean square and correlation), the obtained values of which indicate the adequacy of the reconstructed object scattering function.

4.5. On the basis of the performed numerical assessment, a significant degree of stability of the main feature-forming factors of ACI to variations in the distance and observation angles was established.

References

1. Akyildiz, I.F., Pompili, D., Melodia, T.: Underwater acoustic sensor networks: research challenges. Ad Hoc Netw. **3**(3), 257–279 (2005)

2. Collier, J.S., Humber, S.R.: Time-lapse side-scan sonar imaging of bleached coral reefs: A case study from the seychelles. Remote Sens. Environ. **108**(4), 339–356 (2007)
3. Stewart, W.I., Chu, D., Malik, S., Lerner, S., Singh, H.: Quantitative seafloor characterization using a bathymetric sidescan sonar. IEEE J. Ocean. Eng. **19**(4), 599–610 (1994)
4. LeHardy, P.K., Larsen, J.: Deepwater synthetic aperture sonar and the search for MH370. In: Proc. OCEANS MTS/IEEE Washington, p. 1–4 (2015)
5. Dobeck, G.J.: Fusing sonar images for mine detection and classification. Proc. SPIE **3710**, 602–614 (1999)
6. Ammari, H.M.: Introduction. In: Ammari, H.M. (ed.) Mission-Oriented Sensor Networks and Systems: Art and Science. SSDC, vol. 163, pp. 1–8. Springer, Cham (2019). https://doi.org/10.1007/978-3-319-91146-5_1
7. Ocean Networks Canada. Ocean observing systems. http://www.oceannetworks.ca/innovation-centre/smart-ocean-systems/ocean-observing-systems (2017)
8. Khan, A., Jenkins, L.: Undersea wireless sensor network for ocean pollution prevention. In: Proceedings of the 3rd International Conference on Communication Systems Software and Middleware and Workshops (COMSWARE), p. 2–8 (2008)
9. Cayirci, E., et al.: Wireless sensor networks for underwater survelliance systems. Ad Hoc Netw. **4**(4), 431–446 (2006)
10. Palazov, A., et al.: Black Sea Observing System. Front. Mar. Sci. **6**, 315 (2019). https://doi.org/10.3389/fmars.2019.00315
11. Stanev, E., He, Y., Grayek, S., Boetius, A.: Oxygen dynamics in the Black Sea as seen by Argo profiling floats. Geophys. Res. Lett. **40**, 3085–3090 (2013). https://doi.org/10.1002/grl.50606
12. N. I. Minaieva Sonar detection as an effective Method of Discovery and study of underwater archaeological Objects. Archeology and long history of Ukraine. (in Ukraine). **4** (17), 193–197. ISSN 2227–495 (2015)
13. Hvozdeva, I.M.: Teoria akusticheskih koherentnih izobrazenyj [Text]. Odessa. Nauka i technika. (in Ukraine). ISBN 978–966–1552–70–7 (2015)
14. System Technologies for Modeling of Complex Processes, Monograph edited by prof A.I. Mikhalyov [Text]. Dnipro. NMAU. (in Ukraine) (2016). ISBN 978–966–2596–19–9
15. Verlan, A.F., Myrhorod, V.F., Hvozdeva, I.M.: Modelirovanie processov formirovanija i obrabotki akusticheskih koherentnih izobrazenyj [Text]. Electronic modeling. Kyiv. IPME. (in Ukraine). **26**(5), 111–118 (2004). ISSN: 0204–3572

5G and Beyond: Perspectives of 6G Development

Andriy Luntovskyy(✉)

IT Department, BA Dresden University of Cooperation Education, Saxon Study Academy, Hans-Grundig-Street 25, 01307 Dresden, Germany
Andriy.Luntovskyy@ba-sachsen.de

Abstract. The chapter is devoted to analyzing 6G driving factors, features, and potentials in the mid-term. Further development will concern better coding, new antenna, and cell structures. A special meaning has the use of network softwarization based on further Network Function Virtualization (NFV), Artificial Intelligence (AI), and advanced cyber-security approaches like Blockchain (BC).

Keywords: 5G · 6G · RAT · Slicing · Heterogenous cells · Spectral efficiency · Post-Shannon

1 State-of-the-Art

The 6G architecture generally looks like the ones of the former generations. A typical architecture of 5G and Beyond is depicted in Fig. 1. From the regularly sized cells through hierarchical cells to the heterogenous cell structures are used [1–4].

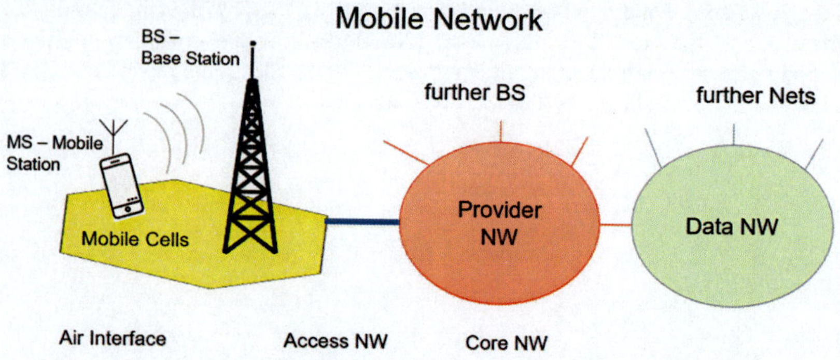

Fig. 1. Typical architecture of 5G and beyond.

The providers create, test, and steadily integrate advanced methods for radio and fixed channel planning (Fig. 2). These new methods [1, 5] are combined with the classical ones between Tx and Rx as follows: ADC/DAC, source/channel coding/decoding, signal spectrum spreading/de-spreading, impulse forming, modulation/demodulation, antenna gaining, and link budget.

M. Klymash et al. (Eds.): TCSET 2022, LNEE 965, pp. 396–408, 2023.
https://doi.org/10.1007/978-3-031-24963-1_23

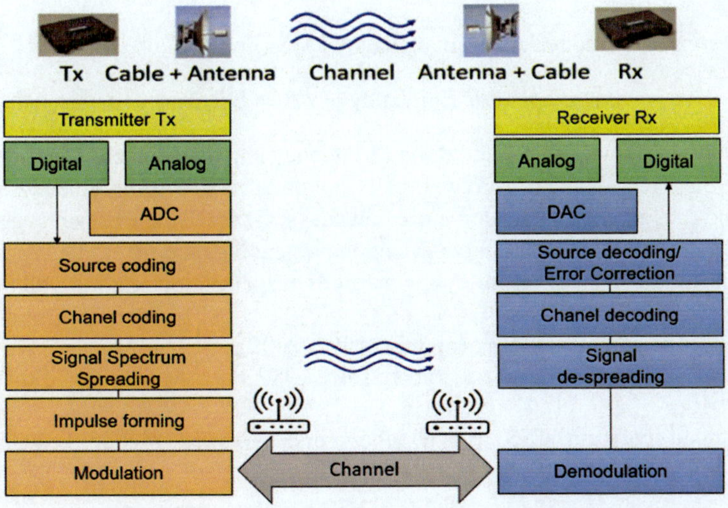

Fig. 2. Channel planning: message source and message drain [5].

5G and Beyond, as the predecessor of 6G, uses so-called "5G NEW RADIO" with the following specifics and advanced performance approaches as follows:

- Higher data rates - up to 20 Gbit/s per user (bandwidth allocation, massive MIMO),
- Tiny round-trip times = latencies (through network slicing, $\Delta = 1$ ms),
- New frequency spectrum: bandwidth B (mmWave),
- Energy and Power Efficiency (PE) for Narrow Band IoT,
- Higher Spectral efficiency (SE).

The following relations (1)–(3) based on the classical relations by Nyquist-Shannon (1948) can be used to describe the corresponding channel capacities. Channel capacity serves as the limit for DR for practical devices and networks.

Taken into account that:

DR = Data Rate < C = Channel capacity, B = Bandwidth, S = Number of signal levels by a selected coding scheme [5–8]:

$$DR = 2B\ ld\ S, \tag{1}$$

$$DR = B\ ld\ (1 + SNR),$$

$$DR \approx B\ /3\ SNR_{dB},$$

where SNR = $P_{signal}/\ P_{noise}$ = Signal-to-Noise-Ratio, sometimes measured in dB,

SNR_{dB} = 10lgSNR.

$$Path\ Losses = PL = PTx/Px = a\,f^2\ d^2, \tag{2}$$

where a – const, f – frequency, d – distance by LOS (line of sight).

$$Spectral\ Efficiency = SE = DR/B. \tag{3}$$

The value PL depends on f and d, i.e., frequency and distance correspondently. It is one of the classical results of Maxwell electrodynamics. Spectral efficiency SE for 5G and Beyond can reach 4 bps/Hz and more nowadays. Generally, the previous generations of mobile radio 3G till 5G are characterized via a steadily spectral efficiency SE growth based on the deployment of the following methods of spectrum spreading (refer to Fig. 3):

- Sequential Frequency Hopping (approx. 1991–2000),
- Orthogonal-assigned subcarriers: known since 1980, but a growing standing role since 2001–2010,
- Chirp spread spectrum (CSS) uses wideband linear frequency modulated chirp pulses to encode information: 1940 – meanwhile.

Some of the examined methods are known since 1940; some of the methods have been awaiting to the ASICs progress in XXI century [4, 5].

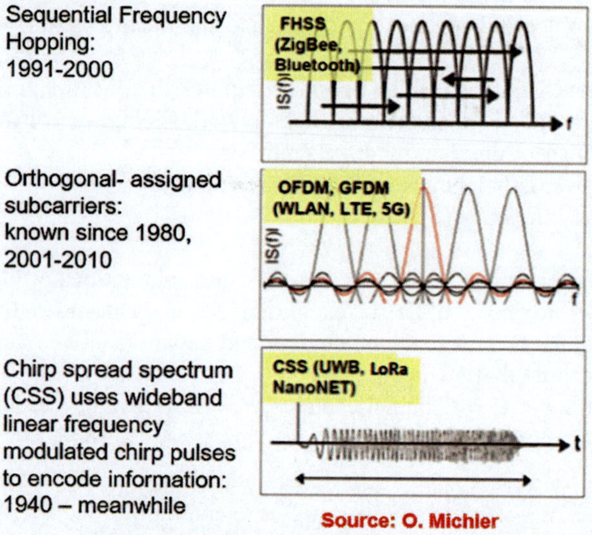

Fig. 3. Spectral efficiency growth [5].

2 Increasing Spectral Efficiency

The growth of spectral efficiency SE by 6G seems to lead to approaching and overcoming known Shannon's limits [5–8]. This effect was described in the literature as so-called "Post-Shannon". Let us analyze the new features and factors of 5G and Beyond, which partially enable so-called "Post-Shannon" effect:

1. New radio access technologies (RATs), refer to next Section.
2. Dense heterogenous cell structures.
3. Advanced spectrum spreading.
4. New modulation schemes, e.g., OFDM.
5. Massive 3D-MIMO Antennas and Beamforming.

As the following essential factors can be considered:

- Softwarization (NFV),
- Functionality slicing,
- Cybersecurity and protocol optimization (AI, Blockchain).

The cell sizes are variable (femto, pico, micro, macro, giga = non-terrestrial cells), and their structures are heterogeneous. They belong as a rule to different small and large providers and are combined with the former cell hierarchy (beginning from 3G). The beamforming instead of isotropic power radiation is used. This leads to better QoS/ QoE in the mobile devices (refer to Fig. 4). Massive 3D-MIMO Antennas enable higher DR and better SE for the dedicated licensed bandwidth B (refer to (1)–(3)). Let us consider so-called RATs.

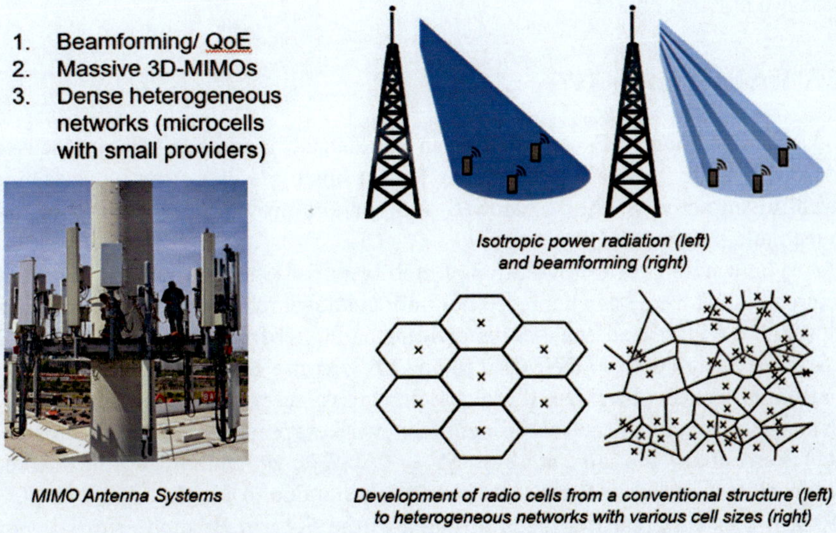

Fig. 4. Transformation of cell hierarchy to heterogeneous cells [2].

The advanced QAM constellation is deployed. A large number of symbols 2**x is used for a higher SE with a sufficiently large SNR. Some examples of even x-Bit-QAM constellations with binary symbols are presented below (Table 1):

Table 1. x-Bit-QAM constellations

Bit number	Type	Symbol number	SNR requirements	LOS	Deployment
6 bits	64-QAM	64 symbols	>24 dB	+	DVB-C, DVB-T, Mobile
8 bits	256-QAM	256 symbols	More	+	DVB-C, Mobile 4G, 5G micro and macro cells
10 bits	1024-QAM	1024 symbols	More	+	DVB, Mobile 5G and Beyond, small cells
12 bits	4096-QAM	4096 symbols	36 dB	+	DVB-C2, ITU-T G.hn

The 4096-QAM is nowadays the largest QAM constellation, which currently being considered, which is just detectable under the best conditions: LOS and SNR of 36 dB to extensive channel coding.

3 5G RAT: What is It?

The inspiration around 5G and Beyond in last decade is understandably focused on exciting new radio access technologies (RATs) since 5G is currently very much a research-driven activity. The researchers and vendors provide a revolution of current radio technologies for 6G [9].

Based on watching of the evolution of mobile network generations, we can consider that increased DR have been achieved not only via novel radio modulation schemes but also through the increased channel bandwidths on higher frequency bands. The 3G-5G with **Spread Spectrum (FHSS, Chirp)** and **OFDM** are based on the approved ideas (that are significantly older than it looks at first sight), refer previous sections.

In fact, the capacity of each of the significant wireless predecessors **(LTE, HSPA, and 802.11)** is essentially the same at **about SE = 4 bps/Hz**, which indicates the differences in achievable data rates for these technologies are not due to the underlying RAT.

What it means is that the focus of much of the 5G and Beyond efforts is around achieving higher DR under the use of intelligent mathematical apparatus to exploit spatial diversity using variants of **MIMO**. As **new modulation schemes** may improve the spectral efficiency SE under constrained conditions, they are unlikely to be capable of delivering ubiquitous coverage at 5G and Beyond data rates (x-QAM).

4 5G Structures and Base Stations

5G spectrum provides the following frequency bands which are being separated into two different frequency ranges [5, 9].

- Frequency Range 1 (FR1) includes sub-6GHz frequency bands, some of which are bands traditionally used by previous standards but have been extended to cover potential new spectrum offerings from 410 MHz to 7125 MHz.
- Frequency Range 2 (FR2) includes frequency bands from 24.25 GHz to 52.6 GHz. Bands in this mmWave range have a shorter-range d but higher available DR than bands in the FR1.

The future 6G will include Microwave, IR, and UWB with 300 GHz till 300 THz, and even VLC with 400 and 800 THz (visible light).

Table 2 shows the base station names for mobile radio technology generations:

Table 2. Base station names [9].

Generation	Radio technology	Base station name
2G	GSM	BTS (Base Transceiver Station)
3G	UMTS	NodeB
4G	LTE	eNB, Evolved NodeB

5G Base stations for regular and small cells are shown in Fig. 5. BS names and types with so-called slicing concept are as follows:

- 5G gNodeB Base Station,
- 5G gNodeB Small Cell Base Station.

The 5G Base Station uses New Radio (NR) technology and is referred to as a gNodeB (gNb). Base Station gNodeB uses Software Defined Radio (SDR) with various MIMO options, including 2×2, 4×4, and 8×8, as well as Massive MIMO options for higher capacities. Options for 5G deployment in Sub-6GHz bands as well as Millimeter Wave (mmWave) bands are available [1, 9].

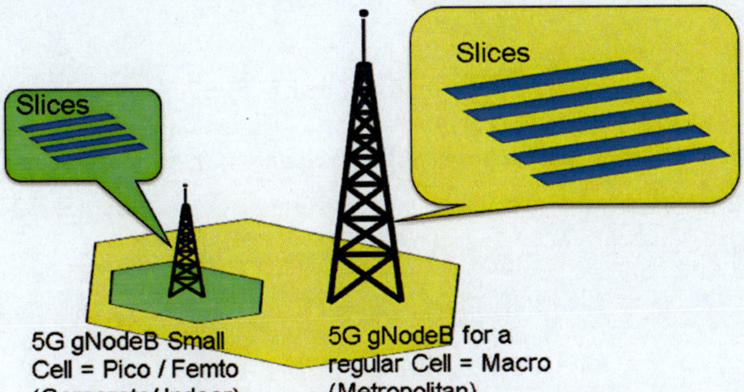

Fig. 5. 5G Base Stations for regular and small cells [9].

The 5G gNodeB Small Cell Base Station supports CableFree 5G Small Cell as outdoor version with 2 × 2 MIMO.

5 New Features of 5G and Beyond: Slicing and Cloud Analytics

5G systems are designed in such a way that virtual networks can be sliced on an as-a-service basis [4, 10, 11]. These services can be scaled up and down quickly and easily. Each network slice can be customized to provide the necessary components for the architecture it requires. For example, 10% of 5G resources can be reserved exclusively for IoT devices. Additionally, each slice is isolated and comprises the device, access, transport, and core network, thereby increasing reliability and network security (by infradata.com). So-called 5G Slicing is used (Fig. 6).

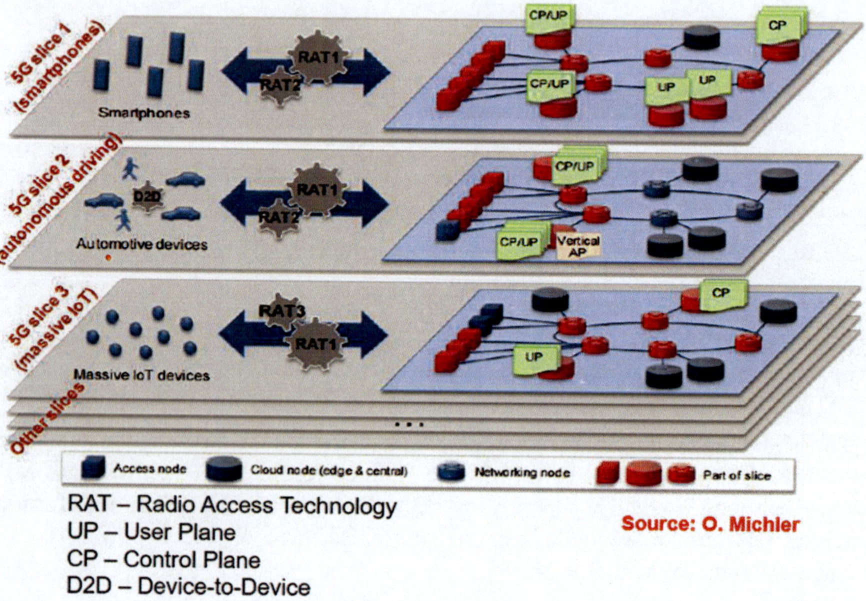

Fig. 6. 5G slicing.

The cloud analytics for 5G and Beyond is shown in Fig. 7. A special meaning obtains network softwarization [2, 4, 10, 11] based on NFV, artificial intelligence (AI), and advanced cyber-security approaches like Blockchain (BC), which will be consecutive implemented in 6G.

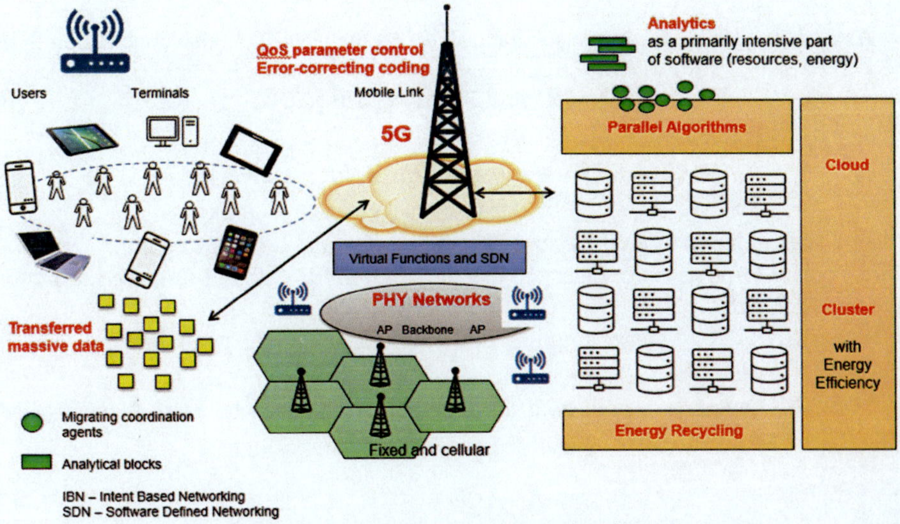

Fig. 7. Cloud Analytics for 5G and Beyond.

6 MIMO and Space Multiplex

Generally, nxn - MIMO antennas work with line of sight (LOS) conditions with polarization and reflection [4, 10, 11]:

- Two antennas – different antenna polarization possible,
- More antennas – further reflection mechanisms required.

The number K of data streams that can be transmitted in parallel, depending on the power of the transmitter Tx/ receiver Rx units for MIMO must be compared with SISO. Herewith, maximal DR increases by the factor K:

$$DR2 = DR1 * K, \quad (4)$$

where DR1 – data rate under the use of SISO, DR2 – data rate under the use of MIMO. At least K as coefficient of DR growth depends in this case on:

$$K \min = NTx * NRx \quad (5)$$

or for 2D structures nxn:

$$K \min = n * *2$$

or even:

$$K = n * *3$$

for 3D structures, n x n x n, e.g., with 5G and 6G networks, where {NTx, NRx, n, m, l} are the dimensions of the used antenna arrays.

As a consequence, for Nyquist-Shannon theorems (1948), K also depends on SNR:

$$DR1 = B \min(2ld\ S, ld(1 + SNR)), \tag{6}$$

$$DR2 = (m * n * 1)B \min(2ld\ S, ld(1 + SNR')),$$

where S – the defined by OSI Layer 1–2 number of signal levels by coding,
SNR – signal-to-noise ratio under the use of SISO, SNR' – signal-to-noise ratio under the use of MIMO,
SNR' >> SNR due to power amplification of antennas (antenna gain factors),

3D MIMO antenna system with the factor m x n x l can be used. It means SNR growth leads to growing of K and DR:

$$K \succ\succ m * n * 1. \tag{7}$$

The DR is herewith increasing and strives for channel capacity C.

7 Towards 6G

Co-existence and synergies of the current transmission technologies that are oriented towards 6G are shown in Fig. 8. Such transmission technologies are complementarity regarding the following features and provide the above-mentioned synergies [3–5]:

- Communication direction,
- Type of information and communication,
- Data rates,
- Frequency ranges,
- Channel planning,
- Mobility properties,
- Supply cell sizes,
- Data integrity and privacy.

So-called enabling technologies, which are used for the further development of 6G, are depicted in Fig. 9. Into account are taken: LiFi, UWB together with WiFi, BT, ZigBee, LoRa WAN and 5G. Future 6G networks include, moreover, better than 5G + Sensing via so-called:

- Active Localization:

The distance x and their derivations x' and x" (velocity and acceleration) can be determined via level, signal propagation time, and phase (GSM, GPS, GALILEO).

- Sensing = Passive Localization:

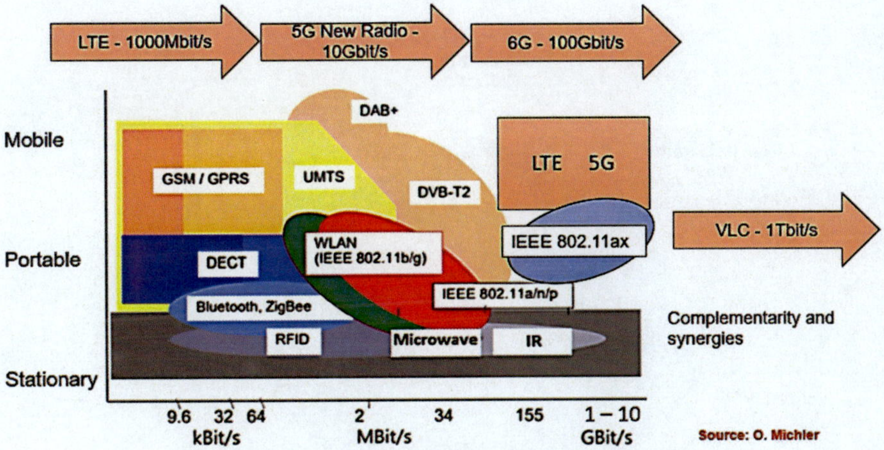

Fig. 8. Transmission Technologies towards 6G [5].

Fig. 9. Towards 6G: Enabling Technologies (based on [3–5]).

The distance x and their derivations x' and x" can be determined via evaluation of the propagation delays due to reflections on the objects (channel impulse response is required: radar techniques, WLANp, VLC, UWB, LiFi).

Let us provide the comparison 5G vs. 6G. The comparison 5G vs. 6G is depicted in Fig. 10. The following factors are herewith compared: Peak Data Rate, User Experienced Data Rate (DR), Energy Efficiency, Spectral Efficiency (SE), Air Latency (Delay), Connection Density, Reliability. The factors are placed on the diagram shown in logarithmic scaling and demonstrate the 6G advantages.

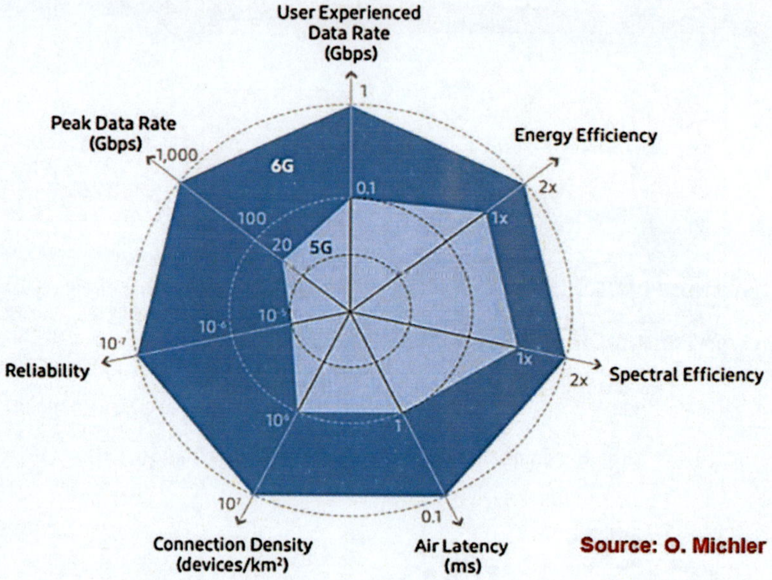

Fig. 10. Comparison 5G vs. 6G (based on [5]).

8 "Post-Shannon"

This topic is one of the most discussed in the last decade—the state-of-the-art attempts to break the classical Shannon limits (1948). So, let us consider that in detail [6–8].

Channel Capacity C and Power Efficiency/ Spectral Efficiency limits are depicted in Fig. 11. The diagram shows the difference between DR and C in practical devices and networks and un unattainable. The researchers speak about "Post-Shannon" and the possibility of a C-limit breakthrough (https://www.gaussianwaves.com/).

A radio system should be able to provide an acceptable bit-error rate (BER) and performance at the lowest possible power level. Often, this performance is charted in terms of BER vs. PE; refer to (1)–(7).

Power Efficiency:

$$PE = Eb/No, \tag{8}$$

Spectral Efficiency:

$$SE = DR/B,$$

Channel Power:

$$P = Eb * DR,$$

where C = Channel capacity, Eb = Signal energy per bit, No = Noise power spectral density per bit.

$$SE = DR/B \prec ld(1 + SNR) = ld(1 + PE * SE) = C/B. \tag{9}$$

Then:

$$PE \succ (2 * *SE - 1)/SE,$$

i.e. Reverse Function:

$$x(y) = (2 * *y - 1)/y \tag{10}$$

for Fig. 11 is used. Here: C = B ld(1 + SNR) vs DR = Bit-per-Second.

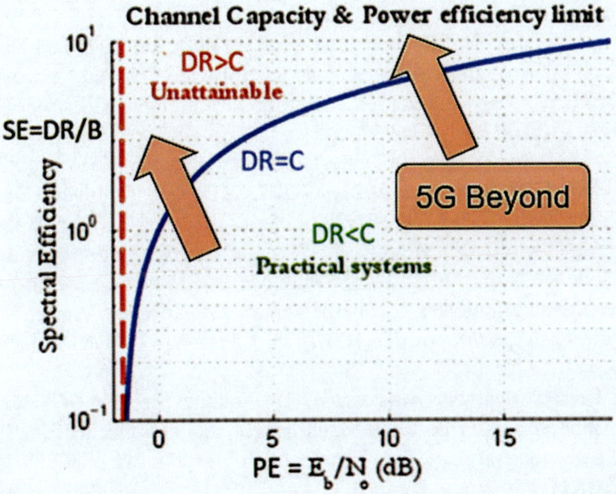

Fig. 11. Channel Capacity and Power Efficiency Limit (based on: https://www.gaussianwaves.com/).

9 Conclusions and Outlook

1. Based on 5G any Beyond IoT radio transmission systems are steadily developed.
2. Some of them strive for very high DR, compatible with the theoretical limits for radio channel capacities C (so-called "Post-Shannon").
3. A special meaning has the use of network softwarization based on further Network Function Virtualization (NFV), Artificial Intelligence (AI), and advanced cyber-security approaches like Blockchain (BC).
4. Future 6G networks include, moreover, better than 5G + Sensing via so-called Sensing like Passive Localization, when the distance x and their derivations x' and x" can be determined via evaluation of the propagation delays due to reflections on the objects (channel impulse response is required: radar techniques, WLANp, VLC, UWB, LiFi).

Acknowledgment. The author's great acknowledgments belong to colleagues from the BA Dresden University of Coop. Education, the Dresden University of Technology, the NTUU KPI Igor Sikorsky in Kyiv, and the Lviv Polytechnical National University, especially to A.Haensel, L.Zipfel, T.Zobjack, I.Melnyk, O.Michler, A.Schill, T.Maksymyuk, M.Klymash, M.Beshley for useful inspiration and challenges by fulfilling this work.

References

1. 5G Labs Germany in Dresden. https://5glab.de/
2. Luntovskyy, A., Beshley, M., Klymash, M. (eds.): Future Intent-Based Networking: on the QoS Robust and Energy Efficient Heterogeneous Software-Defined Networks, in LNEE 831, 28 chapters, XXI, 530 pages, 1st ed (2022). Springer, Heidelberg. (ISBN 978-3-030-92433-1). https://doi.org/10.1007/978-3-030-92435-5
3. Luntovskyy, A., Guetter, D. (eds.): Highly-Distributed Systems: IoT, Robotics, Mobile Apps, Energy Efficiency, Security, 1st ed. 2022, XXXII, 321 pages. Springer, Cham (2022). ISBN-10: 3030928284, ISBN: 978-3-030-92828-5. https://doi.org/10.1007/978-3-030-92829-2
4. Luntovskyy, A., Guetter, D., Melnyk, I.: Planung und Optimierung von Rechnernetzen: Methoden, Modelle, Tools für Entwurf, Diagnose und Management im Lebenszyklus von drahtgebundenen und drahtlosen Rechnernetzen, p. 435. Springer/Vieweg + Teubner Verlag, Wiesbaden (2011). https://doi.org/10.1007/978-3-8348-8242-4. ISBN 978-3-8348-1458-6, (in German)
5. Michler, O.: Perspektiven der FuE sowie Anwendungsbsp. für 5G, 6G und zukünftige Technologie-Generationen: Überblick/Machbarkeit - technologischer Einstieg, ITVS@TUD, Dresden, SiSax Arbeitskreistreffen Funktechnologien, Vortrag, AKFT@Silicon Saxony (2022). (in German)
6. Gaussian Waves: Signal Processing for Communication Systems. https://www.gaussianwaves.com/
7. Berrou, C., Glavieux, A., Thitimajshima, P.: Near Shannon limit error-correcting coding and decoding: Turbocodes. In: IEEE International Conference on Communication, (ICC 1993), vol. 2, pp. 1064–1070 (1993)
8. Post-Shannon: Shannon limit on power efficiency - demystified – Gaussian Waves. https://www.gaussianwaves.com/2019/11/shannons-limit-on-power-efficiency/
9. 5G RAT. https://www.5gradio.com/
10. Luntovskyy, A., Spillner, J.: Architectural Transformations in Network Services and Distributed Systems: Current technologies, standards and research results in advanced (mobile) networks, p. 344. Springer, Wiesbaden (2017). https://doi.org/10.1007/978-3-658-14842-3. ISBN: 9783658148409 https://www.springer.com/gp/book/9783658148409#otherversion=9783658148423
11. Luntovskyy, A., Guetter, D.: Moderne Rechnernetze – Lehrbuch: Protokolle, Standards und Apps in kombinierten drahtgebundenen, mobilen und drahtlosen Netzwerken, 481 S. + 265 Abb., vol.1. Springer, Wiesbaden (2020). ISBN: 978–3–658–25616–6 (2020). 1.Band. (Geleitwort A.Schill) - in German, "Modern Computer Networks - Handbook". https://doi.org/10.1007/978-3-658-25619-7

Method and Means of Synthesis of Barker-Like Codes for Data Transmission Coding

Oleg Riznyk, Ivan Tsmots, and Vasyl Teslyuk(✉)

Lviv Polytechnic National University, Lviv, Ukraine
{oleh.y.riznyk,ivan.h.tsmots,vasyl.m.teslyuk}@lpnu.ua

Abstract. Noise immunity is one of the most important characteristics of modern data transmission systems. Increasing noise immunity at fixed data rates is an urgent problem, for example, for wireless networks. Increasing the noise immunity of data transmission is achieved by increasing the length and power of the noise-tolerant barker-like code sequence used to transmit a single message. Advantages of these barker-like sequences, such as high noise immunity to high-power narrowband interference, code separation capability, transmission secrecy, high resistance to multi-beam propagation, high resolution in navigation measurements, will be widely used in communication and radio systems. An advanced method for the synthesis of noise-tolerant barker-like code sequences using ideal ring bundles. An improved method for quickly finding such noise-tolerant barque-like code sequences that are able to find and correct errors to the greatest extent according to the length of the obtained code sequence. Implemented algorithm for fast finding of noise-tolerant barker-like code sequences, which are able to find and correct errors in the largest amount according to the length of the obtained barker-like code sequence. A simulation model of noise-tolerant barker-like coding with the use of ideal ring bundles has been developed. The software implementation of the simulation model of noise-tolerant barker-like coding on finding and correcting errors in the obtained noise-tolerant barker-like code sequences is carried out. The proposed noise-tolerant barker-like code sequences are of practical value, as the resulting barker-like code sequence contains up to 50% and corrects up to 25% of distorted characters from the length of the noise-tolerant barker-like code sequence.

Keywords: Barker-like Noise-Tolerant Code Sequence · Ideal Ring Bundle

1 Introduction

1.1 The Current State of the Problem

Due to the continuous increase in the amount of information transmitted by communication channels, noise-tolerant encoding and decoding technologies in real time are becoming increasingly important. Analysis of existing channels shows that the frequency of organized threats during data transmission is much higher than natural. Therefore, it is necessary to protect information from both unauthorized access and distortion. One

M. Klymash et al. (Eds.): TCSET 2022, LNEE 965, pp. 409–427, 2023.
https://doi.org/10.1007/978-3-031-24963-1_24

method of protection is to encrypt information using noise-tolerant codes. Interference-resistant coding of information should be understood as a form of its transformation, in which a certain number of errors are detected and corrected. Increasing the noise immunity of the transmission is achieved by increasing the length of the code sequence used to transmit a single message. The use of noise-tolerant encoding during data transmission increases the transmission time. This increases the energy consumption for transmitting a single message. If the data transmission time remains the same, it will expand the signal spectrum by reducing the duration of one symbol and increase the bandwidth allocated to one data channel.

Interference-resistant coding systems are most often used in relay points, where the modulated signal is demodulated and, if necessary, decoded. As a result, the signal is "cleared" of noise and interference, and each repeater transmits the signal without the accumulation of noise and interference in the transmission line. Thus, noise and interference are taken into account only between two adjacent repeaters, which is the basis of data transmission over long distances. Transmission of the modulated signal without the use of noise-tolerant coding leads to the accumulation of errors, distortions and does not ensure the identity of the signal received with the transmitter signal.

1.2 The Urgency of Solving the Problem

The task of regular synthesis of interference-resistant codes with specific characteristics of wireless communication systems is becoming especially important today.

Statement of the research task. It is necessary to develop an algorithm for quickly finding such noise-tolerant code sequences that are able to find and correct errors in the largest number depending on the length of the obtained code sequence.

The object of the study is code sequences that are resistant to interference.

The subject of research is a method of synthesis of noise-tolerant code sequences.

The purpose of the work is to improve the method of synthesis and to develop a simulation model of noise-tolerant coding.

To achieve this goal it is necessary to solve the following *research tasks*:

- to improve the method of synthesis of interference-resistant code sequences with the use of ideal ring bundles;
- develop a simulated model of inclined coding using ideal ring bundles;
- practical implementation of the algorithm for finding and correcting errors of the obtained code sequences, resistant to interference.

A scientific novelty is an improved method for the synthesis of noise-tolerant code sequences.

Practical significance. The proposed noise-tolerant code sequences have practical value, as in the obtained code sequence correct up to 50% and correct up to 25% of distorted characters from the length of the noise-tolerant code sequence.

2 Analysis of Literature Sources

In [1, 2, 4] it was shown that the algorithms for the synthesis of interference-resistant code sequences are based on the principle of redundancy, which allows to find and correct errors due to the specific structure of these sequences. Typically, error-correcting codes are based on cyclic code sequences, all combinations of which can be obtained by cyclically shifting one or more code combinations.

The idea of constructing such codes is based on the use of polynomials that are not specified in the binary number field and are completely divisible only by themselves or by one.

Analysis of works [1, 4] shows that the Bose-Chowdhury-Hawkingham (BCH) codes are widely used for noise-tolerant coding. To decipher them, the Burleckamp algorithm and its main modifications are used, which provide correction of all weight errors $W \leq W_o$, where $W_o + 1 = d_o$. However, the complexity of this algorithm of the order of n2log2n operations requires approximately (5–20) n bits of memory.

In [6, 7] it was shown that the main type of block codes for noise control are non-binary Reed-Solomon (RS) codes. The length of the set is determined by the output field 2l, where l is the length of the set. The decryption of these codes is also based on the Burlackamp algorithm [1]. Despite the research of many researchers, the complexity of the operations performed by the decoder remains very high. Thus, when implementing a PC decoder for code (31, 15), it is necessary to perform about 17 operations for each bit of the decoded message, even if some operations are performed in parallel and performed in binary fields by tabular methods.

In [5, 7, 8] cascade codes are investigated, which provide high probability values at a fairly high level of interference and moderate decoding complexity.

In [9] it was shown that the Viterbi algorithm is widely used in data transmission techniques using convolutional codes that have better characteristics compared to block. However, the complexity of implementing the Viterbi algorithm increases exponentially with the length of the code.

In [8], classes of codes are considered, for which the complexity of decoding increases not exponentially with increasing length, but much more slowly, for example, as a square of length.

In [9–12] the issue of synthesis of noise-tolerant codes with the use of ideal ring bundles is considered. The disadvantage of existing methods and algorithms for the synthesis of noise-tolerant codes using ideal ring bundles is the overriding coding techniques, which slows down the search for optimized code sequences.

The current state of the problem. Due to the continuous increase in the amount of information transmitted by communication channels, noise-tolerant encoding and decoding technologies in real time are becoming increasingly important. Analysis of existing channels shows that the frequency of organized threats during data transmission is much higher than natural. Therefore, it is necessary to protect information from both unauthorized access and distortion. One method of protection is to encrypt information using noise-tolerant codes. Interference-resistant coding of information should be understood as a form of its transformation, in which a certain number of errors are detected and corrected. Increasing the noise immunity of the transmission is achieved by increasing the length of the code sequence used to transmit a single message. The use of noise-tolerant encoding during data transmission increases the transmission time. This increases the energy consumption for transmitting a single message. If the data transmission time remains the same, it will expand the signal spectrum by reducing the duration of one symbol and increase the bandwidth allocated to one data channel. Interference-resistant coding systems are most often used in relay points, where the modulated signal is demodulated and, if necessary, decoded. As a result, the signal is "cleared" of noise and interference, and each repeater transmits the signal without the accumulation of noise and interference in the transmission line. Thus, noise and interference are taken into account only between two adjacent repeaters, which is the basis of data transmission over long distances. Transmission of the modulated signal without the use of noise-tolerant coding leads to the accumulation of errors, distortions and does not ensure the identity of the signal received with the transmitter signal.

3 Research Results

3.1 Improving the Method of Synthesis of Noise-Tolerant Code Sequences Using Ideal Ring Bundles

The greatest effect of information protection is achieved by combining all applied methods and measures into one comprehensive information protection mechanism. It should be noted that information protection should take place simultaneously at three hardware levels (sketchbooks, tokens, cryptographic methods implemented at the software level), software (anti-virus software, data archiving, mandatory identification, encryption) and organizational (control over compliance with all hardware). And software protection methods in full).

However, most experts are inclined to believe that a special noise-protected coding node (implemented in software or hardware) is considered the most reliable method, because it protects the information itself, rather than access to it. The effectiveness of this approach is explained by the fact that the attacker not only needs access to information in the form of a notebook or password, but also needs to know how he needs the information that is encrypted.

The interference-resistant code sequence consists of a set of zeros and ones by which symbols are converted for further reception and transmission. The main features of these noise-tolerant code sequences are improved error detection and correction characteristics. The noise immunity of the system depends on such code sequences. For the same lengths, the properties of interference-resistant code sequences may be different. Interference immunity will depend on both the length of these code sequences and other characteristics, such as cross-correlation of sequences. This means that the choice of the best noise-tolerant code sequence is reduced to the algorithm for finding such code sequences.

Known methods of code sequence conversion do not fully improve the encoding/decoding system. Therefore, one of the important tasks is to study effective models for improving the encoding/decoding of information on such indicators as data rate, resistance to interference in the code sequence, ease of correction and error detection. Such models may be non-equidistant lines of code consisting of integers or strings of numbers defined as their values and the values of all consecutive sums of adjacent elements.

To solve the problem, consider non-equidistant code sequences. By non-equidistant code sequences we mean sequences in which the allowed code combinations form sequences with different distances between ones and zeros.

These non-equidistant code sequences have certain advantages over other interference-resistant sequences. This is easy to find and correct errors on the reception side of the data, because the appearance of the symbol "1" and/or the symbol "0" in the form of an obstacle indicates an error, because the number of allowed distances has changed.

An error is not detected only when the number of error codes is equal to or exceeds the code distance. If distorted characters appear in this non-equidistant code sequence, they will be detected, which contributes to the high security of the non-equidistant sequence.

The problem of improving the noise immunity characteristics of non-equidistant code sequences should be solved on the basis of the use of ideal ring bundles (IRB).

An ideal numerical group is a group in which all numbers exhaust values proportional to the elements of a natural series at a given number of repetitions of each element of this series.

A non-equidistant code sequence is a sequence of $K_n = (k_1, k_2, \ldots, k_i, \ldots, k_n)$ elements, where all possible adjacent elements in the form of ring sums exactly r times exhaust the values of all numbers in the natural series 1, 2,…, $S_n = n(n-1)/r + 1$.

Based on the definitions, we construct a table of ring sums of the model of non-equidistant code sequence K_n (Table 1).

Table 1. Ring amounts of noise-tolerant non-equidistant code sequence are allowed.

p_j	q_j					
	1	2	l-1	l	n-1	n
1	k_1	$\sum_{i=1}^{2} k_i$	$\sum_{i=1}^{l-1} k_i$	$\sum_{i=1}^{l} k_i$	$\sum_{i=1}^{n-1} k_i$	$\sum_{i=1}^{n} k_i$
2	$\sum_{i=1}^{n} k_i$	k_2	$\sum_{i=2}^{l-1} k_i$	$\sum_{i=2}^{l} k_i$	$\sum_{i=2}^{n-1} k_i$	$\sum_{i=2}^{n} k_i$
l-1	$\sum_{i=l-1}^{n} k_i + \sum_{i=1}^{1} k_i$	$\sum_{i=l-1}^{n} k_i + \sum_{i=1}^{2} k_i$	k_{l-1}	$\sum_{i=l-1}^{l} k_i$	$\sum_{i=l-1}^{n-1} k_i$	$\sum_{i=l-1}^{n} k_i$
l	$\sum_{i=1}^{n} k_i + \sum_{i=1}^{1} k_i$	$\sum_{i=1}^{n} k_i + \sum_{i=1}^{2} k_i$	$\sum_{i=1}^{n} k_i$	k_l	$\sum_{i=l}^{n-1} k_i$	$\sum_{i=l}^{n} k_i$
n-1	$\sum_{i=n-1}^{n} k_i + \sum_{i=1}^{1} k_i$	$\sum_{i=n-1}^{n} k_i + \sum_{i=1}^{2} k_i$	$\sum_{i=n-1}^{n} k_i + \sum_{i=1}^{l-1} k_i$	$\sum_{i=n-1}^{n} k_i + \sum_{i=1}^{l} k_i$	k_{n-1}	$\sum_{i=n-1}^{n} k_i$
n	$\sum_{i=n}^{n} k_i + \sum_{i=1}^{1} k_i$	$\sum_{i=n}^{n} k_i + \sum_{i=1}^{2} k_i$	$\sum_{i=n}^{n} k_i + \sum_{i=1}^{l-1} k_i$	$\sum_{i=n}^{n} k_i + \sum_{i=1}^{l} k_i$	$\sum_{i=1}^{n} k_i$	k_n

The total number of all ring weights of non-equidistant elements of the code sequence having different values:

$$S_n = n(n-1)/r + 1. \tag{1}$$

At values of $p_j = 1$, $q_j = \text{n}$, as well as at values of $p_j \neq 1$, $q_j = p_j - 1$, the ring sums of the weights of the elements of the non-equidistant code sequence are equal to S_n.

The number S_n^r of ring sums on the sequence of weights of the elements is determined by thc following dependence:

$$S_n^r = n(n-1)/r. \tag{2}$$

Consider the construction of an obstacle-resistant code sequence. As an example, take a non-equidistant code sequence constructed in accordance with the weights of the elements of the IRB of the 8th order of the 4th multiplicity:

$$1, 1, 1, 2, 2, 1, 3, 4.$$

Since these values of the weights of the elements are numbers of the bunch of the 8th order of the 4th multiplicity, each of these numbers of the natural series from 1 to $n(n-1)/r = 15$ will be represented by four different ways, and the number of all ways is the number of numbers obtained.

For a non-equidistant code sequence with n number of packet elements, our algorithm makes it possible to encode any number from 1 to $S_n = n(n-1)/r + 1$.

The synthesized noise-tolerant code sequence based on IRB weights is shown in Table 2.

Each numerical combination of non-equidistant code sequence, resistant to interference, corresponds to a set of ones and zeros, which is based on the scales of the elements of the IRB according to the following principle: the weight of the IRB 1 is 1, 2 - 10, 3 - 100, 4 - 1000 and so on. We obtain lines of noise-tolerant code sequence based on IRB scales. In general, the code combination 1 is encoded by the first line of Table. 1. Code combination 2 is encoded by the second line of Table. 1, etc., code combination 15 - the fifteenth line of Table. 1.

The number of different $S_n^r(S_n^r - 1)/2$ code sequences contains exactly r of n single characters in the corresponding digits, which follows from the properties of the IRB. The other n-r symbols of any two non-equidistant code sequences differ from the symbols represented in the same bits.

Therefore, the minimum code distance of any noise-tolerant non-equidistant code sequence constructed using IRB will be defined as the ratio of the order and multiplicity of IRB:

$$d_{\min} = 2(n-r). \tag{3}$$

To increase the number of valid combinations of non-equidistant code sequences using IRB, we create a mirror-stable non-equidistant code sequence with IRB weights, in which we replace 1 and 0 with encoding.

Table 2. Noise-tolerant code sequence based on IRB weights of order $n = 8$ multiplicity $r = 4$: 1, 1, 1, 2, 2, 1, 3, 4.

1	1	1	1	0	1	0	1	1	0	0	1	0	0	0
0	1	1	1	1	0	1	0	1	1	0	0	1	0	0
0	0	1	1	1	1	0	1	0	1	1	0	0	1	0
0	0	0	1	1	1	1	0	1	0	1	1	0	0	1
1	0	0	0	1	1	1	1	0	1	0	1	1	0	0
0	1	0	0	0	1	1	1	1	0	1	0	1	1	0
0	0	1	0	0	0	1	1	1	1	0	1	0	1	1
1	0	0	1	0	0	0	1	1	1	1	0	1	0	1
1	1	0	0	1	0	0	0	1	1	1	1	0	1	0
0	1	1	0	0	1	0	0	0	1	1	1	1	0	1
1	0	1	1	0	0	1	0	0	0	1	1	1	1	0
0	1	0	1	1	0	0	1	0	0	0	1	1	1	1
1	0	1	0	1	1	0	0	1	0	0	0	1	1	1
1	1	0	1	0	1	1	0	0	1	0	0	0	1	1
1	1	1	0	1	0	1	1	0	0	1	0	0	0	1

Each numerical combination of mirror non-equidistant code sequence, resistant to interference, corresponds to a set of ones and zeros, based on the scales of the elements of the IRB on the following principle: the weight of the IRB 1 is 0, 2 - 01, 3 - 001, 4 - 0001 and so on. Implemented mirror, noise-tolerant, non-equidistant code sequences with IRB weights illustrates Table. 3. In general, the code combination 1 is encoded by the first line of Table 3. Code combination 2 is encoded by the second line of Table 3, etc., code combination 15 - the fifteenth line of Table 3.

Permissible number of noise-tolerant main and mirror code sequences:

$$P = 2S_n^r. \tag{4}$$

The number of errors t_I detected by a noise-tolerant code sequence is determined by the minimum code distance d_{min}:

$$t_1 \leq d_{min} - 1. \tag{5}$$

The number of errors t_2, which are corrected using a noise-tolerant code sequence, is determined by the number of detected errors t_I:

$$T_2 \leq (t_1 - 1)/2. \tag{6}$$

Let us establish the relation through n and r, which specifies the number of errors t_I that can be detected by interference-resistant code sequence:

$$t_1 \leq 2(n - r) - 1. \tag{7}$$

Table 3. Mirror noise-tolerant code sequence based on IRB scales of order $n = 8$ multiplicity $r = 4$: 1, 1, 1, 2, 2, 1, 3, 4

0	0	0	0	1	0	1	0	0	0	1	0	0	0	1
1	0	0	0	0	1	0	1	0	0	0	1	0	0	0
0	1	0	0	0	0	1	0	1	0	0	0	1	0	0
0	0	1	0	0	0	0	1	0	1	0	0	0	1	0
0	0	0	1	0	0	0	0	1	0	1	0	0	0	1
1	0	0	0	1	0	0	0	0	1	0	1	0	0	0
0	1	0	0	0	1	0	0	0	0	1	0	1	0	0
0	0	1	0	0	0	1	0	0	0	0	1	0	1	0
0	0	0	1	0	0	0	1	0	0	0	0	1	0	1
1	0	0	0	1	0	0	0	1	0	0	0	0	1	0
0	1	0	0	0	1	0	0	0	1	0	0	0	0	1
1	0	1	0	0	0	1	0	0	0	1	0	0	0	0
0	1	0	1	0	0	0	1	0	0	0	1	0	0	0
0	0	1	0	1	0	0	0	1	0	0	0	1	0	0
0	0	0	1	0	1	0	0	0	1	0	0	0	1	0

Let us determine the dependence through n and r of the number of errors t_2 that can be corrected for the noise-tolerant code sequence:

$$t_2 \leq n - r - 1. \tag{8}$$

The minimum code distance for an interference-resistant code sequence is defined as:

$$d_{1,2} = S_n^r - 2(n - r). \tag{9}$$

Find the dependencies to determine the number of errors that can be found using a code sequence resistant to interference:

$$t_1 \leq \begin{cases} 2(n - r) - 1, & \textit{if } S_n^r \geq 4(n - r), \\ S_n^r - 2(n - r) - 1, & \text{otherwise.} \end{cases} \tag{10}$$

Find the dependence to determine the number of errors that can be corrected using a code sequence resistant to interference:

$$t_2 \leq \begin{cases} n - r - 1, & \textit{if } S_n^r \geq 4(n - r), \\ \frac{S_n^r - 2(n-r+1)}{2}, & \textit{otherwise.} \end{cases} \tag{11}$$

We find the optimal relationship between the values of the parameters n and r in terms of the best corrective power of the noise-tolerant code sequence. The noise immunity of this code sequence increases with increasing value of the difference $l = n - r$.

The largest value of l will be provided:

$$S_n^r = 2n. \tag{12}$$

Here is the relationship between the parameters n and r, in which the noise-tolerant code sequence maximally detects and corrects the largest number of errors:

$$L = \begin{cases} n/2, & n - \text{pair} \\ (n-1)/2, & n - \text{odd}. \end{cases} \tag{13}$$

Obtained from ideal ring bundles noise-tolerant code sequences can find up to n – 1 and correct errors to $n/2$–1 at even values of n and find up to n and correct to (n – 1)/2 errors at odd values of n provided that the number of allowable combinations of code sequences based on the bundle is theoretically doubled due to the introduction of noise-tolerant mirror code sequences.

3.2 Development of a Simulation Model of Noise-Tolerant Coding

An algorithm for simulating a noise-tolerant encoding using non-equidistant combinatorial configurations grouped based on the concept of an ideal ring bundle and a representation of a number encoded in ASCII format, and an implemented method of coding using an ideal ring bundle.

The algorithm of the simulation model of noise-tolerant coding includes the following steps:

- enter code parameters and select the ideal ring bundle;
- selection of a code combination with the help of an ideal ring bundle;
- generation of noise-tolerant code with the selected number.

A block diagram of the algorithm of the simulation model of noise-protected coding using non-equidistant combinatorial configurations is developed, which is shown in Fig. 1.

The block diagram of the algorithm for developing a simulation model of noise-tolerant coding using combinatorial non-equidistant configurations consists of the following blocks:

- block «Selection of IRB code parameters» according to the value of code parameters we choose the ideal ring bundle;
- block «Input of information for noise-tolerant coding» input of the necessary parameters of the safety code with IRB restrictions;
- block «If the code power is greater than the amount of IRB» compares the sum of the IRB elements with the entered value. If the value entered is greater than the sum of the IRB elements, you must re-enter the number to select another IRB;
- block «Calculation of code positions for the selected IRB» in the IRB are the corresponding elements and the corresponding sums of elements corresponding to the selected number;

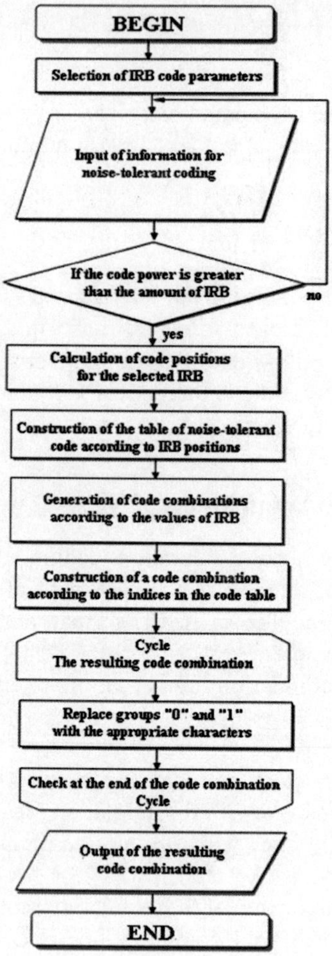

Fig. 1. Block diagram of the algorithm for developing a simulation model of noise-tolerant coding using non-equidistant combinatorial configurations.

- block «Construction of the table of noise-tolerant code according to IRB positions» the table of the found combinations of IRB is formed;
- block «Generation of code combinations according to the values of IRB» - a table of code combinations is formed, where each element of IRB corresponds to its unique code;
- block «Construction of a code combination according to the indices in the code table» the result of the generated noise-tolerant code is created by combining two four-bit codes;
- block «Processing cycle of the received combination» replace groups "0" and "1" with the corresponding groups when searching and correcting errors;
- block «Output of the resulting code combination» saves the encoding results in a file.

As an example, consider the calculation of noise-protected codes built on the basis of IRB with the parameters:

1) $N = 6, R = 1$;
2) $N = 15, R = 7$;
3) $N = 16, R = 8$.

Since the sum of IRB $S_n = 31$, the length of the code combinations and the power of the code in all three cases is the same. The maximum number of errors to be detected or corrected when executing the first of the detected codes is 9 and 4. In each other case, the IRB allows you to generate a code that can detect up to 15 and correct up to 7 errors. Thus, the last two results are much better than the previous one. Therefore, in general, any IRB can be used to build a noise-tolerant code. However, it is most appropriate to use IRB-generated codes, the parameters of which are related to relation (13).

The simulation model of noise-tolerant coding using non-equidistant combinatorial configurations is developed in the Borland Delphi 7 programming language [13–15]. The developed simulation model works perfectly with a wide range of Windows 7–11 operating systems. Encoding of noise-tolerant sequences, modules of formation of random interferences, modules of decoding of noise-resistant sequences (Fig. 2).

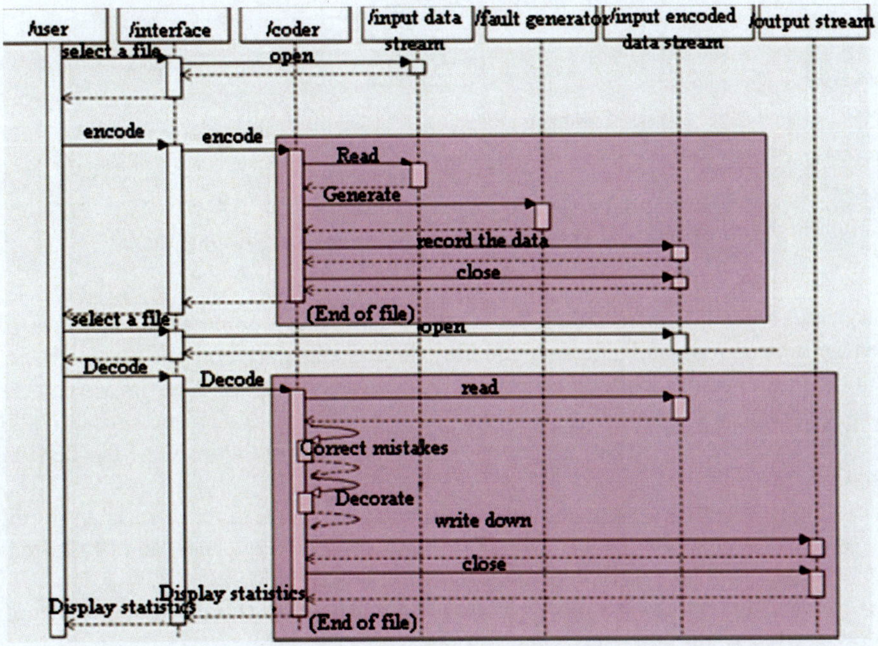

Fig. 2. Diagram of software components of the data encoding process.

As can be seen from the flowchart of the data encoding process, the first step for the user is to select a file and open the file read stream. Next is the file encoding command. Although there is no end to the file, we are reading a block of data. Then we randomly generate interferences and store them in the original file stream. Then close the streams.

Here is a block diagram of the process of decoding the file, shown in Fig. 3.

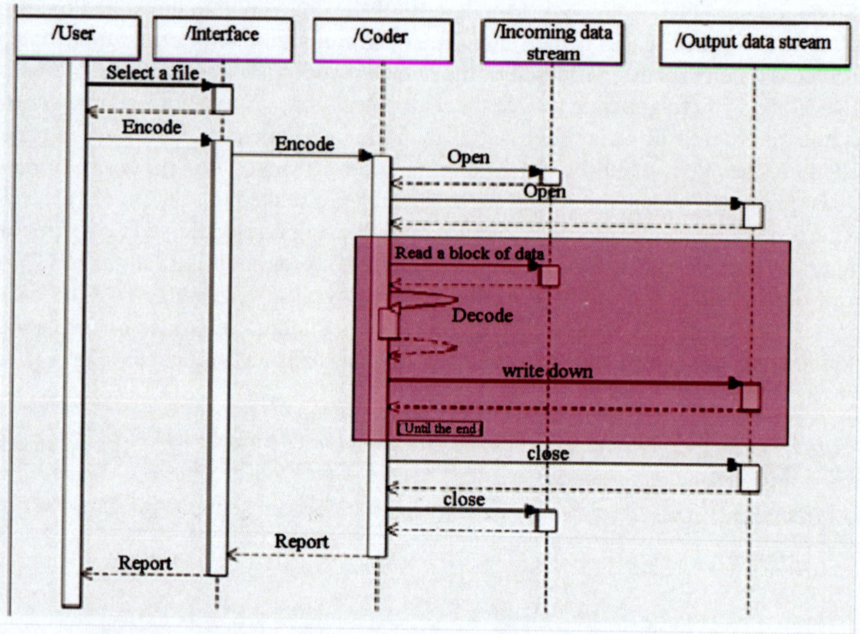

Fig. 3. Diagram of software components of the data decoding process.

We will begin the process of decoding the data from the command to open the file. Then we issue a command to decode the file. The encoded data from the file will be read in blocks to the end of the file. The encoder then checks for errors. If errors are detected, the encoder corrects them. The next step is to decode the block data and save it to a file.

After decoding, we display statistical information about the result of the decoding operation.

Program sequence diagrams show the operation of interference-resistant coding software using non-equidistant interference-resistant code sequences, and illustrate the possibilities of detecting and correcting accidental interference.

The simulation model of noise-tolerant coding using non-equidistant combinatorial configurations has the following characteristics:

- the input data will be the parameters of a non-equidistant code sequence of order n of multiplicity r;
- provides encoding and decoding with finding and correcting errors in the amount of up to t_2.

The program has an easy to use and intuitive interface. The program window after running the program (Fig. 4).

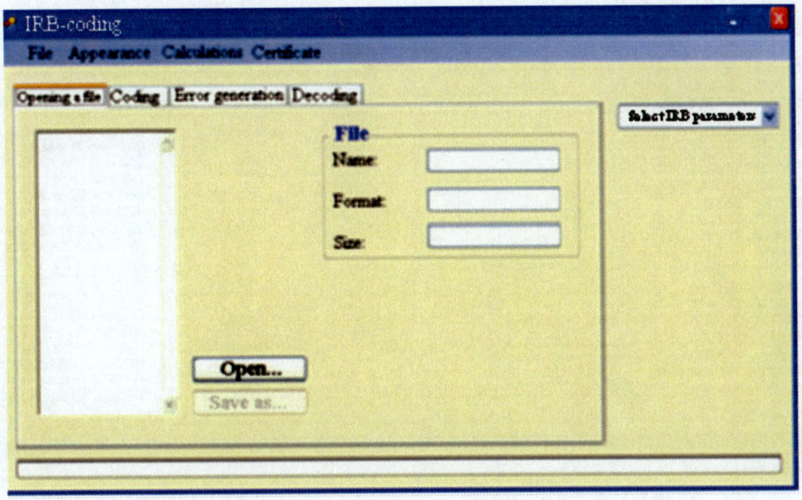

Fig. 4. The program window after starting the program.

The main elements in the form of a window (Fig. 5):

- IRB parameters: order *n*, multiplicity *r*;
- button labeled <Open file> allows you to select the desired file for encoding;
- window named <File> shows the name, format and size in bytes for the selected file;
- button labeled <Coding> encodes the selected file;
- button labeled <Error Generation> generates random interference;
- the button labeled <Decode> decodes the file with random interference.

After entering the IRB parameters and opening the file, the calculation of the code parameters appears depending on the selected IRB (Fig. 6).

In Fig. 6, the following calculations of the code parameters depending on the selected IRB:

- minimum code distance;
- the number of errors detected;
- the number of errors that are corrected;
- length of the input block;
- length of the coded block;
- degree of redundancy.

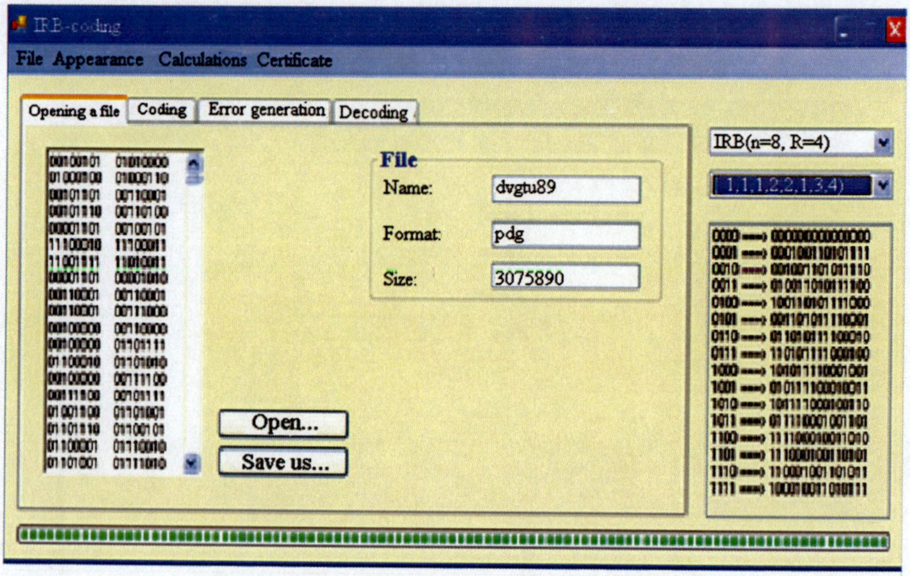

Fig. 5. The main elements in the form of a window.

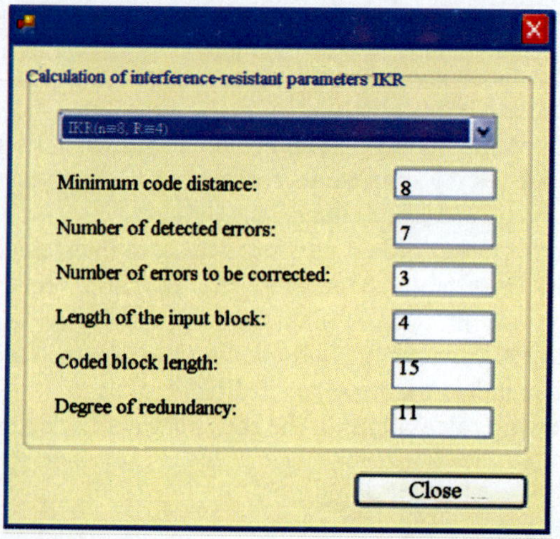

Fig. 6. Calculation of code parameters depending on the selected IRB.

4 Analysis of the Obtained Results

Let's compare with the most well-known noise-resistant Bose-Chowdhury-Hawkingham (BCH) codes, which detect and correct the maximum number of errors. Calculate the relationship between the number of information i and the number k of correction symbols in the code combination of length n, which can correct at least s errors, is based on the following expressions:

$$n = i + k; \tag{14}$$

$$2k \geq n + 1; \tag{15}$$

$$2i \leq 2n/(n + 1); \tag{16}$$

$$s = \mathrm{ent}(d-1)/2, \tag{17}$$

where d is the minimum code distance.

For comparative analysis of BCH codes and tested codes we use the following relations:

$$N = 2n-1; \tag{18}$$

$$t = 2n - 2-1; \tag{19}$$

$$P = 2n + 1; \tag{20}$$

$$N* = 2n - 1; \tag{21}$$

$$t* = (N * -3)/4; \tag{22}$$

$$P* = 2(N * +1), \tag{23}$$

where N i N* – the length of the code sequences;

t i t* – the number of errors that are corrected;

P i P* – the power of the code sequences of the BCH and IRB, respectively.

Formulas (18)–(20) correspond to the ratio of the parameters of the BCH codes [3], and (21)–(23) - the parameters of the IRB codes, which provide maximum noise immunity to the length of the code sequences [4].

Analysis of the results of comparison of these codes, carried out on the basis of formulas (18) (23), shows that at a fixed length of the code sequence the maximum achievable resistance of IRB codes to interference is not worse than BCH codes at the same power of these sequences. A certain advantage of the BCH code is the possibility of increasing the power (20), but due to the loss of interference correction. Unlike BCH codes, the synthesis of IRB codes with a high level of noise immunity does not require complex calculations, and the length of code sequences is determined by a linear relationship (21), which allows the construction of code sequences with a high level of noise protection.

The results of the encoding and decoding with finding and correcting all errors using interference-resistant non-equidistant code sequences are shown in Fig. 7, 8, 9 and 10.

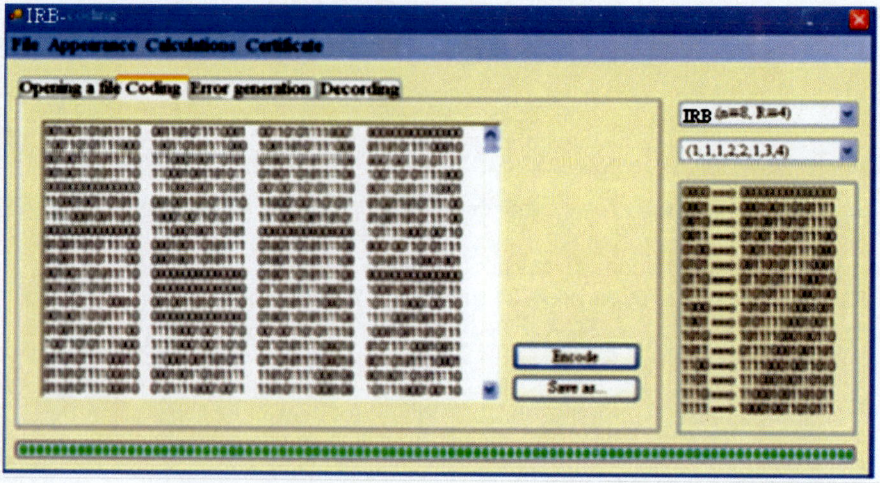

Fig. 7. Encoding the file according to certain rules.

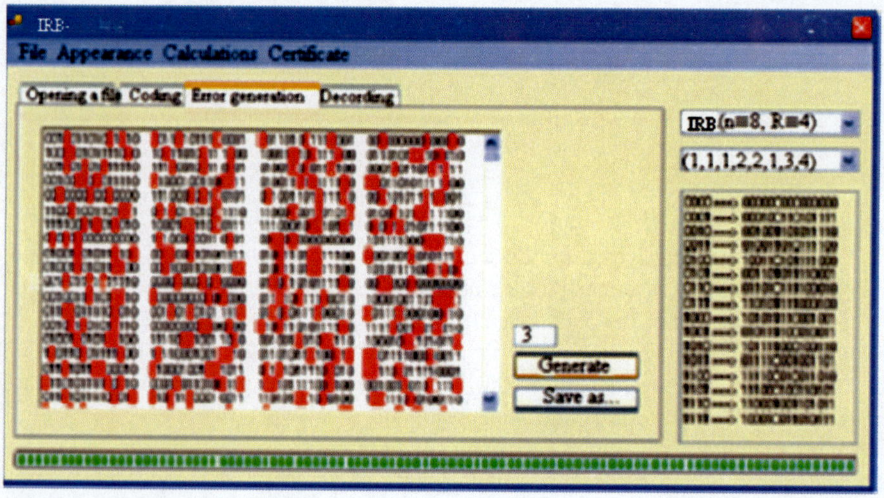

Fig. 8. Artificial generation of interference to demonstrate the operation of the program.

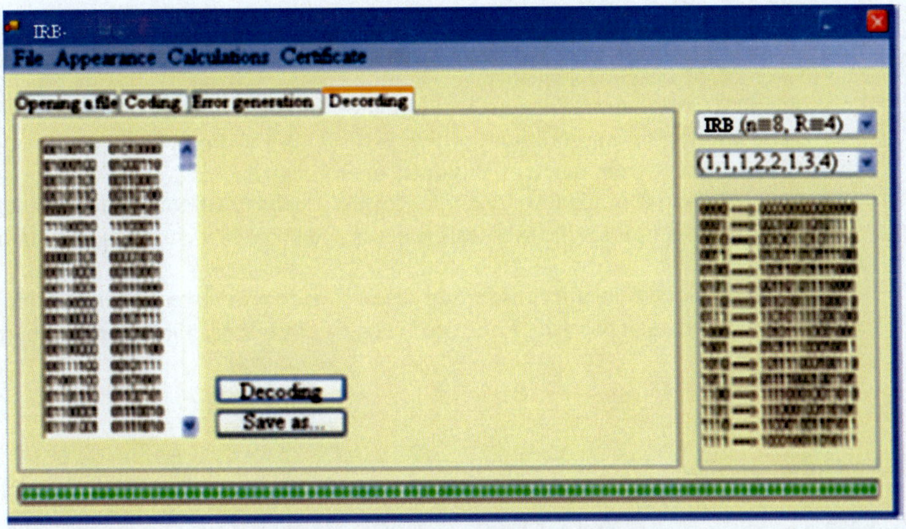

Fig. 9. Decode the file for viewing.

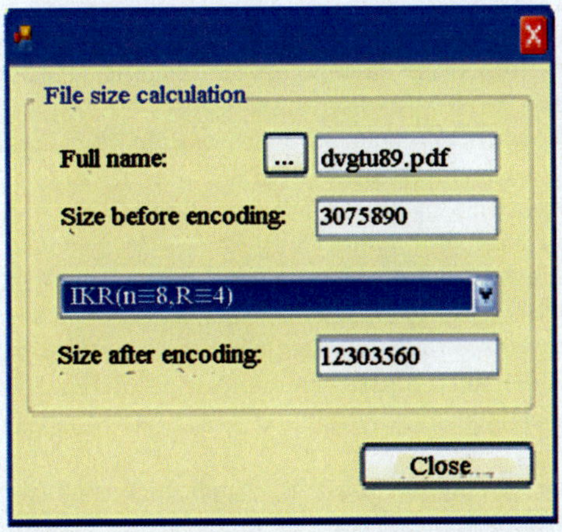

Fig. 10. Calculation of file size depending on the selected IRB.

Analyzing the above experimental data obtained using the developed model, we see that the results are fully consistent with the theory of non-equidistant code sequences, resistant to interference.

5 Conclusions

1. It is shown that to eliminate interference in the transmission of the code sequence by wireless communication channels, it is advisable to use noise-tolerant encoding.
2. Improved the method of synthesis of noise-tolerant code sequences, which through the use of ideal ring bundles provided increased noise immunity of the obtained

sequences, detecting up to 50% and correcting up to 25% of distorted characters from the length of the code sequence.

3. Developed in a programming language simulation model of noise-tolerant coding using non-equidistant configurations, which focuses on low computing power and works flawlessly in a wide range of Windows operating systems.
4. It is shown that the use of ideal ring bundles for the synthesis of noise-tolerant code sequences greatly simplifies the synthesis process and ensures its implementation in real time.
5. It is established that the prospect of further research is to reduce redundancy, which can be achieved through the use of mirrored sequences of codes that are resistant to interference.

References

1. Tsmots, I., Rabyk, V., Riznyk, O., Kynash, Y.: Method of synthesis and practical realization of quasi-barker codes. In: 2019 IEEE 14th International Conference on Computer Sciences and Information Technologies (CSIT), Lviv, Ukraine, pp. 76–79 (2019). https://doi.org/10.1109/STC-CSIT.2019.8929882
2. Ahmad, J., Akula, A., Mulaveesala, R., Sardana, H.K.: Barker-coded thermal wave imaging for non-destructive testing and evaluation of steel material. IEEE Sens. J. **19**(2), 735–742 (2019). https://doi.org/10.1109/JSEN.2018.2877726
3. Fu, J., Ning, G.: Barker coded excitation using pseudo chirp carrier with pulse compression filter for ultrasound imaging. In: BIBE 2018; International Conference on Biological Information and Biomedical Engineering, Shanghai, China, pp. 1–5 (2018)
4. Wang, M., Cong, S., Zhang, S.: Pseudo Chirp-Barker-Golay coded excitation in ultrasound imaging. In: 2018 Chinese Control and Decision Conference (CCDC), Shenyang, pp. 4035–4039 (2018). https://doi.org/10.1109/CCDC.2018.8407824
5. Wang, S., He, P.: Research on low intercepting radar waveform based on LFM and barker code composite modulation. In: 2018 International Conference on Sensor Networks and Signal Processing (SNSP), Xi'an, China, pp. 297–301 (2018). https://doi.org/10.1109/SNSP.2018.00064
6. Xia, S., Li, Z., Jiang, C., Wang, S., Wang, K.: Application of pulse compression technology in electromagnetic ultrasonic thickness measurement. In: 2018 IEEE Far East NDT New Technology & Application Forum (FENDT), Xiamen, China, pp. 37–41 (2018). https://doi.org/10.1109/FENDT.2018.8681975
7. Kovalchuk, A., Izonin, I., Gregush, M.M., Riznyk, O.: An efficient image encryption scheme using projective transformations. In: 10th International Conference on Emerging Ubiquitous Systems and Pervasive Networks, EUSPN 2019 and 9th International Conference on Current and Future Trends of Information and Communication Technologies in Healthcare, ICTH 2019, Affiliated Workshops, Coimbra, Portugal, 4–7 November 2019, code 156993, vol. 160, pp. 584–589 (2019)
8. Riznik, O., Yurchak, I., Vdovenko, E., Korchagina, A.: Model of stegosystem images on the basis of pseudonoise codes. In: 2010 Proceedings of VIth International Conference on Perspective Technologies and Methods in MEMS Design, Lviv, pp. 51–52 (2010)
9. Riznyk, V., Riznyk, O., Balych, B., Parubchak, V.: "Information encoding method of combinatorial optimization. In: 2006 International Conference - Modern Problems of Radio Engineering, Telecommunications, and Computer Science, Lviv-Slavsko, pp. 357–357 (2006). https://doi.org/10.1109/TCSET.2006.4404550

10. Wang, S., Zhang, L., Lu, Z., Zhang, H., Yang, Z., Yu, X.: Photonic generation of barker-code phase-coded terahertz signals. In: 2021 46th International Conference on Infrared, Millimeter and Terahertz Waves (IRMMW-THz), pp. 1–2 (2021). https://doi.org/10.1109/IRMMW-THz50926.2021.9567054
11. Coxson, G.E., Russo, J.C.: Search for generalized MPS and barker sequences. In: 2022 IEEE Radar Conference (RadarConf22), pp. 1–6 (2022). https://doi.org/10.1109/RadarConf2248738.2022.9764290
12. Matsuyuki, S., Tsuneda, A.: A study on aperiodic auto-correlation properties of concatenated codes by barker sequences and NFSR sequences. In: 2018 International Conference on Information and Communication Technology Convergence (ICTC), pp. 664–666 (2018). https://doi.org/10.1109/ICTC.2018.8539367
13. Tsmots, I., Teslyuk, V., Teslyuk, T., Lukashchuk, Y.: The method and simulation model of element base selection for protection system synthesis and data transmission. Int. J. Sensors Wirel. Commun. Control **11**(5), 518–530 (2021). https://doi.org/10.2174/2210327910999201022194630
14. Batyuk, A., Voityshyn, V., Verhun, V.: Software architecture design of the real-time processes monitoring platform. In: Proceedings of the IEEE Second International Conference on Data Stream Mining & Processing, DSMP 2018, Lviv, Ukraine, pp. 98–101 (2018). https://doi.org/10.1109/DSMP.2018.8478589
15. Tsmots, I., Rabyk, V., Berezky, O., Lukaschuk, Y., Teslyuk, V.: Development of modules of neuro-like cryptographic encryption and decryption of data and their implementation on FPGA. In: 2021 IEEE 16th International Conference on the Experience of Designing and Application of CAD Systems (CADSM), pp. 53–57 (2021). https://doi.org/10.1109/CADSM52681.2021.9385228

Processing Marker Arrays of Clustered Transformants for Image Segments

Vladimir Barannik[1](✉), Andrii Krasnorutsky[2], Valeriy Barannik[3], Yurii Babenko[4], Sergii Shulgin[3], Oleksandr Chernenko[5], Oleksandr Slobodyanyuk[6], and Mariia Bondarchuk[7]

[1] V. N. Karazin, Kharkiv National University, Kharkiv, Ukraine
vvbar.off@gmail.com
[2] Ivan Kozhedub Kharkiv National Air Force University, Kharkiv, Ukraine
[3] Kharkov National University of Radio Electronics, Kharkiv, Ukraine
[4] Taras Shevchenko National University of Kyiv, Kyiv, Ukraine
[5] National University of Civil Defense of Ukraine, Kharkiv, Ukraine
[6] Kamianets-Podilskyi Ivan Ohiienko National University, Kamianets-Podilskyi, Ukraine
[7] Lviv Polytechnic National University, Lviv, Ukraine

Abstract. An innovative technological approach to the preservation of the information component of the video data of the aerospace segment of remote video service has been provided. The imbalance between the reliability of the received video image and the efficiency of its delivery via onboard data transmission channels has been substantiated. To solve this imbalance, the technology of reducing the informative intensity of sufficiently informative image segments has been proposed. The essence of this technology is based on taking into account the redundancy of the binary structure of the transformant discrete-cosine transformation of image segments. The technology involves the restructuring of transformant components into several clusters. In each cluster, the probability of the appearance of components relative to their location in the original transformant is expected to increase. The methodology of revealing regularities in the internal binary structure of the transformant component of the image segment on a quantitative basis has been constructed. The stages of identification of transformant components in relation to the cluster structure with the use of marker values has been substantiated. Based on the structural characteristics of the binary representation of the clustered component, a method of forming markers and a marker array of clustered transformants of the image segment has been developed. It has been substantiated that in order to reduce the volumes of clustered transformants it is necessary to reduce the volumes of marker arrays. The strategy of coding the marker array of clustered transformant components that make up the structural cluster is based on the detection of chains of markers that have the same values. This approach reduces the structural description of the linearized marker array of clustered transformants by reducing the number of repeating markers. A method of forming a sequence of two-component corteges containing a length of a cluster trend that precedes the corresponding marker has been developed. This approach allows you to identify structural patterns in the arrays of markers due to the presence of cluster trends. It has been substantiated that there is a relationship between the components of corteges, which is conditioned by the positions of markers in a linearized array

M. Klymash et al. (Eds.): TCSET 2022, LNEE 965, pp. 428–443, 2023.
https://doi.org/10.1007/978-3-031-24963-1_25

of markers, and hence by the correlated lengths of marker chains. For a structural approach to cortege processing, a method of domain formation for a marker matrix has been developed, which takes into account the correlation of marker chain lengths.

Keywords: Video image · Compression · Information intensity · Clustered component · Marker arrays · Marker chain · Marker chain length · Marker array domain

1 Introduction

The study of the reasons for the constant growth of video streams coming to telecommunications networks is associated with the integration of video services of mobile applications in society. The slow bandwidth of mobile video service platforms leads to a significant reduction in their efficiency due to time delays in video delivery. The adaptation of the intensity of video streams to the bandwidth of mobile applications of telecommunications networks is associated with the use of technological platforms aimed at reducing them [1–5].

Existing platforms for reducing the information intensity of video data have significant disadvantages: there is a possibility of reducing time delays in the delivery of video information, but with the disruption of image quality. On the contrary, if the image quality (semantic component) is preserved, the information intensity of video data cannot be reduced to a sufficient level, and in the end the time of video information transmission increases [6–10].

The reason here is the imbalance between the quality of the received video image and the time delays of its delivery. This affects the efficiency of video services using telecommunications networks [11–15]. Therefore, ensuring the reduction of information intensity of video data of mobile platforms of telecommunication networks with given quality is an urgent scientific and applied task.

To solve this problem, the technology of reducing informative intensity of image segments has been proposed [16–19]. The essence of this technology is based on taking into account the redundancy of a binary structure of the transformant discrete-cosine transformation of sufficiently informative image segments [20–22]. The technology involves restructuring of the transformant components into several clusters. In each cluster, the probability of appearance of the components relative to their location in the original transformant is expected to increase [23–26].

To identify the transformant components relative to the cluster structure, it is necessary to create marker values that represent service information [27–29]. The task of the marker is aimed at addressing the components of the clustered transformant in relation to the affiliation of the latter to the corresponding cluster. A marker is an indicator of the affiliation of a particular component of a transformant to one of the structured clusters. It is proposed to use a structural characteristic of binary representation as a marker value, as the number consists of series of units which each of the components of the clustered transformant consists of [30, 31]. Each transformant of the segment will correspond to a marker array of clustered components of this transformant [32, 33]. This approach creates the conditions for further reduction of the structural redundancy of the code representation of the image segment due to the redistribution of transformant components to the level of clusters within which the data will be encoded [34, 35]. In turn, the marker array has a certain amount of service information, which negatively affects the total amount of information intensity that enters communication channels of telecommunications networks. Therefore, the aim of the article is to develop a technology for processing marker arrays of clustered transformants of sufficiently informative image segments.

2 Development of Technology for Labeling Transformant Components of Sufficiently Informative Image Segments

In order to decide on the unambiguous affiliation of the components $y'^{(k,\ell)}_{i,j}$ to the relevant cluster, the concept of labeling the components of the transformant to be processed is introduced. That is, each component $y'^{(k,\ell)}_{i,j}$ is assigned a corresponding marker $m^{(k,\ell)}_{i,j}$ (Fig. 1).

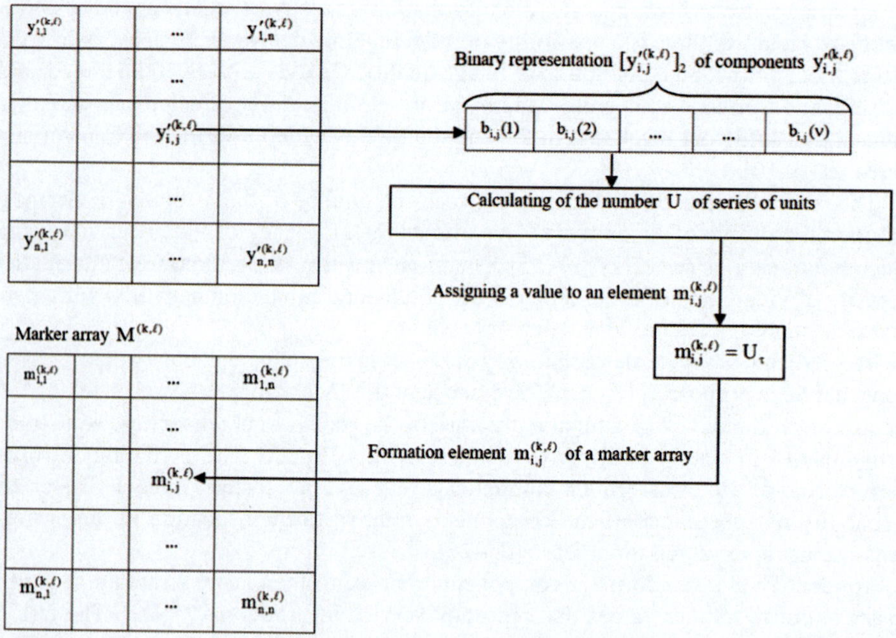

Fig. 1. Block diagram of the process of assigning a corresponding marker to a clustered component.

The scheme of the binary structure $y'^{(k,\ell)}_{i,j}$ is described by the expression:

$$y'^{(k,\ell)}_{i,j} = \{b_{i,j}(1); \ldots b_{i,j}(8); \ldots; b_{i,j}(\nu)\}, \tag{1}$$

where $b_{i,j}(\nu)$ – current binary components.

It is the analysis of the binary structure $y'^{(k,\ell)}_{i,j}$ of the transformant component that underlies the construction of the marker $m^{(k,\ell)}_{i,j}$ of this component. That is, a marker is service information on the availability and U presence of a series of units of which the binary structure $y'^{(k,\ell)}_{i,j}$ consists. The formation of the marker $m^{(k,\ell)}_{i,j}$ values occurs by counting the number of series of units present in the binary space $[y'^{(k,\ell)}_{i,j}]_2$ of each clustered component.

The elements of the marker $M^{(k,\ell)}$ array of clustered transformants are the markers $m^{(k,\ell)}_{i,j}$. Since the number of the markers $m^{(k,\ell)}_{i,j}$ will be equal to the number of the components $y'^{(k,\ell)}_{i,j}$ of the transformant $Y'(\alpha,\beta)^{(k,\ell)}$, the dimension of the marker matrix $M^{(k,\ell)}$ will coincide with the dimension of the transformant $Y'(\alpha,\beta)^{(k,\ell)}$.

It is known that each marker $m^{(k,\ell)}_{i,j}$ has its own numerical value. Therefore, the number of binary digits on the representation of the marker $[m^{(k,\ell)}_{i,j}]_2$ affects the total volume $V(M^{(k,\ell)})$ of the formed marker array $M^{(k,\ell)}$.

The comparative characteristic of the volume $V(M^{(k,\ell)})$ of the formed marker array and the coded clustered transformant $V_o(Y'(\alpha,\beta)^{(k,\ell)})$ for different compression coefficients K_{co}, allows us to state that if the compression coefficient increases, the volumes of the coded transformant decrease. At the same time, the volume $V(M^{(k,\ell)})$ of the formed marker array of clustered $y'^{(k,\ell)}_{i,j}$ transformant components remains unchanged and will exceed the volume $V_o(Y'(\alpha,\beta)^{(k,\ell)})$. The volumes $V(M^{(k,\ell)})$ of the marker array of clustered components are about 37% of the total volume $V(Y'(\alpha,\beta)^{(k,\ell)})$ of this transformant.

There is an imbalance: on the one hand, to compress the transformants $Y'(\alpha, \beta)^{(k,\ell)}$ the restructuring of the information space of these transformants by forming the clusters $Cls^{(\tau)}(\alpha, \beta)^{(k,\ell)}$ based on the structural feature of the number of binary series is used. On the other hand, to identify clustered $y'^{(k,\ell)}_{i,j}$ component transformants relative to the cluster structure, the marker values $m^{(k,\ell)}_{i,j}$, formed in the marker arrays $M^{(k,\ell)}$ are used, but the volumes $V(M^{(k,\ell)})$ of marker arrays may exceed the volumes $V_o(Y'(\alpha,\beta)^{(k,\ell)})$ of coded transformants. This increases the total amount of encoded image representation.

To resolve this imbalance, it is necessary to develop a technology that reduces the volume of the formed marker $M^{(k,\ell)}$ array of clustered components $y'^{(k,\ell)}_{i,j}$ of the transformant $Y'(\alpha, \beta)^{(k,\ell)}$.

3 Development of Technology for Processing Marker Arrays of Clustered Transformants of Sufficiently Informative Image Segments

The strategy of coding the array of markers $y'^{(k,\ell)}_{i,j}$ of the transformant $Y'(\alpha, \beta)^{(k,\ell)}$ of which the structural cluster $Cls^{(\tau)}(\alpha, \beta)^{(k,\ell)}$ consists is developed taking into account such properties as:

1) the number of structural clusters in the spectral space of discrete-cosine transformation (DCT) transformants is limited. They are due to the following aspects:
- limiting the number of ν binary bits (bits) used to describe the binary structure $[y'^{(k,\ell)}_{i,j}]_2$ of the values of the transformant components $Y'(\alpha,\beta)^{(k,\ell)}$ DCT;
- concentration of the main number of values of the transformant component $y'^{(k,\ell)}_{i,j}$ of DCT near the medium frequencies and zero level of high-frequency components;
- the ability to redistribute values of the components of one cluster to other structural clusters;
- the number of clusters will be further reduced in case of previous use of the transformant quantization process. This takes into account the fact that high-frequency components have a less significant effect on the visual perception of images than medium-frequency and low-frequency components;

2) *the presence in the dirufiant component of chains* $y'^{(k,\ell)}_{i,j}$ *belonging to one cluster of zigzag bypass*. As a result, it is possible to identify sequences of the markers $m^{(k,\ell)}_{i,j}$ that have the same values;
3) the distribution of values within clusters will be characterized by its non-uniformity. This is due to the fact that:
- nominal power of clusters has non-uniform distribution;
- as a result of quantization processes of the transformant components $y'^{(k,\ell)}_{i,j}$ and their redistribution by other clusters $Cls^{(\tau)}(\alpha,\beta)^{(k,\ell)}$ an additional increase in the real level of non-uniformity is achieved.

Taking into account the presented properties of marker arrays, it is proposed to use the following approaches for their compact representation:

1) to identify sequences of the same values of cluster markers in the direction of the "zigzag bypass";
2) values of the markers are proposed to be encoded with one of the statistical codes;
3) a probability-statistical model for coding cluster markers is proposed to be formed by one of the two approaches:
- using the nominal non-uniform distribution of capacities of structural clusters;
- summing up the probabilities of the components included in them.

In this case, if the entropy estimates of the distribution of cluster marker values for neighboring transformants differ slightly from the initial one, it is proposed to use the probability-statistical model of the previous marker array.

For sufficiently informative segments, the characteristics of the realistic probability-statistical model of structural cluster markers will approach the nominal characteristics. On the contrary, in case of the segments with low information content, for the characteristics of the realistic probability-statistical model of markers of structural clusters there will be a tendency to increase the probability of markers with low values.

Formulation of the problem of coding.

It is necessary to create such technology of coding of low-range markers of structural clusters for which reduction of a bit volume under the conditions of complete exclusion of information losses will be provided; exclusion of significant impact on the loss of quality or distortion in the processing of transformants; the complexity of the implementation of encoding and decoding processes should be linearly dependent on the size of the marker arrays.

Let us consider the operation of the algorithm for forming the array $M^{(k,\ell)}$ of the markers $m_{i,j}^{(k,\ell)}$ of the cluster $Cls^{(\tau)}(\alpha, \beta)^{(k,\ell)}$ of the clustered transformant $Y'(\alpha,\beta)^{(k,\ell)}$.

Linearization of the marker array.

Execution of the "zigzag bypass" algorithm in order to linearize the elements $m_{i,j}^{(k,\ell)}$ of the matrix $M^{(k,\ell)}$ of markers of clustered transformants $Y'(\alpha,\beta)^{(k,\ell)}$ of the image segment:

$$F_{lin} : M^{(k,\ell)} = \{ m_{i,j}^{(k,\ell)}\}, i = \overline{1, n}; j = \overline{1, n} \longrightarrow LM^{(k,\ell)} = \{m_w^{(k,\ell)}\}, w = \overline{1, n^2}, \quad (2)$$

where F_{lin} – is the function of the elements of the matrix of markers;

$LM^{(k,\ell)}$ – is a linearized matrix of markers;

(α,β) – are coordinates of the segment in the image;

(k,ℓ) – are coordinates of transformants in the segment;

(i,j) – are coordinates of the element in the matrix of markers;

$m_w^{(k,\ell)}$ – is the element of the linearized matrix;

(w) – are coordinates of the elements $m_w^{(k,\ell)}$;

$(n \times n)$ – is the dimension of the matrix $M^{(k,\ell)}$ of markers.

We construct the elements $m_w^{(k,\ell)}$ of a linearized matrix in one line. Here, the two-coordinate elements $m_{i,j}^{(k,\ell)}$ of the marker matrix $M^{(k,\ell)}$ are transformed into the elements $m_w^{(k,\ell)}$ with one coordinate (w).

After the linearization of the elements $m_{i,j}^{(k,\ell)}$ of the matrix $M^{(k,\ell)}$ of markers of the clustered transformants, we obtain the following sequence:

$$LM^{(k,\ell)} = \{m_1^{(k,\ell)}, m_2^{(k,\ell)}, \ldots, m_w^{(k,\ell)}, \ldots, m_{n^2}^{(k,\ell)}\}.$$

The presented structural and statistical properties, as well as the properties due to the psycho-visual features of the perception of video images by the visual system determine the conditions for creating new methods of coding transformants. In this regard, one of the stages of processing the arrays $M^{(k,\ell)}$ of markers of structural clusters is to identify statistical and structural-statistical patterns.

In the beginning, the detection of sequences and the calculation of lengths $Le(g)^{(k,\ell)}$ of the chain $Ch(g)^{(k,\ell)}$ consisting of the markers $m_{i,j}^{(k,\ell)}$ having the same values are performed, i.e. the trend properties of each structural cluster relative to the DCT transformant are evaluated.

Formation of marker chains.

The analysis of the adjacent elements $m_w^{(k,\ell)}$ of the linearized $LM^{(k,\ell)}$ matrix of markers is performed and the marker chains $Ch(g)^{(k,\ell)}$ are built according to the rule: the elements $m_w^{(k,\ell)}$ must be next to each other and have the same values $m_w^{(k,\ell)} = m_{w+1}^{(k,\ell)}$.

$$F_{chain} : \{m_w^{(k,\ell)}\}, w = \overline{w_1, w_2} \overset{samevalue}{\longrightarrow} Ch(g)^{(k,\ell)}, \quad (3)$$

where F_{chain} – is the detection function of marker chains;

$Ch(g)^{(k,\ell)}$ – marker chain;

(g) – coordinates of the marker chain;

$\overline{w_1, w_2}$ – elements of the beginning and end of a sequence of markers having the same values.

A sequence of marker chains is constructed $Ch(g)^{(k,\ell)}$, $g = \overline{1, G}$. Their number will be equal to G. Each chain $Ch(g)^{(k,\ell)}$ consists of the elements $m_w^{(k,\ell)}$, i.e.: $Ch(g)^{(k,\ell)} = \{m_w^{(k,\ell)}\}$, $w = \overline{w_1, w_2}$.

Let us consider an example of the formation of the marker chains $Ch(g)^{(k,\ell)}$.

Formation of the length of marker chains.

The length $Le(g)^{(k,\ell)}$ of the marker chain $Ch(g)^{(k,\ell)}$ is determined by counting the number of the elements $m_w^{(k,\ell)}$ (having the same values) in each formed chain $Ch(g)^{(k,\ell)}$

$$Le(g)^{(k,\ell)} = F_{length}(Ch(g)^{(k,\ell)}), \tag{4}$$

where F_{length} – is the function of the decision rules for counting the number of elements $m_w^{(k,\ell)}$ in each chain $Ch(g)^{(k,\ell)}$ of the sequence $LM^{(k,\ell)}$.

Each marker chain $Ch(g)^{(k,\ell)}$ has a corresponding length $Le(g)^{(k,\ell)}$, equal to the number of adjacent equivalent elements $m_w^{(k,\ell)}$, that make up the chain $Ch(g)^{(k,\ell)}$.

After combining adjacent equivalent $m_w^{(k,\ell)}$ elements of the matrix $LM^{(k,\ell)}$ of markers, they become the informative $m_g^{(k,\ell)}$ marker.

The informative marker $m_g^{(k,\ell)}$ has coordinates (*g*) in a linearized $LM^{(k,\ell)}$ matrix of markers and is equal to the numerical value of the corresponding $m_w^{(k,\ell)}$ marker from which it was formed.

As a result of the detection of chains of markers having identical values, we will receive the following description of the linearized marker array:

$$\begin{aligned} LM^{(k,\ell)} = \{&(Le(1)^{(k,\ell)}; m_1^{(k,\ell)}), (Le(2)^{(k,\ell)}; m_2^{(k,\ell)}), (Le(3)^{(k,\ell)}; m_3^{(k,\ell)}), \\ &\ldots, (Le(g)^{(k,\ell)}; m_g^{(k,\ell)}), \ldots, (Le(G)^{(k,\ell)}; m_G^{(k,\ell)})\}, g = \overline{1, G} \end{aligned},$$

where $m_g^{(k,\ell)} \in \{m_w^{(k,\ell)}\}$, $w = \overline{w_1, w_2}$.

This approach reduces the structural description of the linearized marker array of the clustered transformant. Here the number of repeating markers is reduced.

Formation of two-component cortege for a matrix of markers.

There is a classification placement of the value of the chain $Ch(g)^{(k,\ell)}$ length $Le(g)^{(k,\ell)}$ of markers along with the corresponding equivalent elements $m_g^{(k,\ell)}$ for the formation of the cortege $Kr(g)^{(k,\ell)}$.

This transformation is given by the expression:

$$Kr(g)^{(k,\ell)} = F_{cortege}(Ch(g)^{(k,\ell)}) = \{Le(g)^{(k,\ell)}; m_g^{(k,\ell)}\}, \tag{5}$$

where $F_{cortege}$ – the function of decision-making rules on the conformity of the length $Le(g)^{(k,\ell)}$ of the chain of markers to the particular element $m_g^{(k,\ell)}$;

$m_g^{(k,\ell)}$ – markers having the same values;

$Kr(g)^{(k,\ell)}$ – cortege of marker matrix.

(g) – the coordinates of the marker chain will be equal to the coordinates of the corteges. Accordingly, their number will be equal to the number G of marker chains.

Each two-component cortege $Kr(g)^{(k,\ell)}$ has: the elements with odd coordinates (odd position) is the length $Le(g)^{(k,\ell)}$ of the chain $Ch(g)^{(k,\ell)}$ of the marker matrix, the elements with even coordinates (even position) are the values of the elements $m_g^{(k,\ell)}$, corresponding to this chain. That is, the elements with even coordinates correspond to the $m_g^{(k,\ell)}$ value of the information marker.

As a result of the formed two-component cortege $Kr(g)^{(k,\ell)}$ we obtain the following description of the linearized marker array:

$$M^{(k,\ell)} \rightarrow LM^{(k,\ell)} = \{(Le(g)^{(k,\ell)}; m_g^{(k,\ell)})\}_{g=\overline{1,n_G}},$$

where g – is the position of the cortege as part of a linearized marker array, where $g = \overline{1, n_G}$. Accordingly $n_G \leq n^2 - 1$, where n – is the dimension of the marker array; n_G – the number of corteges.

The two-component cortege $\{Le(g)^{(k,\ell)}; m_g^{(k,\ell)}\}$ contains the length $Le(g)^{(k,\ell)}$ of the ***cluster trend***, that precedes the corresponding marker $m_g^{(k,\ell)}$. As a result of such process a sequence of two-component corteges of an array of markers of structural classes is formed. The length of such a sequence is denoted as n_G, $G = \overline{1, n_G}$.

Constructing a sequence of a *two-component cortege* allows us to identify structural patterns in the arrays of markers due to the presence of cluster trends. In the absence of elimination of psycho-visual redundancy, the length of *cluster trends* allows us to assess the following. After identifying cluster trends and describing their values, the increase in the length of cluster trends, especially in the area of their low values, indicates the level of informative content and coherence of the segments. Accordingly, it should be understood that the area of the segment with high coherence is uninformative.

It should be understood that after identifying cluster trends and describing their values by their length, the statistical properties of structural cluster markers change. And here the distribution of probabilities of their occurrence becomes more uniform.

To increase the level of non-uniformity (decrease the corresponding value of entropy) it is proposed to organize such transformations as:

- quantization of component values;
- redistribution of the DCP transformant components between adjacent structural clusters.

Such operations are carried out taking into account:

- the presence of the quality of interleaving the components of neighboring clusters with each other, i.e. the components of different clusters have similar values;
- the presence of psycho-visual redundancy for the video signal and the possibility of its elimination in the spectral space of the DCT transformant. For example, there is a pattern that the higher the level of coherence of the segment, the higher the level of energy concentration in the low-frequency components. That is, the amount of energy of the output video fragment, which will be contained (concentrated) in a limited number of low-frequency components, increases. This leads to an increase in the

number of the chain lengths of markers with low values. Therefore, nominally, the load distribution on structural clusters is shifted from the center towards the smaller values of their markers, i.e. towards the structural clusters, which will contain sequences with a smaller value of the number of series of units.

Let us consider the processing of two-component corteges. There is a relationship between the components of the corteges. First, this is due to the positions of the $m_w^{(k,\ell)}$ markers in the linearized array of markers $L\,M^{(k,\ell)}$. Hence, the shorter chain length will correspond to the most important information markers. Secondly, linearized markers $m_w^{(k,\ell)}$ of clustered components $y'^{(k,\ell)}_{i,j}$ in connection with the processes of transformation and quantization will be characterized by a non-uniform law of distribution of their values. The markers with lower values are more likely to appear. Such properties determine the stage of processing of transformants, which lies in constructing of a code structure for the vector of two-component corteges. Here, coding technologies are divided into two approaches:

1. For the first approach, dynamic static coding is organized, where contextual modeling is used. Contextual modeling is necessary for adapting the law of distribution of probabilities of appearance of $m_w^{(k,\ell)}$ markers in the array $L\,M^{(k,\ell)}$. This approach reduces the number of bits spent on the information component. But at the same time it requires additional bit costs for the representation of the service component, which contains information about the possibilities of distribution of markers. In addition, the disadvantage of this approach is the increase in processing delays associated with the enumeration of the probabilities of markers.
2. The second approach is related to the statistical coding with fixed tables (static implementation of the statistical codec). This reduces processing delays and bit costs for the service component. But on the other hand, the adaptability of the statistical model to the variable characteristics of the transformants decreases. This leads to an increase in the length of the code of the information component of the code structure.

To eliminate these shortcomings, it is proposed to consider a structural approach to the processing of corteges.

Formation of domains for a matrix of markers.

We find the corteges which have identical length $Le(g)^{(k,\ell)}$ of a chain from all space of corteges $Sp\{Kr(g)^{(k,\ell)}\}$, $g = \overline{1, G}$.

To identify the corteges $Kr(\alpha)^{(k,\ell)}$ with the same length of a marker chain, in each cortege analysis is performed for the equivalence of odd elements, i.e. the length $Le(g)^{(k,\ell)}$ of each marker chain $Ch(g)^{(k,\ell)}$, for example:

$$Le(1)^{(k,\ell)} = Le(3)^{(k,\ell)} = \ldots = Le(g)^{(k,\ell)}.$$

This is specified by the expression:

$$\begin{aligned} Se(Le(g)^{(k,\ell)}) &= \{Kr(1)^{(k,\ell)}, \ldots, Kr(a)^{(k,\ell)}, \ldots, Kr(\Psi_{Le(g)})^{(k,\ell)}\} \\ &= \{Le(g)^{(k,\ell)}; m_1^{(k,\ell)}\}, \ldots, \{Le(g)^{(k,\ell)}; m_a^{(k,\ell)}\}, \ldots, \{Le(g)^{(k,\ell)}; m_{\Psi_{Le(g)}}^{(k,\ell)}\}, \end{aligned} \quad (6)$$

where $Se(Le(g)^{(k,\ell)})$ – is the sequence of corteges having the same length Le(g) of a marker chain;

$\Psi_{Le(g)}$ – the number of corteges with the same length $Le(g)$ of a marker chain.

In the right part the expression (6) there are corteges containing the same length $Le(g)^{(k,\ell)}$ of a marker chain. Here $m^{(k,\ell)}_{\Psi_{Le(g)}}$ – is the last sequence marker $Se(Le(g)^{(k,\ell)})$.

To combine corteges with the same length of a marker chain, the concept of a domain is used.

The domain $D(\omega)^{(k,\ell)}$ of the corteges is a totality of the corteges $Kr(\alpha)^{(k,\ell)}$, $\alpha = \overline{1, \Psi_{Le(g)}}$ that have the same marker chain length.

The construction of domains is given by the expression:

$$D(\omega)^{(k,\ell)} = F_{domen}\,(Kr(a)^{(k,\ell)}, a = \overline{1, \Psi_{Le(g)}}), \tag{7}$$

where F_{domen} – is the function of the decisive rules on the formation of the domain $D(\omega)^{(k,\ell)}$ with an index ω, which is formed by those very corteges that have the same value $Le(g)^{(k,\ell)}$ of the chain length sampling from the space of two-component corteges. The domain $D(\omega)^{(k,\ell)}$ indexing takes value $\omega = \overline{1, \Omega}$.

Let us consider the method of coding domains of corteges, taking into account the peculiarities of the formation of marker arrays and the detection of marker chains.

Coding the structure of the domain of corteges.

Each domain $D(\omega)^{(k,\ell)}$ is proposed to be reformatted as follows: the first element is a fixed length $Le(g)^{(k,\ell)}$ of a marker chain; the first element has a sequence of markers corresponding to this length.

That is:

$$D(\omega)^{(k,\ell)} = \{Le(g)^{(k,\ell)}; \{m_1^{(k,\ell)}, \ldots, m_\alpha^{(k,\ell)}, \ldots, m^{(k,\ell)}_{\Psi_{Le(g)}}\}\}. \tag{8}$$

Therefore, the matrix $M^{(k,\ell)}$ of markers of the clustered transformant $Y'(\alpha,\beta)^{(k,\ell)}$ consists of a set $Va\{D(\&)^{(k,\ell)}\}$ of cortege domains. The number of such domains, taking into account the peculiarities of the formation of marker arrays and the detection of marker chains will be equal to Ω.

That is:

$$M^{(k,\ell)} = Va\{D(\omega)^{(k,\ell)}\} = \{D(1)^{(k,\ell)}, \ldots, D(\omega)^{(k,\ell)}, \ldots, D(\Omega)^{(k,\ell)}\}. \tag{9}$$

The transformation which lies in creation of the domain can be called the technology of reclusterization of the marker array $M^{(k,\ell)}$ by the length of marker chains.

Reclusterization of the marker $M^{(k,\ell)}$ array by the length of marker chains suggests that the sequence of informative $m_g^{(k,\ell)}$ markers of such an array will be characterized by an uneven distribution law. Therefore, such a sequence, respectively, is encoded by element-by-element statistical code (for example, one of the modifications of the Huffman code) or one of the block codes.

Formation of a codogram of a matrix of markers of clustered transformants of an image segment.

It is known that the reformatted matrix $M^{(k,\ell)}$ of markers of the clustered image segment transformants $Y'(\alpha,\beta)^{(k,\ell)}$ now consists of a plurality of the domains $Va\{D(\omega)^{(k,\ell)}\}$.

Therefore, you need to build codograms for each of the domains $D(\omega)^{(k,\ell)}$, that have appropriate indexing ω. That is, the coding of the marker matrix $M^{(k,\ell)}$ is reduced to the coding of the domains $D(\omega)^{(k,\ell)}$ of corteges.

The encoding of domains of corteges is specified by the expression:

$$F_{coding} : D(\omega)^{(k,\ell)} \overset{coding}{\longrightarrow} E(\omega)^{(k,\ell)}, \tag{10}$$

where F_{coding} – is the function of the decisive rules for the formation of the codogram $E(\omega)^{(k,\ell)}$ of domains of corteges.

The first element of the domain (corresponding to the length $Le(\omega)^{(k,\ell)}$ of the marker chain) is encoded by the uniform binary code BC_{un}. Subsequent domain elements (the sequence of markers $m_{\alpha}^{(k,\ell)}, \alpha = \overline{1, \Psi_{\omega}}$) are encoded by the Huffman code BC_{huf}.

This is given by the expression:

$$E(\omega)^{(k,\ell)} = \{BC_{un}^{(\&)}; BC_{huf}^{(\&)}\}. \tag{11}$$

where $E(\omega)^{(k,\ell)}$ – is the codogram of cortege domains,

ω – domain indexing; takes value $\omega = \overline{1, \Omega}$.

After coding the domains $D(\omega)^{(k,\ell)}$, the codogram of the market $M^{(k,\ell)}$ matrix is formed. This transformation is given by the expression:

$$F_{codogram} : M^{(k,\ell)} \overset{coding}{\longrightarrow} E(M^{(k,\ell)}), \tag{12}$$

where $F_{codogram}$ – is the function of decision-making rules on the formation of the codogram $E(M^{(k,\ell)})$ of the marker matrix.

In turn, the codogram $E(M^{(k,\ell)})$ consists of a sequence (array) of codograms $Ar^{(k,\ell)}$ of cartridge domains. Therefore, the codogram of the matrix $M^{(k,\ell)}$ of the markers of the clustered transformants $Y'(\alpha,\beta)^{(k,\ell)}$ of the image segment is given by the expression:

$$E(M^{(k,\ell)}) = Ar^{(k,\ell)} = \{\{BC_{un}^{(1)}; BC_{huf}^{(1)}\}, \ldots, \{BC_{un}^{(\omega)}; BC_{huf}^{(\omega)}\}, \ldots, \{BC_{un}^{(\Omega)}; BC_{huf}^{(\Omega)}\}\}. \tag{13}$$

Thus, it is the codogram $E(M^{(k,\ell)})$ of the formed matrix of markers $M^{(k,\ell)}$ of the clustered transformant $Y'(\alpha,\beta)^{(k,\ell)}$ that is submitted for further processing concerning the stage of effective syntactic representation of the formed clusters.

Structural and functional diagram of the technology of processing marker arrays of clustered transformants of sufficiently informative image segments is given in Fig. 2.

4 Evaluation of the Effectiveness of the Developed Technology

It should be noted that the developed technology of processing marker arrays of clustered transformants of sufficiently informative image segments allows one to create the conditions for reducing the volume of the codogram $V(E(M^{(k,\ell)}))$ of the marker array. This leads to the prospect of reducing the total amount of encoded image representation.

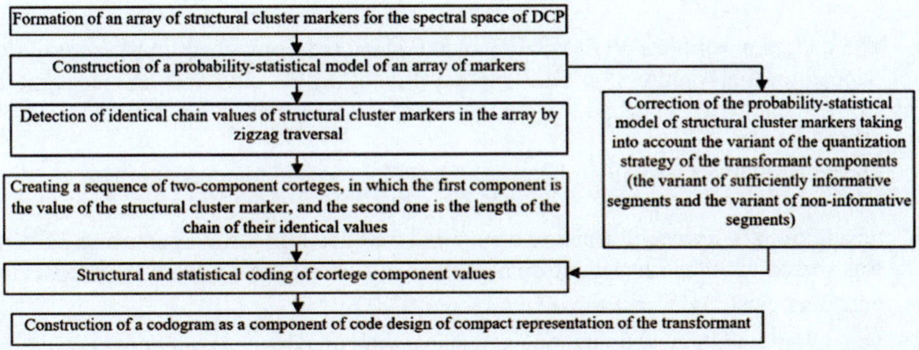

Fig. 2. Structural and functional diagram of the technology of processing marker arrays of clustered transformants of sufficiently informative image segments.

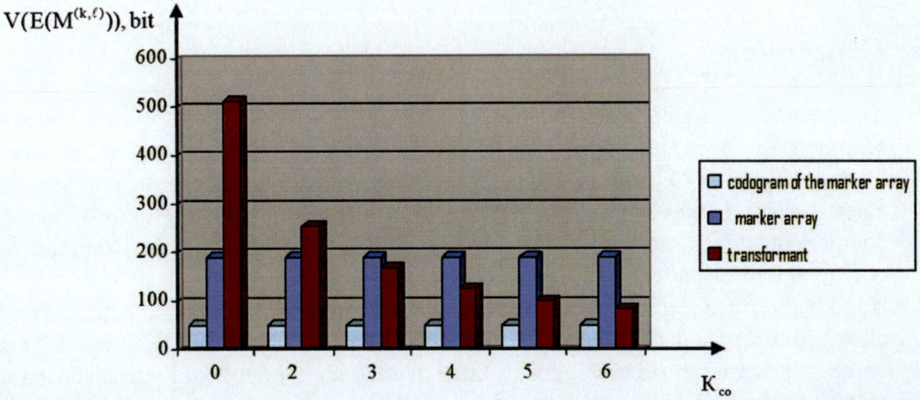

Fig. 3. The diagram of the dependence of the volume $V(M^{(k,\ell)})$ of the formed marker array of the volume of the codogram $V(E(M^{(k,\ell)}))$ of the marker array and the coded clustered transformant $V_{co}(Y'(\alpha,\beta)^{(k,\ell)})$ for different compression ratios.

Figure 3 shows the results of the experimental estimation of the volume of the codogram of the marker array of the clustered transformant of a sufficiently informative image segment. A comparative characteristic of the volume of the codogram $V(E(M^{(k,\ell)}))$ marker array of the clustered image segment transformant and the coded clustered transformant $V_{co}(Y'(\alpha,\beta)^{(k,\ell)})$ for different compression ratios is given.

The analysis of Fig. 3 suggests that the developed technology for processing the marker array transformants of the clustered transformant of sufficiently informative image segments, has jmnme $V(M^{(k,\ell)})$ of the formed marker array 3.7 times.

Obviously, provided the compression ratio K_{co} increases and the volume $V_{co}(Y'(\alpha,\beta)^{(k,\ell)})$ of the coded transformant decreases, the volume of the marker array codogram $E(M^{(k,\ell)})$ remains unchanged and will be much smaller than the volume $V_{co}(Y'(\alpha,\beta)^{(k,\ell)})$.

The codgram volumes $V(E(M^{(k,\ell)}))$ of the clustered component marker array do not depend on the compression ratio applied directly to the clustered image segment transformant and will not exceed 50% of the total coded clustered transformant volume $V_{co}(Y'(\alpha,\beta)^{(k,\ell)})$.

The technology of coding of the marker array has certain features, namely:

1) no additional service information is required for encoding the marker arrays $M^{(k,\ell)}$. It is shared with the service information that is evaluated and used in the process of encoding clustered transformant components $y'^{(k,\ell)}_{i,j}$;
2) you do not need to use prefix codes to encode the marker chain lengths $Le(\omega)^{(k,\ell)}$. This eliminates the variation of increasing the length of the codogram to ensure mutually unique decoding on the receiving side, i.e. to implement the quality of the prefix codograms.

5 Conclusions

1. The method of reduction of informative intensity of sufficiently informative image segments has been developed. The method involves restructuring of the transformant components of discrete-cosine transformation of image segments into several clusters. This makes it possible to increase the probability of the appearance of components relative to their location in the original transformant. Therefore, due to coding, the information intensity of video data is reduced.
2. A method of forming marker arrays (based on the analysis of the binary structure of clustered components of the image segment transformants) has been developed. This makes it possible to (unambiguously) identify the components of the transformant relative to the cluster structure using marker values.
3. The technology of processing marker arrays of clustered transformants of sufficiently informative image segments (on the basis of revealing regularities in internal binary structure of components on a quantitative basis) has been developed.

The technology has the following components:

3.1. The method of detecting sequences and determining the lengths of chains consisting of markers with the same values. This allows us to estimate the trend properties of each structural cluster relative to the transformant of discrete-cosine transformation of image segments;
3.2. The method of forming two-component corteges for a matrix of markers containing a chain length. This allows us to identify structural patterns in the arrays of markers due to the presence of cluster trends;
3.3. The method of forming domains for a marker matrix that takes into account the correlation of chain lengths. This allows you to reclassify the marker array by the length of the chains. Therefore, the sequence of informative markers will be characterized by an uneven distribution law and, accordingly, creates conditions for coding with element-by-element statistical code or one of the block codes;

3.4 The method of encoding the sequence of domains of the marker matrix. This enables us to say that prefix codes do not need to be used to encode marker chain lengths. Therefore, this eliminates the variation of increasing the length of the codogram to ensure mutually unique decoding on the receiving side.

This approach makes it possible to reduce the amount of generated marker arrays, which affects the overall information intensity of video data coming to telecommunications networks.

4. The developed technology for processing the marker array transformants of the clustered transformants of sufficiently informative image segments, has the ability to reduce the volume of the formed marker array 3.7 times. The coded marker array does not exceed the threshold value from the total amount of coded clustered transformants by 15%. The proposed technology reduces the delay time for mobile applications of telecommunications networks by up to 10%.

Scientific novelty.
For the first time the technology of processing a marker array of the clustered transformant of sufficiently informative image segments has been developed. The basic differences are that the technology allows one to identify structural patterns in the marker arrays of clustered components due to the presence of cluster trends and to assess the trend properties of each structural cluster relative to the transformants of discrete-cosine transformation. This provides an effective syntactic description of the markers of the clustered components of the segment transformant.

This approach creates the conditions for improving the efficiency of the wireless segment of telecommunications networks, namely the preservation of key semantic information and reduce time delays in the delivery of video data. Thus, positive conditions are provided, such as reducing the response time of the control system related to the efficiency of video delivery.

References

1. JPEG Privacy & Security Abstract and Executive Summary (2015). https://jpeg.org/items/20150910_privacy_security_summary.html. Accessed 7 April 2021
2. Barannik, V., Sidchenko, S., Barannik, D.: Technology for protecting video information resources in the info-communication space. In: IEEE 2nd International Conference on Advanced Trends in Information Theory (IEEE ATIT 2020), pp. 29–33 (2020). https://doi.org/10.1109/ATIT50783.2020.9349324
3. Barannik, V., Sidchenko, S., Barannik, N., Barannik, V.: Development of the method for encoding service data in cryptocompression image representation systems. Eastern-Eur. J. Enterp. Technol. **3**(9(111)), 112–124 (2017)
4. DSTU 7624:2014: Informatsiini tekhnolohii. Kryptohrafichnyi zakhyst informatsii. Alhorytm symetrychnoho blokovoho peretvorennia [Information Technology. Cryptographic protection of information. Symmetric block transformation algorithm]. Ministry of Economic Development of Ukraine, 39 p. (2015)
5. Data Encryption Standard (DES), Federal Information Processing Standards Publication 46–3, 26 p. (1999)

6. DSTU GOST 28147:2009: Systema obrobky informatsii. Zakhyst kryptohrafichnyi. Alhorytm kryptohrafichnoho peretvorennia (HOST 28147–89) [Information processing system. Cryptographic protection. Cryptographic transformation algorithm (GOST 28147–89)], State Committee for Technical Regulation and Consumer Policy (Derzhspozhivstandart) of Ukraine, 20 p. (2008)
7. Rivest, R.L., Shamir, A., Adleman, L.M.: A method for obtaining digital signatures and public-key cryptosystems. Commun. ACM **21**(2), 120–126 (1978). https://doi.org/10.1145/359340.359342
8. Barannik, V., Babenko, Yu., Kulitsa, O., Barannik, V., Khimenko, A., Matviichuk Yudina, O.: Significant microsegment transformants encoding method to increase the availability of video information resource. In: IEEE 2nd International Conference on Advanced Trends in Information Theory (IEEE ATIT 2020), pp. 52–56 (2020). https://doi.org/10.1109/ATIT50783.2020.9349256
9. Chen, T.-H., Wu, Ch.-S.: Efficient multi-secret image sharing based on Boolean operation. Sig. Process. **91**(1), 90–97 (2011). https://doi.org/10.1016/j.sigpro.2010.06.012
10. Barannik, V., Shulgin, S., Krasnorutsky, A., Slobodyanyuk, O., Gurzhii, P., Korolyova, N.: Methodological fundamentals of deciphering coding of aerophotography segments on special equipment of unmanned Complex. In: IEEE 2 nd International Conference on Advanced Trends in Information Theory (IEEE ATIT 2020), pp. 38–43 (2020). https://doi.org/10.1109/ATIT50783.2020.9349257
11. Li, F., Krivenko, S., Lukin, V.: Two-step providing of desired quality in lossy image compression by SPIHT. Radioelektronni i komp'uterni sistemi – Radioelectronic and computer systems **2**(94), 22–32 (2020). https://doi.org/10.32620/reks.2020.2.02
12. Ji, Sh., Tong, X., Zhang, M.: Image encryption schemes for JPEG and GIF formats based on 3D baker with compound chaotic sequence generator (2012). https://arxiv.org/abs/1208.0999. Accessed 7 April 2021
13. Belikova, N., Lekakh, A., Dovbenko, O., Dodukh, O.: Method of increasing the capacity of information threat detection filters in modern information and communication systems. In: IEEE 3rd International Conference Advanced Information and Communications Technologies, pp. 426–429 (2019). https://doi.org/10.1109/AIACT.2019.8847754
14. Naor, M., Shamir, A.: Visual cryptography. In: Proceedings of the Advances in Cryptology. EUROCRYPT 1994. LNCS, vol. 950, pp. 1–12. Springer, Heidelberg (1995). https://doi.org/10.1007/bfb0053419
15. Wu, Yu., Agaian, S., Noonan, J.: Sudoku associated two dimensional bijections for image scrambling. IEEE Trans. Multimedia, 30 (2012). https://arxiv.org/abs/1207.5856v1. Accessed 7 April 2021
16. Tsai, Ch.-L., Chen, Ch.-J., Hsu, W.-L.: Multi-morphological image data hiding based on the application of Rubik's cubic algorithm. In: IEEE International Carnahan Conference on Security Technology (ICCST), pp. 135–139 (2012). https://doi.org/10.1109/CCST.2012.6393548
17. Wong, K.W.: Image encryption using chaotic maps. Intell. Comput. Based Chaos **184**, 333–354 (2009). https://doi.org/10.1007/978-3-540-95972-4_16
18. Cheng, P., Yang, H., Wei, P., Zhang, W.: A fast image encryption algorithm based on chaotic map and lookup table. Nonlinear Dyn. **79**(3), 2121–2131 (2015).https://doi.org/10.1007/s11071-014-1798-y
19. Guesmi, R., Farah, M.A.B., Kachouri, A., Samet, M.: A novel chaos-based image encryption using DNA sequence operation and Secure Hash Algorithm SHA-2. Nonlinear Dyn. **83**(3), 1123–1136 (2015). https://doi.org/10.1007/s11071-015-2392-7

20. Barannik, V., Barannik, V.: Binomial-Polyadic binary data encoding by quantity of series of ones. In: 15th IEEE International Conference on Modern Problems of Radio Engineering, Telecommunications and Computer Science (TCSET 2020), pp. 775–780 (2020). https://doi.org/10.1109/TCSET49122.2020.235540
21. Kurihara, K., Watanabe, O., Kiya, H.: An encryption-then-compression system for JPEG XR standard. In: IEEE International Symposium on Broadband Multimedia Systems and Broadcasting (BMSB), pp. 1–5 (2016). https://doi.org/10.1109/BMSB.2016.7521997
22. Sharma, R., Bollavarapu, S.: Data security using compression and cryptography techniques. Int. J. Comput. Appl. **117**(14), 15–18 (2015). https://doi.org/10.5120/20621-3342
23. Zhou, J., Liu, X., Au, O.C., Tang, Y.Y.: Designing an efficient image encryption-then-compression system via prediction error clustering and random permutation. IEEE Trans. Inf. Forensics Secur. **9**(1), 39–50 (2014). https://doi.org/10.1109/TIFS.2013.2291625
24. Barannik, V.: The method of increasing accessibility of the dynamic video information resource. In: Barannik, V., Shulgin, S. (EDS.) 13th International Conference TCSET IEEE Modern Problems of Radio Engineering, Telecommunications and Computer Science, pp. 621–623 (2016)
25. Information technology – JPEG 2000 image coding system: Secure JPEG 2000, International Standard ISO/IEC 15444–8, ITU-T Recommendation T.807, 108 p. (2007)
26. Barannik, V.: Method coding efficiency segments for information technology processing video. In: Barannik, V., Tarasenko, D. (eds.) IEEE Problems of Infocommunications, Science and Technology: 4th International Scientific-Practical Conference PIC S&T, pp. 551–555 (2017)
27. Farajallah, M.: Chaos-based crypto and joint crypto-compression systems for images and videos (2015). https://hal.archives-ouvertes.fr/tel-01179610. Accessed 7 June 2021
28. Wong, K., Tanaka, K.: DCT based scalable scrambling method with reversible data hiding functionality. In: 4th International Symposium on Communications, Control and Signal Processing (ISCCSP), pp. 1–4 (2010). https://doi.org/10.1109/ISCCSP.2010.5463307
29. Yang, Y., Zhu, B.B., Li, S., Yu, N.: Efficient and syntax-compliant JPEG 2000 encryption preserving original fine granularity of scalability. EURASIP J. Inf. Secur. **2007**(1), 1–13 (2007). https://doi.org/10.1155/2007/56365
30. Watanabe, O., Uchida, A., Fukuhara, T., Kiya, H.: An encryption-then-compression system for JPEG 2000 standard. In: IEEE International Conference on Acoustics, Speech and Signal Processing (ICASSP), pp. 1226–1230 (2015). https://doi.org/10.1109/ICASSP.2015.7178165
31. Minemura, K., Moayed, Z., Wong, K., Qi, X., Tanaka, K.: JPEG image scrambling without expansion in bitstream size. In: 19 th IEEE International Conference on Image Processing, pp. 261–264 (2012). https://doi.org/10.1109/ICIP.2012.6466845
32. Phatak, A.: A non-format compliant scalable RSA-based JPEG encryption algorithm. International Journal of Image, Graphics and Signal Processing **8**(6), 64–71 (2016). https://doi.org/10.5815/ijigsp.2016.06.08
33. Kurihara, K, Watanabe, O., Kiya, H.: An encryption-then-compression system for JPEG XR standard. In: IEEE International Symposium on Broadband Multimedia Systems and Broadcasting (BMSB), pp. 1–5 (2016). https://doi.org/10.1109/BMSB.2016.7521997
34. Komolov, D., Zhurbynskyy, D., Kulitsa, O.: Selective method for hiding of video information resource in telecommunication systems based on encryption of energy-significant blocks of reference I-frame. In: 1st International Conference on Advanced Information and Communication Technologies (AICT 2015), pp. 80–83 (2015)
35. Wu, Y., Noonan, J.P., Agaian, S.: NPCR and UACI randomness tests for image encryption. Cyber J. Multi. J. Sci. Technol. J. Sel. Areas Telecommun. (JSAT) **2**, 31–38 (2011). https://doi.org/10.4236/jss.2015.33005

A Method of Scrambling for the System of Cryptocompression of Codograms Service Components

Vladimir Barannik[1](✉), Serhii Sidchenko[2], Dmitriy Barannik[3], Mykhailo Babenko[4], Volodymyr Manakov[3], Oleh Kulitsa[5], Vasyl Kryshtal[5], Vitalii Kolesnyk[6], and Oleh Kuzmin[7]

[1] V. N. Karazin, Kharkiv National University, Kharkiv, Ukraine
vvbar.off@gmail.com
[2] Ivan Kozhedub Kharkiv National Air Force University, Kharkiv, Ukraine
[3] Kharkov National University of Radio Electronics, Kharkov, Ukraine
[4] Dniprovsk State Technical University, Kamianske, Ukraine
[5] National University of Civil Defense of Ukraine, Kharkiv, Ukraine
[6] State Research Institute of Armament and Military Equipment Testing and Certification, Chernihiv, Ukraine
[7] Lviv Polytechnic National University, Lviv, Ukraine

Abstract. A method of scrambling for a service components system in crypto-compression codograms formed under the condition of discarding the pixel brightness values in RGB space least significant been developed. The difference between this method and the known ones is that, what before performing scrambling transformations, the service data merging presented in a reduced dynamic range into 8-bit combined elements is organized. At the permutation stage of the combined 8-bit data, not only the original 7-bit elements of the service components' values location changing organized, but also their values change too. This allows to increase known permutation transformations cryptographic characteristics. The developed method provides: availability of video data increasing by further reducing the image crypto-compression representation; cryptographic stability increasing by the elements of service data system values changing, breaking the correlation between the elements and changing the frequency of pixels. Scrambling transformations based on permutation tables applied to the service components in cryptocompression codograms system, ensure the stability of the image visual information to errors in codograms that occur in the communication channel. This is despite the fact that cryptocompression codograms are original images concise representation.

Keywords: Cryptocompression encryption · Information protection · Scrambling · Encryption · Permutation · Compression · Privacy · Image

1 Introduction

Crisis management systems need to ensure transmitted video data confidentiality while maintaining information specified quality and without availability reducing. However,

M. Klymash et al. (Eds.): TCSET 2022, LNEE 965, pp. 444–459, 2023.
https://doi.org/10.1007/978-3-031-24963-1_26

there is a problem due to the fact that the video data confidentiality can be organized either by the video data availability while maintaining a given quality, or by reducing useful information amount to maintain a given availability.

The aim of the article is to develop a scrambling method for the cryptocompression codograms service components system to ensure video data cryptographic stability while maintaining the information specified quality without reducing its availability.

The cryptocompression coding peculiarity is that the service components on the one hand determine the information component corresponding codes correct length, and on the other hand determine the source image elements number that formed these codes and allows to correctly decode their values. Service components that are changed based on scrambling or data encryption transformations in the crypto-compression decoding process cannot be correctly compared with the codes of information components and, accordingly, are erroneous for them. In addition, incorrect service components do not allow you to properly decode the image. Moreover, in the cryptocompression decoding process based on erroneous service components, errors accumulate. This is due to the fact that erroneous service cryptocompression codograms components when first stage service components intermediate values decoding also form erroneous values. They, in turn, incorrectly decode the reconstructed elements that do not match the original values.

The cryptocompression codograms service components cryptographic transformation organizing process will be considered in general, which depends on the presenting protected data method and their transformation without choosing a specific transformation algorithm organization. The cryptocompression codograms service components can be represented in the original dynamic range and in the reduced dynamic range due to the rejection of the least significant digit in the elements of the original images in the process of their cryptocompression coding. In the first case, when the service components are presented in the original dynamic range, to store the brightness value of one element is allocated 8 bits (1 byte). In the second case, to store the brightness value of one element in a reduced dynamic range requires $(8 - n_{LSB})$ bit, where n_{LSB} – the number of discarded lower bits in the image element. To ensure image reconstruction quality crypto-compression transformations minimal loss, proposed to discard one least significant digit of the original image elements. Therefore, to store the one element brightness value of the service components in the reduced dynamic range is allocated 7 bits.

2 Problem Statement

The most Existing approaches to image confidentiality analysis has shown that the most common cryptographic protection methods are based on data scrambling and encrypting. They are used in the following approaches:

– application of cryptographic transformations to different video data representations [1–5];

– based on the compression and cryptographic transformations phased implementation [6];

– based on steganographic methods [7–16].

– visual cryptography technologies based on image elements combining and scattering [17–21];

– schemes of uncompressed images elements scrambling [22–27];

– perceptual encryption schemes before performing compression transformation [28–31];

– application of cryptographic transformations at different stages of compression technologies [32–42];

Scrambling methods are less stable from the point of view of information protection. Most often, they are organized on the basis of permutation operations that change data location within the processed blocks or the whole image. However, permutations do not change data being processed. Organization permutations within the small dimensions source data blocks does not ensure the images confidentiality. Due to the video data perception psychovisual features, large objects in such scrambled images can be seen. To eliminate this shortcoming, permutations are used for data larger sizes blocks up to the size of images, as well as permutations of the processed blocks within the entire frame and/or between frames. Scrambling transformations based on permutation tables applied to open video data, increase the visual information of the image errors resistance in scrambled images fragments that occur as a result of communication channel errors [43]. However, scrambling transformations in modern compression technologies neutralize this beneficial effect. This is due to the fact that errors in compressed data are very significant.

Encryption methods provide more robust data protection. They change the data themselves without changing their location. A small exception to the location change position in block data processing is byte and/or bit permutations within the block. However, when processing data in encryption blocks or semi-blocks ranging in size from 32 to 256 bits, such permutations without using other cryptographic primitives are negligible because they involve only 4 to 32 bytes of information (8-bit processed data elements). Such permutations in encryption algorithms are designed to organize the avalanche effect when scattering and mixing data. Encryption methods are critical to errors in encrypted data. Even a single error in the block of encrypted data leads to its incorrect decryption. Existing approaches to ensuring the video data confidentiality are characterized by the next problematic shortcomings as:

– ensuring the video data confidentiality without compression technologies using leads to their significant availability reduction;

– ensuring the images confidentiality using compression technologies after and/or between data compression process stages actually based on the encryption and compression functionality separation. This also reduces the video data availability;

– there are no indeterminate characteristics in cryptographic transformations. This can affect the level of cryptocurrency especially when using scrambling transformations.

To eliminate these problematic shortcomings, approaches to crypto-compression image coding have been developed, which provide encryption and compression technologies combination [44–50]. Formed images cryptocompression codograms consist of informational and service components. The information component, which contains information about image elements original values, is an encrypted representation that cannot be properly decoded without service components presence. The service components contain information about the image identified structural characteristics. It is a key element for information component decoding. In the open form from the service component elements, you can reconstruct the shape of the reduced size image.

3 Main Part

Cryptocompression coding is organized for each individual image plane A by the dimension of elements $M \times N$, where M – number of lines in the image, and N – number of columns.

The conceptual method of floating cryptocompression coding of images in a differentiated basis without loss of information quality organizes a three-stage scheme for the formation of cryptocompression codograms [46, 47]. The first two stages provide simultaneous formation of code structures of information components that do not require additional confidentiality, and key elements as service components.

At the first stage of processing, the formation of the information component from the elements of the original image is organized with the formation of intermediate two-dimensional matrices of service components.

To form service components the plane A is divided into identical blocks $A^{(\gamma;\chi)}$, where γ – coordinate of the block in the image vertically, χ – coordinate horizontally. Blocks $A^{(\gamma;\chi)}$ have dimension of elements $m \times n$, where m- number of rows in the block, and n – number of columns. Blocks $A^{(\gamma;\chi)}$ are two-dimensional arrays of elements $a_{i,j}^{(\gamma,\chi)}$ where $\gamma = \overline{1, [\frac{M}{m}]}\ \chi = \overline{1, [\frac{N}{n}]}, i = \overline{1, m}, j = \overline{1, n}$.

For each block in the i-line $A^{(\gamma;\chi)} = \{a_{i,j}^{(\gamma,\chi)}\}$ direction, the maximum $\lambda_i^{(\gamma,\chi)}$ and minimum values $\mu_i^{(\gamma,\chi)}$ of the elements $a_{i,j}^{(\gamma,\chi)}$ are determined:

$$\lambda_i^{(\gamma,\chi)} = \max_{1 \le j \le n} (a_{i,j}^{(\gamma,\chi)}) \tag{1}$$

$$\mu_i^{(\gamma,\chi)} = \min_{1 \le j \le n} (a_{i,j}^{(\gamma,\chi)}) \tag{2}$$

At the first stage to form the information component E matrix of the image plane $A = \{a_{i,j}^{(\gamma,\chi)}\}$ is transformed into vector representation $A = \{a_\tau\}$, where $\tau = \overline{1, M \cdot N}$. Transformation of two-dimensional matrices of service components $\Lambda = \{\lambda_i^{(\gamma,\chi)}\}$ and $\Theta = \{\mu_i^{(\gamma,\chi)}\}$ into vector representation can be arranged:

– with an extension of the dimensionality of the resulting vectors to the size of the vector representation of the video data. It is organized by repeating once n each element $\lambda_i^{(\gamma,\chi)}$ of,$\mu_i^{(\gamma,\chi)}$. Then the vectors $\Lambda' = \{\lambda'_\tau\}$ and $\Theta' = \{\mu'_\tau\}$, $\tau = \overline{1, M \cdot N}$ are formed;

– without organising the dimension expansion of the service component vectors. Then the vectors $\Lambda = \{\lambda_{m \cdot [\frac{\tau-1}{m \cdot n}] + \tau - m \cdot [\frac{\tau-1}{m}]}\}$ and $\Theta = \{\mu_{m \cdot [\frac{\tau-1}{m \cdot n}] + \tau - m \cdot [\frac{\tau-1}{m}]}\}$, $\tau = \overline{1, M \cdot N}$ are formed.

The formation of the code value E_α for vector data representation is given by the following expressions [45, 46]:

$$E_\alpha = \sum_{\tau=\tau(0)_\alpha}^{\tau(0)_\alpha + \Psi_\alpha - 1} \left\langle (a_\tau - \mu'_\tau) \cdot W_\tau \right\rangle = \sum_{\tau=\tau(0)_\alpha}^{\tau(0)_\alpha + \Psi_\alpha - 1} \left\langle (a_\tau - \mu_{m \cdot [\frac{\tau-1}{m \cdot n}] + \tau - m \cdot [\frac{\tau-1}{m}]}) \cdot W_\tau \right\rangle, \tag{3}$$

$$W_\tau = \begin{cases} \prod_{\xi=\tau+1}^{\tau(0)_\alpha+\Psi_\alpha-1} (\lambda'_\xi + 1 - \mu'_\xi) = \prod_{\xi=\tau+1}^{\tau(0)_\alpha+\Psi_\alpha-1} (\lambda_{m\cdot[\frac{\xi-1}{m\cdot n}]+\xi-m\cdot[\frac{\xi-1}{m}]} + 1 - \mu_{m\cdot[\frac{\xi-1}{m\cdot n}]+\xi-m\cdot[\frac{\xi-1}{m}]}), \\ \qquad \to \tau < \tau(0)_\alpha + \Psi_\alpha - 1; \\ 1, \\ \qquad \to \tau = \tau(0)_\alpha + \Psi_\alpha - 1, \end{cases} \tag{4}$$

where $\tau \in [\tau(0)_\alpha; \tau(0)_\alpha + \Psi_\alpha - 1]$ and $\tau(0)_\alpha + \Psi_\alpha - 1 \leq M \cdot N$,

α – is the sequence number of codeword value of the E_α to be generated;

τ, ξ – are the linear vector coordinates that determine the position of the data to be processed during the encoding process;

$\tau(0)_\alpha$ – is the start coordinate of the element a_τ in vector form from which the codeword value to be generated starts;

Ψ_α – floating (non-deterministic) number of elements a_τ involved in forming the code value E_α of the information component, which depends on the values of the data being processed;

W_τ – the weighting factor for the τ-element a_τ, which is the product of the base elements λ'_ξ following it, taking into account the reduction of their dynamic ranges by μ'_ξ.

The code values E_α form the information component of the cryptocompression codogram $E = \{E_\alpha\}$.

In the second stage, additional cryptocompression encoding of matrices is organized based on the use of the technological core of the cryptocompression data conversion system described by the expressions (3) and (4) [47].

The third coding stage ensures the confidentiality of the service components of cryptocompression codegrams based on scrambling and encryption methods [49].

Consider the cryptocompression codograms service components transformations scrambling process organizing.

Service components rearrangements can be organized both without taking into account the change in their dynamic range, and taking into account this change.

In the first case, it is assumed that the data in the original dynamic range are subject to permutation, i.e. only a change in their location within the scrambling unit or within all service components. Thus, the original dynamic range service components permutation involves normal byte permutations organization. Reduced dynamic range service components permutation involves elements of length $(8 - n_{LSB})$ bit permutations organization. For service components with one rejected lower digit, a permutation of 7-bit elements is organized. This permutation organization provides a between service elements components correlations rupture (although the elements themselves do not change) and the gap of service components compliance with the information components codes.

The second option for transforms scrambling organizing involves combining service data presented in a reduced dynamic range with length $(8 - n_{LSB})$ bit each at $n_{LSB} > 0$ in 8-bit merged elements. This is due to the fact that, as a rule, data is processed and stored in 8-bit form. In fact, each service component element contains unfilled (receiving zero values) senior bits, the number of which is equal to n_{LSB}. Since to ensure image reconstruction quality minimal loss in cryptocompression transformations it was proposed to discard one least significant digit of original image elements, the organization of combining service data process in the reduced dynamic range into 8-bit data will be considered from the standpoint of 7-bit data. The combined 8-bit data is subsequently subjected to matrix permutations byte vector or within the scrambling unit or within all service components. Suppose you need to form a sequence of 8-bit data d_{i_d}, where i_d is the formed sequence serial number, for the organization of permutations over them, $i_d = \overline{1, i_{d_{\max}}}$, $i_{d_{\max}}$ is the generated 8-bit element maximum sequence number. This uses a sequence of 7-bit service components a_{i_a}, where i_a is the service components serial number elements, $i_a = \overline{1, i_{a_{\max}}}$, $i_{a_{\max}}$ is the maximum of service components element serial number. Maximum serial number $i_{a_{\max}}$ can be determined by the formula:

$$i_{a_{\max}} = \frac{M \cdot N}{m \cdot n} \tag{5}$$

The latter will generate an 8-bit element $d_{i_{d_{\max}}}$ with serial number $i_{d_{\max}}$, which is determined according to the maximum sequence number $i_{a_{\max}}$ value (defined by expression (5)) for service data a_{i_a} in the reduced dynamic range and the reduction of the dynamic range n_{LSB} level based on the formula:

$$i_{d_{\max}} = \frac{(8 - n_{LSB}) \cdot M \cdot N}{8 \cdot m \cdot n} = \frac{(8 - n_{LSB})}{8} \cdot i_{a_{\max}} \tag{6}$$

When combining 7-bit values of crypto-compression codograms service components, when one lower-order bit is rejected $n_{LSB} = 1$, expression (6) will take the next form:

$$i_{d_{\max}} = \frac{7 \cdot M \cdot N}{8 \cdot m \cdot n} = \frac{7}{8} \cdot i_{a_{\max}}. \tag{7}$$

This value will correspond to:

– the total number of generated 8-bit elements that are subject to further cryptographic transformation;

– the required table of permutation or the corresponding matrix form vector organization dimensions;

– the total service data amount in bytes for the processed color plane.

The last generated 8-bit element $d_{i_{d_{\max}}}$ must be completely filled, i.e. in accordance with expression (7) the condition must be met:

$$i_{d_{\max}} = [\frac{7}{8} \cdot i_{a_{\max}}] . \frac{7}{8} \cdot i_{a_{\max}} = [\frac{7}{8} \cdot i_{a_{\max}}] \tag{8}$$

If condition (8) is not met, the remaining bits are randomly filled with zero and/or single bit values.

Condition (8) is fulfilled at the minimum value $i_{a_{\max}} = 8$. In this case, the value $i_{d_{\max}}$ based on expression (7) will be equal to 7. Condition (8) will be satisfied for all values $i_{a_{\max}}$ multiples of eight. Accordingly, it can be concluded that every 7 consecutive 8-bit pooled data will be formed from 8 consecutive 7-bit values of service components.

Combining of cryptocompression codograms 7-bit a_{i_a} service components values scheme into 8-bit data for their further cryptographic transformation presented in Fig. 1

for the first seven merger options. In the future, they will be repeated cyclically. The variable is written in the data blocks on the diagram a_{i_a}, which participates in the association, and through a comma the number of its digits.

The combination of bit degits can be organized on division and multiplication of values by 2 mathematical operations basis or the using operations of bit of cyclic shift to the left *shl* and to the right *shr*. In this chapter framework we will use bit arithmetic.

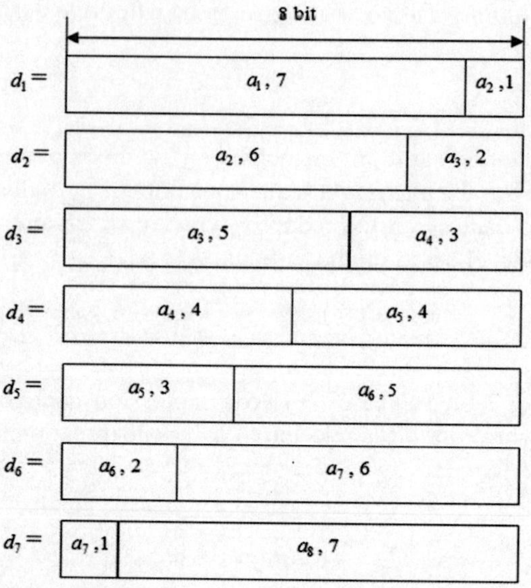

Fig. 1. The scheme of service components 7-bit values into 8-bit combined data combining.

Formation 8-bit merged data d_{i_d} based on cyclic shift to the left *shl* and to the right *shr* bit operations combined to the operation of service components 7-bit values bit addition modulo 2 for the combination presented in Fig. 1 first seven options, is organized using the expressions:

$$d_1 = shl_1 a_1 \oplus shr_6 a_2 \tag{9}$$

$$d_2 = shl_2 a_2 \oplus shr_5 a_3 \tag{10}$$

$$d_3 = shl_3 a_3 \oplus shr_4 a_4 \tag{11}$$

$$d_4 = shl_4 a_4 \oplus shr_3 a_5 \tag{12}$$

$$d_5 = shl_5 a_5 \oplus shr_2 a_6 \tag{13}$$

$$d_6 = shl_6 a_6 \oplus shr_1 a_7 \tag{14}$$

$$d_7 = shl_7 a_7 \oplus a_8 \tag{15}$$

Here $\oplus$ is a bit operation "excludes OR" (addition by module 2).

Cyclic shift to the left *shl* and to the right *shr* bit operations describes the next example. Numeric $204 = 11001100_2$ as a result of cyclic shift on 2 bits to the left $shl_2 204$ operations turns into meaning $48 = 00110000_2$. And as a result of cyclic shift on 2 bits to the right $shr_2 204$ operations will be a number $51 = 00110011_2$.

Expressions (9) - (15) for the based on bit operations 8-bituniteddata formation in the general case will take the form:

$$d_{i_d} = \begin{cases} shl_{i_d - [\frac{i_d}{7}] \cdot 7} a_{i_d + [\frac{i_d}{7}]} \oplus shr_{7 - \left(i_d - [\frac{i_d}{7}] \cdot 7\right)} a_{i_d + [\frac{i_d}{7}] + 1}, & \rightarrow \quad [\frac{i_d}{7}] \neq \frac{i_d}{7}; \\ shl_7 a_{i_d + [\frac{i_d}{7}] - 1} \oplus a_{i_d + [\frac{i_d}{7}]}, & \rightarrow \quad [\frac{i_d}{7}] = \frac{i_d}{7}. \end{cases} \tag{16}$$

From 7-bit service components a_{i_a} of 8-bit combined data d_{i_d} formation scheme analysis presented in Fig. 1 and described by expressions (9) - (16), it is seen that:

– each of combined data d_{i_d} byte for further cryptographic transformation consists of bits belonging to two 7-bit service components standing side by side, a_{i_a} and a_{i_a+1};

– in the only every first and every seventh 8-bit merged element d_{i_d} formation of all bit bits are involved, respectively, every first and every eighth 7-bit service components element a_{i_a}. In the all other 8-bit merged elements d_{i_d} formation (namely, from the second to the sixth) there are separate bit digits of 7-bit service component elements, standing side by side, a_{i_a} and a_{i_a+1};

– every 7 consecutive 8-bit combined data d_{i_d} formation used for further cryptographic transformation, with fully filled bit bits organized on 8 service component consecutive elements a_{i_a} basis, of which all bit values are used;

– the total length of 7 consecutive 8-bit combined data d_{i_d} with fully filled bit bits is 56 bits.

After uniting everyone of $i_{a_{max}}$ 7-bit service components a_{i_a} in 8-bit combined data d_{i_d}, the latter undergo cryptographic transformation based on vector or matrix permutation. As a permutation transformation result, the 8-bit merged data location changed d_{i_d}. And since they consist two service component a_{i_a} and a_{i_a+1} adjacent 7-bit elements bit values, this permutation actually leads to the 7-bit elements division into parts and has its own characteristics:

– when 8-bit combined data d_{i_d} every first and seventh element forming there are situations when all of the service components a_{i_a} 7-bit element bits fall into one 8-bit element d_{i_d} (this is every first and every eighth 7-bit element a_{i_a} service components when dividing them into groups of 8 elements). Such data permutation only leads to a change in their location;

– every service components from groups of 8 adjacent elements second – seventh 7-bit element a_{i_a} bits fall into two different adjacent 8-bit merged elements d_{i_d}. It is these 7-bit elements a_{i_a} service components by changing the location of 8-bit aggregated data d_{i_d} are separated and the coordinates of their individual bit subsequences are separated from each other. The service components individual bits of 7-bit elements a_{i_a} position within 8-bit aggregate data d_{i_d} does not change. Service components of 7-bit elements a_{i_a}

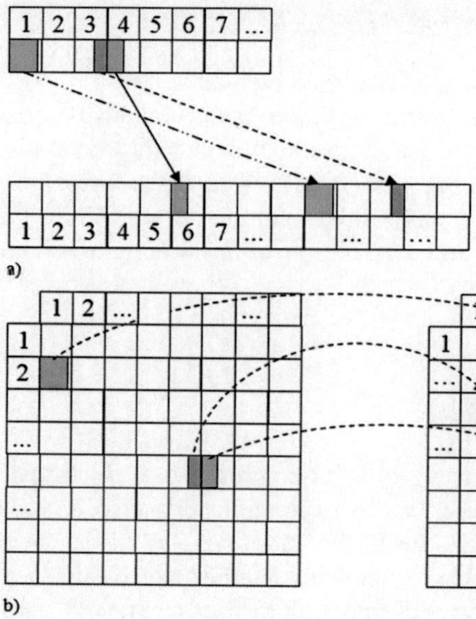

Fig. 2. Examples of the service components element separation by changing the combined data in the process of permutation location: a - permutation vector, b - matrix permutation.

separation by 8-bit aggregated data d_{i_d} in the process of vector and matrix permutation location changing examples are presented in Fig. 2.

From the point of cryptography view, this approach combines 7-bit elements into 8-bit data with the byte permutation subsequent organization performs cryptocompression codograms service components scattering and mixing functions.

At the 7-bit elements $a^{\bullet}_{i_a}$ service components reconstruction stage without the inverse scrambling transformation organization or its organization on the non-authentic permutation basis is not only incorrect recovery the 8-bit data $d^{\bullet}_{i_d} \neq d_{i_d}$ location, but the service components elements themselves can not be restored, i.e. $a^{\bullet}_{i_a} \neq a_{i_a}$.

This cryptocompression codograms service components scrambling option organizing based on the elements under the condition of combining 7-bit service data into 8-bit combined views permutation is more advantageous than using the option without combining data.

The win is as follows:

– provides scrambled service components cryptographic stability increasing by mixing and scattering data at the stage of combining them into 8-bit views with the subsequent organization of byte vector or matrix permutation.

In contrast to the standard approach, organizing the elements location change, provides an additional change in their values.

Moreover, when using the standard approach based on the elements in the original dynamic range permutations, the crypto-compression codograms crypto-stability determined only by the crypto-stability of the scrambling transformation used (not higher than this transformation). The second option, when the permutation is organized in conjunction with the 7-bit elements into 8-bit views combination, actually provides an increasing in cryptographic strength;

– for the scrambling transformations organization within the whole service components elements set requires the permutation tables of smaller dimensions formation.

This will increase the permutation table formation efficiency while maintaining their cryptographic stability (key parameters and permutation table formation algorithm remain unchanged), as well as increase the transformation permutation efficiency itself by processing less data.

All this also helps to increase the availability of video information.

The dimension of the permutation matrix for the case of its organization for elements in the original dynamic range exceeds the variant of its organization for 7-bit service components, combined into 8-bit data, by 12.5%.

Service components 7-bit elements $a^{\bullet}_{i_a}$ reconstruction from the combined 8-bit data $d^{\bullet}_{i_d}$ (which may or may not be authenticated) organized on the inverse transformations basis using bit arithmetic operations in the general case using the expression:

$$a^{\bullet}_{i_a} = \begin{cases} shl_1 d^{\bullet}_{[\frac{i_a}{8}]\cdot 7+1}, \\ \qquad \to [\frac{i_a}{8}] \neq \frac{i_a}{8}, i_a - [\frac{i_a}{8}]\cdot 8 = 1; \\ shl_{8-\left(i_a-[\frac{i_a}{8}]\cdot 8\right)} d^{\bullet}_{[\frac{i_a}{8}]\cdot 7+(i_a-[\frac{i_a}{7}]\cdot 8)-1} \oplus shr_{i_a-[\frac{i_a}{8}]\cdot 8} d^{\bullet}_{[\frac{i_a}{8}]\cdot 7+(i_a-[\frac{i_a}{7}]\cdot 8)}, \\ \qquad \to [\frac{i_a}{8}] \neq \frac{i_a}{8}, i_a - [\frac{i_a}{8}]\cdot 8 \neq 1; \\ shr_1\left(shl_1 d^{\bullet}_{[\frac{i_a}{8}]\cdot 7}\right), \\ \qquad \to [\frac{i_a}{8}] = \frac{i_a}{8}. \end{cases} \tag{17}$$

If at the 7-bit elements $a^{\bullet}_{i_a}$ service components reconstruction stage, an authenticated descramble transform scheme was organized (authenticated key was used) and combined 8-bit data $d^{\bullet}_{i_d}$ not subject to modification, i.e. $d^{\bullet}_{i_d} = d_{i_d}$, then the service components reconstructed values $a^{\bullet}_{i_a}$ will coincide with the initial a_{i_a}, i.e. equality will be fulfilled $a^{\bullet}_{i_a} = a_{i_a}$.

4 Experiments

Experimental studies were conducted to assess images in RGB space privacy in the cryptocompression encoding scheme quality by reducing the output values dynamic range by discarding one least significant digit with service components combined into 8-bit data subsequent scrambling. The evaluation was performed on the basis of comparison of the original images with reconstructed images recovered from cryptocompression codograms with scrambled service components. For example, the article presents the processing only a few test images of different saturation degrees results. The obtained results are typical for most of the processed video data.

The following approaches were used to assess quality:

– visual assessment of the original and corresponding scrambled images quality;

– quantify of image processing quality using the standard deviation RSME (mean squared error) indicators, the peak ratio of signal to noise PSNR (Peak Signal-to-Noise Ratio) and the correlation coefficient;

– Number of Pixel Changing Rate (NPCR), which is most often used to assess the image encryption quality;

– evaluation scattering and mixing pixels in scrambled data quality by constructing correlation histograms between elements in scrambled images and histograms of the elements in scrambled images distribution;

– assessment of the additional data compression possibility based on using ZIP and RAR archivers.

In the experiment, logistic mapping was used as the basic scrambling transformation [51]. The results of the quality evaluation of video data confidentiality ensuring based on proposed solutions are presented in Fig. 3, 4 and in Table 1.

Fig. 3. Examples of scrambled images visualization.

Table 1. The results of scrambled images quality evaluation.

Test image	Processing quality indicators			
	RSME	PSNR, dB	Correlation coefficient	NPCR,%
Baboon	88.94	9.15	–0.0030	99.5832
Lena	87.02	9.34	0.0760	99.5519
Airport area	82.20	9.83	0.0029	99.5518

The following conclusions can be drawn from the obtained results analysis:

– visual image quality assessment (Fig. 3) shows their complete destruction.Scrambled different images representations have become almost similar to each other. The dominant individual original images background completely destroyed;

– values of RSME, PSNR and correlation coefficient between scrambled and the original images (Table 1) fully confirm the video data complete destruction visual assessment data. For all image types, the RSME value is above 80, the PSNR is below 10 dB. The correlation coefficients values for most images are in the region of 0, although for some video data the correlation coefficient values can be observed in the region up to 0.1;

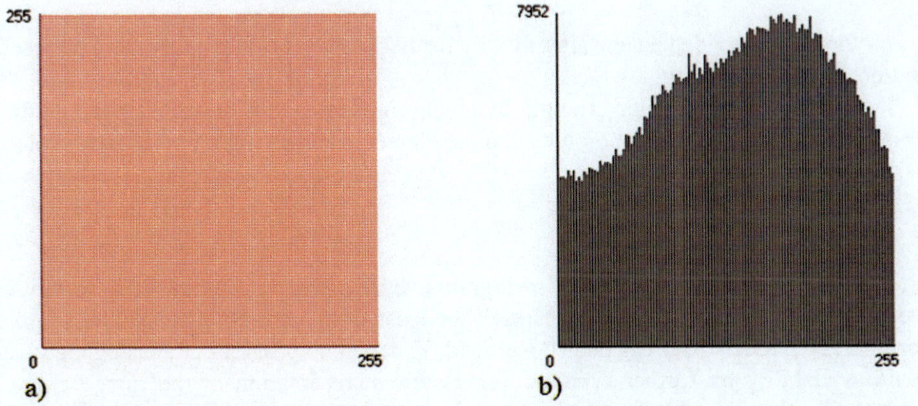

Fig. 4. Example of histograms for scrambled test images content assessing: a - correlation between elements; b - distribution of elements.

– the number of variable pixels, NPCR (Table 1) for all images is above the theoretical threshold of 99.5341% [52], which indicates a encrypted data high resistance to differential attacks;

– correlation histograms between elements for different images coincide regardless of their content (Fig. 4, a). A square image was formed on the histograms, which significantly changed the original video data histograms. Due to the fact that the data are reconstructed in a reduced range due to the lower digit rejection, as in the original video data in the histogram in Fig. 4, squared pixels are arranged one by one in the vertical and horizontal directions. The histogram in the form of a square indicates a completely destroyed correlation between adjacent elements;

– elements distribution histograms for different images relative to the original video data have changed greatly, there has been number of elements significant alignment (Fig. 4, b). Due to the fact that the data are processed to reduce their dynamic range, the samples histograms are arranged one by one. In fact, there are only elements with even values. The histograms show the element's number changes and there are no options when there is no paired brightness value or the number of these values is small. Data change values only every six 7-bit service components values out of eight, while two values (first and eighth) remain unchanged, on the other hand.

Additional compression of video data cryptocompression codograms evaluation using ZIP and RAR archivers showed that the codograms size is not further reduced. For the most part, on the contrary, there is a slight increase in their volume due to the formation of ZIP and RAR archivers of their service information at the archive beginning. This indicates the redundancy in the generated cryptocompression codograms absence and the correlations between elements elimination.

The rejection of one minor sign, which is organized in the crypto-compression transformation scheme without information quality loss, on the one hand, reduces the reconstructed data quality. Although, this reduction in quality is insignificant. For all images, the value of RSME is 0.71, PSNR is above 51 dB, and the correlation coefficient is 0.9999. On the other hand, it provides:

– increase the video data availability by further image crypto-compression representation amount reducing;

– increase cryptographic strength by changing elements of the service data system values, breaking the correlation between the elements and changing pixels frequency.

5 Conclusions

A method of crypto-compression codograms service components system scrambling formed under the condition of discarding the least significant bit in the values of pixel brightness in RGB space has been developed. The method provides increasing the video data availability by further crypto-compression representation of the image amount reducing; increasing cryptographic strength by changing the service data system elements values, breaking the correlation between the elements and changing the frequency of pixels.

The scientific novelty. A method of crypto-compression codograms service components system scrambling formed under the condition of discarding the least significant bit in the values of pixel brightness in RGB space has been developed. The difference between this method and the known ones is that, before performing scrambling transformations, the merging of service data presented in a reduced dynamic range into 8-bit combined elements is organized. At the stage of permutation of the combined 8-bit data, not only the change of the location of the values of the original 7-bit elements of the service components is organized, but also the change of their values. This allows to increase the cryptographic characteristics of known permutation transformations.

The practical significance. Created a scrambling the system of service components in crypto-compression codograms software implementation method, which provides:

– protected cryptocompression code construction formation with scrambled service components. Reconstructed images by unauthenticated users are completely destroyed and all scrambled images are not visually different from each other;

– for all scrambled images provides a significant reduction in their quality compared to the original video data. The quality scores for such images are as follows: RSME is above 80, PSNR is below 10 dB. The most images correlation coefficients are in the region of 0, although for some video data correlation coefficient values can be observed in the region of up to 0.1. The number of pixels that change, NPCR for all images is above the theoretical threshold of 99.5341%..

Prospects for further research to create consistent methods of adaptive scrambling of video information, optimization of their software implementations, as well as experimental study of the proposed approaches on a larger set of practical problems of different nature and dimension.

References

1. Schneier, B.: Applied Cryptography: Protocols, Algorithms, and Source Code in C / B, p. 784. Wiley, Schneier, New York (2015)
2. Announcing the ADVANCED ENCRYPTION STANDARD (AES). Federal Information Processing Standards Publication, 197 (2001)

3. DSTU 7624: 2014 Information Technology. Cryptographic protection of information. Symmetric block transformation algorithm. Corrections. — [Valid from 01.03.2016]. — Official publ., Kyiv : UkrREC, 228 p. (2016)
4. DSTUGOST 28147: 2009 Information Processing System. Cryptographic protection. Algorithm of cryptographic transformation. [Not valid from 01.01.2022]. Official publ. Kyiv : UkrREC, 228 p. (2016)
5. Rivest, R., Shamir, A., Adleman, L.: A method for obtaining digital signatures and public-key cryptosystems. Commun. ACM **2**(21) 120–126 (1978)
6. Sharma, R., Bollavarapu, S.: Data Security using compression and cryptography techniques. Int. J. Comput. Appli. **117**(14), 15–18 (2015). https://doi.org/10.5120/20621-3342
7. Barannik, D.: Stegano-compression coding in a non-equalible positional base. In: IEEE Advanced Trends in Information Theory : 2nd International Conference ATIT 2020, pp. 83–86. https://doi.org/10.1109/ATIT50783.2020.9349328
8. Barannik, V., Barannik, D., Fustii, V., Parkhomenko, M.L.: Evaluation of effectiveness of masking methods of aerial photographs. In: IEEE Advanced Information and Communications Technologies: 3rd International Conference AICT, pp. 415–418 (2019)
9. Barannik, V., Barannik, N., Ryabukha, Y., Barannik, D.: Indirect steganographic embedding method based on modifications of the basis of the polyadic system. In: Modern Problems of Radio Engineering, Telecommunications and Computer Science: 15th IEEE International Conference TCSET, pp. 699–702 (2020)
10. Barannik, V.: Binomial-polyadic binary data encoding by quantity of series of ones. In: IEEE Modern Problems of Radio Engineering, Telecommunications and Computer Science: 15th IEEE International Conference TCSET 2020, pp. 775–780 (2020)
11. Vladimir, B., Ryabukha, Y., Tverdokhleb, V., Dmitr, B.: Methodological basis for constructing a method for compressing transformers bit representation, based on non-equilibrium positional encoding. In: Advanced Information and Communication Technologies: 2nd International Conference AICT 2017, pp.188–192 (2017)
12. Barannik V., Krasnoruckiy. A., Khakhanova, A.: The positional structural-weight coding of the binary view of transformants. In: East-West Design & Test Symposium : International Conference, EWDTS 2013, p. 4 (2013)
13. Barannik, V., Ryabukha, Y., Kulitsa, O.: The method for improving security of the remote video information resource on the basis of intellectual processing of video frames in the telecommunication systems. Telecommun. Radio Eng. **76**(9), 785–797 (2017)
14. Barannik, V., Barannik, V., Havrylov, D., Sorokun, A.: Development second and third phase of the selective frame processing method. In: Advanced Information and Communications Technologies : 3rd International Conference AICT, Lviv-slavske 2019, pp. 54–57 (2019)
15. Barannik, V., Shulgin , S.: The method of increasing accessibility of the dynamic video information resource. In: IEEE Modern Problems of Radio Engineering, Telecommunications and Computer Science: 13th International Conference TCSET, 2016, pp. 621–623 (2016)
16. Barannik, V., Tarasenko, D.: Method coding efficiency segments for information technology processing video. In: IEEE Problems of Infocommunications, Science and Technology: 4th International Scientific-Practical Conference PIC S&T 2017, pp. 551–555 (2017)
17. Chen, C., Wu, W.: A secure Boolean-based multi-secret image sharing scheme. J. Syst. Softw. **92**, 107–114 (2014)
18. Chen, C., Wu, W.: Efficient multi-secret image sharing based on Boolean operation. Signal Process. **91**(1), 90–97 (2011)
19. Deshmukh, M., Nain, N., Ahmed, M.: Multi secret image sharing scheme using boolean XOR and modular arithmetic. In: IEEE Advanced Information Networking and Applications: 30th International Conference, AINA 2016, pp. 690–697 (2016)
20. Naor, M., Shamir, A.: Visual cryptography. In: Advances in Cryptology EUROCRYPT 1994. LNCS, vol. 950, pp. 1–12 (1995). https://doi.org/10.1007/bfb0053419

21. Yang, Ch.-N., Cai, S.-R.: Enhanced Boolean-based multi secret image sharing scheme. J. Syst. Softw. **116**, 22–34 (2016)
22. Barannik, V., Babenko, Y., Kulitsa, O., et al.: Significant microsegment transformants encoding method to increase the availability of video information resource. IIn: EEE Advanced Trends in Information Theory: 2nd International Conference ATIT 2020, pp. 52–56 (2020)
23. Ramakrishnan, S.: Cryptographic and Information Security Approaches for Images and Videos, p. 962. CRC Press, Ramakrishna (2018)
24. Tsai, Ch.-L., Chen, Ch.-J., Hsu, W.-L.: Multi-morphological image data hiding based on the application of Rubik's cubic algorithm. IEEE Int. Carnahan Conf. Sec. Technol. ICCST 2012, pp. 135–139 (2012). https://doi.org/10.1109/CCST.2012.6393548
25. Wong K.-W. Image encryption using chaotic maps / K.-W. Wong // Intelligent Computing Based on Chaos, 2009, vol. 184, pp. 333–354
26. Yu, W., Agaian, S., Noonan , J.: Sudoku Associated Two Dimensional Bijections for Image Scrambling. Available via arXiv preprint. arXiv: 1207.5856v1
27. Zhou, Y., Panetta., K., Agaian, S.: Image encryption using P-Fibonacci transformation and decomposition. CPL. Chen. Optics Commun. **285**(5), 594–608 (2012)
28. Kurihara K., Shiota, S., Kiya , H.: An encryption-then-compression system for JPEG standard. PIn: icture Coding Symposium, PCS 2015, pp. 119–123 (2015)
29. Kurihara, K., Watanabe, O., Kiya, H.: An encryption-then-compression system for JPEG XR standard. In: IEEE Broadband Multimedia Systems and Broadcasting: International Symposium, BMSB 2016, pp. 1–5 (2016)
30. Watanabe, O., Uchida, A., Fukuhara, T., Kiya, H: Encryption-then-Compression, A., system for: standard. In: IEEE Acoustics, Speech and Signal Processing: International Conference ICASSP 2015, pp. 1226–1230 (2000)
31. Zhou, J., Liu, X., Au, O.C., Tang, Y.Y.: Designing an efficient image encryption-then-compression system via prediction error clustering and random permutation. IEEE Trans. Inf. Forens. Sec. **9**(1), 39–50 (2021)
32. Auer, S., Bliem, A., Engel, D., et al.: Bitstream-based JPEG encryption in real-time. Int. J. Digital Crime Foren (2013). https://doi.org/10.4018/jdcf.2013070101
33. Dufaux, F., Ebrahimi, T.: Toward a Secure JPEG Applications of Digital Image Processing XXIX, vol. 6312 (2006)
34. Honda, T., Murakami, Y., Yanagihara, Y., et al.: Hierarchical image-scrambling method with scramble-level controllability for privacy protection. In: IEEE Midwest Symposium on Circuits and Systems : 56th International Symposium MWSCAS 2013, pp. 1371–1374 (2013). https://doi.org/10.1109/MWSCAS.2013.6674911
35. Information technology – JPEG 2000 image coding system: Secure JPEG 2000. International Standard ISO/IEC 15444–8; ITU-T Recommendation T.807 2007, 108 p. (2007)
36. Ji, S., Tong, X., Zhang, Z.: Image encryption schemes for JPEG and GIF formats based on 3D baker with compound chaotic sequence generator. Available via arXiv preprint. arXiv: 1208.0999. (2012)
37. Kobayashi, H., Kiya , H.: Bitstream-Based JPEG Image Encryption with File-Size Preserving. In: IEEE Consumer Electronics : 7th Global Conference GCCE, 2018, pp. 1–4 (2018) https://doi.org/10.1109/gcce.2018.8574605
38. JPEG image scrambling without expansion in bitstream size / [K. Minemura, Z. Moayed, K. Wong and all] // Image Processing : 19th IEEE International Conference, 2012, pp. 261–264 DOI: https://doi.org/10.1109/ICIP.2012.6466845
39. Phatak, A.: A non-format compliant scalable RSA-based JPEG encryption algorithm. Int. J. Image, Graph. Signal Process. **8**(6), 64–71 (2016). https://doi.org/10.5815/ijigsp.2016.06.08
40. Wong, K., Tanaka, K.: DCT based scalable scrambling method with reversible data hiding functionality. In: Communications, Control and Signal Processing: 4th International Symposium, ISCCSP 2010, 4 p. (2010)

41. Yang, Y., Zhu, B., Li, S., Yu, N.: Efficient and syntax-compliant JPEG 2000 encryption preserving original fine granularity of scalability. EURASIP J. Inf. Sec. **2007**, 13 (2007)
42. Yuan, L., Korshunov, L., Ebrahimi, T.: Secure JPEG Scrambling enabling Privacy in Photo Sharing. In: IEEE Automatic Face and Gesture Recognition: 11th International Conference and Workshops, FG 2015, 6 p. (2015)
43. Gschwandtner, M., Uhl, A., Wild , P.: Transmission error and compression robustness of 2D chaotic map image encryption schemes. EURASIP J. Inf. Sec. 16 (2007)
44. Barannik, V.: Technology for protecting video information resources in the info-communication. In: Proceedings of IEEE 2nd International Conference 2020, pp. 29–33 (2020). https://doi.org/10.1109/ATIT50783.2020.9349324
45. Alimpiev, A., Barannik, V., Sidchenko, S.: The method of cryptocompression presentation of videoinformation resources in a generalized structurally positioned space. Telecommun. Radio Eng. **76**(4), 521–534 (2017). https://doi.org/10.1615/TelecomRadEng.v76.i6.60
46. Barannik, V., Sidchenko, S., Barannik, N., Barannik, V.: Development of the method for encoding service data in cryptocompression image representation systems. Eastern-Euro. J. Enterprise Technol. **3**(9)(111), 103–115 (2021). https://doi.org/10.15587/1729-4061.2021.235521
47. Barannik, V., Sidchenko, S., Barannik, D., Shulgin, S., Barannik, V., Datsun, A.: Devising a conceptual method for generating cryptocompression codograms of images without loss of information quality. Eastern-Europ. J. Enterprise Technol. **4**(2)(112), 6–17 (2021). https://doi.org/10.15587/1729-4061.2021.237359
48. Barannik, V., Sidchenko, S., Barannik, N., Barannik, D., Shulgin, S.: Methods for decoding informational codes of cryptocompression codegrams to improve information security In: EUR Workshop Proceedings (CEUR-WS.org), Aachen, Germany, 07 August 2021, vol. 2923, pp. 143–152 (2021). CEUR-WS.org/Vol-2923/paper16.pdf
49. Barannik, V., Sidchenko, S., Barannik, N., Khimenko, A.: The method of masking overhead compaction in video compression systems. Radioelectronic Comput. Syst. **2**, 51–63 (2021). https://doi.org/10.32620/reks.2021.2.05
50. Barannik, V., Sidchenko, S., Barannik, D., Barannik, V., Hurzhii, I., Babenko, Y.: Evaluating of the resistance of crypto-compression image, codograms, to errors in the data transmission channel. Adv. Trends Inf. Theory (ATIT: Proceedings of IEEE 3rd Internernational Conference on Kyiv 2021, pp. 52–56 (2021). https://doi.org/10.1109/ATIT54053.2021.9678774
51. May, R.: Simple mathematical models with very complicated dynamics. Nature **261**(5560), 459–467 (1976)
52. Wu, Y., Noonan, J.P., Agaian, S.: NPCR and UACI Randomness Tests for Image Encryption. Cyber J. Multidis. J. Sci. Technol. J. Selected Areas Telecommun. (JSAT), 31–38 (2011)

Smart Ergonomic Lighting for Circadian Rhythm

Yehor Zheliazkov[1,2](✉) and Julia Jamnenko[1,3]

[1] Igor Sikorsky Kyiv Polytechnic Institute, Kyiv, Ukraine
yzheliazkov2@gmail.com
[2] Technical University of Dresden, Dresden, Germany
[3] Technical University of Munich, Munich, Germany
i.yamnenko@tum.de

Abstract. Comfortable level is an important element that has considerable implications on circadian energy and the perception of lit environments. Light has been defined only in terms of the human visual system, not the circadian system. Light source and systems development should consider the needs of both the visual and non-visual systems. Discussed are the lighting characteristics impacting these two systems and the implications for designing light for various healthcare and medical applications. Do not knowing comfort level it is impossible to build the certain determination system and to measure human circadian rhythm.

Keywords: Circadian lighting · Ergonomic lighting · Visual comfort · Intelligent lighting system · Unified Glare Rating

1 Introduction

People can often hear different phrases such as "Do not bend your back!", "Do not cross one leg over the other", "Hold your PC for a certain distance from your eyes" and so on. There are also a lot of different aspects, which people need to take into account in the workplace. There is also certain knowledge such as "Ergonomic".

Ergonomics (or "human factors" as it is referred to in North America) is a branch of science that aims to learn about human abilities and limitations, and then apply this learning to improve people's interaction with products, systems and environments [1].

Ergonomics is essential for a modern working place. Most of the companies and employees today are more educated on office well-being, a key component of an ergonomic workplace, which is also often forgotten, is lighting.

Ergonomic lighting is crucial for workplace well-being, and it plays an important role in preventing Computer Vision Syndrome (CVS). According to a 2014 survey, 60% of adults between the ages of 20 and 50 have experienced CVS symptoms. These symptoms include blurred vision, headaches, dry eyes and some another illnesses.

Improper lighting, whether it's too little or too much lighting, contributes significantly to digital eye strain. Appropriate lighting, without shadows or glare, can reduce eye fatigue and headaches. Thus, if you often experience dry eyes, blurred vision, or

M. Klymash et al. (Eds.): TCSET 2022, LNEE 965, pp. 460–494, 2023.
https://doi.org/10.1007/978-3-031-24963-1_27

even headaches after working on your computer, optimizing the lighting ergonomics in your office or at your desk can help ease these symptoms [2].

Lighting systems allow to adjust the individual characteristics and human healthy demands. Particularly, during adjusting the lighting parameters, it is possible to create a positive effect on a human circadian rhythms. The violation of these rhythms is an important problem during the durable operation of computers or laptops [3].

In this article ergonomic lighting will be describe through the connection between ergonomics and circadian lighting. Circadian lighting means lighting that coincides with our circadian system for the entire day. Outdoor lighting changes color throughout the day from a warm light color to a cool daylight color and again to a warm color [4]. It is important to pay attention that circadian lighting isn't limited to light bulbs and fixtures, it includes electronic devices as well.

2 Motivation and Demarcation

Nowadays, advanced wireless and mobile protocols, such as 5G, LTE Cat-M, NB-IoT, IEEE 802.11p/pWLAN/pWi-Fi, Bluetooth, ZigBee, 6LoWPAN, LoRA WAN, LiFi etc. are used for data transfer between smart things in IoT (Internet of Things), Smart Home and Smart Office, self-driving cars and cloud analytics to support all of them.

Emerging Networking (EmN) appears nowadays spontaneously and accompanies us in our everyday life as well as supports modern industrial processes, ongoing digitalization of workflows, and multiple Smart Office and Smart Ergonomics scenarios.

Embedded Systems integration, modern electronic devices, and IoT deployment require an interdisciplinary approaches as multiple technologies come together to the networked control algorithms, automation routines, and monitored workflows.

EmN can support so-called Networked Ergonomics based on appropriate lighting, HVAC monitoring (Heating, Ventilation, Air-Conditioning), aimed at the optimization of working environments and office communication, i.e. aimed to further Digitalization.

Smart Lightning is an important issue for networking environments are increasing used in our everyday, office, industry, education and entertainment.

Comfortable level is an essential element that has considerable implications on circadian energy and the perception of lit environments. The networked distributed applications, Web apps and Wearable Data Analysis Platform are widely used to support optimal and health-aware lighting for circadian rhythm.

3 Theoretical Part of Circadian and Comfort Lighting

3.1 Ergonomic Lighting

Lighting ergonomics means the relationship between an individual light acceptation and the source of lighting. It has a huge impact on productivity, workplace well-being and preventing CVS.

There are lot of offices that have overhead lighting fixtures as the primary source of light; this causes glare or shadows. The situation becomes even worse if there is insufficient illumination or flickering. During the working in an office with poor lighting for extended periods can be reason of improper contrast (low light from the room and bright light from the computer screen), low level of lighting distribution, and wrong colour temperature, which causes eye strain and fatigue [2].

As were mentioned above, lighting ergonomics is an important factor for office designing. Poor lighting can be caused by low level of productivity, high employee error rates, eye fatigue, low focusing, inability to match and select correct colors, headaches, low worker morale, general sickness and another.

In Fig. 1 there is one of the examples how do people need to act during the working process.

Fig. 1. Correct sitting position during the work with a PC.

During the working process in a workstation with low level of lighting can significantly affect your performance. Plus, the wrong color temperature can affect your mood and emotions, affecting your bottom line. Therefore, people need to improve the lighting ergonomics in their office to maintain a healthy work environment, as well as to boost their productivity.

Common and Recommended Light Levels Indoors. The outdoor light level is approximately 10,000 lx on a clear day. In the building, in the area closest to windows, the light level may be reduced to approximately 1,000 lx [5]. In the middle area its may be as low as 25–50 lx. Additional lighting equipment is often necessary to compensate the low levels.

Earlier it was common with light levels in the range 100 - 300 lx for normal activities. Today the light level is more common in the range 500–1000 lx - depending on activity. For precision and detailed works, the light level may even approach 1500–2000 lx.

The Table 1 below is a guide for recommended light level in different workspaces [6].

Table 1. Light level in different workspaces.

Workspace	Brightness (cd/m^2)	Illumination (lux, lumen/m^2)
Corridors	350–500	50–250
Meeting room	350–800	150–2500
Board room	400–1000	200–700
Reception area	500–1600	250–1000
Office desk area	350–800	200–500
Artium	700–2500	400–5000

Unified Glare Rating in a Room. The following list provides a brief overview of several methods to quantify disability glare.

The method of UGR is the most useful in conjunction with lighting software, and the GR and luminance measures are more practical for measurement in a field application. Unified Glare Rating – this model was developed by the international commission on illumination, CIE, for applications of interior lighting [CIE 117-1995 - Discomfort Glare in Interior Lighting] [7]. UGR is based on a measurement of the luminance of a fixture for a specified direction of line of sight.

The Unified Glare Rating calculations are calculated in accordance with the CIE document CIE 117-1995 - Discomfort Glare in Interior Lighting. The shielding effect of the observer's eyebrows and forehead are considered in this calculation. The CIE definition of Discomfort Glare is "glare that causes discomfort without necessarily impairing the vision of objects". Glare which impairs the vision of objects is termed Disability Glare and is not covered in this metric.

UGR values generally range from 10 to 30 where a high value indicates significant discomfort glare, and a low value indicates little discomfort glare. Electric lighting systems producing UGR values of 10 or less are assumed to produce no discomfort [8, 9].

One UGR unit represents the least detectable step in discomfort glare evaluation, and three UGR units represent an acceptability step in glare criteria. Average UGR value range from 10-13-16-19-22-25-28.

The relationship between calculated UGR value and Hopkinson's discomfort glare criteria is as soon as in Table 2.

Table 2. Hopkinson's discomfort glare criteria.

UGR	10	13	16	19
Discomfort Glare Criterion	Imperceptible	Just perceptible	Perceptible	Just acceptable
Illuminance, lx	300	460	708	917
UGR	22	25	28	
Discomfort Glare Criterion	Unacceptable	Just uncomfortable	Uncomfortable	
Illuminance, lx	1678	2585	3980	

Analyzing this diagram, you can see that at low color temperatures, low light is required for objects in the room, and at high color temperatures, to make a person feel comfortable, the light should be increased by 5–200 times. Since the illuminance is proportional to the magnitude of the luminous flux, at high color temperatures it is necessary to provide a luminous flux much higher than for low color temperatures.

The quality of lighting with the use of LED lighting fixtures is significantly influenced by their high brightness, so the urgent task of this work is to study the uncomfortable and dazzling brightness of these fixtures using modern CIE recommendations.

The ability to predict the degree of discomfort from the brightness of light sources is useful for users, as it is possible to avoid. Manufacturers will be able to develop products and classify the suitability of their products for different applications.

UGR is determined by the ratio of the brightness of the light sources and the brightness of the background:

$$UGR = 8\lg\left(\frac{0.25}{L_B}\sum_{i=1}^{N}\frac{L_i^2 \cdot \omega_i}{p_i^2}\right), \tag{1}$$

where L_B is the background brightness (cd/m^2); L_i is the brightness of the luminous parts of each luminaire in the direction of the observer's eye (cd/m^2), which can be also defined as $L_i = \pi \cdot E_{ind}$, E_{ind} is the vertical indirect illuminance at the observer's eye $L_B = E_{ind}/\pi$; ω_i is the solid angle of the luminous parts of each luminaire at the observer's eye (sr); p_i is the Guth position index for each luminaire (displacement from the line of sight). The Guth position index is based on two angles: α = angle from vertical of the plane containing the source and the line of sight in degrees and β = angle between the line of sight and the line from the observer to the source. Schematic image was represented in Fig. 2.

The Guth position index is expressed as:

$$p_i = \exp\left[(35.2 - 0.31889\alpha - 1.22e^{-\frac{2\alpha}{9}})10^{-3}\beta + (21 + 0.26667\alpha - 0.0029663\alpha^2)10^{-5}\beta^2\right]. \tag{2}$$

According to the information about climate chamber, construction parameters of the walls, floor, ceiling and lamps position are constant. This means, that Guth position index is also constant [10].

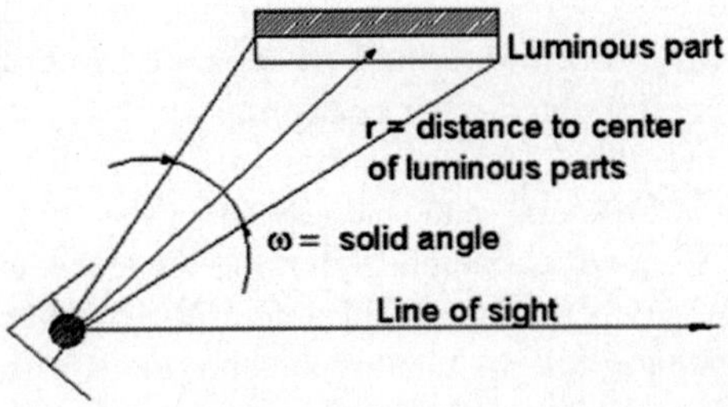

Fig. 2. Representation to acception of Unified Glare Ratings.

To provide a pleasing visual experience and constant high performance of the employees, there are some standards, which should be satisfied.

The artificial lighting installation needs to fulfil the requirements according to DS/EN 12464–1 [11] for workplaces in Denmark. Some of the requirements for healthcare spaces relevant for this project are presented in Table 3. The values for the rooms for general use, wards and examination rooms are the most important for the case studies described in this report.

Table 3. Light requirements for chosen types of health care premises – wards, maternity wards.

Type of area, task or activity	Em	UGR_L	U_0	R_a
General lighting (illuminance at floor level)	100	19	0.40	80
Reading lighting	300	19	0.70	80
Simple examinations	300	19	0.60	80
Examinations and treatment	1000	19	0.70	90
Night light, observation lighting	5	-	-	80
Bathrooms and toilets for patients	200	22	0.40	80

In DS 12464-1 [11], the recommended illuminance is given by the maintained illuminance (E_m) which considers possible light loss from the luminaires over time. In the table, the maximum value for The Unified Glare Rating limit (UGR_L), the minimum illuminance uniformity (U_0) and the minimum value for color rendering index (R_a) is shown as well.

For evaluating potential glare issues in the patient rooms, the Daylight Glare Probability (DGP) is used to measure glare under daylight conditions, and the Unified Glare Ratio (UGR) is used to assess glare under the electric lighting conditions.

3.2 Circadian Rhythm and Circadian Lighting

"Circadian rhythms" [12] mean changes in physiological, biochemical and behavioral functions of the body. These rhythms are synchronized with the natural light dynamics of the surrounding world ("day and night") by means of internal molecular "hours", surround temperature, humidity and other conditions.

Circadian rhythms are also 24-h cycles that are part of the body's internal clock, running in the background to carry out essential functions and processes. One of the most important and well-known circadian rhythms is the sleep-wake cycle.

Human organism can adopt to the certain time of the day and the different systems of the body follow circadian rhythms that are synchronized with a master clock in the brain. This master clock is directly influenced by environmental cues, especially light, therefore circadian rhythms are tied to the cycle of day and night.

As properly aligned, a circadian rhythm can promote consistent and restorative sleep. But if this circadian rhythm is thrown off, it can create significant sleeping problems for the organism, including insomnia. Earlier researches are also revealing that circadian rhythms play an integral role in diverse aspects of physical and mental health.

Certain list of circadian rhythm with a certain time and characteristic of human organism was represented in Fig. 3. This shows certain period of time, when people have highest or lowest either blood pressure, and body temperature; time of deepest sleep and so on.

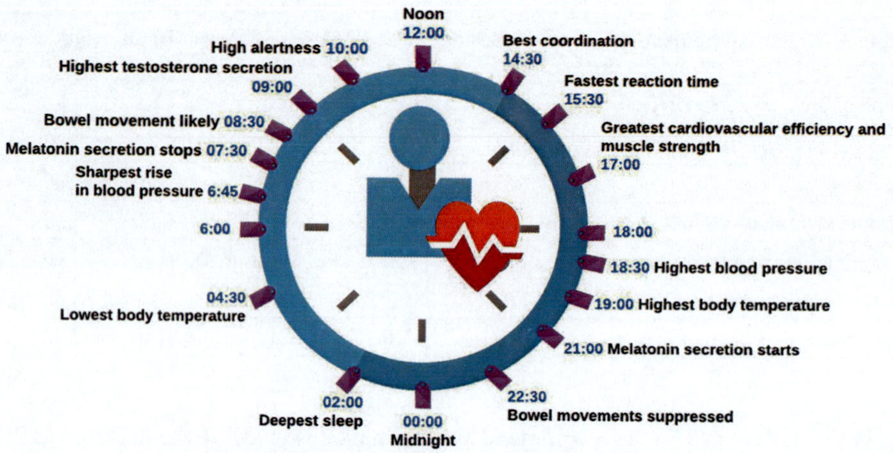

Fig. 3. Circadian rhythm during the different period of day.

The terms circadian and integrative lighting are now often used interchangeably [13]. Throughout this report, we have chosen to use the term circadian lighting, as the project was funded for studying circadian lighting, lighting that is intended to support the natural sleep and wake cycle of humans that is normally entrained by the presence and absence of naturally varying daylight.

Research has indicated that light affects on the human healthy both of non-visual and our visual systems and that electric light can make impact on circadian rhythm [14]. Circadian lighting is the concept that electric light can be used to support human healthy by minimizing the effect of electric light on the human circadian rhythm. Research workers have discovered that long-term exposure to certain wavelengths of blue light at a specific intensity can have either positive or negative impact on melatonin production. The conception of using lighting to influence human circadian rhythm is a relatively new one in the lighting industry: all aspects of this exciting new topic are still being studied for a long time, and the current data is still in its infancy. Nowadays, there are three main electric light approaches to implementing a circadian lighting system: intensity tuning and stimulus tuning. Certain representation demonstrates Fig. 4.

Fig. 4. Perception of lighting of different color temperature.

Intensity tuning is the most familiar and cost-effective solution to circadian lighting. Fixtures maintain a fixed correlated color temperature (CCT) while the intensity (brightness) of the fixture is adjusted, through a controlled dimming system, to correlate with time of day. Light fixtures are set to a lower intensity in the early morning, transition to a higher intensity as the day progresses, and reduce to a lower intensity in the evening.

Color tuning involves changing the light intensity and CCT to mimic the daytime/nighttime cycle. We experience cooler color temperatures (ranging from 4000 K up to about 10,000 K) when the sun is highest in the sky and people are typically most alert during the day. Therefore, cooler CCTs are used in spaces and during times when it's appropriate to promote alertness and attention. Warmer color temperatures (ranging from <2700 K to 3500 K) represent daylight hours when the sun is rising and setting when people are falling asleep or waking up. Circadian lighting systems are set to adjust based on the CCT we typically observe at any given time of the day.

Stimulus tuning is lighting technology that replaces the "blue light hazzard" with "good blue" light wavelengths. This circadian lighting approach more closely mimics the daylight natural spectrum. Stimulus tuning light fixtures can be programmed for reducing blue light wavelengths during the evening/nighttime hours to limit melatonin suppression without changing the CCT. This lighting approach is similar to color tuning and is the most effective when paired with intensity tuning [15].

The consequences of circadian rhythm disturbances are reduced sleep quality, drowsiness, low work productivity and bad mood. The effect of radiation on the human body includes visual perception, influence on the psychophysiological and emotional state. At the same time, there is damage to the elements of the eye and skin, as well as a number of other negative factors for physical health (photobiological danger).

3.3 Comfort Level in the Building

Most of all visual comfort can be usually defined through a set of criteria based on the level of light in the room, the balance of contrasts, CCT and the absence or presence of glare [16].

There are actually many different definitions of visual comfort level and what people should measure as part of it.

The most widely used and accepted metrics tend to focus on:

- The measure or the quantity of natural light over the year: the potential of the certain building and its location to provide enough daylight to occupants. This is typically explored through daylight autonomy or through the useful daylight illuminance;
- Lighting distribution as perception: whether a space will be too highly contrasting or too bright;
- Another aspects: there are another conceptions that don't have a metric but on which academics widely agree. The quality (through its spectral) of a view is also a consideration but is hard to quantify; however, the first methodologies are starting to appear.

As were mentioned above in subtitle "Ergonomic lighting" of this section, certain role plays coefficient UGR.

During our research, office conditions will be accept. Actually, in cellular offices, good conditions have been observed in general and giving the users full responsibility for controlling their own indoor climate during occupied hours works very well. The occupants have demonstrated a sparing use of supplementary mechanical cooling/heating equipment. They appear to have a preference for using windows and clothing to modify conditions in mild mid-season weather. However, most occupants appear to have upper and lower "tolerance" limits, beyond which active intervention will be applied if the opportunity is available [17].

The control strategy should determine both time and rate control. It should also determine different control modes in relation to different weather conditions. The actual control strategy should reflect the demands of the building owner, the needs of the users and the requirements in standards and regulations [17].

In the literature, user comfort has been addressed mainly based on thermal or/and visual comfort metrics as control variables including: indoor temperature, outdoor temperature, solar radiation, sky condition, task illuminance, Daylight Glare Index (DGI), Daylight Glare Probability (DGP), Daylight Autonomy (DA), spatial Daylight Autonomy (sDA) and Annual Sun Exposure, Useful Daylight Illuminance (UDI), vertical eye illuminance (Ev), glare frequency, and so forth. But, to ensure a comfortable indoor condition, designers need to choose an appropriate control strategy based on climatic variations and user visual/thermal demands. In most cases, researchers automated the shading system when it exceeded a certain threshold by utilizing a built-in control algorithm to keep user's comfort in acceptable range [18].

4 Methods of Researching

4.1 Climate Chamber at the Technical University of Dresden

For experiment was used climate chamber at the Technical university of Dresden.

The physical behavior of the climate room is extremely flexible. The climate room at the TU Dresden (Fig. 5) has an inner dimension of $l \times w \times h = (5 \times 4 \times 2.5)\text{m}^3$ [19].

The interior surfaces are divided into 73 individually tempered surface areas and the connected ventilation system can be controlled by temperature, humidity and volume flow. In order not to exert any significant influence on the measurement results, the ventilation system only serves to supply the subjects with fresh air during the test.

Activities are currently underway at various scientific institutions that focus on refining the local comfort models and on checking the global comfort models.

Fig. 5. Climate room at the TU Dresden outside.

The surface temperatures of the walls, ceiling and floor can be regulated in a range of T = 10 … 50 °C. Air volume flow maximal, actually maximal volumetric flow per unit time 600 m^3/h.

In order to define important for measuring parameters of lighting there were used optical devices. There are digital color temperature sensor and illuminance device FLAD23CCT with ALMEMO D6-connector [20], which were used during the experiment, are in Fig. 6.

Certain technical dates for measuring device are represented in Table 4.

Color temperature and illuminance are determined as a means to plot and evaluate lighting systems.

Digital color temperature sensor with "TrueColorSensorchip" and integrated signal processor "The TrueColorSensorchip" (3 sensors on 1 chip) detects - separately - each

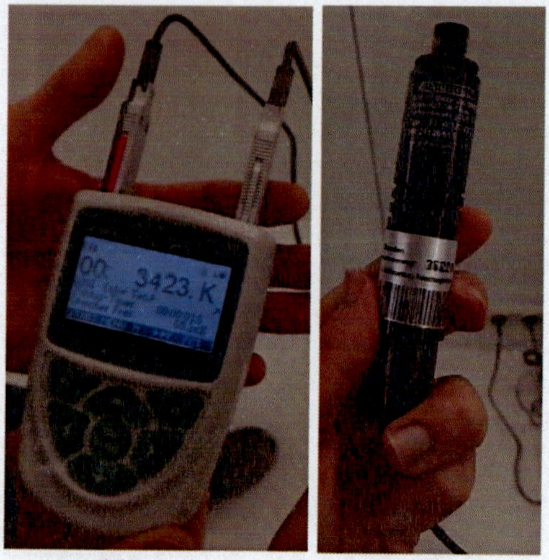

(a) (b)

Fig. 6. ALMEMO D6 plug (a) with a digital sensor for illuminance and color temperature FLAD23CCT (b).

of the three colors - red, green, blue (RGB). The respective sensitivities of these 3 color sensors are adapted to the standard spectral curves as per CIE and DIN. On the basis of these RGB values the computer calculates the color point within the RGB range in terms of coordinates X and Y and determines the correlated color temperature (CCT) in Kelvin.

Compact sensor, particularly suitable for mobile applications. The display shows simultaneously both this color temperature and the illuminance in a range of lux (lx) or kilolux (klx).

During the experiment test people are sitting in the middle of the room and accepting the light as represented in Fig. 5 There is represented front view (a), side view (b) and plan view (c) with an area dimension and lamps, each of the lamps numbered, has a

Table 4. Technical dates for measuring devices.

Spectral sensitivity	380 nm–720 nm
Correlated Color Temperature (CCT)	54…30 000 K (in a range of 120 lx… 170 klx)
Illuminance V-Lambda:	10…65,000 lx (factory settings) or 0.05…170.00 kLx
Room temperature conditions	−10… + 40 C°
Refresh rate:	1.5 s for all channels
Settling time:	3 s (when operating the data logger in sleep mode, a sleep delay of 3 s must be programmed)
Supply voltage	6…13 V DC

certain dimension and Guth position index (p_1, p_2, p_3, p_4) [21]. Lighting in Fig. 7 is represented in case of arrows.

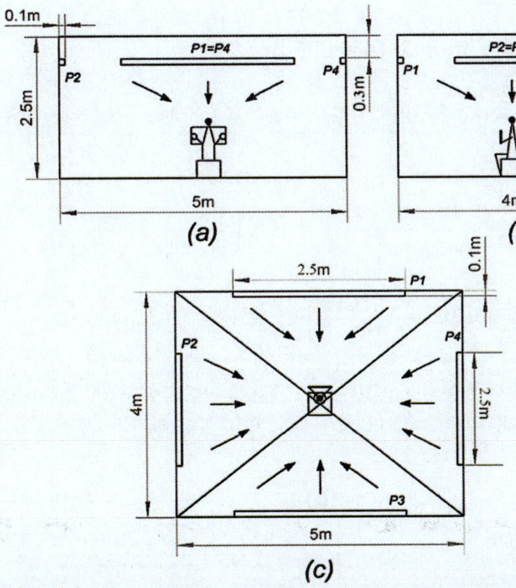

Fig. 7. Schematic representation of elevation and person, which is sitting in the middle of the room: a) front view; b) side view; c) plan view.

Person in Fig. 7 can be represented with a big scale as in Fig. 8. According to easy math calculation for test person, angle $\alpha = 90° = \frac{\pi}{2}$ and can be defined according to formula:

$$\cos\beta = \frac{Line\_of\_sight}{Lline\ from\ the\ observer\ to\ the\ source} = \frac{b}{c}, \quad (3)$$

where line from the observer to the source can be defined according to Pifagor formula:

- for c, p_3 and p_1:

$$c = \sqrt{a^2 + b^2} = |a = 2.5\,\text{m} - (0.3\,\text{m} + 1.23\,\text{m}) = 0.97| = \sqrt{2.5^2 + 2^2} = 3.2\,\text{m};$$

$$\cos\beta = \frac{2\,\text{m}}{3.2\,\text{m}} = 0.625 \Rightarrow \beta = 0.895\,\text{rad} = 51.3°.$$

- for p_2 and p_4:

$$c = \sqrt{a^2 + b^2} = |a = 2.5\,\text{m} - (0.3\,\text{m} + 1.23\,\text{m}) = 0.97| = \sqrt{2.5^2 + 2^2} = 3.2\,\text{m};$$

$$\cos\beta = \frac{2.5\,\text{m}}{3.2\,\text{m}} = 0.78125 \Rightarrow \beta = 0.674\,\text{rad} = 38.6^\circ.$$

According to formula (2), $p_1 = p_3 = 1.031$ and $p_2 = p_4 = 1.0232$. There is the situation when only p_1 is using in final formula for UGR:

$$UGR = 8\lg\left(\frac{0.25}{L_B}\left(\frac{L_1^2\cdot\omega_1}{p_1^2} + \frac{L_2^2\cdot\omega_2}{p_2^2} + \frac{L_3^2\cdot\omega_3}{p_3^2} + \frac{L_4^2\cdot\omega_4}{p_4^2}\right)\right),$$

$$UGR = 8\lg\left(\frac{(\pi\cdot E_{ind})^2\cdot\omega_1}{4\cdot L_B\cdot p_1^2}\right) = 8\lg\left(\frac{1.12(\pi\cdot E_{ind})^2}{4\cdot\frac{E_{ind}}{\pi}\cdot 1.031^2}\right),$$

$$UGR = 8\lg\left(\frac{34.73E_{ind}}{4.252}\right) = 8\lg(8.17E_{ind}). \tag{4}$$

This means, that illuminance E_{ind} for our experiment can be found in range of $E_{ind} = [0.4272lx...0.9044lx]$ for $UGR = 10...16$ appropriately.

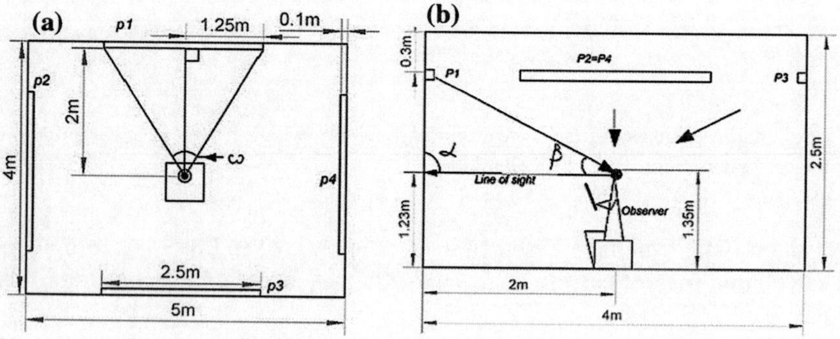

Fig. 8. Detailed schematic representation of the test person a) plan view; b) side view.

The brightness change function is expressed through the normalization of the illuminance function by the maximum illuminance by 7000K and the multiplication by the maximum brightness of the intelligent lighting system:

$$L = (0.0002\cdot CCT - 0.2587)L_{\max}, \tag{5}$$

where CCT is correlated color temperature of the emitting light, $L_{\max}$ – the maximum brightness that can be provided by an intelligent lighting system at a color temperature of 7000 K. This equation also describes approximation of brightness and lighting according to comfort [22].

There were received dependences of color temperature during the day that color temperature is 5580 K in the period from 9:30 to 16:30 and color temperature does not change. In the morning at 5 o'clock. 30 min and in the evening at 8 p.m. 30 min it is 4359 K and accordingly slowly increases from this value or decreases to it. Based on these data, the function of changing the color temperature of the emitting light of the intelligent lighting system during the day was implemented in the system operation algorithm:

$$CCT = \begin{cases} 4359, \; if \; 20:30 < t \leq 24 \; or \; 00 < t \leq 5:30 \\ 9.1385 \cdot t^3 - 291.52 \cdot t^2 + 3096.8 \cdot t - 5375.4, \; if \; 5:30 < t \leq 10:00 \\ 5580, \; if \; 10:00 < t \leq 15:30 \\ -9.1385 \cdot t^3 + 421.28 \cdot t^2 - 6470.4 \cdot t + 38689, \; if \; 15:30 < t \leq 20:30 \end{cases} . \quad (6)$$

Certain dependence demonstrates graphic in Fig. 9.

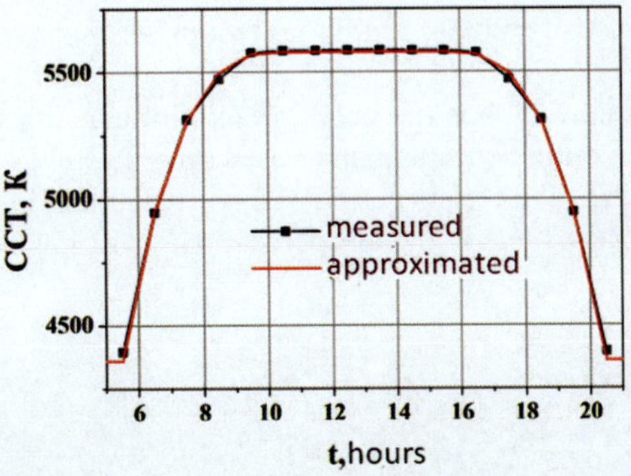

Fig. 9. The function of changing the color temperature intelligent lighting system during the day.

4.2 Experiment Task and Participants

One of the tasks for measuring is defining of human comfortable level. There were prepared special conditions in order to define impact on human comfortable level through the lighting and to confirm some theoretical moments between lighting parameters and comfort parameter. It also confirms basically ergonomics moments.

Participants. There are five persons were called for an experiment, especially boys and girls. Each of them has been interviewed before experiment about different biological and personal parameters: age, sex, how did people get to the labor (bicycle, walking, car or something else), height, weight, first imagination of temperature feeling (hot, warm, cold) in the climate chamber.

4.3 Realization

During the experiments, both Correlated Color Temperature (CCT) and illuminance are only variable parameters. Another parameters such as room and surface temperature, airflow are constants.

The influence of brightness on human sensation has been extensively studied in the past and is also established in various standards and regulations. In order to be able to assess the influence of color temperature on the thermal sensation of humans, tests are currently being carried out on test persons in the Combined Energy Lab of the TU Dresden [10]. A uniform operative room temperature of 22 °C is set in the indoor climate room. One result could be that at the selected operative room temperature of 22 °C, the thermal sensation of the test persons is influenced by the choice of color temperature - "warm" light could provide a higher thermal sensation than "cold" light. If this correlation can be shown, this obviously has a very great potential for the operation of a heating system or its energy consumption as well as CO2 emissions.

With the help of LED lighting, the color temperature can be varied in the range of T = 2600… 4400 K and brightness can be varied in the range of 79…250 lx. This results in a three-stage experiment with color temperatures T1 = 2600 K, T2 = 3500 K and T3 = 4400 K with constant ambient conditions (room perimeter surfaces and supply air have 22 °C).

The color temperature is thus the only variable influencing thermal comfort. The test person answers an online questionnaire based on DIN EN ISO 28802 every 5 min. Experiment passes during the certain period of day-time: either in the morning, or in the afternoon. Figure 10 shows one of the test person, who was involved into voluntarily.

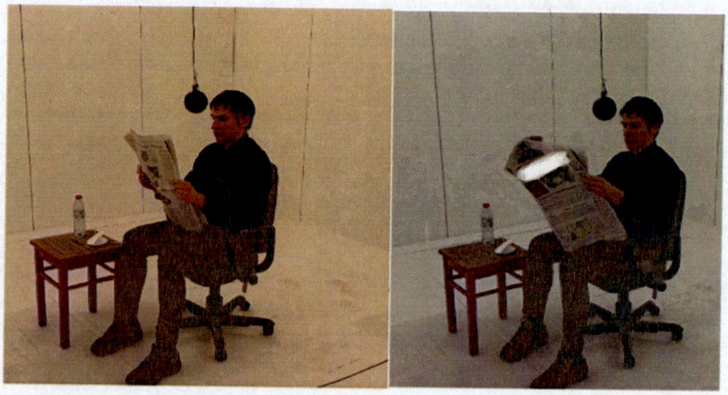

Fig. 10. The Experimental setup, left T1 = 2600 K, right T3 = 4400 K.

There are some more conditions:

- Test person must not use any other devices such as computers, mobile phones, tablet or any other lighting devices;
- Each test person has a right to take along either book or newspaper;
- Each test person has a right to take along a bottle of water and drink it.

5 Experimental Results

There were different participants, who have already taken part during the experiment. Certain information about biological and personal parameters, which were mentioned in Sect. 3.2, are shown in Table 5.

Table 5. Parameters of the test persons.

Spectral sensitivity	380 nm–720 nm
Correlated Color Temperature (CCT)	54…30 000 K (in a range of 120 lx… 170 klx)
Illuminance V-Lambda:	10…65,000 lx (factory settings) or 0.05…170.00 kLx
Room temperature conditions	−10… + 40 C°
Refresh rate	1.5 s for all channels
Settling time	3 s (when operating the data logger in sleep mode, a sleep delay of 3 s must be programmed)
Supply voltage	6…13 V DC

During the each position there people must sitting in the middle of the room.

As it was mentioned above, each test person had to answer online questionnaire every 5 min. Here are some approximate results of experiment in Table 4.

During the each position there people must sitting in the middle of the room (Table 6).

Table 6. Experimental middle measuring results based on questionnaire of one of the statistical person.

Questionnaire of one of the statistical person							
Time	Period	Phase	CCT, K	Brightness, lx	Feeling of room temp.	Did you feel any changes	Do you agree with temp.
9:23		Clear all Data	4400	222			
9:46		Test realization	4172	81	Something warm	Cooling	Yes
9:55	0:00	Phase 1	4133	85	Neutral	Cooling	Yes
10:15	0:20	Start of Pause	4138	84	Neutral	Cooling	Yes
10:20	0:25	Phase 2	3346	69	Something cold	Cooling	Yes
10:40	0:45	Start of Pause	2600	78	Neutral	No changes	Yes
10:45	0:50	Phase 3	2600	85	Neutral	No changes	Yes
11:05	1:10	Finish	2600	88	Neutral	No changes	Yes

Simple explanation of Kruithof curves can be defined as two curves:

- hyperbolic function for bluish or cold color temperature:

$$y = \sqrt{10(x - 2000)}. \tag{7}$$

- parabolic function for reddish or warm color temperature:

$$y = \frac{(x - 2000)^2}{1538.46}. \tag{8}$$

Final curves are in Fig. 11. An enormous in Fig. 11(a) there is approximation of Kruithof curves with a square of cold and warm color temperature. At the same time there was received an experimental measuring results based on questionnaire of the statistical persons in Fig. 11(b). Square in the center of graphic is "Comfort working zone", which is limited by four direct lines ($CCT_1 = 4000\,\text{K}$, $CCT_2 = 5500\,\text{K}$, $E_{V1} = 200\,lx$, $E_{V2} = 500\,lx$) and shows the most pleasant zone for a work.

Graphics in Fig. 11 were built in software environment MATLAB according to formulas (6) and (7). There are two forms of the graphic, which show us zones of both cold and warm temperature:

```
clf;
x = [0000:0.1:5000]';
y = ((x-2000).^2)/1538.46;
  h = area (x, y);
  fill(x,y,[0.9290 0.6940 0.1250])
hold on
   y1=sqrt(10*(x-2000));
area(x,y1,'FaceColor','b','FaceAlpha',.3,
'EdgeAlpha',.3)
hold on
```

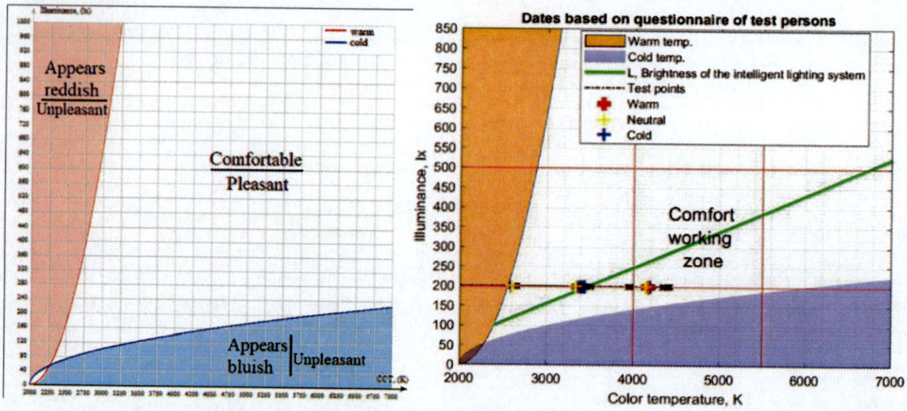

Fig. 11. Practical realization of Kruithof curves with an: a) warm and cold color temperatures; b) results based on experiments with test persons.

According to the certain researches, standards and recommendations, there is a useful for working process range of color temperature T = 4000 … 5500 K. It helps for stimulation of the brain, for feeling fit. It is recommended for offices and studying rooms. At the same time it is not recommended color temperature T = 6500 K and more, where are still working people for a very long time.

Low color temperature (2000K–3500K) corresponds to morning or evening light. It helps to relax and will be successfully used in the bedroom, children's room or other recreation areas. But do not forget that the color temperature can affect the perception of interior colors. This light is suitable for a room decorated in warm colors. Warm light is recommended for lamps and lighting fixtures used for reading, because during the warm lighting we have lower eye-strains.

As the result, according to Eq. (5), different graphics for a different period of times were represented in Fig. 10. Green line and/or green point shows preferred value for comfort level.

These graphics were built by software environment MATLAB. Green line is describe by code:

```
   x8 = [2391:0.1:5000]';
y8 = (0.0002*x8-0.2587)*460;
plot(x8,y8,'green-','LineWidth',2)
hold on
```

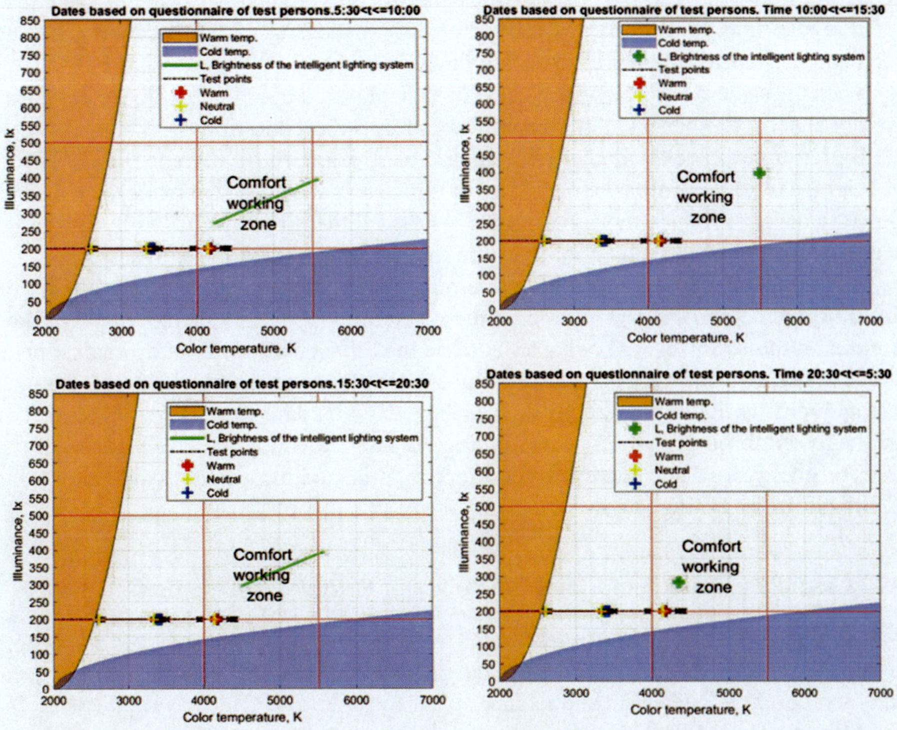

Fig. 12. Dates based on questionnaire of test persons during the different period of day.

6 Methods of Researching

6.1 Algorithm of Device. Functional Scheme and Research Methodology

There is an algorithm scheme of adaptive device based on determination human circadian rhythm day represented in Fig. 11.

First of all our control system was programmed to the initial state, which means almost-real life operating conditions and factory settings for average person.

Afterward system check if bracelet connected to the hand of human. If yes, bracelet takes physical dates about test-person and then takes into account information about lighting parameters, an enormous brightness, illuminance, color temperature, time of day season of year (summer, spring, winter or autumn). If the watch is not connected to human arm, it takes only lighting parameters into account.

According to the certain requirements for algorithms and operation, there was built the block diagram of the device Fig. 11. It demonstrates which main components of the device are here and how do network nodes will be interconnected. Structure scheme can be changed according to host of functions and or software and hardware of system.

It is also important to pay attention not only on season, but also country. Because each country has a different climate and daylight period.

Then the system defines: which value of each photometric quantity do we have. Later system check, whether mentioned parameters fulfill our conditions.

System must check, whether photometric parameters fulfill our conditions. If value of brightness and/or colour temperature is more than we need, control systems send a signal in order to reduce the signal. If value of brightness and/or CCT is lower than we need, control systems send an information in order to amplify the signal.

This approach allows the designer to simulate the entire number of design alternatives (brute-force method) that would be a time-consuming approach using typical building simulation tools; however, it helps to increase the robustness of the study against uncertainties. The research process includes three main stages as shown in Fig. 12 from collecting inputs until deriving outputs that will be explained he outputs are post-processed and analyzed through Python programming language to study the implications of variables on energy load and occupant's comfort, relations within variables and outputs, and main findings [23] (Fig. 14).

6.2 Circadian Rhythm of Blood Pressure and Regression Equation

We use bracelet. It can be also fitness watch as shown in the example of Fig. 13 is adopted in this study to collect the following biodata from the participated subjects; heart rate, body temperature, and skin conductance.

Almost all health and fitness companies use optical sensors in their wearable devices for non-invasive measurements of fitness parameters. These optical sensors use the principle of photoplethysmography, which has become the standard for portable heart rate sensors and which is able to replace uncomfortable electrocardiogram (ECG) chest straps for many physical activities. In contrast to ECG, which is typically measured by multiple electrodes at the chest, a photoplethysmographic (PPG) sensor can be placed at a single site on the body with good blood circulation, such as the earlobe, fingertip, or wrist,

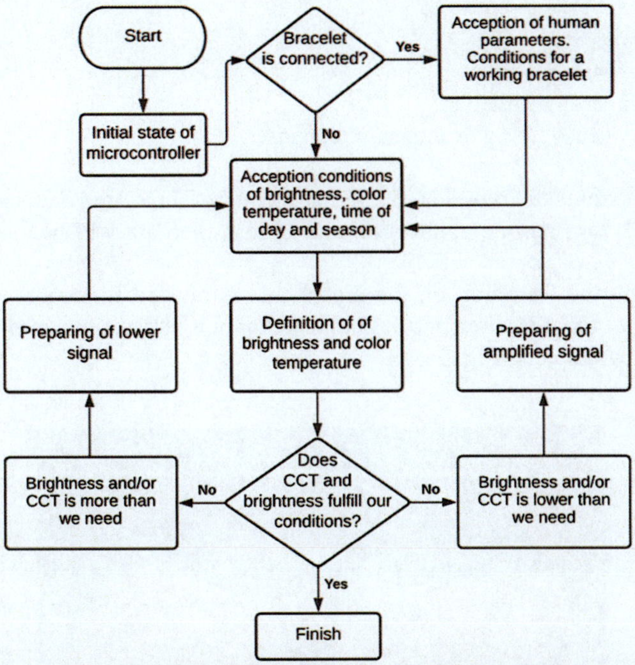

Fig. 13. Algorithm scheme of adaptive device based on determination human circadian rhythm day.

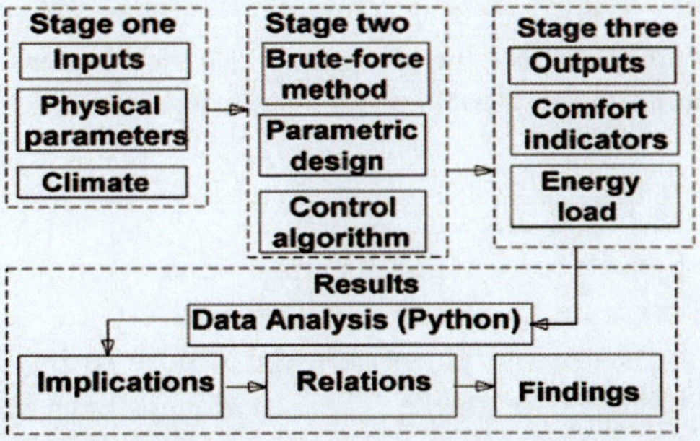

Fig. 14. Research methodology of adaptive device based on human circadian rhythms.

making it easy to wear and collect measurement data throughout the day. In addition, a PPG sensor can be cost-effectively integrated into a wearable, and any video camera of a smartphone can be transformed into a PPG sensor [24, 25].

Photoplethysmography has great potential to provide new health insights, to measure additional physiological parameters of the body, and to reveal health risks.

This watch contains five sensors:

- Optical blood flow sensor;

- 3D accelerometer;
- Body temperature reading;
- Ambient temperature reading;
- Conductivity sensor for galvanic skin response.

This watch stores the bio-data in internal memory. The stored data can be exported from the watch to a mobile application and in turn the data will be available via the cloud.

However, further research on the algorithms is needed to improve the accuracy and reliability of the measurements and to develop new estimation methods for further physiological parameters (Fig. 15).

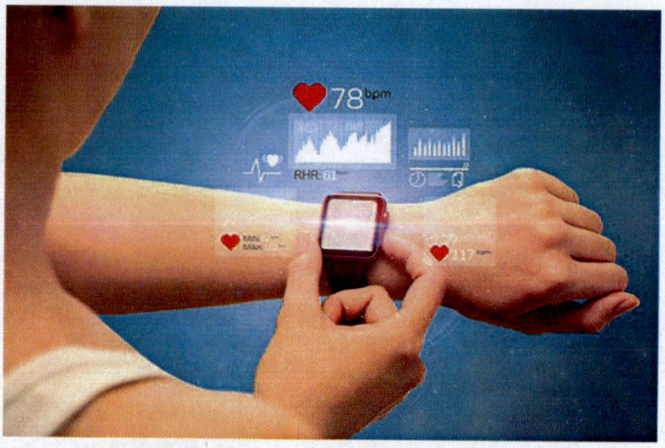

Fig. 15. Smart watches and other wearables designed with optical biosensors can provide a wealth of health insights via heart-rate and blood-oxygen monitoring.

Circadian rhythm of blood pressure is also a topic that has been widely explored. In different day, blood pressure variation has many characters. Several papers pointed out that normal blood pressure in nighttime will drop by 10–20% of the daytime blood pressure. And, if the nighttime blood pressure drops little or increases, the population of this group has the higher risk of stroke. These evidences have not difficult to find that the stroke risk is more dependent on the blood pressure variation of the day and night [26].

At present, the cuff-less method of blood pressure measurement has features, such as sleeveless, non-invasive, non-inductive and wearable device. The principle of blood pressure is estimated by the negative correlation between blood pressure and Pulse Transit Time (PTT) [27]. PTT decreases while blood pressure rises, and vice versa. And also show about distance that PTT divided by time was called Pulse Wave Velocity (PWV). The value of PTT is measured by Electrocardiogram (ECG) and Photoplethysmogram (PPG).

The relationship between PTT, PWV and BP about the physiology are important factors for arterial wall thickness, blood elastic and blood vessel diameter. Pressure waves produced at the heart propagate through the arteries at a certain velocity known as the pulse-wave velocity [26], which depends on the elastic properties of arteries and blood:

$$PWV = \frac{L}{PTT} = \sqrt{\frac{S \cdot E_0 \cdot e^{\alpha_b \cdot BP}}{\rho \cdot d}}. \tag{9}$$

where $L =$ blood vessel length, $S =$ vascular wall thickness, $d =$ vessel diameter, $\rho =$ blood density, $E_0 =$ elastic modulus, $\alpha_b =$ blood vessel modulus, $P =$ blood pressure.

To process the PPG and ABP signals is under the same procedure including to select the common feature points, such as the peak value of the signal, the valley value and the maximum slope point. The ABP signal will take the peak as the value of the systolic blood pressure.

Pulse transit time, heart rate, pulse width and pulse wave interval were put into the blood pressure regression model as corrections. And using correction S wave in ECG as the PTT start time which is shown in Fig. 16 [26, 27].

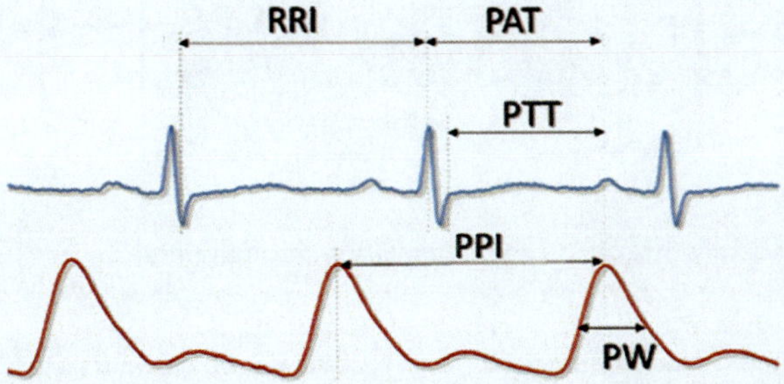

Fig. 16. Characteristic parameters of blood pressure regressive analysis model.

Blood pressure regression model used in the past shown in Eq. 9. A simplified equation that modifies the Eq. 9 to Eq. 10 shown in below.

$$\begin{aligned} BP = {} & A \cdot PAT + b \cdot HR + c \cdot PW + d \cdot AMP + \\ & e \cdot PPI \cdot \Delta PAT + g \cdot \Delta HR + h \cdot \Delta AMP + i \cdot \Delta PPI + j \end{aligned} \tag{10}$$

$$BP = A \cdot PAT[n] + b \cdot HR[n] + c \cdot PW[n] + d \cdot PPI[n] + e. \tag{11}$$

PPI is a sum of diastole and systole signals.

This database has more than one ECG signal with different leads, and some data include continuous arterial blood pressure (ABP) signal and PPG signal. Each data has different data length with the sampling rate in 125 Hz. ECG, PPG and ABP signals were used to verify the performance for the different cuffless BP regressive models.

The signal contains two important components such as DC and AC. The DC component is due to constant reflective matter such as skin, liquid and solid parts, muscle, bone, venous blood and certain list of other parts. If a body is at rest and motion is less of a factor, the AC component comprises mainly of reflected light from the pulsation of artery blood. The AC component depends on heart rate and artery thickness, with more reflected or transmitted light in systolic (pump) than the dystopic (relaxation).

During the period of systolic phase, blood is pumped from the heart, which increases atrial blood pressure. The growth in blood pressure expands the arteries and leads to an increase in atrial blood volume. This increase in blood causes an increase in light absorption. Blood pressure drops during the phase of diastolic and that is why certain changes can be found through the absorption of light. Figure 17 shows the diastolic trough and systolic peaks caused by the beating heart [24].

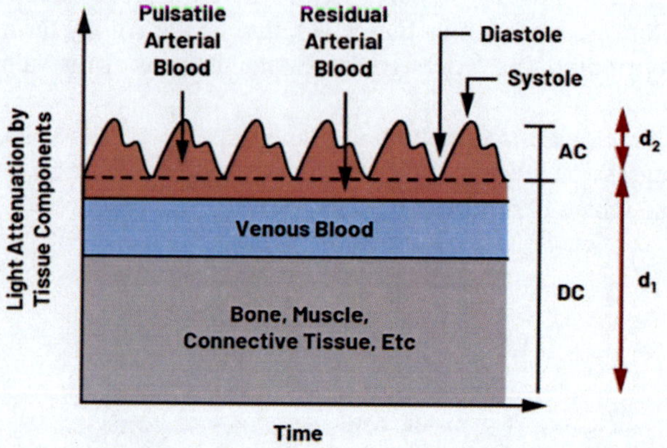

Fig. 17. Light attenuation through tissue.

By computing AC and DC from a PPG signal, we are able to determine the change in absorption of light in atrial blood caused by blood pumping from the heart, with no contribution from other tissue.

The Beer-Lambert law explains that light decays exponentially when travelling though absorptive material. This can be used to determine the level of oxygenated hemoglobin to total hemoglobin.

The intensity of light absorbed at the diastole and systole are related by:

$$I_{systole} = I_{diastole} \cdot \exp(-\alpha \cdot l) \text{ or } I = I_0 \cdot e^{-\alpha \cdot l}. \tag{12}$$

After the logarithmic transformation was received formula

$$l = [\ln(I_0) - \ln(I)]/\alpha. \tag{13}$$

In this case α is a coefficient, which shows human blood absorption during the certain lighting wavelength. Common process of transmission intensity changing according to the Beer-Lambert law [28] is represented in Fig. 18.

As the example, it is represented PPG signal, which can be obtained using a transmissive or reflective LED and photodiode (PD) configuration (Fig. 19) [29]. A transmissive configuration measures the nonabsorbed light passed through a part of the body. This configuration is best suited to areas such as the finger and earlobe where measurement benefits from the capillary density of these body locations, which make the measurements more stable, repetitive, and less sensitive to variations in placement. Transmissive configurations achieve a 40 dB to 60 dB increase in the perfusion index.

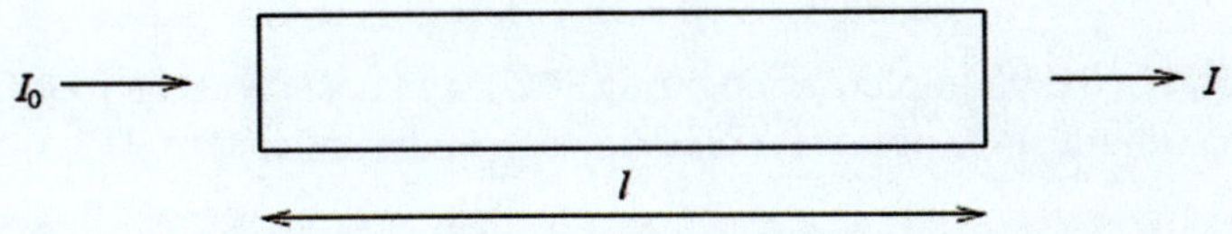

Fig. 18. Process of intensity transmission and changing through artery.

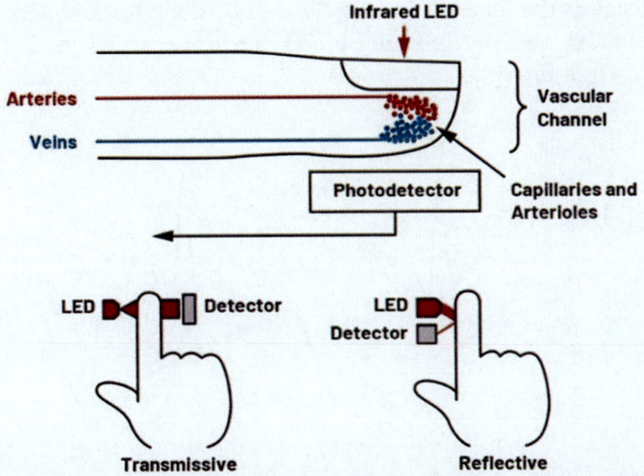

Fig. 19. LED-PD configuration.artery for lighting transmission.

Reflective PPG configurations are chosen when the PD and LED must be placed next to each other for practicality, such as with wrist- or chest-worn devices.

According to Eqs. (8), (10), (11):

$$\frac{L}{PTT} = \sqrt{\frac{S \cdot E_0 \cdot e^{\alpha_b BP}}{\rho \cdot d}} \Rightarrow \frac{L^2}{PTT^2} = \frac{S \cdot E_0 \cdot e^{\alpha_b BP}}{\rho \cdot d} \Rightarrow ln\left(\frac{L}{PTT}\right)^2$$
$$= ln\left(\frac{S \cdot E_0 \cdot e^{\alpha_b BP}}{\rho \cdot d}\right) \Rightarrow$$
$$2[ln(L) - ln(PTT)] = ln\left(\frac{S \cdot E_0}{\rho \cdot d}\right) + \alpha_b \cdot BP. \tag{14}$$

In this case $[ln(L) - ln(PTT)]$ can be changed as $ln(PWV)$, blood vessel length can be defined as $L = d + 2S$ there is also length $L = l = [ln(I_0) - ln(I)]/\alpha$

$$\Rightarrow 2ln(L) - 2ln(PTT) = ln\left(\frac{\frac{L-d}{2} \cdot E_0}{\rho \cdot d}\right) + \alpha_b \cdot BP \Rightarrow$$
$$2ln(L) - 2ln(PTT) = ln\left(\frac{LE_0}{2\rho \cdot d} - \frac{E_0}{2\rho}\right) + \alpha_b \cdot BP. \tag{15}$$

PTT is typically measured indirectly through a related quantity known as Pulse Arrival Time (PAT). PAT is calculated as the delay between the R peak of ECG and a particular point in the photoplethysmogram (PPG) signal, such as the foot (PATf), peak (PATp) or maximum slope point (PATs) (Fig. 20). PAT is related to PTT as follows

$$PAT = PEP + PTT, \tag{16}$$

where PEP is the Pre-Ejection Period. PEP represents the isovolumetric contraction time of the heart, which is the time it takes for the myocardium to raise enough pressure to open the aortic valve and start pushing blood out of the ventricle [30]. Only PTT is related to BP through Eq. (8).

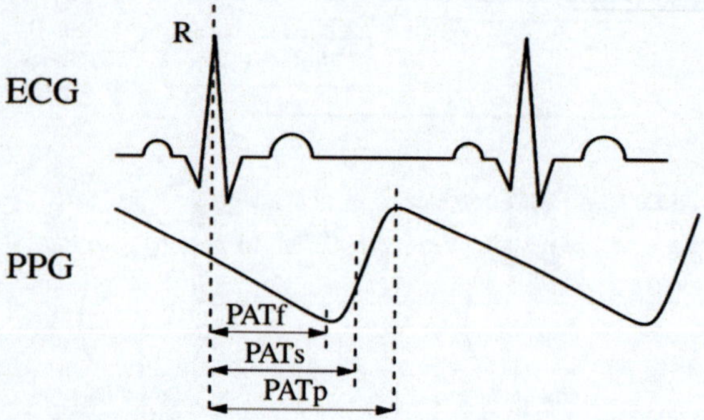

Fig. 20. PAT measured between R peak of ECG and a particular point of PPG.

After the series of transformation can be received final equation, which demonstrates logical series "light – photopletysmogram – blood pressure". According to Eqs. (10) and (13):

$$\begin{aligned} &\frac{2}{\alpha_b} ln\left(\frac{L}{PTT}\right)^2 - \frac{1}{\alpha_b} ln\left(\frac{(L-d)\cdot E_0}{2\rho\cdot d}\right) \\ &= A\cdot PAT[n] + b\cdot HR[n] + c\cdot PW[n] + d\cdot PPI[n] + e \end{aligned} \tag{17}$$

According to definition of intensity, $L = \ln(I/I_0)/\alpha$:

$$\begin{aligned} &\frac{2}{\alpha_b} ln\left(\frac{\ln(\frac{I}{I_0})}{\alpha\cdot PTT}\right)^2 - \frac{1}{\alpha_b} ln\left(\frac{(\frac{1}{\alpha}\ln(\frac{I}{I_0}) - d)\cdot E_0}{2\rho\cdot d}\right) \\ &= A\cdot PAT[n] + b\cdot HR[n] + c\cdot PW[n] + d\cdot PPI[n] + e. \end{aligned} \tag{18}$$

Absolute black body is a physical body that absorbs all incident electromagnetic radiation, regardless of frequency or angle of incidence. Radiation intensity analization of absolute black body depends on the temperature and frequency, which defines by Planck constant:

$$I_{n\,\max} = \frac{2h\nu^3}{c^2}\frac{1}{e^{h\nu/kT} - 1} = \frac{2hc^3}{\lambda^3 c^2}\frac{1}{e^{hc/\lambda kT} - 1} = \frac{2hc}{\lambda^3}\frac{1}{e^{hc/\lambda kT} - 1}. \tag{19}$$

Planck constant $h = 6.63 \cdot 10^{-34}[Joule \cdot se]$; $c = 310^8[\text{m/se}]$ – lighting velocity in vacuum.

[m/sec]; λ_n – wavelength of lighting $[nm]$.

Radiation intensity can be found from the formula

$$I_{n0} = \frac{2hc}{\lambda_n^3} \frac{1}{e^{hc/\lambda_n kT} - 1} \Rightarrow \ln(2hc) - ln(I_{n0}\lambda_n^3) = ln(e^{hc/\lambda_n kT} - 1),$$

$$ln(e^{hc/\lambda_n kT_n} - 1) = -56.18 - ln(I_{n0}) - 3ln(\lambda_n). \tag{20}$$

Color temperature definition of the absolute black body temperature T_n is a constant for the certain conditions λ_n.

$$\ln(\frac{I}{I_0}) = ln(\frac{\frac{2hc}{\lambda_n^3}\frac{1}{e^{hc/\lambda_n kT_{n-1}}}}{\frac{2hc}{\lambda_0^3}\frac{1}{e^{hc/\lambda kT_{-1}}}}) = ln(\frac{\lambda_0^3(e^{hc/\lambda kT} - 1)}{\lambda_n^3(e^{hc/\lambda_n kT_n} - 1)}). \tag{21}$$

An enormous, there were received two important Eqs. (18) and (21), which can show us best explanation of logical connection "electronic connection – physical connection – lighting connection". It is background for lighting devices in order to adjust human parameters with a certain physical parameters during the certain period of time (location of the country, month, day, hours, minutes, seconds).

7 Wearable Data Analysis

7.1 Wearable Data Analysis Platform (WDAP)

Schematic of the Wearable Data Analysis Platform (WDAP), which analyzes user-submitted data, represented in Fig. 21.

In order to facilitate the future analysis of data from a wide range of settings, there were created a Wearable Data Analysis Platform (WDAP), which can collect and process data anonymously submitted by the public through the Social Rhythms app, available free through the App Store (iOS) and the Google Play Store (Android). An application programed interface server transmits anonymous user-submitted data to a processing server, which analyzes the data by using our method [31].

7.2 Flowchart of Mobile Application Communication with Wearable Device

Our data-system can be interpreted into mobile Application as in Fig. 22.

The mobile Application once opened first checks if the wearable device is turned on, if not turned on, it requests the user to turn on the Bluetooth. Once the Bluetooth is turned on, it allows the user to search for nearby wearable devices and to connect. If a wearable device already exists and the Bluetooth is turned on, it tries to connect automatically to the device; if the device is turned on then connection will be established; otherwise, the Application automatically exits.

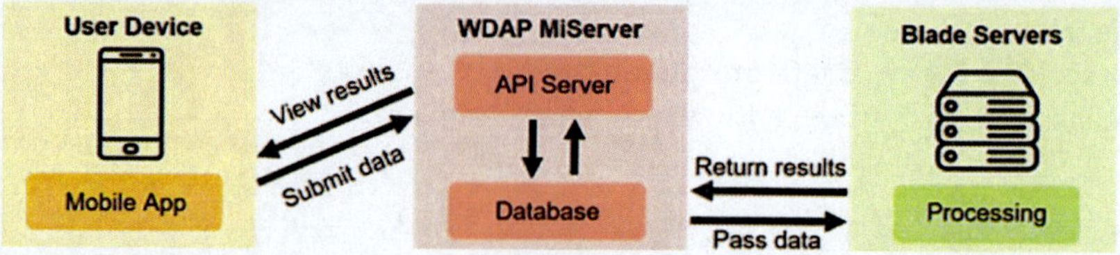

Fig. 21. Schematic of the Wearable Data Analysis Platform (WDAP), which analyzes user-submitted data.

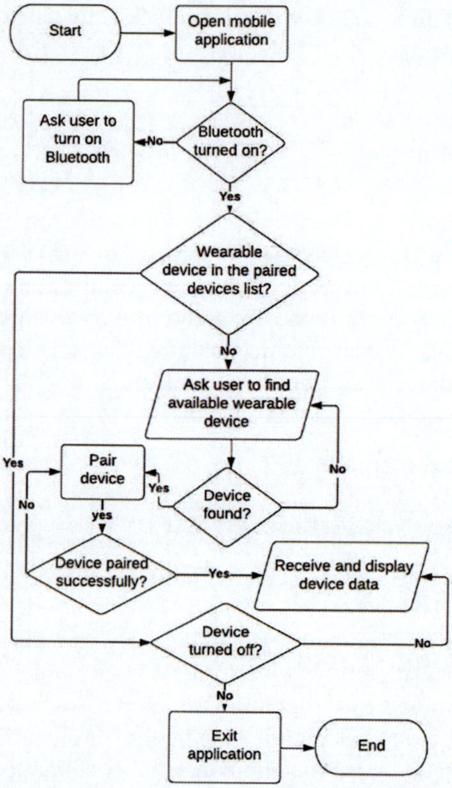

Fig. 22. Flowchart showing communication between mobile Application and any wearable device.

Nevertheless, as were mentioned above, there was used fitness device for our research.

The Remote monitoring of Human-Machine-Interface (HMI).

HMI system first requests for permission from the device it wants to monitor by querying an API, awterwards gets response through another API for the response. It should be noted that the practical details of this process is hidden from the user. The user only clicks a button to get fitness update and gets either a failed response if permission was denied, or a response containing the fitness data if permission was granted [32].

Figure 23 below provides a graphical illustration of this process. Subsequent sections will discuss more details on how this network is implemented.

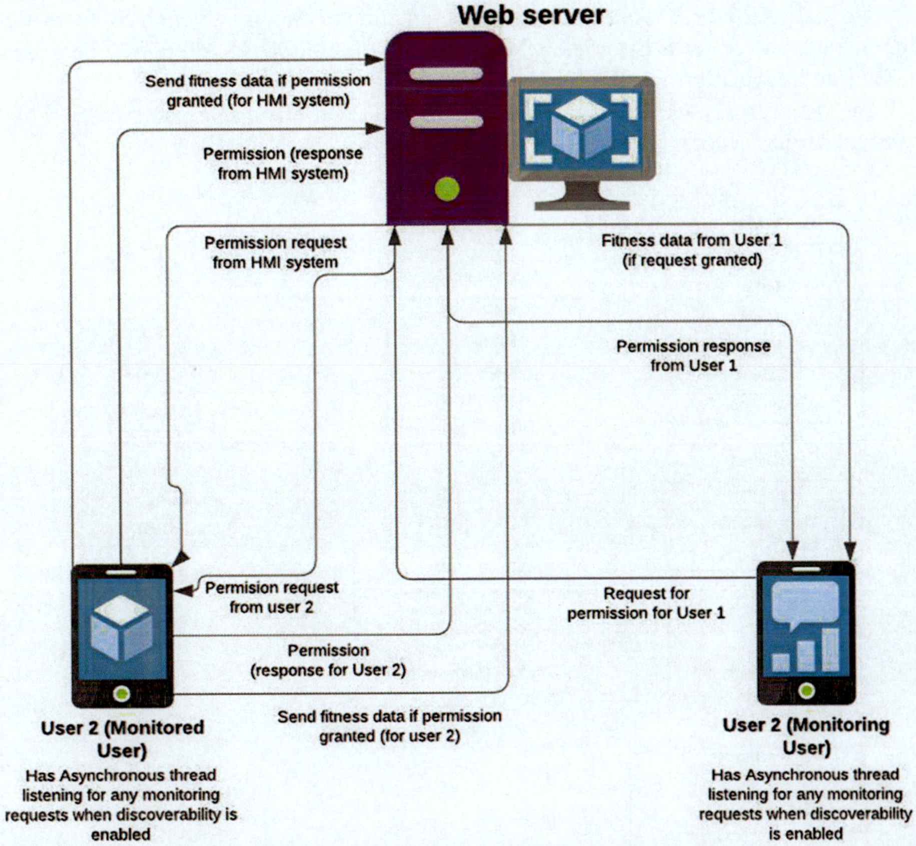

Fig. 23. Remote/Inter-user monitoring Network Structure.

Asynchronous programming is used to implement this feature; it is a form of parallel programming in which a thread or more run independent of the primary thread that runs the entire Application.

It is a powerful feature and is supported by certain programming language, which is used for developing the Android Application for the fitness device. Asynchronous threads are used in this project for communication to and from the webserver, which is an essential part of the Remote/Inter-user monitoring network.

It allows the creation of different exercise routines, assigns those routines to a person by date. In Fig. 24 there is approximate representation of database, which could be used either for a one or for a list of people. This seems like it could potentially be a large amount of information to store for Each Person and complicated to do so.

The diagram shows list of lighting, biophysical and electronic parameters, which were taken into account.

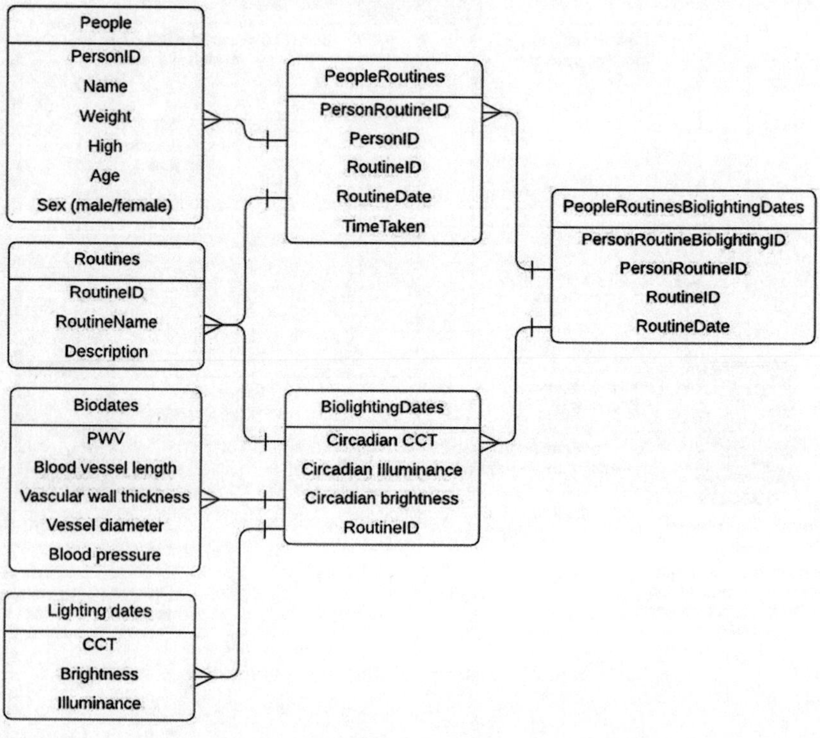

Fig. 24. Approximate representation of logical combination between test person and circadian parameters.

For each table, determine a correct unique key that will be immutable and unique for all records. The investigations have not been fully completed yet. The data obtained can then be used to develop a correlation needs.

8 Wearable Technical Realization and Structure of Electrical Scheme

According to requirements in Sect. 5 for our device, certain structure scheme is represented in Fig. 25.

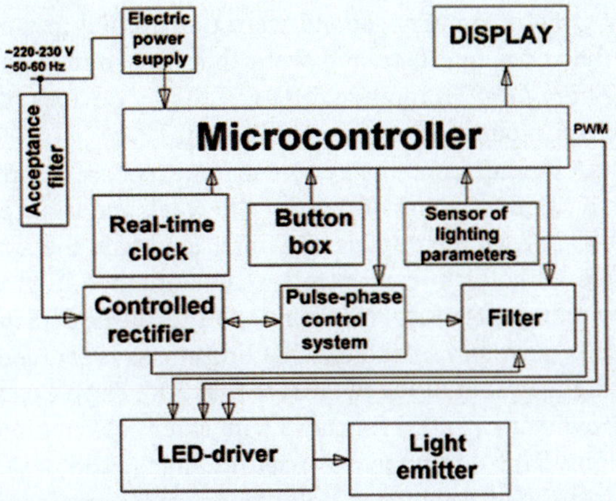

Fig. 25. Structure scheme of adaptive device based on determination of human circadian rhythm day.

The Algorithm of device. Functional scheme and research methodology structure scheme of adaptive device based on determination human circadian rhythm day.

Microcontroller is a "brain" and main part of the device. It contains firmware and system data. It is responsible for buttons and sensor data processing and calculating, wireless connection with mobile phone, tablet PC, personal computer and so on through either Bluetooth or Wi-Fi. It is also responsible for maintaining by peripherals: LED-driver, commutators, display, pulse-phase control system. The series of this functions can be realized by ESP8266 [18, 33, 34].

Another part, which plays certain role, is Electric power supply block. This one provides working process for device, transforms voltage supply to the voltage range, which is enough for photodiodes, LED-driver, microcontroller and other peripherals.

Light emitter. In case of this thing, we use LED-string. An enormous we can use either RGB or bicolor string. Main task of this component is shining function with a list of important lighting parameters, which can regulate circadian rhythms. The list of those parameters was described above and can be regulated. LED-driver is an element, which can maintain various types of the LED-strings: either RGB, or bicolor. Another types of emitting strings are also possible.

Dates are being generated every second were transmitted to a microcontroller unit ESP8266 through the serial interface and were then transmitted through either Wi-Fi or Bluetooth Low Energy (BLE) module (HC-05) to any devices, which can be a data logger – smartphone, computer, data tablet and so on.

Real-time clock. A module contains a power source to store information after turning off the main power, a calendar, a list of days of the week and timekeeping algorithms.

It is a module DS3231, which contains information about the time and date, as well as the day of the week. It helps microcontroller to be oriented in time [35, 36].

Display. Display represents useful information for the user. It displays the state of the system, which mode is currently selected, what brightness is set, and so on. In this case was chosen organic light emitter diode (OLED) SSD1306. People can control the system with a phone, but the display is used for more convenient orientation in the settings.

A block of buttons. This component is used to change the basic parameters of the system quickly, such as mode, brightness, color temperature or certain settings according to requirements. Users can also make these settings with a smartphone and other devices. If you need to change one of the basic settings quickly, it will be much easier to do it with the button panel.

Stabilized power supply for LEDs [37]. The supply voltage U1 is flowing through the input filter to the primary transformer coil. The voltage of the secondary transformer coil U2 is coming to the uncontrolled bridge-circuit rectifier $VD_3 \div VD_6$. The voltage U3 is flowing to the controlled voltage boost rectifier VS1, VS_2, VD_5, VD_6. The input voltage of the filter is the sum of the voltages of the two rectifiers. The stabilizer is a closed-loop system of automatic regulation of the output voltage supplied to the RGB-strings or matrix of super-bright LEDs $HL_{11} \div HL_{nk}$. This system maintains by ESP8266 and includes list of important electrical parts: measuring and reference elements, two signals of which are going to the comparison device. The amplified differential control error signal is coming to the pulse-phase control system. In the pulse-phase control system, the error signal converted into thyristor control pulses, synchronized by the supply voltage.

As the result, certain electronic scheme was constrained and represented in Fig. 26.

Color light-to-digital converter with IR-filter is also one of the main parts of the device, which plays role of "bridge" between lighting and electrical parameters. The TCS3472 device provides a digital return of red, green, blue (RGB), and clear light sensing values. An IR blocking filter, integrated on-chip and localized to the color sensing photodiodes, minimizes the IR spectral component of the incoming light and allows color measurements to be made accurately. The high sensitivity, wide dynamic range, and IR blocking filter make the TCS3472 an ideal color sensor solution for use under varying lighting conditions and through attenuating materials.

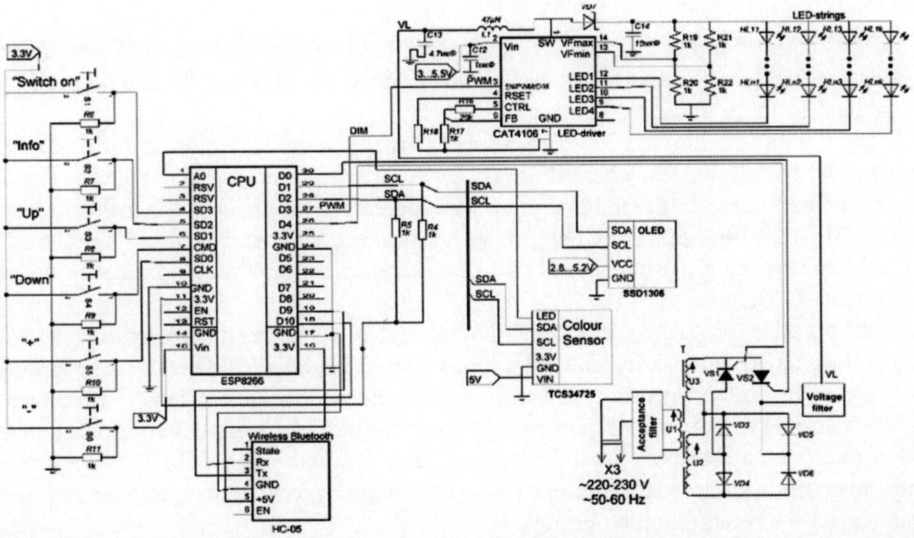

Fig. 26. Electrical scheme of adaptive device based on determination of human circadian rhythm day.

The TCS3472 color sensor has a wide range of applications including RGB LED backlight control, solid-state lighting, health/fitness products, industrial process controls and medical diagnostic equipment. In addition, the IR blocking filter enables the TCS3472 to perform ambient light sensing. Ambient light sensing is widely used in display-based products such as cell phones, notebooks, and TVs to sense the lighting environment and enable automatic display brightness for optimal viewing and power savings.

9 Conclusions

The optimal CCT is more beneficial than increased illuminance in moderate ambient indoor lighting, as it provides better lighting comfort.

This work is dedicated to the development of "smart lighting in the building" systems that create a positive impact on the physiological properties of human. By applying IoT and sensor technology we designed and implemented a smart lighting system, which can dynamically control lighting environment inside a room.

During the projecting of lighting systems, the choice of color temperature and spectral composition of sources should be carefully considered in order to avoid disruption of natural circadian rhythms or, if necessary, to increase the efficiency of human physiological parameters (for example, daytime or nighttime working period). Particularly, the beginning of the color temperature is in the numerical range of 4000–5500 K.

There was defined original and practical zone from "Kruithof curve", which forms square "Comfort working zone", which is limited by four direct lines (,,,) and shows the most pleasant zone for a work.

Restriction of direct brightness is performed by the unified indicator of discomfort UGR (Unified Glare Rating), adopted in international practice to take into account the combined effect of all lamps in the room.

The combined indicator of discomfort is an international criterion for assessing uncomfortable brightness, which causes unpleasant sensations due to the uneven distribution of brightness in the field of view. The indicator is without a dimensional value from 10 to 30, which cannot be determined by photometric methods, it can only be calculated approximately. But it needs more calculations and methodic in order to define coefficient UGR.

The proposed electronic scheme of lighting control allows regulating lighting parameters - brightness, intensivity, and color temperature. Experimental result in the laboratory of the Technical University Dresden confirm theoretical conclusions. The research in laboratory is still continues, and research plan for the future implies consideration of other microclimate parameters together with an individual parameters with a test person, an enormous: circadian rhythms based on humidify, room temperature and human parameters for an ergonomic lighting.

There were defined conception, which helps to define logical connection "electronic connection – physical connection – lighting connection". Due to adjust human parameters with a certain physical parameters during the certain period of time, there were explained direct connection through the lighting parameters, such as color temperature, brightness and lighting wavelength, but certain connection needs more researching.

Acknowledgement. This research is funded by the Technical University of Munich - Institute for Advanced Study.

Authors are thankful to the Technical University of Dresden, an enormous to the Faculty of mechanical and science engineering, Chair of building energy systems and heat supply for possibility to make some measurements and experiments in the climate chamber, which is a part of the Combined Energy Lab 2.0.

References

1. Dohrmann consulting "What is Ergonomics?". https://www.ergonomics.com.au/what-is-ergonomics/
2. Autonomous "5 Ways to Improve Lighting Ergonomics and Stay Productive". https://www.autonomous.ai/ourblog/5-ways-to-improve-lighting-ergonomics-and-stay-productive
3. Suzer, O.K.: The effects of correlated colour temperature on wayfinding: a study in a virtual airport environment. In: 6th International Conference on Spatial Cognition (ICSC 2015). Vol. 16, issue 1 Supplement, Italy, Rome, p. S72 (2015)
4. SMUD Energy Education and Technology Center "Circadian Lighting and You", pp. 1–23 (2017)
5. The Engineering ToolBox "Illuminance - Recommended Light Level". https://www.engineeringtoolbox.com/light-level-rooms-d_708.html
6. NEC Display Solutions Europe GmbH "How Display brightness affects successful implementation of Digital Signage applications", München, Germany, pp. 1–6 (2015)
7. International Commission on Illumination "CIE 117: Discomfort Glare in Interior Lighting". ISBN-9783900734701, pp. 1–39 (1995)

8. Vasilieva, J., Lyashenko, O.: Mark of Unified Glare Rating (UGR), projected in DIALUX. In: O.M. Beketov National University of Urban Economy in Kharkiv. Conference "Metrology and devices", Kharkiv, Ukraine, pp. 30–34 (2017). ISSN 2307-2180
9. Waldorf, J., Zvei, V.: Lighting division "position paper UGR method - application and limits", Frankfurt am Main, Germany, October 2021
10. Calculations UGR Concepts. https://docs.agi32.com/AGi32/Content/adding_calculation_points/Calculations_UGR_Concepts.htm
11. Danish Standards Association, "DS/EN 12464-1:2011 Light and lighting - Lighting of workplaces - Part 1: Indoor workplaces", Danish Standards Association, Charlottenlund (2016)
12. Zheliazkov, Y., Seifert, J., Yamnenko, J., Beyer, M., Schinke, L.: System of adaptive lighting in the building based on determination of human circadian rhythms. In: 2022 IEEE. Proceedings of 16th International Conference on Advanced Trends in Radioelectronics, Telecommunications and Computer Engineering (TCSET). Lviv-Slavske, Ukraine, 22–26 February 2022
13. Osterhaus, W., Erhardtsen, I., Gkaintatzi-Masouti, M., Gert Nielsen, K., Dobos, F.: Case studies circadian lighting, Aarhus University, AFRY and IVE, 100 p. (2020)
14. The lighting practice "What is Circadian Lighting?" https://www.thelightingpractice.com/what-is-circadian-lighting/
15. Zheliazkov, Y.: Vplyv dzherel synoho svitla v osvitlyuvalnykh ustanovkakh na tsyrkadni rytmy lyudyny [The effect of blue light source in a lighting installations on human circadian rhythms]. Microsystems, Electronics and Acoustics, vol. 26 No. 2, 237533-1, August 2021. https://doi.org/10.20535/2523-4455.mea.237533
16. Liu, A., Tuzikas, A., Arturas, Z., Vitta, P., Vaicekauskas, R., Shur, M.S.: Cultural preferences to color quality of illumination of different objects, 7 p., May 2013. https://doi.org/10.48550/arXiv.1305.6308
17. Heiselberg, P.: "Principles of hybrid ventilation", Hybrid Ventilation Centre, Aalborg University, Aalborg, Denmark, p. 73 (2002). ISSN 1395–7953 R0207
18. Tabadkani, A., Roetzel, A., Li, H.X., Tsangrassoulis, A., Attia, S.: Analysis of the impact of automatic shading control scenarios on occupant's comfort and energy load. Appl. Energy **294**, 21 (2021). https://doi.org/10.1016/j.apenergy.2021.116904
19. Seifert, J., Schinke, L., Buchheim, A., Beyer, M.: The new climate room for transient investigations of thermal comfort. The REHVA Eur. HVCA J. **54**(1) (2017)
20. Optical radiation "Digital sensor for color temperature and illuminance FLAD23CCT with ALMEMO® D6 plug", October 2021. https://www.mrclab.com/digital-sensor-for-color-temperature-and-illuminance
21. Shuster, J.: Addressing glare in solid-state lighting. White Paper WP528001EN, Eaton, 3 January 2017
22. Kornaga, V.: Methods and ways of construction of intelectual lighting systems on the basis of synthesis of white light with the specific spectrum, Kyiv (2017). UDC: 628.931
23. Seifert, J.: Grundlagen der Wärmephysiologie, ITM InnoTech Medien GmbH, p. 160 (2019). ISBN: 978-3-96143-082-6
24. Vulcan, R.S., André, S., Bruyneel, M.: Photoplethysmography in normal and pathological sleep. Sensors **21**, 2928 (2021). https://doi.org/10.3390/s210929
25. Schäck, D.-I.T.: Photoplethysmography-Based Biomedical Signal Processing, Darmstadt, p. 121 (2019)
26. Chen, P.-Y., Ting, H.-J., Chen, M.-F., Lin, W.-C., Lin, K.-P.: Blood pressure variation trend analysis based on model study. In: Lin, K.-P., Magjarevic, R., de Carvalho, P. (eds.) ICBHI 2019. IP, vol. 74, pp. 115–121. Springer, Cham (2020). https://doi.org/10.1007/978-3-030-30636-6_17

27. Mukkamala, R., Hahn, J., Inan, O.T., Mestha, L.K., Kim, C., Hakan, T.: Toward ubiquitous blood pressure monitoring via pulse transit time: theory and practice. IEEE Trans. Biomed. Eng. **62**, 1879–1901 (2015)
28. Zheliazkov, Y.: Bachelor work, Digital meter of blood oxygen saturation, Kyiv (2018)
29. Finnerty, R.: How to design a better pulse oximeter: specifications. Analog Devices, American College of Surgeons – ACS, p. 5, 11 August 2021
30. Cattivelli, F.S., Garudadri, H.: Noninvasive Cuffless estimation of blood pressure from pulse arrival time and heart rate with adaptive calibration. In: 2009 Sixth International Workshop on Wearable and Implantable Body Sensor Networks, pp. 114–119 (2009). https://doi.org/10.1109/BSN.2009.35
31. Bowman, C., et al.: A method for characterizing daily physiology from widely used wearables. Cell Rep. Methods **1**(4), 100058 (2021). https://doi.org/10.1016/j.crmeth.2021.100058
32. Akpan, A., Aldabbagh, A.: Remote body fitness monitoring system with inter-user/multi-user tracking software applications and social distancing warning sensor. In: 2020 2nd International Conference on Electrical, Control and Instrumentation Engineering (ICECIE), pp. 1–8 (2020). https://doi.org/10.1109/ICECIE50279.2020.9309607
33. George, D.P., Sokolovsky, P.: MicroPython Documentation Release 1.9.4. MIT License (MIT), p. 146 (2018)
34. Dopieralski, R.: Micropython on ESP8266 Workshop Documentation Release 1.0, MIT License (MIT), p. 27 (2017)
35. Zinchenko, A.: Dyplomnyy projekt. Kontroler osvitlennya ta mikroklimatu [Diploma project. Lighting and microclimate controller], Kyiv Polytechnic Institute named after Igor Sikorsky, Faculty of electronic, Kyiv, p. 100 (2021)
36. Fahad, E.: ESP8266 Bluetooth: HC05 Bluetooth interfacing with Nodemcu ESP8266 (2021). https://www.electroniclinic.com/esp8266-bluetooth-hc05-bluetooth-interfacing-with-nodemcu-esp8266
37. Belousov, D.: Stabilizovane dzherelo zhyvlennya svitlodiodnykh svitylnykiv [Stabilized power supply of LED lamps]. Lighting technic and electricity, Kharkiv, Ukraine, vol. 2, pp. 6–10 (2015). ISSN 2079- 424X. http://nbuv.gov.ua/UJRN/svitteh_2015_2_3
38. May, P., Ehrlich, H.-C., Steinke, T.: ZIB structure prediction pipeline: composing a complex biological workflow through web services. In: Nagel, W. E., Walter, W. V., Lehner, W. (eds.) Euro-Par 2006. LNCS, vol. 4128, pp. 1148–1158. Springer, Heidelberg (2006). https://doi.org/10.1007/11823285_121
39. Foster, I., Kesselman, C.: The Grid: Blueprint for a New Computing Infrastructure. Morgan Kaufmann, San Francisco (1999)
40. Czajkowski, K., Fitzgerald, S., Foster, I., Kesselman, C.: Grid information services for distributed resource sharing. In: 10th IEEE International Symposium on High Performance Distributed Computing, pp. 181–184. IEEE Press, New York (2001). https://doi.org/10.1109/HPDC.2001.945188
41. Foster, I., Kesselman, C., Nick, J., Tuecke, S.: The physiology of the grid: an open grid services architecture for distributed systems integration. Technical report, Global Grid Forum (2002)
42. National Center for Biotechnology Information. http://www.ncbi.nlm.nih.gov

The Approach to Assessment of Technical Condition of Microprocessor Systems that Are Implemented on Integrated Circuits with a Programmable Structure

Serhii Shtanenko[1], Yurii Samokhvalov[2], Serhii Toliupa[2](✉), and Oleksiy Silko[1]

[1] Military Institute of Telecommunication and Information Technologies named after the Heroes of Kruty, Moskovska Str. 45/1, Kyiv, Ukraine
sh_sergei@ukr.net

[2] Taras Shevchenko National University of Kyiv, Volodymyrs'ka Str. 64/13, Kyiv, Ukraine
yu1953@ukr.net, tolupa@i.ua

Abstract. The chapter considers basic methods of testing digital devices implemented on the technology of programmable logic device (PLD) and it also reveals their advantages and disadvantages. It analyzes the existing approaches to diagnostics of microprocessor systems, which are based on integrated circuits with programmable structure. It is suggested to use a service processor as a diagnostic device, to be an integrated element of multiprocessor system. Such a processor allows to carry out diagnostics of microprocessor system and to perform operative restoration of the system functioning by means of its reconfiguration.

Keywords: Microprocessor system · Programmable logic device · Digital device diagnostics

1 Introduction

A programmable logic integrated circuit (integrated circuit with a programmable structure, programmable logic device, PLD) is an electronic component used to create digital integrated circuits, the classification of which is shown in Fig. 1. Unlike conventional digital microcircuits, the logic of PLD operation is set by programming with special tools: programmers and software [1–3].

A significant advantage of PLDs is their versatility and the fact that they can be quickly programmed to perform the functions of virtually any digital device.

According to the authors [4–7] it is the use of PLDs as an element base for building modern microprocessor systems today is a promising direction in terms of designing highly reliable, fault-tolerant (survivable) complex technical systems operating under adverse conditions, including cyber attacks.

In the design, manufacture and subsequent operation of such complex technical devices as PLDs, various types of deviations such as design errors, faults (defects) in

M. Klymash et al. (Eds.): TCSET 2022, LNEE 965, pp. 495–508, 2023.
https://doi.org/10.1007/978-3-031-24963-1_28

manufacture are inevitable and lead to failures of system components and sub-assemblies in the course of operation. Most of these deviations, such as design errors, have little effect on the performance of given functions, but some deviations, such as malfunctions (defects) and failures of elements and components of the system in operation, lead to the malfunction of the PLDs, resulting in its inappropriate use. One of the most common ways to detect such deviations is through test and functional diagnostics.

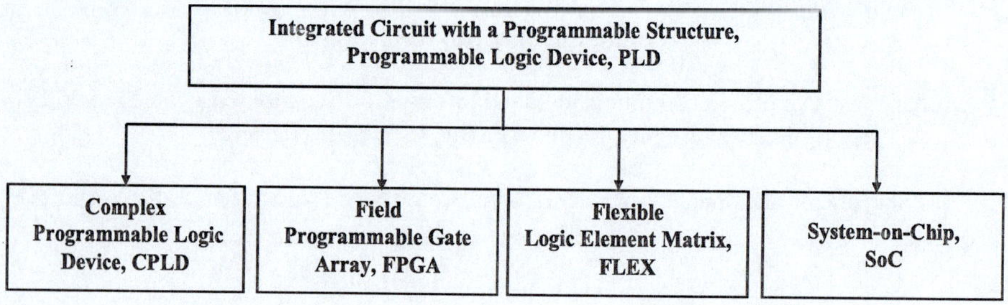

Fig. 1. Classification programmable logic devices, PLD.

Thus, according to [8–10] test diagnostics consists of applying a series of test input signals to the inputs of an integrated circuit (IC) and observing the corresponding sequence of output signals.

Functional testing, on the other hand, is used to verify that a digital device correctly implements its intended functions, without regard to information about the logic structure of the circuit. The best known types of functional testing are random and exhaustive testing.

In addition, it should be noted that test and functional diagnostics methods currently do not provide complete diagnostic information about the technical state of the IC. And considering the fact that ICs with programmable structure form a great competition to standard ICs, the issues related to the need to apply existing and develop new methods of PLDs diagnostics are relevant.

2 Analysis of Recent Studies and Publications

Nowadays a large number of studies are devoted to the testing of digital devices. For example, in [11] the problem of low efficiency of known deterministic test construction methods for digital circuits is considered, the main reason being uncontrolled enumeration of all signal combinations, which provide manifestation of a fault at the output of a defective element. A focused search method for test construction is proposed as a preference.

In [12], a cubic digital circuit analysis technique for test generation and quality assessment is presented. A cubic primitive model fault handling method and a test generation method for constant faults using cubic fault list coverage to construct an activation path are proposed.

In [13], a new test and diagnosis method based on bitstream back-reading is presented which includes a bitstream parsing algorithm as well as a corresponding error diagnosis and diagnosis algorithm based on bitstream back-reading for PLD.

The paper [14] proposed to test digital devices using genetic algorithms, which is an adaptive search method based on selection of the best elements in the population, i.e. logic elements on an integrated circuit. At the same time, the efficiency of this algorithm is significantly influenced by the right choice of coding method of individuals, as well as the right choice of parameters.

However, it should be noted that in the reviewed works, testing of digital circuits, as well as assessing the technical state of digital devices (serviceable, workable and properly functioning) is considered from the perspective that these digital devices have a rigid structure, in contrast to the PLD, where the structure is reconfigurable in nature. It should be noted that testing of digital integrated circuits with a programmable structure does not depend on the functioning of the final, user-created digital device [15].

Given the above, the purpose of this article is to justify the approach to the diagnosis of microprocessor system in the basis of PLD, which is a further development of self-diagnostic method, the use of which will improve survivability and consequently the reliability of complex technical systems functioning under adverse effects, including cyber attacks.

3 Testing Methods for Digital Devices in an PLD Basis

Programmable logic integrated circuits are a promising architecture for digital integrated circuits which are nowadays increasingly used as typical elements of functioning complex digital systems with a high degree of integration.

One of the advantages of PLD is that the operating logic is not defining in manufacture but is set by programming, making it possible to simplify the process of synthesising complex digital circuits. A second advantage is that the same regularity of the structure enables a more efficient use of the of the crystal area, thereby increasing the degree of integration. And finally, the third advantage is that having the important property of reconfigurability, PLD can be a hardware and software basis for the design of highly reliable, fault-tolerant microprocessor systems, thereby increasing the survivability and consequently the reliability of complex technical systems.

Thus, today the studies [15] describe the testing methods for first generation of PLD (PAL –Programmable Array Logic, GAL – Genetic Array Logic) with a combinational type structure quite fully.

However, the advent of PLD with structures based on programmable logic blocks-macro cells connected to external pins and internal buses (CPLD – Complex Programmable Logic Device) and CLB blocks (Configurable Logic Block) with LUTs (Look Up Table) with boolean functions from multiple arguments (FPGA – Field Programmable Gate Array) – has raised challenges for this type of PLD.

In addition, given the fact that these structures form the basis for an entirely new direction in microelectronics – universal microprocessor systems on crystal (System-on-Chip – SoC), there are challenges in testing not just digital devices, but microprocessor systems on an PLD basis.

Let us consider the most effective methods for testing PLDs with CPLD and FPGA type structures [16]:

Built-In Self-Test (BIST) is a self-diagnostic circuitry method. This method consists in the implementation of a specialized circuit (device) at the firmware level of the PLD that enables internal self-testing. Figure 2 shows the structure of a self-testing device consisting of a test signal generator (TPG), a response analyzer (ORA) and control logic.

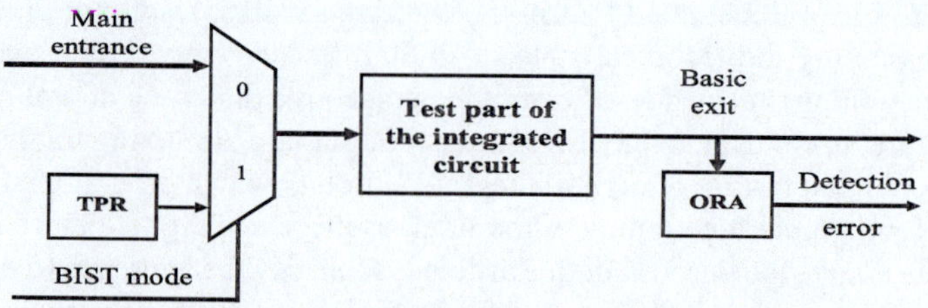

Fig. 2. Standard schema architecture self-diagnosis (BIST).

The essence of this method is to apply pseudo-random test sets to the input of the device under test and then compare the resultant value with the reference value in a signature analyzer. Implementation of the algorithm based on this method implies using a certain percentage of PLD internal logic, while the entire amount of internal logic needs to be tested when performing a functional test. Therefore this method is most suitable for testing internal PLD blocks.

The method of repeatable test sets (ILA-Iterative Logic Arrays). The essence of this method is that the configuration logic blocks (CLBs) that make up the PLD are configured in such a way that the output signals of the previous CBL block are the input signals of the next block, thus forming a test chain. This method is optimal for functioning inspection of the internal logic of a PLD. One of the advantages of the method is its high observability, allowing the location of the deviation to be determined. The main disadvantage of the method is the need of specialized test equipment as well as the low test speed compared to BIST.

The method of reading the output responses via the embedded interfaces. This method is used to perform functional control of the embedded PLD blocks. The advantages of this method are its high speed and the ability to perform functional tests without using expensive equipment. The main drawback is that this method is only applicable to next-generation PLD with a dedicated hardware interface, JTAG (Joint Test Action Group) based on IEEE standard 1149.1. In this case, testing the entire PLD is quite time-consuming and consists of combining this method with the method of built-in self-diagnostic circuits via configuration firmware.

Thus, it can be concluded that the most effective method is the one based on the BIST-Built-In Self-Test technique, i.e. built-in self-test. On this basis, further consideration of microprocessor-based PLD systems in terms of diagnostics will be based on the self-diagnostic method.

4 Approach to Diagnostics of Microprocessor Systems

The microprocessor system as an object of diagnostics is a complex computing structure with a bus organization (Fig. 3).

It consists of four main groups of ICs – microprocessor, memory, I/O controllers and object communication controllers. Each of these functional subsystems is in turn quite complex in terms of diagnostics.

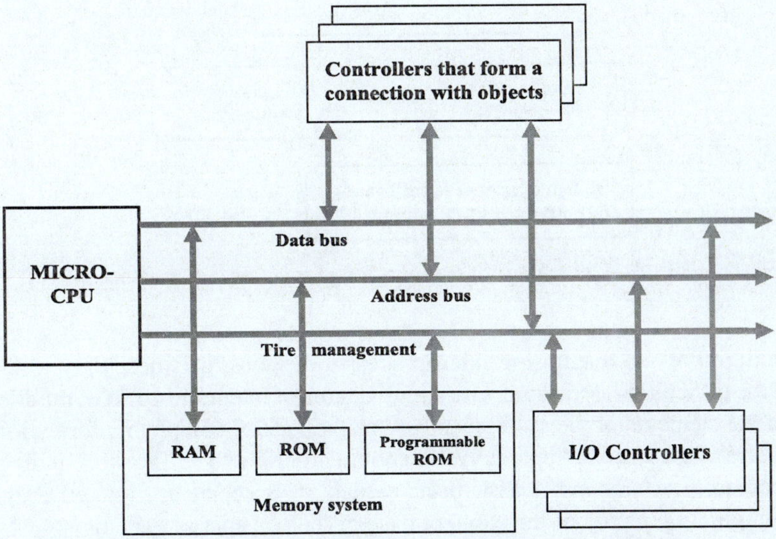

Fig. 3. Structural diagram of the microprocessor system.

Therefore, a decomposition approach is used when organizing test and functional diagnostics of microprocessor systems. In this approach separate functional devices act as an object of diagnostics: arithmetic-logic device, processor, operational-memory device, input-output device. Figure 4, 5 are generalized schemes of the process of diagnostics of microprocessor systems, which consist of the diagnosing object, diagnostics means and troubleshooting system, i.e. the system of operative restoration of correct functioning.

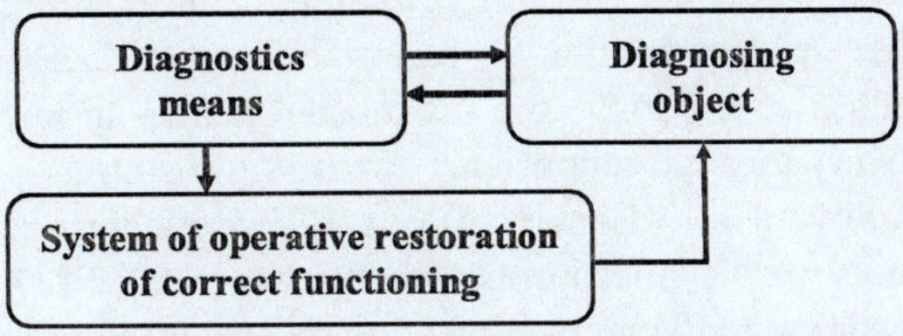

Fig. 4. The scheme of organization of test diagnostics.

The scheme of organization of test diagnostics (Fig. 4) differs from the scheme of organization of functional diagnostics (Fig. 5) by the possibility of supplying the diagnosed object with specially organized test influences, while in the process of functional diagnostics only the working influence is used. In this case, the application of these schemes in microprocessor systems solves the problems of rapid detection of errors and search for fault locations.

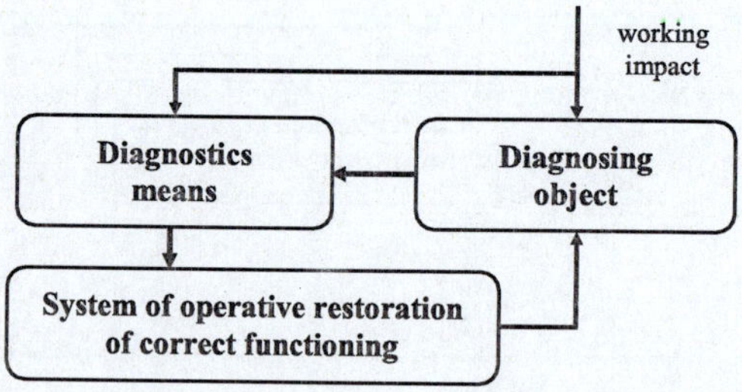

Fig. 5. The scheme of organization of functional diagnostics.

In addition, when diagnosing microprocessor systems, it is necessary to take into account the difficulties associated with a high degree of integration of ICs, ramified links between the elements of the system, as well as the lack of complete information about the structure of the system. Separately, we should highlight the difficulties of diagnosing microprocessor systems in the PLD basis, which are associated, as noted above, with the reconfiguration nature of the element base of the considered systems.

For these reasons, according to [17, 18], existing methods (constructive and abstract) of constructing diagnostic tests for microprocessor systems on standard ICs (fault function table method, existing path method, *D*-algorithm, equivalent normal form method, boolean differentiation, register transfer models, Petri network model, etc.) cannot always be effective in diagnosing microprocessor systems on PLD basis. Besides, according to [19] the use of external means of control and diagnostics in microprocessor systems is usually unjustified. This is due to the fact that interference in the microprocessor system leads to its complexity and reduces both performance and reliability of the system.

Based on the above, to determine the technical state of the microprocessor system in the PLD basis (serviceable, operable and properly functioning), as well as diagnose faults in it, it seems appropriate to use the principle of interaction between processors by introducing a service processor into, for example, a multiprocessor system [20]. The main function of such a processor is the diagnosis of the multiprocessor system and rapid automatic recovery through reconfiguration of the system.

Thus, further improvement of survivability and consequently reliability, as well as enhancing functionality of the microprocessor system in the PLD basis is associated with the use of hardware redundancy, by introducing special hardware that implements the surveillance function, i.e. using the principle of interaction.

Consider the most common interaction models ($P-$, $B-$, $R-$, $K-$, models) that describe diagnosis and control in multi-machine systems [19].

Thus, $P-$ model (Prepararto-Metze-Chien) is based on the representation of the diagnostic process by interaction of pairs of modules, the consequence of which is to obtain a set of diagnostic results [21]. The diagnostic model is represented as a diagnostic graph, where modules are vertices and inter-module links are arcs.

In this model, the module ϑ_i controls the module ϑ_j. The verification of the technical condition of the module ϑ_i by the module ϑ_j using the test sequence h_{ij} is called an elementary verification, and as a result is denoted r_{ij}. The set of tests h_{ij} forms the set J, and the set of elementary checks r_{ij} forms the set R, which is called a syndrome. The entire set J of tests is carried out in one diagnostic cycle. The result of an elementary test is marked with "0" if the test "passed" and "1" otherwise.

$B-$ model (Barsi-Grandoni-Maestrini) differs from the P model only in that a control result $r_{ij} = 0$ is considered invalid if there are simultaneous failures in the elements ϑ_i and ϑ_j. The use of the B model results in a smaller number of acceptable system operability states [22].

$R-$ model (Russel-Kime) describes the testing of one element by several other elements, either by hardware or software procedures [23]. In systems described by such models, n elements are assigned a set of $F_0 = (f_1, \ \ldots, \ f_n)$ admissible single faults f_i, for which a set of $J = (h_1, \ \ldots, \ h_p)$ tests $j = 1, \ \ldots, \ p$ is given. It is assumed that each test is complete for one fault from the set F_0.

Finally, the K- model (Makam-Avizienis) highlights the following types of faults [24]:

$A = a_1, \ \ldots, \ a_{|A|}$ – a set of atomic faults;

$M = m_1, \ \ldots, \ m_{|M|}$ – a set of macro faults;

$P = p_1, \ \ldots, \ p_{|P|}$ – a set of group defects, where $|A|, \ |M|, \ |P|$ is the power of the sets A, M, P respectively.

The set of atomic faults A describes faults at the lowest logical level of the system. They can be described by a list of functional elements or by permanent faults. Group faults of the set M represent physically or logically replaceable parts of the system, which determine the depth of the fault finding in the system. Macro faults are the consequence of the occurrence of the subset of atomic faults from the fault set A.

There are certain relationships between the fault sets. In particular, the function $f_{ap} : A \to P$ is introduced and displays the relationship r_{am} between A and M and r_{mp} between M and P. The function f_{ap} assigns one and only one group fault to each atomic fault. A group fault is considered to have occurred if an atomic fault occurs, i.e. $\forall a_i \in A, \exists p_i \in P, f_{ap}(a_i) = p_j$.

The mapping r_{am} describes the relation-tA and M, whereby the macro-fault m_j occurs whenever an atomic fault occurs a_i. In many cases, the mapping r_{am} can be represented by the function f_{am}. The relationship between the elements of the set M and P is that if m_j is a defect from a group defect, then at least one of the defects is a consequence of the macro-fault.

The elements of the sets A, M and P are single faults of the system. To define fault sets for each of these sets A', M' and P' are defined as subsets of all elements from A, M, P i.e. $M' \subseteq C(M)$, $A' \subseteq C(A)$, $P' \subseteq C(P)$ respectively.

For the sets A', M', P' define the functions f_{AP} and f_{AM} as follows:

$$f_{AP} : A' \to C(P), \forall A^i \in A', f_{AP}(A^i) = P^j = \{p_k | p_k = f_{ap}(a_i), a_i \in A^i\}$$

$$f_{AM} : A' \to C(M), \forall A^i \in A', f_{AM}(A^i) = M^j = \{m_k | a_k r_{am} m_k, a_i \in A^i\}$$

A generalized structure of links between sets is shown in Fig. 6.

The diagnostics in the system are performed with atomic tests u_i of the set $U = \{u_1, ..., u_i, ..., u_{|U|}\}$ And macro test h_j of the set $T = \{h_1, ..., h_j, ..., h_{|T|}\}$, as well as input sequences of the set $J = \{J_1, ..., J_{|J|}\}$. An atomic test consists of observing a single output at a point in time when an input sequence is applied to the circuit input. A macro test is a set of atomic tests.

Thus, comparing the models with each other, it can be noted that *P*- and *B*-models are mainly focused on diagnostics using the results of interaction of elements between themselves. In *R*- and *K*-models, when searching for fault (defects), the functional structure of the diagnostic object, diagnostic attributes and capabilities of software tools are taken into account to a greater extent.

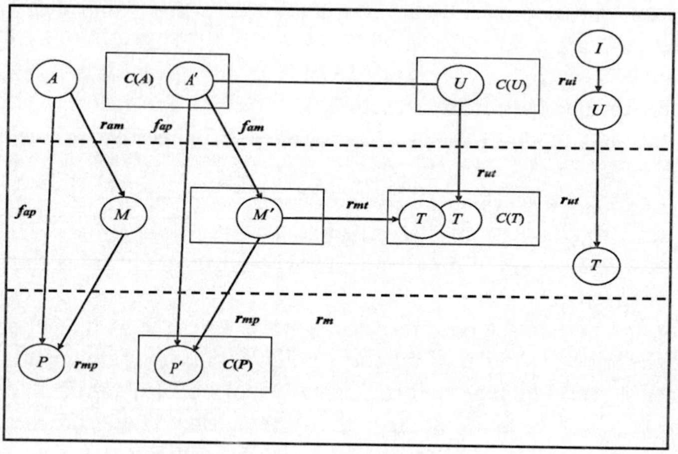

Fig. 6. Relationship in *K*-model.

In addition, the implementation of these interaction models makes it possible in the presence of diagnostic information about the technical state of the microprocessor system to decide on its reconfiguration, i.e. the reallocation of tasks (functions) by reprogramming the PLD.

5 Task Redistribution in a Multiprocessor System

Let us consider one of the approaches to increasing reliability of multiprocessor systems based on task redistribution in case of system modules failures. In this case we will use the notion of state systems. By state $s(t)$ of the system at an arbitrary moment t it is understood to be the set of states of all its microprocessors at that moment, i.e. $s(t) =< \sigma_1, \ldots \sigma_n >$, where $\sigma_i \in \{0, 1\}$, $\sigma_i = 0$, if the microprocessor is in the state of operability, $\sigma_i = 1$, if the MP is in the state of failure ($i = 1, ..., n$). Assuming that at the moment $t = 0$ is the initial state of the system $s^0 =< 00 \ldots 0 >$.

Let $\aleph = \{M_1, ..., M_n\}$ is the set of all MPs of the system, the number of which is n, and $\Omega^0 = \{\Re_1, ..., \Re_L\}$ is the set of all the tasks that the system is capable to solve while in the state of s^0, i.e. for solving of which the system has appropriate software and hardware available. We believe that for the state s^0 an initial distribution of tasks among all MPs is given so that with each M_i a subset of the tasks matches $\Omega_i^0 = \left\{\Re_i^1, ..., \Re_\mu^i, ..., \Re_{mi}^i\right\}$ which it is capable of solving if the system is in a state of s^0. Thus, at any time t, which is in the system of the state s^0, the process of solving some subset of problems is implemented, the number of which let it be equal $r_0(t)$.

If there are any failures or transitions to the system state $s_\nu(\nu = 1 ..., 2^n - 1;\ s_\nu \neq s^0)$, the system may lose the ability to solve any of the problems of the set Ω^0. Let $\Omega^\nu = \left\{\Re_1, ..., \Re_{L_\nu}\right\}$ is the set of all tasks, which the system is capable of solving, while in the state of s_ν, and the number of which equals L_ν; $\Omega_i^\nu = \Re_1^{i\nu}, ..., \Re_{\mu'}^{i\nu}, ..., \Re_{m_{i\nu}}^{i\nu}$ is a subset of the tasks that M_i is able to solve if the system is in a state of s_ν. The problems of a subset Ω_i^ν will be called module tasks M_i in the state s_ν.

Let $\aleph_f^\nu = \left\{M_f^\nu\right\}$ and $\aleph_r^\nu = \left\{M_r^\nu\right\}$ are the sets of failed and serviceable MPs, corresponding to the state of s_ν; $A_{\nu f}^0$ is the set of all tasks that in the state s^0 are performed in modules $M_r^\nu \in \aleph_r^\nu$.

If the system, while turning to the state s_ν, makes the redistribution of tasks, then in relation to the tasks of the set $A_{\nu f}^0$, the following decisions can be applied:

a_f – all tasks of the set $A_{\nu f}^0$, that are stored in the system (i.e. transmitted to perform a workable MPM set);

b_f – all tasks of the set $A_{\nu f}^0$, that ceased to be performed in the system (i.e. are discarded);

c_f – a part of the set $A_{\nu f}^0$, that is retained in the system, the others are discarded.

At the same time, in relation to the set $A_{\nu f}^0$, one of the following decisions can be applied:

a_r – all-task-sets of $A_{\nu r}^0$, that are retained in the system (i.e. either continue to be carried out in their own MPMs or transferred to other functioning MPs);

b_r – all-task-sets of $A_{\nu r}^0$, that are discarded;

a_r – a part of the tasks of the set $A_{\nu r}^0$ is discarded, the rest is retained in the system;

The set of $\prod$ possible decisions that can be taken when reallocation of tasks in the state s_ν, is defined by the set of all possible combinations of the above solutions for subset problems $A_{\nu r}^U$ and $A_{\nu r}^0$, namely:

$$\prod = \left\{(a_f a_r),\ (a_f b_r),\ (a_f c_r),\ (b_j a_r),\ (b_f b_r),\ (b_f c_r),\ (c_f a_r),\ (c_f b_r),\ (c_f c_r)\right\}.$$

Thus, there are currently two approaches to organizing task reallocation in a system in case of module failures. The first approach, static, assumes that a subset $S = \{s_\nu\}$ of system states, for which task reallocation is necessary, is known and it is required to find a set $\Gamma = \{G_\nu\}$ of task distribution such that each distribution G_ν, corresponding to state s_ν, satisfies the requirements to the chosen indicators of system reliability. At that, optimal distribution G_ν for all states $s_\nu \in S$ are found when designing the system, and each module when building the system is provided with resources necessary to perform tasks both in the state s_0, and in each of the states $s_\nu \in S$ according to the distributions

G_ν. When the system is switched to the state $s_\nu \in S$ and a failure is detected in all the operable modules of the system, the tasks corresponding to the distribution G_ν are started.

The second approach, called dynamic approach, is based on the fact that the task of finding the optimal allocation G_ν for the state s_ν is solved each time (with the help of a special solver, i.e. service processor) when the system enters the state s_ν and its result may depend on the previous states of the system. In general, in both static and dynamic approaches, obtaining the optimal allocation may require task reallocation of both failed and operable modules.

6 Static Approach to Organizing Task Reallocation

Consider a static approach to organizing task reallocation, assuming that only failed modules are reallocated when the system enters the state s_ν [26].

For each state $s_\nu \in S$ find the distribution G_ν, satisfying the conditions:

$$\Phi_\nu = \max;\ C_\nu \leq C_{\nu\ addit},\ T_i^\nu \leq T_{i\ addit},\ i = 1,\ \ldots,\ g_\nu,$$

where C_V is the cost required to be able to transfer tasks of failed modules to operable modules in the state s_ν; T_i^ν is the average time to service the application in the module M_i in the state s_ν; $C_{\nu\ addit}, T_{i\ addit}^\nu$ are permissible values of the specified indicators for the state s_ν.

Let the task $\Re_j$, executing at the initial distribution in the module M_i, denote $\Re_{ji}$. Let the initial allocations G_0 is given by the matrix $\|\beta_{ji},\|j = 1,\ \ldots,\ L,\ i = 1,\ \ldots,\ n$. The element β_{ji} is the weight of the task $\Re_j$, performed at the initial allocation in the module M_i, defined as $\beta_{ji} = \sum\limits_{\varphi_{ji}^s \in F_{ji}} \rho_{ji}^s$, where $F_{ji} = \left\{\varphi_{ji}^s\right\}$ is the set of technological processes which requires the task $\Re_{ji}$; ρ_{ji}^s is the weight of technological processes φ_{ji}^s, $s = 1,\ \ldots,\ h_{ji}, h_{ji} = |F_{ji}|$. Let $D_\nu^f = \{M_l\}, D_\nu^r = \{M_i\}$ are the sets of failures respectively and still workable modules for the state s_ν; A_ν^f is the set of all the tasks, that are performed in the modules $M_l \in D_\nu^f$ in the state s^0.

Then the component of the system efficiency value in the state s_ν, determined by the performance of the operable modules in this state $\Re_{ji} \in A_\nu^f$, can be expressed as.

$$\Phi_\nu^f = \sum_{\Re_{jl} \in A_\nu^f} b_{jl}\beta_{il},$$

where $b_{jl} = 1$, if in the state s_ν the task $\Re_{jl}$ is performed by one of the operable modules $b_{jl} = 0$ otherwise.

Let $C_{(jl)i}$ be the costs required to implement the transfer of the task $\Re_{jl}$ ($\Re_{jl} \in A_\nu^f$) to some module M_i, where $M_i \in D_\nu^r$, the values $C_{(jl)i}$, independent of l, are given by the matrix $\|C_{ji}.\|$

Then the cost required to implement the transfer of a fixed task $\Re_{jl}$ to any of the workable models in the state s_ν,is.

$$C_{\nu(jl)} = \sum_{M_i \in D_\nu^r} d_{jl}^i C_{(jl)i},$$

where $d_{jl}^i = 1$, if the task $\Re_{jl}$ is transferred to the module M_i,$d_{jl}^i = 0$ otherwise.

If a module M_l fails, its task $\Re_{jl}$ can either be discarded, i.e. not passed to any of the capable modules, or passed to only one of the modules in the set D_ν^r. In the first case $d_{jl}^i = 0$ for all $M_i \in D_\nu^r$ and $b_{jl} = 0$, in the second case only one of the values $d_{jl}^i = 1$ and hence $b_{jl} = 1$. Therefore,

$$b_{jl} = \sum_{i=1}^{g_\nu} d_{jl}^i$$

where $d_{jl}^i \in \{0, 1\}$ and $b_{jl} \in \{0, 1\}$ are valid.

The total cost required to implement the reallocation of all tasks $\Re_{jl} \in A_\nu^f$, is expressed as.

$$C_\nu = \sum_{\Re_{jl} \in A_\nu^f} C_{\nu(jl)} = \sum_{\Re_{jl} \in A_\nu^f} \sum_{M_i \in D_\nu^r} d_{jl}^i C_{(jl)i}.$$

The average service time for an application in the module M_i in the state s_ν is determined by the expression.

$$T_i^\nu = T_i^0 + \sum_{\Re_{jl} \in A_\nu^f} d_{jl}^i \Delta T_{(jl)i},$$

where $\Delta T_{(jl)i}$ is the increment of the average service time of a serviceable module M_i when a task is transferred to it $\Re_{jl}$, where $\Re_{jl} \in A_\nu^f$; values $\Delta T_{(jl)i}$, independent of l, are given by the matrix $\|\Delta T_{jl}\|$.

Thus, finding the optimal allocation of C_ν tasks in the system for a fixed state s_ν is reduced to solving an integer linear programming problem, which consists of finding a set of values of the variables d_{jl}^i, where $d_{jl}^i \in \{0, 1\}$, at which the maximum value of the functions is achieved

$$\Phi_\nu^f = \sum_{i=1}^{g_\nu} \sum_{\Re_{jl} \in A_\nu^f} d_{jl}^i \beta_{jl}$$

with the following restrictions:

$$\left.\begin{array}{l} C_\nu = \sum\limits_{i=1}^{g_\nu} \sum\limits_{\Re_{jl} \in A_\nu^f} d_{jl}^i C_{(jl)i} \le C_{\nu\ addit}\ ; \\ T_i^\nu = T_i^o + \sum\limits_{\Re_{jl} \in A_\nu^f} d_{jl}^i \Delta T_{(jl)i} \le T_{i\ addit}^\nu,\ i = 1,\ 2,\ \ldots,\ g_\nu \end{array}\right\}$$

This problem can be solved by any of the known methods of integer linear programming.

Once the set $\Gamma = \{G_\nu\}$ of optimal task distributions for a given subset $S = \{s_\nu\}$ of system states is found, the set of tasks that it should be able to solve in the state s^0 and in all states $s_\nu \in S$ should be found for each module of the system and the total costs required to implement the resulting set $\Gamma = \{G_\nu\}$ of distributions, the average time to serve a request in each module and the system performance for each state of the given set $S = \{s_\nu\}$ should be calculated.

The optimal distribution of tasks found for each state $s_\nu \in S$ $(\nu = 1, \ldots, r)$ we describe by the matrix $\left\| A^\nu_{jl} \right\|$, whose rows correspond to tasks $\Re_j$ $(j = 1, \ldots, L)$, and columns to modules M_i $(i = 1, \ldots, n)$, A^ν_{ji} are some subset of tasks defined below. For the columns corresponding to workable modules M_i in state s_ν, $A^\nu_{ji} = \Re_{jl} \cup \left\{\Re_{jl}\right\}^{\nu i}$, where $\Re_{ji}$ is the task performed in M_i in state s^0, $\left\{\Re_{jl}\right\}^{\nu i}$ is the set $\Re_{jl}$ transferred to M_i in state s_ν.

For the columns corresponding to the failures M_i, $A^\nu_{ji} = \emptyset$. Let us construct the resulting matrix $\left\| A_{ji} \right\|$, where $A_{ji} = \bigcup\limits_{\nu=1}^{r} A^\nu_{ji}$.

The matrix $\left\| A_{ji} \right\|$ defines for each module M_i, $(i = 1, \ldots, n)$, the set of those tasks for which it should have the necessary resources in order for each of the corresponding state of the system to achieve the maximum value of its efficiency under given constraints on module performance and additional costs. This matrix can be used to calculate the above-mentioned system performance.

To summarize all above mentioned, it can be stated that the implementation of the principle of interaction between processors by introducing a service processor in the PLD-based multiprocessor system makes it possible not only to assess the technical state of the system, but also to decide on the automatic recovery of the system by reprogramming the PLD.

7 Conclusion

The problem of diagnostics of microprocessor-based PLD systems, as a basis for building complex technical systems operating under adverse conditions, including cyber-attacks, is considered. An analysis of existing methods of testing integrated circuits with programmable structure is considered, their advantages and disadvantages are noted. It was proposed to use a service processor as a diagnostic device, as a integral element of the multiprocessor system. This processor will not only collect diagnostic information, but also decide on the reconfiguration of the system to restore its functioning automatically by reprogramming the PLD.

In general, the proposed approach to the diagnosis of microprocessor systems in the basis of PLD is a further development of self-diagnostic method. If a complex technical system is provided with the property of adaptation to changing operating conditions, then such a system can be considered as a system with elements of artificial intelligence.

References

1. Tahoori, M.B., Mitra, S.: Computer-aided design of integrated circuits and systems **23**(2) (2005) 261. https://doi.org/10.1109/TCAD.2003.822112
2. Tischler, M.: Programmable logic devices in circuit development. ATZ Worldw **107**, 20–21 (2005). https://doi.org/10.1007/BF03224785
3. Samokhvalov, Y., Toliupa, S., Buchyk, S., Shtanenko, S.: Design of robotic systems in the basis of cad intel quartus prime, 2021. In: IEEE 3rd International Conference on Advanced Trends in Information Theory (ATIT), 2021, p. 179 – 183, https://doi.org/10.1109/ATIT54053.2021.9678857
4. Joost, R., Salomon, R.: Advantages of FPGA-based multiprocessor systems in industrial applications. In: 31st Annual Conference of IEEE Industrial Electronics Society, 2005. IECON p. 6 (2005) https://doi.org/10.1109/IECON.2005.1568946
5. Mousavi, M., De, S., Pourshaghaghi, H.R., Corporaal, H.: Fault Tolerant FPGAs: where to spend the effort?. In: 2019 22nd Euromicro Conference on Digital System Design (DSD), pp. 651–654 (2019). https://doi.org/10.1109/DSD.2019.00103
6. Rapid System Prototyping with FPGAs: Accelerating the Design Process. R. C. Cofer, Benjamin F. Harding. Elsevier, p. 6 (2006)
7. Jun, S., et al.: A cost-efficient software based router and traffic generator for simulation and testing of IP network. Electronics (Switzerland) **9**(1) (2020)
8. Shtanenko, S., Samokhvalov, Y., Toliupa, S., Silko, O.: Increasing survivability of technological systems based on the technology of programmable logic device. In: Conference: Information Technology and Implementation (IT&I-2021), Dec 01–03, 2021, Kyiv, Ukraine At: CEUR Workshop Proceedings (Vol-3132), p. 237 – 245 (2022)
9. Beshley, M., Seliuchenko, M., Panchenko, O., Zyuzko, O., Kahalo, I.: Experimental performance analysis of software-defined network switch and controller. In: Paper presented at the 14th International Conference on Advanced Trends in Radioelectronics, Telecommunications and Computer Engineering, TCSET 2018 - Proceedings, 2018-Apr 282–286. https://doi.org/10.1109/TCSET.2018.8336203 Retrieved from www.scopus.com
10. Simons, R.: "Strategy Execution Module 11: Using Diagnostic and Interactive Control Systems. Harvard Business School Module Note 117–111, November 2016. (Revised December 2016.)
11. Shalobanov, S.V., Shalobanov, S.S.: Diagnostics of Automatic Control Systems Using Trial Deviations of Model Parameters and Binary Diagnostic Signs. 530–534. (2020). https://doi.org/10.1109/RusAutoCon49822.2020.9208057
12. Shvachko, A.A., Prokopenko, T.N., Yanakova, E.S., Kozlov, V.K.: Digital circuit testing methods: an analysis of modern problems. In: 2021 IEEE Conference of Russian Young Researchers in Electrical and Electronic Engineering (ElConRus), 2021, pp. 2739–2742 (2021). https://doi.org/10.1109/ElConRus51938.2021.9396341
13. Hahanov, V.I., Babich, À.V., Hyduke, S.M.: Test generation and fault simulation methods on the basis of cubic algebra for digital devices. In: Proceedings of the Euromicro Symposium on Digital Systems Design DSD2001. Warsaw, Poland. September, 4–6, pp. 228 – 235 (2001)
14. Ruan, A., et al.: A bitstream readback-based automatic functional test and diagnosis method for Xilinx FPGAs. Microelectron. Reliab. **54**(8), 1627–1635 (2014)
15. Gorodilov, A., Morozenko, V.: Genetic algorithm for finding the key's length and cryptanalysis of the permutation cipher. Inform. Theor. Appl. **15**(1) 94–99 (2008)
16. Aksenova, G.P.: Compaction of test response at self-testing in the programmable logic matrices. Autom. Remote Contr. **74**, 264–274 (2013). https://doi.org/10.1134/S0005117913020082

17. Stroud, C., Nall, J., Lashinsky, M., Abramovici, M. (2002). BIST-based diagnosis of FPGA interconnects. In: IEEE International Test Conference (TC), pp. 618 – 627
18. Koong, C.-S., et al.: Automatic testing environment for multi-core embedded software-ATEMES. J. Syst. Softw. **85**, 43–60 (2012). https://doi.org/10.1016/j.jss.2011.08.030
19. Obukhova, K., Zhuravska, I., Burenko, V.: Diagnostics of power consumption of a mobile device multi-core processor with detail of each core utilization. In: 2020 IEEE 15th International Conference on Advanced Trends in Radioelectronics, Telecommunications and Computer Engineering (TCSET), pp. 368–372, (2020) https://doi.org/10.1109/TCSET49122.2020.235456
20. Shinghal, D., Dinesh, C.: Design and analysis of a fault tolerant microprocessor based on triple modular redundancy using VHDL. Int. J. Adv. Eng. Technol. **1**(1), 21–27 (2011). ISSN: 2231–1963
21. Principles of High-Performance Processor Design For High Performance Computing, Deep Neural Networks and Data Science. Springer, Cham (2021). https://doi.org/10.1007/978-3-030-76871-3
22. Preparato, F.P.: On the Connection Assignment Problem of Diagnosable Systems. In: F. P. Preparato, G. Metze, R. J. Chien. IEEE Trans. **6** 848–854 (1967)
23. Barsi, T., Crandoni, T., Maestrini, P. A.: Theory of Diagnosability of Digital Systems. IEEE Trans. **6** 585–593 (1976)
24. Russel, J., Kime, C.: Sistem Fault Diagnosability Without Repair. IEEE Trans. **24**(11), 1078–1089 (1975)
25. Kime, C.K.: An abstract model for digital system diagnosis. IEEE Trans. **28**(8) 754–767 (1979)
26. Aghaee, F., Dehkordi, N.M., Bayati, N., Hajizadeh, A.: Distributed control methods and impact of communication failure in AC microgrids: a comparative review. Electronics **8**, 1265 (2019). https://doi.org/10.3390/electronics8111265
27. Kheiralla, A., Siddig, O., Mokhtar, A., Esameldeen, M., Abdalla, O.: Design and Development of a Low Cost Programmable Logic Controller Workbench for Education Purposes (2007).

Robust Approach to the Signal and Data Processing

Ihor Prokopenko(✉)

National Aviation University, Lubomyr Huzar Ave., 1, Kyiv 03058, Ukraine
igorprok48@gmail.com

Abstract. The chapter is devoted to the development of the theory of robust methods of signal and data processing construction of space distributed information-measuring systems in their distortion by non-Gaussian impulse interference and failures of the recording equipment, purposed for EmN (Emerging Networking). The problem of robust estimation of regression parameters of a random process is considered, the recurrent algorithm of gradient search of the solution of likelihood equations for non-Gaussian perturbations with density of distribution of Tukey type is synthesized and their efficiency is investigated.

Keywords: Robustness · Maximum likelihood methods · Non-Gaussian pulse interference · Regression parameters estimation

1 Introduction

Distributed information and measurement systems (DIMS) in EmN (Emerging Networking) deployment for various purposes, such as radar systems for monitoring of the meteorological situation in large areas or radar airspace surveillance systems (European air traffic control radar network) have become widely used. These systems receive information from many spatially-distributed sensors and interact with the control center and with each other using telecommunications computer networks. The distributed structure of DIMS creates the tasks of primary processing of sensor signals (RADAR), which consist in measuring information parameters (spatial coordinates of object), compressing data streams (tracking) transmitted by communication channels to the center of complex processing and decision-making. These systems operate under conditions of interference created by different sources and changed over time. Algorithms for signal and data processing in such systems must be noise-tolerant, i. e. robust. In this paper the problem of synthesis of robust algorithms of signal and data processing on the basis of use of statistical model of noise of distribution with the complex distribution law of Tukey type is investigated.

The problem of providing the stability of signal processing algorithms operation is known since 1960s. The scope of signal and data processing methods for networking purposes and EmN is growing. It includes multiple distributed and networked applications like radar, biomedical diagnostic, communications, meteorology, reliability etc.

M. Klymash et al. (Eds.): TCSET 2022, LNEE 965, pp. 509–522, 2023.
https://doi.org/10.1007/978-3-031-24963-1_29

Signal processing methods are usually based on the Gaussian model of probability density distribution. However, the real distribution may differ significantly from the Gaussian one. This leads to a decrease in the reliability of statistical inference - an increase in the probability of errors in decision-making and a decrease in the accuracy of measuring information parameters. Particularly harmful are large deviations from the Gaussian model on the "tails" of the distribution - large emissions that occur due to the action of impulse interference of natural or artificial origin.

The theoretical foundations of a robust approach in statistics were laid down by the works of Huber [1], Hampel [2, 13], Hogg [3] in the 60–70 years of the last century. The introduction of a robust approach to the synthesis of radar signal processing algorithms should be traced back to the 1970s. These are the works of E. Korniliev [4], H. Rohling [5], I. Prokopenko [6–11], O. Zelensky [12]. Recently, the flow of work on robust signal and data processing techniques has increased significantly [13–25].

The approach proposed by Huber in [1] is based on the spline model of the distribution, the central part of which is described by the Gaussian curve $f_0(x)$, and the "tails" by the Laplace distribution.

$$f(x) = \begin{cases} (1-p)f_0(x)exp\,(k(x-x_0)), & x \le x_0, \\ (1-p)f_0(x), & x_0 < x < x_1, \\ (1-p)f_0(x)exp\,(-k(x-x_0), & x \ge x_1, \end{cases} \quad , \tag{1}$$

where p is probability of pollution, that described deviation of base Gaussian distribution by interference.

Herewith:

$$1 - p = \int_{x_0}^{x_1} f_0(x)dx,$$

Value of k is connected with p by the following equation:

$$1 - p + (f_0(x_0) + f_0(x_1))/k = 1/(1-p).$$

This distribution minimizes the Fisher information for the parameter to be estimated from the sample $x_1, .., x_n$. The estimation of the parameter is sought by the method of maximum likelihood as one that maximizes the likelihood function, ie provides minimal losses at the worst "pollution" of the sample by the interference with the Laplace distribution. This approach allows us to synthesize robust algorithms in many tasks [4–11]. But its disadvantage is the spline structure of the distribution density (1), which leads to the spline structure of nonlinear converters of sample values $x_1, .., x_n$ and complicates the adaptation process when changing the noise parameters, in particular when changing the degree of "pollution" of the sample p.

This work uses a more convenient model to synthesis of the robust signal processing procedures - a composite distribution, which corresponds to the sampling algorithm shown in Fig. 1. We use Tukey type model of signal and interference mixture probability density function and maximum likelihood method. The robust algorithm for estimation of regression model parameters on the background of mixed Gaussian noise and impulse interference will be synthesized.

This work uses a more convenient model to synthesis of the robust signal processing procedures - a composite distribution, which corresponds to the sampling algorithm shown in Fig. 1. We use Tukey type model of signal and interference mixture probability density function and maximum likelihood method. The robust algorithm for estimation of regression model parameters on the background of mixed Gaussian noise and impulse interference will be synthesized.

2 Mathematical Model of Signal Distribution

The set of distributions that to some extent differs from the Gaussian one can be given by composite distributions of the Tukey type

$$f(x) = \frac{(1-p)}{\sqrt{2\pi\sigma}} exp\left(-\frac{(x-a)^2}{2\sigma^2}\right) + p\lambda exp(-\lambda|x-a|), \tag{2}$$

where a, σ^2 - the mathematical expectation and variance of the Gaussian distribution, λ - the parameter of the scale of the Laplace distribution, p - the parameter that determines the degree of deviation from the Gaussian model.

Distribution (2) is a mixture of Gaussian distribution and Laplace distribution, which is formed by the algorithm shown in Fig. 1.

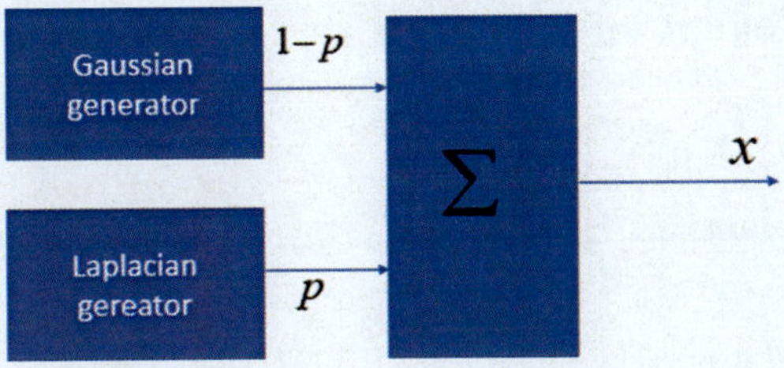

Fig. 1. Random variable modeling algorithm with distribution.

The sequence of random numbers, according to the algorithm Fig. 1, is formed by adding to the Gaussian numbers generated by the Gaussian generator, pulses whose amplitude is distributed according to Laplace's law, and the probability of such a pulse during one sampling interval is equal to p. In this case, the output of the mixture sensor with a probability of (1-p) will be present Gaussian noise and with a probability of p - a mixture of Gaussian noise and impulse noise. This is consistent with the physical mechanism of formation of a mixture of Gaussian noise and impulse interference in radio channels and radio signal processing equipment. In Fig. 2 shows the densities of the distribution (1) and (2) by some parameters p, $\lambda = 1$, $\sigma = 1$ and the implementation of sequence of random numbers with distributions (2).

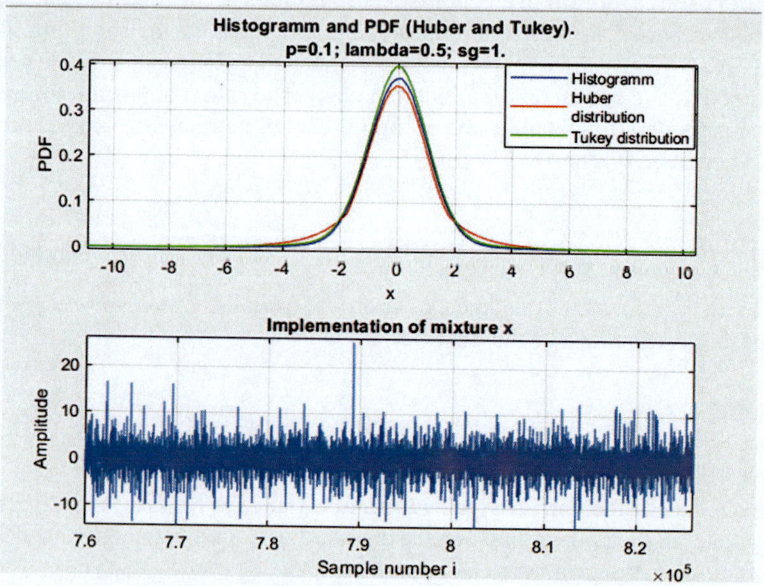

Fig. 2. Non-Gaussian probability distributions with heavy "tails" and implementation of Gaussian noise and Laplace pulse interference with parameters $p = 0.1$; $\lambda = 0.5$; $\sigma = 1$.

3 Problem Statement of Research

Assume that for $x_1, \ldots x_n$ - sampling of a mixture of signal and interference with one-dimensional distribution (2) and vector parameter $\bar{a} = \{a_1, .., a_n\}$. The task is to estimate the vector of parameters $\Theta = \{\bar{a}_i, \ \sigma, \ \lambda, \ p\}$. To solve this problem, we write the likelihood function

$$\varphi(x_1, \ldots, x_n, \theta) = \prod\nolimits_{i=1}^{n} f(x_i, \theta) = \prod\nolimits_{i=1}^{n} \frac{(1-p)}{\sqrt{2\pi\sigma^2}} exp\left(-\frac{(x_i - a_i)^2}{2\sigma^2}\right) + p\lambda exp(-\lambda|x_i - a_i|). \quad (3)$$

Depending on the a priori information about the values of the distribution parameters (3), the problem has a corresponding degree of complexity. In the general case, if all the parameters of distribution (3) are unknown, then the maximum likelihood method should be used to look for such a vector of estimates of unknown parameters, at which the likelihood function (3) reaches the maximum value.

$$\theta^* = (\varphi(x_1, \ldots, x_n, \theta)). \quad (4)$$

Analytically, this problem is extremely difficult to solve with a large dimension of the vector of unknown parameters. For this purpose can use the Monte Carlo method, which theoretically solves the problem of finding the global extremum of a function of many variables with a certain accuracy that depends on the number of tests. To achieve the desired accuracy in a multidimensional problem, this number can be very large.

The function maximum of many variables can be found by the gradient search method.

We consider the problem of estimating the parameters of a polynomial trend of a random process with a non-Gaussian one-dimensional distribution (2), the mathematical expectation of which changes over time.

Estimation of polynomial regression parameters under the action of Gaussian uncorrelated noise and impulse clutter.

Let the observed process with discrete time be described by the equation

$$x(t_i) = \sum_{k=0}^{m} a_k t_i^k + \omega_i, i = \overline{1, n}. \quad (5)$$

where $a_0, .., a_m$ - parameters of regression, ω_i, $i = \overline{1, n}$- samples of non-Gaussian interference of distribution (2).

Likelihood function according to (3) takes the form

$$\varphi(x_1, \ldots, x_n, t_1, \ldots, t_n\theta) = \prod_{i=1}^{n} \frac{(1-p)}{\sqrt{2\pi\sigma}} exp\left(-\frac{\left(x_i - \sum_{k=0}^{m} a_k t_i^k\right)^2}{2\sigma^2}\right) + p\lambda exp\left(-\lambda a\left|x_i - \sum_{k=0}^{m} a_k t_i^k\right|\right) \quad (6)$$

and its logarithm

$$log(\varphi(x_1, \ldots, x_n, t_1, \ldots, t_n\theta)) = \sum_{i=1}^{n} log((1-p)exp\left(-\frac{\left(x_i - \sum_{k=0}^{m} a_k t_i^k\right)^2}{2\sigma^2}\right) + p\lambda exp\left(-\lambda a\left|x_i - \sum_{k=0}^{m} a_k t_i^k\right|\right)). \quad (7)$$

Let the parameters of the noise distribution be known. Consider the problem of estimating regression parameters $a_0, .., a_m$. . To solve it, according to the method of maximum likelihood, we must find the maximum of the logarithm of the likelihood function (6) in the range of possible values of the parameters $a_0, .., a_m$.

Next, we will use the notation for a random variable normalized by mathematical expectation

$$z(t_i) = x_i - \sum_{k=0}^{m} a_k t_i^k. \quad (8)$$

We will use two methods of finding the extremum of the function of many variables - the method of gradient descent and the proposed method of compressive coatings.

Gradient descent method. This method involves finding partial derivatives of function (7) and constructing an iterative procedure of descent to the extremum point on the gradient.

$$\begin{pmatrix} a_{0,i+1} \\ \vdots \\ a_{m,i+1} \end{pmatrix} = \begin{pmatrix} a_{0,i} \\ \vdots \\ a_{1,i} \end{pmatrix} + \begin{pmatrix} \lambda_{0,i} \\ \vdots \\ \lambda_{1,i} \end{pmatrix} \times \begin{pmatrix} g_1(a_{0,i}, .., a_{m,i}) \\ \vdots \\ g_m(a_{0,i}, .., a_{m,i}) \end{pmatrix} \quad (9)$$

where $g_1(a_{0,i}, .., a_{m,i}), .., g_m(a_{0,i}, .., a_{m,i})$ - partial derivatives at i-step, expressions for which have the form:

$$g_k(\overline{x}, a_{0,i},.., a_{m,i}) = \frac{d}{da_k}\log\left(\varphi(x_1,..,x_n,t_1,..,t_n,x_{0,i},..,x_{m,i})\right) =$$

$$= \sum_{i=1}^{n} \frac{\frac{(1-p)}{\sqrt{2\pi\sigma^2}}(z_i(t_i))t_i^k)\exp(-\frac{(z_i(t_i))^2}{2\sigma^2}) + pt_i^k\lambda^2\text{sign}(z_i(t_i))\exp(-\lambda|z_i(t_i)|)}{\frac{(1-p)}{\sqrt{2\pi\sigma}}\exp\left(-\frac{(z_i(t_i))^2}{2\sigma^2}\right) + p\lambda\exp(-\lambda|z_i(t_i)|)} \quad (10)$$

$k = \overline{0, m}.$

Expression (10) can be transformed into a form

$$g_k(\underline{x}, a_0, .., a_m) =$$

$$= \sum_{i=1}^{n} \frac{z_i(t_i))t_i^k + p\sqrt{2\pi}\sigma^2\lambda^2 t_i^k sign(z_i(t_i))exp\left(\frac{z_i(t_i)^2}{2\sigma^2} - \lambda|z_i(t_i)|\right)/(1-p))}{1 + \frac{(1-p)}{\sqrt{2\pi}\sigma}exp\left(-\frac{(z_i(t_i))^2}{2\sigma^2}\right) + p\lambda exp(-\lambda|z_i(t_i)|)} \quad (11)$$

$$k = \underline{0, m}.$$

Graphs of the random sequence implementation function with probability distribution (2), logarithm of the likelihood function (6) and its derivatives for linear regression (m = 1) are shown in Fig. 1–3.

The derivative value fields $g_0(\bar{x}, a_{0,i}, a_{1,i})$, calculated on the grid of values of the standard deviation of Gaussian noise σ and the estimation of the regression coefficient a_0^* for the two values of the Laplace distribution parameter λ are shown in Fig. 4.

The block diagram of the recurrent algorithm for estimating the regression coefficients is shown in Fig. 7.

4 The Results of Calculations

An important element of the gradient procedure is the change of factors for derivatives depending on the desired rate of coincidence and the accuracy of the evaluation result. The following algorithm was used in this work.

The following initial values of the coefficients were accepted: $\lambda_0 = 1$; $\lambda_1 = 0.05$. These values did not change until the signs of the derivatives began to change regularly. After fixing the change of the sign, the corresponding coefficient was multiplied by 0.85 and the procedure was continued until the next change of the sign of the derivative.

A prerequisite for the stability of the efficiency characteristics of robust algorithms is their nonlinear nature.

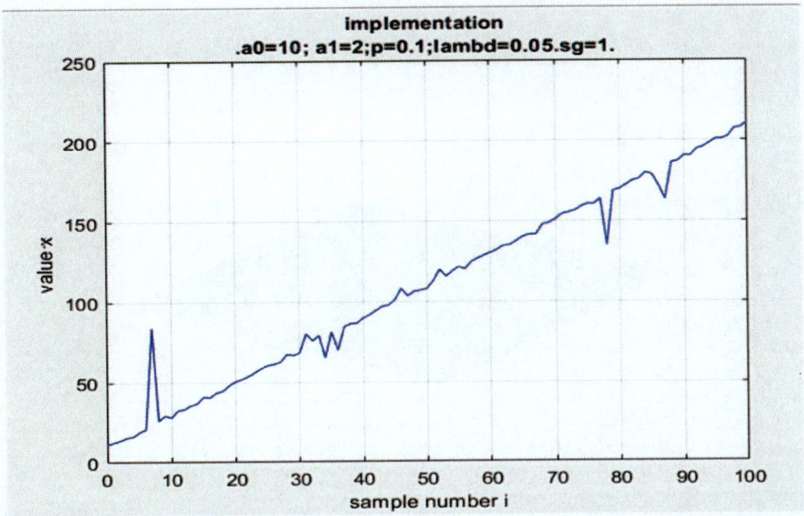

(a)

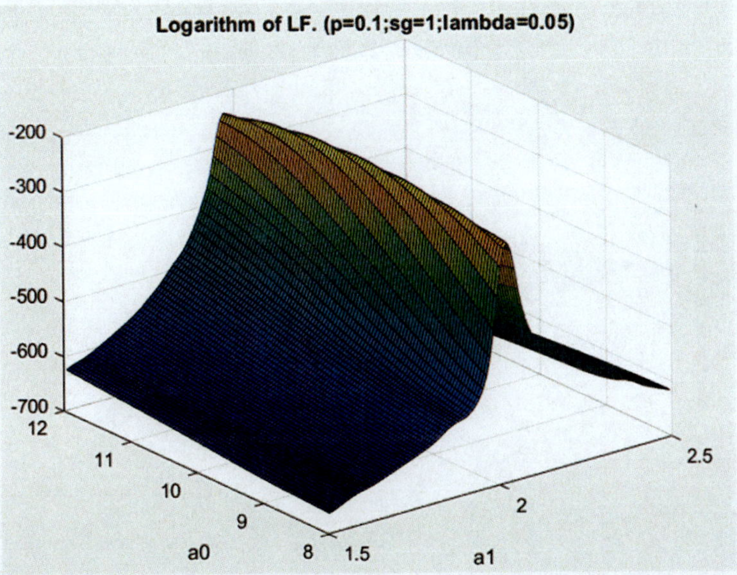

(b)

Fig. 3. a) The implementation of random sequence of mixture line (a0 = 10; a1 = 2), Gaussian uncorrelated noise ($\sigma = 1$)and pulse noise with the distribution of pulse amplitudes according to Laplace's law ($\lambda = 0.05$) and probability of defeat p = 0.1; b) logarithm of the likelihood function.

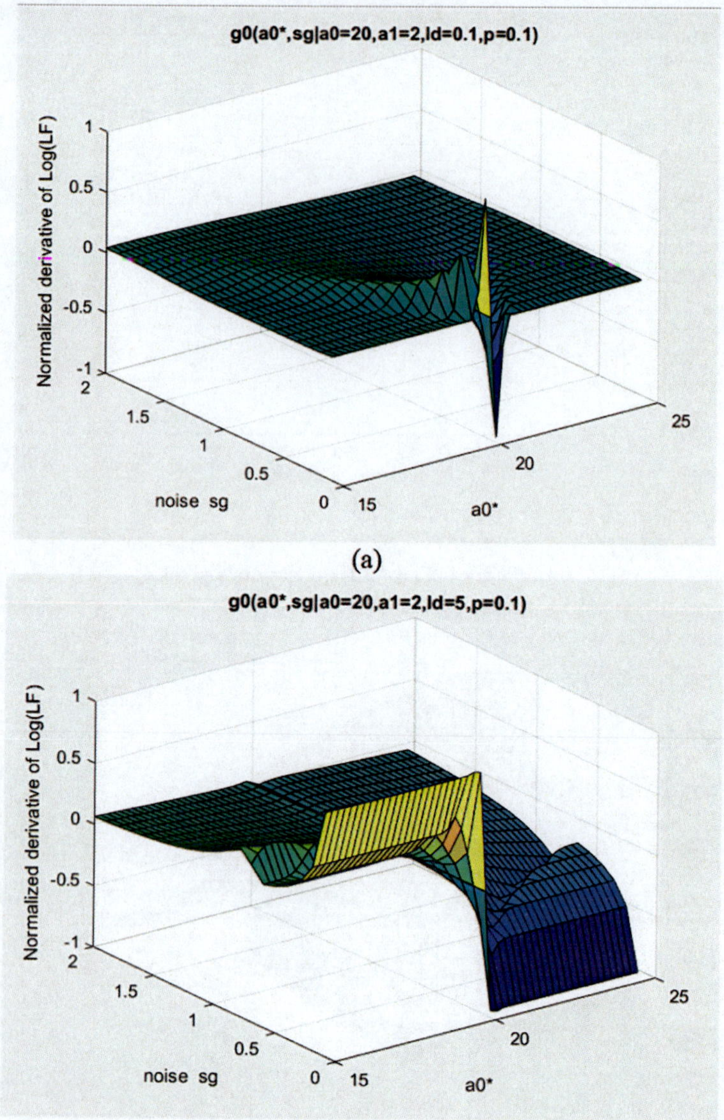

(b)

Fig. 4. The derivative value fields $g_0 = (a_0^*, \sigma | a_0 = 20; a_1 = 2; p = 0.1)$ as function of two parameters - the standard deviation of Gaussian noise σ and Laplace distribution parameter λ: a) $\lambda = 0.1$; b) $\lambda = 5$.

As can be seen from the above characteristics (Fig. 5, 6), the working area of the algorithm lies in a limited range, which is determined by the dispersion of Gaussian noise. With a small dispersion ($\sigma = 0.1$), the linear section of the characteristics of the channels of adjustment of derivatives is relatively small (Fig. 5b), with increasing noise dispersion, the linear section expands. (Fig. 5a). If the deviations from the trajectory

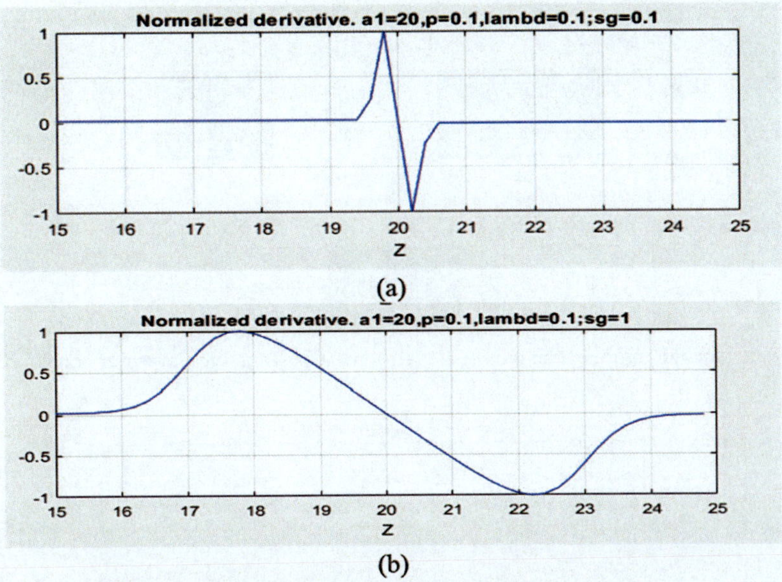

Fig. 5. The field intersection (Fig. 4a): a) for $\sigma = 0.1$; b) $\sigma = 1$.

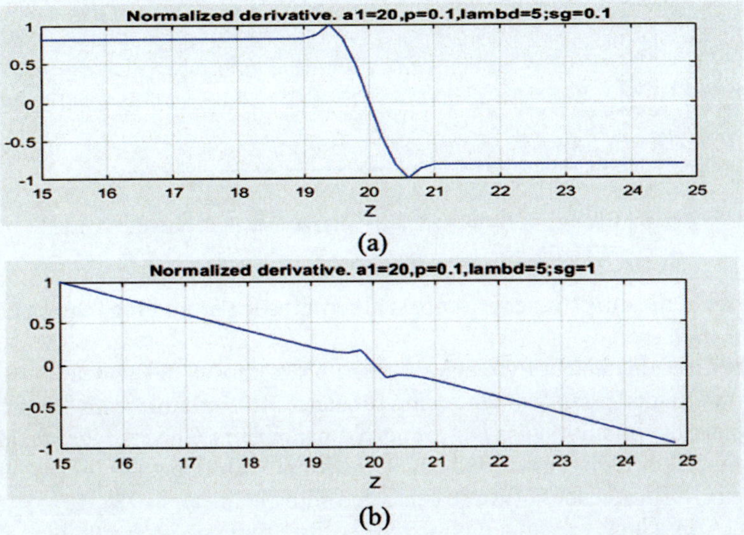

Fig. 6. The field intersection (Fig. 4b): a) for $\sigma = 0.1$; b) $\sigma = 1$.

go beyond this zone, they are suppressed and the adjustment of the channels of the derivatives is reduced or stopped altogether.

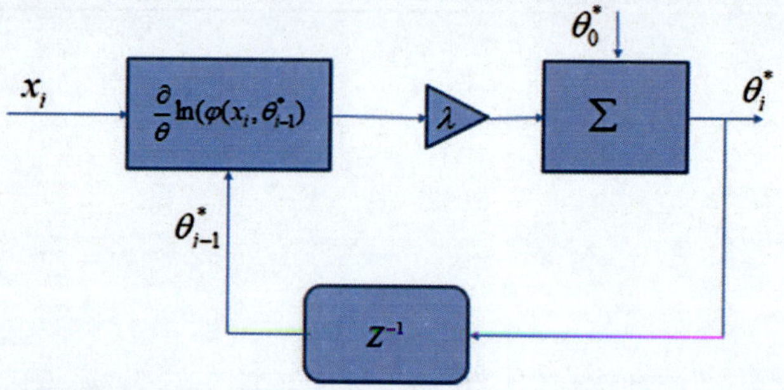

Fig. 7. The block diagram of the recurrent algorithm for estimating the regression coefficients.

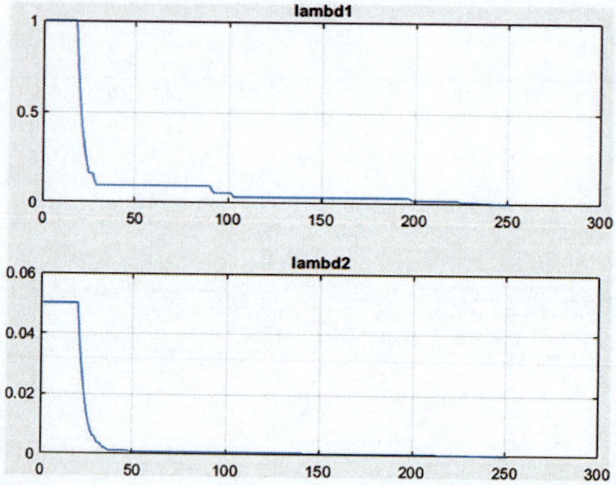

Fig. 8. Change of factors at partial derivatives in iterative process.

Analysis of the effectiveness of robust algorithms for estimating regression parameters.

To study the efficiency of the algorithms, a program was developed on MATLAB, which allows to model random regression sequences in a mixture with Gaussian noise and non-Gaussian impulse noise (4) with given parameters. The study of the efficiency of processing algorithms was carried out by the Monte Carlo method. According to this method, M = 400 sequences (4) were generated with parameters a0 = 10, a1 = 2 lasting 100 samples, to which was added a mixture of Gaussian uncorrelated noise and pulse sequence with Laplace distribution of pulse amplitudes and pulse probability p. Each sequence was processed according to a gradient algorithm (7–9). The noise distribution parameters were considered known and their values were introduced into the formula of the regression parameter estimation algorithm.

The influence of failures on the efficiency of robust algorithms is investigated in the work. Failures in surveillance systems occur due to various factors. These can be power surges in the power supply system, leading to failures of the recording equipment, can be a interference that leads to overloading of the system of trajectory (secondary) processing of radar targets, external interference in the data channel, packet loss in computer networks due to DOS attacks etc.

Failure leads to the replacement of the j-th reference of sequence (4) by zero. Robust algorithms can significantly increase the reliability and accuracy of estimating the informative parameters contained in the registered sequences of a mixture of signals and data.

An example of the measurement's implementation of the trajectory coordinates of the object with failures is shown in Fig. 9.

Arrays of regression coefficient estimates $a_{01}^*, \ldots a_{0M}^*$ and $a_{11}^*, \ldots a_{1M}^*$ were used to calculate the mathematical expectations of the estimates and their standard deviations. The results of the calculations are shown in Fig. 10.

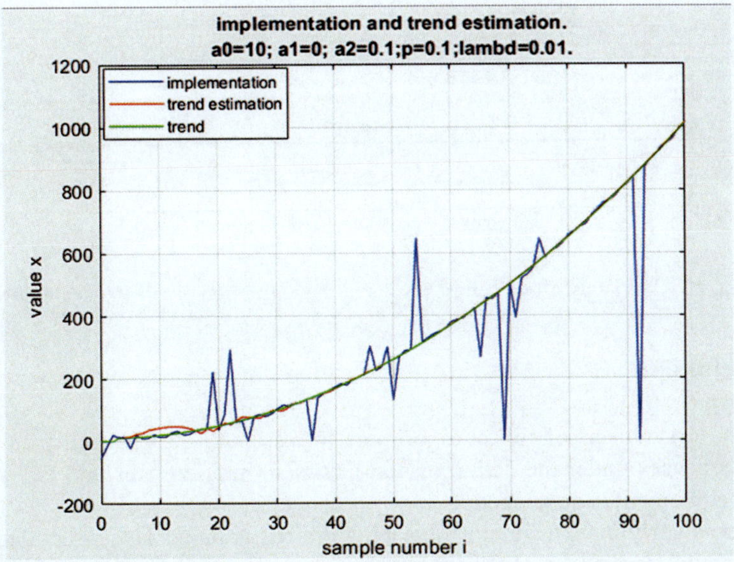

Fig. 9. Example of random process with second order trend and failures implementation and trend estimation.

Figure 7 shows the study results of the estimation effectiveness of linear regression parameters under the action of non-Gaussian impulse noise (blue curves) and the results of estimation the parameters of linear regression in case of failures (red curves). The top two graphs show the mathematical expectation and the standard deviation of the coefficient estimate, respectively, the value of which is 10. The third and fourth graphs show the mathematical expectation and the standard deviation of the coefficient estimate, respectively. (The value of the coefficient a1 = 2).

The proposed algorithm allows to obtain unbiased estimates of regression coefficients in complex noise conditions, as well as in case failures at registration.

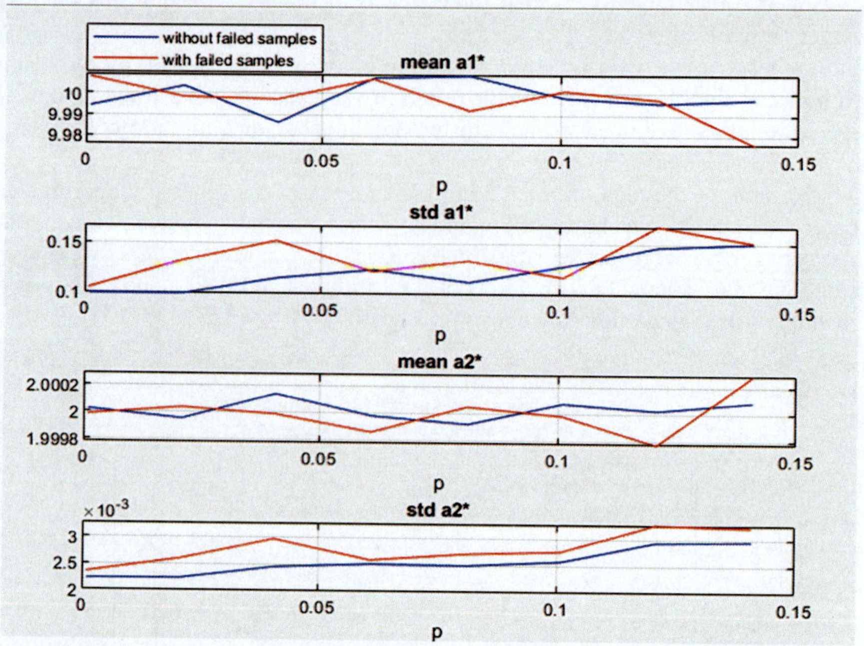

Fig. 10. The study effectiveness results of the regression parameters estimates.

5 Conclusion

The theory of synthesis of robust algorithms for estimating regression parameters of random sequences under the action of non-Gaussian interferences and failures in the recording equipment is developed.

A robust algorithm for estimating linear regression parameters is synthesized, expressions for partial derivatives of the likelihood function are obtained, and the efficiency of the algorithm for estimating regression parameters by the gradient descent method is investigated.

The proposed robust methods of signal and data processing provide high accuracy and noise immunity from impulse interference and failures.

References

1. Huber, P., Ronchetti, E.: Robust Statistics. In: 2nd Edition, A John Wiley and Sons, Inc., Publication, p. 380 (2009). https://doi.org/10.1002/9780470434697
2. Hampel, F.: A general qualitative definition of robustness. Ann. Math. Stat. **42**, 1887–1896 (1971). https://doi.org/10.1214/aoms/1177693054
3. Hogg, R.: Introduction to noise-immune estimation, in: Stable statistical methods of data evaluation. Moscow, Mechanical Engineering. (1984)
4. Korniliev, E., Prokopenko, I., Chuprin, V.: Stable algorithms in automated information processing systems. Kyiv, Technika, p. 223 (1989). (in Russian)

5. Rohling, H.: Radar CFAR Thresholding in Clutter and Multiple Target Situations. IEEE Trans. Aerospace Electron. Syst. **Aes-19**(4), 608–621 (1983) https://doi.org/10.1109/TAES.1983.309350
6. Omelchuk, I., Prokopenko, I., Chyrka, I.: Multichannel target speed estimation by a colocated Doppler-pulse MIMO radar. In: International Conference Radio Electronics & Info Communications (UkrMiCo), pp. 11–16 (Sept.2016). https://doi.org/10.1109/UkrMiCo.2016.7739601
7. Prokopenko, I., Osipchuk, A.: Robust algorithm for estimation the phase of a harmonic signal on the background of Gaussian noise and impulse noise. In: Problems of development of the global system of communication, navigation, surveillance and air traffic management, CNS/ATM, p. 68 (November 21–23, 2016). (in Ukrainian).
8. Prokopenko, I., Omelchuk, I., Chirka, J., Yanovsky, F.: Detection of Markovian signals on the background of Markovian interferences. Prior uncertainty case. In: The International Society for Optical Engineering (SPIE), Proceedings, pp. 1–4 (2009) https://doi.org/10.1117/12.838049
9. Prokopenko, I.: Detection of a harmonic signal in a mixture with narrowband interference. In: 22nd International Microwave and Radar Conference (MIKON-2018), Poznań, Poland, Proceedings, pp. 614–617. (May 15–17) (2018). https://doi.org/10.23919/MIKON.2018.8405303
10. Prokopenko, I.: Robust methods and algorithms of signal processing. In: IEEE Microwaves, Radar and Remote Sensing Symposium (MRRS-2017), 29–31, Kyiv, Ukraine, Proceedings, pp. 71–74 (2017). https://doi.org/10.1109/MRRS.2017.8075029
11. Prokopenko, I.: Statistical Synthesis of Robust Signal Detection Algorithms under Conditions of Aprioristic Uncertainty. In: Cybernetics and Information Technologies, vol. 15, no. 7, Special Issue on Information Fusion, Sofia. (2015). https://doi.org/10.1515/cait-2015-0085
12. Lukin, V., Melnik, V., Pogrebniak, A., Zelensky, A., Saarinen, K., Astola, J.: Digital adaptive robust algorithms for radar image filtering. J. Electron. Imag. **5**(3), 410–421 (1996). https://doi.org/10.1117/12.240715
13. Hampel, F.: Robust Estimation: A Condensed Partial Survey. Z. Wahr. Verw. Geb. **27**, 87–104 (1973). https://doi.org/10.1007/BF00536619
14. Kassam, S., Poor, H.: Robust Signal Processing for Communication Systems. IEEE Commun. Mag. **21** 20–28 (1983). https://doi.org/10.1109/MCOM.1983.1091322
15. Hajek, J., Sidak, Z.: Theory of Rank Tests, 2rd ed., Academic Press, (1999). https://doi.org/10.1016/B978-012642350-1/50021-7
16. Kendall, M., Stuart, A.: Statistical inference and connections. In: Inference and Relationship, Published by International Statistical Institute, vol. II, pp. 167–169.(1976)
17. Lukin, K.: Contributions to Electromagnetic Theory, Radar and Communication Technologies. In: (by Dr. Henning F. Harmuth). 9-th International Conference on Ultrawideband and Ultrashort Impulse Signals (UWBUSIS 2018), Proceedings (2018)
18. Bhatta, N., Priya, M.: RADAR and its Applications. In: International Conference on Novel Issues and Challenges in Science & Engineering, (NICSE-16), at: Noorul Islam University, Kumaracoil, Thuckalay, Tamilnadu, India, vol. IJCTA, 9(28), pp. 1–9, © International Science Press (2016)
19. Thayaparan, T., Wernik, C.: Noise Radar Technology Basics Defence, R&D Canada–Ottawa, Technical Memorandum DRDC Ottawa TM, p. 266 (2006)
20. Ershov, A.: Stable methods for estimating parameters. Overview. Autom. Telemech. **8**, 66–100 (1978)
21. Omelchuk, I., Chyrka, I.: A Closed-form arma-based ml-estimator of a single-tone frequency. Circ. Syst. Signal Process. **37**(8), 3441–3456 (2017). https://doi.org/10.1007/s00034-017-0714-3

22. Sinharay, S., Puhan, G., Haberman, S.: Reporting Diagnostic Scores: Temptations, Pitfalls, and Some Solutions. ETS Princeton, New Jersey, p. 41 (2009)
23. Levin, B.R.: Theoretical Foundations of Statistical Radio Engineering, book 1. Moscow: Sovetskoye radio, p. 752 (1969). (in Russian)
24. Solomentsev, O., Zaliskyi, M., Herasymenko, T., Petrova, Y.: Data Processing Method for Deterioration Detection During Radio Equipment Operation. In: Microwave Theory and Techniques in Wireless Communications (MTTW), pp. 1–4, (2019). https://doi.org/10.1109/MTTW.2019.8897232
25. Zaliskyi, M., Solomentsev, O., Kozhokhina, O., Herasymenko, T.: Reliability parameters estimation for radioelectronic equipment in case of change-point. In: Signal Processing Symposium (SPSympo), pp. 1–4 (2017). https://doi.org/10.1109/SPS.2017.8053676

Automation of Vehicle Control Systems and Emerging Networking

Yuriy Zaychenko[1], Serhii Melnykov[2], Galib Hamidov[4], Aydin Gasanov[3](✉), and Petro Malezhyk[3]

[1] National Technical University of Ukraine "Igor Sikorsky Kyiv Polytechnic Institute", Peremohy Avenue 37, Kyiv 03056, Ukraine
zaychenkoyuri@ukr.net

[2] International Research and Training Center for Information Technologies and Systems of the National Academy of Sciences of Ukraine and Ministry of Education and Science of Ukraine, Acad. Glushkov Avenue 40, Kyiv 03187, Ukraine
psmail@i.ua

[3] National Pedagogical Drahomanov University, Pyrohova Street 9, Kyiv 020000, Ukraine
{a.s.hasanov,p.m.malezhyk}@npu.edu.ua

[4] Azerishiq JSC Electrical Distribution Grid Company, Baki Narimanov District, Kazimzade-20, 1008 Baku, Azerbaijan

Abstract. The issues of organization of intelligent vehicle control systems, including those using the capabilities of modern and advanced computer networks, are considered. Directions for the organization of automated and intelligent control systems are shown. The main stages of automation of control systems for dynamic objects, the evaluation of the effectiveness of control functioning are determined. As a result of the review, the characteristic levels of development of the intelligence of control systems are presented, the features of emerging networking are highlighted, and the effectiveness of the above ideas for solving practical problems is shown.

Keywords: Organization of control system · Automation · Intellectualization · Man-machine systems · Emerging networking

1 Introduction

The purpose of the review is to show the directions for organizing intelligent control (including remote control) systems, as well as the need and usefulness of a system review that takes into account: the requirements of the target task, the characteristics of the external environment, the means used to obtain and process information, executive controls, functional characteristics and work experience of the human operator.

The idea of consideration is to reduce the options for choosing target situational solutions from among the possible states of the environment and the control object. This requires a preliminary analysis of the relationship between the goals of management and the used information characteristics of the selected environment and object (composition

M. Klymash et al. (Eds.): TCSET 2022, LNEE 965, pp. 523–545, 2023.
https://doi.org/10.1007/978-3-031-24963-1_30

and dimension of characteristics). It also requires the initial support structure of management, which can be improved and changed in the process of practical implementation of management. These representations can be used when searching for solutions to a large number of practical problems. They are especially important when searching for the control of dynamic objects under conditions of significant uncertainty and the need to make urgent decisions in critical situations. Management of land, sea, air and space transport falls within this scope. Further, for concretization, attention is drawn to unmanned aerial vehicles and cars. An example of automating the control of an aircraft in a turbulent atmosphere in the event of a malfunction of the engine is considered. When controlling aerodynamic surfaces and actuators, there is a local distribution of control resources in a certain network structure. If it is necessary to control or manage the movement of vehicles remotely, it is required to use local and global network resources.

The development over the past decade of technical devices for processing and presenting information, and especially means of communication (in particular computer networks) between objects with different locations, creates new possibilities for controlling moving objects. Unmanned vehicles are widely used and put into practice, which is especially evident in the development of unmanned aerial vehicles and cars [1–3]. The use of modern computer networks makes it possible to expand the scope of unmanned vehicles by obtaining up-to-date information over the network, the ability to control the state of an object and remotely control an object in case of difficult situations. The efficiency of control systems can be improved by using cloud databases and knowledge about the characteristics of the environment, traffic participants and the current situation on the routes. This requires an increase in the speed characteristics of information transfer in the network, the use of updated data transfer protocols, the possibility of quickly changing the network configuration, and the rational distribution of controls in the network nodes, taking into account time delays [4]. The intellectualization of computer networks implies the possibility of adapting their structure and functions to the current requirements for the efficiency and reliability of driving vehicles of various classes.

Vehicle control systems must support different communication standards, as well as cover different devices and networks. In addition, such systems must be equipped with tracking, interaction and control functions that are not limited to one vehicle, but fit into a vast transport complex, including adapting to traffic control systems.

In addition to the convenience for the driver of the car, equipping cars with sensors opens up new opportunities for improving road safety. Using 5G networks, cars will be able to communicate with each other and make instant decisions about what to do in a given situation based on information received from other vehicles on the road. For example, a car could send a signal about its hard braking, so that a car that is threatened with a collision because of this can also brake hard in automatic mode. In 4G networks, it is impossible to implement such a service, since the signal delay is too high to drive a car in real time [5].

5G technology, due to faster response, will become preferred for broadband streaming multimedia, sending diagnostic information to a service center, use in V2X (Vehicle-to-Everything) systems - a system for exchanging data between a car and other road infrastructure objects, for example, with cars, road signs, markings, traffic lights; the infrastructure must also be connected to the Internet).

Many cars use radar for collision avoidance, while self-driving cars rely on cameras, radar, and lidar to navigate predetermined routes. The technology will allow drivers and self-driving cars to not only avoid collisions, but also send and receive all types of data to help balance city traffic and save fuel.

6G wireless networks with data transfer rates up to terabytes per second (100 times faster than 5G networks) will be the basis of the future Internet of Everything (IoE).

Existing mobile communication systems perform one main function - wireless communication. 6G networks will simultaneously perform functions such as communication, computing, control, localization and reading.

The combination of 6G networks and AI technologies to improve the network has become a trend since 5G networks. Intelligence will become an essential feature of the 6G network.

2 The Main Steps of the Automation of Control Systems for Dynamic Objects

Intelligent technology, intelligent control, intellectualization - concepts are perceived in most cases intuitively. When there is a transition from a simple target functioning of the system to processes that have intellectual properties, there are various rather subjective opinions. They are gradually changing towards investing in the concept of intelligence the ability to find solutions to increasingly complex tasks. From the point of view of the natural evolution of biological organisms, elements of intellectual and adaptive functions can be seen at different stages of their development. The formation of higher manifestations of intelligence occurs with an increase in the levels of generalization and accumulation of knowledge about the environment, the construction of models of the interaction of objects with the outside world [6].

The currently rapidly developing methods for constructing neural-like networks of deep learning, methods for clustering fuzzy sets, combining neural network methods with classical methods of the control theory, methods for working with big data [6, 7]. When using various methods for creating and carrying out the automation of practical systems for managing dynamic objects, certain stages can be distinguished.

At the first stage of preliminary processing of the received information, the main target characteristics of the current situation and their dimensions are identified, the time intervals necessary for the formation of effective control (discreteness of processes) are determined. For this, elements of data filtering, learning opportunities and determining the statistical characteristics of the environment can be involved. At this stage a control object model and a environment model are creating. It is necessary to build models of optimal accuracy for target tasks. At this stage, the limiting capabilities of the control object as a whole are studied; the problems of stabilizing individual characteristics and the creation of automatic subsystems for them are solved.

The second stage is usually devoted to the creation of a basic control structure that is stable and reliable in operation. The structure has the logic of actions for the main set of realizable situations, the logic of testing the main nodes and the logic of using backup and autonomous adaptive devices. At this stage of constructing the initial structure for automated or automatic control of an object, the relationship between the input data and the required control goal is determined. The patterns of accumulated practical experience in managing a specific object (stable dependencies that give the desired result), heuristic assumptions can be used.

The third stage is the construction of means for representing control information and means for monitoring the functioning of the system. It is especially important to take into account the target requirements of the human operator, knowledge of its functional characteristics and use of experience with typical control systems of a certain class. In addition, opportunities are created for an extended analysis of situations and mutual adaptation of a person and a control system.This stage of verification, refinement and adaptation of control structures, the characteristics used to control processes and make operational decisions are adjusted to the preferences of a particular operator.

The fourth stage aims to provide the control system with the means of interaction with the human operator in the language of natural ideas and concepts, the possibility of implementing creative elements of the control process. At this stage of expanding the possibilities of organizing effective management with realistic models of the object and environment are created, additional information is used from databases and knowledge about the current state of the environment. Long-term management forecasts are obtained and global decisions are made. In the future, the use of models of the surrounding world and the interaction of a person with a system in natural languages can be considered.

The presented ideas are connected with the emergence and development of natural evolutionary intelligence, but they also allow the possibility of a hybrid combination with other approaches. The intelligence of the control system is related to the efficiency of its functioning, but there is not always an unambiguous correspondence. The overall assessment of the effectiveness of the created system and the measure of its intellectualization is relative and can be obtained by conducting research on the functioning of systems built by different methods in real environments or by computer simulation. A qualitative assessment of the intelligence of a control system implies the presence of a certain structure (blocks for adapting parameters, storing actions and results, building environment and object models, etc.). A quantitative assessment of intelligence can be expressed by the number of microsituations in which a solution to the target task is found within a limited given time. Time is a critical factor and affects the overall effectiveness of the practical functioning of the system.

A computer network for managing moving objects implies the operation of a control system with a number of features associated with the distribution of control links. When ensuring the intellectualization of a computer control network, it is natural to take into account the general patterns of intellectualization of control systems and the specifics of the network operation [4, 8].

3 Methodological Foundations of Intellectualization of Control Systems

The idea of consideration is based on the selection of a certain target space for the functioning of an object in an environment, which may contain conditionally specified sets of macrosituations, situations and microsituations [9]. The definition of these sets or subsets is quite subjective and emphasizes the hierarchy of levels of the management process. Usually, a macrosituation is the most general statement of the control problem; if possible and with practical experience in obtaining intermediate solutions, it is divided into separate situations. Situations have their own internal goal of management, which serves as a stage in achieving the main goal of management. Situations are structured into sets of microsituations in which elementary control actions are formed in the current time intervals. The value of these intervals is chosen from the conditions for the possibility of reliably achieving the control goal in a finite number of control steps. Obtaining the necessary control in each microsituation takes a certain time interval. A compromise is required between the time to determine the solution and the accuracy and reliability of achieving the goal.

The intellectualization of control systems and networks is aimed at transferring and expanding the total set of initial microsituations to the final area, where the control task is completed. In practice, it is usually not possible to obtain positive solutions in the entire range of initial values due to the presence of noise components of various nature in information data, restrictions on the computational and energy resources of the system, and restrictions on the time to achieve the required solutions. Intellectualization of control processes, including full automation of individual links, basically expands the initial area from which the required result is achieved. However, especially time constraints may reduce or undo the expansion of the initial area and less intelligent but fast solutions will be required. Thus, the intelligence and overall efficiency of the control system may not always have the same direction. On the evolutionary path of development of natural intelligence, the efficiency of using relatively simple control structures is also preserved.

The use of the laws of evolutionary development of biological systems for the development of technical devices has been carried out throughout the history of mankind. The development and improvement of the intellectual functions of vehicle control systems is no exception. It is proposed to use the basic natural patterns of intellectual development to create technical systems (without rejecting the possibility of using existing effective technical solutions in hybrid systems). When creating such technical systems, special attention is paid to the selection and processing of the necessary information to solve a specific target task and further use of the data obtained.

The key points in solving complex practical problems are taking into account and matching the used characteristics of the environment and the object with the target requirements of the task, taking into account the possibilities of obtaining information limited in accuracy from sensors in the presence of noise, and also taking into account the discreteness of time intervals. As an example, the problem of controlling the launch of an aircraft to a given final region from the region of initial states is considered below. Accounting for these requirements allows us to conduct a microsituational analysis and find operational control, obtain areas of attainability of target values for various control algorithms, and draw conclusions on the effectiveness of the algorithms used.

There are a large number of works aimed at describing the intellectual properties of biological systems and their implementation in technical devices [6, 10, 11]. Especially popular and successful at present is the use of analogues of neural networks. The practical implementation of intelligent systems is carried out in a wide variety of areas of human activity, now it is fundamental for unmanned transport systems [12, 13]. Difficulties in the implementation of intelligent systems largely lie in the high dimensionality of input data, the presence of noise components, the uncertainty of internal and external parameters of the control process, as well as the lack of effective reference control solutions and structures that can be upgraded. A systematic evolutionary consideration of the formation of the intellectual properties of biological organisms helps to overcome these difficulties in the practice of creating intelligent technical devices.

The search for the solutions required to achieve the goal in selected situations and microsituations. Taking into account the evolutionary development of biological systems, we will single out the main directions for their improvement and development in terms of increasing the level of intellectual capabilities. In tasks solved by the simplest biological organisms, the similarity of intellectual properties can manifest itself in the specifics of subsystems for receiving and processing external information (intelligent sensors). A further increase in the level of intelligence is associated with the construction of models of the environment and models of the possible behavior of an object in an environment with an assessment of the result.

When organizing intelligent control systems, let's pay attention to topical methodological issues of extracting and processing input information, also taking into account the aspects of building distributed control structures and databases, and problemas of predicting decisions in non-stationary environments [7].

We can assume that if the system achieves a certain goal by its actions more efficiently than in the absence of activity, then there is a certain level of intelligence.

As an example of a possible idea of the gradation of the levels of development of the intellectual properties of automatic and automated control systems, the following levels are presented.

1. The level of intellectualization, at which the processing of input information is carried out in one control channel (such as a specialized target sensor or a directive control system).
2. The level of intellectualization, at which work is carried out with several information channels, preliminary processing and selection of data significant for achieving the goal is carried out.
3. The level of intellectualization at which the construction of information converters is carried out, which take into account the stable patterns of human operator activity (such as logical inference or behavior stereotypes).
4. The level of intellectualization, at which adaptation to the specifics of perception and search for a solution of a particular operator is carried out, there is an expansion of the possibilities for choosing alternative information transformations.
5. The level of intellectualization at which control takes into account an extended set of databases and knowledge about the object and the environment. The adequacy of functioning is realized on the presentation of situations by a human operator and the formation of solutions for the technical part of the system in a certain subject area of control.
6. The level of intellectualization, which ensures the adequacy of the functioning of the control system and the person according to the main information data (and outside a separate subject area of control), coordinates the interaction of the human operator with the technical part of the control system in most situations.

7. The level of intellectualization, at which the systems have the properties of creative development and improvement.

4 Features of the Organization of Distributed Remote Control Systems

The functioning of modern computer networks is based on packet transmission and switching methods, which are implemented on the basis of converting any type of information into digital sequences, followed by dividing this sequence into packets equipped with all the necessary information for their identification, routing, and error correction. This approach makes it possible to transmit all types of information in a single information channel, to use various ways and various means for transmitting information, to use universal means of switching that are invariant with respect to all types of information. In this context, the computing environment can be considered as some kind of universal environment for the dissemination of information.

Let us consider some key issues of the influence of network delays on the functioning of a distributed control system using heterogeneous existing local and global network structures. Issues of compression, coding, encryption and increasing the reliability of network information require separate consideration.

The possibility of using existing network protocols and channels was analyzed in order to develop recommendations for organizing information transfer in distributed network structures, and the influence of network delays on the quality of remote control of a mobile object was evaluated. As an example of a practical problem of remote control of a moving object, the problem of trajectory control of the movement of an aircraft (working out system control commands: Traffic alert and Collision Avoidance System -TCAS 2) was considered.

Considering the operation of communication protocols at the network level, one should analyze the delays that occur during the transmission of information packets over the network. Significant types of delays are nodal processing delay, waiting delay, transmission delay, and propagation delay.

Node processing delays are related to the operation of the router. The latency depends on the number of packets in the router's queue and can vary significantly from router to router as the packet travels from sender to receiver.

Transmission delay is the total time required for a router to transmit a packet over a link. This time is proportional to the link speed and packet size. Transmission delays can be the most significant type of delay for low speed connections.

Propagation delay is the time required to transmit a bit of information over a communication link. This time is directly proportional to the length of the communication line and depends on the physical properties of the transmission medium. In modern fiber optic networks, the speed of signal propagation is close to the speed of light. Propagation delay is significant for satellite links and can exceed all other types of delays.

The wait delay is the most complex kind of delay, and it's also the only kind of delay that can have different values for different packets. There are three main factors that affect the amount of waiting delay: the frequency of receiving packets, the transmission rate of the output communication line, and the law of distribution of received packets in time.

To reduce network delays, priority processing of synchronous traffic packets is used. However, even with priority processing, synchronous traffic packets can be delayed in communication devices, since they can usually only implement processing algorithms with relative priorities, not absolute ones. At the same time, in segments of shared media, a high-priority packet must wait not only for the completion of the transmission of a low-priority packet, but also for the release of the shared environment. Some newer protocols provide a mechanism for prioritizing media access. Prioritizing access to a shared environment greatly reduces delays in packet delivery to the destination host.

For the transmission of asynchronous type of traffic, it is most advantageous to use long packets, since in this case, less overhead information is transmitted, which increases the utilization of the communication channel. The only thing that limits the maximum packet size is the presence of errors in the communication channel. Large packets cannot be used to transmit synchronous type traffic due to the fact that the packet transmission time is directly proportional to its length, and this increases the delay in the transmission of synchronous type traffic.

The number of intermediate nodes in the network significantly affects the delay in the transmission of synchronous type traffic by subscribers. The transmission of asynchronous traffic with a large number of intermediate nodes (which is typical for global networks) does not significantly affect the quality of services provided.

When transmitting real-time data over a network, typical delay ranges when using heterogeneous equipment can be: coding - 20–45 ms; process delay when entering the network - 15–25 ms; network access - 1–8 ms; network transmission delay - 20–200 ms; exit from the network - 1–8 ms; process delay when leaving the network - 15–25 ms; compensating buffer - 20–225 ms; decompression - 15 ms. The total typical data transmission delay is 100–550 ms.

The level of development of network technologies makes it possible to use computer networks with packet data transmission for the implementation of distributed systems for the intelligent control of aircraft. In this case, the aircraft control system is located far from the aircraft, data exchange between the aircraft and the control system is carried out via WAN networks. In this scheme, WAN networks are also used to obtain information about oncoming aircraft, weather conditions, etc.

The main parameters affecting the quality of operation of such a distributed aircraft control system are the delay value that shifts data packets in time when they are transferred to the aircraft control system, and the delay value that shifts control command packets in time when they are transmitted to the aircraft. The impact of these delays is particularly impacted by those air traffic control processes that require real-time control.

During the simulation, the analysis of aircraft motion was based on the numerical solution in the time domain of systems of differential equations describing the motion of aircraft under given initial conditions with a data sampling rate of 50 ms. The delay of data packets in the navigation computer network during their transfer to the aircraft control system varied in the range of 50–550 ms. The delay of packets of control commands during their transmission from the control system to the aircraft during the simulation also varied in the range of 50–550 ms, i.e. the total delay in the transmission of data packets and command packets varied in the range of 100–1100 ms.

During the simulation, the operation of three aircraft control systems was studied: the first is a test (locally located without packet transmission over a computer network) aircraft control system; the second is the aircraft control system distributed in the computer network, similar to the test one, but with delays in the transmission of data packets and packets of control commands over the network; the third is the system built taking into account the developed original method for high-speed network control processes [8].

The results of the research have shown that the quality of the work of the second aircraft control system distributed in the computer network is unsatisfactory even if there are total delays in the transmission of data packets and command packets within 200–300 ms, and with total delays in the transmission of data packets and command packets greater than 400 ms, this control system does not achieve the control objectives.

At the same time, the results of research into the operation of the third distributed aircraft control system showed that this system provides control quality comparable to the quality of the test local aircraft control system in the presence of the most typical for real networks total network delays in the transmission of data packets and command packets in the range of 100–700 ms.

In addition, the third aircraft control system made it possible, with a slight decrease in quality, to still achieve the goals of controlling the movement of the aircraft along a given route with a variable flight altitude profile and prevent dangerous approaches of the aircraft when executing commands from the collision avoidance system (TCAS 2) for the entire range of initial traffic conditions and an extended range of delays in the transmission of data packets and command packets within 700–1100 ms.

When building efficient distributed remote control systems (which can use computer networks for their implementation), such analytical stages should be distinguished [4, 8].

Statement of the problem of remote control, determination of requirements for the functional characteristics of the object and control system.

Formation of time requirements for functional agents that are allocated as part of the control system.

Determining the characteristics of the existing network structure that can be used to solve this remote control task.

Determination of time delays when using different protocols of network information exchange.

Distribution of control information in terms of importance with respect to control goals and allowable delays in its transmission.

Defining the specification of data transfer protocols for different types of information.

Determination of the expediency of using different channels and routes of information transmission.

Setting priorities for the significance of data and the sequence of their transmission in information structures.

Creation of structures that provide reliable encryption, duplication or redundancy of transmitted data.

Functional optimization of the control of a remote object due to additional knowledge about the characteristics of the object and the environment.

Introduction to network structures of new information devices for receiving and processing information.

Creating a network infrastructure that evolves with the addition of new features.

Refinement of information exchange protocols in the network and organization of new information entry channels.

Based on the analysis, the distribution structure of the functional modules of the control system and the object is determined depending on the possible delay intervals of the control signals.

5 Organization of Control System of an Aircraft in Emergency Conditions

Taking into account the stated patterns of building goal-achieving systems with elements of intelligent functions and the features of distributed control, models of the movement of an aircraft in the air and aircraft control algorithms were developed.

The construction of the aircraft dynamics was carried out with the aim of further modeling different flight modes and the operation of the control system, for this a complete system of equations of the spatial movement of the aircraft was used.

The movement of the aircraft occurs under the influence of aerodynamic forces and moments, forces and moments from engines and gravitational forces. When determining the movement of the aircraft, it is generally necessary to solve the following tasks [14]:

- find the corner and linear speeds of the aircraft, due to the action of the listed forces and moments on it;
- determine the angles of orientation of the aircraft relative to the flow that runs on it, and the axes of the coordinates associated with the ground;
- Determine the movement of the aircraft relative to the Earth.

Each of these private tasks is responsible for its own group of differential equations. The model uses quite accurate and complete equations of the aircraft for calculations and allows you to get reliable results when modeling the operation of the control system.

It is believed that the plane is an absolutely tough body.

The mass and moments of the inertia of the aircraft are considered unchanged throughout the time of the flight modeling and meet the initial state of equilibrium flight.

It is believed that the configuration of the aircraft has a plane of symmetry and that the masses are distributed symmetrically in relation to this plane.

To represent the model of movement of an unstable aircraft in a three-dimensional pro-space, the development of algorithms for stabilizing the aircraft, taking into account cross-country relations in the expanded range of changes in parameters, was also carried out. Original models of a stochastic turbulent environment have been added and a map of the area with a designation of prohibited and dangerous areas of movement was created.

When performing the movements of the aircraft in space, it is necessary to transfer an aircraft from the field of initial or current states to a given area of the final states. The area of final states can be an arbitrarily specified area of space, the exit area for planting along the glyssade, the alignment area or the area of the strip touch. When performing maneuvers, the presence of various kinds of disturbing influences, both external (wind, turbulence of atmosphere, etc.) and internal (inaccuracies of the work of computers and sensors) are assumed. To build the trajectories of a given movement in the target area, it is necessary to take into account the most significant information characteristics of the control and environmental object and be able to constantly adjust the trajectory due to the influence of random influences. Given the general provisions indicated above, we will consider an example of organizing the control of the aircraft at the stage of landing in emergency conditions and the action of random wind disturbances.

The movement of the aircraft at the planting stage is the most complex task of aircraft, so a lot of attention is paid to the automation of the aircraft planting. The aircraft planting system for the engine failure is designed as an integral part of the integrated complex of control of complex dynamic objects, which includes a program that models the movement of the aircraft in three -dimensional space.

Planting a aircraft in emergency mode when engine failure puts forward the complementary requirements for organizing the functioning of an interconnected pilot system - automatic control system - aircraft. Such systems are focused on active interaction with a human operator in an extreme situation. This can be formulated as a mutual coordination of the circuit and the structure of the automated part of the control system and the activity of the human operator. A similar consideration of the system as a holistic human-machine system can serve as the basis for determining the general management structure and for optimizing the resulting structural and functional solutions.

When creating the system, it was accepted that the human-machine system for managing the object in the critical mode of functioning for an effective and reliable achievement of management goals requires accounting for such factors [9, 15]:

- characteristics of the activity of man operator, features of the processes of perception and processing by a person of information and the development of managing actions;
- the flexible structure of the laid algorithms and the consistency of the structures of algorithms with stable stereotypes of the human operator;
- the possibility of prompt switching of control functions between a person and an automaton;
- Accounting and compensation for disturbances of various origin, dynamic analysis of the situation, correction and synthesis of trajectories.

The strategy of synthesis of landing trajectories is based on the functioning of the system: the pilot - the automatic control system - the aircraft - the environment as a game, where the ergatic system seeks to implement goal behavior, and the environment introduces unpredictable perturbations. The action of the ergatical system is aimed at minimizing environmental actions and maximizing the ultimate goal.

You can give a formal description of the emerging task within the general game theory. We introduce the following designations: $y \in Y$ - a three -dimensional vector of spatial coordinates of the aircraft, Y is a set of all possible provisions of the aircraft in space; $p \in P$ - vector of plane movement parameters (speed, roll, pitch and yaw angles, angular velocities, course angle and angle of inclination of the trajectory, thrust of engines, etc.), P is a set of possible sets of plane motion parameters; $u \in U$ - a vector that sets the position of the aircraft control (control vector), U is a set of possible control vectors; $v \in V$ is a three-dimensional vector, the coordinates of which are wind speeds on the axis of spatial coordinates, V is a set of all possible vectors v; $T = [0, t]$ is the time interval at which the planting task is solved. Also, we introduce $X = Y \times P$ - the set of phase states of the aircraft and $x = (y, p)$ - a vector that describes the phase condition of the aircraft. Suppose g $X \times U \times V \times T \rightarrow X$ the law of the plane movement in space. The aircraft movement will be described as:

$$dx/dt = g(x, u, v, t), \tag{1}$$

where dx/dt is a derivative of aircraft state vector, $g(x, u, v, t)$ is the function of the specified parameters.

As a 1st player strategy, we consider the reflection of $\sigma_u : X \times U \times V \rightarrow U$ - the law of managing the plane at the stage of landing from the set Σ_u of the permissible management laws.

As a 2nd player's strategy, we will consider the reflection of $\sigma_v : Y \times T \rightarrow V$ - a windy environment in the area of landing from the set Σ_v of possible settings.

We denote the $X_0 \subset X$ - the set of possible initial phase states of the aircraft, and $X_1 \subset X$ - the set of phase states of the aircraft, for which one can speak about successful landing (the set of advantages of the 1st player).

Then, with a given $x_0 \in X_0$ and strategy of the 2nd player $\sigma_v \in \Sigma_v$, the task must be solved to carry out successful landing:

-find the strategy $\sigma_u \in \Sigma_u$, the restriction is taken into account

$$x \in X_l \tag{2}$$

and the criterion of optimality

$$|L_0 - L| \rightarrow min, \tag{3}$$

where $L_0 = L_0(x_0, v_0)$ is the optimal trajectory length (in terms of consumption of energy supply); L is the the actual length of the trajectory depending on the length of the aircraft approach for landing.

Next, let $D(x_0, \sigma_v) \subset \Sigma_u$ - the area of acceptable solutions of the problem of mathematical programming under the initial condition $x0 = (y0, p0)$ and a given strategy σ_v.

Then $Q(p_0, \sigma_v) \subset Y$ is an area in which the 1st player wins with the given starting parameters of p_0 and a given strategy of σ_v (the area of the airspace from which successful landing is possible in the air situation).

The formal description of the used windshield model in the landing area or the set of permissible strategies of the 2nd player Σ_v looks like this:

$$\Sigma_v = \{(w(h,t) \cdot cos[\Psi_w(h,t)], w(h,t) \cdot sin[\Psi_w(h,t)], 0)\}, \quad (4)$$

where $\Psi_w(h,t)$ is the azimuth of the wind vector; $w(h, t)$ is a wind vector module. The azimuth and module of wind speed vector are functions depending on the height of h and variable time t and are represented in the form of a constant component, depending on h and variable component, depending on t:

$$w(h,t) = w^l(h) + w^2(t), \quad (5)$$

$$\Psi_w(h,t) = \Psi_w^1(h) + \Psi_w^2(t.)$$

The variable component is used to model gust wind in the runway area with a sharp change in its speed and direction.

The permanent component of wind speed is set as follows:

$$w^l(h) = \begin{cases} -0,005h + 150, 16000 \le h < 20000 \\ 70, 12000 \le h < 16000 \\ 0,005h + 10, 0 \le h < 12000 \end{cases} \quad (6)$$

When solving the described problem, the set of landing strategies Σ_u is the following. There are a number of consecutive maneuvers that must be carried out for landing without a power plant. The first is the reversal and exit to the trajectory of the movement in the direction of the runway axis. The second is the exit of the aircraft to the axis of the runway extension and movement along it to the glide path. The third is the movement along the glide path (landing event). The fourth is landing with a runway. Each of the stages of planting is characterized by a certain set of stereotypes of the pilot. Taking into account the requirements for the ergatic control system of the object in the critical mode of functioning, the scheme of synthesis of the landing trajectory must implement these maneuvers.

The following are considered as heuristic criteria for the construction of the traffic. At the heart of the first criterion is the requirement of compliance with the length of the trajectory of its optimal length in the sense of the possibility of implementation, taking into account the disturbing actions of the environment and possible errors in management. The second heuristic criterion takes into account the desire of the pilot to get the simplest kind of trajectory of further movement. The latter is expressed in reducing the number of maneuverable trajectory sections and the desire to perform them in minimal intervals.

The action of an ergatic control system should be aimed at minimizing the action of the environment and maximizing the achievement of the ultimate goal. Minimizing the action of the environment involves taking into account the ergatic system of flexibility of the intended trajectory of movement and the possibility of its constant adjustment (carrying out feedback on the results of control). Microsituations consideration in determining the control actions allows you to realize these capabilities.

The microsituation approach involves the synthesis of the optimal trajectory, including all or part of these stages of landing, at any time before the aircraft pre-landing maneuver, and in the process of pre-landing maneuver and approach to landing. The automatic control system must be able to synthesize the optimal trajectory for an arbitrary microsituation, characterized by the location of the aircraft and the runway, as well as the parameters of the aircraft and environmental disturbances. Such a construction of the control system allows to compensate for various control errors and the impact of wind actions. The ability to change the trajectory is especially important in situations where the energy reserve of the aircraft is significantly limited and, accordingly, limited room for maneuver to return to the former optimal trajectory. No less important is the fact that such a structure of the automatic control system allows at any time to switch controls, allows a person to take control "on himself", and then return control to automation. When implementing such an approach, it is important to ensure the smooth transition from one trajectory to another, ie small changes in the microsituation should correspond to a small change in the position of the trajectory.

The synthesis of the landing trajectory follows the following scheme (Fig. 1).

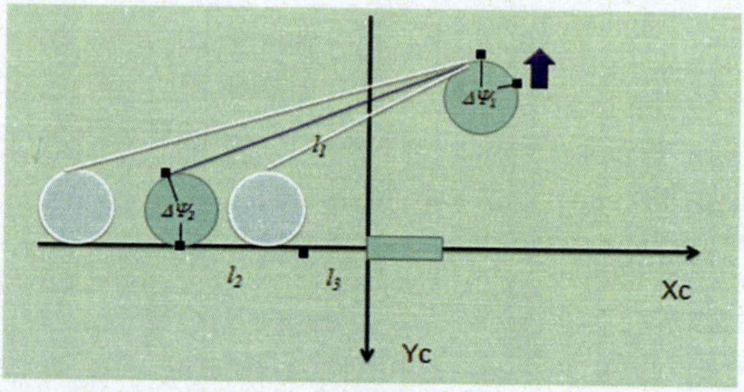

Fig. 1. View of the positions of the predicted trajectory of the aircraft in the horizontal plane (Xc, Yc). Primary reversal with radius *r* by angle $\Delta\Psi_1$. Flight with a constant course to the intersection with a conditional extension of the runway axis (l_1). The second turn at an angle $\Delta\Psi_2$ with access to the landing course. Movement with landing azimuth and exit to the glide path (l_2). Movement along the glide path (l_3) and landing.

The length of the projection of the trajectory on the horizontal plane will generally be

$$L = \Delta\Psi_1 \cdot r + l_1 + \Delta\Psi_2 \cdot r + l_2 + l_3, \tag{7}$$

where l_1, l_2 - respectively the lengths of the first and second rectilinear sections of the trajectory;
l_3- length of the glide path;
r is the radius of the turning cylinder.

The calculated length L must correspond to the average possible length of the trajectory L_0, which can be estimated based on the average aerodynamic quality of the glider K. The estimate L_0 is as follows:

$$L_0 = (h - h_3) \cdot K + l_3, \tag{8}$$

where h_3 is the height of the exit to the glide path;
h is the current altitude of the aircraft.

The optimal landing trajectory should have the following property:

$$|L - L_0| \rightarrow min. \quad (9)$$

The parameter on which the optimization is performed is the position of the turning cylinder on the extension of the runway axis, which is determined by the length of the line l_2.. To synthesize the optimal trajectory, the optimization problem must be solved:

$$l_2 = argmin|L - L_0|. \quad (10)$$

This problem is a nonlinear programming problem, and the function, the extremum of which is sought, is significantly non-convex in a number of practical situations. To overcome the problems associated with non-convexity, the following strategy of finding the global extremum has proved its effectiveness in the experimental implementation of the algorithm. The primary interval is divided into n ($10 < n < 20$) subintervals, and after finding the values of the function that is minimized, at the ends of the subintervals the extremum is searched by a gradient method with automatic step selection on the selected subinterval. Such a strategy, while remaining acceptable in terms of the speed of finding a solution, ensured finding a global minimum for a virtually complete set of situations that arise in this task.

When calculating it is necessary to take into account the possibility of choosing the direction of the primary turn. When choosing a maneuver option, the desire to minimize the total turning angle $\Delta\Psi_1 + \Delta\Psi_2$. Seems optimal. Thus it is necessary to analyze separately situations in which the primary turn can be executed only in the direction coinciding with the direction of the second turn.

In the control channel of longitudinal movement at this stage should be provided stabilization of the angle of inclination of the trajectory Θ, the estimated value of which can be obtained from the following relationship:

$$ctg\Theta = (L - l_3)/(h - h_3). \quad (11)$$

Although such an estimate of the required trajectory angle does not take into account additional energy losses when making turns, the estimation error is comparable to the errors caused by the environment and control accuracy, and is compensated by constant refinement of the optimal trajectory and its parameters.

After the aircraft unfolds and moves along the conditional extension of the runway axis, the trajectory synthesis scheme changes. The type of trajectory in the projection on the horizontal plane is significantly simplified. This allows the pilot to pay more attention to control in the longitudinal channel when controlling the approach to the landing in automatic mode and facilitates the task if necessary, switching the control "on itself". In the lateral motion control channel, guidance is performed on a point running along the extension of the runway axis to the runway section. The distance from the aiming point to the runway cut is selected in proportion to the distance of the aircraft from the runway. This ensures a smooth correction of the course of the aircraft and access to the landing course, even with the initial significant deviation from the runway axis after the turn.

The implementation of the proposed structure of the automatic control system assumes the presence of two-channel control of the aircraft. Aircraft course control (aircraft roll) is the lateral control channel, and the aircraft pitch angle control (vertical overload) is the longitudinal control channel. Such two-channel control can be implemented both by a human pilot and by autopilot. Since at each stage the main is only one control channel, and the second plays the role of an additional control channel, such a division facilitates the human pilot as part of an automated control system and is rational in terms of the principle of human-machine compatibility.

The scheme of the landing trajectory with a movable turning cylinder on the extension of the runway axis provides the necessary smooth transition from one trajectory to another. In addition, such a scheme allows you to dynamically take into account the effects of wind, atmospheric conditions and other environmental factors, as their full consideration at the initial stage of trajectory planning is hardly possible. The estimate of the average possible length of the trajectory L_0 is calculated based on the average aerodynamic quality of the glider K, which significantly depends on the aircraft speed, direction and wind speed and changes dynamically upon receipt of the relevant information.

Developed and optimized in terms of the composition and type of control information, the interface for the interaction of the pilot with the control system.

6 Evaluation of Control Effectiveness

An estimate of the effectiveness of solutions in the general case should be obtained over the entire range of admissible initial values. Usually this area is quite large and a simple enumeration of all points or options requires considerable time and computational resources. When taking into account the system methodology for specific implementations of the control process as a whole, it is possible to reduce the number of options for determining the effectiveness of the solutions under study by taking into account the continuity of the control process and the patterns of system interaction. The continuity, as well as the stability of the nature of management processes, suggests that a change in external conditions that affect the change in the state of the object causes a relatively small change in the final state. Usually these values are not more than some certain value within one given microsituation.

Examples of reachability areas (Fig. 2–4) for evaluating the effectiveness of control algorithms for a transport aircraft (the arrow indicates the initial course of the aircraft, the acute angle of the triangle indicates the direction of the wind).

Developed and optimized in terms of the composition and type of control information, the interfaces for the interaction of the pilot with the control system. The creation of interfaces for the interaction of the pilot with the control system was based on providing him with basic and additional information (augmented reality) for the reliable performance of target tasks. Microsituational analysis and synthesis was also used to increase the options for choosing control decisions based on the preferences of a particular human operator. Examples of the created interfaces are shown at Fig. 5, 6.

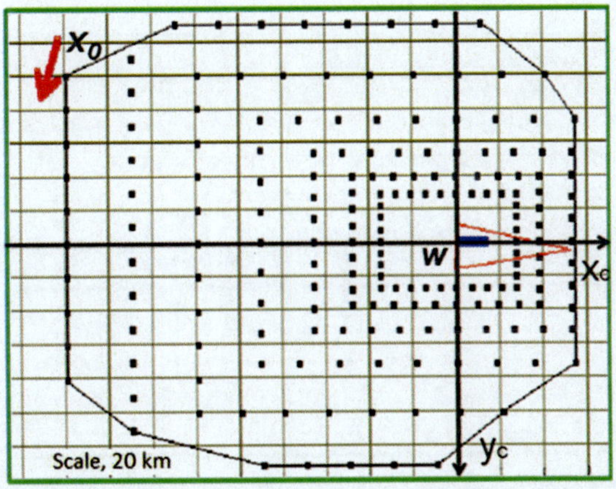

Fig. 2. The area of the initial positions of the aircraft, from where a successful landing is carried out. High-altitude echelon - 5000 m. Fair wind (Xo - aircraft initial state, W - runway wind state, Xc, Yc coordinate directions).

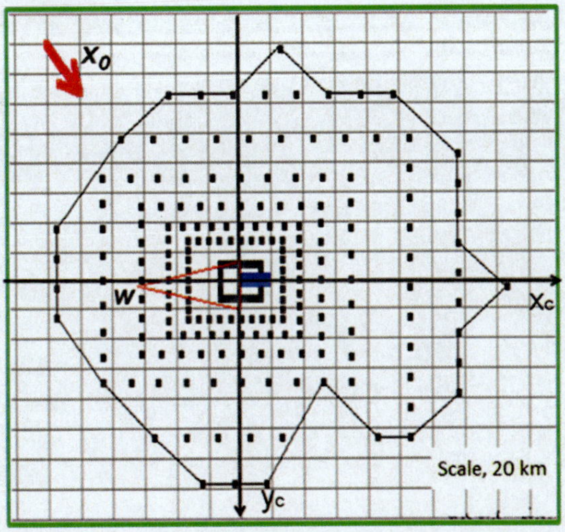

Fig. 3. The area of the initial positions of the aircraft, from where a successful landing is carried out. High-altitude echelon - 5000 m. Head wind (Xo - aircraft initial state, W - runway wind state, Xc, Yc coordinate directions).

Figure 7, 8 shows the interfaces of the pilot and navigator when solving the problem of laying safe routes for the movement of controlled objects and avoiding dangerous encounters. On the left side of the interface screens, there is a microsituational analysis of the current situation (divided into cells by coordinates, course or speed maneuvers) for the selected object relative to other traffic participants. The resulting slice of the current situation allows the human operator to make an reasonable decision on control, taking into account his preferences.

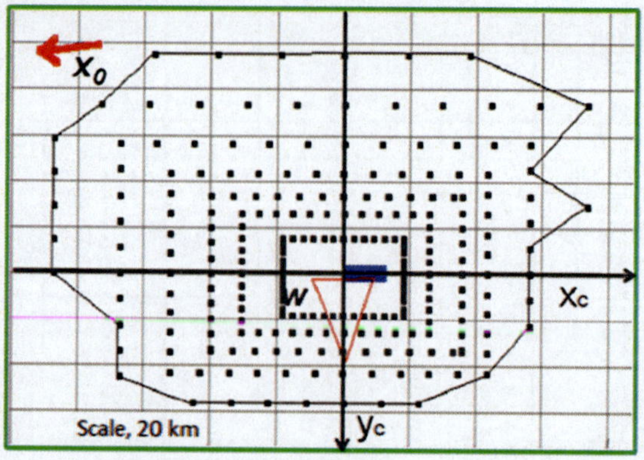

Fig. 4. The area of the initial positions of the aircraft, from where a successful landing is carried out. High-altitude echelon - 3000 m. Side wind (Xo - aircraft initial state, W - runway wind state, Xc, Yc coordinate directions).

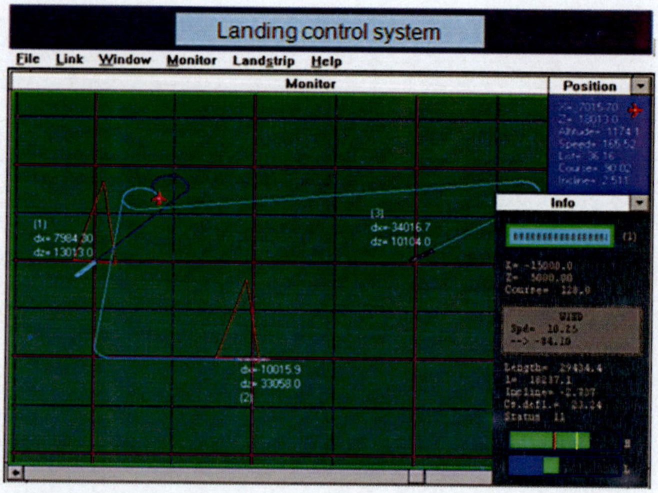

Fig. 5. Pilot's interface during an emergency landing in the presence of several airports in the accessible vicinity.

Computer simulation of the developed control algorithms for an emergency landing with a realistic model of the aircraft and the environment made it possible to evaluate the effectiveness of the final goal in a difficult navigation environment. To carry out an emergency landing of a transport aircraft, it is possible to obtain a significant increase in the size of the region of initial states (up to 2.5 times), from which a positive result is achieved in comparison with known control algorithms along an initially specified trajectory.

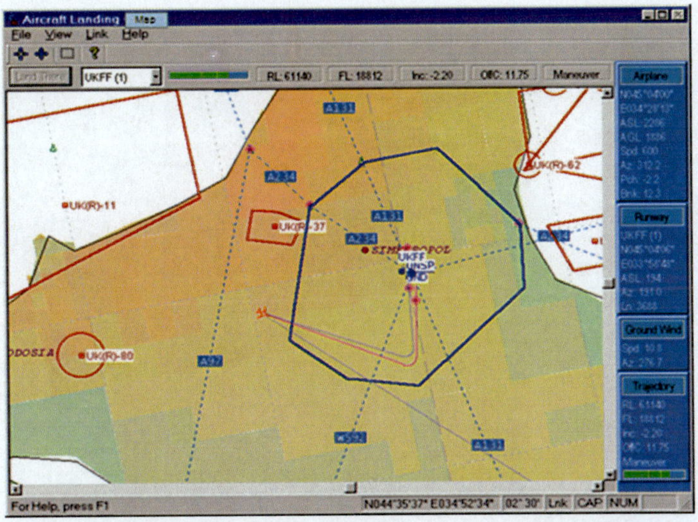

Fig. 6. Pilot interface using an additionally developed terrain map with special restricted zones and signs.

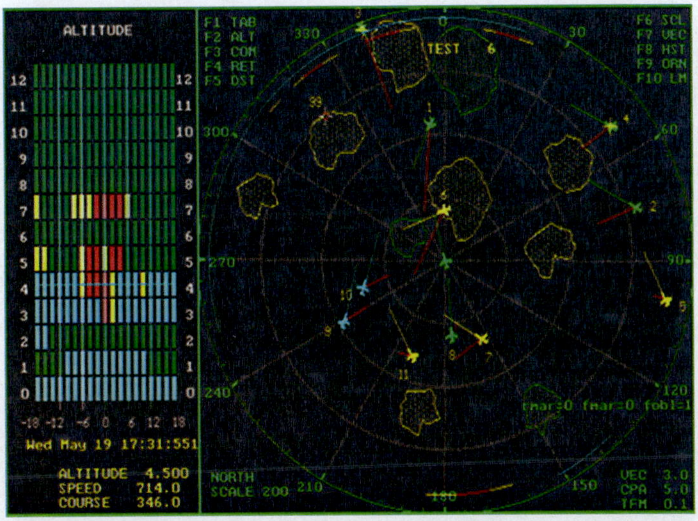

Fig. 7. Pilot interface to prevent dangerous collisions in the presence of a group of aircraft at different flight levels and cloud masses that make it difficult to observe.

The simulation of the aircraft motion and the construction of the areas of initial conditions, from which a positive result is achieved, was carried out under the control of the developed algorithms in automatic mode. The pilot or operator can be in the cockpit of the aircraft or in the control center and control the process of performing maneuvers. In

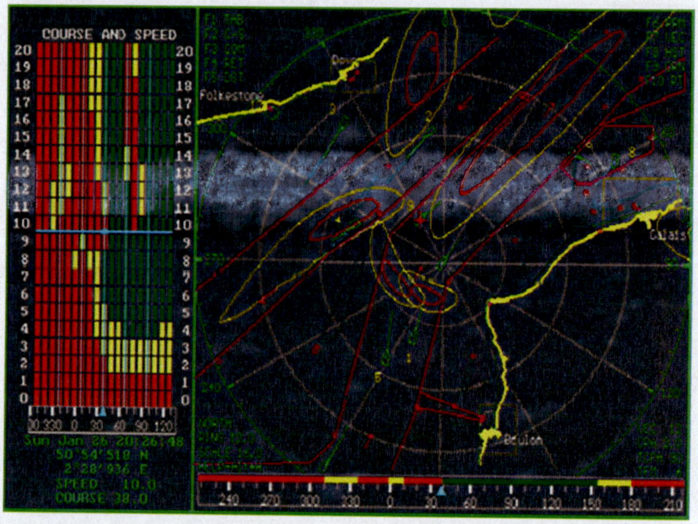

Fig. 8. Built on the basis of microsituation analysis, the interface of the navigator to control the vessel in areas of heavy traffic and the need to avoid collisions.

difficult situations, the operator can take control. With remote management, the requirements for the network through which control and management take place are quite high. However, due to the previously mentioned information optimization of the developed algorithms and user interfaces, the solution of problems arising at various stages of aircraft landing in the existing network architecture (3G, 4G) is achieved. Expanding the list of tasks for remote control of mobile objects (for example, the tasks of combat use of aircraft or driving a car in urban infrastructure) requires the use of 5G and 6G technologies and greater intellectualization of communication and control processes.

7 Directions for Advanced Research

The experience of creating control systems and complexes capable of operating in automatic mode and effectively interacting with a person in critical situations makes it possible to identify promising areas of research and development.

- Creation of systems that contain basic semantic concepts, databases of data and knowledge, objects and their interaction (selected or general model of the world).
- Creation of basic elements and processors that have the properties of adaptation, learning and retraining for organizing parallel data processing in stochastic environments.
- Creation of architecture of interaction of modern computing facilities, basic elements and stochastic environments.
- Development of the structure and individual components of integrated and distributed control systems.
- Achieving automation of the common use of heterogeneous distributed databases and knowledge.

- Achievement of automatic and automated remote control of dynamic objects in real time.
- Strengthening the intellectual capabilities of control systems and decision support systems using distributed human and machine resources.
- Construction of hybrid human-machine intelligent complexes and systems for processing databases and knowledge.

These are quite general directions, which in modern conditions acquire special significance. The development of multiprocessor computing tools and cloud services, the accumulation and availability of large databases on control objects and the environment using remote access provide new opportunities for creating systems with signs of intelligence [7]. Of particular note is the development of neural structures and computing technologies such as deep learning convolutional networks [6, 16, 17]. The use of available databases for different types of objects for training neural-like systems has made it possible to obtain results of automatic object recognition that are superior to the results of human object recognition. Based on such technologies, autonomous transport control systems, machine translation systems for different languages, systems for searching objects in an image and generating their descriptions, systems for autonomous dialogues in certain areas of activity, systems for synthesizing stylized images and many other developments are also being created.

8 Conclusion

In the process of organizing an efficiently functioning human-machine control system using computer networks, the main stages of building algorithms and control structures can be distinguished. These general provisions on the organization of intelligent control systems receive expanded possibilities for implementation in modern technical and information conditions. The correct discretization of the processes of obtaining information, the processes of information transformations and management processes must be consistent with the target task of the system and interaction with a person. A detailed consideration of control processes on relatively small intervals (microsituations) correlates well with the development of modern neural-like systems such as multilayer convolutional networks.

The emergence of new communication capabilities (5G, 6G networks) and computing such as a tensor processor and cloud services that provide access to such equipment allow a greater degree of intellectualization of information processes and the construction of intelligent control systems. Such control systems will have an optimal distribution of functional blocks at the control object and in the nodes of the global network, will make fuller use of virtual and augmented reality tools, and will allow efficient control of dynamic objects located at large distances from the control center.

The practical results of the work are focused on the control of moving dynamic objects, in particular, an aircraft in critical flight modes.

The removal of the aircraft into a given end area with a given state characteristics in the presence of significant external disturbances and restrictions on management resources was considered. A computer three-dimensional model of the aircraft, environment and control algorithms has been built, which has a high measure of similarity with the real functioning of practical systems.

The developed management algorithms correlate with the experience of the pilots on the management of aircraft in conditions of strong disturbances of different nature and in the presence of significant uncertainty of information data. Algorithms allow at each moment the adoption of the management decision to build the optimal target trajectory based on the current characteristics of the state of the controlled object and the environment.

The use of methods of microsituational analysis and synthesis allows the automation of various spheres of human practical activity, especially where there is a significant data uncertainty, a shortage of decision-making.

The results obtained are aimed at developing information technologists for the purposeful systems and are of great practical significance for solving a wide set of tasks for driving vehicles of various classes in difficult and emergency conditions.

References

1. Shaoshan, L.: Engineering autonomous vehicles and robots: the DragonFly modular-based approach. Hoboken, NJ : John Wiley & Sons, Inc. p. 203 (2020)
2. Beard, R.W., McLain, T.W.: Small Unmanned Aircraft: Theory and Practice. Princeton Univ. Press. Princeton. NJ. p. 320 (2012)
3. Mira-Bonnardel, S., Antonialli, F., Attias, D. (eds.): The robomobility revolution of urban public transport. TREP, Springer, Cham (2021). https://doi.org/10.1007/978-3-030-72976-9
4. Melnykov, S.V.: Zakonomernosti organizatsii slozhnykh ergaticheskikh sistem i postroyeniye struktury kompleksov udalennogo upravleniya dinamicheskimi ob"yektami [Principles of organization of the complex ergatic systems and creation of structure of remote control dynamic objects]. Cybernetics and Computer Engineering. № 168, pp. 70–79 (2012) (In Russian).
5. Meyer, G., Beiker, S. (eds.): AUVSI 2018. LNM, Springer, Cham (2019). https://doi.org/10.1007/978-3-319-94896-6
6. Zgurovsky, M., Zaychenko, Y.: The Fundamentals of Computational Intelligence: System Approach Studies in Computational Intelligence SCI 652 Springer Cham (2017). https://doi.org/10.1007/978-3-319-35162-9
7. Zgurovsky, M., Zaychenko, Y.: Big Data: Conceptual Analysis and Applications Studies in Big Data SBD 58 Springer Cham (2020). https://doi.org/10.1007/978-3-030-14298-8
8. Pavlova, S.V., Bogachuk, Yu.P., Melnykov, S.V., Gospodarchuk, O.Yu.: Organizatsiya effektivnogo obmena dannymi v setevykh sistemakh upravleniya dinamicheskimi ob"yektami [Organization of effective exchange data in network control system for dynamic objects]. Cybernetics and Computer Engineering. № 177, pp. 43–53. (2014) (In Russian).
9. Melnykov, S.V.: Zakonomernosti organizatsii upravleniya v ergaticheskikh sistemakh [Principles of organization in the ergatic control systems]. Cybernetics and Computer Engineering. № 155. pp. 3–15. (2007) (In Russian).
10. Tegmark, M.: Life 3.0: Being Human in the Age of Artificial Intelligence, p.380 Penguin. (2018)
11. Yao, M.: Applied Artificial Intelligence: A Handbook for Business Leaders. Topbots, 246 p. (2018)

12. Shaoshan, L., Liyun, L., Jie, T., Shuang W., Jean-Luc, G.: Creating autonomous vehicle systems. San Rafael. California. Morgan & Claypool. p.186 (2018)
13. Hugh, L., Bo, Z.: Formation control of multiple autonomous vehicle systems. Hoboken, NJ. Wiley. p. 271 (2018)
14. Bushgens, G.S., Studnev, R.V.: Aerodinamika samoleta. Dinamika prodol'nogo i bokovogo dvizheniya [Aerodynamics of airplane. Dynamics of longitudinal and lateral motion]. Moscow. Mashinostroenie, p. 372 (1979). (In Russian)
15. Pavlov, V.V.: Sintez strategiy v cheloveko–mashinnykh sistemakh [A synthesis of strategies is in the man-machine systems]. Kyiv: Vyscha Shkola. p. 162 (1989). (in Russian).
16. Wilkins, N.: Artificial Intelligence: An Essential Beginner's Guide to AI, Machine Learning, Robotics, The Internet of Things, Neural Networks, Deep Learning, Reinforcement Learning, and Our Future. Bravex Publications, p. 112 (2019)
17. Goodfellow, I., Bengio, Y., Courville, A.: Deep learning. Cambridge. MA. USA. The MIT Press, p. 800 (2016)

Metaverse of Things in 6G Era: An Emerging Fusion of IoT, XR, Edge AI and Blockchain Technologies

Taras Maksymyuk[1,2(✉)], Juraj Gazda[3], Bohdan Shubyn[1], Oleksandr Karpin[2], Oleg Kapshii[2], Oksana Urikova[1], El-Mehdi Amhoud[4], Madhusanka Liyanage[5,6], Minho Jo[7], and Mischa Dohler[8]

[1] Lviv Polytechnic National University, 79013 Lviv, Ukraine
{taras.a.maksymiuk,bohdan.p.shubyn,oksana.m.urikova}@lpnu.ua
[2] Infineon Technologies, 79000 Lviv, Ukraine
{oleksandr.karpin,oleg.kapshii}@infineon.com
[3] Technical University of Kosice, 04001 Kosice, Slovakia
juraj.gazda@tuke.sk
[4] Mohammed VI Polytechnic University, 43150 Ben Guerir, Morocco
ElMehdi.AMHOUD@um6p.ma
[5] University College Dublin, Belfield, Dublin 4, Ireland
madhusanka@ucd.ie
[6] University of Oulu, 90014 Oulu, Finland
madhusanka.liyanage@oulu.fi
[7] Korea University, 30019 Sejong Metropolitan City, Republic of Korea
minhojo@korea.ac.kr
[8] Advanced Technology Group, Ericsson Inc., Santa Clara, CA 95054, USA
mischa.dohler@ericsson.com

Abstract. An emerging future networks will be driven by the deep integration of the Internet of Things (IoT), artificial intelligence (AI) and blockchain technologies. With the enhanced capabilities of the 5G mobile networks, currently we are observing the growing interest to the metaverse and extended reality (XR) applications. Considering the demand for remote engagement solutions, in this chapter we describe the new emerging paradigm called Metaverse of Things (MoT), which aims to synchronize humans, things and digital entities across real and virtual worlds. We have outlined a state-of-the-art of the XR, IoT, edge computing, AI and blockchain, and explained how their combination can enable the MoT development. Finally, we describe the decentralized service management workflow in 6G network considering the synchronized transmission of the multiple data flows, which are transmitted through multiple devices with different quality requirements, but are associated with the single MoT application.

Keywords: XR · IoT · Edge AI · Metaverse · Blockchain · NFT · 6G

M. Klymash et al. (Eds.): TCSET 2022, LNEE 965, pp. 546–564, 2023.
https://doi.org/10.1007/978-3-031-24963-1_31

1 Introduction

The post-pandemic world has undoubtedly changed the lifestyle of the humanity by favouring the remote work and education over the real connection of the people. This has tremendously boosted the popularity of various online platforms and services, which are essential to perform social and business activities. The fusion of Internet of Things (IoT), artificial intelligence (AI) and 5G technologies, which we are currently observing, has shaped the modern business and social processes [1]. A proliferation of industrial IoT (IIoT) and smart city applications made a valuable impact on the economy through the digitization of the manufacturing cycles, supply chains, and public services [2,3]. Despite the benefits of full interconnection between machines and sensors within the digital economy, there are still blank spaces left in the processes, which require critical thinking and creativity of humans. Even the most advanced AI solutions are not yet capable of human-level comprehensive intelligence to take important critical actions in industrial processes [4]. Therefore, a so-called metaverse concept has emerged to address the issue of synergistic collaboration among humans, machines and intelligent agents in various domains [5]. The key idea of metaverse is to create a human-centric solution, which can facilitate a transition to the Internet of Everything and empower humanity with new means of interaction and engagement [6,7]. Metaverse is envisioned as a subsequent step in the Internet evolution, which provide a transition from the flat screens of smartphones or tablets to the immersive remote engagement of users in a virtual copy of the real world, including sounds, images, touches and other senses [8,9].

Technically, metaverse is the further evolution of the virtual and augmented reality (VR/AR) applications, which has extensively incorporated the advancements of the artificial intelligence and blockchain technologies to deliver a new unique experience in gaming and various social applications [10–12]. The use of blockchain allows to achieve an essential level of decentralization where all users interact without a central authority by using smart contracts and non-fungible tokens (NFTs) [13]. Correspondingly, the most advanced means of visualization and sensing along with the latest achievements of AI and edge computing are shaping the future unprecedented immersive user experience [14].

With the more extensive proliferation of the metaverses, the data volumes transmitted in 5G and 6G mobile networks will exponentially grow. Considering the need of delivering the fully realistic VR experience in most gaming and entertainment applications, we can assume that 6G will become a network for VR, just like 5G became a network for ultra high-definition video streaming [15].

Nevertheless, there is still other side of the metaverse "coin", which assumes a combination of AR experience, IoT and digital twins [16]. Such a concept will provide many new unimaginable use cases based on a large variety of the connected devices such as AR-glasses, earbuds, smart-watches and sensors. With such a variety of the connected devices the existing service models should be reconsidered from conventional single-flow services to the multi-flow services through multiple end-devices. The important aspect of the multi-flow service provisioning is the ability to synchronize several data flows from separate devices considering the difference of their data rates and latency constraints.

In [7] authors have defined a new paradigm called *Metaverse of Things (MoT), which interconnects humans, physical objects and other entities with their digital twins across real and virtual world.* MoT enables context-aware AR-experience, that allows to interact with "things" in the virtual domain and trigger real actions in the physical domain, and vice versa. One of the most feasible future applications of MoT is the digital factory, which is composed of the multiple synchronized digital twins of industrial robots and other equipment. MoT will transform the industrial IoT to the collaborative environment of humans and machines within real and virtual domains as envisioned by the Industry 5.0 paradigm [17,18].

A wide adoption of MoT requires a synergy among multiple technologies such as 6G, AI, IoT and AR/VR, blockchain, edge computing and multi-sensing [19]. In this chapter, we review the current state-of-the-art solutions which enable the future development of the MoT applications and discuss main constraints and future challenges. We also describe the possible architecture of the future mobile network, which can support deterministic and time-sensitive transmission of multiple data flows with different QoS requirements, by using the AI and blockchain technologies. Among many research works in this area, currently there are not any comprehensive surveys that address the problem of decentralized service provision over multiple data flows, which are associated with a single metaverse application.

The main contributions of this chapter are as follows:

1. We review the current state-of-the-art of the key enabling technologies for MoT applications.
2. We describe a potential 6G network architecture empowered by AI and blockchain technologies.
3. We explain an approach for adaptive and flexible service selection for any type of device in the decentralized multi-operator network environment.

The remainder of this chapter is organized as follows. Section 2 covers a review of the state-of-the-art enabling solutions for the MoT implementation. Section 3 discusses the general 6G mobile network architecture and flexible MoT service selection in the decentralized network environment. Section 4 concludes the paper.

2 A Review of the State-of-the-Art Technologies for Metaverse of Things

In this section, we review the state-of-the-art achievements of the key enabling technologies for the MoT such as VR/AR/XR, IoT, edge computing, AI and blockchain.

2.1 Immersive Engagement Through VR/AR/XR Technologies

VR has been around for quite a long time and got its momentum mostly through gaming and entertainment industry [20]. VR is typically associated with the head-mounted displays, which are used to immerse users into the rendered virtual 3D world, while "blinding" him temporary in the real world [21]. Most of the conventional VR applications are independent and represent their own finite virtual 3D world, where user can be engaged through his smartphone, head-mounted device and control joysticks. VR is considered as one of the main pillars of the metaverse due to its capability to reflect spatial 360°C view corresponding to the user movements [22]. However, due to the lack of user interaction with the surrounding real world, VR use in MoT applications is limited to the specific and mostly static use cases.

AR is an alternative solution, which in contrary to VR does not disconnect user from the real world, but overlays the virtual vision right on top of the real view [23]. One of the most popular examples of the AR is the PockemonGo game, where popular fantastic creatures were rendered right on the smartphone's screen on top of the camera view [24]. The key benefit of AR is the full understanding of the surrounding environment on the user side, which allows high-mobility use cases. Implementation of AR is less clear comparing to VR, because of the additional degrees of freedom. VR is all the time related to the complex world simulation, involving large data processing, 3D graphics rendering and high throughput requirements. AR experience can range from showing simple textual information over the camera view to the full color interactive user interface rendered in the AR-glasses. Therefore, the requirements to the network significantly vary depends on the actual AR application. It is worth noting that AR technologies are not limited to the glasses, and can be also used with any other setup of screen and camera, which significantly extends its range of feasible MoT applications [25].

Both VR and AR have their target specific use cases in the future applications. Therefore, the key conceptual challenge is to develop the extended reality (XR) experience, where parts of AR and VR will harmoniously coexist and naturally align with each other [26]. XR aims to synchronize real and virtual reality by synergistic combination of VR's immersive 3D experience and AR's capability to perceive the real environment. Such a combination allows users to interact with virtual objects integrated in the real environment [27]. In our vision, XR is the key point where metaverse experience should be started from, to make a significant impact on the economy and social life. XR allows to visually spread metaverse over the entire real world rather than keep it within the virtual non-overlapping universe, where user can temporary "jump" in. Therefore, further in this chapter we will refer to XR as to the complete "all-in-one" solution for interfacing the metaverse.

2.2 Pervasive Sensing via Internet of Things

The concept of IoT has appeared long time ago from the initial idea of enabling the machines and devices with communication capabilities. IoT devel-

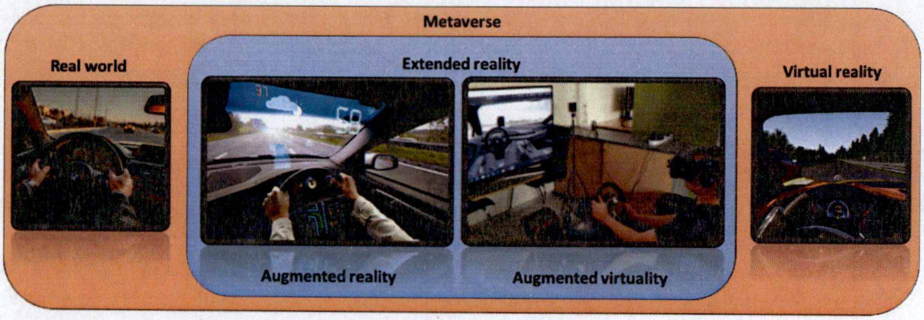

Fig. 1. The relation between VR, AR and XR applications for a driving scenario.

opment has started from the industrial and telemetry applications, where simple sensors/actuators were able to send/receive information in the Internet [28]. Recently, the concept of IoT has been significantly enhanced due to several supportive factors. First, the tremendous progress of semiconductors has enabled more powerful computing and storage capabilities within small form factor, limited battery and for a lower price. Second, the Internet itself became more effective for the data transmission with low latency and high throughput requirements. Finally, the growing capabilities of data processing and AI allow to develop many interesting IoT applications, which haven't been possible previously [6,29].

Thus, so far we have observed the rapid growth of the connected IoT devices in the world, which are spreading for a wide range of different applications. The number of IoT devices has already surpassed the size of the human population by nearly three times. The most common use cases of IoT are the industrial automation, home automation, smart cities and healthcare [30,31]. In the context of the metaverse, we can consider IoT as the low-latency sensing layer which can make the virtual objects alive according to the concept of digital twins [32]. The further evolution of IoT sensing capabilities, has a direct impact on the complexity and completeness of the new appearing digital twins by opening new dimensions of pervasive sensing and augmentation of environmental state. By combination of IoT and XR, we can augment visual and sensory information from the real world and translate it in real-time to the virtual world and digital twins in the metaverse. Correspondingly, IoT and XR can operate in the opposite direction by translating virtual actions from the virtual domain of the MoT, to the actual IoT actuators and controllers in the real world.

2.3 Ubiquitous Edge Computing

Nowadays, data is essentially an "electricity of mind", which drives the evolution and revolutions in the area of information technologies [33]. The overall volume of data processing is tremendous and exponentially growing with each new connected IoT device. Moreover, apart from the number of users, there is

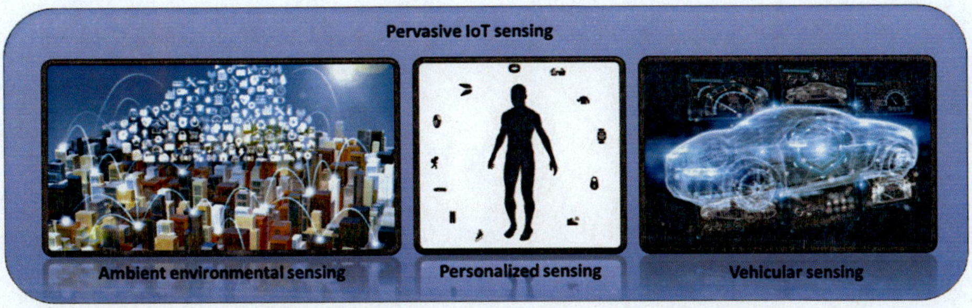

Fig. 2. Applications of pervasive IoT sensing.

also a sustainable trend of the growing content size due to the higher resolution of cameras, more detailed high density sensor measurements, and 3D XR representation required by the MoT. It is obvious that energy-constrained and low-capable IoT devices are not able to handle the large data processing tasks, which can be associated with MoT. The traditional way to overcome this limitation is to use cloud computing, which offloads the computational tasks from devices to the large data centers [34]. The common problem, which we are facing in most cloud computing applications is the high latency of data transmission due to the bottlenecks of underlying mobile and transport network infrastructure [35].

Edge computing has emerged as a promising solution to tackle the latency problem of cloud computing [36]. The core idea of edge computing is coming form the natural data distribution over the territory. In real life, most of the data are relevant only for specific regional domain such as city, enterprise, factory, etc. Hence, offloading such data to the remotely located data centers is not reasonable due to the large round-trip latency of the data transmission, which negatively affects the overall performance of the data processing. Edge computing solves this problem by a network of distributed servers, which are located at the edge of the logical data domains in order to bring the necessary computational power as close as possible to the users in terms of network latency and throughput. Recently, edge computing has been closely tightened with the 5G mobile network infrastructure, by following the same hierarchical structure as 5G cells [37]. As result, edge servers are usually collocated with the base stations and access points that allows to reduce distance between data source and computational power to the single wireless link with high throughput and low latency [38] (Fig. 3).

One of the most prominent direction of edge computing is the full application streaming, where literary nothing will be computed on the end device, but streamed from the edge cloud [39]. Such applications are currently getting momentum due to the continuous efforts of developers to fit the best possible experience to the limited capabilities of the mobile devices. Notwithstanding the fact that mobile devices becoming more powerful with each new generation, the complexity of applications is still increasing at a much faster rate. With the XR applications of metaverse the complexity of graphical rendering of a 3D scenery

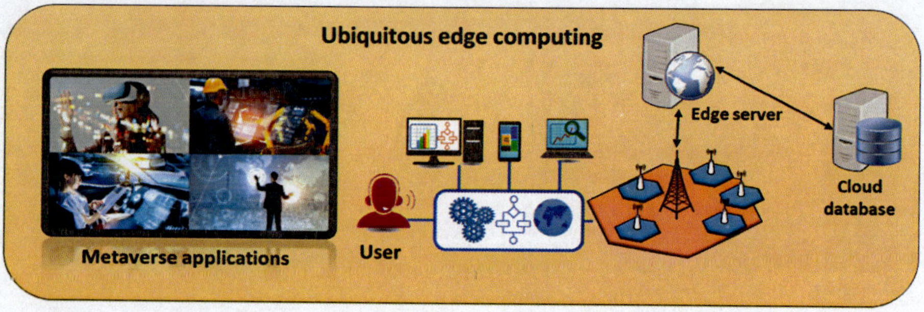

Fig. 3. General example of ubiquitous edge computing for metaverse applications.

and computation of the large number of parameters is exceeding all imaginable capabilities of modern smartphones and XR headsets. Moreover, there is not much room for trade-offs between quality and performance of real-time applications, because any attempt to reduce the computational workload will definitely end up with the poor user experience. Therefore, we envision that most of computational workload of the MoT applications, will be handled by the edge servers of different hierarchical levels, and all graphical experience will be transmitted directly to the end device such XR headset, only for visualisation [40]. Note that IoT sensors, mentioned in the previous subsection, will play a pivotal role in the MoT by transferring the important information about user orientation and haptic actions within the real environment to the edge server [41]. Correspondingly, edge servers will render the exactly needed spherical sector of the whole 3D visual scenery, so that precious network resources will not be wasted for the transmission of a redundant graphical content. Thus, ubiquitous edge computing is an important technology, which can empower the MoT with required computational capacity for new advances applications.

2.4 Comprehensive Intelligence

AI in general is considered as a "silver bullet" for most of modern informational applications. Although its wide adoption is still questionable, there are already several directions where AI is common, such as computer vision, patterns recognition, natural language processing and many others [42]. The MoT involves nearly all aspects where AI can be leveraged. First of all, XR applications will rely on the precise segmentation of environment and recognition of the anchor objects to render the corresponding information on top of that [43]. Second, AI can solve problems with pervasive sensing by recognition of the certain patterns among data from numerous sensors and translate them to the corresponding representations in the metaverse [44]. Finally, AI has improved so far in the generative tasks, which can be used to generate virtual 3D worlds with realistic objects and avatars for the MoT applications [45,46].

Nevertheless, an important factor, which should be considered, is that there is not any single AI algorithm to solve all problems. Typically, each specific problem requires dedicated neural network architecture depend on the target MoT application. All these models should be orchestrated somehow among the edge servers, in order to ensure their ubiquitous availability for the users [47]. Therefore, apart of the numerous AI models based on dense, convolutional and recurrent neural networks, generative adversarial networks, autoencoders and others, MoT requires a comprehensive development and operation framework to control and manage different AI models across all available edge servers [48].

2.5 Decentralized Trust via Blockchain

Since the beginning of the metaverse trend, blockchain has been considered as the best solution to merge the virtual worlds with our reality [49]. Blockchain is the distributed ledger technology, which is widely considered as the enabler of decentralized trust between different parties, without the central authority [50]. Traditionally, blockchain has been considered as the solution for peer-to-peer financial transactions. Nevertheless, currently we witness the trend of blockchain adoptions in many other domains, such as healthcare, industrial IoT, metaverse and many others [51–53]. In the metaverse context, the decentralized consensus used in blockchain can solve the security concerns of the large number of IoT devices, which are not secured properly due to low cost and constrained performance [54–56].

In addition to aforementioned use cases, recently blockchain has extended its functionality by introducing ERC-721 Non-Fungible Token (NFT) standard [57]. Unlike commonly adopted tokens, which are more like digital versions of the conventional currencies, the NFTs have the feature of uniqueness, like real world assets. In other words, fungible tokens are the money, which are equal one to another, i.e. 1 dollar is equal to any other 1 dollar. Instead, the NFTs is more like a precious asset, which is unique, i.e. Van Gogh's picture which exist only in the one authentic copy. Therefore, NFTs are gained a lot of interest in the metaverse, because user is able to buy and sell virtual items through blockchain, and make sure that his ownership rights will be preserved across real and virtual world as long as he keeps the corresponding NFT in his crypto-wallet [58].

Currently, we observe the rise of metaverse internal marketplaces of virtual goods. The most known example is Decentraland, which represents the pieces of virtual land by NFTs and sold them to the users. Another example in Nikeland, where users can buy unique sport clothing and shoes of the corresponding brand and wear their virtual avatars in the metaverse. There are also numerous other similar example of marketplaces mostly related to the gaming and entertainment industry [58]. Therefore, we believe that blockchain and NFTs can significantly empower the MoT, by providing a legal and secure ground to the digitization of any real or virtual assets.

3 Metaverse of Things in the Decentralized 6G Mobile Network Environment

In this section, we describe the novel decentralized 6G mobile network architecture and its main operational principles, which enable the development of the multi-flow MoT applications (Fig. 4).

Fig. 4. Relation among the key components of the Metaverse of Things.

3.1 The Evolution of the Mobile Networks to Meet the MoT Requirements

MoT applications impose more strict requirements to the throughput, reliability and latency of the 5G mobile network infrastructure. Modern 5G networks provide the key technological solutions to cover most of the MoT use cases, such as enhanced mobile broadband (eMBB), massive machine type communication (mMTC) and ultra-reliable and low latency communication (URLLC). These solutions have enabled the major breakthrough of the IoT applications so far by replacing most of the traditional cable infrastructure and simplifying the automation process of many industrial applications [59].

To manage the data flows with different requirements, as mentioned above, the 3GPP has defined a network slicing procedure in TS 29.531 [60]. Network slicing provides an isolated virtual network infrastructure with target QoS parameters, i.e. for mMTC, URLLC or eMBB services. Recently, many research works have appeared to further investigate the different approaches, which can improve the performance of the IoT applications. In [59], authors investigate the impact of different physical and medium-access control configurations. Their findings indicate that reserving the dedicated bandwidth for the eMBB and URLLC applications allows to improve the overall efficiency of these services in terms of throughput and latency. In other research [61], authors propose an alternative approach for simultaneous support of eMBB and URRLC services by an explicit prioritization method. Proposed approach allows to isolate both domains and provides stable performance characteristics of URLLC applications, while eMBB traffic experiences dynamic variation of application performance. For the industrial IoT applications, authors in [62] comprehensively studied the eMBB and URLLC performance in the isolated 5G scenarios.

The requirements for MoT differ from those for the industrial IoT. The goal of IoT is to synchronize physical things with their digital twins in the Internet. On the other hand, MoT aims to synchronize physical and virtual realities, and redefine the meaning of a "thing", which can refer not only to real connected instance, but also to the unique virtual objects, which do not exist in the real world [7]. For example, in the industrial domain we need synchronize robots, production lines, sensors and actuators with their corresponding virtual copies represented as NFTs [7]. Despite all the advancements of 5G with respect to the previous generations, it has yet not been designed for the challenges, which metaverse applications will impose. Therefore, it is necessary to reconsider the current adopted practices of 5G deployment and operation, and make the corresponding improvements in the 6G standard.

3.2 Decentralized 6G Mobile Network Architecture for the Metaverse of Things

The true power of the metaverse is in its decentralized nature and joint administration by all users through blockchain. However, the 5G networks are mostly centralized in terms of deployment and management of the data flows. Whereas

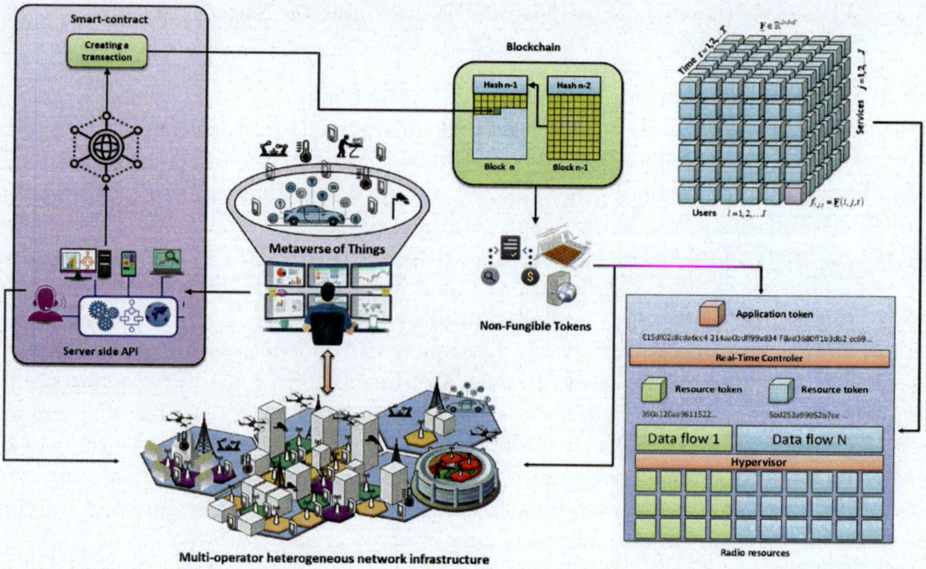

Fig. 5. The decentralized 6G architecture for the Metaverse of Things

conventional IoT and 5G services can be somehow differentiated between eMBB, URLLC and mMTC slices, the MoT will often require some combination of them within a single application. This can be done by a synchronized transmission of data flows with different QoS parameters [7]. Moreover, in some previous works it is assumed that these data flows can be transmitted by different operators [63], and even by multiple devices [7]. To be clear, we consider not only conventional smartphones and tablets, but also any other devices such as sensors, cameras, actuators, XR headsets, smart watches, etc. In the MoT context, one user application can involve several devices simultaneously. For example, let's consider an AR navigation. Such use case involves eMBB data transmission for AR view rendering, which is synchronized with URLLC data transmission from wearable sensors to determine the spatial orientation of the user, and supplied with mMTC data from ambient sensors and actuators to determine the precise localization of the user. Moreover, those data flows can be transmitted by different devices, and by different operators.

Summarizing our previous research [7,63,64], here we derive our vision for managing the MoT applications in the decentralized 6G mobile network based on the multi-flow service provision and blockchain. In the decentralized 6G network, each device has a unique blockchain identifier and all service-level agreements are secured by smart contracts and NFTs [63] (Fig. 5).

As shown in the Fig. 5, the MoT applications will be composed from multiple data flows, which can be represented as a sparse matrix of 3 dimensions: devices, services and operators. This matrix reflects the association among devices, service types and operators in quasi-real time:

$$\underline{\mathbf{F}} = \begin{bmatrix} f_{ij11} & f_{ij12} & \cdots & f_{ij1K} \\ f_{ij21} & f_{ij22} & \cdots & f_{ij2K} \\ \vdots & \vdots & \ddots & \vdots \\ f_{ijT1} & f_{ijT2} & \cdots & f_{ijTK} \end{bmatrix} \in \mathbb{R}^{I \times J \times T \times K}, \tag{1}$$

where the smallest data flow can be represented by a connection of the $i - th$ UE with the $j - th$ service to the $k - th$ MNO at the time interval t.

The matrix of data flows (1) defines the corresponding scheduling of radio resources by each operator [65–67]. The underlying blockchain infrastructure is needed to track the corresponding records of spectrum allocation and service level agreements. In order to ensure the continuous availability of resources for each type of service, we consider the forward market of radio and computing resources represented by NFTs. Thus, each user will be able to reserve the corresponding resources in advance by purchasing the NFT of corresponding MoT applications. If an application is composed by multiple data flows and multiple devices, a smart contract will automatically split the whole data flow into multiple data flows with different service requirements, and secure the corresponding radio and computational resources by NFTs. Although, current use case is quite complicated for real-time operation, over time it can converge to the sustainable statistical pattern where users will get the same services from the same operators for the specific time intervals and locations.

3.3 Adaptive Service Selection for MoT Applications in the 6G Mobile Network

The typical MoT application will be composed by the multiple data flows with different types of services. Some services may be more suitable for one operator, while other ones can be better provided by other operator. Within this uncertainty, the selection of the corresponding operators and services becomes more complicated. Moreover, service level agreements are typically connected to the service price, which can influence the final decision of the user. To tackle the challenge of explicit calculation of the quality of experience (QoE) based on measurable parameters of service quality, in [63,68,69] the utility function for weighted QoE optimization is defined based on the price and multiple quality parameters of the service:

$$u(q,p) = \left[\mu\left(q - q_{\min}\right) + (1-\mu)\left(p_{\max} - p\right)\right], \tag{2}$$

where q is the integral QoE value of the data flow, p is the service price for the data flow normalized by time and determined by the algorithm presented in [70], q_{min} is the lowest quality acceptable by the user, and p_{max} is the highest service price that the user can pay. An additional parameter $\mu \in (0,1)$ is used to adjust the user's trade-off between price and quality.

An integral quality parameter q is defined as a weighted sum of all other parameters [7]:

$$\begin{gathered} q = R \cdot \beta^{(R)} + T \cdot \beta^{(T)} + L \cdot \beta^{(L)} \text{ and} \\ \beta^{(R)} + \beta^{(T)} + \beta^{(L)} = 1, \end{gathered} \tag{3}$$

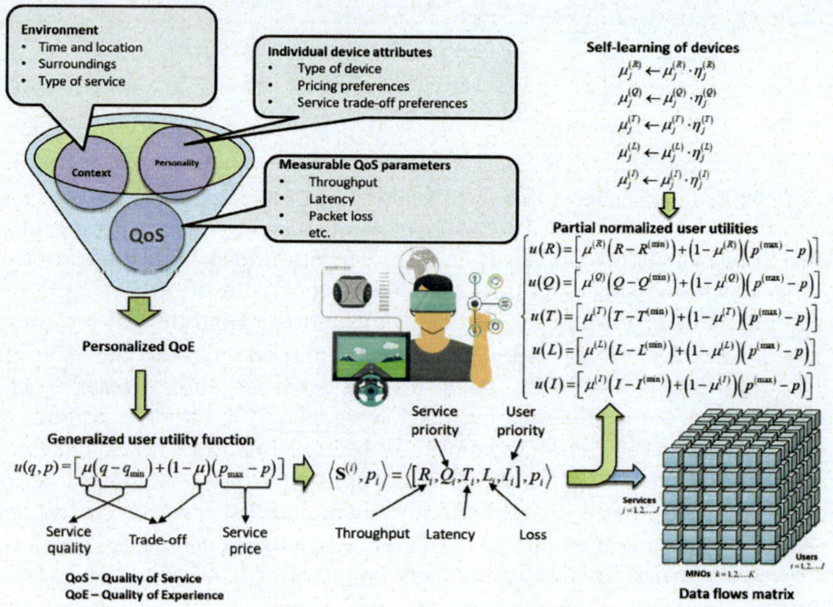

Fig. 6. QoE management workflow for MoT services in 6G

where R, T, L - are the throughput, latency and packet loss, respectively, as defined in 3GPP TS 23.501 [71]. By tweaking the corresponding values of $\beta^{(R)}, \beta^{(T)}, \beta^{(L)}$, each device can adapt the balance between all parameters and define its own quality preferences. The whole adaptive QoE management workflow is depicted in Fig. 6.

MoT immersive experience requires precise synchronization of all data flows with different requirements, such as tactile data, video data, and data from ambient sensors. In [7], this problem is resolved by using NFTs, which represent multi-flow applications as single entities in the blockchain, which can be purchased by the user from multiple operators simultaneously.

The translation from single service NFT to multiple data flows NFTs is done by blockchain smart contract, which orchestrate them according to service requirements, target operators and different devices (XR headsets, smartwatches, sensors, wearables, etc.).

Each device choose the service with the highest utility function among all available bids for each type of service j at each discrete time interval t:

$$f^{(k)} = \max_{\forall i,j,t} \underline{\mathbf{U}}\,(i,j,k,t)\,. \tag{4}$$

After execution of a smart contract, each selected operator performs the scheduling of data flows within the available spectrum and computational capacity of edge servers. On the other hand, each user receives a NFT associated with dedicated network resources, and the corresponding amount of money is secured

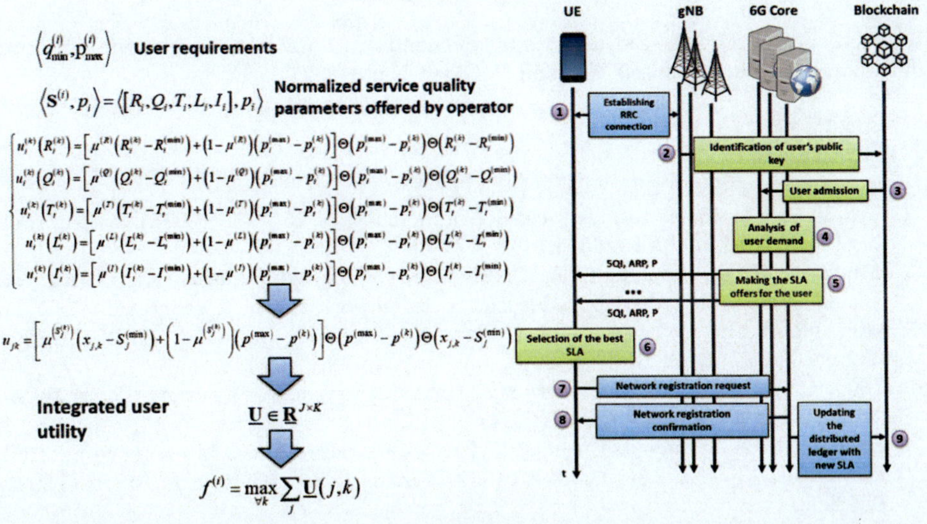

Fig. 7. Protocol diagram of adaptive service selection by end-user.

by a smart contract until successful confirmation of service provision, and then transferred to the corresponding operators (Fig. 7).

4 Conclusion

In this chapter, we have provided an overview of the state-of-the-art solutions and our developments on the MoT implementation in 6G era. The key enabling technologies such as XR, IoT, edge computing, AI and blockchain have been described in the context of their synergistic combination for the MoT. In particular, we have explained the main challenge and available solutions for synchronized multi-flow data transmission through multiple sensing and visualization devices. The key idea is reserve the required spectrum and computational capacity via blockchain by tokenizing them as NFTs. The advantage of the proposed solution is in the automated management of the tokenized resources by the smart contracts, which can precisely adjust the QoE parameters for each application by tweaking the corresponding QoS parameters of individual data flows. In our future research, we will provide a more insights to the feasible future MoT applications in the context of the state-of-the-art capabilities of AI, blockchain and 6G technologies.

Acknowledgement. This research was supported by the Ukrainian government project #0120U100674 "Designing the novel decentralized mobile network based on blockchain architecture and artificial intelligence for 5G/6G development in Ukraine", by the Slovak Research and Development Agency, project number APVV-18-0214 and APVV-21-0318, by the Scientific Grant Agency of the Ministry of Education, science, research and sport of the Slovakia under the contract: 1/0268/19 and by the Academy of Finland project "6Genesis Flagship" (Grant No. 318927).

References

1. Yarali, A.: Artificial Intelligence, 5G, and IoT, pp. 251–268 (2022). https://doi.org/10.1002/9781119685265.ch14
2. Medhane, D.V., Sangaiah, A.K., Hossain, M.S., Muhammad, G., Wang, J.: Blockchain-enabled distributed security framework for next-generation IoT: an edge cloud and software-defined network-integrated approach. IEEE Internet Things J. **7**(7), 6143–6149 (2020)
3. Da Xu, L., He, W., Li, S.: Internet of things in industries: a survey. IEEE Trans. Industr. Inf. **10**(4), 2233–2243 (2014)
4. Brooks, R.: A human in the loop: AI won't surpass human intelligence anytime soon. IEEE Spectr. **58**(10), 48–49 (2021). https://doi.org/10.1109/MSPEC.2021.9563963
5. Park, S.M., Kim, Y.G.: A metaverse: taxonomy, components, applications, and open challenges. IEEE Access **10**, 4209–4251 (2022). https://doi.org/10.1109/ACCESS.2021.3140175
6. Dohler, M., et al.: Internet of skills, where robotics meets AI, 5G and the tactile internet. In: 2017 European Conference on Networks and Communications (EuCNC), pp. 1–5. IEEE (2017)
7. Maksymyuk, T., Gazda, J., Bugár, G., Gazda, V., Liyanage, M., Dohler, M.: Blockchain-empowered service management for the decentralized metaverse of things. IEEE Access **10**, 99025–99037 (2022). https://doi.org/10.1109/ACCESS.2022.3205739
8. Nadir, Z., Taleb, T., Flinck, H., Bouachir, O., Bagaa, M.: Immersive services over 5G and beyond mobile systems. IEEE Netw. **35**(6), 299–306 (2021). https://doi.org/10.1109/MNET.121.2100172
9. Gao, Y., Wei, X., Chen, J., Zhou, L.: Toward immersive experience: evaluation for interactive network services. IEEE Netw. **36**(1), 144–150 (2022). https://doi.org/10.1109/MNET.121.2100323
10. Jot, J.M., Audfray, R., Hertensteiner, M., Schmidt, B.: Rendering spatial sound for interoperable experiences in the audio metaverse. In: 2021 Immersive and 3D Audio: From Architecture to Automotive (I3DA), pp. 1–15 (2021). https://doi.org/10.1109/I3DA48870.2021.9610971
11. Yoon, K., Kim, S.K., Jeong, S.P., Choi, J.H.: Interfacing cyber and physical worlds: introduction to IEEE 2888 Standards. In: 2021 IEEE International Conference on Intelligent Reality (ICIR), pp. 49–50 (2021). https://doi.org/10.1109/ICIR51845.2021.00016
12. Kim, S.S., Dohler, M., Dasgupta, P.: The Internet of Skills: use of fifth-generation telecommunications, haptics and artificial intelligence in robotic surgery. BJU Int. **122**(3), 356–358 (2018)
13. Mouris, D., Tsoutsos, N.G.: NFTs for 3D models: sustaining ownership in industry 4.0. IEEE Consum. Electron. Mag., 1 (2022). https://doi.org/10.1109/MCE.2022.3164221
14. Wang, X., Han, Y., Wang, C., Zhao, Q., Chen, X., Chen, M.: In-edge AI: intelligentizing mobile edge computing, caching and communication by federated learning. IEEE Netw. **33**(5), 156–165 (2019). https://doi.org/10.1109/MNET.2019.1800286

15. Fantacci, R., Picano, B.: Edge-based virtual reality over 6G terahertz channels. IEEE Netw. **35**(5), 28–33 (2021). https://doi.org/10.1109/MNET.101.2100023
16. Mihai, S., Yaqoob, M., et al.: Digital twins: a survey on enabling technologies, challenges, trends and future prospects. IEEE Commun. Surv. Tut. **24**, 2255–2291 (2022). https://doi.org/10.1109/COMST.2022.3208773
17. Groshev, M., Guimarães, C., Martín-Pérez, J., de la Oliva, A.: Toward intelligent cyber-physical systems: digital twin meets artificial intelligence. IEEE Commun. Mag. **59**(8), 14–20 (2021). https://doi.org/10.1109/MCOM.001.2001237
18. Fraga-Lamas, P., Varela-Barbeito, J., Fernández-Caramés, T.M.: Next generation auto-identification and traceability technologies for industry 5.0: a methodology and practical use case for the shipbuilding industry. IEEE Access **9**, 140700–140730 (2021). https://doi.org/10.1109/ACCESS.2021.3119775
19. Kumar, T., et al.: BlockEdge: blockchain-edge framework for industrial IoT networks. IEEE Access **8**, 154166–154185 (2020). https://doi.org/10.1109/ACCESS.2020.3017891
20. Menin, A., Torchelsen, R., Nedel, L.: An analysis of VR technology used in immersive simulations with a serious game perspective. IEEE Comput. Graph. Appl. **38**(2), 57–73 (2018). https://doi.org/10.1109/MCG.2018.021951633
21. Hossan, M.T., Chowdhury, M.Z., Shahjalal, M., Jang, Y.M.: Human bond communication with head-mounted displays: scope, challenges, solutions, and applications. IEEE Commun. Mag. **57**(2), 26–32 (2019). https://doi.org/10.1109/MCOM.2018.1800527
22. Climent, M.M., Soler-Vilageliu, O., Vila, I.F., Langa, S.F.: Vr360 subtitling: requirements, technology and user experience. IEEE Access **9**, 2819–2838 (2021). https://doi.org/10.1109/ACCESS.2020.3047377
23. Morín, D.G., Pérez, P., Armada, A.G.: Toward the distributed implementation of immersive augmented reality architectures on 5G networks. IEEE Commun. Mag. **60**(2), 46–52 (2022). https://doi.org/10.1109/MCOM.001.2100225
24. Druga, M.: Pokemon go: where VR and AR have gone since its inception. IEEE Potentials **37**(1), 23–26 (2018). https://doi.org/10.1109/MPOT.2017.2675498
25. Qiao, X., Ren, P., Dustdar, S., Liu, L., Ma, H., Chen, J.: Web AR: a promising future for mobile augmented reality-state of the art, challenges, and insights. Proc. IEEE **107**(4), 651–666 (2019). https://doi.org/10.1109/JPROC.2019.2895105
26. Huzaifa, M., et al.: ILLIXR: an open testbed to enable extended reality systems research. IEEE Micro **42**(4), 97–106 (2022). https://doi.org/10.1109/MM.2022.3161018
27. Hu, F., Deng, Y., Zhou, H., Jung, T.H., Chae, C.B., Aghvami, A.H.: A vision of an XR-aided teleoperation system toward 5G/B5G. IEEE Commun. Mag. **59**(1), 34–40 (2021). https://doi.org/10.1109/MCOM.001.2000581
28. Cheng, J., Chen, W., Tao, F., Lin, C.L.: Industrial IoT in 5G environment towards smart manufacturing. J. Ind. Inf. Integr. **10**, 10–19 (2018). https://doi.org/10.1016/j.jii.2018.04.001. https://www.sciencedirect.com/science/article/pii/S2452414X18300049
29. Wang, Y., Yang, S., Ren, X., Zhao, P., Zhao, C., Yang, X.: IndustEdge: a time-sensitive networking enabled edge-cloud collaborative intelligent platform for smart industry. IEEE Trans. Ind. Inf. **18**, 2386–2398 (2021)
30. Lema, M.A., et al.: 5G case study of Internet of Skills: slicing the human senses. In: 2017 European Conference on Networks and Communications (EuCNC), pp. 1–6. IEEE (2017)
31. Kubendiran, M., Singh, S., Sangaiah, A.K.: Enhanced security framework for e-health systems using blockchain. J. Inf. Process. Syst. **15**(2), 239–250 (2019)

32. Simsek, M., Aijaz, A., Dohler, M., Sachs, J., Fettweis, G.: 5G-enabled tactile internet. IEEE J. Sel. Areas Commun. **34**(3), 460–473 (2016)
33. Hernandez-Ramos, J.L., Geneiatakis, D., Kounelis, I., Steri, G., Nai Fovino, I.: Toward a data-driven society: a technological perspective on the development of cybersecurity and data-protection policies. IEEE Secur. Priv. **18**(1), 28–38 (2020). https://doi.org/10.1109/MSEC.2019.2939728
34. Zhou, B., Dastjerdi, A.V., Calheiros, R.N., Srirama, S.N., Buyya, R.: mCloud: a context-aware offloading framework for heterogeneous mobile cloud. IEEE Trans. Serv. Comput. **10**(5), 797–810 (2017). https://doi.org/10.1109/TSC.2015.2511002
35. Maksymyuk, T., Dumych, S., Krasko, O., Jo, M.: Software defined optical switching for cloud computing transport systems. In: Proceedings of the 9th International Conference on Ubiquitous Information Management and Communication, IMCOM 2015, New York, NY, USA (2015). https://doi.org/10.1145/2701126.2701232
36. Heo, J., Bhardwaj, K., Gavrilovska, A.: Poster: Enabling flexible edge-assisted XR. In: 2021 IEEE/ACM Symposium on Edge Computing (SEC), pp. 465–467 (2021). https://doi.org/10.1145/3453142.3491408
37. Šlapak, E., Gazda, J., Guo, W., Maksymyuk, T., Dohler, M.: Cost-effective resource allocation for multitier mobile edge computing in 5G mobile networks. IEEE Access **9**, 28658–28672 (2021). https://doi.org/10.1109/ACCESS.2021.3059029
38. Tran, T.X., Hajisami, A., Pandey, P., Pompili, D.: Collaborative mobile edge computing in 5G networks: new paradigms, scenarios, and challenges. IEEE Commun. Mag. **55**(4), 54–61 (2017)
39. Gao, Y., Wei, X., Kang, B., Chen, J.: Edge intelligence empowered cross-modal streaming transmission. IEEE Netw. **35**(2), 236–243 (2021). https://doi.org/10.1109/MNET.011.2000474
40. Magro, M., Bugeja, K., Spina, S., Debattista, K.: Cloud-based dynamic GI for shared VR experiences. IEEE Comput. Graph. Appl. **40**(5), 10–25 (2020). https://doi.org/10.1109/MCG.2020.3002474
41. Qiao, Y., Zheng, Q., Lin, Y., Fang, Y., Xu, Y., Zhao, T.: Haptic communication: toward 5G tactile internet. In: 2020 Cross Strait Radio Science & Wireless Technology Conference (CSRSWTC), pp. 1–3. IEEE (2020)
42. Li, X.H., et al.: A survey of data-driven and knowledge-aware explainable AI. IEEE Trans. Knowl. Data Eng. **34**(1), 29–49 (2022). https://doi.org/10.1109/TKDE.2020.2983930
43. Ko, J.G., Lee, S., Lee, S., Lee, J.: Lightweight deep learning based intelligent mobile augmented reality. In: 2021 IEEE International Conference on Consumer Electronics-Asia (ICCE-Asia), pp. 1–3 (2021). https://doi.org/10.1109/ICCE-Asia53811.2021.9641972
44. Mohammadi, M., Al-Fuqaha, A., Sorour, S., Guizani, M.: Deep learning for IoT big data and streaming analytics: a survey. IEEE Commun. Surv. Tut. **20**(4), 2923–2960 (2018). https://doi.org/10.1109/COMST.2018.2844341
45. Yang, C., et al.: BodyGAN: general-purpose controllable neural human body generation. In: 2022 IEEE/CVF Conference on Computer Vision and Pattern Recognition (CVPR), pp. 7723–7732 (2022). https://doi.org/10.1109/CVPR52688.2022.00758
46. Segal, Y., Hadar, O.: Constructing a skeleton database and enriching it using a generative adversarial network (gan) simulator to assess human movement. In: 2022 IEEE 38th International Conference on Data Engineering (ICDE), pp. 3226–3229 (2022). https://doi.org/10.1109/ICDE53745.2022.00304

47. Wu, Y.: Cloud-edge orchestration for the internet of things: architecture and ai-powered data processing. IEEE Internet Things J. **8**(16), 12792–12805 (2021). https://doi.org/10.1109/JIOT.2020.3014845
48. Chang, L., et al.: 6G-enabled edge ai for metaverse: challenges, methods, and future research directions. arXiv preprint arXiv:2204.06192 (2022)
49. Yang, Q., Zhao, Y., Huang, H., Xiong, Z., Kang, J., Zheng, Z.: Fusing blockchain and AI with metaverse: a survey. IEEE Open J. Comput. Soc. **3**, 122–136 (2022). https://doi.org/10.1109/OJCS.2022.3188249
50. Li, X., Russell, P., Mladin, C., Wang, C.: Blockchain-enabled applications in next-generation wireless systems: challenges and opportunities. IEEE Wirel. Commun. **28**(2), 86–95 (2021). https://doi.org/10.1109/MWC.001.2000455
51. Hasselgren, A., Kralevska, K., Gligoroski, D., Pedersen, S.A., Faxvaag, A.: Blockchain in healthcare and health sciences-a scoping review. Int. J. Med. Inf. **134**, 104040 (2020)
52. Taylor, P.J., Dargahi, T., Dehghantanha, A., Parizi, R.M., Choo, K.K.R.: A systematic literature review of blockchain cyber security. Digit. Commun. Netw. **6**(2), 147–156 (2020)
53. Nguyen, C.T., Hoang, D.T., Nguyen, D.N., Dutkiewicz, E.: Metachain: A novel blockchain-based framework for metaverse applications. In: 2022 IEEE 95th Vehicular Technology Conference (VTC2022-Spring), pp. 1–5 (2022). https://doi.org/10.1109/VTC2022-Spring54318.2022.9860983
54. Huang, J., Kong, L., Chen, G., Wu, M.Y., Liu, X., Zeng, P.: Towards secure industrial IoT: blockchain system with credit-based consensus mechanism. IEEE Trans. Industr. Inf. **15**(6), 3680–3689 (2019)
55. Cai, X., et al.: A sharding scheme based many-objective optimization algorithm for enhancing security in blockchain-enabled industrial internet of things. IEEE Trans. Industr. Inf. **17**, 7650–7658 (2021)
56. Han, R., Gramoli, V., Xu, X.: Evaluating blockchains for IoT. In: 2018 9Th IFIP International Conference on New Technologies, Mobility and Security (NTMS), pp. 1–5. IEEE (2018)
57. Galal, H.S., Youssef, A.M.: Aegis: privacy-preserving market for non-fungible tokens. IEEE Trans. Netw. Sci. Eng. **10**, 92–102 (2022). https://doi.org/10.1109/TNSE.2022.3205428
58. Mouris, D., Tsoutsos, N.G.: NFTs for 3D models: sustaining ownership in industry 4.0. IEEE Consum. Electron. Mag., 1 (2022). https://doi.org/10.1109/MCE.2022.3164221
59. Mhedhbi, M., Morcos, M., Galindo-Serrano, A., Elayoubi, S.E.: Performance evaluation of 5G radio configurations for industry 4.0. In: 2019 International Conference on Wireless and Mobile Computing, Networking and Communications (WiMob), pp. 1–6. IEEE (2019)
60. Kukliński, S., et al.: A reference architecture for network slicing. In: 2018 4th IEEE Conference on Network Softwarization and Workshops (NetSoft), pp. 217–221. IEEE (2018)
61. Markoval, E., Moltchanov, D., Pirmagomedov, R., Ivanova, D., Koucheryavy, Y., Samouylov, K.: Priority-based coexistence of eMBB and URLLC traffic in industrial 5G NR deployments. In: 2020 12th International Congress on Ultra Modern Telecommunications and Control Systems and Workshops (ICUMT), pp. 1–6. IEEE (2020)
62. Chettri, L., Bera, R.: A comprehensive survey on Internet of Things (IoT) toward 5G wireless systems. IEEE Internet Things J. **7**(1), 16–32 (2019)

63. Maksymyuk, T., Volosin, M., Gazda, J., Liyanage, M.: Blockchain-based decentralized service provisioning in local 6G mobile networks. In: Proceedings of the 19th ACM Conference on Embedded Networked Sensor Systems, pp. 516–519 (2021)
64. Maksymyuk, T., et al.: Blockchain-empowered framework for decentralized network management in 6G. IEEE Commun. Mag. **58**(9), 86–92 (2020). https://doi.org/10.1109/MCOM.001.2000175
65. Afolabi, I., Bagaa, M., Boumezer, W., Taleb, T.: Toward a real deployment of network services orchestration and configuration convergence framework for 5G network slices. IEEE Netw. **35**(1), 242–250 (2021). https://doi.org/10.1109/MNET.011.2000146
66. Viswanathan, H., Mogensen, P.E.: Communications in the 6G era. IEEE Access **8**, 57063–57074 (2020). https://doi.org/10.1109/ACCESS.2020.2981745
67. Maksymyuk, T., Pelishok, V.: The LTE channel transmission rate increasing. In: Proceedings of International Conference on Modern Problem of Radio Engineering, Telecommunications and Computer Science, pp. 251–252 (2012)
68. Xing, Y., Chandramouli, R., Cordeiro, C.: Price dynamics in competitive agile spectrum access markets. IEEE J. Sel. Areas Commun. **25**(3), 613–621 (2007)
69. Bugár, G., Vološin, M., Maksymyuk, T., Zausinová, J., Gazda, V., Horváth, D., Gazda, J.: Techno-economic framework for dynamic operator selection in a multi-tier heterogeneous network. Ad Hoc Netw. **97**, 102007 (2020). https://doi.org/10.1016/j.adhoc.2019.102007
70. Gazda, J., Bugár, G., Vološin, M., Drotár, P., Horváth, D., Gazda, V.: Dynamic spectrum leasing and retail pricing using an experimental economy. Comput. Netw. **121**, 173–184 (2017)
71. Saboorian, T., Xiang, A., Thiébaut, L.: Network slicing and 3GPP service and systems aspects (SA) standard. IEEE Software Defined Networks, IEEE Softwarization, vol. 7 (2017)

Data Management in IoT: From Sensor Data to Intelligent Applications

Volodymyr Vasyutynskyy(✉) and Andriy Luntovskyy

BA Dresden Univ. of Coop. Education (Saxonian Study Academy), Hans-Grundig-Str. 25, 01307 Dresden, Germany
vasyut@gmail.com

Abstract. Emerging networks allow to connect the smart devices and applications efficiently and everywhere, opening a lot of opportunities and use cases for getting and using data. The resulting huge amount of heterogeneous data leads to the new challenges on the data management, like data storage, processing, privacy, lifecycle etc. This work gives and overview of these challenges and the current solutions like cloud computing and data mesh.

Keywords: Data management · IoT · Cloud computing · Data mesh

1 Networked Applications and Challenges

Emerging Networking appears nowadays spontaneous and accompanies us in our everyday life as well as supports modern industry processes, ongoing digitalization of workflows and multiple smart office scenarios that can be summarized under the term IoT (Internet of the Things). It provides flexible network monitoring, performs network analytics and demonstrates full advantage of Machine Learning (ML) too. The emerging networking solutions enable better process automation and proactive problem solving as well as decentralized operation of the network infrastructure. To be used efficiently, the data from emerging networks must be combined with other data sources, which may be [1–4]:

- Data from sensors and production capacities (Smart Manufacturing),
- Navigation data from GPS and other navigation systems, e.g. Galileo,
- Data from GIS (Geographic Information Systems),
- Traffic and telematics data (IoT),
- Robotics data (navigation, sensing, measurements),
- QoS management data and traffic data from mobile providers,
- Power and renewable energy stations (data for smart metering),
- Data from scientific simulations and parallel computing (clusters, grids),
- Imaging data of patients from CT and MRT (Computer and Magnetic Resonance Tomography) in healthcare,
- Marketing information from cookies, social networks and Web2.0 semantic networks,

M. Klymash et al. (Eds.): TCSET 2022, LNEE 965, pp. 565–573, 2023.
https://doi.org/10.1007/978-3-031-24963-1_32

- Collected personal data of citizens and much more,
- Industrial applications, including home automation, patient health data, M2M and robotic data, business intelligence, pharmacological research, networking, and experimental data.

The real power of this data unleashes in the intelligent integration and processing of it, which enables more **optimized processes and completely new applications** like:

- Industrial applications (also known as Industrial IoT or Industry 4.0): condition monitoring, adaptive maintenance, predictive repair, security monitoring, process optimisation. They often rely on real-time data to quickly react on the changing conditions, but also include the batch data processing e.g., for ML scenarios.
- Business networks: logistics networks optimization, emergent events reaction.
- Smart cities: traffic optimization, environment monitoring, e-democracy, energy efficiency.
- Smart homes: energy consumption optimization, ambient assistant living.
- Biomedicine: biomedical sensor integration, health monitoring with wearables.

Data management is the process which takes the overall available data and refines it down to important information, which typically includes data acquisition, preprocessing, integration, analysis, governance and lifecycle support. The IoT applications imply different requirements on the data management, including real-time aspects (like latency or reaction times), security, availability, costs. There is no one-for-all approach for data management, but rather different data management patterns must be combined for each use. Along with that, there are main trends of using the cloud computing and organizing the data architectures according to the principles of data mesh, which every IoT data specialist should know and which are described in the next chapters.

2 From Devices to Cloud

Networked sensors and devices can send data to each other directly for rather simple sensing, analytical and control tasks like condition monitoring or automation control, which often require real-time communication. The edge devices can filter data to reduce bandwidth, ensure performance and security with low costs, as well as serve as the gateways to the cloud applications.

In the last decade, the massive data from IoT applications were processed with Big Data technological stack [5], which is based for example on Apache Spark for distributed data processing. Although still a valid approach, administrating and scaling the big data stack on its own requires a lot of expertise and infrastructure costs. This led to delegating of these tasks to the cloud in the last years.

Cloud computing significantly facilitates the data management by providing cost-efficient infrastructure with centralized data management and is one of the main trends of the last decade also in the area of networked devices and systems.

Typically, the devices are connected to the cloud over the cloud gateways as shown in Fig. 1.

Moving of data and data management to cloud brings the following **advantages**:

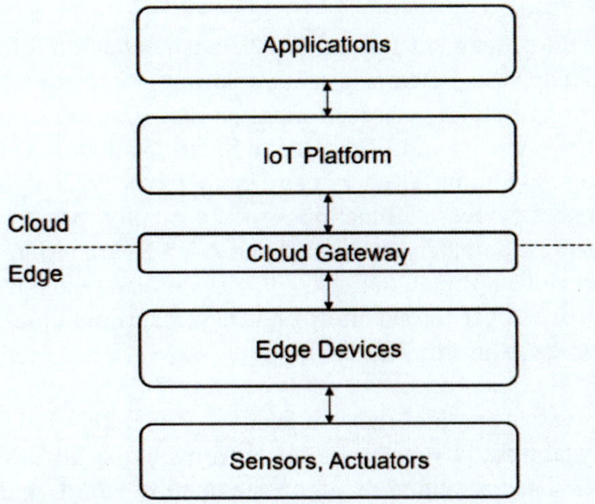

Fig. 1. Architecture of cloud-enabled IoT systems.

- Reducing the need for advanced computing resources at the client side. The clients can use the infrastructure of the cloud without a need to maintain their own servers.
- Scalability and on-demand usage. The additional resources in the cloud like CPUs or data storage can be added quickly when needed for computational-heavy tasks like machine learning model trainings and reduced if not needed anymore.
- Access from everywhere. Data gets safer since persisted with replicas by the cloud provider.
- IT support is reduced. The clients can use the standard cloud realizations for security, access, analytics, deployment, configuration, monitoring, lifecycle management, user management, authentication, metering, billing, compliance etc.
- No need to buy software licences but can flexibly use the services when needed.

IoT platforms in the cloud significantly reduce the efforts for developing and running the data management applications by offering the common services as a part of the platform. The providers of IoT device and data management in cloud are among others Amazon IoT Device Management and IoT Management [6], Google Cloud IoT [7], Microsoft Azure IoT [8] or SAP IoT [9]. These platforms support device management (including onboarding using device provisioning, configuration and organizing into groups, lifecycle management and connection management), data storage, analysis and data processing (filtering, data mining, stream processing, ML training and inference etc.) and exposure for further usages.

A great advantage of using cloud solutions is combining the device data with further types of data like business and technical documents, financial numbers, clickstreams etc. These may be not only the company-internal data, but also services and data from 3rd party providers like weather or geographical information.

Moreover, having data in the cloud allows using of further intelligent services to create the new applications quicker, cp. Amazon Machine Learning Services [10]. The examples of such **services** are:

- Text recognition and document extraction. The business and information documents can be automatically interpreted to provide relevant information like business entities, technical configurations, price information etc. Examples are SAP Document Information Extraction [11] and IBM Watson Smart Document Discovery [12].
- Image and video recognition. The services can be trained to analyze camera pictures and videos to detect device malfunction, product quality, persons, traffic jams etc. Examples are Azure Computer Vision [13] and AWS Rekognition [14].
- Digital assistants and conversational AI. The services can recognize and react on the voice commands or analyze the situation, or, vice versa, communicate with the human users. One example is Amazon Lex [15].

To avoid lock-in to one provider, the companies can use multi-cloud data management frameworks like Databricks [16] or business platforms like SAP Business Technology Platform (BTP [17]. They are often based on the open standards and frameworks like Delta Lake [18] for storage and Apache Spark [19] for data processing pipelines.

3 From Data Lake to Data Mesh

The ease of putting big volumes of heterogeneous data into the big storages (also called data lakes, no matter if in cloud or in an own server infrastructure) facilitated by developments shown in the previous section leads to further challenges for data management of this data:

- Different formats and data quality. The data may include structured (databases and data warehouses), semi-structured (XML and JSON documents) and unstructured information (CSV, images, videos etc.).
- The data has different meaning (semantics) and thus requires deep understanding and transformation for their interpreting.
- Different teams are working on same data with different purposes.

Consequently, traditional centralized approaches of bringing all data into one Big Data platform (typically a data warehouse or Spark-based platform) have led to big monolithic solutions which fail in providing quick solutions for constantly changing requirements on the data usage [20]. The central teams responsible for such a central platform are not familiar with application specifics with all implying communication and integration issues.

Thus, a shift of the paradigm to "*considering domains as the first-class concern, applying platform thinking to create self-serve data infrastructure, and treating data as a product*" [20] is happening, which results in the idea of **Data Mesh** which supports *decentralization* and *distribution of responsibility* to people or teams following the principles of DDD (Domain-Driven Design, see [21]).

In data meshes (Fig. 1), the central team maintains only the infrastructure of the platform, whereas the teams structured along the DDD boundaries care about all aspects of data management (data ingestion, data storage, conceptual modeling, processing pipelines, exposure of data). The data is exposed to any other potential usages via clearly defined APIs as *data products*. Correspondingly, other teams can easily use the data products, integrate them and offer new services and data products. This may also include the customers and partners outside of the company, which can consume the data products of a company, or serve as data providers for it.

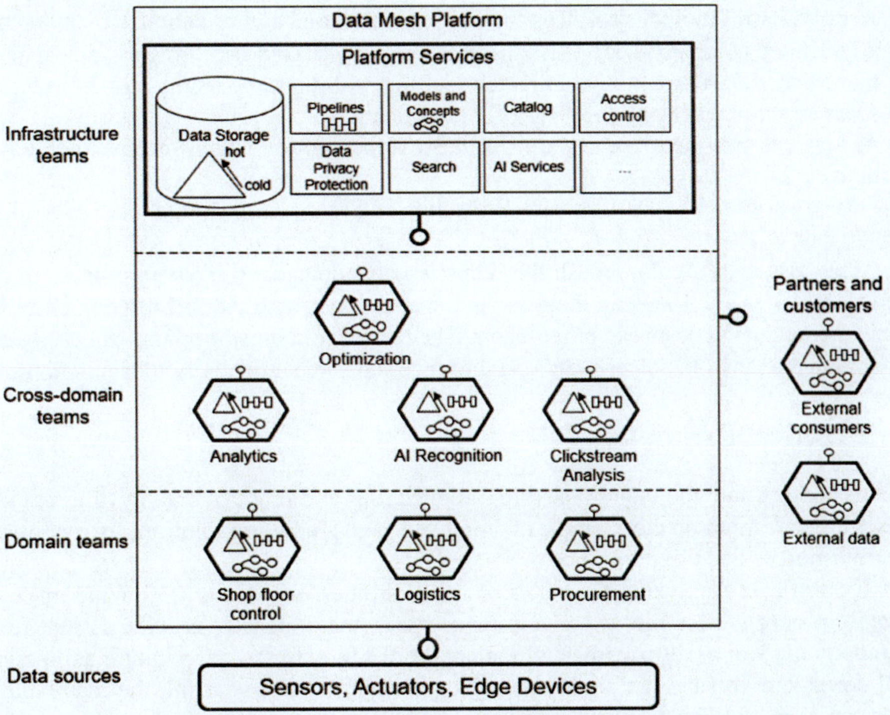

Fig. 2. Data mesh architecture (adapted based on [20]).

Let's investigate the typical data mesh architecture in detail. The central **platform services** include the typical services that can be reused by the teams:

- Infrastructure for running and orchestrating the pipelines. Here, the usual big data tools like Apache Spark can be used. Along with that, the teams may still be free in selection of their pipeline tools to match their needs and available expertise, as the pipelines remain the implementation detail of the team and not visible outside of it.
- Models and concepts functionality provide uniform formats for describing the models (like OpenAPI for APIs or JsonSchema for data models). This service may contain also shared vocabularies, like common terms, units of measure or currencies. To combine the metadata of the data products, knowledge graphs can serve as a semantic glue, see [22], as realized in Anzo platform [23]. The knowledge graphs can integrate the industry ontologies for better reuse and integration with external systems, see for example [24].

- Catalog service provides the overview of available data artefacts (data products, data types, models etc. along with their descriptions and lineage information) and allows to quickly find the necessary ones. As a best practice, it should provide semantically rich descriptions, including not only the models, but also how and in which context they can be used.
- Access control includes services for authentication and authorization, which are ideally defined company-wide. Also, here, the cloud platform functionality may be used like Microsoft Azure Active Directory. Along with that, the teams may introduce a separate set of roles and privileges for their domain.
- AI services provide reusable services like text recognition which we have mentioned in Sect. 2.
- Further services like Data Privacy Protection or Search simplify the reuse as well.

As a result, in the data mesh the domain teams delegate the common tasks to the central services and focus on their domain-specific data with the full responsibility for their data and its exposure to other teams. The teams from other domains can easily use their data thanks to the clearly defined interfaces and semantically rich data models.

4 Practical Examples

Let us demonstrate the cloud-based IoT applications and reuse of data by different use cases in a data mesh on the example of monitoring and prediction solutions for renewable energy sources like wind turbines or solar plants.

Example 1: Analyzing the states of wind turbines, see demo [25]. Wind turbines require proper lubrication to prevent unnecessary and expensive repairs, so that fully synthetic oils are used to withstand the conditions in extreme temperature and potential corrosion. To save the significant costs of changing the gear oil, the maintaining companies would like to run the early anomaly detection and troubleshooting to precisely evaluate the operational health and for example better predict the timepoint of oil change. For that, they aggregate several variants of data points like the oil temperature of the turbine. The solution is based on SAP Data Intelligence [26], which allows to integrate all data sources, implemented data processing pipelines and in the end develop machine learning models capable of reliably predicting the operational health. As result, the companies can constantly evaluate the wind turbine stability based on time series data obtained from the IoT infrastructure by conducting inference with periodically retrained ML models.

From the viewpoint of data mesh, the team maintaining this use case is focusing on the domain of analyzing the wind turbine data with the time series processing methods and physical modelling, offering to other domains higher-value data of the wind turbine state.

Example 2: Financial decision support for sustainable energies. The data from the previous example can be combined with other types of data to realize more comprehensive scenarios like Kaiserwetter Energy Asset Management system called ARISTOTELES [27]. It aggregates, structures, and correlates technical and financial data from wind farms and solar parks to support decision making, supporting all stakeholders like energy providers, maintaining companies, investors and analysts. The platform enriches the information about the facilities with data from third-party sources, such as meteorology or the electricity market.

It is built on SAP Business Technology Platform and uses a number of further SAP solutions, including SAP Internet of Things [28] for data collection in the cloud, and SAP Analytics Cloud [29] for visualization and analysis. This set of tools supports the reuse of the data from different scenarios in the same application, building a basis for the data mesh with a much more comprehensive functionality, which is still manageable as each team cares about its domain.

From the viewpoint of data mesh, this use case realizes the cross-domain scenario, which uses the data from the domains of wind turbine and solar monitoring, financial and weather services. Each of domains adds additional value to the data, claiming further on the data pyramid.

5 Conclusions and Outlook

The current technical developments in Emerging Networks and IoT have led to further democratization of data management. The factors leading to that are:

- lowered costs for sensors, edge devices, data storage and processing capabilities in the cloud that allow to collect more data and to process them more efficiently,
- the cloud platforms and corresponding gateways, which allow to bring the data from different sources together in a uniform and safety way,
- out-of-the-box available and highly adjustable cloud services and open source frameworks, which lower the entry level for providing the comprehensive analysis scenarios and new services,
- new organizational principles like data mesh which support complex scenarios by exposing the data as data products intended for reuse.

These developments are also necessary to support the tremendous growth of the connected devices, the number of which is estimated to reach 29 billion by 2022, of which around 18 billion will be related to IoT [30]. This development will require a much higher grade of automation of all the data management tasks, following the developments we have sketched in this paper.

We see the following trends in the data management for data from Emerging Networks and IoT in the coming years:

- Preprocessing (like filtering and simple consolidation and inference scenarios) in edge devices. An example is computer vision, where the video information is processed directly in the cameras to extract most important features and communicate only them to reduce the required bandwidths.
- Advanced networking with bigger bandwidths and lower latencies like 5G and soon 6G will enable further real-time scenarios in robotics or autonomous vehicles.
- Development of much more easily configurable and adjustable systems with reusable services (e.g., from cloud providers or multi-cloud business platforms) and data (e.g., from the so-called data marts) to further democratize the realization of new scenarios.

- Still, the services and platforms must further improve the basic features like security, integration and performance.
- Combination of data and AI in even more comprehensive scenarios.
- New business models with reusable service and data, like highly cooperative models from logistics area.

References

1. Ulema, M.: Big Data and telecommunications telecom analytics. In: Tutorial at International IEEE Conference on BlackSeaCom, Varna, Bulgaria (2016)
2. Luntovskyy, A., Spillner, J.: Architectural transformations in network services and distributed systems: service vision. Case Studies. Springer Nature, **344** (2017). ISBN: 9–783–6581–484–09
3. Globa, L., Svetsynska, I., Luntovskyy, A.: Case studies on big data. J. Theoret. Appl. Comput. Sci., JTACS, Polish Acad. Sci. Gdansk (2), 10 (2016). ISSN 2299–2634
4. Konys, A., Rogoza, W.: Big data and ontologies. In: Talk at ACS International Conference 2016 in Międzyzdroje, p. 3 (October 2016)
5. Luntovskyy, A., Globa, L., Shubin, B.: From big data to smart data: the most effective approaches for dataaAnalytics. In: Advances in Information and Communication Technology and Systems, pp. 23–40. Springer (2020)
6. Amazon IoT Device Management and IoT Management (https://aws.amazon.com/iot-analytics)
7. Google Cloud IoT [https://cloud.google.com/solutions/iot]
8. Microsoft Azure IoT [https://azure.microsoft.com/en-us/overview/iot/]
9. SAP IoT [https://www.sap.com/products/iot-data-services.html]
10. Amazon Machine Learning Services [https://aws.amazon.com/free/machine-learning/]
11. SAP Document Information Extraction [https://help.sap.com/docs/DOCUMENT_INFORMATION_EXTRACTION]
12. IBM Watson Smart Document Discovery [https://cloud.ibm.com/docs/discovery?topic=discovery-sdu]
13. Azure Computer Vision [https://azure.microsoft.com/en-us/services/cognitive-services/computer-vision/#overview]
14. AWS Rekognition [https://aws.amazon.com/rekognition/]
15. Amazon Lex [https://aws.amazon.com/lex]
16. Databricks [https://databricks.com/]
17. SAP Business Technology Platform (BTP [https://www.sap.com/germany/products/business-technology-platform.html]
18. Delta Lake [https://delta.io/]
19. Apache Spark [https://spark.apache.org/]
20. Dehghani, Z.h.: Data Mesh: Delivering Data-Driven Value at Scale. O'Reilly (2022)
21. Vernon, V.: Implementing Domain-Driven Design. Addison Wesley (2013)
22. Strengholt, P.: Data Management at Scale. O'Reilly (2020)
23. Anzo platform [https://cambridgesemantics.com/anzo-platform/]
24. Ontologies for Industry 4.0 (2019). https://www.cambridge.org/core/journals/knowledge-engineering-review/article/ontologies-for-industry-40/BF86BB5310356D642C82470D67974804
25. Online demo of SAP Wind Turbines Energy Data Intelligence [https://sapdemostore.com/sap/bc/ui5_ui5/sap/yunifiedstore/index.html#/scenario/16609]

26. SAP Data Intelligence [https://www.sap.com/products/data-intelligence.html]
27. Kaiserwetter Looks Ahead with SAP Artificial Intelligence to Support Investors in Renewable Energy. ASUG News and Views (2020) [https://www.asug.com/insights/kaiserwetter-looks-ahead-with-sap-artificial-intelligence-to-support-investors-in-renewable-energy]
28. SAP Internet of Things [https://www.sap.com/products/intelligent-technologies/iot.html]
29. SAP Analytics Cloud [https://www.sap.com/products/cloud-analytics.html]
30. Paolo Collela. Ushering in a better connected future, Ericsson (2017). [https://www.ericsson.com/en/about-us/company-facts/ericsson-worldwide/india/authored-articles/ushering-in-a-better-connected-future]

A Neuron Network Learning Algorithm for the Recognition of Fractal Image and the Restoration of Their Quality

Ostap Yunak(✉), Bohdan Strykhaliuk, Mikhailo Klymash, Yuliia Pyrih, and Olha Shpur

Lviv Polytechnic National University, S. Bandery Str., 12, Lviv 79013, Ukraine
ostap.yunak@gmail.com

Abstract. The chapter discusses the results of a neural network learning algorithm for recognizing fractal images of the "Fractal dust" type (Kantor's set). This algorithm allows you to restore a fractal image with low quality. A generated set of matrices of the randomized system of iterative functions (RSIF), and fractals built on these matrices were used for training the neural network Mathematical formulas and schemes for calculating and constructing a fractal image and recognition are derived. This algorithm consists of finding RSIF for an existing fractal with different image quality. A learnt neural network makes it possible to compress an existing fractal image to a small set of coefficients. With their help you can reproduce an image of any format (resolution), as well as improve (distort) the quality of the image. The use of the algorithm in the future will make it possible to develop an automated library which can be integrated for graphic editors, data protection, information compression and improvement of display quality.

Keywords: Neural network · Fractal · Image recognition · Data compression · Randomized system of iterative functions (RSIF)

1 Introduction

The use of neural networks for object recognition is an urgent task in the field of information technologies. Object recognition is a complex task because it requires a large collection of input and output data.

Recognizing fractal images is an even more difficult task because the construction of fractals requires large computing power. In the case of using RSIF, the task is somewhat simplified, there is no need to store a large amount of data in the memory.

Therefore, this algorithm is convenient to use in computers with limited resources. The construction of fractals using RSIF includes the calculation at each step of the coordinates of one point to display it on the screen [1].

In this case it is necessary to calculate the number of operations which directly depends on the number of pixels of the fractal image. RSIF makes it possible to change the quality of the image which in its turn makes it possible to teach the neural network to recognize a fractal with different quality and restore it.

M. Klymash et al. (Eds.): TCSET 2022, LNEE 965, pp. 574–584, 2023.
https://doi.org/10.1007/978-3-031-24963-1_33

2 Basic Material Presentation

To learn a neural model that would be able to recognize fractal images and transform them into a system of iterative RSIF functions, input and output data are required. The input data will be fractal images (image pixels), the output data will be RSIF (see Fig. 1).

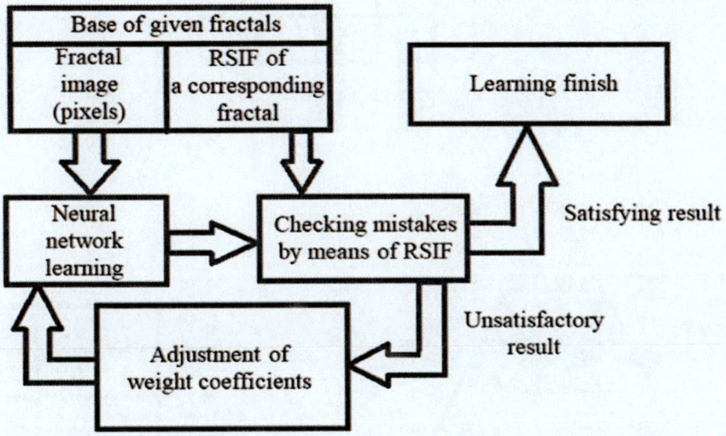

Fig. 1. Algorithm of learning a neural network.

To learn a neural network, you need to create a database of fractals. The article [3] describes the forward and reverse algorithm for constructing fractals and converting them into RSIF. This algorithm will give you a possibility to form a system of iterative functions of RSIF (1) and use it to construct a fractal image:

$$\left[\begin{array}{l} \begin{cases} x_n = x_a - \frac{x_a - x_{n-1}}{k_{Xa}} \\ y_n = y_a - \frac{y_a - y_{n-1}}{k_{Ya}} \end{cases} \\ \\ \ldots \\ \\ \begin{cases} x_n = x_b - \frac{x_b - x_{n-1}}{k_{Xb}} \\ y_n = y_b - \frac{y_b - y_{n-1}}{k_{Yb}} \end{cases} \end{array}\right., \tag{1}$$

where (x_n, y_n) - the current coordinates of the point that builds the fractal step b step;

$(x_a, y_a) \ldots (x_b, y_b)$ - coordinates of rosaries forming the first iterations of the fractal [3];

k_{Xa}, k_Y_a... k_{Xb}, k_{Yb}- coefficients of proportionality of the corresponding figures of the first iteration [3].

We can write RSIF in the form of a matrix (X, Y, K_X, K_Y) presented in (see Fig. 2):

With the help of RSIF (3) and the algorithm presented in the article [3], we can build any fractal (see Fig. 3–4):

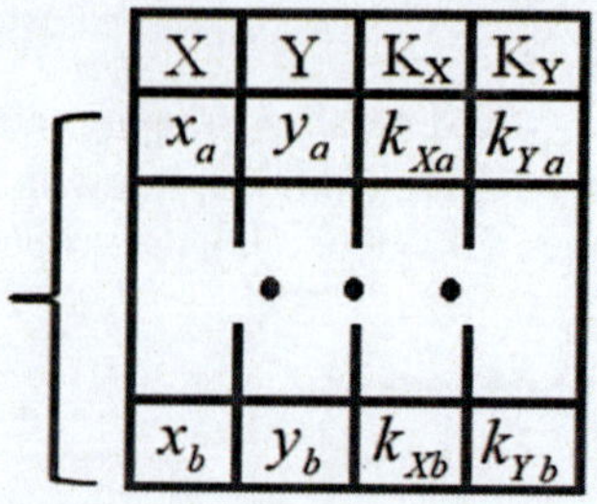

Fig. 2. Representation of RSIF in the form of a matrix of numbers (where n is the number of figures of the first iteration).

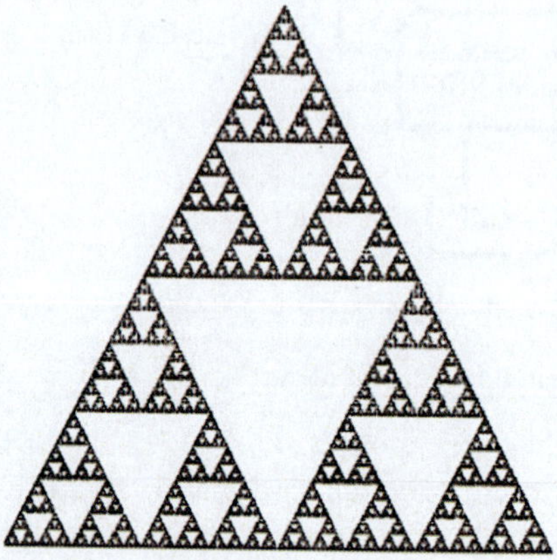

n = 3

X	Y	Kx	Ky
300	0	0,5	0,5
0	600	0,5	0,5
600	600	0,5	0,5

Fig. 3. Serpinski's triangle (the result of the algorithm).

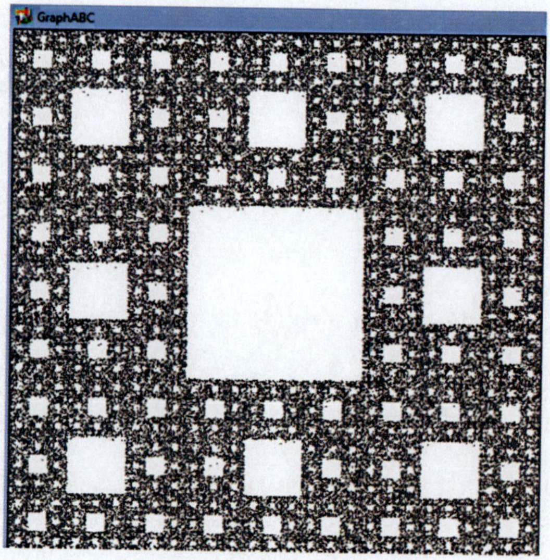

n = 8

X	Y	Kx	Ky
0	0	0,33	0,33
300	0	0,33	0,33
600	0	0,33	0,33
0	300	0,33	0,33
600	300	0,33	0,33
0	600	0,33	0,33
300	600	0,33	0,33
600	600	0,33	0,33

Fig. 4. Sierpinski's square (the result of the algorithm).

As we can see from the previous results (see Fig. 3, Fig. 4), the RSIF matrix mostly depends on the image resolution. It becomes necessary to set (generate) the dimensions of the image L_X, L_Y.

Since our neural model can be learnt, let's start with a limited range of resolution up to 1024 px by 1024 px (2):

$$\begin{aligned} 1 < L_X < 1024,\ L_X = Random(0..1024), \\ 1 < L_Y < 1024,\ L_Y = Random(0..1024). \end{aligned} \tag{2}$$

The next step is the need to set (generate) the number of figures of the first iteration. Since our neural model can be learnt, to begin with, we will set a limited range of generation of figures of the first iteration to 100 (3):

$$1 < n < 100, n = Random(0..100). \tag{3}$$

The next step will be to determine the parameters of the iterative functions X_i, Y_i, $K_{X\,i}$, $K_{Y\,i}$ to build the RSIF matrix by means of the randomized Algorithm (4–7):

$$X_i = Random(0..L_X), \tag{4}$$

$$Y_i = Random(0..Y_X), \tag{5}$$

$$K_{X\,i} = Random(0..1), \tag{6}$$

$$K_{Y\,i} = Random(0..1). \tag{7}$$

The next step is to determine the number of image pixels N. We will use the following formula (8) for this:

$$p = \left(\frac{\sum_{i=1}^{n} (K_{Xi}) \cdot \sum_{i=1}^{n} (K_{Yi})}{n^2} \right); N = L_X \cdot L_Y \cdot \left(\sum_{i=1}^{n} \left(\frac{1}{K_{Xi} \cdot K_{Yi}} \right) \right)^{\log_p (L_X \cdot L_Y)}. \tag{8}$$

All parameters for constructing RSIF (2–8) are defined (2–8), the scheme of the algorithm of the generator of RSIF matrices and the constructed corresponding fractals is shown in Fig. 5.

The given construction algorithm does not use recursive functions and the entry a cycle into a cycle. This fact allows not to spend large computing power and it is quite optimized.

Therefore, this algorithm is convenient to use on computers with limited resources for the construction of fractals using RSIF. It includes the calculation of the coordinates of one point at each step to display it on the screen.

The number of operations of the algorithm on a straight line depends on the number of pixels of the fractal image (8).

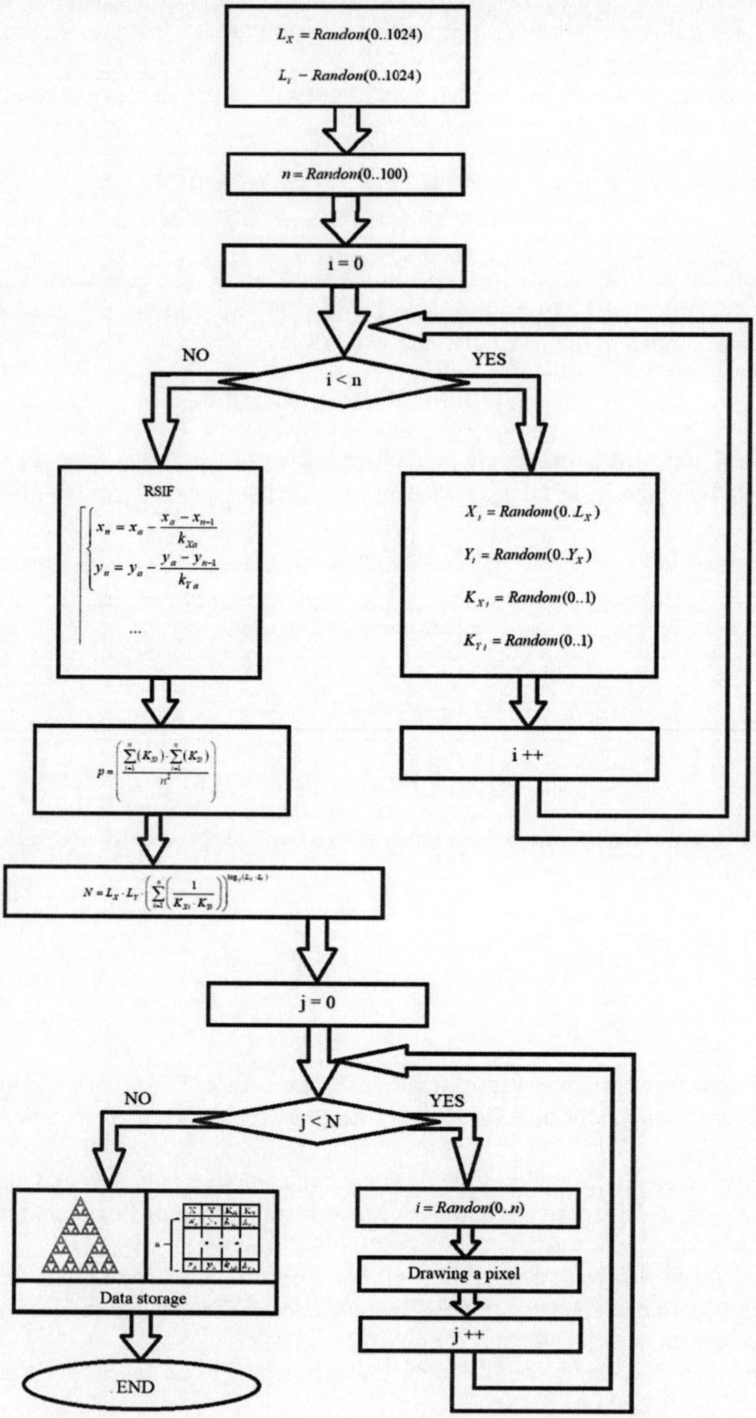

Fig. 5. Algorithm of action of the fractal generator and RSIF.

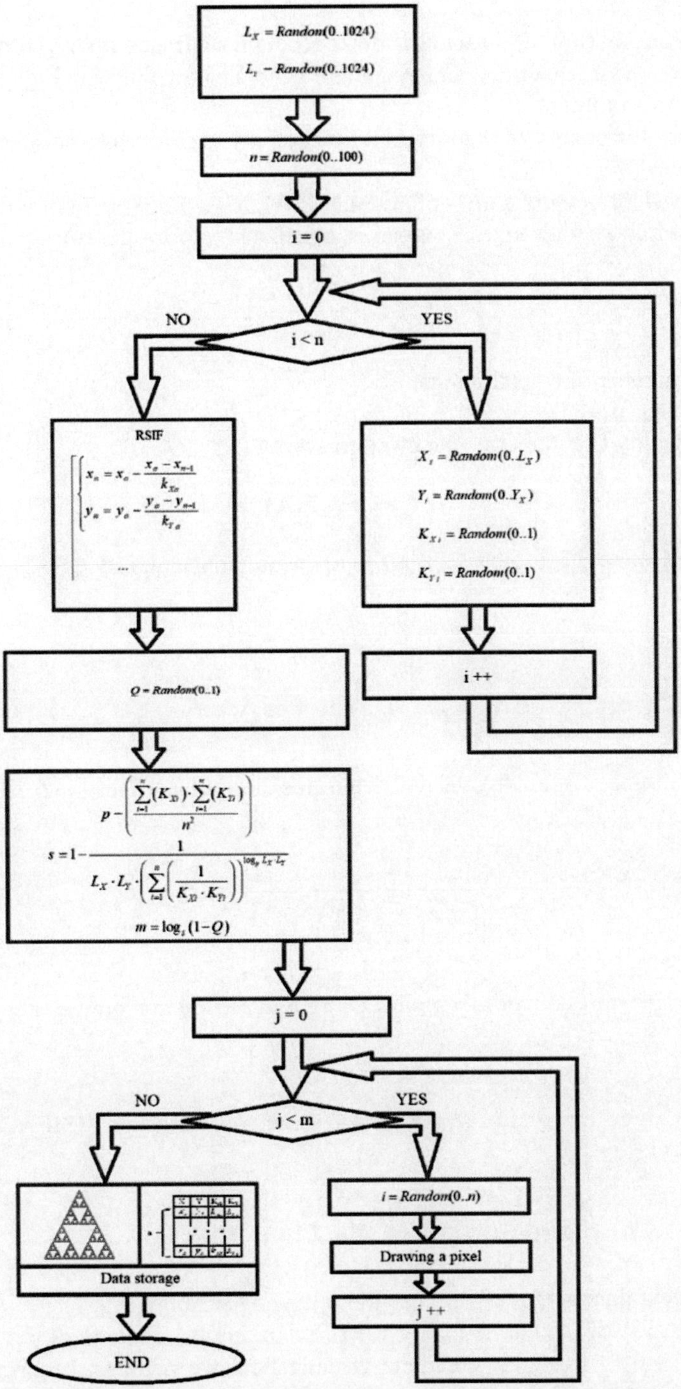

Fig. 6. Algorithm of action of the fractal generator and RSIF with different image quality.

This algorithm (Fig. 5) does not allow to recognize a fractal image with low quality, this is due to the fact that the fractal generator generates fractals with high quality. Let's try to solve this problem.

To reduce the quality of the image, it is necessary to reduce the number of pixels of the fractal image.

Let's consider the probability of an event P(A) that a point does not fall into a given place on the fractal, with m trials, it can be calculated by formula (9):

$$P(A) = \left(\frac{N_f - 1}{N_f}\right)^m, \tag{9}$$

where N_f - number of fractal points;

m - number of trials.

Image quality Q is the opposite event to event P(A) (10):

$$Q = 1 - \mathrm{P(A)}. \tag{10}$$

Let's determine the number of trials m to ensure image quality Q:

$$Q = 1 - \left(\frac{N_f - 1}{N_f}\right)^m, \tag{11}$$

$$m = \log_{1-\frac{1}{N_f}} 1 - Q. \tag{12}$$

By substituting formula (8), we will determine the number of tests m for a rectangular fractal image:

$$s = 1 - \frac{1}{L_X \cdot L_Y \cdot \left(\sum_{i=1}^{n}\left(\frac{1}{K_{Xi} \cdot K_{Yi}}\right)\right)^{\log_p L_X \cdot L_Y}}; \; m = \log_s(1 - Q). \tag{13}$$

To determine the quality of the image Q, we will use the randomized algorithm (14):

$$Q = Random(0..1). \tag{14}$$

Now the algorithm of the fractal image generator based on RSIF will have the following form (Fig. 6).

3 Software Implementation of the Algorithm

To train the neural network, we will use the following scheme (Fig. 7).

We will use WEB technologies for the software implementation of the neural network. Namely, for writing the code and visualization, we will use the Javascript programming language and the JavaScript library brain.js. The program implementation scheme is shown in Fig. 8.

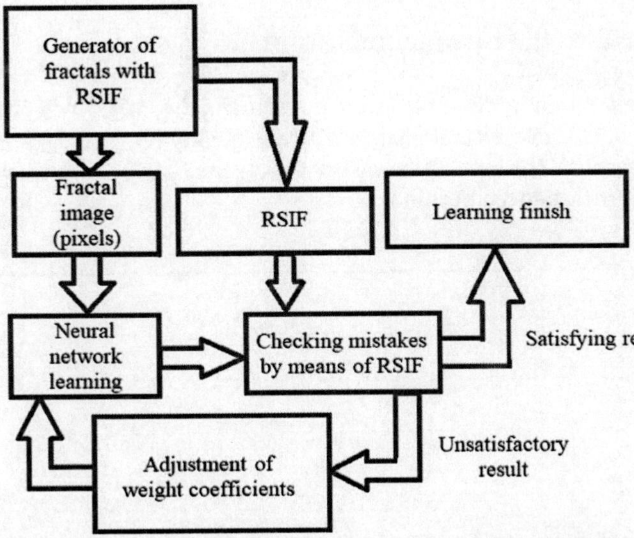

Fig. 7. Algorithm for learning a neural network using a generator of fractal images with RSIF.

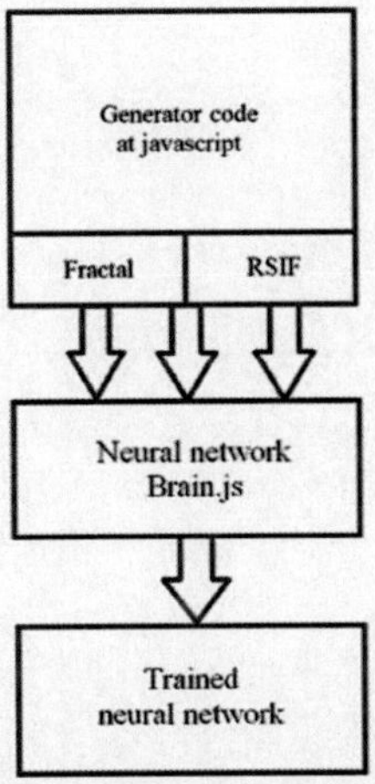

Fig. 8. Software implementation of a neural network based on the Brain.js library.

4 Results of Practical Implementation

After writing the program code and learning the neural network, we will check the work results. To do this, let's take fractal images with low quality (Q < 0.15) (see Fig. 9), and pass this image through our learned neural network (see Fig. 10). Then we will get RSIF and then build our fractal (see Fig. 11).

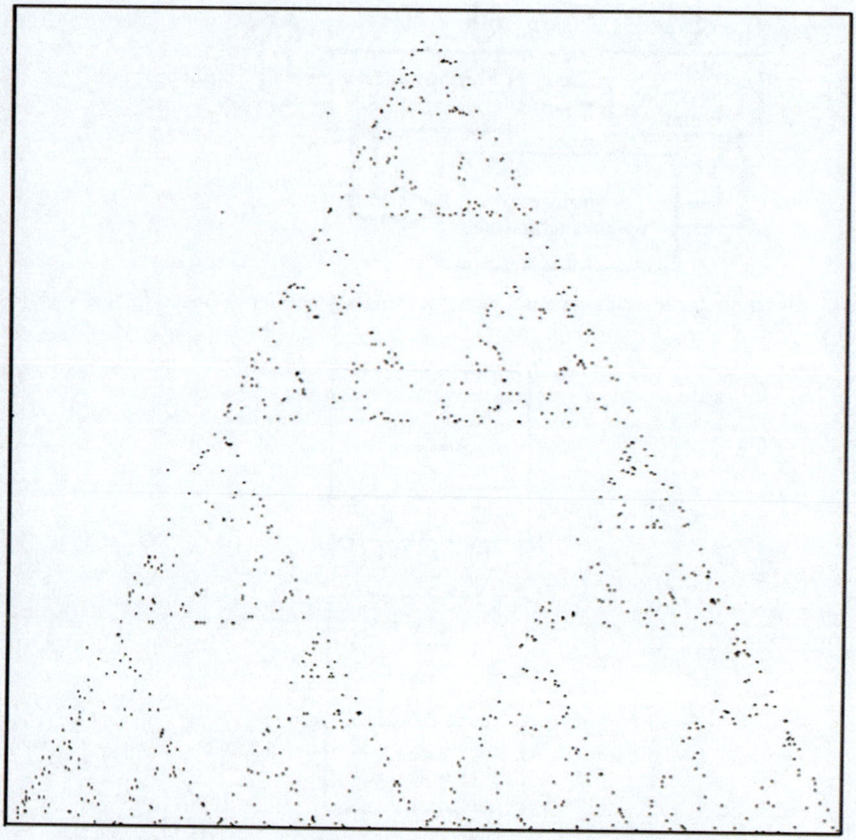

Fig. 9. Fractal image (Serpinski's triangle), with quality Q < 0.15.

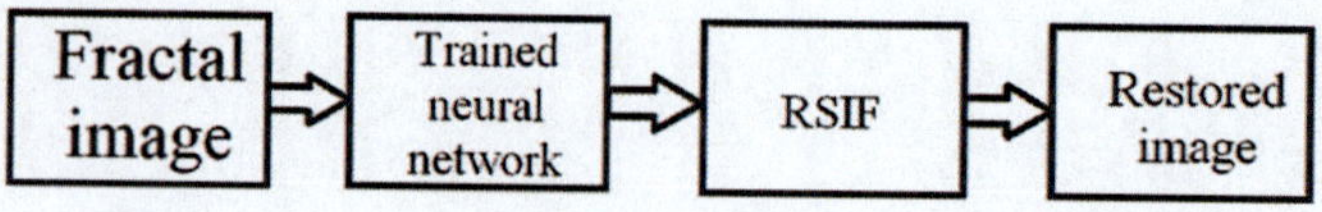

Fig. 10. Scheme of action of the trained neural network.

Fig. 11. Results of action of the trained neural network (Serpinski's triangle with quality Q = 0.95).

5 Conclusions

Object recognition is one of the urgent tasks in the field of information technologies. Recognizing self-similar objects (fractal structures) is one of the difficult tasks, this is due to the fact that it is very difficult, to collect (form) data (fractals) since the construction of fractal structures requires significant information calculations and resources. RSIF was used to solve this problem, since there is no need to store a large amount of data in the memory. The suggested construction algorithm does not use recursive functions and the entry of a cycle into a cycle. It allows not to spend a lot of computing power and it is rather optimized.

Therefore, this algorithm is convenient to use on computers with limited resources. The construction of fractals using RSIF includes the calculation of the coordinates of one point at each step to display it on the screen. The number of operations of the algorithm on a straight line depends on the number of pixels of the fractal image.

The results of the algorithm showed that the learned neural network qualitatively recognizes fractal images and converts them into RSIF. The resulting RSIF matrix allows you to improve the quality of the fractal image. It transforms raster graphics into vector graphics, which in its turn will allow you to change the size of the image without the loss of the quality. This algorithm also reduces the size of the image to a set of coefficients.

In the future, this neural network can be integrated into pattern recognition systems and graphic editors as a library.

References

1. Al-, W.F.H.: Deterministic algorithm for constructing fractal attractors of iterated function systems. Eur. J. Sci. Res. **134**, 121–131 (2015)
2. Mandelbrot, B.B. The Fractal Geometry of Nature, W.H. Freeman & Company: New York, NY, USA, 1999
3. Yunak, O.: Algorithm forming randomized system of iterative functions by based cantor structure. In: Yunak, O., Shpur, O., Strykhaliuk, B., Klymash, M., (eds.) Information and Communication Technologies, Electronic Engineering, vol. 1 (2), pp. 71–80 (2021)
4. Mandelbrot, B.B. The Fractal Geometry of Nature: Echo Point Books & Media, LLC, 490p. (2021) ISBN-10: 1648370403
5. Yunak, O., Peleschak, B., et al.: Transformation of an image of a fractal structure of the "Fractal Dust" type (Kantor's set) into a randomized system of iterative functions. In: XII International Scientific and Practical Conference "Latest Achievements in European Science – 2016", Sofia "Byal GRAD-BG" OOD, vol. 13, 90 p. (2016). (in Ukrainian)
6. Mandelbrot, B.B. Fractals: Form, Chance and Dimension, Echo Point Books & Media; Reprint ed., 656 p. (2020)
7. Falconer, K.: Fractal Geometry: Mathematical Foundations and Applications, 3rd edn., 400 p. (2014) ISBN-10: 111994239X
8. The Mandelbrot Set and Beyond, 308 pages, Springer, New York (2004) ISBN: 0–387–20158–0
9. Gutzwiller, M.C., Mandelbrot, B.B., Evertsz, C.J.G., et al.: Fractals and Chaos: The Mandelbrot Set and Beyond. Springer New York (2010) ISBN: 1441918973
10. Falconer, Z.Z., Falconer, K.: Techniques in Fractal Geometry. Wiley & Sons, Incorporated, 274 p. (1997) ISBN: 0471957240
11. Massopust, P.R.: Fractal Functions, Fractal Surfaces, and Wavelets. Elsevier Science & Technology, 383 p. (1995) ISBN:0124788408

Advances in Data Reduction Techniques to Solve Power Spectrum Estimation Problems for Emerging Wireless Networks

Andriy Lozynsky[1(✉)], Igor Romanyshyn[1], Bohdan Rusyn[1(✉)], Mykola Beshley[2], Mykhailo Medvetskyi[2], and Danylo Ivantyshyn[2]

[1] Karpenko Physico-Mechanical Institute of the NAS of Ukraine, Naukova Str. 5, Lviv 79060, Ukraine
romanyshyn@ipm.lviv.ua, b.rusyn.prof@gmail.com
[2] Lviv Polytechnic National University, Lviv 79013, Ukraine
mykola.i.beshlei@lpnu.ua

Abstract. New trends in the development of wireless networks are due to the rapid growth of the speed and computing power of software and hardware digital signal processing. The increase in the rate of information transfer leads to a decrease in the duration of processing signals, which means that finding faster and more efficient methods of spectrum signal processing for future 5G/6G networks becomes relevant. The generally accepted methods of spectral analysis based on harmonic basis functions (Fourier transform) become inefficient since their implementation increases either the cumbersomeness of the spectral representation or the reproduction error due to the use of weight windows. Moreover, the application of non-harmonic basis functions when implementing the discrete wavelet transform is even less effective than the discrete Fourier transform because of the complexity of eliminating temporal redundancy. The chapter proposes state-of-the-art techniques to reduce redundancy in signal spectrum estimation tasks in order to improve the efficiency of computing resources for these new 5G/6G wireless systems. Namely, a new technique for estimating the autocorrelation function (and power spectrum) of a stationary random signal for finite sampling at different variants of nonuniform data sampling is proposed. The peculiarity of the technique is the orientation to minimize the number of sampled data by reducing the redundancy. In the long term, it allows to offload significantly the wireless communication channels and digital signal processing systems with high noise immunity. The concepts of reduction ruler are introduced to estimate the autocorrelation function and power spectrum. The ways of its construction according to the known Golomb's ruler are described and analyzed, as well as the convenient ways of application in practice based on irregular sampling with the switching of the sampling frequency with two close frequencies. The results of numerical modeling of estimation accuracy based on the calculation of the standard deviation, which, as a rule, increases when the number of data decreases, are given and illustrate the efficiency of estimation of different data selection options. The proposed solution will reduce the amount of computation in the implementation of digital signal processing for emerging wireless networks.

M. Klymash et al. (Eds.): TCSET 2022, LNEE 965, pp. 585–601, 2023.
https://doi.org/10.1007/978-3-031-24963-1_34

Keywords: Stationary random signal · Autocorrelation function estimation · Power spectrum · Nonuniform sampling · Data reduction · Estimation variance

1 Introduction

In the field of wireless communication systems, there has been significant progress in recent years, including an increase in bandwidth. Thus, the emerging 5G/6G networks will play an increasingly important role in our lives [1–3]. The advances in wireless communication technology are constantly evolving based on signal processing research. In digital signal processing and analysis first of all we pay attention to mathematical expectations, dispersion, signal shape and its frequency spectrum. In the general case, signal sampling is carried out taking into account the frequency bandwidth of the signal, on the basis of which according to Nyquist criterion the sampling frequency is selected. Compliance with the Nyquist criterion gives a theoretically valid lower estimate of the required sampling rate. Usually higher values are used, which enable a certain data redundancy. With limited bandwidth of the data channel (or data storage), this redundancy is tried to be minimized by approaching the sampling rate to its theoretical limit - the Nyquist frequency [4, 5].

In some cases when we deal with random signals, only their amplitude spectrum is informative. Discrete Fourier transform of received signal realization allows us to estimate a complex signal spectrum, i.e. amplitude and phase of each harmonic. Knowing this information we can reconstruct the signal in time domain [6, 7]. Note, that for a signal with noise nature its shape is not informative. Accordingly, the phase part of the spectrum will not be informative, but only the amplitude part. It turns out that the sampling of a noise signal with Nyquist frequency for the task of estimating its amplitude spectrum is redundant. At the same time it cannot be lower in principle, because in this case spectrum distortions occur. Therefore it is reasonable to consider the question - is it possible to find such samples in the signal, which can be removed without losses in the amplitude spectrum.

The problem of data preprocessing, estimation of spectral densities, correlation matrices, and the use of these characteristics for solving applied problems are considered in the paper [8]. The accuracy of the estimations is determined on the basis of their dispersion.

The problem of finite time sample length is encountered when estimating the correlation function [9]. Analytical expressions for the variance of the autocorrelation function estimation by first- to fourth-order moments were obtained.

Tables for determining the sample size, based on the required accuracy and stochasticity characteristics of the process, have been developed for practical use [10].

Estimates of power spectrum and correlation function are known for different measurement schemes: when exactly one measurement is made at each moment and when a random number of measurements is made at each moment [11]. Unbiased estimates of the power spectrum in the form of a first-order spline were obtained for these measurement schemes. The unbiased estimates are substantiated. It is shown that the variance of the estimates behaves asymptotically as 1/T, where T is the observation time.

Problems when it is necessary to analyze signals at insufficient quantity of samples, absence of separate data are known [12]. One of the ways of the decision of these problems consists in reduction of samples to a uniform grid by means of interpolation. The reduction to a uniform grid requires the calculation of many values of a signal, which do not give informativity, but serve only for realization of the chosen way of the analysis. Therefore, it is important to be able to estimate what minimum number of calculated samples is sufficient in this case.

When estimating spectral power on the basis of Fourier transform one faces with non-uniformity of counts in astronomical researches [13–15]. Power spectrum is used in materials science to estimate surface roughness [16, 17].

Works using the Holomb ruler are closely related to this study. Among them [18] - to maximize the "bad" lags in the correlation functions of the received signal, as well as to reduce the density of "bad" lags close to the absolute minimum. Note that Golomb rulers have found wide application in cryptography, in particular, in data encryption problems [19].

The theory and applications of signal processing are concerned with the identification, modeling and exploitation of patterns and structures in signal processes. The signals of surveillance are usually corrupted, incomplete and noisy, and therefore reduction of noise and data redundancy, elimination of channel corruption and replacement of lost samples are important parts of the signal processing system for future emerging wireless networks [20–22]. This chapter discusses emerging trends and research directions for reducing redundancy in the spectral analysis process for emerging wireless technologies.

2 Problem Statement

The amplitude spectrum of a signal uniformly discretized with a frequency meeting the Nyquist criterion can be obtained in several ways. One of them, which is widely used in practice, is by means of discrete Fourier transform. In this case, the intermediate result will be a complex spectrum mutually unambiguously related to the signal realization in the time domain. We will use the amplitude spectrum from it, and discard the redundant information - the phase spectrum. On the basis of the amplitude spectrum alone it is impossible to uniquely reconstruct the shape of the signal, but the amount of data needed to store information is halved based on its symmetry with respect to zero frequency.

Another way is based on the Wiener - Khinchin theorem, which connects by Fourier transform the amplitude spectrum of the signal and its correlation function [23]. The correlation function contains as many values as the number of harmonics in the spectrum, i.e., half as many samples of the signal in the time domain. Each value of the correlation function is the expectation of products of pairs of signal values taken with the appropriate offset. Recalling that a random variable belonging to a certain general population can serve as a correct unbiased estimate of the expectation of this population, we can conclude that in the limiting case we only need to have such samples of signal realizations whose pairs cover all the landslides necessary to calculate the correlation function, taking into account the known recommendations for pruning its "tail". In the optimal case (in the sense of minimizing the number of signal samples) it is enough that each shift is formed by only one pair of samples. Such a sample of signal values can be obtained relying on a perfect Golomb's ruler, that is, one in which all possible pairs are present and none is repeated.

Let us consider the problem from the other side. Suppose that the signal spectrum lies in a band of *n*/2 harmonics whose amplitude and phase are quantized and can take any of k_a and k_p values, respectively. Applying Fourier transforms to a particular set of amplitudes and phases, we obtain a signal in the time domain. Taking into account the one-to-one correspondence between the spectrum and the signal, the number of different signals that can be formed in this way is $k_a^{n/2}k_p^{n/2}$. Since by convention only the amplitude spectrum is informative in our case, the set of signals is divided into $k_a^{n/2}$ equivalence classes, each of which contains $k_p^{n/2}$ signals. Uniform sampling of signal by Nyquist frequency allows unambiguously to identify each of these signals and their respective amplitude and phase spectra, which is redundant, because we are interested only in the class to which the signal belongs. That is, such sampling is redundant by a factor of $k_p^{n/2}$. Since a simple lowering of the sampling frequency will lead to ambiguity, it is clear that elimination of redundancy is possible only with non-uniform sampling. As a variant the work considers a way of elimination of some samples of the signal, selected by Nyquist frequency.

Let's estimate a small number of samples of a signal, necessary for receiving its amplitude range, assuming that the range of a signal lies in a band of *n*/2 harmonics. This means that the correlation function of such a signal has *n*/2 values, formed by pairs of samples of the signal, taken through increasing sequential offsets from 1 to *n*/2. The number of signal values m must be such that $C_m^2 \geq n/2$. In case of equality there will be no redundancy, otherwise the data has some redundancy.

3 Ways to Solve the Problem by Using a Reduction Ruler

For the convenience of the following exposition, let us introduce the notion of a reduction ruler - a subsample of a sequence of numbers of uniform sampling.

An ascending sequence of m integral integers $x_1, x_2, ...x_m$ whose pairs $\langle x_i x_j \rangle$ form a continuous series of distances $l = x_j - x_i$ between them from 1 to *L*, (possibly with repetitions) we call a reduction ruler ${}_L^mX$ of order m of length *L*. We denote by ${}_L^mX_i$ its *i*-th element. A useful parameter of a reduction ruler can be the reduction coefficient, defined as the ratio $\frac{L+1}{m}$. Reduction rulers, the elements of which are obtained by a shift by the same number will be considered equivalent.

An example of a reduction ruler is the perfect Golomb ruler [24]. For it there is no redundancy, and the reduction coefficient takes the maximum value. The set of consecutive $m = L + 1$ numbers also belongs to the class of reduction rulers. It is easy to see that for them the reduction coefficient is equal to one, and the redundancy is maximal.

Before considering several practical ways to construct reductive rulers, let us formulate and prove two useful theorems.

Theorem 1. Let there be two equivalent reduction rulers *A* and *B* of order m and length *L*, and the ruler *B* is shifted by distance $D = 2L + 1$, that is ${}_L^mB_i = {}_L^mA_i + D$. Then the concatenation of *A* and *B* will also be a reduction ruler of order $2m$ and length $3L + 1$.

Proof of Theorem 1. Consider an arbitrary pair of elements $\langle a_i a_j \rangle$ of reductive ruler A and its corresponding pair of elements $\langle b_i b_j \rangle$ of reductive ruler B, the distance between the elements in each pair l. Then the pairs $\langle a_i b_j \rangle$ and $\langle b_i a_j \rangle$ will correspond to distances $D+l$ and $D-l$ respectively. Since by definition the initial reduction rulers A and B form a continuous series of distances from 1 to L, concatenation of A and B gives all pairs of numbers with distances from $D-L$ to $D+L$, and if $D-L = L+1$ (hence $D = 2L+1$), then considering the length of B, the length of the resulting ruler will be $3L+1$, and its order will be the sum of the orders of A and B, that is $2m$.

The theorem is proved.

Theorem 2. Suppose there are three equivalent reduction rulers A, B and C of order m and length L, and ruler B is shifted by distance $2L+1$ and ruler C by distance $6L+3$. Then the concatenation of A, B and C is also a reduction ruler of order $3m$ and length $7L+3$.

Proof of Theorem 2. It follows from Theorem 1 that the concatenation of reduction rulers A and B forms a reduction ruler of length $3L+1$, so the smallest distance which the concatenation of rulers C to A and B must provide is $3L+2$. Given that by definition the elements of reduction rulers are ordered in ascending order, of course this distance must be formed by a pair $\langle b_m c_1 \rangle$, hence the shift between rulers B and C given the length of one ruler is $D_{BC} = 4L+2$. Since ruler B is shifted by $2L+1$, the shift of ruler C is $6L+3$. Then, based on similar considerations as in the proof of Theorem 1, we see that the pairs formed by elements of reduction rulers B and C overlap away from $3L+2$ to $5L+2$, and the pairs formed by elements of reduction rulers A and C overlap away from $5L+3$ to $7L+3$. This means that such a concatenation of A, B and C forms a reduction ruler of length $7L+3$.

The theorem is proved.

Remark 1. Thinking similarly as in the proof of Theorem 2, it can be extended to the cases of concatenation of more equivalent reduction rulers by increasing the offset of each subsequent one by $4L+2$. The length of the resulting reduction ruler will increase by the same amount, and its order will increase by m.

Remark 2. Theorems 1 and 2 consider the cases of maximum offsets between the original reduction rulers. If necessary, the displacement can be reduced up to their complete overlapping. In case of partial overlapping, the elements of the last ruler that become smaller than the last element of the previous ruler are discarded with a corresponding decrease in the order of the resulting ruler.

Remark 3. It is clear that at addition of a reduction ruler with a certain shift the number of repeated pairs increases, as all pairs with recoil from 1 to L repeat the number of times corresponding to multiplicity, and pairs with recoil equal to shift between rulers repeat the number of times corresponding to the order of the initial ruler.

The application of the above theorems (together with extending remarks 1 and 2) makes it possible to form reduction rulers of desired length on the basis of a much smaller reduction ruler.

Many different ways of constructing reduction rulers can be proposed. Based on the needs of solving a particular problem, they can be optimized according to the selected criteria, which can be, for example, minimization of redundancy, which will certainly be reflected in an increase in the variance of the estimate. Or, for the selected reduction coefficient, we will achieve the minimum variance. Let us consider some practical ways of constructing reduction rulers.

4 Methods of Constructing Reduction Rulers

Method 1 - Series of Consecutive Numbers. As mentioned above, a series of consecutive numbers from 1 to m is a reduction ruler (we can say trivial) of order m and length $L = m - 1$. For it, the reduction factor $r = 1$ (if repeated many times, it can be larger). Figure 1a, 1b shows a sample forming a ruler of order 3 of length 2 and a graph of pair recurrence for it. This ruler is taken as the basis for threefold repetition (see Theorem 2). This made it possible to form a reduction ruler of order 9 of length 17. The results obtained are shown in Figs. 2a, 2b.

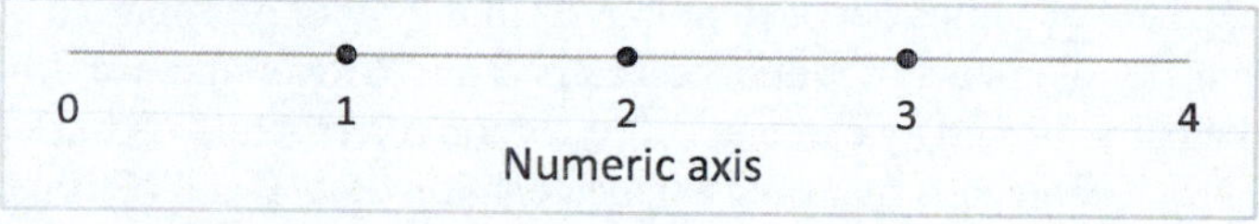

(a)

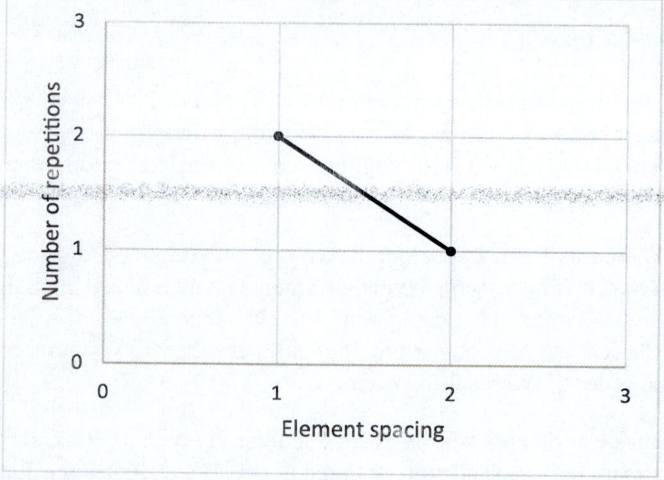

(b)

Fig. 1. Example of a reduction ruler: a) - reduction ruler formed by three consecutive numbers, its order 3, length 2; b) - graph of the number of pair recurrence for it.

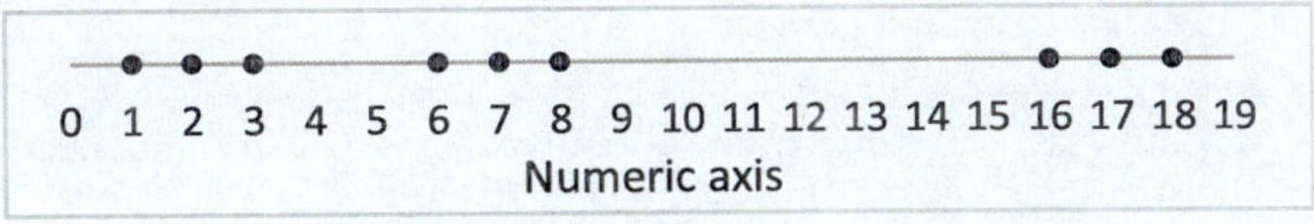

(a)

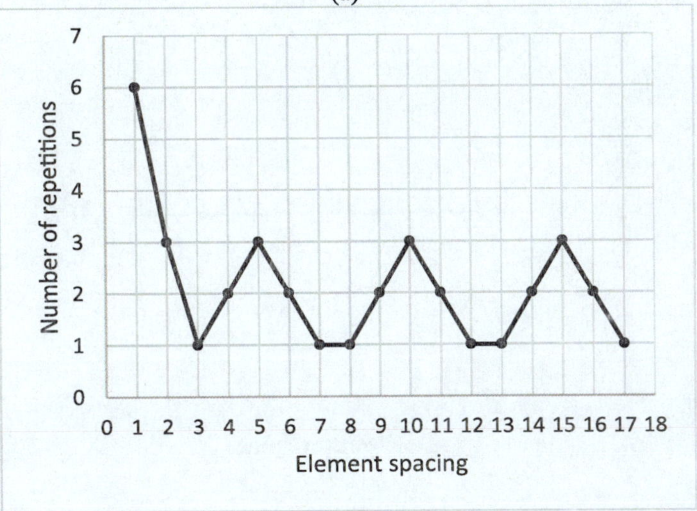

(b)

Fig. 2. Example of a reduction ruler: a) - a reduction ruler formed by threefold repetition of the ruler shown in Fig. 1, its order 9, length 17; b) - a graph of the number of repeating pairs for it.

In Fig. 2b, 4 maxima are clearly observed.

The first, for a pair with a distance between elements equal to 1, is a threefold increase in the number of repetitions of the corresponding pair in the original ruler, the second, a distance of 5, corresponds to the offset between the first and second groups of elements of the reduction ruler, the third, a distance of 10, between the second and third groups, and the fourth, a distance of 15, between the first and third groups. Figure 2b serves as a good illustration of the above remark 3 to Theorems 1 and 2.

Method 2 - Golomb Rulers. A Golomb ruler, a set of integral integers arranged as divisions on an imaginary ruler in such a way that the distance between any two divisions is unique, can be used to construct reduction rulers.

This means that values of distance between elements of Holomb's ruler do not occur more than once. But some values may be missing, which is not allowed for a reductive ruler.

Thus, if you use a Golomb ruler as a reduction ruler, its (reduction ruler) length can be less than the distance between the outermost elements, that is, than the length of the Golomb ruler. And only for a perfect Golomb ruler will these lengths be equal. Note that perfect Holomb rulers of order higher than 4 do not exist.

Figures 3, 4 show the reduction rulers formed on the basis of a perfect Golomb ruler of order 4 and the graphs of pair recurrence for them. In Fig. 3 - without repeats, and in Fig. 4 - its threefold repetition using Theorem 2.

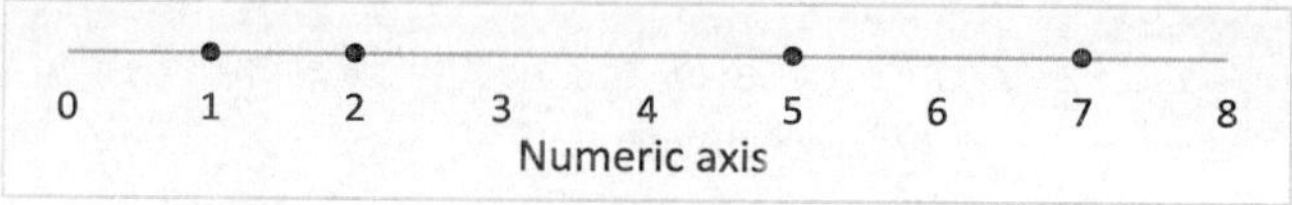

(a)

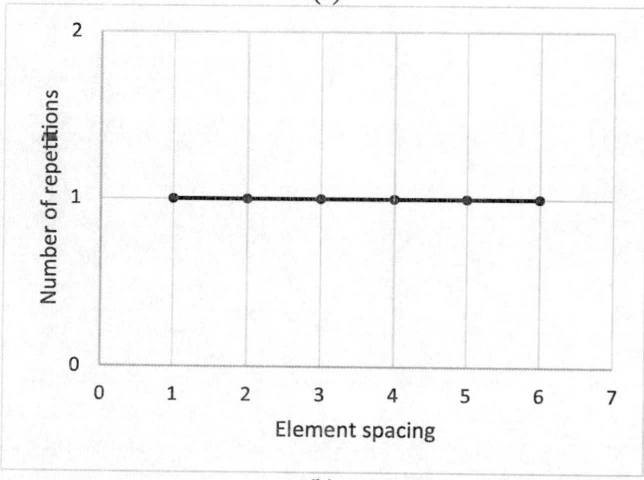

(b)

Fig. 3. Example of a reduction ruler: a) - reduction ruler corresponding to a perfect Golomb ruler of order 4; b) - graph of the number of pair recurrence for it.

The figures show how the use of Golomb's ruler is more efficient for constructing reduction rulers (in the sense of less redundancy and, consequently, more reduction coefficient - greater length at a given order) than using a set of consecutive numbers - the vast majority of pairs occur only once. The maxima (at positions corresponding to note 3 to the theorems, they cannot be eliminated) have a minimal width in this case: one. At the same time, it should be noted that Golomb's rulers are somewhat inconvenient for practical implementation, especially for hardware implementation, since they require a tabular assignment of the positions of the elements of the generated reduction ruler.

Below we will consider two other methods of construction of reduction rulers, which are more convenient for practical implementation.

Method 3 - with Switching of Sampling Frequency. The method consists in sampling k elements with frequency corresponding to the Nyquist frequency, and further, by the end of the sequence with frequency k times lower. The required length of the ruler is easily achieved by shortening the last interval to the required value. It is not difficult to show that the minimum order of m for a given length L (maximum reduction factor) is achieved by choosing k close to the square root of the length. To do this, for given k and the length of the ruler L, find its order and equate the derivative of k to zero. The order of the ruler $m = k + {}^{L}\!/_{k} - 1$, the derivative $\frac{dm}{dk} = 1 - {}^{L}\!/_{k^2}$, whence $k \cong \sqrt{L}$ and the minimal order of $m \cong 2\sqrt{L} - 1$ (the approximate equal sign because both k and m must be integers, and k can be either greater or less than the exact value of the root, and m will remain the same as a result of the equality of the derivative $\frac{dm}{dk}$ to zero at that point).

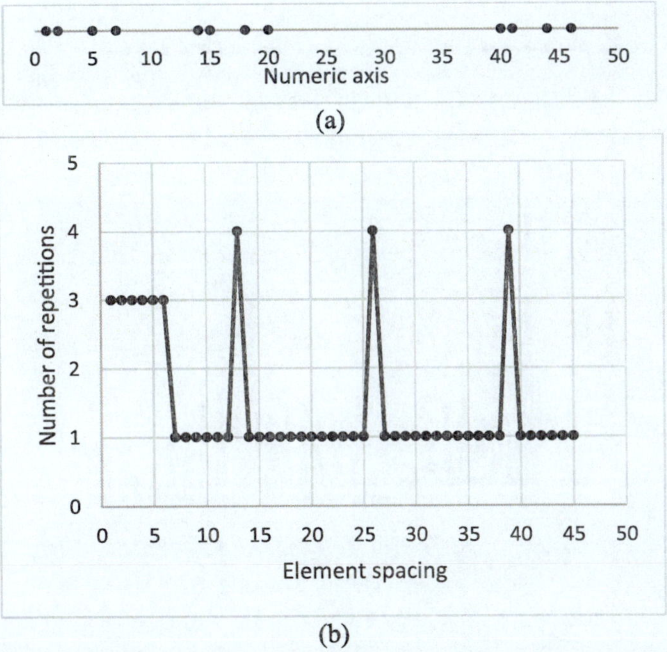

Fig. 4. Example of a reduction ruler: a) - a reduction ruler formed by threefold repetition of the ruler shown in Fig. 3, its order 12, length 45; b) - graph of the number of repeating pairs for it.

Given that the length of the reduction ruler (without taking into account the truncation of the correlation function) must not be less than half the sample length of the Nyquist frequency n and cannot exceed n, we obtain an estimate of the maximum reduction factor $\frac{n/2+1}{\sqrt[2]{n/2-1}} < r_{\max} < \frac{n+1}{2\sqrt{n-1}}$, or, given that usually $n > 1$, $\frac{1}{2}\sqrt{\frac{n}{2}} < r_{\max} < \frac{1}{2}\sqrt{n}$.

Figure 5a shows a reduction ruler formed by twice (according to Theorem 1) repeating a ruler of $k = 4$, $m = 12$, $L = 36$. Its order is 24, and its length is 106. The graph of the number of repeating pairs for it is shown in Fig. 5b.

Method 4 - Sampling with Two Close Frequencies. To implement the method, we sample simultaneously at two close frequencies, significantly lower than the Nyquist frequency. If the first frequency is chosen k times lower than the Nyquist frequency, then the second frequency is chosen $k + 1$ times lower (Fig. 6a, k = 6). Both subsamples are uniform and can be continued continuously. The reduction ruler is obtained by combining both subsamples (Fig. 6b). The resulting sequence of distances between adjacent samples is periodic with a period $T = k(k + 1)$. It is easy to see that one period contains all pairs with distances from 1 to $2k$, that is, the length of such a reduction ruler is $L = 2k$, and adding a period increases its length by the value T. That is, if a reduction ruler consists of N such periods (the distance between the outermost elements of the ruler is then $NT - k$), then its length is $L = (N - 1)T + 2k$ and its order is $m = 2Nk$. In general, the length of such a reduction ruler is less than the distance between its outermost elements by $k(k - 2)$ and only for $k = 2$ they coincide. Note that the last $k - 2$ elements of

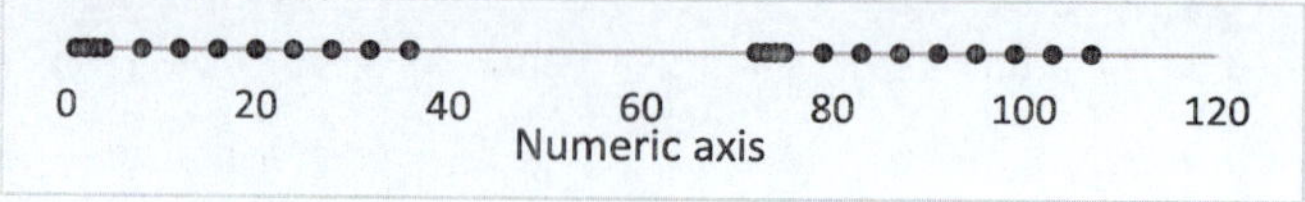

(a)

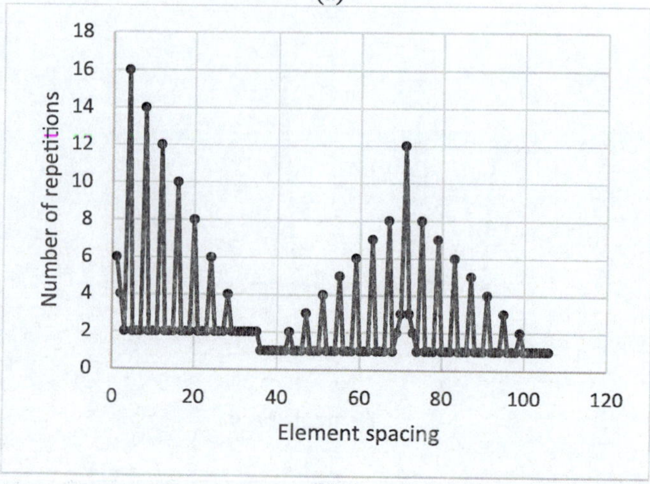

(b)

Fig. 5. Example of a reduction ruler: a) - a reduction ruler formed by repeating twice the ruler of $k = 4$, $m = 12$, $L = 36$, its order 24, length 106; b) - a graph of the number of repeating pairs for it.

the reduction ruler obtained by the method under consideration are redundant and can be eliminated without reducing its length. Then its order will be $m = (2N - 1)k + 2$, and the distance between the extreme elements will be $(2N - 1)\frac{T}{2} + \frac{k}{2}$ if k is even, or $(2N - 1)\frac{T}{2} + \frac{k+1}{2}$ if k is odd. Such a sequence contains $N - 1$ full periods and ends with a half-period plus one element, which is convenient for hardware implementation of data sampling cycles, because a pause of almost half-period duration is formed before the start of the next cycle, which can be used, for example, for data transmission. The value of k, at which for the chosen length of the reduction ruler, the distance between the outermost elements, its order will be minimal, can be found by selection (starting from lower values of N), since the decoupling of the given equations in integers leads to non-monotone, jump-like changes in values. The reduction ruler obtained by sampling at close frequencies for one period T has limited application, because its reduction coefficient $r = 1 + {}^{1}/_{2k}$ is close to 1 (for reduced to a half-period - $r = 2 - \frac{3}{k+2}$, the reduction coefficient is less than 2). Tangible values of the reduction coefficient are obtained only for a reduction ruler consisting of several periods - $r = \frac{(N-1)T+2k+1}{(2N-1)k+2}$, it grows rapidly at small N and asymptotically approaches $\frac{k+1}{2}$ at large N.

The reduction ruler constructed in this way can serve as the basis for constructing larger rulers by means of repetitions (Theorems 1, 2), but it should be borne in mind that its important property, continuity of samples, is lost in this case. Figure 6b shows a reduction ruler formed using the described sampling method by two adjacent frequencies at $k = 6$, that is, one frequency is 6 times lower than the Nyquist frequency, and the second frequency is 7 times lower.

The subsamples forming the reduction ruler are shown in Fig. 6a. The figure clearly shows two sampling periods ($N = 2, T = 42$) and one element of the next period added at the end (just to be more illustrative). The order of the ruler $m = 25$, its length $L = 54$, and the distance between the outermost elements 84. Figure 6c shows a graph of the number of repeating pairs for it.

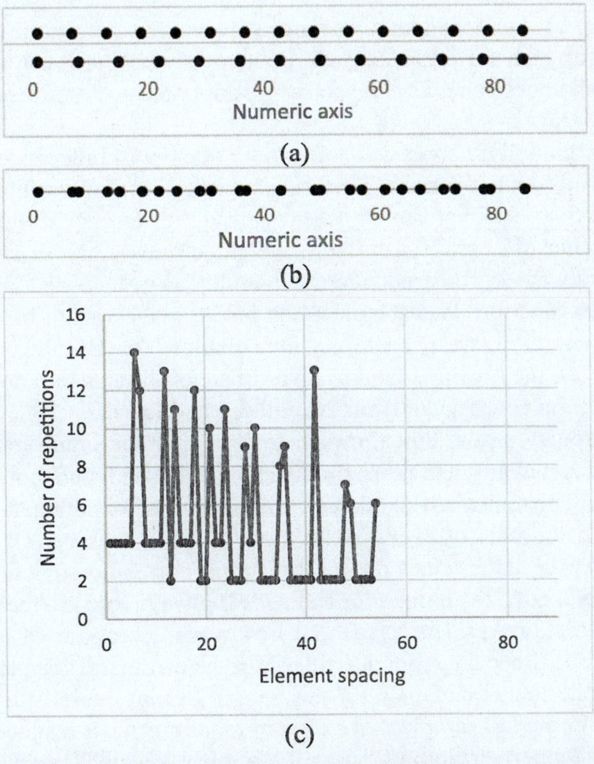

Fig. 6. Reduction ruler formed by sampling two close frequencies (6 and 7 times less than the Nyquist frequency, $k = 6$). a) - components of the subsample; b) - combining the subsamples into a reduction ruler, its order 25, length 54. c) - graph of the number of pair repeatability for it.

5 Estimation Standard Deviation. Numerical Modeling of Estimation Accuracy Based on the Calculation of the Standard Deviation

As mentioned above, reducing the sample redundancy leads to an increase in the variance of the estimate of the correlation function counts and, as a result, to an increase in the standard deviation of the estimate of the amplitude spectrum of the signal.

In the general case, the variance of the correlation function estimation for each lag depends on the specific type of spectrum with other parameters unchanged, so studying the effect of the data reduction proper on the standard deviation of the spectrum estimation in the general case is quite a difficult task. Under such conditions, it is convenient to consider a class of signals having a rectangular amplitude spectrum, the same at all frequencies - "white" noise. For such signals, selected in the time domain by the Nyquist frequency, it is characteristic that all samples are interdependent.

In this case, all their paired products that serve to calculate the values of the correlation function will also be interdependent. Therefore, the variance of the correlation function count estimate for each particular lag will be inversely proportional to the number of products of signal count pairs with the corresponding distance between them, that is, to the number of averages.

Thus, the correlation function will consist of a number of independent random variables, each with its own variance. Since each sample of the amplitude spectrum is obtained using a Fourier transform with all values of the correlation function taken into account, the variance of its estimate will be a weighted sum of their variances. In order to compare different ways to construct a reduction ruler, as well as the effect of truncation of the correlation function, let us express the result as the square root of the sum of variances and pronormalize it by a similar value obtained for the full Nyquist frequency sample. We will use the obtained ratio as an estimate of the standard deviation ratio for the amplitude spectra constructed from the reduced and full data.

Consider a reduction ruler that allows us to obtain the minimum standard deviation of the result for a signal as white noise for given lengths of full and reduced samples (or reduction factor). The minimum standard deviation with other parameters unchanged is achieved when the number of averages for each lag other than zero is the same. And those lags, which exceed the selected truncation of the correlation function, do not affect the result. For lag "zero", the number of averages is always equal to the ruler order, and for other lags at least one less, because there are always fewer pairs of samples than the samples themselves. Since the reduction ruler is based on the full sample by eliminating some elements, the number of pairs of any length cannot exceed the corresponding number of pairs for the full sample. Such a set of requirements is achieved only in some exceptional cases for perfect Golomb rulers, the number of which is significantly limited.

Therefore, we will consider a certain hypothetical reduction ruler that satisfies the above requirements. It should be borne in mind that since there are no practical ways to construct such a ruler, the estimates obtained will be underestimated, but nevertheless give an idea of the characteristic properties (consequences) of the application of reduction.

Figure 7 shows the dependences of the ratio of the standard deviations of the full and reduced samples R on the ratio of their lengths $^n/_m$ for the hypothetical reduction ruler at different values of the truncation correlation function ($^n/_2$, $^n/_3$, $^n/_4$, $^n/_{10}$).

The calculations were performed for the length of the full sample $n = 1000$ elements. For large and somewhat smaller values of n the dependences differ insignificantly, the difference becomes perceptible only for $n < 100$.

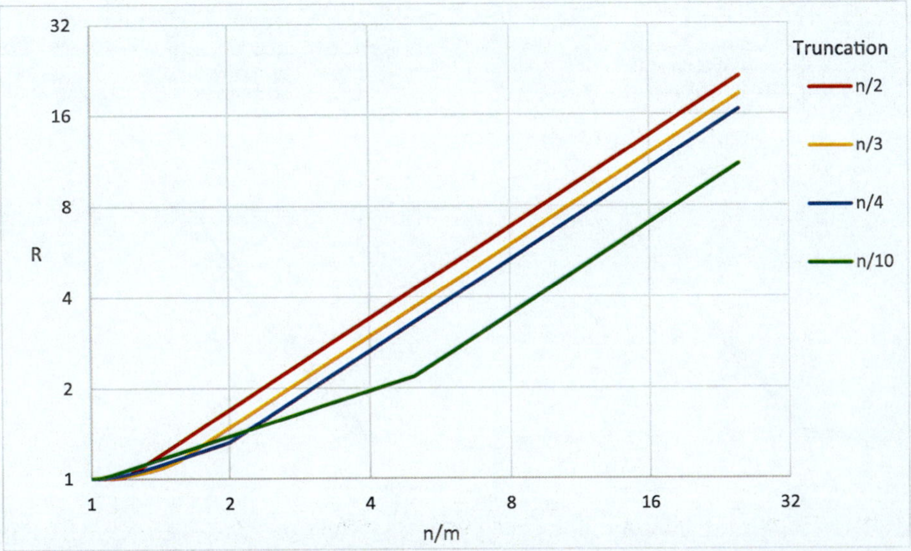

Fig. 7. Dependences of the ratio of standard deviations of the full and reduced samples R on the ratio of their lengths $^n/_m$ for the hypothetical reduction ruler at different truncations of the correlation function ($^n/_2$, $^n/_3$, $^n/_4$, $^n/_{10}$, full sample of items $n = 1000$, length of reduced sample m).

The dependences shown in Fig. 7 dependences have a characteristic "scrap", which is well observed on the graph for truncation of correlation function $^n/_{10}$. It is caused by the fact that at smaller reduction coefficients the total number of pairs formed by elements of the reduction ruler is quite sufficient to meet the requirements described above, "extra" pairs are placed along the point of truncation of correlation function and do not affect the result, but further they become insufficient, the maximum possible values of pair repetition cannot be realized, so the graph of dependence becomes steep.

Figure 8 shows the results of numerous experiments with different reduction rulers, for which the methods described above have been used for plotting.

The results obtained illustrate the possibilities of constructing such rulers, which allow to achieve low values of the standard deviation of the estimate of the amplitude spectrum of the signal at a given reduction $^n/_m$ - the ratio of the lengths of the full ($n = 1000$) and reduced samples. The truncation of the correlation function is chosen $^n/_2$. For comparison, the same Limit line corresponding to the $^n/_2$ line in Figs. 8a-8d is plotted in Figs. 7.

Figure 8 illustrates the possibilities of data reduction when using reduction rulers derived from a series of consecutive numbers.

Graph #1.1 is a single solid section of length m numbers, graphs #1.2 and #1.3 are two and three repetitions of a reduced length section respectively, offset by distances somewhat smaller than those derived from Theorems 1 and 2.

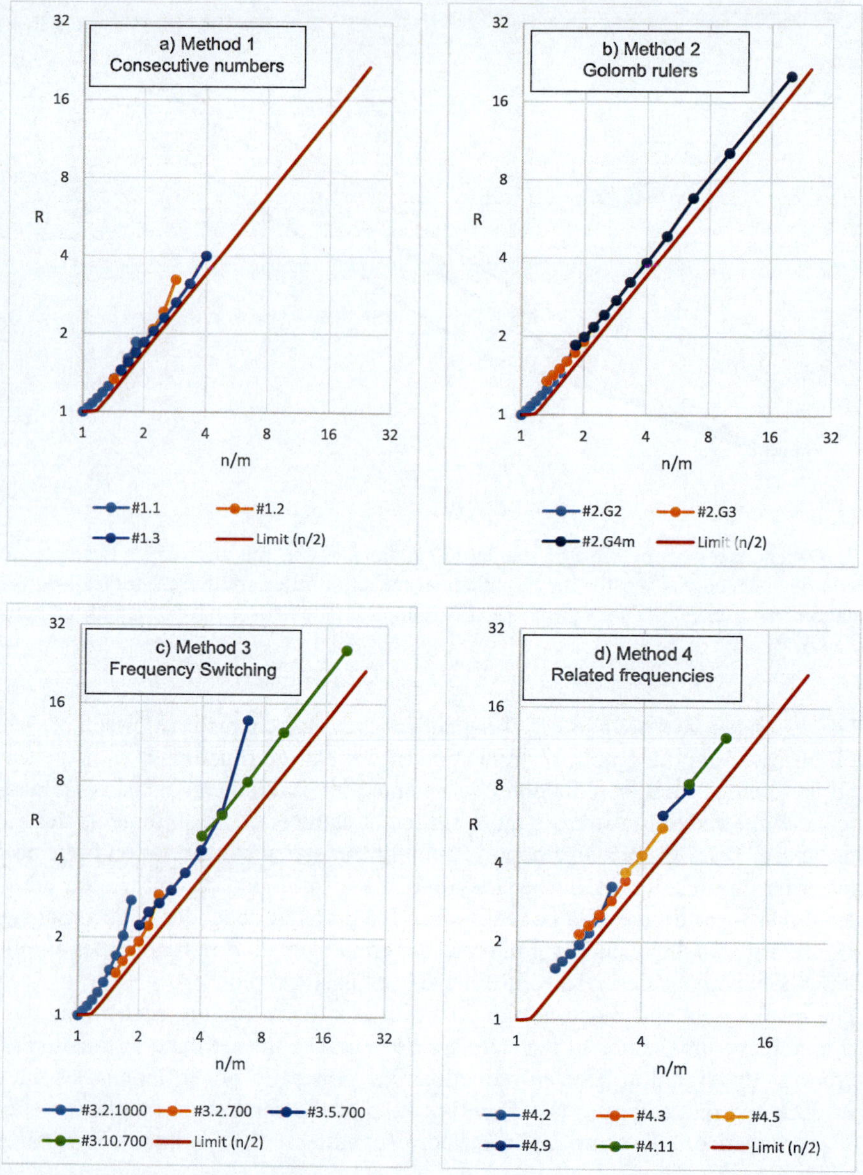

Fig. 8. Results of numerous experiments with reduction rulers constructed using the described methods. R - ratio of standard deviations of amplitude spectra estimation for full and reduced samples, ratio of sample lengths - $^{n}/_{m}$.

These distances were chosen to provide smaller values of standard deviation.

Figure 8b shows the results for the reduction rulers derived from perfect Golomb rulers of orders 2, 3, and 4 (plots #2.G2, #2.G3, and #2G4m, respectively). The required length of the reduction ruler was achieved by recursive application of Theorems 1 and

2, i.e., if the obtained intermediate reduction ruler was not long enough, it was taken as the basis for the subsequent step of the increase. Only for #2G4m was the offset value in the last ramping step chosen to obtain the minimum value of the standard deviation.

Figure 8c corresponds to the method of constructing a reduction ruler with switching sampling frequency. For graphs #3.2.1000 and #3.2.700, k was chosen to be 2, with ruler lengths of 1000 and 700, respectively. Graph #3.5.700 was obtained at $k = 5$, and graph #3.10.700 was obtained at $k = 10$. For these, the constructed ruler length was 700.

The results obtained for the reduction rulers plotted in the contiguous frequency method are shown in Fig. 8d. Graphs #4.2, #4.3, #4.5, #4.9 and #4.11 differ in the values of k - 2, 3, 5, 9 and 11, respectively.

Figure 8 allows us to draw a practical conclusion that with a good choice of a reduction ruler how many times the amount of data decreases, the standard deviation of the result will increase approximately by the same factor. In this case, as can be seen from Fig. 8, the best results can be achieved with the construction of reduction rulers based on Golomb rulers - both a wide range of reduction (up to 20 times) and a relatively low standard deviation of the result. If you take a series of consecutive numbers as a basis, good results are achieved only for much smaller values of reduction, larger values of reduction are difficult to obtain. Methods with switching frequencies and adjacent frequencies allow to obtain a wider reduction range, but with a large standard deviation.

6 Conclusion

It has been noted that in the spectral analysis of random signals, when only the amplitude spectrum is informative, the uniform sampling with the frequency corresponding to the Nyquist criterion is significantly redundant. If a certain level of growth of the spectrum estimation variance is acceptable, the sample redundancy can be reduced without loss of informativity by reducing its volume. In this case, the discretization cannot remain uniform. One of the variants of non-uniform sampling is elimination of certain samples from the uniform sampling, i.e. reduction of the full sample corresponding to discretization of the signal with Nyquist frequency. The list of remaining samples is specified by the reduction ruler. Perfect Golomb rulers can act as boundary reduction rulers. But their practical application is limited by the fact that their number (and length) is limited by the fourth order. Ways of increasing the reduction rulers to the required size have been proposed, and the corresponding theorems have been formulated and proved. This extends the range of data reduction to the values necessary in practical applications. Several ways of constructing reduction rulers are considered and their simulation for broadband signals is performed. The results allow us to estimate the growth of the standard deviation caused by data reduction.

Further research will be related to the application of reduction rulers for the estimation of the mutual correlation function of signals, as well as the noise-resistant coding of signals.

References

1. Beshley, H., Beshley, M., Medvetskyi, M., Pyrih, J.: QoS-aware optimal radio resource allocation method for machine-type communications in 5G LTE and beyond cellular networks. Wireless Commun. Mobile Comput. **2021**, 9966366–1–9966366–18 (2021)
2. Klymash, M., Savchuk, R., Pozdnyakov, P., Beshley, M.: The researching and modeling of structures of mobile networks for providing of multiservice radio access. In: Proceedings of International Conference on Modern Problem of Radio Engineering, Telecommunications and Computer Science, pp. 281–282 (2012)
3. Masiuk, A., Beshley, M., Lavriv, O., Deschynskiy, Y.: Common radio resource management model for heterogeneous cellular networks. In: 2016 13th International Conference on Modern Problems of Radio Engineering, Telecommunications and Computer Science (TCSET), pp. 661–663 (2016) https://doi.org/10.1109/TCSET.2016.7452145
4. Pan, Z., Wang, J., Weng, Y.: Digital signal processing techniques in Nyquist-WDM transmission systems. In: 2015 14th International Conference on Optical Communications and Networks (ICOCN), pp. 1–3 (2015) https://doi.org/10.1109/ICOCN.2015.7203653
5. van Wyk, M.A., Ping, L., Chen, G.: Multivaluedness in networks: shannon's noisy-channel coding theorem. IEEE Trans. Circuits Syst. II Express Briefs **68**(10), 3234–3235 (2021c). https://doi.org/10.1109/TCSII.2021.3074925
6. Candan, C., Kutay, M.A., Ozaktas, H.M.: The discrete fractional Fourier transform. IEEE Trans. Signal Process. **48**(5), 1329–1337 (2000). https://doi.org/10.1109/78.839980
7. Cooley, J., Lewis, P., Welch, P.: The finite Fourier transform. IEEE Trans. Audio Electroacoust. **17**(2), 77–85 (1969). https://doi.org/10.1109/TAU.1969.1162036
8. Zheng, L., Hu, W., Min, Y.: Raw wind data preprocessing: a data-mining approach. IEEE Trans. Sustainable Energy **6**(1), 11–19 (2015). https://doi.org/10.1109/TSTE.2014.2355837
9. Hanson, J.A., Haw Yang, H.: A general statistical test for correlation in a finite-length time series. J. of Chemical Phys. **128**, 214101 (2008)
10. Bujang, M.A., Baharum, N.: Sample size guideline for correlation analysis. World J. of Soc. Sci. Res. 3(1), 37–46 (2016)
11. Ma, J., Tao, R., Li, Y., Kang, X.: Fractional Power Spectrum and Fractional Correlation Estimations for Nonuniform Sampling. IEEE Signal Process. Lett. **27**, 930–934 (2020). https://doi.org/10.1109/LSP.2020.2997561
12. Chen, B.-W., Wang, J,-C.: Incomplete data analysis/computer science. Appli. Pattern Recogn.15 (2021). https://doi.org/10.5772/Intechopen.94068
13. Singh, P., Gupta, A., Joshi, S.D.: General parameterized fourier transform: a unified framework for the fourier, Laplace, Mellin and Z Transforms. IEEE Trans. Signal Process. **70**, 1295–1309 (2022). https://doi.org/10.1109/TSP.2022.3152607
14. Kuhn J.R.: Recovering spectral information from unevenly sampled data: two mashine-efficient solutions. The Astronomical J. 87(2), 196–202 (1982)
15. Deeming, T.J.: Fourier analysis with unequally-spaced data. Astrophysics Space Sci. **36**, 137–158 (1975)
16. Persson, B.N.J., Albohr, O., Tartaglino, U., Volokitin, A.I., Tosatti, E.: On the nature of surface roughness with application to contact mechanics, sealing, rubber friction and adhesion. Topical Review. J. Phys. Condens. Matter. **17**, P.R1-R82 (2005). https://doi.org/10.1088/0953-8984/17/1/R01
17. Janeczko, D.J.: Power Spectrum Standard for Surface Roughness: Part I. In: Proceedings of SPIE 1165, Scatter from Optical Components, (2 January 1990). https://doi.org/10.1117/12.962847
18. Berngardt, O., Voronov, A., Grkovich, K.: Optimal signals of Golomb rules class for spectral measurements at EKB Super DARN radar: Theory and experiment. Radio Sci., 50(6), 486–500 (2015). 10/1002/2014RS005589

19. Wolfram, S.: A new kind of science. –Illinois: Stephen Wolfram, LLC. (2002)
20. M. Beshley, M. Seliuchenko, O. Panchenko, O. Zyuzko and I. Kahalo, "Experimental performance analysis of software-defined network switch and controller," 2018 14th International Conference on Advanced Trends in Radioelecrtronics, Telecommunications and Computer Engineering (TCSET), 2018, pp. 282–286, https://doi.org/10.1109/TCSET.2018.8336203.V
21. Romanchuk, Beshley, M., Polishuk, A., Seliuchenko, M.: Method for processing multiservice traffic in network node based on adaptive management of buffer resource. In: 2018 14th International Conference on Advanced Trends in Radioelecrtronics, Telecommunications and Computer Engineering (TCSET), pp. 1118–1122 (2018). https://doi.org/10.1109/TCSET.2018.8336390
22. Kryvinska, N., Kaczor, S., Strauss, C., Gregus, M.: Servitization strategies and product-service-systems. IEEE World Congress Serv. **2014**, 254–260 (2014). https://doi.org/10.1109/SERVICES.2014.52
23. Jin, R.-B., Shimizu, R.: Extended Wiener-Khinchin theorem for quantum spectral analysis. Optica **5**, 93–98 (2018)
24. Golomb, S.W.: How to number a graph in graph. In: Read, R.C., (ed.) Theory and Computing, pp. 23–37. Academic Pres, New York (1972)

Investigation of High-Speed Methods for Determining the Equilibrium State of a Network Based on the Principle of Maximum Entropy

Ruslan Politanskyi[1](✉), Andrii Samila[1], Leonid Politanskyi[1], Vitalii Vlasenko[1], Valentin Popa[2], Yuriy Bobalo[3], and Ihor Tchaikovsky[3]

[1] Yuriy Fedkovych Chernivtsi National University, Kotsyubynsky 2, Chernivtsi 58012, Ukraine
polrusl@ukr.net
[2] Stefan Cel Mare University, Str. Universitatii 13, 720229 Suceava, Romania
[3] Lviv Polytechnic National University, Bandery St. 12, Lviv 79013, Ukraine

Abstract. The chapter considers the possibility of applying the entropy method to determine the most probable state of the network, which has restrictive conditions for the total traffic generated and received in each network node. Additionally, it is assumed that the network is closed, i.e. the sum of all generated information flows is equal to the sum of all received flows. Possibilities of application of the method in networks with several planes of decentralization are analyzed. It is explained how to build a model of a decentralized network in the form of a two-dimensional sparse matrix of information flows, which is structured in a special way. Thus, the task of analyzing a decentralized network is to determine the optimal entropy value of the system represented by such a matrix. A method for estimating the range of entropy values of a decentralized network by analyzing significantly un-balanced networks with a significant number of zero values in the matrix of in-formation flows is proposed. Calculations for elementary configurations of a network represented by a flat graph and a decentralized network with the least topological complexity are given. A huge advantage of this method is the much lower cost of computing power to analyze the state of the network, even if the network has a complex decentralized structure.

Keywords: Entropy maximum principle · Decentralized networks · Equilibrium state of the system

1 Introduction

The concept of thermodynamic equilibrium can often be applied to problems of determining the most probable state of macrosystems of different physical nature or artificial origin. The concept of thermodynamic equilibrium is often transferred to the task of determining the most probable state of macrosystems. According to this approach, the most used and informative concept is the entropy of the system. Depending on the problems to be solved, different types of entropy can be used. Despite the fact that this

M. Klymash et al. (Eds.): TCSET 2022, LNEE 965, pp. 602–614, 2023.
https://doi.org/10.1007/978-3-031-24963-1_35

concept has been used in science for a long time, today the study of entropy is used by researchers in the most diverse fields of science and technology: finance [1], social systems [2], telecommunication networks [3, 4], quantum transmission and processing systems information [5], technical and natural systems [6, 7]. This is a very important branch of modern science, which is associated with a significant increase in the complexity and scale of the studied systems.

The article considers the algorithm for determining the equilibrium state, which is based on the concept of entropy and is applied to telecommunication systems. It is assumed that the most probable state of the network is the state in which the entropy reaches its maximum value under the condition of equality of input and output information flows in each network node.

The second law of thermodynamics postulates that all processes in a system over which no external forces perform work, and in which only heat exchange can take place, all processes flow in the direction of increasing the entropy of this system. If we transfer this law to non-thermodynamic systems, which can be in many states, we can say that the most probable state is the state with maximum entropy, if no work is done to increase the order of the whole system.

Such work in the case of a telecommunications network is routing algorithms operating in it, connecting a large number of new subscribers, entering the market of new operators, upgrading equipment, improving software algorithms for network management and more. But even in this case, if the network is sufficiently developed, the deviation of information flows in the network from some equilibrium still have certain limits, unless there is a radical transformation of the network structure.

High-speed algorithms for determining the equilibrium state of a telecommunications system are of particular importance due to the increasing complexity of systems due to the introduction of new communication standards, especially modern 5G and 6G standards. One of the features of such standards is the need for decentralized distribution of infrastructure and frequency resources of the network among a significant number of players in the telecommunications services market, including small and medium-sized businesses. This approach is made possible by the latest technologies related to the sharing of frequency resources, including methods of artificial intelligence and software defined radio [8, 9], blockchain technology [10] and the search for new methods of signal generation and modulation [11, 12]. All this leads to a global trend in the development of telecommunications technologies in the direction of their convergence [13, 14].

The use of services by the subscriber in 6G networks will be mostly connected with the cyber-physical world based on a large number of factors for perception. This means that the network must provide all or almost the entire context of the end users. In addition, the services that will be provided to subscribers will be combined directly in the devices of subscribers from a large number of information flows transmitted by different communication networks and operators. Therefore, in the short term, the current centralized model of construction and operation of network infrastructure will inevitably change to a more flexible decentralized model with an open market of radio frequency resource and network infrastructure. The decentralized model will stimulate the emergence of a large number of micro-operators who will be narrowly focused on addressing certain needs of the economy and society.

To formalize the decentralized model, it is convenient to use the matrices of information flows operating in each plane of decentralization. The boundary conditions of such a problem are the values of the total information flows of each network node. This model can be closer to determining the physical parameters of the network, if the structure of the matrix meets the condition that direct information exchange can occur only between nodes of a single plane. A priori, it can be a plane of infrastructure. However, with the development of telecommunications technology, there may be means of communication that use the direct exchange of information between individual subscribers (point-to-point connection). Therefore, the task of determining the modelling of the network, which should take into account the restrictions on the existence of flows only between nodes that are physically capable of. This task is a generalization of the partial case when information exchange can occur between all nodes of the network.

2 The Matrix of Information Flows and the Entropy of the Network Represented as a Flat Graph

Consider the simulation of a telecommunications network formed by a finite number of nodes l: $\{n_1, n_2, \ldots, n_l\}$, , and which is represented by a flat graph. One of the most detailed characteristics of the network is the matrix of information flows (1):

$$A = \begin{Vmatrix} \rho_{11} \ldots \rho_{1l} \\ \ldots \ldots \; .. \\ \rho_{l1} \ldots \rho_{ll} \end{Vmatrix}. \tag{1}$$

Each element of this matrix determines the intensity of traffic directed from one network node to another. Based on the content of the elements of the matrix (1), each of them is non-negative:

$$\rho_{ij} \geq 0, 1 \leq i \leq l, 1 \leq j \leq l. \tag{2}$$

Consider further the boundary conditions of the problem, which limit the set of possible values of the correspondence matrix. Part of the network nodes can generate traffic (sources of information flows), and part – to absorb it (receivers of information flows). Each node can be both a source and a receiver of information flow. The sum of all threads generated in the node s_i is defined as the sum of the values of the elements of the correspondence matrix that belong to the corresponding row of the matrix:

$$s_i = \sum\nolimits_{j=1}^{l} \rho_{ij}. \tag{3}$$

The sum of the received in each node threads d_i is equal to the sum of the values of the flow matrix that belong to the appropriate column of the matrix:

$$d_i = \sum\nolimits_{j=1}^{l} \rho_{ji} \tag{4}$$

One of the formulations of the problem is the condition of network closure. It is obvious that in this case the value of the matrix of information flows is subject to the condition of equality of the total intensity of all generated and all received flows:

$$\sum_{i=1}^{l} s_i = \sum_{i=1}^{l} d_i. \tag{5}$$

Obviously, the set of solutions to this problem is quite powerful. The approximate upper power limit of this set N can be obtained from the well-known problem of dividing some numbers s_i or d_i by a given number of terms l, which is defined as the product of binomial coefficients:

$$N = max\left\{\prod_{i=1}^{l} C_{s_i+l-1}^{s_i}, \prod_{i=1}^{l} C_{d_i+l-1}^{d_i}\right\}. \tag{6}$$

Consider the problem of determining the matrix of information flows, which provides the maximum value of entropy. The entropy to be maximized is a function of matrix elements, and it is defined by the following expression [15]:

$$H(\| \rho_{ij} \|) = \sum_{i=1}^{l} \sum_{j=1}^{l} \rho_{ij} \cdot in\frac{\nu_{ij}}{\rho_{ij}} \rightarrow max, \tag{7}$$

where ν_{ij} is weighting value of the flow matrix $\rho_{ij} \neq 0$. If the value of some information flow $\rho_{i'j'} = 0$, then in expression (6) it is not taken into account:

$$h(0) = 0, \tag{8}$$

where $h(\rho_{ij})$ is term in the expression for entropy, formed by the information flow ρ_{ij}.

If all the branches that make up the network have the same technical characteristics, then the value of entropy takes the form:

$$H(\| \rho_{ij} \|) = \sum_{i=1}^{l} \sum_{j=1}^{l} \rho_{ij} \cdot in\frac{1}{\rho_{ij}} \rightarrow max. \tag{9}$$

The algorithm for solving the optimization problem (7, 9) with boundary conditions (2–5) is described in detail in [15]. According to the specified algorithm in the first step of the problem all values of the matrix of flows are assigned arbitrary nonzero values. For example, you can specify the same values for each element of the matrix of information flows:

$$\rho_{ij} = 1/l^2. \tag{10}$$

To evaluate the method, consider the solutions of the problem for the most primitive element of the network, which consists of only two nodes.

The flow matrix for such a problem has the following form:

$$A = \left\| \begin{matrix} \rho_{11} & \rho_{12} \\ \rho_{21} & \rho_{22} \end{matrix} \right\|. \tag{11}$$

Both nodes in Can be both receivers and sources of information flows, the values of which are known $\{s_1, d_1, s_2, d_2\}$.

Then the values of the flow matrix satisfy the system of equations:

$$\begin{cases} s_1 = \varrho_{11} + \rho_{12} \\ s_2 = \varrho_{21} + \rho_{22} \\ d_1 = \varrho_{11} + \rho_{21} \\ d_2 = \varrho_{12} + \rho_{22} \\ s_1 + s_2 = d_1 + d_2 \end{cases} . \quad (12)$$

For such a problem, the search for the optimal value for entropy is reduced to determining only one element of the matrix (11), let it be an element ϱ_{11}, which we denote as x. All other elements of the matrix (11) can be expressed in x:

$$\begin{cases} s_1 = x + \rho_{12} \\ s_2 = \varrho_{21} + \rho_{22} \\ d_1 = x + \rho_{21} \\ d_2 = \varrho_{12} + \rho_{22} \\ s_1 + s_2 = d_1 + d_2 \end{cases} \Rightarrow \begin{cases} \rho_{12} = s_1 - x \\ \rho_{22} = s_2 - \varrho_{21} \\ \rho_{21} = d_1 - x \\ \rho_{22} = d_2 - \varrho_{12} \\ s_1 + s_2 = d_1 + d_2 \end{cases} \Rightarrow \begin{cases} \rho_{12} = s_1 - x \\ \rho_{22} = s_2 - d_1 + x \\ \rho_{21} = d_1 - x \end{cases} . \quad (13)$$

From the formed system (13) and proceeding from the condition of nonzero flows (2) it is possible to deduce the restrictions imposed on values x:

$$\begin{cases} x \geq 0 \\ x \geq d_1 - s_2 \\ x \leq s_1 \\ x \leq d_1 \end{cases} . \quad (14)$$

The entropy of such a matrix is a function of one variable x:

$$H(\|\rho_{ij}\|) = H(x), \quad (15)$$

and it is defined by an expression:

$$H(x) = -\begin{bmatrix} x \cdot ln(x) + (s_1 - x) \cdot ln(s_1 - x) + \\ +(s_2 - d_1 + x) \cdot ln(s_2 - d_1 + x) + \\ +(d_1 - x) \cdot ln(d_1 - x) \end{bmatrix} . \quad (16)$$

Consider an example of a network by specifying the values of flows in nodes: $s = (99, 90)$, $d = (128, 61)$. The set of x values is determined by a two-way inequality:

$$38 \leq x \leq 99. \quad (17)$$

The graphical dependence of $H(x)$ is shown in Fig. 1.

It is obvious that the function described by expression (16) has a single maximum that can be found by any numerical methods, including the use of standard functions of mathematical packages of computer data processing. Calculated to the nearest 0.01 using

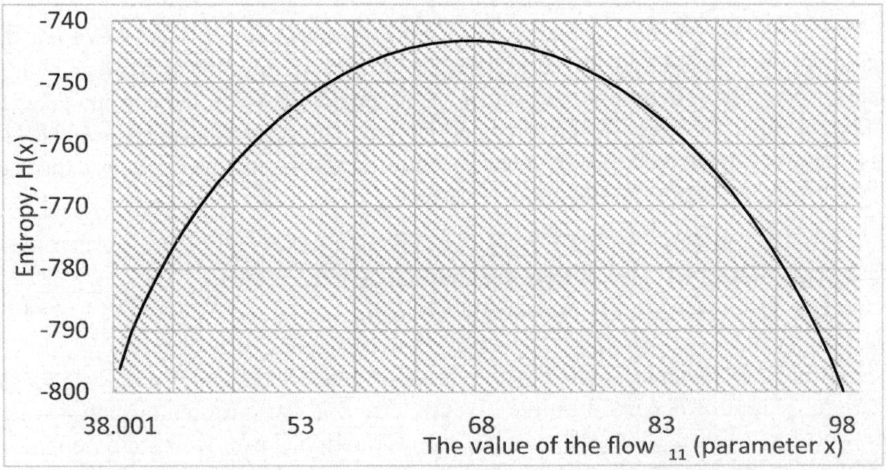

Fig. 1. Entropy of the simplest topological element of the network.

standard SciLab environment functions, the maximum entropy value is $H_{max} = -741.03$ for the value of $x = 67.05$.

Most methods for determining the maximum value for a transcendental function, which is also a logarithmic function in expression (16), are iterative methods. For fast-converging methods, the number of iterations at the accuracy of the argument 0.01 is approximately equal to 3. In addition, for fast-converging methods, the process of determining each iteration requires the calculation of the first and even the second derivative. For other, slower methods, such as splitting a segment in half, you need twice as many iterations.

We apply the method of determining the optimal distribution of flows in the network, which is determined by the principle of maximum entropy [14]. To do this, choose arbitrary values of the elements of the matrix (11). For example, we can set the initial values of the matrix depending on the indices of the matrix element:

$$\rho_{ij}^{0} = \frac{i \cdot j}{l^2}. \tag{18}$$

The process of recalculating the values of the elements of the matrix of information flows provides partial fulfilment of the boundary conditions of the task set for the received information flows:

$$\rho_{ij}^{'} = \rho_{ij}^{0} \cdot s_i \cdot \left[\sum\nolimits_{j \in D} \rho_{ij}^{0}\right]^{-1}. \tag{19}$$

After calculating the values of the matrix of flows with formulas (18) you need to recalculate the values of these coefficients so as to take into account the boundary conditions of the problem for the generated information flows:

$$\rho_{ij} = \rho_{ij}^{'} \cdot d_j \cdot \left[\sum\nolimits_{i \in S} \rho_{ij}^{'}\right]^{-1}. \tag{20}$$

The advantage of this method is that it allows determining the values of the flow matrix without a long iterative process, because only one stage of the matrix recalculation makes it possible to determine the values of the matrix at which the entropy takes the maximum value. In this case, the values of the flow matrix calculated by formulas (18–20) coincide with the solutions obtained by optimizing the function given by expression (16) with any degree of calculation accuracy.

3 Application of the Entropy Maximum Method to the Decentralized Network Model

The existing model can be applied to decentralized networks in which nodes are located in several planes of decentralization. To fully describe the information flows in such a network, you can use several matrices of information flows, which are defined for each plane of the decentralized telecommunications system [16]. This model makes it possible to determine the actual amount of traffic that arises in the process of information exchange between any nodes of the decentralized system, and the result of its application is a detailed scheme of information exchange in terms of all possible subjects and objects of the telecommunications system.

The model of information exchange in a decentralized network can be reduced to a two-dimensional matrix of information flows. The elements of such a matrix, in addition to the two lower indices denoting the numbers of the nodes within each plane, also contain two upper indices denoting the numbers of the planes to which the nodes belong:

$$\left\| \rho_{i,j}^{k,l} \right\|. \tag{21}$$

where k is the number of the plane, in which the node i is located; l is the number of the plane, in which the node j is located. In most real systems, information exchange can occur only between neighbouring planes of decentralization ("plane of subscribers-> plane of operators", "plane of operators-> plane of infrastructure", etc. [16]), so, as a rule,$l = k + 1$.

In such a matrix, the assignment of the information flow with nodes of some pair of planes is given by grouping the corresponding elements in the matrix structure.

Another important approximation of the model is the lack of information flows between nodes that are in the same plane of the decentralized network, except for the plane of the infrastructure, where there is an exchange of information flows at the physical level. This means that the appropriate elements of the matrix have zero values. Thus, the consolidated matrix of information flows of the decentralized network is sparse. The boundary conditions of the problem determine the flows that each node generates or receives.

For example, consider a system that has the lowest topological complexity among systems that cannot be described by a two-dimensional graph. This network has only two planes of decentralization, each of which contains two nodes (see Fig. 2).

In our proposed model, the matrix of information flows has the form of marix with dimension 4×4. Next, we take into account that the information exchange between the nodes of one of the planes (plane 1) does not occur. Another limitation is the zero

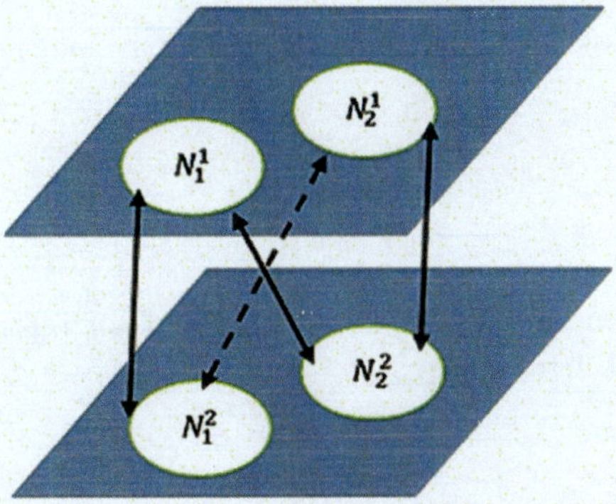

Fig. 2. Entropy of the simplest topological element of the network.

equality of the diagonal elements of the matrix, because in a real telecommunications system, the existence of loop routes is most likely a shortcoming of the routing algorithm or a fault at the network design stage. Given these limitations, the matrix of information flows is as follows:

$$\begin{Vmatrix} 0 & 0 & \rho_{11}^{12} & \rho_{12}^{12} \\ 0 & 0 & \rho_{21}^{12} & \rho_{22}^{12} \\ \rho_{11}^{21} & \rho_{12}^{21} & 0 & \rho_{12}^{22} \\ \rho_{21}^{21} & \rho_{22}^{21} & \rho_{21}^{22} & 0 \end{Vmatrix}. \tag{22}$$

Next, it is also necessary to specify the values of incoming and outgoing traffic in each node as mentioned in (3) and (4) with $l = 4$.

$$\begin{cases} S = \{s_1^1, s_2^1, s_1^2, s_2^2\} \\ D = \{d_1^1, d_2^1, d_1^2, d_2^2\} \end{cases}. \tag{23}$$

Thus, the system of equations describing the network has the form:

$$\begin{cases} s_1^1 = \rho_{11}^{12} + \rho_{12}^{12} \\ s_2^1 = \rho_{21}^{12} + \rho_{22}^{12} \\ s_1^2 = \rho_{11}^{21} + \rho_{12}^{21} + \rho_{12}^{22} \\ s_2^2 = \rho_{21}^{21} + \rho_{22}^{21} + \rho_{21}^{22} \end{cases}, \begin{cases} d_1^1 = \rho_{11}^{21} + \rho_{21}^{21} \\ d_2^1 = \rho_{12}^{21} + \rho_{22}^{21} \\ d_1^2 = \rho_{11}^{12} + \rho_{21}^{12} + \rho_{21}^{22} \\ d_2^2 = \rho_{12}^{12} + \rho_{22}^{12} + \rho_{12}^{22} \end{cases}. \tag{24}$$

Another condition is the condition of system closure, i.e. there is equality of sums of all input and output information flows:

$$s_1^1 + s_2^1 + s_1^2 + s_2^2 = d_1^1 + d_2^1 + d_1^2 + d_2^2 = s. \tag{25}$$

Let us analyse the system (24) with constraint given by expressions (25). Determine the number of independent variables that uniquely define the matrix of the network, based on the following considerations: the matrix includes 10 non-zero values, they satisfy eight equations, one of which is dependent because the network is closed. Therefore, there are three independent variables, and they uniquely define the matrix of information flows in the network:

$$\begin{cases} \rho_{12}^{12} = x_1 \\ \rho_{11}^{12} = x_2 \\ \rho_{21}^{21} = x_3 \end{cases}. \tag{26}$$

Next, we use the condition of non-negativity of all information flows and find restrictions on the set of their values:

$$\begin{cases} 0 \le x_1 \le s_1^1 \\ 0 \le x_2 \le s_2^1 \\ 0 \le x_3 \le d_1^1 \\ (d_1^1 + d_2^2 - s_1^2) \le (x_1 + x_2 + x_3) \le d_2^1 + d_1^1 + d_2^2 - s_1^2 \\ x_1 + x_2 \ge s_1^1 + s_2^1 - d_1^2 \end{cases}. \tag{27}$$

Therefore, the entropy of the flow matrix (22) is determined by the function of three independent variables with constraints on the values of the variables given by the system (27):

$$H\left(\left\|\rho_{kl}^{ij}\right\|\right) = \sum\nolimits_{i,j,k,l} \rho_{kl}^{ij} \cdot ln\frac{1}{\rho_{kl}^{ij}} = H(x_1, x_2, x_3). \tag{28}$$

4 Numerical Examples of Calculating the Entropy of a Decentralized Multi-level Network

Consider an example of determining the optimal state of the simplest multi-level network (see Fig. 2) with a given vectors of input and output flows in a network nodes:

$$\begin{cases} S = \{30, 55, 107, 208\} \\ D = \{88, 49, 150, 113\} \end{cases}. \tag{29}$$

The optimal distribution of information flows, determined by the entropy method, is as following:

$$\begin{Vmatrix} 0 & 0 & 13.05 & 16.95 \\ 0 & 0 & 23.93 & 31.07 \\ 27.01 & 15.04 & 0 & 64.96 \\ 61.00 & 33.97 & 113.03 & 0 \end{Vmatrix}. \tag{30}$$

The values in (30) are given to two decimal places, the maximum entropy value is –1570.3, and it is determined over 10 iterations.

The maximum value of the entropy is possible to calculate via function $H(x_1, x_2, x_3)$ in formula (23) by equating to zero the first partial derivatives of this function for each of the independent variables:

$$\begin{cases} \frac{\partial H(x_1, x_2, x_3)}{\partial x_1} = 0 \\ \frac{\partial H(x_1, x_2, x_3)}{\partial x_2} = 0 \\ \frac{\partial H(x_1, x_2, x_3)}{\partial x_3} = 0 \end{cases}. \tag{31}$$

However, the analytical method becomes too cumbersome for larger dimensional matrices.

Interestingly, the value of entropy on the set of all matrices of information flows deviates slightly from the average value. Thus, for this example, the maximum (optimal) value of entropy is –1570.3, its minimum value is –1723.0, and the average value is –1602.6. Thus, the deviation of entropy from the mean value in the direction of decrease does not exceed 2%, and in the direction of the largest (optimal) value does not exceed 7.5%. This is very important for the study of non-equilibrium processes in telecommunications networks, but a detailed statistical analysis of the entropy range calculated for possible matrices of information flows requires fundamentally new computational methods or the use of high-performance computing systems. This also means that most network states are close to the optimal state.

Consider one more example of modelling a telecommunications system with three planes of decentralization (see Fig. 3).

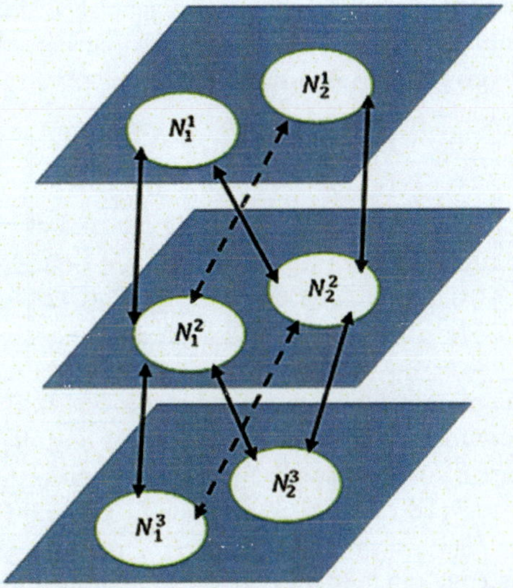

Fig. 3. The simplest network with three planes of decentralization.

Assume that the direct exchange of information flows is possible only between nodes of the third plane, and between all nodes located in adjacent planes of decentralization. Then the matrix of information flows is a sparse matrix, the elements in it are grouped on three planes, and it has the form:

$$\left\| \begin{matrix} & pl1 & & pl2 & & pl3 & \\ pl1 & 0 & 0 & \rho_{11}^{12} & \rho_{12}^{12} & 0 & 0 \\ & 0 & 0 & \rho_{21}^{12} & \rho_{22}^{12} & 0 & 0 \\ pl2 & \rho_{11}^{21} & \rho_{12}^{21} & 0 & 0 & \rho_{11}^{23} & \rho_{12}^{23} \\ & \rho_{21}^{21} & \rho_{22}^{21} & 0 & 0 & \rho_{21}^{23} & \rho_{22}^{23} \\ pl3 & 0 & 0 & \rho_{11}^{32} & \rho_{12}^{32} & 0 & \rho_{12}^{33} \\ & 0 & 0 & \rho_{21}^{32} & \rho_{22}^{32} & \rho_{21}^{33} & 0 \end{matrix} \right\|. \tag{32}$$

Let the given vectors of input and output flows in network nodes:

$$\begin{cases} S = \{40, 51, 51, 42, 32, 48\} \\ D = \{17, 25, 91, 32, 48, 51\} \end{cases}. \tag{33}$$

Calculated with an accuracy of two decimal places according to the algorithm for determining the maximum entropy, the elements of the matrix for the three-level network with flows in each node given by (33), are given below:

$$\left\| \begin{matrix} 0 & 0 & 29.59 & 10.41 & 0 & 0 \\ 0 & 0 & 37.73 & 13.27 & 0 & 0 \\ 9.29 & 13.67 & 0 & 0 & 11.51 & 16.53 \\ 7.65 & 11.25 & 0 & 0 & 9.48 & 13.62 \\ 0 & 0 & 8.33 & 2.93 & 0 & 20.74 \\ 0 & 0 & 15.57 & 5.47 & 26.96 & 0 \end{matrix} \right\|. \tag{34}$$

The maximum value of entropy for this network is –751.08.

The computational complexity of the problem of analysing all possible network states for this system is too great for a personal computer, because the number of independent variables increases to 7 (18 variables in the matrix (32), and 11 linearly independent equations), thereby $H = H(x_1, x_2, x_3, x_4, x_5, x_6, x_7)$. Therefore, various approximations should be used to estimate the range of the entropy. One of the most obvious approximations is the state of the network, in which as many node pairs as possible are not involved in the information exchange, and the corresponding elements of the matrix have zero values. Of course, this is rare in practice, but formal conditions such as the total flows in the nodes and the structure of the flow matrix are also fulfilled in this state. Based on these considerations, consider the example of a substantially non-equilibrium network with a maximum possible number of zero values, which a priori has a low entropy value:

$$\left\| \begin{matrix} 0 & 0 & 40 & 0 & 0 & 0 \\ 0 & 0 & 51 & 0 & 0 & 0 \\ 0 & 0 & 0 & 0 & 0 & 51 \\ 17 & 25 & 0 & 0 & 0 & 0 \\ 0 & 0 & 0 & 32 & 0 & 0 \\ 0 & 0 & 0 & 0 & 48 & 0 \end{matrix} \right\|. \tag{35}$$

The value of the entropy of the network in this state is –973.87 and can be assumed as its smallest value:

$$-973.87 \leq H(x_1, x_2, x_3, x_4, x_5, x_6, x_7) \leq -751.08. \tag{36}$$

Based on this estimate, it can be concluded that the entropy of the network can vary within no more than 29% of its optimal value.

Therefore, the entropy can serve as an aggregate quantitative measure of the deviation of the network from its equilibrium state. This allows you to predict the behaviour of the network in extreme conditions of connecting a large group of new subscribers, the emergence of new players in the market of telecommunications services, equipment upgrades, the introduction of new algorithms, etc.

5 Conclusions

The study of the possibility of applying the method of network optimization, which is based on determining the maximum value of the function of information flows in the network, which in appearance resembles entropy. By numerical simulation for the least complex topological elements of the network, it is established that the method is effective for networks in which information flows are exchanged between each pair of nodes. It matches the solutions obtained by the method of determining the maximum of the function with any predetermined accuracy, and requires a much smaller number of computational operations.

The chapter also investigates the possibility of applying the optimization method, based on the principle of maximum entropy, for a network with nodes between which there is no direct information exchange. It is shown that the introduction of zero values into the matrix of information flows leads to significant restrictions on the possible values of the elements of the matrix. It is also shown that the method of finding the optimal values of information flows based on the principle of maximum entropy can be applied to such networks as well.

Thus, the algorithms of fast finding of the optimal configuration of telecommunication networks based on the concept of entropy (entropy method), which can be applied to networks with several planes of decentralization, are considered in the work. In this case, information flows can be displayed in the form of a two-dimensional sparse matrix, where they are grouped accordingly. The application of the entropy model complements and expands existing methods, algorithms and models for building decentralized networks. An important advantage of the method is that it does not require the use of significant computing power compared to other methods of finding the optimal values of information flows in networks.

References

1. Cao, H., Lin, F., Li, Y., Wu, Y.: Information flow network of international exchange rates and influence of currencies. Entropy **23**(12), 1696 (2021)
2. Chen, T.-L., Chou, E.P., Fushing, H.: Categorical nature of major factor selection via information theoretic measurements. Entropy **23**(12), 1684 (2021)

3. Liu, Y., Zheng, W.-M., Liu, S., Chai, Q.-W.: Gaussian-based adaptive fish migration optimization applied to optimization localization error of mobile sensor networks. Entropy **24**(8), 1109 (2022)
4. Liu, G., et al.: A novel epidemic model base on pulse charging in wireless rechargeable sensor networks. Entropy **24**(2), 302 (2022)
5. Park, J.: Estimating non-gaussianity of a quantum state by measuring orthogonal quadratures. Entropy **24**(2), 289 (2022)
6. Bormashenko, E.: Magnetic entropic forces emerging in the system of elementary magnets exposed to the magnetic field. Entropy **24**(2), 299 (2022)
7. Klymash, M., Beshley, M., Luntovskyy, A. (eds.): Future Intent-Based Networking. LNEE, vol. 831. Springer, Cham (2022). https://doi.org/10.1007/978-3-030-92435-5
8. Politanskyi, R., Klymash, M.: Application of Artificial Intelligence in Cognitive Radio for Planning Distribution of Frequency Channel. In: 3rd International Conference on Advanced Information and Communications Technologies (AICT), pp. 390–394. IEEE, Lviv, Ukraine (2019)
9. Lavric, A., Petrariu, A.I., Coca, E., Popa, V.: Lora traffic generator based on software defined radio technology for lora modulation orthogonality analysis: Empirical and experimental evaluation. Sensors (Switzerland) **20**(15), 1–19, 4123 (2020)
10. Maksymyuk, T., Andrushchak, V., Dumych, S., Shubyn, B., Bugar, G., Gazda, J.: Blockchain based network functions virtualization for 5G network slicing. Acta Electrotechnica et Informatica **20**(4), 54–59 (2020)
11. Veryga, A., Politanskyi, R., Lesinskyi, V., Ruda, T.: Analysis of using of fractal signals for noise immune information transmission system. In: 15th International Conference on Advanced Trends in Radioelectronics, Telecommunications and Computer Engineering (TCSET), pp.162–165. IEEE, Lviv-Slavske, Ukraine (2020)
12. Politanskyi, R., Lesinsky, V., Kachur, V., Vistak, M.: Theoretical studies of the information capacity of complex signals generated by the method of their decomposition in two-dimensional space. In: 16th International Conference on Advanced Trends in Radioelectronics, Telecommunications and Computer Engineering (TCSET), pp. 162–165. IEEE, Lviv-Slavske, Ukraine (2022)
13. Jo, M., Maksymyuk, T., Batista R.L., Maciel, T,F., Almeida, A.L.F., Klymash, M.: A Survey of Converging Solutions for Heterogeneous Mobile Networks. IEEE Wireless Commun. Lett. **21**(8):54–62 (2014)
14. Luntovskyy, A.: Planning Paradigms for IoT Systems. In: 16th International Conference on Advanced Trends in Radioelectronics, Telecommunications and Computer Engineering (TCSET), pp. 162–165. IEEE, Lviv-Slavske, Ukraine (2022)
15. Bregman, L.: Proof of the convergence of Sheleikhovskii;s method for a problem with transportation constraints. Comput. Math. Math. Phys. **7**(1), 191–204 (1967)
16. Maksymyuk, T., et al.: Blockchain-empowered framework for decentralized network management in 6G. IEEE Commun. Mag. **58**(9), 86–92 (2020)

Elliptical Polarized Radio Wave Decay in Land Communication Links

Yehuda Ben-Shimol[1(✉)], Nathan Blaunstein[1], and Christos Christodoulou[2]

[1] Ben Gurion University of the Negev, Beer Sheva, Israel
benshimo@bgu.ac.il
[2] The University of New Mexico, Albuquerque, NM, USA
christos@unm.edu
https://in.bgu.ac.il/Pages/default.aspx

Abstract. The topic of depolarization in terrestrial wireless communication links has been poorly researched even for technologies beyond 4G. The main source of knowledge to this field are the Stokes parameters and various experimental results. Here we present an innovative analytical approach based on the multi-parametric stochastic model, as a combination of vision on radio propagation above built-up terrain and statistical description of main features of the terrain: buildings overlay profile, their density and characteristics, elevation of the transmitter and receiver antennas with respect to obstructions surrounding, etc. The proposed approach provides a sufficient way of predicting depolarization decay of radio waves and allows reducing polarization losses in both-sides terminal antennas, thus achieving cheaper data transmission inside wireless networks. This paper makes use of a wave's total intensity expression and derives probability distributions for parameters that describe the spatial polarization ellipse, energy distribution along and across directions of propagation, and changes in the shape and spatial orientation of the elliptically polarized wave depending of the terrain features. These equations are based on a new physical vision on the Stokes parameters and on the test results accounting for usually used real data of mixed-residential and sub-urban/urban radio communication links.

Keywords: Depolarization · Incoherent component of intensity · Stoke's parameters · Urban area

1 Introduction

Roughly speaking, *depolarization* is the deformation in the polarization ellipse due to either diffraction or reflection from obstacles, or propagation in space between buildings. In a land communication link full of obstacles, e.g. buildings, the absorbed signal might be polarized utterly differently, compared to its original form, as depicted in Fig. 1.

M. Klymash et al. (Eds.): TCSET 2022, LNEE 965, pp. 615–633, 2023.
https://doi.org/10.1007/978-3-031-24963-1_36

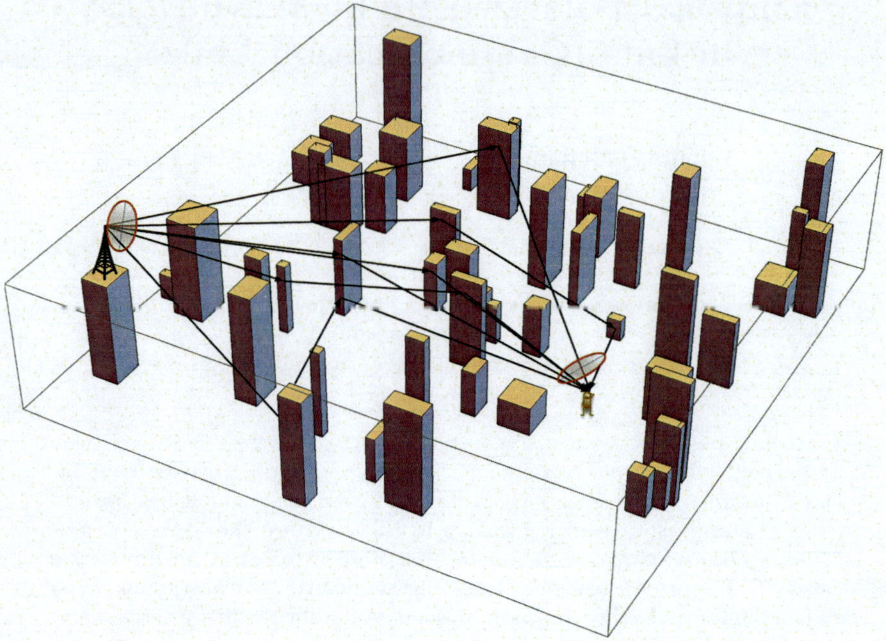

Fig. 1. A circularly polarized electromagnetic wave depolarizing after hitting objects between antennas

Modern cellular technologies (e.g., 5G and beyond) use multibeam antennas, for which, the distribution of energy in the space domain is an important issue. However, the effect of depolarization leads to energy losses of the arriving signals which result in losses of data. Therefore, depolarization becomes important for link budget design in such systems. Using circular or elliptically polarized antennas do not mitigate the problem of depolarization, especially in built-up environments.

Stoke's parameters were used in link budget design of 2G and 3G cellular networks. However, Stoke's based methods cannot be adapted to modern technologies, such as 5G and beyond. The goal of the present work is to present a careful analysis of depolarization to help with the design of link-budget in various built-up environments for modern technologies. To achieve this goal, we combine Stoke's parameters with the stochastic multi-parametric model [1,2]. The combined analysis allows the prediction of depolarization parameters with relation to the parameters of the built-up environment for the non-line-of-sight (NLOS, or incoherent) component of the received signal.

In more detail, four main characteristic variations of a diffracted/reflected wave may be discussed. The wave might variate either in the semi-axes lengths of the polarization ellipse, an angle of rotation, inclination or in azimuth angle [3–8,8–10].

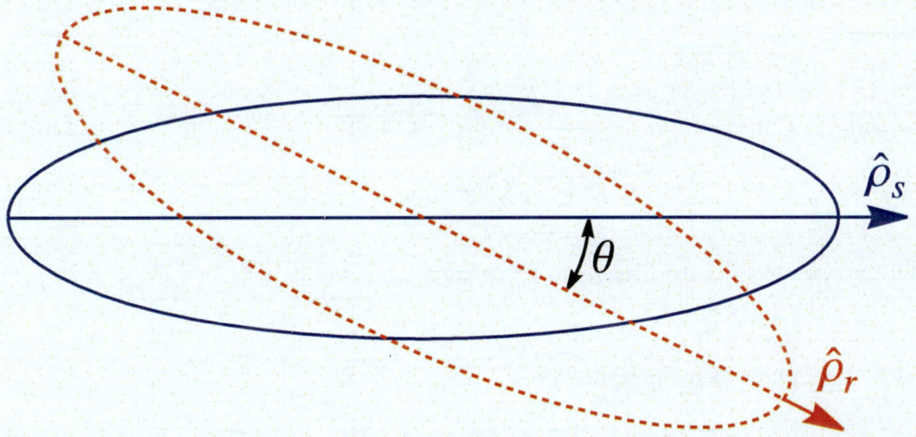

Fig. 2. Polarization unit vectors (blue - source, red - rotated) and the angle of mismatch (Color figure online).

Considering polarization losses, the most important parameter is the angle of rotation of the polarization ellipse, as shown in Fig. 2. Polarization loss is the mismatch between the orientations of the arriving wave and the receiving antenna [10].

This mismatch angle, the ratio between the polarization ellipse's axes and the inclination angle, can be calculated using the analytical expressions presented in this paper. Since no details of the orientation of a receiver antenna are provided, we assume the receiving antenna is perpendicular to the ground, and the angle of mismatch is denoted as β. The corresponding PDF (will be defined below) is more important characteristic of the polarization ellipse's spatial orientation and rotation in space, that is, the polarization mismatch. In our further discussions, we put the main attention on investigating the PDF of the inclination angle β. Other independent PDFs, presented also below, are related to the shape and dimensions of the polarized ellipse and describe changes in its shape and dimensions in the space domain.

1.1 Polarization Model by Stokes Parameters

Following [8,10] one can write an expression for the electric part of the electromagnetic wave in space, using the quadrature components $\mathbf{S}$ and $\mathbf{C}$ and the relations between them. The corresponding electric field in the orthogonal system of coordinates may be represented by the cosine and sine components:

$$\mathbf{E}(t) = \mathbf{S}\sin(\omega t) + \mathbf{C}\cdot\cos(\omega t) \tag{1}$$

where

$$\mathbf{S} = -\sum_{i=1}^{3} \mathbf{E}_i \sin(\varphi_i) \tag{2}$$

$$\mathbf{C} = \sum_{i=1}^{3} \mathbf{E}_i \cos(\varphi_i) \tag{3}$$

The normal vector in direction of propagation is

$$\mathbf{N} = \mathbf{C} \times \mathbf{S} \tag{4}$$

with the square of its magnitude

$$N^2 = C^2 S^2 - (\mathbf{C} \cdot \mathbf{S})^2 \tag{5}$$

Considering an elliptical polarization and using the Pauli spin matrices noted in [3–5] and expressions (2), (3), the Stokes parameters can be rewritten as:

$$I = \sum_{i=1}^{3} (C_i^2 + S_i^2) \tag{6}$$

$$Q = I - 2I_3 / \sin^2(\theta) \tag{7}$$

$$U = 2\left(N_1 \cos(\varphi) + N_2 \sin(\varphi)\right) / \sin(\theta) \tag{8}$$

$$V = 2\,|\mathbf{S} \times \mathbf{C}| = 2N \tag{9}$$

Here $I_i = C_i^2 + S_i^2$, $i = 1, 2, 3$ and θ and φ are the azimuthal and polar angles of a spherical coordinate system. Using expressions (6)–(9), one can introduce a joint probability density function of all six quadrature components of the polarization ellipse in a product form of a of six one-dimensional normal density probability functions. We assume that all the relevant parameters are normally distributed [5,8–10]:

$$\omega(\mathbf{C}, \mathbf{S}) = \prod_{i=1}^{3} \frac{1}{2\pi\sigma_i^2} \exp\left(-\frac{C_i^2 + S_i^2}{2\sigma_i^2}\right) \tag{10}$$

In addition, the polarization ellipse can be described by using its normal vector $\mathbf{N}$, and the polar angles $\varphi \in [0, 2\pi]$ and $\theta \in [0, \pi]$ are fixing the position of $\mathbf{N}$ with respect to the coordinate axes. The projections of $\mathbf{N}$, namely, $\mathbf{N}_1, \mathbf{N}_2, \mathbf{N}_3$, are related to the angles θ, φ by

$$\varphi = \tan^{-1}(N_2/N_1) \tag{11}$$

$$\theta = \tan^{-1}(N_3/N_1) \tag{12}$$

The ratio between the minor axis to the major axes of the polarization ellipse is described by the parameter R:

$$R = \frac{2N}{I + \sqrt{I^2 - 4N^2}} \tag{13}$$

A more commonly used form of (13) is the ratio

$$p = \frac{1 - R^2}{1 + R^2} \tag{14}$$

Finally, we can rewrite (10) as a function of the parameters I, p, θ, β and φ, using the absolute value of the determinant of the Jacobean matrix for the change of variables. This representation fully describes the spatial polarization ellipse:

$$\omega(I, p, \theta, \beta, \varphi) = \frac{I^2 p \sin(\theta)}{(2\pi)^3} \prod_{i=1}^{3} \frac{1}{2\sigma_i} \exp\left(-\frac{I_i}{2\sigma_i^2}\right) \tag{15}$$

Expression (15) is less applicable to the investigation of wireless terrestrial communication links, in the sense that it does not account for any channel characteristics. However, $\omega(I, p, \theta, \beta, \varphi)$ depends on the signals intensity I, which, in turn, can be described by the channel's characteristics.

In expression (15), σ_i^2 $(i = 1, 2, 3)$, describe the redistribution of energy of an elliptically polarized wave propagating in various environments. According to the 2-D geometry of propagation of elliptically polarized radio wave in built-up terrain environment, as shown in Fig. 3, we indicate $\sigma_1^2 = \sigma_\parallel^2$ and $\sigma_2^2 = \sigma_\perp^2$, directing the first one along the semi-axis of ellipse along wave propagation, and the second one – in the semi-axis normal to direction of wave propagation, following the geometry sketched in Fig. 4. Such notations give us a physically vivid picture on the energy distribution inside the polarization ellipse of radio wave propagating in various built-up environments. The rotation in space of the polarization ellipse provides information on the polarization mismatch and losses, occurring during the propagation of elliptically polarized waves in built-up areas. The relations between the parameters of the polarization ellipse and the properties of the built-up terrain are discussed in the following paragraph.

Subsequent paragraphs, however, are indented.

2 Modified Terrestrial Stochastic Multiparametric Model

Blaunstein et al. [1,2] present an expression for the wave's average total intensity as a function of channel's characteristics. This expression for the average total energy is further divided into two parts, denoted as the coherent part of wave intensity, I_{co}, and its incoherent part, I_{inc}. I_{co} is a deterministic value, that is less affected by the communication link, compared to the incoherent part.

$$\langle I_{co} \rangle = \exp\left(-\gamma_0 d \frac{\bar{h} - z_1}{z_2 - z_1}\right) \cdot \left(\frac{\sin(k z_1 z_2 / d)}{2\pi d}\right)^2 \tag{16}$$

$$\langle I_{inc} \rangle = \frac{\Gamma}{8\pi} \cdot \frac{\lambda \ell_h}{\lambda^2 + (2\pi \ell_h \bar{L} \gamma_0)^2} \cdot \frac{\lambda \ell_v}{\lambda^2 + \left(2\pi \ell_v \gamma_0 (\bar{h} - z_1)\right)^2} \cdot \frac{z_2 - \bar{h}}{d^3} \tag{17}$$

and the total intensity is

$$\langle I_{tot} \rangle = \langle I_{co} \rangle + \langle I_{inc} \rangle \tag{18}$$

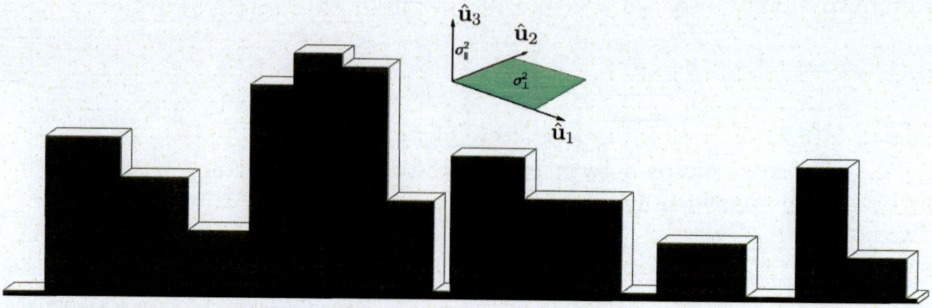

Fig. 3. Terrestrial link profile and the energy variances

The meaning of the various parameters in expressions (16)–(18) are given in Table 1 (together with values used for the numerical investigations).

The incoherent part can be further decomposed into two additional expressions, which describe the energy variances in two orthogonal planes: parallel ($\sigma_\parallel^2$) and perpendicular ($\sigma_\perp^2$) [11], that is,

$$\langle I_{inc} \rangle = \sigma_\perp^2 \cdot \sigma_\parallel^2 \tag{19}$$

Figure 3 illustrates a scenario for the decomposition of $\langle I_{inc} \rangle$ into parallel and perpendicular components with relation to a reference system of Cartesian coordinates. In the case presented in Fig. 3, $\sigma_\perp^2$ is the energy variance in the plane defined by the unit vectors $\hat{\mathbf{u}}_1$, $\hat{\mathbf{u}}_2$, and $\sigma_\parallel^2$ is the energy variance in the direction parallel to $\hat{\mathbf{u}}_3$. Figure 4 illustrates an additional explanation for the meaning of the energy variances in parallel and perpendicular to the wave's plane for an arbitrary Cartesian system of coordinates $(\hat{\mathbf{u}}_1, \hat{\mathbf{u}}_2, \hat{\mathbf{u}}_3)$, and the tip of the electrical field is rotating in a plane normal to the direction of propagation $\hat{\mathbf{n}} = \mathbf{N}/|\mathbf{N}|$ (the purple plane where the polarization ellipse is placed in Fig. 4).

Since the two energy variances are defined orthogonal, we can finally achieve their expressions by

$$\sigma_\parallel^2 = \frac{\lambda \ell_v}{\lambda^2 + \left(2\pi \ell_v \gamma_0 (\bar{h} - z_1)\right)^2} \frac{z_2 - \bar{h}}{d^2} \tag{20}$$

$$\sigma_\perp^2 = \frac{\Gamma}{8\pi} \cdot \frac{\lambda \ell_h}{\lambda^2 + (2\pi \ell_h \gamma_0 \bar{L})^2} \cdot \frac{1}{d} \tag{21}$$

For better consideration of the built-up terrain, we introduce a new function $P_h(z)$, a complementary cumulative distribution function that determines the probability of an event for which $z < \bar{h}$, and $\bar{h}$ is the average building height. Presenting the buildings height profile inside the city layer by [2,11,12]:

$$P_h(z) = H(h_1 - z) + H(z - h_1) H(h_2 - z) \left[\frac{h_2 - z}{h_2 - h_1} \right]^n \tag{22}$$

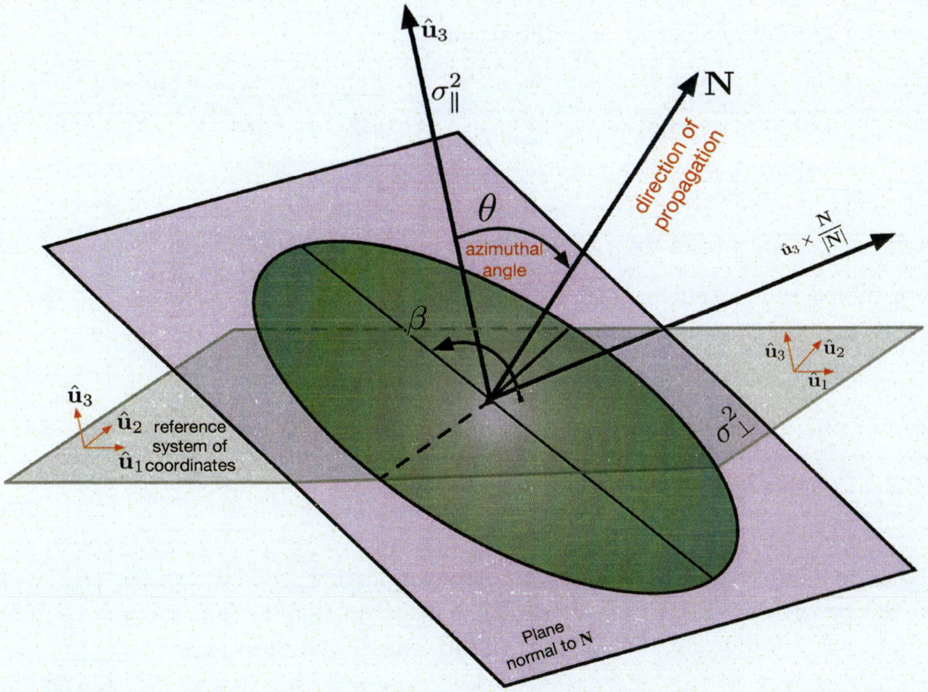

Fig. 4. Energy variances in parallel and perpendicular to the normal in direction of propagation.

Here h_1, h_2 are the minimum and maximum of building heights, respectively, $0 < z < h_2$ is the antenna height, and $n > 0$ is an indicator of the ratio between the number of low buildings to the number of high buildings. The function $H(\cdot)$ is the Heaviside unit step function.

Using (22), a function representing the built-up profile for the area between two antennas can be represented as

$$F(z_1, z_2) = \int_{z_1}^{z_2} P_h(z)dz \tag{23}$$

Expression (23), denoted in [2,11] as the "shadow function", represents the probability of a signal to reach a client that is present in the shadow of obstacles in a given terrain. Using (23), we can rewrite (20), (21) to achieve the average signal intensity in suburban areas for a single-ray scattering and diffraction:

$$\sigma_{\parallel}^2 = \frac{\lambda \ell_v}{\lambda^2 + (2\pi \ell_v \gamma_0 F(z_1, z_2))^2} \frac{\sqrt{\frac{\lambda d}{4\pi^3} + (z_2 - \bar{h})^2}}{d^2} \tag{24}$$

$$\sigma_{\perp}^2 = \frac{\Gamma \lambda}{8\pi \sqrt{\lambda^2 + \left(2\pi \ell_h \gamma_0 \bar{L}\right)^2}} \frac{1}{d} \tag{25}$$

and for two ray scattering and diffraction

$$\sigma_{\parallel}^2 = \frac{\Gamma\lambda\ell_v^2}{3\pi\left[\lambda^2 + (2\pi\ell_v\gamma_0 F(z_1,z_2))^2\right]^2}\frac{\frac{\lambda d}{4\pi^3} + (z_2 - \bar{h})^2}{d^2} \tag{26}$$

$$\sigma_{\perp}^2 = \frac{\Gamma\lambda}{8\pi\sqrt{\lambda^2 + (2\pi\ell_h\gamma_0\bar{L})^2}}\frac{1}{d} \tag{27}$$

For mixed residential terrains, in which most of the obstacles are about of the same height, we can simplify (24) and (25) to achieve:

$$\sigma_{\parallel}^2 = \frac{\lambda\ell_v}{\lambda^2 + (2\pi\ell_v\gamma_0(\bar{h} - z_1))^2}\frac{\sqrt{\frac{\lambda d}{4\pi^3} + (z_2 - \bar{h})^2}}{d^2} \tag{28}$$

$$\sigma_{\perp}^2 = \frac{\Gamma}{8\pi}\frac{\lambda\ell_h}{\lambda^2 + (2\pi\ell_h\gamma_0\bar{L})^2}\frac{1}{d} \tag{29}$$

To better understand how a channel affects the EM wave, we present the ratio of the energy variance in parallel to that in perpendicular directions, as shown from Figs. 3 and 4. For the single-ray scattering and diffraction:

$$\frac{\sigma_{\perp}^2}{\sigma_{\parallel}^2} = \frac{\Gamma}{8\pi\ell_v}\frac{\lambda^2 + (2\pi\ell_v\gamma_0 F(z_1,z_2))^2}{\sqrt{\lambda^2 + (2\pi\ell_h\gamma_0\bar{L})^2}}\frac{d}{\sqrt{\frac{\lambda d}{4\pi^3} + (z_2 - \bar{h})^2}} \tag{30}$$

For the two-ray scattering and diffraction:

$$\frac{\sigma_{\perp}^2}{\sigma_{\parallel}^2} = \frac{8}{3\ell_v^2}\frac{\left[\lambda^2 + (2\pi\ell_v\gamma_0 F(z_1,z_2))^2\right]^2}{\sqrt{\lambda^2 + (2\pi\ell_h\gamma_0\bar{L})^2}}\frac{d}{\frac{\lambda d}{4\pi^3} + (z_2 - \bar{h})^2} \tag{31}$$

Notice that every additional ray at the multipath is further attenuated in around 15 dB [1,2,11]. Thus, for practical use, it is enough to analyze up to two rays. For the mixed residential areas:

$$\frac{\sigma_{\perp}^2}{\sigma_{\parallel}^2} = \frac{\Gamma\ell_h}{8\pi\ell_v}\frac{\lambda^2 + (2\pi\ell_v\gamma_0(\bar{h} - z_1))^2}{\lambda^2 + (2\pi\ell_h\gamma_0\bar{L})^2}\frac{d}{\sqrt{\frac{\lambda d}{4\pi^3} + (z_2 - \bar{h})^2}} \tag{32}$$

Finally, from [9,10], we present the modified form of expression (15) as:

$$\omega(I,p,\theta,\beta,\varphi) = \frac{I^2 p\sin(\theta)}{8(2\pi)^2\sigma_{\parallel}^2\sigma_{\perp}^2}\cdot\exp\left[-\frac{I}{2}\left(\frac{1-\xi\sin^2(\theta)}{\sigma_{\perp}^2} - \frac{\xi\sin^2(\theta)}{\sigma_{\parallel}^2}\right)\right] \tag{33}$$

where $\xi = (1 - p\cos 2\beta)/2$, $I > 0$, $0 \leq p \leq 1$, $0 \leq \beta \leq \pi/2$, $0 \leq \varphi \leq 2\pi$, $0 \leq \theta \leq \pi/2$. Graphical illustration of β and θ is given in Fig. 4. Finally, after integrating (33) over each parameter, we achieve the one-dimensional probability density functions for the parameters that describe the polarization ellipse in space [12]:

$$\omega(\theta) = \frac{\Delta p \sin(\theta)}{2\left[\cos^2(\theta) + \Delta \sin^2(\theta)\right]^{3/2}} \tag{34}$$

$$\omega(\beta) = \Delta \frac{(1+\eta)^2(1+\eta_0)}{2\pi(\eta-\eta_0)^2}\left[(\eta - 3\eta_0)\sqrt{\eta}\tan^{-1}\sqrt{\eta} + 2\eta_0^{3/2}\tan^{-1}\sqrt{\eta_0} + \frac{\eta(\eta-\eta_0)}{1+\eta}\right] \tag{35}$$

$$\omega(R) = \frac{8G(1-R)^2}{(1+R^2)^3}\frac{1+\eta_0^2/2}{1-(\eta_0 p)^2} \tag{36}$$

where R was already defined by (15) and

$$\Delta = \sigma_{\perp}^2/\sigma_{\parallel}^2 \tag{37}$$

$$\eta_0 = \frac{1-\Delta}{1+\Delta} \tag{38}$$

$$\eta = \frac{\eta_0(1-\cos(2\beta))}{1+\cos(2\beta)} \tag{39}$$

$$G = \frac{2(1-\eta_0^2)\eta_0^2}{\eta_0^4 - 2(1-\eta_0^2)\ln(1-\eta_0^2)} \tag{40}$$

Note that expression (35) is the most relevant for the prediction of the polarization loss as explained earlier.

3 Numerical Analysis of Probability Functions and Polarization Ellipse Energetic Parameters for Various Built-Up Environments

Figure 5 illustrates variants of elevation of the BS and MS antennas with respect to the buildings' overlay profile. As will be clear in our analysis, situations of both terminal antennas change significantly the physical process of radio propagation in built-up communication link, and therefore, redistribution of energy inside the polarization ellipse. Usually, there are three possible situations, as illustrated in Fig. 5:

- Case (a) - Both antennas, receiver and transmitter, are placed above the rooftop level (red transmitter antenna).
- Case (b) - One of the antennas is higher than the rooftop level, and the second is lower than the rooftop level (blue transmitter antenna).
- Case (c) - Both antennas are below the rooftop level (magenta transmitter antenna).

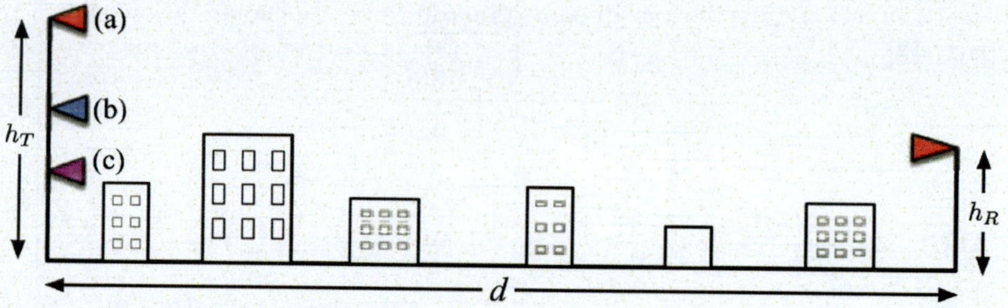

Fig. 5. Main scenarios of BS and MS antennas elevation in built-up areas: transmitter antenna is (a) above rooftops, (b) same level as roof tops and (c) below rooftops.

In line-of-sight (LOS) conditions (Fig. 5.a), the coherent part of the signal intensity prevails and the incoherent part becomes irrelevant, or too small. Its effects increase with the decrease in the base station antenna height (for limited lower elevated MS antenna) to the average buildings' height (Fig. 5.b).

In the second scenario (Fig. 5.b) only buildings surrounding the MS and BS antenna give sufficient impact on the incoherent component of the elliptically polarized wave, due to multipath stochastic phenomena of multiple scattering and multiple diffractions. Moreover, the diffractive deterministic component may exceed scattering (the randomly distributed component).

In the last case (Fig. 5.c), the multipath interference, due to multiple scattering from all obstructions lining the range between the BS and MS antennas, becomes predominant in the incoherent part of wave intensity, and the diffractive part effects are weaker. In this case of NLOS, the coherent component of wave intensity becomes insufficient. Considering these situations with BS and MS antennas allows us to understand and explain the further numerical analysis and results.

The relevant expressions for this analysis are (30)–(32) and (34)–(36). One needs first to find the relevant values for the energy variances $\sigma_{\parallel}^2$, $\sigma_{\perp}^2$, and then substitute the results into the desired probability functions and ratio between the energetic components of the elliptically polarized wave propagating in the built-up communication link.

It is especially difficult to achieve "representative" results due to the large number of different parameters and the vast amount of their combinations. The following simulations we will make use of the values given in Table 1, for the frequencies $f = 0.9, 1.8, 2.4, 5.2$ GHz. For the mixed residential zones, we will use the two-ray model described by (32), as it is more realistic compared to the one-ray model and for the suburban areas, we will use expression (31).

3.1 Mixed-Residential Areas

Here we analyze the operational parameters of the polarization ellipse for the case of mixed residential areas, with the parameters presented in Table 1. Figure 6

Table 1. Parameter values for mixed residential and suburban zones.

Parameter	Meaning	Mixed residential	Suburban
z_1 [m]	Receiver height	4	7
z_2 [m]	Transmitter height	35	45
d [km]	Distance	0.5–5	0.2–2
$h_{\min}$ [m]	Minimum buildings height	5	15
$h_{\max}$ [m]	Maximum buildings height	10	30
$\bar{L}$ [m]	Average length	20–40	50–60
ℓ_h	Horizontal coefficient	1	2
ℓ_v	Vertical coefficient	1	2
ν [km^{-2}]	Areal building density	30–40	50–60
γ_0 [km^{-1}]	Obstacles line density	0.4, 0.7, 1	1.6, 2, 2.3
Γ	Reflection coefficient	0.7 (stone)	0.2 (glass)
n	Obstacle height index	1, 10	0.1, 1, 20

shows the value of the ratio Δ (in logarithmic scale), defined in (37), as function of distance between the antennas, and the signal's frequency, with $\gamma_0 = 0.4\,\text{km}^{-1}$, $n = 1$ (i.e., equal number of low and high buildings) and the average building height is $\bar{h} = 7.5\,\text{m}$. The frequencies range from 900 MHz to 5.2 GHz. As is depicted in Fig. 6, near the transmitting antenna (for small ranges between terminal antennas) $\Delta \ll 1$, i.e., most of the energy of $\langle I_{inc}\rangle$ is gathered in $\sigma_{\|}^2$, while for large distances from the transmitter, this behavior is reversed. This behavior becomes less dominant as the carrier frequency increases. Following Fig. 6 (bottom) the increase in Δ is also less dominant as the line density of buildings increases and also for an increase in n (i.e., the ratio between the number of small to high buildings).

An interesting question is how Δ changes as a function of the antenna height. Using expression (31), the maximal Δ is achieved at $z_2 = (h_2 + h_1 n)/(1+n)$, i.e., the average buildings height according to (22). For the specific values $h_1 = 5\,\text{m}$, $h_2 = 10\,\text{m}$, $n = 1$, used for calculating Fig. 7, the maximal value of Δ is obtained at $z_2 = 7.5\,\text{m}$. Although, (31) is rather simple for determining the values of z_2 for which $\Delta > 1$ (or $\Delta < 1$), the large number of parameters involved leads to cumbersome expressions. Therefore, for the rest of this discussion refer only to the specific parameters used for the calculations of Fig. 7, which shows the depolarization phenomena occurring in the mixed residential areas for three values of γ_0 in two distances of 500 and 1000 m between the antennas. The height of the lower antenna is $z_1 = 4\,\text{m}$ (below the lowest building) and the height of the BS antenna, z_2, changes from 2 to 50 m. The results presented in Fig. 7 show that the maximal value of $\Delta > 1$ is achieved for $z_2 = 7.5\,\text{m}$ (i.e., antenna height is equal to the average buildings height). Raising z_2 above the average buildings height, decreases Δ, meaning that the wave energy in the parallel plane, with respect to direction $\mathbf{N} = \mathbf{k}$ of wave propagation, decreases comparing to that lying in the vertical plane to direction of wave propagation. This decrease is gradual, and the value of z_2 for which $\Delta = 1$ is frequency dependent, in general,

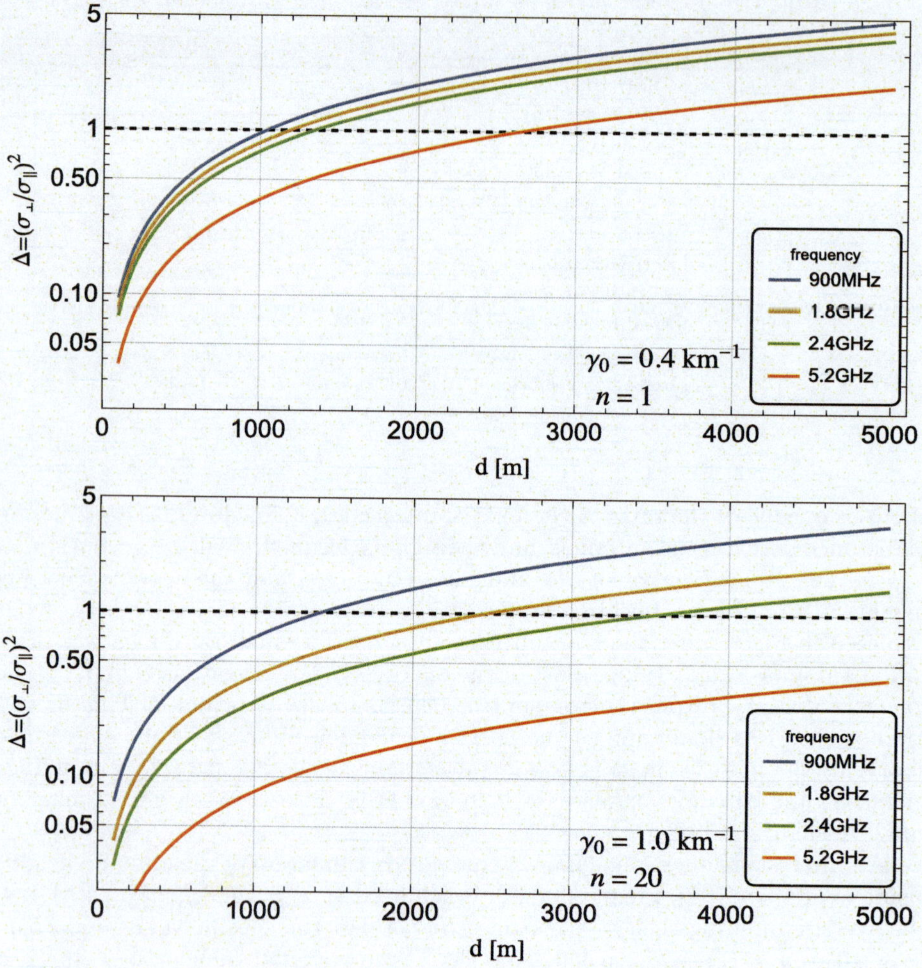

Fig. 6. The ratio $\Delta = \sigma_\perp^2/\sigma_\|^2$ as a function of distance and frequency in mixed residential area with $\gamma_0 = 0.4\,\text{km}^{-1}$ (top) and $\gamma_0 = 1\,\text{km}^{-1}$(bottom)

larger for smaller carrier frequencies. For example, for $\gamma_0 = 0.4$, $\Delta = 1$ for $z_2 = 20.7\,\text{m}$ and $12.7\,\text{m}$ for frequencies of 900 MHz and 5.2 GHz, respectively (see the top left subfigure of Fig. 7).

Increasing z_2 higher than the value for which $\Delta = 1$, Δ becomes less than one. Physically, this phenomenon can be explained as follows: the denominator of Δ is responsible for the distribution of the wave energy caused by multiple scattering from houses and other obstructions (hills, trees, etc.), whereas the denominator of Δ is responsible for distribution of the wave energy caused by multiple diffractions from tops of houses and other obstructions located in mixed-residential areas. So, we can state, following the presented illustrations, that

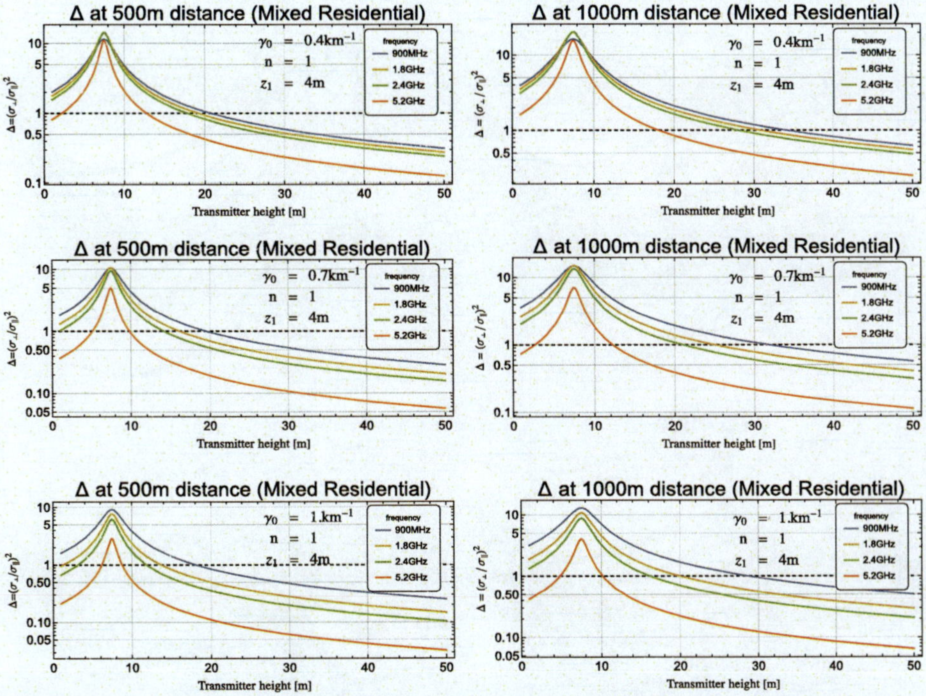

Fig. 7. Δ vs. z_2 in mixed residential areas, for three representative line densities of buildings. Left column, distance of 500 m between the antennas, right column distance of 1 km.

with increase of the heights of the transmitter antenna, in the first Fresnel zone the incoming rays arrive after multiple diffraction from houses tops, and this effect is prevailed under the effects of multiple scattering from houses walls and their roughness's features. But, when both antennas are low enough, comparing to houses heights, conversely, effects of multi-scattering prevail the effects of multiple diffraction, and Δ still exceeds one. We find important to emphasize this effect for mobile communication when both subscribers' antennas are lower than houses heights. With the increase in frequency from 0.9 GHz to 5.2 GHz, all the features mentioned above become vividly observed. The same tendency is observed with increase of distance between the two terminal antennas, from 500 m (the left column of Fig. 7) to 1 km (the right column of Fig. 7). Only the area surrounding the maximal value of Δ widens.

Next, using (35) and the parameters used for computations of Fig. 6, we calculate the normalized probability function, for the angle of inclination β (see illustration of β in Fig. 4) for the same four common carrier frequencies. For all carrier frequencies and for short distance from the transmitting antenna the inclination angle β is well centered around $\pi/2$. However, as the distance from the antenna increases, $w(\beta)$ widens and it gets more difficult to tell about the

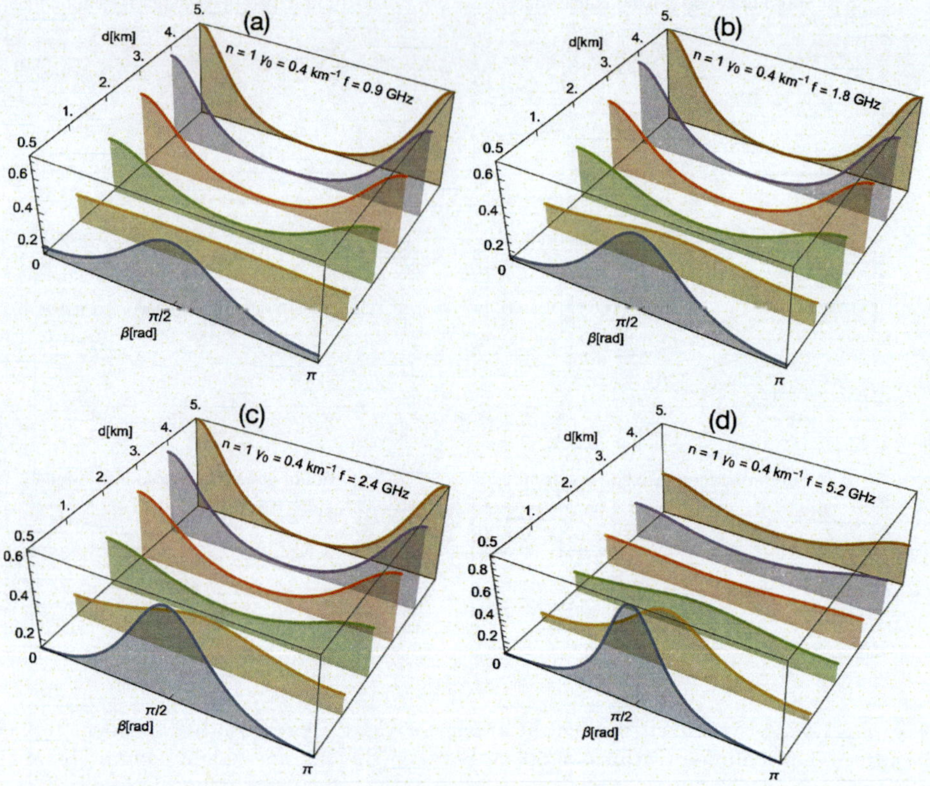

Fig. 8. The PDF $w(\beta)$ in mixed-residential areas for various distances with $n = 10$ (more low buildings), and $\gamma_0 = 0.4\,\text{km}^{-1}$. (a) $f = 900\,\text{MHz}$, (b) $f = 1.8\,\text{GHz}$, (c) $f = 2.4\,\text{GHz}$, (d) $f = 5.2\,\text{GHz}$.

probable value of the wave's polarization ellipse rotation angle β. Note that an ellipse is equal to itself if it is rotated 180°C, therefore, at far distances from the antenna it is more probable to find β around small values (again, notice that $w(\beta) = w(\pi - \beta)$). As the carrier frequency increases, the transition of $w(\beta)$ from a bell shape to more flat distribution (with the increase of distance) is more moderate as can be noticed when comparing Figs. 8(a) and (d).

3.2 Suburban and Urban Areas

Suburban/urban areas are identified by higher and larger buildings and denser building layout. Using (31) for two typical cases, Fig. 9 shows the behavior of Δ in this environment. The top figure refers to the case where the number of low building is approximately equal to the number of high buildings (i.e., $n = 1$)

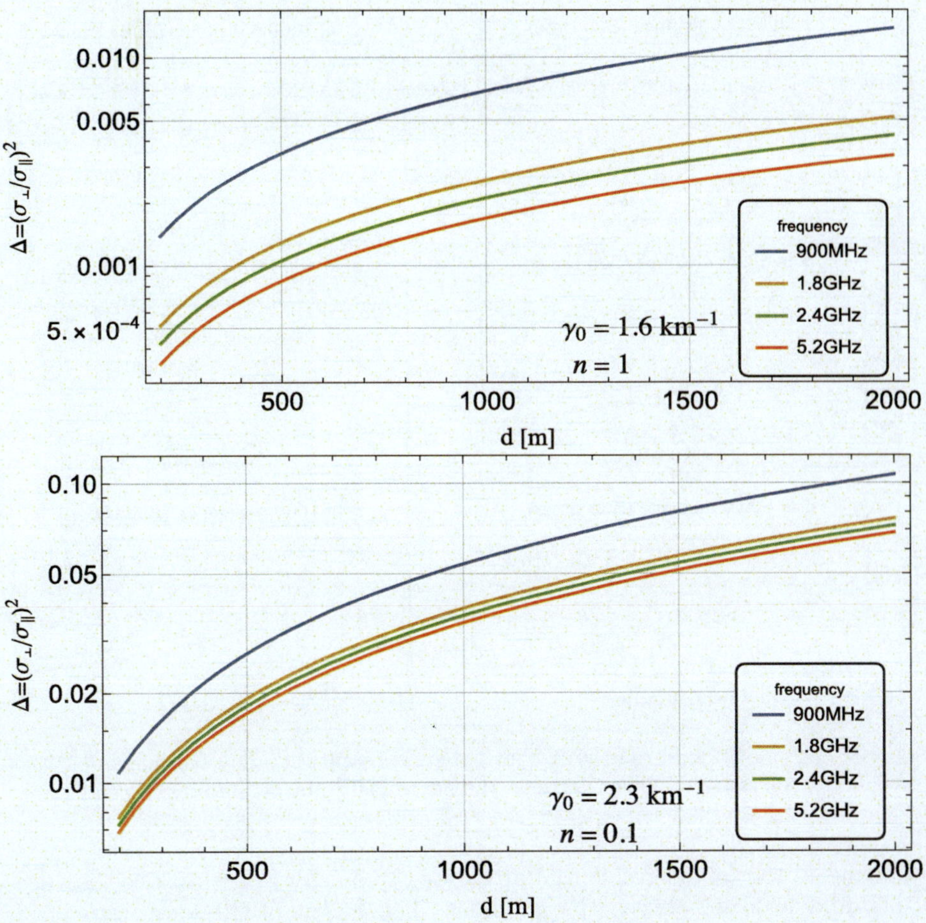

Fig. 9. The ratio $\Delta = \sigma_{\perp}^2/\sigma_{\parallel}^2$ as a function of distance and frequency in suburban area

and the line density of the buildings is relatively small. The bottom figure refers to a case where there are 10 times more high buildings ($n = 0.1$) and higher line density ($\gamma_0 = 2.3\,\text{km}^{-1}$). For a fixed line density Δ is higher (for the same distance) as n increases, as expected. Δ also increases with the increase in the distance. For fixed distance and n, Δ increases with the increase in line density (again, as expected).

Again, we consider the effect of the antenna heights on Δ. Using expression (32) the maximal Δ value is achieved at $z_2 = (h_2 + h_1 n)/(n + 1)$, i.e., the average buildings height according to (22). For the specific values $h_1 = 15\,\text{m}$, $h_2 = 30\,\text{m}$, $n = 1$, used for calculating Fig. 10, the maximal value of Δ is obtained at $z_2 = 22.5\,\text{m}$. For the rest of this discussion refer only to specific parameters used for the calculations of Fig. 10, which shows the depolarization phenomena

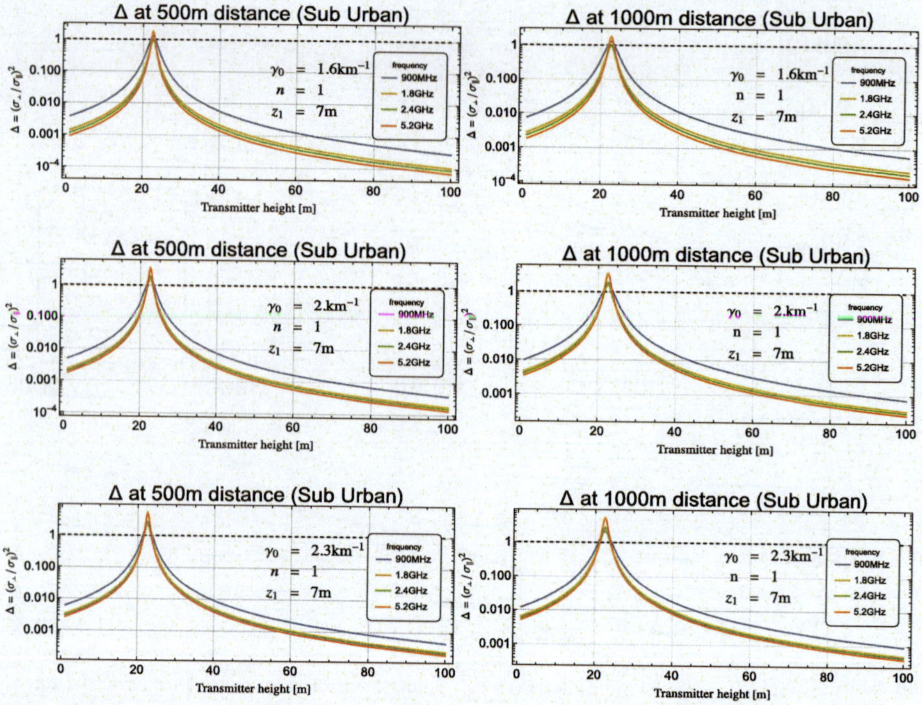

Fig. 10. Δ vs. z_2 in suburban areas, for three representative line buildings densities. Left column, distance of 500 m between the antennas, right column distance of 1 km.

occurring in the mixed residential areas for three values of γ_0 in two distances of 500 and 1000 m between the antennas. The height of the lower antenna is $z_1 = 7\,\text{m}$ (below the lowest building) and the height of the BS antenna, z_2, changes from 2 m to 100 m. The results presented in Fig. 10 show that:

Maximal value of Δ (only slightly above 1) is achieved for $z_2 = 22.5\,\text{m}$ (i.e., antenna height is equal to the average buildings height). Raising z_2 above the average buildings height decreases Δ, that is, the wave energy in the parallel plane, with respect to direction $\mathbf{N} = \mathbf{k}$ of the wave propagation, decreases comparing to that lying in the plane vertical to the direction of propagation.

In the range of $2\,\text{m} < z_2 < 22\,\text{m}$ and $z_2 > 23\,\text{m}$ the ratio Δ becomes less than one. Physically, this phenomenon can be explained by the following. In the ratio Δ, given in (32), the dominator is responsible for distribution of the wave energy caused by multiple scattering from buildings and their roughness (balcony, windows, etc.), whereas the denominator is responsible for distribution of the wave energy caused by multiple diffraction from rooftops of buildings. So, we can state, following the presented illustrations in Fig. 10, that with the increase of the heights of the transmitter antenna, in the first Fresnel zone, incoming rays are after multiple diffraction from roof tops, and this effect is

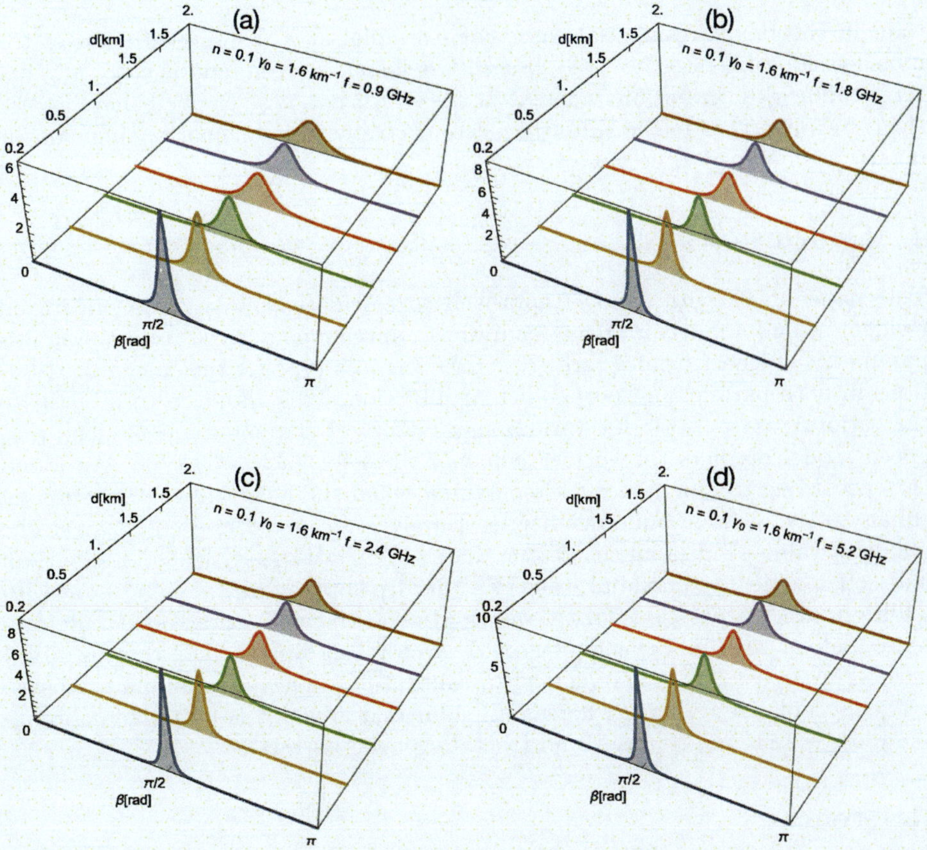

Fig. 11. The PDF $w(\beta)$ in suburban areas for various distances with $n = 1$ (equal number of low and high buildings), and $\gamma_0 = 1.6\,\mathrm{km}^{-1}$. (a) $f = 900\,\mathrm{MHz}$, (b) $f = 1.8\,\mathrm{GHz}$, (c) $f = 2.4\,\mathrm{GHz}$, (d) $f = 5.2\,\mathrm{GHz}$.

prevailed under effects of multiple scattering from buildings' walls and their roughness. The same tendency is observed, when both antennas are low enough (i.e., $z_2 < 22\,\mathrm{m}$, $z_1 = 7\,\mathrm{m}$) comparing to buildings' heights, and Δ becomes less than one, an effect important to emphasize for mobile communication when both subscribers' antennas are lower comparing the buildings rooftop heights. With increasing carrier frequency from 0.9 GHz to 5.2 GHz, all above mentioned features become vividly observed. The same tendency is observed with increase of distance between the two terminal antennas, from 500 m (left column of Fig. 10) to 1 km (right column of Fig. 10). Again, for the suburban case, we calculate the normalized probability function $w(\beta)$, for the angle of inclination β (see Fig. 4) for the same four common carrier frequency values. Figure 11 shows that for all given frequencies, β is centered around $\pi/2$. Although there is a broadening of the distribution $w(\beta)$ and the distance increases, this effect is much smaller than in the mixed-residential case. For example, at a distance of 2 km in the mixed-residential case the $w(\beta)$ losses its bell-shaped form and is close to a flat (i.e., uniform) distribution, whereas in the sub-urban case the bell-shape is still clear, meaning that in the sub-urban case we expect the inclination angle of the polarization ellipse to be close to $\pi/2$.

4 Conclusions

This paper deals with a novel approach to calculate depolarization in mixed residential and suburban areas. Analytical expressions derived from the multi-parametric stochastic model are given for each area type for the energy distributions in both parallel and perpendicular direction. This allows us to investigate the ratio $\Delta = (\sigma_{\perp}/\sigma_{\|})^2$ for various parameters of the environment. The proposed model predicts the correct behavior of Δ for each area type. We found that according to our model, Δ is maximal when the height of the BS antenna equals to the average buildings height. However, in mixed residential areas the maximal value of Δ is much greater than 1, i.e., $\sigma_{\perp}^2 > {\sigma_{\|}}^2$, while in suburban and urban areas the maximal value of Δ hardly approaches 1, i.e., $\sigma_{\perp}^2 \approx {\sigma_{\|}}^2$. In addition, exact expressions for the distribution of the inclination angle β are presented, along with its dependency on Δ. As a result we conclude that in mixed residential areas, it is recommended to calibrate the receiver antenna to absorb vertically polarized waves, whereas, in suburban areas, it is better to calibrate the receiver antenna to absorb horizontally polarized waves.

References

1. Blaunstein, N.: Prediction of cellular characteristics for various urban environments. IEEE Antennas Propag. Mag. **41**(6), 135–145 (1999)
2. Blaunstein, N., Katz, D., Censor, D., Freedman, A., Matityahu, I., Gur-Arie, I.: Prediction of loss characteristics in built-up areas with various buildings' overlay profiles. IEEE Antennas Propag. Mag. **43**(6), 181–191 (2001)
3. McMaster, W.H.: Polarization and the Stokes parameters. Am. J. Phys. **22**(6), 351–362 (1954)
4. McMaster, W.H.: Matrix representation of polarization. Rev. Mod. Phys. **33**(1), 8 (1961)
5. Morgan, M., Evans, V.: Synthesis and Analysis of Polarized Ellipses. Wiley (1961)
6. Zhang, J., Pan, C., Pei, F., Liu, G., Cheng, X.: Three-dimensional fading channel models: a survey of elevation angle research. IEEE Commun. Mag. **52**(6), 218–226 (2014). https://doi.org/10.1109/MCOM.2014.6829967
7. Bitler, Y.: A two-channel vertically spaced UHF diversity antenna system. In: Proceedings of Microwave Mobile Radio Symposium, pp. 13–15 (1973)
8. Burrows, C.R., Hunt, L.E., Decino, A.: Ultra-short-wave propagation: mobile urban transmission characteristics*. Bell Syst. Tech. J. **14**(2), 253–272 (1935). https://doi.org/10.1002/j.1538-7305.1935.tb00415.x
9. Huang, C.X., Zhang, J., Cheng, Q., Cui, T.J.: Polarization modulation for wireless communications based on metasurfaces. Adv. Func. Mater. **31**(36), 2103379 (2021). https://doi.org/10.1002/adfm.202103379

10. Dai, H., Wang, X., Xie, H., Xiao, S., Luo, J.: Spatial Polarization Characteristics of Radar Antenna. Analysis, Measurement and Anti-jamming Application. Springer, Singapore (2019). https://doi.org/10.1007/978-981-10-8794-3
11. Blaunstein, N., Christodoulou, C.G.: Radio Propagation and Adaptive Antennas for Wireless Communication Networks: Terrestrial, Atmospheric, and Ionospheric, 2nd edn. Wiley (2014)
12. Ben-Shimol, Y., Blaunstein, N., Christodoulou, C.: Depolarization effects of radio wave propagation in various land built-up environments. In: 2014 IEEE Antennas and Propagation Society International Symposium (APSURSI), pp. 955–956. IEEE (2014)

Fault Identification in Linear Dynamic Systems by the Method of Locally Optimal Separate Estimation

Andrii Volovyk[1(✉)], Vasyl Kychak[1], Alexander Osadchuk[1], and Bohdan Zhurakovskyi[2]

[1] Vinnytsia National Technical University, Khmelnytske Shose Street 95, 21021 Vinnytsia, Ukraine
voland@vntu.edu.ua

[2] National Technical University of Ukraine "Igor Sikorsky Kyiv Polytechnic Institute", Peremohy Avenue 37, Kyiv 03056, Ukraine

Abstract. In this article, the synthesis of two-stage structure recursive filter is carried out. The filter is resistant to unknown disturbances and is able to simultaneously evaluate both the state of the system and the malfunction has arisen in it. The presented synthesis method is based on the assumption that there is no a priori information about the dynamics of operating faults and disturbances. The provided synthesis method is based on assumption of the lack prior data about fault dynamics and perturbations. In this case, it is assumed that faults affect both the state of the system and the output variables, and disturbances affect only the state variables. Synthesis of the recursive filter was made in two options. In the first case, it was assumed that the distribution matrix of faults in the observation channel is a full rank matrix, then the result obtained was generalized to the case of an arbitrary rank matrix. The proposed filters are locally optimal in the sense of forming separate unbiased fault estimates and states with minimal variance. For these purposes, a matrix version of the Lagrange multiplier method was used. The operability of the considered method was tested on the example of an aircraft landing system. A comparative analysis of the qualitative variables of the proposed filter with the corresponding variables of the standard Kalman filter has been carried out. Solving this task is very important to improving the stability of operation and reliability of different communication equipment in the emerging networks, both local and global, both fixed and wireless.

Keywords: Decomposition · Kalman filter · Separate estimation · Lagrange multiplier method · Two-cascade estimation algorithm

1 Introduction

The modern state of telecommunication technologies can be characterized by the rapid acceleration of the network convergence processes, which represents mutual penetration of networks for different purposes based on the use of common components and the

M. Klymash et al. (Eds.): TCSET 2022, LNEE 965, pp. 634–651, 2023.
https://doi.org/10.1007/978-3-031-24963-1_37

unification of functionality. The result of such convergence is the formation of overly complex physical systems. In turn, the development of telecommunications has reached a level at which qualitative changes in approaches to the construction of telecommunication networks must inevitably begin, while the quality of the functioning of networks, their reliability and safety always stay on the cutting edge of technologies.

In practice, there are often cases when the dynamics of complex physical systems will experience sudden changes. This, as a rule, leads to a deterioration in their quality indicators. In general, such changes can be classified as malfunctions or failures. Faults manifest themselves in the form of deviations from the regulated norms, and failures - in the form of an abnormal development of the process, due to a sharp change in the parameters of the system or its structure. In this regard, the system may become unable to perform the task assigned to it. To maintain the stable functionality of the system, the basic provisions of the fault-tolerant control theory are often used, based on a very simple idea - compensation for the influence of faults due to hard-ware-architectural or functional redundancy [1]. According to the well-known distribution theorem, strictly proven only for linear systems, the general problem of fault-tolerant control can be divided into two relatively independently solved problems - the problem of the state system estimation and the control problem using the obtained estimation results. In this work, attention is focused exclusively on the estimation problem associated with the detection of faults and their recognition by applying appropriate model-based methods. Recently, several basic approaches to solving this problem have been developed, for example, the method of parity relations, the detecting filters method, the method of observers with uncertain inputs, the generalized likelihood ratio, etc. Their characteristic features were considered in a number of review papers [2–7]. Several monographs and manuals have been published devoted to certain aspects of fault diagnostics in linear dynamic systems [1, 7, 8], and applied aspects were partially considered in works [7, 9]. To date, de facto, two concepts have already been formed regarding the solution of problems of estimating the state of linear systems in the presence of faults and uncontrolled disturbances. The first one is based on the idea of expanding the state vector of a nominal system by introducing an additional unknown vector input into its mathematical model, associated with the influence of existing faults and disturbances. However, this concept provides for the presence of an a priori given model of an unknown input. The optimal solution of the estimation problem is guaranteed by the extended Kalman filter, and with a large number of faults and disturbances taken into account, the dimension of the extended Kalman filter becomes much larger than the dimension of the nominal system. In order to save computational resources, Friedland [10] suggested approximating the extended Kalman filter by a parallel autonomously operating cascade structure of lower dimension, which turned out to be only quasi-optimal in terms of equivalence to the extended Kalman filter outputs. The application of the basic provisions of work [10] to the stochastic type of faults and disturbances encountered difficulties of an exclusively practical nature [11–13]. Today, the main efforts are aimed at finding methods for approximating the extended Kalman filter, which combine acceptable accuracy with fairly simple (for practical applications) restrictions [14, 15]. For example, for this purpose, an adaptive version of the two-stage Friedland filter was developed in [16] and its stability was analyzed [17].

An alternative concept is based on the assumption of complete absence of a priori information regarding the dynamic properties of unknown inputs. Kitanidis [18] was the first to start solving this problem in order to obtain linear unbiased estimators with minimal generalized variance by introducing constraints that were imposed on the structure of the system under study. In [20], the authors generalized the results of [18], applying a parametric approach to obtain optimal estimates. A little later, in [21], an optimal filter with a minimum dispersion was obtained, which took into account the problem of degradation of the accuracy characteristics inherent in Kitanidis filters. In [22], the problem of detecting faults and their localization was solved using a geometric approach, while generating difference signals with directional properties.

In the present work, methods of locally optimal separate filtering solve the problem of fault diagnostics in linear dynamic systems in the presence of perturbations of an indefinite structure. The study is based on the main provisions of [18] in order to reduce the restrictions that contribute to the divergence of the filtration process.

2 Problem Statement

Let us assume that the model of a system with accompanying faults and disturbances can be described by a system of equations:

$$s(k+1) = \boldsymbol{\Sigma}(k+1,k)\boldsymbol{s}(k) + \boldsymbol{\Gamma}(k+1,k)\boldsymbol{u}(k) + \boldsymbol{\Psi}(k+1,k)\boldsymbol{f}(k) + \boldsymbol{\Omega}(k+1,k)\boldsymbol{d}(k) + \boldsymbol{w}(k) \tag{1}$$

$$\boldsymbol{y}(k) = \boldsymbol{H}(k)\boldsymbol{s}(k) + \boldsymbol{F}(k)\boldsymbol{f}(k) + \boldsymbol{v}(k) \tag{2}$$

where $\boldsymbol{s}(k) \in \mathcal{R}^n$– is the system state vector, $\boldsymbol{y}(k) \in \mathcal{R}^m$– is the observation vector, $\boldsymbol{u}(k) \in \mathcal{R}^r$– is the precisely known input action vector, $\boldsymbol{f}(k) \in \mathcal{R}^p$– is the additively acting fault vector, and $\boldsymbol{d}(k) \in \mathcal{R}^q$ – is the perturbation vector. White noise sequences $\boldsymbol{w}(k)$, $\boldsymbol{v}(k)$ are of Gaussian White noise, not correlated, have zero means and given covariance matrices, $\boldsymbol{Q}(k) \geq 0$, $\boldsymbol{R}(k) \geq 0$, respectively. A priori, it is assumed that the structure of the perturbation vector $\boldsymbol{d}(k)$ is indefinite, and may not even have a probabilistic description, and for it to be isolated from the state system vector, a decomposition procedure must be performed, which requires a number of additional restrictions, namely:

a) $[\boldsymbol{H}(k),\ \boldsymbol{\Sigma}(k+1,k)]$ – renewable pair; b) $n > m \geq p+q$; c) $rank[\boldsymbol{F}(k)] = p$;

$$\text{d) } rank[\boldsymbol{H}(k)\boldsymbol{\Omega}(k,k-1)] = rank[\boldsymbol{\Omega}(k,k-1)] = q. \tag{3}$$

He assumes that the system matrices $\boldsymbol{\Sigma}(k+1,k)$, $\boldsymbol{\Gamma}(k+1,k)$, $\boldsymbol{\Psi}(k+1,k)$, $\boldsymbol{\Omega}(k+1,k)$, $\boldsymbol{H}(k)$ and $\boldsymbol{F}(k)$ are known and have the corresponding dimensions. The initial system state $\boldsymbol{s}(0)$ is a Gaussian random variable, which is not correlated with noise processes $\boldsymbol{w}(k)$, $\boldsymbol{v}(k)$, has a given vector of average values $\mathcal{M}\{\boldsymbol{s}(0)\} = \bar{s}_0$ and the covariance matrix $\mathcal{M}\{[\boldsymbol{s}(0) - \bar{s}_0]\}[\boldsymbol{s}(0) - \bar{s}_0]^T = \boldsymbol{P}_s(0)$. Here the symbol $\mathcal{M}\{\cdot\}$ will mean mathematical expectation. Taking into account the Eq. (1)–(2), it is easy to see that faults simultaneously affect both the system state and the output variables, and the perturbation affects only the system state. Taking into account the type of the introduced restrictions and relying on the fundamental type results of [18, 19], it can be argued that the conditions for dividing the resulting filter into two components are satisfied. The first of them evaluates the system state vector, while ignoring the influence of disturbances, and the second, for the same conditions, estimates only mal-functions. For a given split filter structure, the following standard relations are valid [19]:

$$\hat{s}\left({}^{k}/_{k}\right) = \hat{s}\left({}^{k}/_{k-1}\right) + K_s(k)\left[y(k) - H(k)\hat{s}\left({}^{k}/_{k-1}\right)\right] \tag{4}$$

$$\hat{s}\left({}^{k}/_{k}\right) = \hat{s}\left({}^{k}/_{k-1}\right) + K_s(k)\left[y(k) - H(k)\hat{s}\left({}^{k}/_{k-1}\right)\right] \tag{5}$$

$$\hat{s}\left({}^{k}/_{k}\right) = \hat{s}\left({}^{k}/_{k-1}\right) + K_s(k)\left[y(k) - H(k)\hat{s}\left({}^{k}/_{k-1}\right)\right] \tag{6}$$

where (6) describes the unifying extrapolator function. Thus, the task of synthesis is reduced to the choice of such values of the transmission matrices of separated filters $K_s(k) \in \mathcal{R}^{n\times m}$ and $K_f(k) \in \mathcal{R}^{p\times m}$, that would meet the selected quality criteria:

– absence of shifts in estimates of state vector $\hat{s}\left({}^{k}/_{k}\right)$ and faults $\hat{f}\left({}^{k}/_{k}\right)$;

$$\mathcal{M}\{\tilde{s}(k)\} \triangleq \mathcal{M}\left\{s(k) - \hat{s}\left({}^{k}/_{k}\right)\right\} = 0;\ \mathcal{M}\{\tilde{f}(k)\} \triangleq \mathcal{M}\left\{\left[f(k) - \hat{f}\left({}^{k}/_{k}\right)\right]\right\} = 0 \tag{7}$$

– minimal mean squared errors in estimates $\hat{f}\left({}^{k}/_{k}\right)$;

– minimal trace of the error covariance matrix in the estimates of the state vector $\hat{s}\left({}^{k}/_{k}\right)$, subject to the validity of restrictions (7), that is

$$\min\ tr\mathcal{M}\left\{\tilde{s}(k)\tilde{s}^T(k)\right\} = \min\ tr\left[P_s\left({}^{k}/_{k}\right)\right]. \tag{8}$$

3 Synthesis of Locally Optimal Filters of Separate Estimation

Let us apply the filter (4)–(6) to systems t (1)–(2) and determine the residual of observations $r(k) \triangleq y(k) - H(k)\hat{s}\left({}^{k}/_{k-1}\right)$ subject to constraint (3 c). It is easy to show that the difference $r(k)$ can be reduced to the form:

$$r(k) \triangleq y(k) - H(k)\hat{s}\left({}^{k}/_{k-1}\right) = F(k)f(k) + H(k)\Psi(k, k-1)d(k-1) + e(k); \tag{9}$$

where

$$e(k) = H(k)\tilde{s}^*\left({}^{k}/_{k-1}\right) + v(k); \tag{10}$$

$$\tilde{s}^*\left({}^{k}/_{k-1}\right) = \Sigma(k, k-1)\tilde{s}(k-1) + \Psi(k, k-1)\tilde{f}(k-1) + w(k-1), \tag{11}$$

and expressions for estimation errors should be presented as:

$$\tilde{f}(k) \triangleq \left[f(k) - \hat{f}\left({}^{k}/_{k}\right)\right] = \left[I - K_f(k)F(k)\right]f(k) - K_f(k)H(k)\Omega(k, k-1)d(k-1) - K_f(k)e(k); \tag{12}$$

$$\tilde{s}(k) \triangleq \left[s(k) - \hat{s}\left(^{k}/_{k}\right)\right] = [\boldsymbol{I} - \boldsymbol{K}_s(k)\boldsymbol{H}(k)]\tilde{\boldsymbol{s}}^*\left(^{k}/_{k-1}\right) - \boldsymbol{K}_s(k)\boldsymbol{F}(k)\boldsymbol{f}(k) - \\ - [\boldsymbol{K}_s(k)\boldsymbol{H}(k)\boldsymbol{\Omega}(k, k-1) - \boldsymbol{\Omega}(k, k-1)]\boldsymbol{d}(k-1) - \boldsymbol{K}_s(k)\boldsymbol{v}(k). \tag{13}$$

First, consider the constraints that must be met in order to obtain unbiased estimates of the vectors $\hat{s}\left(^{k}/_{k}\right)$ and $\hat{\boldsymbol{f}}\left(^{k}/_{k}\right)$ by criterion (7), and the estimation process itself will be considered in the text below. In accordance with the above requirements, the following algebraic restrictions must be imposed on the matrices $\boldsymbol{K}_s(k)$ and $\boldsymbol{K}_f(k)$:

$$\boldsymbol{K}_f(k)\boldsymbol{G}(k) = \boldsymbol{\Phi}(k), \ \boldsymbol{K}_s(k)\boldsymbol{G}(k) = \boldsymbol{Z}(k), \tag{14}$$

where $\boldsymbol{G}(k) = [\boldsymbol{F}(k) \ \ \boldsymbol{H}(k)\boldsymbol{\Omega}(k, k-1)]$; $\boldsymbol{\Phi}(k) = \left[\boldsymbol{I}_p \ \ \boldsymbol{0}\right]$; $\boldsymbol{Z}(k) = [\boldsymbol{0} \ \ \boldsymbol{\Omega}(k, k-1)]$. (15)

This is easy to verify if you perform the following steps. For example, we restrict ourselves to Eq. (12). In order to avoid shifts, in (12), we equate to zero the expression: $\left[\boldsymbol{I} - \boldsymbol{K}_f(k)\boldsymbol{F}(k)\right]\boldsymbol{f}(k) - \boldsymbol{K}_f(k)\boldsymbol{H}(k)\boldsymbol{\Omega}(k, k-1)\boldsymbol{d}(k-1) = \boldsymbol{0}.$

A series of further, simple calculations gives the desired result:

$$\boldsymbol{K}_f(k)\boldsymbol{F}(k)\boldsymbol{f}(k) + \boldsymbol{K}_f(k)\boldsymbol{H}(k)\boldsymbol{\Omega}(k, k-1)\boldsymbol{d}(k-1) = \boldsymbol{I}_p\boldsymbol{f}(k);$$

$$\left[\boldsymbol{K}_f(k)\boldsymbol{F}(k) \ \ \boldsymbol{K}_f(k)\boldsymbol{H}(k)\boldsymbol{\Omega}(k, k-1)\right]\begin{bmatrix}\boldsymbol{f}(k)\\ \boldsymbol{d}(k-1)\end{bmatrix} = \left[\boldsymbol{I}_p \ \ \boldsymbol{0}\right]\begin{bmatrix}\boldsymbol{f}(k)\\ \boldsymbol{d}(k-1)\end{bmatrix};$$

$$\boldsymbol{K}_f(k)[\boldsymbol{F}(k) \ \ \boldsymbol{H}(k)\boldsymbol{\Omega}(k, k-1)]\begin{bmatrix}\boldsymbol{f}(k)\\ \boldsymbol{d}(k-1)\end{bmatrix} = \left[\boldsymbol{I}_p \ \ \boldsymbol{0}\right]\begin{bmatrix}\boldsymbol{f}(k)\\ \boldsymbol{d}(k-1)\end{bmatrix};$$

$$\boldsymbol{G}(k) \triangleq [\boldsymbol{F}(k) \ \ \boldsymbol{H}(k)\boldsymbol{\Omega}(k, k-1)]; \quad \boldsymbol{\Phi}(k) \triangleq \left[\boldsymbol{I}_p \ \ \boldsymbol{0}\right].$$

Similar actions using (13) allow us to calculate the matrix $\boldsymbol{Z}(k) = [\boldsymbol{0} \ \ \boldsymbol{\Omega}(k, k-1)]$.

If assumptions b, d from (3) are valid, and the matrix is a full rank matrix, then a necessary and sufficient condition for the existence of unbiased estimates and is the requirement that the matrix be a full rank matrix by the columns, that is:

$$rank[\boldsymbol{G}(k)] = rank[\boldsymbol{F}(k) \quad \boldsymbol{H}(k)\boldsymbol{\Omega}(k, k-1)] = p + q. \tag{16}$$

To confirm this, we first rewrite Eq. (14) in the vector form $\begin{bmatrix}\boldsymbol{K}_f(k)\\ \boldsymbol{K}_s(k)\end{bmatrix}\boldsymbol{G}(k) = \begin{bmatrix}\boldsymbol{\Phi}(k)\\ \boldsymbol{Z}(k)\end{bmatrix}$. This inhomogeneous matrix equation has a single solution with respect to $\begin{bmatrix}\boldsymbol{K}_f(k)\\ \boldsymbol{K}_s(k)\end{bmatrix}$, if the augmented matrix (due to the inclusion of the right hand side) has the rank of the matrix $\boldsymbol{G}(k)$ [20], that is $rank\begin{bmatrix}\boldsymbol{\Phi}(k)\\ \boldsymbol{Z}(k)\\ \boldsymbol{G}(k)\end{bmatrix} = rank[\boldsymbol{G}(k)].$

Further, we will expand the matrices $\begin{bmatrix}\boldsymbol{\Phi}(k)\\ \boldsymbol{Z}(k)\end{bmatrix}$ in accordance to the expression (15):

$$rank\begin{bmatrix}\boldsymbol{I}_p & \boldsymbol{0}\\ \boldsymbol{0} & \boldsymbol{\Omega}(k, k-1)\\ \boldsymbol{F}(k) & \boldsymbol{H}(k)\boldsymbol{\Omega}(k, k-1)\end{bmatrix} = rank[\boldsymbol{F}(k) \quad \boldsymbol{H}(k)\boldsymbol{\Omega}(k, k-1)]. \tag{17}$$

The matrix that is on the left side of the equals sign has rank $p + q$. This is easy to check using assumptions (3 b, d), in addition, it is assumed that the rank of the matrix is $\boldsymbol{F}(k)$ equal to p. Since faults and disturbances are independent processes, the requirement (16) remains the condition for the absence of shifts in the estimates $\widehat{\boldsymbol{f}}\left({}^{k}/_{k}\right)$ and $\hat{\boldsymbol{s}}\left({}^{k}/_{k}\right)$.

The Optimal Faults Estimation. Let's perform the transformation in (9) bringing it to the form:

$$\boldsymbol{r}(k) = \boldsymbol{G}(k)\begin{bmatrix} \boldsymbol{f}(k) \\ \boldsymbol{d}(k-1) \end{bmatrix} + \boldsymbol{e}(k) \tag{18}$$

where $\boldsymbol{e}(k)$ in accordance with (10) denoted as $\boldsymbol{e}(k) = \boldsymbol{H}(k)\widetilde{\boldsymbol{s}}^{*}\left({}^{k}/_{k-1}\right) + \boldsymbol{v}(k)$. It is easy to see that the process $\boldsymbol{e}(k)$ doesn't belong to random processes with unit variance. For this reason, it doesn't meet the conditions of the Markov theorem and normality inherent in the innovation process [24], and the use of the least squares method will not guarantee the minimum value of the estimation errors dispersion. However, the extrapolation error covariance matrix has the form

$$\boldsymbol{W}(k) = \mathcal{M}\left\{\boldsymbol{e}(k)\boldsymbol{e}^{T}(k)\right\} = \boldsymbol{H}(k)\overline{\boldsymbol{P}}_{s}\left({}^{k}/_{k-1}\right)\boldsymbol{H}^{T}(k) + \boldsymbol{R}(k), \tag{19}$$

where $\overline{P}_{S}\left({}^{k}/_{k-1}\right) \cong \mathcal{M}\left[\tilde{s}^{*}\left({}^{k}/_{k-1}\right)\tilde{s}^{*T}\left({}^{k}/_{k-1}\right)\right]$. Despite this, it seems possible to obtain the minimum value of the dispersion, if we use the weighted least squares method, while introducing the normalization matrix weight coefficient $\boldsymbol{W}^{-1}(k)$. To do this, it is necessary to assume that the extrapolation error estimate $\widetilde{\boldsymbol{s}}^{*}\left({}^{k}/_{k-1}\right)$ is unbiased, the matrix $\boldsymbol{G}(k)$ has a full rank by columns, and $\boldsymbol{W}(k)$ is positively denoted, then an unbiased fault estimate with minimum variance is guaranteed by the transfer matrix $\boldsymbol{K}_{f}^{*}(k)$ that is given by the expression

$$\boldsymbol{K}_{f}^{*}(k) = \boldsymbol{\Phi}(k)\boldsymbol{G}^{*}(k), \quad \text{where } \boldsymbol{G}^{*}(\mathrm{k}) = \left[\boldsymbol{G}^{\mathrm{T}}(\mathrm{k})\boldsymbol{W}^{-1}(\mathrm{k})\boldsymbol{G}(\mathrm{k})\right]^{-1}\boldsymbol{G}^{\mathrm{T}}(\mathrm{k})\boldsymbol{W}^{-1}(\mathrm{k}). \tag{20}$$

Moreover, if the matrix $\boldsymbol{W}(k)$ is positively denoted and has an inverse, then it is always possible to find such the matrix $\boldsymbol{\Theta}(k) \in \mathcal{R}^{m\times m}$ that is related to the matrix $\boldsymbol{W}(k)$ by the relation $\boldsymbol{\Theta}(k)\boldsymbol{\Theta}^{T}(k) = \boldsymbol{W}(k)$ [22]. Then formula (18) can be written as

$$\boldsymbol{\Theta}^{-1}(k)\boldsymbol{r}(k) = \boldsymbol{\Theta}^{-1}(k)\boldsymbol{G}(k)\begin{bmatrix} \boldsymbol{f}(k) \\ \boldsymbol{d}(k-1) \end{bmatrix} + \boldsymbol{\Theta}^{-1}(k)\boldsymbol{e}(k). \tag{21}$$

Since the matrix $\boldsymbol{G}(k)$ s the full column rank matrix (that is equal to $p + q$) by assumption, then the matrix $\boldsymbol{G}^{T}(k)\boldsymbol{\Theta}^{-1}(k)\boldsymbol{G}(k)$ can be inverse. For this reason, the solution obtained by the weighted least squares method for (8) is equivalent to the least squares solution for (21)

$$\boldsymbol{f}^{*}\left({}^{k}/_{k}\right) = \boldsymbol{\Phi}(k)\left[\boldsymbol{G}^{T}(k)\boldsymbol{W}^{-1}(k)\boldsymbol{G}(k)\right]^{-1}\boldsymbol{G}^{T}(k)\boldsymbol{W}^{-1}(k)\boldsymbol{r}(k). \tag{22}$$

Then the quantity $\boldsymbol{\Theta}^{-1}(k)\boldsymbol{e}(k)$ can be considered as such that it has unit variance and satisfies the Markov conditions and is Gaussian. After all, (22) is an unbiased estimator with a minimum variance. In this case, the fault estimation error can be represented as

$$\widetilde{\boldsymbol{f}}^{*}(k) = -\boldsymbol{K}_{f}^{*}(k)\boldsymbol{e}(k), \tag{23}$$

and using formula (23), one can calculate the covariance matrix of filtering errors in the form:

$$\boldsymbol{P}^{*}_{f}\left({}^{k}/_{k}\right) \triangleq \mathcal{M}\left\{\boldsymbol{f}^{*}(k)\boldsymbol{f}^{*T}(k)\right\} = \boldsymbol{K}^{*}_{f}(k)\boldsymbol{W}(k)\boldsymbol{K}^{*T}_{f}(k) = \boldsymbol{\Phi}(k)\left[\boldsymbol{G}^{T}(k)\boldsymbol{W}^{-1}(k)\boldsymbol{G}(k)\right]^{-1}\boldsymbol{\Phi}^{T}(k). \tag{24}$$

The Optimal State Vector Estimation. In this section, we consider the procedure for correctly choosing the matrix filter gain $\boldsymbol{K}_{s}^{*}(k)$ that minimizes the trace of the covariance matrix of filtering errors in the presence of constraints $\boldsymbol{K}_{s}(k)\boldsymbol{G}(k) = \boldsymbol{Z}(k)$. This guarantees the absence of shifts in the estimates of the system state vector $\boldsymbol{s}^{*}\left({}^{k}/_{k}\right)$. If the matrix $\boldsymbol{G}^{T}(k)\boldsymbol{W}^{-1}(k)\boldsymbol{G}(k)$ is not singular, then the transfer matrix of the filter that generates the estimate $\boldsymbol{s}^{*}\left({}^{k}/_{k}\right)$ with the minimum value of the covariance matrix trace of filtering errors $\boldsymbol{P}_{s}^{*}\left({}^{k}/_{k}\right)$ is defined as

$$\boldsymbol{K}_{s}^{*}(k) = \overline{\boldsymbol{P}}_{s}\left({}^{k}/_{k-1}\right)\boldsymbol{H}^{T}(k)\boldsymbol{W}^{-1}(k)\left[\boldsymbol{I} - \boldsymbol{G}(k)\boldsymbol{G}^{*}(k)\right] + \boldsymbol{Z}(k)\boldsymbol{G}^{*}(k) \tag{25}$$

on condition of restriction observance (14).

To test this assumption turning to formulas (13)–(14), we calculate the filtering errors covariance matrix by the expression:

$$\begin{aligned}\boldsymbol{P}_{s}\left({}^{k}/_{k}\right) \triangleq \mathcal{M}\left\{\widetilde{\boldsymbol{s}}(k)\widetilde{\boldsymbol{s}}^{T}(k)\right\} &= [\boldsymbol{I} - \boldsymbol{K}_{s}(k)\boldsymbol{H}(k)]\overline{\boldsymbol{P}}_{s}\left({}^{k}/_{k-1}\right)[\boldsymbol{I} - \boldsymbol{K}_{s}(k)\boldsymbol{H}(k)]^{T} + \boldsymbol{K}_{s}(k)\boldsymbol{R}(k)\boldsymbol{K}_{s}^{T}(k) = \\ &= \boldsymbol{K}_{s}(k)\boldsymbol{W}(k)\boldsymbol{K}_{s}^{T}(k) - 2\overline{\boldsymbol{P}}_{s}\left({}^{k}/_{k-1}\right)\boldsymbol{H}^{T}(k)\boldsymbol{K}_{s}^{T}(k) + \overline{\boldsymbol{P}}_{s}\left({}^{k}/_{k-1}\right).\end{aligned} \tag{26}$$

In the future, to minimize the trace of this matrix, we will apply the variational methods, for example, the method of Lagrange vector multipliers. To do this, we introduce an auxiliary function, the so-called Lagrange matrix function [23]

$$\begin{aligned}\varphi_{L}(k) = tr&\left\{\boldsymbol{K}_{s}(k)\boldsymbol{W}(k)\boldsymbol{K}_{s}^{T}(k) - 2\overline{\boldsymbol{P}}_{s}\left({}^{k}/_{k-1}\right)\boldsymbol{H}^{T}(k)\boldsymbol{K}_{s}^{T}(k) + \overline{\boldsymbol{P}}_{s}\left({}^{k}/_{k-1}\right)\right\} - \\ &- 2tr\left\{[\boldsymbol{K}_{s}(k)\boldsymbol{G}(k) - \boldsymbol{Z}(k)]\boldsymbol{\Lambda}^{T}(k)\right\}\end{aligned} \tag{27}$$

where $\boldsymbol{\Lambda}(k)$– is the matrix of Lagrange multipliers. Here, the factor 2 is chosen purely for convenience, since it doesn't affect the minimization process in any way. If we calculate the gradient of this function with respect to $\boldsymbol{K}_{s}(k)$ and equate it to zero, then the minimum condition will

$$\boldsymbol{W}(k)\boldsymbol{K}_{s}^{T}(k) - \boldsymbol{H}(k)\overline{\boldsymbol{P}}_{s}\left({}^{k}/_{k-1}\right) - \boldsymbol{G}(k)\boldsymbol{\Lambda}^{T}(k) = \boldsymbol{0}. \tag{28}$$

Now let's combine Eqs. (14) and (28) into a system of vector equations

$$\begin{bmatrix}\boldsymbol{W}(k) & -\boldsymbol{G}(k)\\ \boldsymbol{G}^{T}(k) & 0\end{bmatrix}\begin{bmatrix}\boldsymbol{K}_{s}^{T}(k)\\ \boldsymbol{\Lambda}^{T}(k)\end{bmatrix} = \begin{bmatrix}\boldsymbol{H}(k)\overline{\boldsymbol{P}}_{s}\left({}^{k}/_{k-1}\right)\\ \boldsymbol{Z}^{T}(k)\end{bmatrix}. \tag{29}$$

System (29) has a single solution with respect to $\begin{bmatrix} \boldsymbol{K}_s^T(k) \\ \boldsymbol{\Lambda}^T(k) \end{bmatrix}$ if the block matrix $\begin{bmatrix} \boldsymbol{W}(k) & -\boldsymbol{G}(k) \\ \boldsymbol{G}^T(k) & 0 \end{bmatrix}$ is not singular – it has an inverse. Turning to [25], it is easy to make sure that the condition for the existence of the inverse block matrix is the non-singularity of the matrix $\boldsymbol{G}^T(k)\boldsymbol{W}^{-1}(k)\boldsymbol{G}(k)$. This means that (29) has a single solution.

In a standard way according to formulas (4)–(5) and (11), the filter is corrected in the time domain:

$$\overline{\boldsymbol{P}}_s\left(k/_{k-1}\right) \triangleq \mathcal{M}\left\{\boldsymbol{s}^*\left(k/_{k-1}\right)\boldsymbol{s}^{*T}\left(k/_{k-1}\right)\right\} =$$
$$= \left[\boldsymbol{\Sigma}(k,k-1) \quad \boldsymbol{\Psi}(k,k-1)\right]\begin{bmatrix} \boldsymbol{P}_s^*\left(k-1/_{k-1}\right) & \boldsymbol{P}_{sf}^*\left(k-1/_{k-1}\right) \\ \boldsymbol{P}_{fs}^*\left(k-1/_{k-1}\right) & \boldsymbol{P}_f^*\left(k-1/_{k-1}\right) \end{bmatrix}\begin{bmatrix} \boldsymbol{\Sigma}^T(k,k-1) \\ \boldsymbol{\Psi}^T(k,k-1) \end{bmatrix} + \boldsymbol{Q}(k-1); \tag{30}$$
$$\boldsymbol{P}_{sf}^*\left(k/_k\right) \triangleq \mathcal{M}\left\{\boldsymbol{s}^*\left(k/_k\right)\boldsymbol{f}^*\left(k/_k\right)^T\right\} =$$
$$= \boldsymbol{K}_s^*(k)\boldsymbol{R}(k)\boldsymbol{K}_f^{*T}(k) - \left[\boldsymbol{I} - \boldsymbol{K}_s^*(k)\boldsymbol{H}(k)\right]\overline{\boldsymbol{P}}_s\left(k/_{k-1}\right)\boldsymbol{H}^T(k)\boldsymbol{K}_f^{*T}(k).$$

4 Generalization of the Locally Optimal Estimation for the Case of the Arbitrary Rank Distribution Matrix of Faults

In this section, we consider the case when the distribution matrix of faults in the observation subsystem can have an arbitrary rank in the range from one to *p*, that is, $0 < rank[\boldsymbol{F}(k)] < p$. The extension of the filter functionality is achieved by introducing additional restrictions that are imposed on the system matrices of Eqs. (1)–(2), and optimization is performed on a space of lower dimension, since not all modes can be restored from the observation results. Moreover, non-renewable modes must remain in the persistent modes class, otherwise the filter cannot be implemented in practice. Since the vast majority of mathematical calculations are identical to those given in the previous subsection, the presentation of the material will be brief with references to individual literary sources. If we substitute formulas (10)–(11) into Eqs. (12), then the estimate error fault will be presented as:

$$\begin{aligned}\widetilde{\boldsymbol{f}}(k) &= \left[\boldsymbol{I} - \boldsymbol{K}_f(k)\boldsymbol{F}(k)\right]\boldsymbol{f}(k) - \boldsymbol{K}_f(k)\boldsymbol{H}(k)\boldsymbol{\Omega}(k,k-1)\boldsymbol{d}(k-1) - \boldsymbol{K}_f(k)\left[\boldsymbol{H}(k)\widetilde{\boldsymbol{s}}^*\left(k/_{k-1}\right) + \boldsymbol{v}(k)\right] = \\ &= -\boldsymbol{K}_f(k)\boldsymbol{H}(k)\boldsymbol{\Psi}(k,k-1)\widetilde{\boldsymbol{f}}(k-1) - \boldsymbol{K}_f(k)\boldsymbol{H}(k)\boldsymbol{\Sigma}(k,k-1)\widetilde{\boldsymbol{s}}(k-1) + \\ &+ \left[\boldsymbol{I} - \boldsymbol{K}_f(k)\boldsymbol{F}(k)\right]\boldsymbol{f}(k) - \boldsymbol{K}_f(k)\boldsymbol{H}(k)\boldsymbol{\Omega}(k,k-1)\boldsymbol{d}(k-1) - \boldsymbol{K}_f(k)\boldsymbol{H}(k)\boldsymbol{w}(k-1) - \boldsymbol{K}_f(k)\boldsymbol{v}(k).\end{aligned} \tag{31}$$

Next, we introduce a series of notation, provided there is no shift $\mathcal{M}\{\widetilde{\boldsymbol{s}}(k-1)\} = \boldsymbol{0}$ in the previous estimate:

$$\boldsymbol{\Phi}(k) \triangleq \boldsymbol{K}_f(k)\boldsymbol{F}(k) = \boldsymbol{I}_p - \boldsymbol{\Xi}(k), \tag{32}$$

where $\boldsymbol{\Xi}(k) = \boldsymbol{I} - \boldsymbol{F}^{\#}(k)\boldsymbol{F}(k)$; $\boldsymbol{G}_f(k) \triangleq \boldsymbol{K}_f(k)\boldsymbol{H}(k)\boldsymbol{\Psi}(k,k-1)$;

$G_d(k) \triangleq K_f(k)H(k)\Omega(k, k-1)$; # – symbol denotes the Moore-Penrose pseudo-inversion operation.

Using the method of work [26], we expand the mathematical expectation of the fault estimation error into a functional matrix series:

$$\begin{aligned}
&\mathcal{M}\{\widetilde{f}(k)\} = \Xi(k)f(k) - G_f(k)\Xi(k-1)f(k-1)+\\
&+G_f(k)[G_f(k-1)\Xi(k-2)]f(k-2) + ... + (-1)^k G_f(k) \times ... \times G_f(2)[G_f(1)\Xi(0)]f(0)-\\
&-G_d(k)d(k-1) + G_f(k)G_d(k-1)d(k-2) + ... + (-1)^k G_f(k) \times ... \times G_f(1)G_d(1)d(0), \quad (33)
\end{aligned}$$

and confine ourselves to the first-order approximation, putting all the values $G_f(i)\Xi(i-1)=0;\ G_d(i)=0$ для $i=1,...,k$.

This will mean that the previous values of the fault vectors do not influence the cur-rent value of the mathematical expectation. Then we get $\mathcal{M}\{\widetilde{f}(k)\} = \Xi(k)f(k)$. By analogy with (14)–(15), in order for the fault estimate to have no shift, the transfer matrix must satisfy the imposed constraint:

$$K_f(k)F(k) = \Phi(k); \quad K_f(k)H(k)\Psi(k, k-1)\Xi(k-1) = \mathbf{0}; \quad K_f(k)H(k)\Omega(k, k-1) = \mathbf{0}. \quad (34)$$

In what follows, the system of Eqs. (34) will be represented in matrix form:

$$K_f(k)\overline{G}(k) = \overline{\Phi}(k), \quad (35)$$

where $\overline{G}(k) = [H(k) \quad H(k)\Psi(k, k-1) \quad H(k)\Omega(k, k-1)]$; $\overline{\Phi}(k) = [\Phi(k) \ \mathbf{0} \ \mathbf{0}]$.

Combining (35) with (32) allows us to reduce the transfer matrix $K_f(k)$ to the form

$$K_f^*(k) = \overline{\Phi}(k)\overline{G}^*(k); \quad \overline{G}^*(k) = \left[\overline{G}^T(k)W^{-1}(k)\overline{G}(k)\right]^{\#}\overline{G}^T(k)W^{-1}(k). \quad (36)$$

Consider the state vector estimation error:

$$\begin{aligned}
&\widetilde{s}(k) = [I - K_s(k)H(k)]\widetilde{s}^*(k/k-1) - K_s(k)F(k)f(k)-\\
&-[K_s(k)H(k)\Omega(k, k-1) - \Omega(k, k-1)]d(k-1) - K_s(k)v(k) =\\
&= [I - K_s(k)H(k)]\Sigma(k, k-1)\widetilde{s}(k-1) + [I - K_s(k)H(k)]\Psi(k, k-1)\widetilde{f}(k)-\\
&-[K_s(k)H(k)\Omega(k, k-1) - \Omega(k, k-1)]d(k-1) - K_s(k)F(k)f(k)+\\
&+[I - K_s(k)H(k)]w(k-1) - K_s(k)v(k). \quad (37)
\end{aligned}$$

In order for the estimate of the camp vector to have no shifts, the matrix must satisfy the following introduced restriction:

$$\begin{aligned}
&K_s(k)F(k) = \mathbf{0}; \quad K_s(k)H(k)\Psi(k, k-1)\Xi(k-1) = \Psi(k, k-1)\Xi(k-1);\\
&K_s(k)H(k)\Psi(k, k-1)\Xi(k-1) = \Psi(k, k-1)\Xi(k-1); \quad K_s(k)H(k)\Omega(k, k-1) = \Omega(k, k-1). \quad (38)
\end{aligned}$$

In the combined form, the equivalent notation for (38) is

$$K_s(k)\overline{G}(k) = \overline{Z}(k); \overline{Z}(k) = [\mathbf{0} \quad \Psi(k, k-1)\Xi(k-1) \quad \Omega(k, k-1)]. \quad (39)$$

Table 1. Summary table of separate assessment formulas.

<table>
<tr><td colspan="2">Model of the process under study:
$\boldsymbol{s}(k+1)=\boldsymbol{\Sigma}(k+1,k)\boldsymbol{s}(k)+\boldsymbol{\Gamma}(k+1,k)\boldsymbol{u}(k)+\boldsymbol{\Psi}(k+1,k)\boldsymbol{f}(k)+\boldsymbol{\Omega}(k+1,k)\boldsymbol{d}(k)+\boldsymbol{w}(k)$;</td></tr>
<tr><td colspan="2">Observation channel model: $\boldsymbol{y}(k)=\boldsymbol{H}(k)\boldsymbol{s}(k)+\boldsymbol{F}(k)\boldsymbol{f}(k)+\boldsymbol{v}(k)$.</td></tr>
<tr><td colspan="2">Prior data:
1. $\boldsymbol{s}(k)\in\mathcal{R}^n,\ \boldsymbol{u}(k)\in\mathcal{R}^r,\ \boldsymbol{f}(k)\in\mathcal{R}^p,\ \boldsymbol{d}(k)\in\mathcal{R}^q,\ \boldsymbol{y}(k)\in\mathcal{R}^m$;
2. $\boldsymbol{u}(k)$ – well-known control input;
3. $\boldsymbol{\Sigma}(k+1,k),\ \boldsymbol{\Gamma}(k+1,k),\ \boldsymbol{\Psi}(k+1,k),(k),\ \boldsymbol{\Omega}(k+1,k),\ \boldsymbol{H}(k),\ \boldsymbol{F}(k)$ – given matrices;
4. $\mathcal{M}\{\boldsymbol{s}(0)\}=\bar{s}_0;\ \mathcal{M}\{\boldsymbol{w}(k)\}=\mathcal{M}\{\boldsymbol{v}(k)\}=\boldsymbol{0};\ \mathcal{M}\left\{\left[\boldsymbol{s}(0)-\bar{s}_0\right]\left[\boldsymbol{s}(0)-\bar{s}_0\right]^T\right\}=\boldsymbol{P}_s(0)$;
$\mathcal{M}\left\{\left[\boldsymbol{w}(k)\boldsymbol{w}^T(j)\right]\right\}=\boldsymbol{Q}(k)\boldsymbol{\delta}(k-j);\ \mathcal{M}\left\{\left[\boldsymbol{v}(k)\boldsymbol{v}^T(j)\right]\right\}=\boldsymbol{R}(k)\boldsymbol{\delta}(k-j)$;
$\mathcal{M}\left\{\left[\boldsymbol{s}(0),\boldsymbol{w}^T(k)\right]\right\}=\mathcal{M}\left\{\left[\boldsymbol{s}(0),\boldsymbol{v}^T(k)\right]\right\}=\mathcal{M}\left\{\left[\boldsymbol{w}(k),\boldsymbol{v}^T(j)\right]\right\}=\boldsymbol{0}$;
5. Constraint: a) $\left[\boldsymbol{H}(k),\ \boldsymbol{\Sigma}(k+1,k)\right]$ – observed pair; b) $n>m\geq p+q$;
c) $0<rank\left[\boldsymbol{F}(k)\right]\leq p$; d) $rank\left[\boldsymbol{H}(k)\boldsymbol{\Omega}(k,k-1)\right]=rank\left[\boldsymbol{\Omega}(k,k-1)\right]=q$.</td></tr>
<tr><td colspan="2">Fault evaluation</td></tr>
<tr><td colspan="2">$\boldsymbol{W}(k)\triangleq\boldsymbol{H}(k)\overline{\boldsymbol{P}}_S\left(k/_{k-1}\right)\boldsymbol{H}^T(k)+\boldsymbol{R}(k)$;</td></tr>
<tr><td>$rank\left[\boldsymbol{F}(k)\right]=p$</td><td>$0<rank\left[\boldsymbol{F}(k)\right]<p$</td></tr>
<tr><td>$\boldsymbol{G}(k)=\left[\boldsymbol{F}(k)\ \ \boldsymbol{H}(k)\boldsymbol{\Omega}(k,k-1)\right]$;
$\boldsymbol{\Phi}(k)=\left[\boldsymbol{I}_p\ \ \boldsymbol{0}\right]$;
$\boldsymbol{G}^*(k)=\left[\boldsymbol{G}^T(k)\boldsymbol{W}^{-1}(k)\boldsymbol{G}(k)\right]^{-1}\times$
$\times\,\boldsymbol{G}^T(k)\boldsymbol{W}^{-1}(k)$;
$\boldsymbol{K}_f(k)=\boldsymbol{\Phi}(k)\boldsymbol{G}^*(k)$.</td><td>$\boldsymbol{\Xi}(k-1)=\left[\boldsymbol{I}-\boldsymbol{F}^{\#}(k-1)\boldsymbol{F}(k-1)\right]$;
$\overline{\boldsymbol{G}}(k)=\begin{bmatrix}\boldsymbol{F}(k)\ \ \boldsymbol{H}(k)\boldsymbol{\Psi}(k,k-1)\boldsymbol{\Xi}(k-1)\times\\ \times\,\boldsymbol{H}(k)\boldsymbol{\Omega}(k,k-1)\end{bmatrix}$;
$\boldsymbol{\Phi}(k)=\boldsymbol{I}_p-\boldsymbol{\Xi}(k);\ \overline{\boldsymbol{\Phi}}(k)=\left[\boldsymbol{\Phi}(k)\ \ \boldsymbol{0}\ \ \boldsymbol{0}\right]$;
$\overline{\boldsymbol{G}}^*(k)=\left[\overline{\boldsymbol{G}}^T(k)\boldsymbol{W}^{-1}(k)\overline{\boldsymbol{G}}(k)\right]^{\#}\overline{\boldsymbol{G}}^T(k)\boldsymbol{W}^{-1}(k)$;
$\boldsymbol{K}_f(k)=\overline{\boldsymbol{\Phi}}(k)\overline{\boldsymbol{G}}^*(k)$.</td></tr>
</table>

(*continued*)

Table 1. (*continued*)

$f\left(k/k\right)=K_f(k)\left[y(k)-H(k)\hat{s}\left(k/k-1\right)\right]$; $P_f\left(k/k\right)=K_f(k)W(k)K^T_f(k)$.	
System vector estimation	
$rank\left[F(k)\right]=p$;	$0<rank\left[F(k)\right]<p$
$Z(k)=\left[0 \quad \Omega(k,k-1)\right]$; $K_s(k)=\overline{P}_s\left(k/k-1\right)H^T(k)W^{-1}(k)\times$ $\times\left[I-G(k)G^*(k)\right]+Z(k)G^*(k)$.	$\overline{Z}(k)=\left[0 \quad \Psi(k,k-1)\Xi(k-1) \quad \Omega(k,k-1)\right]$; $K_s(k)=\overline{P}_s\left(k/k-1\right)H^T(k)W^{-1}(k)\left[I-\overline{G}(k)\overline{G}^*(k)\right]+$ $+\overline{Z}(k)\overline{G}^*(k)$;
$\hat{s}\left(k/k\right)=\overline{s}\left(k/k-1\right)+K_s(k)\left[y(k)-H(k)\hat{s}\left(k/k-1\right)\right]$; $P_s\left(k/k\right)=\left[I-K_s(k)H(k)\right]\overline{P}_s\left(k/k-1\right)\left[I-K_s(k)H(k)\right]^T+K_s(k)R(k)K^T_s(k)$; $P_{sf}^*\left(k/k\right)=-\left[I-K_s(k)H(k)\right]\overline{P}_s\left(k/k-1\right)H^T(k)K^T_f(k)+K_s(k)R(k)K^T_f(k)$.	
Time Filter Correction : $\hat{s}\left(k/k-1\right)=\Sigma(k,k-1)\hat{s}\left(k-1/k-1\right)+\Gamma(k,k-1)u(k-1)+\Psi(k,k-1)f\left(k-1/k-1\right)$; $\overline{P}_s\left(k/k-1\right)=\left[\Sigma(k,k-1) \quad \Psi(k,k-1)\right]\begin{bmatrix} P_s\left(k-1/k-1\right) & P_{sf}\left(k-1/k-1\right) \\ P_{fs}\left(k-1/k-1\right) & P_f\left(k-1/k-1\right)\end{bmatrix}\begin{bmatrix}\Sigma^T(k,k-1) \\ \Psi^T(k,k-1)\end{bmatrix}+Q(k-1)$.	

Taking into account the entries (10)–(11) and (37), it is now possible to calculate the error estimation covariance matrix of the state vector by the expression:

$$P_s\left(k/k\right) \triangleq \mathcal{M}\left\{\tilde{s}(k)\tilde{s}^T(k)\right\} = [I - K_s(k)H(k)]\overline{P}_s\left(k/k-1\right)[I - K_s(k)H(k)]^T + $$
$$+ K_s(k)R(k)K_s^T(k) = K_s(k)W(k)K_s^T(k) - 2\overline{P}_s\left(k/k-1\right)H^T(k)K_s^T(k) + \overline{P}_s\left(k/k-1\right). \tag{40}$$

Minimization of the trace $P_s\left(k/k\right)$ is performed by the method of Lagrange multipliers. By analogy with (27)–(29) we get:

$$\begin{bmatrix} W(k) & -\overline{G}(k) \\ \overline{G}^T(k) & 0 \end{bmatrix}\begin{bmatrix} K^{*T}_s(k) \\ \Lambda^T(k) \end{bmatrix} = \begin{bmatrix} H(k)\overline{P}_s\left(k/k-1\right) \\ \overline{Z}^T(k) \end{bmatrix}. \tag{41}$$

If the matrix $\overline{G}^T(k)W^{-1}(k)\overline{G}(k)$ is not singular, the matrix Eq. (41) has a single solution, and the transfer matrix $K_s(k)$ is given by

$$K_s^*(k) = \overline{P}_s\left(k/k-1\right)H^T(k)W^{-1}(k)\left[I - \overline{G}(k)\overline{G}^*(k)\right] + \overline{Z}(k)\overline{G}^*(k). \tag{42}$$

5 Illustrative Example

As a test example, consider the process of an aircraft landing consisting of several stages. In the first of them, the aircraft is guided to the required airport by means of radio guidance equipment. At the second stage, starting from the moment of entering the coverage area of the glide path radio beacon, the pilot directs the aircraft along the selected glide path line at an angle of approximately – 3^0 to the runway. At a height of about 30 m, the final phase begins - alignment. Here, due to the close proximity of the earth, radio beam guidance becomes ineffective. Further planning at an angle of – 30 to the horizon plane also doesn't meet the conditions of flight comfort and safety. Therefore, at the leveling stage, the pilot controls the aircraft in manual mode, focusing on the on-board facilities indications, such as altimeters, and his visual observations. However, in any case, effective aircraft flight control involves the use of an a priori given nominal mathematical model of the aircraft as a control object. In addition, the nominal mathematical model of the control object is a basic element in the construction of dynamic system models that will be affected by disturbances or faults. If we assume that the roll angle at the leveling stage is equal to zero, then the movement of the aircraft is divided into two components: longitudinal and lateral. Further, we confine ourselves to considering only the longitudinal component of the aircraft motion. The small value of the landing trajectory inclination gives grounds to believe that the longitudinal movement of the aircraft during the leveling stage is completely determined by the angles of deflection of the elevators, provided that the pilot maintains the throttle stick in such a position that the airspeed of the aircraft re-mains constant. The small value of the landing trajectory inclination gives reason to believe that the longitudinal movement of the aircraft during the leveling phase is completely determined by the deflection elevator angles, provided that the pilot maintains the thrust handle in such a position that the airspeed of the aircraft remains constant. From the point of view of flight stability and controllability along a given trajectory, the most significant role is played by the short-period component of the longitudinal motion, the linearized equation of which in terms of variable states has the form [29]:

$$\begin{bmatrix} s_1 \\ s_2 \\ s_3 \\ s_4 \end{bmatrix} = \begin{bmatrix} 0 & 1 & 0 & 0 \\ 0 & -\dfrac{1}{T_0} & \dfrac{V_0}{T_0} & 0 \\ 0 & 0 & 0 & 1 \\ 0 & \left(\dfrac{1}{V_0 T_0^2} - \dfrac{2\xi_\vartheta \omega_\vartheta}{V_0 T_0} - \dfrac{\omega^2_\vartheta}{V_0}\right) & \left(\dfrac{1}{T_\vartheta^2} - \dfrac{2\xi_\vartheta \omega_\vartheta}{T_\vartheta} - \omega_\vartheta^2\right) & \left(T_0^{-1} - 2\xi_\vartheta \omega_\vartheta\right) \end{bmatrix} \begin{bmatrix} s_1 \\ s_2 \\ s_3 \\ s_4 \end{bmatrix} + \begin{bmatrix} 0 \\ 0 \\ 0 \\ K\omega_\vartheta^2 T_0 \end{bmatrix} \delta_e. \quad (43)$$

where ξ_ϑ – is the damping coefficient in the pitch channel; ω_ϑ – frequency of free oscillation; V_0 – landing approach speed; K – is the amplification factor of short-period oscillations; T_0 – trajectory time constant; δ_e– angle of deflection of the elevator. The parameters of Eq. (43) are determined by the aircraft design features and depend on the coefficients of aerodynamic forces and moments, which are very complex non-linear functions of many parameters and change with time depending on the flight conditions.

However, the processing of the results of flight and bench tests shows that the vast majority of pilots rate the aircraft as a control object as "good" or "satisfactory" if its design parameters are within the following limits: $\xi_\vartheta = 0{,}5$–$0{,}7$; $\omega_\vartheta = 1$–$3{,}5$ c^{-1}; $K = 0{,}5$–2; $T_0 = 1$–5 c. In accordance with the instructions of work [26], we will dwell on the following values of the above parameters: ξ_ϑ=0.6; T_0=3,5c; T_ϑ=3,1c ;K_ϑ=1,8 c^{-1};V_0=75мс$^{-1}$, that are consistent with the recommendations of other works, for example [27]. As state variables, we choose the altitude, the rate of its change, the pitch angle and the rate of pitch angle change. This choice has the advantage that all state variables are technically measurable. The discrete equivalent of the nominal model (43) supplemented by the perturbation component has the form [29]:

$$s(k+1) = \boldsymbol{\Sigma}(k+1,k)s(k) + \boldsymbol{\Gamma}(k+1,k)\boldsymbol{u}(k) + \boldsymbol{\Omega}(k+1,k)\boldsymbol{d}(k) + \boldsymbol{w}_s(k);$$
$$\boldsymbol{y}(k) = \boldsymbol{H}(k)s(k) + \boldsymbol{v}(k), \tag{44}$$

where

$$\boldsymbol{\Sigma}(k+1,k) = \begin{bmatrix} 1.000 & 0.0249 & 0.0075 & 0.0001 \\ 0 & 0.9920 & 0.6010 & 0.0072 \\ 0 & 0.0001 & 0.9930 & 0.0235 \\ 0 & 0.0073 & -0.5457 & 0.8829 \end{bmatrix}; \boldsymbol{\Gamma}(k+1,k) = \begin{bmatrix} 0 \\ -0.0043 \\ -0.0209 \\ -1.6417 \end{bmatrix};$$

$$\boldsymbol{H}_y(k) = \begin{bmatrix} 1 & 0 & 0 & 0 \\ 0 & 1 & 0 & 0 \\ 0 & 0 & 1 & 0 \\ 0 & 0 & 0 & 1 \end{bmatrix}; \quad \boldsymbol{\Omega} = \begin{bmatrix} 0 & 0 \\ T_0^{-1} & 0 \\ 0 & 0 \\ -\Sigma_{042} & -(2\omega_\vartheta)\left[(V_0 T_0)^{-1} + T_\vartheta^{-1} + 1\right] \end{bmatrix}.$$

$\boldsymbol{\Omega}$ – the perturbation distribution matrix according to the recommendations [9] was under the assumption that the perturbations source is the change of the aircraft speed $V_0 + \Delta V$ under the wind influence and the instability of the damping coefficient $\xi_0 + \Delta\xi$. The state vector components were also perturbed by white Gaussian noise $\boldsymbol{w}_s(k)$ in order to take into account modeling errors and the influence of the turbulent wind component. Observation errors were simulated by introducing white Gaussian noise $\boldsymbol{v}(k)$ that is uncorrelated with $\boldsymbol{w}_s(k)$. The above noises had a zero mean value, and their intensity was determined by covariance matrices $\boldsymbol{Q}_s(k) \triangleq \mathcal{M}\{ww^T\} = 0.0081 * \boldsymbol{I}_{(4)}$; $\boldsymbol{R}(k) \triangleq \mathcal{M}\{vv^T\} = 0.16 * \boldsymbol{I}_{(4)}$. The initial conditions were taken as follows: $\boldsymbol{s}^T(0) = \begin{bmatrix} 0.0 & -3.6 & 0.1 & 0.004 \end{bmatrix}^T$; $\boldsymbol{P}(0/0) = 0.01 * \boldsymbol{I}_{(4)}$; $\boldsymbol{s}^T(0/0) = \begin{bmatrix} 0.0 & 1.9 & -0.1 & 0.004 \end{bmatrix}^T$. Since the component $\boldsymbol{\Gamma}(k+1,k)\boldsymbol{u}(k)$ is a deterministic value, it doesn't affect the essence of the evaluation process. In order to simplify the modeling process, we set it to zero, that is, we temporarily assume that $\boldsymbol{u}(k) = 0$, and in the final results we take into account the influence of the actual steering sequence, based on the requirements for the landing trajectory of the aircraft at the leveling stage. The simulation results for the standard Kalman filter are shown in (Fig. 1 a,b).

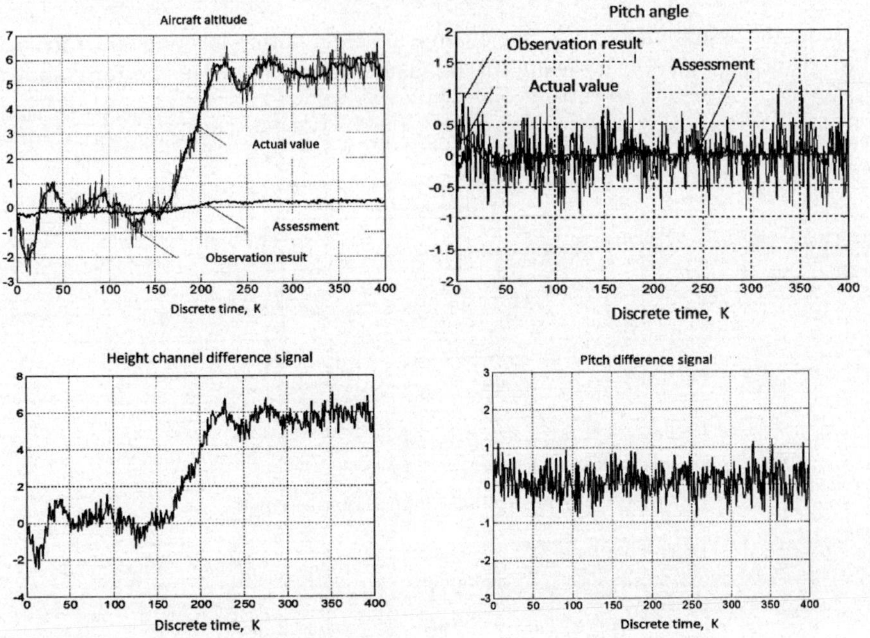

a) b)

Fig. 1. Results of modeling separate properties of the standard Kalman filter: a) – height channel; b) – pitch angle channel.

Visually, it is easy to see that the difference signal in the height channel is not separated from the influence of disturbances, that is, its average value is far from zero. The consequence is the filtering process divergence starting from the disturbance moment (k = 150) and the impossibility of resolving the fault signal against the related disturbances and noise background. To take into account the effect of the control sequence, we will use the list of main requirements and restrictions specific to the landing maneuver at the leveling stage [27]:

1. The curve for the desired change in aircraft altitude during the leveling process has an exponential and linear component, as shown in (Fig. 2). Thanks to this trajectory, it is possible to ensure the safety and convenience of the landing process. We assume that the leveling stage starts from a height of 30 m and lasts 30 s for an air-craft approaching for landing at a speed of 75 m/s. For the convenience of using the graph, the origin of the coordinates is referred to the runway contact point, and the horizontal axis is graduated in units of distance.
2. At the moment of first touching the runway, the vertical component of the descent velocity should not exceed -0.5 m/s, and the pitch angle should be within 0–10 °. During landing, the aircraft attack angle must not exceed the regulated value of -180. Further, we assume that the synthesized control $\boldsymbol{u}(k)$ satisfies the above conditions.

Under the assumptions made, the change in height during leveling under the action of disturbances $\boldsymbol{d}(k)$ and using the standard Kalman filter has the form shown in (Fig. 3).

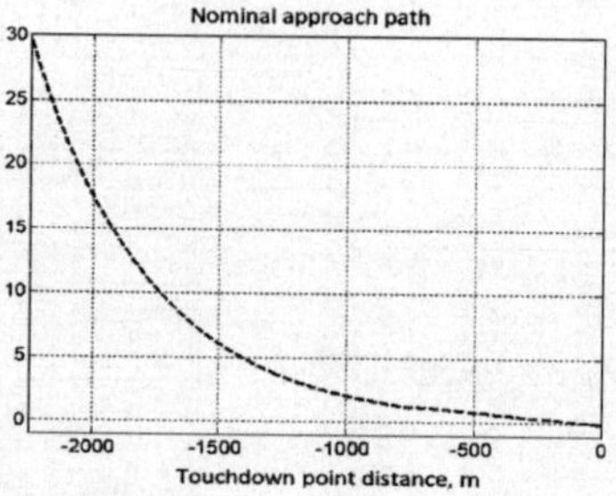

Fig. 2. View of the nominal approach path.

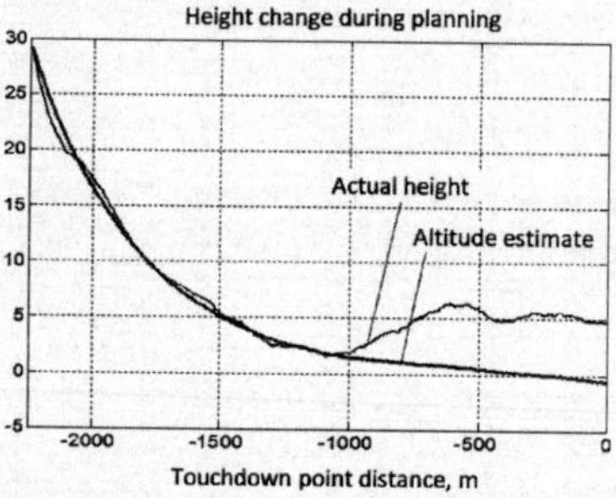

Fig. 3. Aircraft height estimation in the presence of disturbances using standard Kalman filter.

It is easy to see that in this case the actual height is not tracked, and the filtering process diverges. According to previous estimates, the runway contact point shifts to the right by approximately 250–300 m, that can create an emergency situation associated with the aircraft leaving the runway. This is especially critical for airports where, for one reason or another, the runway has a limited length. On (Fig. 4) presents the simulation results for the uncertain inputs filter, synthesized according to the method described in Table 1. The above results show that the height estimate moves towards its actual value only in the asymptotic sense. This will mean that the difference signal of the height channel will not be completely separated from uncontrolled perturbations, that is, only partial decomposition takes place in the time interval $150 < k < 300$. In the future, for $k > 300$, the situation improves. As the estimation process converges, the errors approach zero in the RMS sense, and the difference signal oscillates around the zero value. Also, a similar situation, but in a less pronounced form, takes place in the pitch channel.

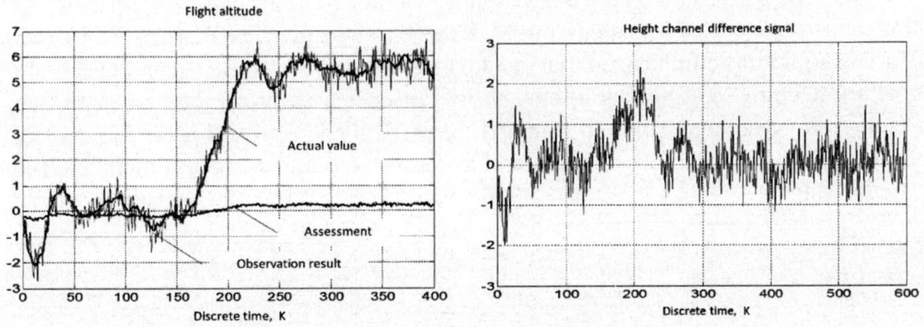

Fig. 4. Height channel estimation results.

The change in the height of the aircraft in the vertical plane, subject to the action of disturbances $\boldsymbol{d}(k)$ and the use of a separate estimation filter, has the form shown in (Fig. 5). It should be noted that the proposed filter, on the whole, adequately reflects the actual change in height, with the exception of the transient process in the area of 1000–500 m, where there is a systematic error. As the estimation process converges, the aircraft crew obtains an objective estimate of the actual altitude and can implement reductions in engine thrust in order to maintain a regulated descent trajectory.

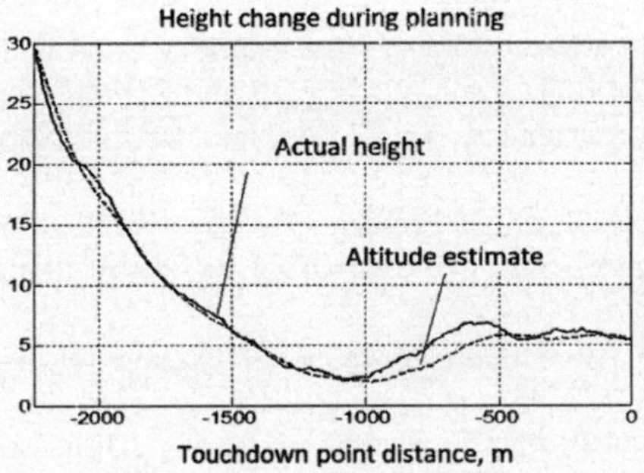

Fig. 5. Height estimation in the presence of disturbances using locally - optimal filter of separate estimation.

6 Conclusion

In the presented work, the solution of the problem of simultaneous separate estimation of faults and the state of a linear dynamic system under the influence of uncontrolled disturbances. An algorithm for recurrent separate filtering is proposed. This algorithm is capable of generating unbiased minimum variance estimates for the case where the fault distribution matrix is a full rank matrix. It is shown that in the case when the rank above the specified matrix remains undefined and is within $0 < rank[\boldsymbol{F}(k)] < p$, optimization remains possible, but in a space of lower dimension, provided that unobserved modes are found in the subspace of persistent modes. Otherwise, the problem of separate estimation has no solution. It is emphasized that the advantages of the proposed separate estimation algorithm become tangibly significant in the absence of a priori data on disturbances or faults, the structure of which remains uncertain. The operability of the method is demonstrated on a meaningful example of an aircraft landing maneuver at the final stage of the landing.

References

1. Isermann, R.: Fault-diagnosis systems. An Introduction from Fault Detection to Fault Tolerance. Springer, Berlin (2006). https://doi.org/10.1007/3-540-30368-5
2. Blanke, M., Kinnaert, M., Lunze, J., Staroswiecki, M.: Diagnosis and Fault Tolerant Control. Springer, Berlin, p. 695 (2016). https://doi.org/10.1007/978-3-662-47943-8
3. Hwang, I., Kim, S., Kim, Y., Seah, C.A.: Survey of fault detection, isolation, and reconfiguration methods. IEEE Trans. Control Syst. Technol. **18**(3), 636–653 (2010)
4. Escobet, T., Bregon, A.B., Puig, P.: Fault Diagnosis of Dynamic Systems Quantitative and Qualitative Approaches. vol. 2019, p. 462. Springer, Cham (2016)
5. Patton, R.J., Uppal, F.J., Lopez-toribio, C.J.: Soft computing approaches to fault diagnosis for dynamic systems: a survey. In: IFAC Proceedings Volumes, vol. 33, no. 11, pp. 303–315 (2000)
6. Gertler, J.: Survey of model-based failure detection and isolation in complex plants. IEEE Contr. Syst. Mag. **8**(6), 3–11 (1988)
7. Willsky, A.S.: Detection of abrupt changes in dynamic systems. In: Basseville, M., Benveniste, A. (eds.) Detection of Abrupt Changes in Signals and Dynamical Systems. Lecture Notes in Control and Information Sciences, vol. 77. Springer, Berlin (1985). https://doi.org/10.1007/BFb0006388
8. Varga, A.: Solving Fault Diagnosis Problems. Linear Synthesis Techniques. Studies in Systems, Decision and Control. Springer, Chams, vol. 84, 394 p. (2017). https://doi.org/10.1007/978-3-319-51559-5
9. Varga, A., Ossmann, D.: LPV model-based robust diagnosis of flight actuator faults. Control Eng. Pract. **31**, 135–147 (2014)
10. Friedland, B.: Treatment of bias in recursive filtering. IEEE Trans. Autom. Contr. **14**(4), 359–367 (1969)
11. Chen, J., Patton, R.: Optimal filtering and robust fault diagnosis of stochastic systems with unknown disturbances. IEE Proc. Control Theor. Appl. **143**, 31–36 (1996)
12. Alouani, A.T., Rice, T.R., Blair, W.D.: Two-stage filter for state estimation in the presence of dynamical stochastic bias. In: Proceedings of the American Control Conference, pp. 1784–1788, Chicago, Ill, USA (1992)
13. Ignagni, M.: Optimal and suboptimal separate-bias Kalman estimators for a stochastic bias. IEEE Trans. Autom. Control **45**(3), 547–551 (2000)
14. Volovyk, A., Kychak, V., Havrilov, D.: Discrete Kalman filter invariant to perturbations. Acta Polytechnica Hungarica **18**(10), 21–41 (2021)
15. Chen, F.C., Hsieh, C.S.: Optimal multistage Kalman estimators. IEEE Trans. Autom. Control **45**(11), 2182–2188 (2000)
16. Kim, K.H., Lee, J.G., Park, C.G.: Adaptive two-stage Kalman filter in the presence of unknown random bias. Int. J. Adapt. Control Signal Process. **20**(7), 305–319 (2006)
17. Kim, K.H., Lee, J.G., Park, C.G., The stability analysis of the adaptive two-stage Kalman filter. Int. J. Adapt. Control Signal Process. **21**(10), 856–870 (2007)

18. Kitanidis, P.K.: Unbiased minimum-variance linear state estimation. Automatica **23**(6), 775–778 (1987)
19. Aoki, M.: Optimization of stochastic systems. Per. s anhl. Nauka M.: – 424 p. (1971). (in Russian)
20. Gillijns, S., Moor, B.: Unbiased minimum-variance input and state estimation for linear discrete-time systems. Automatica **43**(1), 111–116 (2007)
21. Hsieh, C.S.: Optimal minimum-variance filtering for systems with unknown inputs. In: Proceedings of the World Congress on Intelligent Control and Automation (WCICA '06), vol. 1, pp. 1870–1874, Dalian, China (2006)
22. Massoumnia, M.A.: A geometric approach to the synthesis of failure detection filters. IEEE Trans. Automat. Contr. **AC-31**(9), 839–846 (1986)
23. DeRusso, P., Roy, R., Close, C., Desrochers, A.: State Variables for Engineers, 2nd edn., 592 p. Wiley-Interscience, Hoboken (1997)
24. Sage, A., Melsa, J.: Estimation Theory With Applications to Communications and Control, 1st edn, 752 p. McGraw-Hill Book Company, Inc., New York (1971)
25. Hilbe J., Robinson A. Methods of Statistical Model Estimation Chapman and Hall, 1st edn, 255 p. CRC Press, Boca Raton (2019)
26. Hsieh, C.S., Chen, F.C.: Optimal Solution of The Two-Stage Kalman Estimator. In: Proceedings of 34th Conferenceon Decision & Control, New Orleans, LA, pp. 1532–1537 (1995)
27. Biushhens, H.S., Studnev, R.V.: Dynamyka samoleta. Prostranstvennoe dvyzhenye. Mashynostroenye, 320 s. (1983). (in Russian)
28. Ellert, F.J., Merriam III, C.W.: Synthesis of feedback control optimization theory: an example. IEEE Trans. Autom. Control **AC–8**, 89–103 (1963)
29. Volovyk, A., Kychak, V.: Detection filter method in diagnostic problems for linear dynamic systems. Visnyk NTUU KPI Seriia – Radiotekhnika Radioaparato-buduvannia **2021**(84), 30–39 (2021)

Conclusions

Keywords

- 5G and Beyond
- 6G
- Blockchain
- CAPEX and OPEX
- CO_2 Footprint
- Cybersecurity, IDS and IPS, CIDN
- Digitalization and Digital Transformation
- Digital Economy
- Discrete Algorithm
- Energy Efficiency
- Finite State Machine
- IoT and Smart Office
- ML and AI
- Market Efficiency
- Networked Ergonomics
- Neural Networks
- Process Automation
- QoS and QoE
- Reliability and Control
- Rescue and Emergency Ad Hoc Networks
- Robotics
- SDN and NFV
- Self-Driving Cars
- Signal Processing
- Smart Applications
- Smart Economy
- Smart Home

M. Klymash et al. (Eds.): TCSET 2022, LNEE 965, pp. 653–656, 2023.
https://doi.org/10.1007/978-3-031-24963-1

- Smart Office
- Smart Shopping
- Starlink
- Terrestrial and Non-Terrestrial Cells
- UAV and Flying Objects
- Unmanned Vehicles
- Wireless Sensor Networks

1 Reference Data about the Book

The book contains 660 pages, 37 chapters each up to 14–24 pages with the preface, introduction and appeared under own edition in close cooperation of the scientists from multiple countries: Germany, Ukraine, Israel, Switzerland, Slovak Republic, Poland, Czech Republic, South Korea, China, Italy, North Macedonia, Azerbaijan, Kazakhstan, France, Latvia, Greece, Romania, USA, Finland, Morocco, Ireland, United Kingdom.

2 Reader Circles and Target Groups

2.1 Target Group 1

The topics to be addressed are definitely of interest to **Postgraduates**: **Graduated Engineers** and **Professionals** who have been active in the telecommunications industry for a while and want to get to know the emerging and innovative computer network architectures, Digital Transformations and Smart Applications, network function virtualization, advanced network technologies, IoT, energy efficiency and security do. The researchers are also involved in the discussion about usable platforms, protocols and best practices.

2.2 Target Group 2

The future book is suitable for **teaching specialist lecturers and professors**. It addresses a holistic view of QoS and QoE, which become nowadays via Digital Transformations less technically and more socially driven. This includes aspects of energy efficiency and ecology (Green IT), security (Secure Transactions, Blockchain-supported IoT) and comfort (QoE based always on). Each chapter contains a real system example. The book has a summary and an outlook as well as further source recommendations.

2.3 Target Group 3

The future book is suitable for **Undergraduates too**: students of universities and technical high schools can access the work as e-book via so-called Springer links.

3 How to Read It?

The material of the presented book covers the latest research results in, but is not limited to, the following research topics. Therefore, we recommend to study the material on the given sequence through the listed topics for better comprehension:

- EMERGING NETWORKS ACCELERATE DIGITALIZATION
  - Emerging Networking: what is it in the future?
  - Digitalization Age, Digital Transformations, Networked Process Automaton
  - Impacts of Smart Economy and Digitalization on Market Efficiency
  - IoT in Emerging Scenarios
  - Robotics and UAV-Unmanned (Aerial) Vehicles, Self-Driving Cars, Discrete Vehicle Automation Algorithms
  - Functional Stability in Emerging Networks
  - Node Reconfiguration and Advanced Routing
  - Reliability and Control in Emerging Networks
  - Networked Sustainability, QoS and QoE
  - Intelligent Connectivity
  - SDN Protocols
  - Traffic Modelling and Simulation
  - Rescue and Emergency Networking
- SMART APPLICATIONS MEET DIGITALIZATION
  - Advanced Software Approaches and Process Models
  - Smart Applications: Smart Information Systems, Smart Office, Smart Home, Smart Shopping, Smart Building Management
  - Highly Decentralised Networking: Adaptation, Coordination, Highly-Distributed Applications
- 5G AND BEYOND
  - 5G and Beyond: Bird Perspective, AI, ML, and Deep Learning, Blockchain
  - Quo Vadis: 6G/NET-2030 goes terrestrial or satellite-based?
  - Polarization and Radio Wave Decay
  - Towards 6G: SpaceX/Starlink Technology, RAT, "Post-Shannon"
  - SAT Radio Networking

- DATA SECURITY FOR EMERGING NETWORKING
  - Data Security for Emerging Networking
  - Cybersecurity: Honeypotting, Intrusion Detection, and Intrusion Prevention Systems and Networks
  - Cryptography and Steganos for Emerging Networking
  - Potentially Dangerous IoT and Networks
  - Blockchain Techniques
- EMBEDDED SYSTEMS AND NETWORKED AUTOMATION
  - Embedded Systems and Networked Automation
  - Networked Control for Embedded and Electronics
  - Modeling, Simulation, Visualisation in Networks
  - Networked Ergonomics: Lighting, HVAC, Smart Home, Smart Office
  - Networked PHY Modelling and Visualisation
  - Neural Network Architectures.

Acknowledgment

The book was prepared on the latest research results and in close cooperation of scientists from Germany, Ukraine, Israel, Switzerland, Slovak Republic, Poland, Czech Republic, South Korea, China, Italy, North Macedonia, Azerbaijan, Kazakhstan, France, Latvia, Greece, Romania, USA, Finland, Morocco, Ireland, United Kingdom. The Editorial Team is very grateful to the authors from different countries, which enabled this book and to the colleagues from Springer Nature.

Our acknowledgment belongs to Yulia Pyrih and Halyna Beshley for technical support and assistance by book preparation.

M. Klymash et al. (Eds.): TCSET 2022, LNEE 965, p. 657, 2023.
https://doi.org/10.1007/978-3-031-24963-1

Author Index

M. Klymash et al. (Eds.): TCSET 2022, LNEE 965, pp. 659–660, 2023.
https://doi.org/10.1007/978-3-031-24963-1